Mineral Nutrition of Rice

Mineral Nutrition of Rice

Nand Kumar Fageria

CRC Press
Taylor & Francis Group
Boca Raton London New York

CRC Press is an imprint of the
Taylor & Francis Group, an **informa** business

CRC Press
Taylor & Francis Group
6000 Broken Sound Parkway NW, Suite 300
Boca Raton, FL 33487-2742

Version Date: 20130315

International Standard Book Number-13: 978-1-4665-5806-9 (Hardback)

Visit the Taylor & Francis Web site at
http://www.taylorandfrancis.com

and the CRC Press Web site at
http://www.crcpress.com

To my wife Shanti
Her partnership in my life has contributed
significantly to my successful scientific career.

Contents

Preface

Rice is a very fascinating plant because it grows in diverse environmental conditions. It is the main food crop that can be grown on oxidized as well as reduced or flooded soil conditions. Rice is the staple food for more than half of the world's population, and it will continue to be the major food crop in the next century too. It is the third most important cereal crop after wheat and corn, and it is the main food for the people of Asia, Africa, and South America. It is also consumed by large numbers of Asian, African, and South American immigrants living in the United States, Europe, and Australia. As a primary source of food, rice accounts for 35–75% of the calorie intake of more than three billion people.

It is estimated that the world population may be more than nine billion by the year 2050. Most of this increase will be in Asia, Africa, and South America. The current annual growth rate of the global population is 1.22%; this growth rate is six times faster in developing nations than in developed regions. Hence, on these three continents, the demand for rice will be greater than for other crops. To meet this growing demand, rice production should be increased about 60% by the year 2025. This could be made possible by increasing yield per unit area, crop intensity, and the land area planted with rice. However, increase in land area under rice cultivation would be possible only in South America and Africa.

Most rice varieties are grown in lowland or under flooded conditions. This type of rice cultivation uses large amounts of water. In the future, water scarcity may be a serious problem for rice production in most of the rice-growing regions. Efficient utilization of nutrients can improve water use efficiency in crop plants, including rice.

Mineral nutrition encompasses the supply, uptake, and utilization of essential plant nutrients for the growth and development of crop plants. Essential plant nutrients in adequate amounts and with appropriate balance are responsible for 40% yield increase of annual crops under all agro ecosystems, including rice production. However, in modern agricultural practice, fertilizers are among the most expensive factors used to increase crop yields. If fertilizers are not used judiciously, essential plant nutrients may leach downward in the soil profile and may contaminate groundwater, rivers, and lakes. This may result in serious environmental pollution. All 15 chapters in this book provide valuable information on the ecophysiology of rice and the adequate management of essential nutrients for maximizing rice yield. This information may lead to increased yield, reduced production costs, and avoidance of environmental pollution.

Agricultural science has advanced at a very rapid pace over the past decades. I hope this volume will be useful for agricultural scientists, students, and professors, as well as agricultural administrators and extension workers in the fields of agricultural and environmental sciences. The information provided in this book can be utilized in the classroom, in laboratories, and in agricultural research fields in order to keep the environment (including land and water) safe for future generations.

I could not have written such a comprehensive book without the help of many people and I thank all of them. I sincerely thank the National Rice and Bean Research Center of the Brazilian Enterprise for Agricultural Research (EMBRAPA) for providing necessary support to write this book and also its staff members for their cooperation, ensuring a friendly atmosphere and a favorable academic environment to write this book. I extend my gratitude to the Brazilian Scientific and Technological Research Council (CNPq) for providing financial support for my many research projects; the results from several of these projects are included in the chapters of this book. I thank the staff at the Taylor & Francis Group/CRC Press, especially Randy Brehm and David Fausel, for their excellent handling of numerous issues and their dedication toward producing a high-quality book.

Finally, I thank my wife Shanti, daughter Savita, daughter-in-law Neera, sons Rajesh and Satya Pal, and grandchildren Anjit, Maia, and Sofia for their patience, encouragement, and understanding, without which I could not have possibly found the time and energy required to write this book.

Nand Kumar Fageria
National Rice and Bean Research Center of EMBRAPA
Santo Antônio de Goiás, Brazil

About the Author

Nand Kumar Fageria, doctor of science in agronomy, has been the senior research soil scientist at the National Rice and Bean Research Center, Empresa Brasileira de Pesquisa Agropecuária (EMBRAPA), since 1975. Dr. Fageria is a nationally and internationally recognized expert in the area of mineral nutrition for crop plants and has been a research fellow and ad hoc consultant to the Brazilian Scientific and Technological Research Council (CNPq), since 1989. In 1975, Dr. Fageria was the first to identify zinc deficiency in upland rice grown on Brazilian Oxisols. He has developed crop genotype screening techniques for aluminum and salinity tolerance and nitrogen, phosphorus, potassium, and zinc use efficiency under controlled and field conditions. He established adequate soil acidity indices such as pH, base saturation, Al saturation, and Ca, Mg, and K saturation for dry beans grown in conservation or no-tillage systems on Brazilian Oxisols. He also determined adequate and toxic levels of micronutrients in soil and plant tissues of upland rice, corn, soybeans, dry beans, and wheat grown on Brazilian Oxisols. Dr. Fageria determined adequate rates of N, P, and K for lowland and upland rice grown on Brazilian lowland soils, locally known as "Varzea," and for Oxisols of the Cerrado region. He also screened large numbers of tropical legume cover crops for acidity tolerance and for N, P, and micronutrient use efficiency. Dr. Fageria characterized the chemical and physical properties of Varzea soils of several Brazilian states, and it has been helpful in better fertility management of these soils for sustainable crop production. Dr. Fageria also determined adequate rate and sources of P and acidity indices for soybeans grown on Brazilian Oxisols. His studies have been published in scientific papers, technical bulletins, book chapters, and congress or symposium proceedings.

Dr. Fageria is the author/coauthor of 12 books and more than 320 scientific journal articles, book chapters, review articles, and technical bulletins. His three books, *The Use of Nutrients in Crop Plants* (published in 2009); *Growth and Mineral Nutrition of Field Crops*, Third Edition (published in 2011); and *The Role of Plant Roots in Crop Production* (published in 2013), have been bestsellers for CRC Press. Dr. Fageria has written several review articles for *Advances in Agronomy*, a well-established and highly regarded serial publication, on nutrient management, enhancing nutrient use efficiency in crop plants, ameliorating soil acidity by liming on tropical acidic soils for sustainable crop production, and the role of mineral nutrition on root growth of crop. He has been an invited speaker to several national and international congresses, symposiums, and workshops. Dr. Fageria is a member of the editorial board of the *Journal of Plant Nutrition* and the *Brazilian Journal of Plant Physiology*. Since 1990, he has also served as a member of an international steering committee of symposiums on plant–soil interactions at low pH. He is an active member of the American Society of Agronomy and the Soil Science Society of America.

1 Ecophysiology of Rice

1.1 INTRODUCTION

Rice (*Oryza sativa* L.) is an important food crop for a large portion of the world's population, and it is the staple food for the people of Asia, Latin America, and Africa. Rice is cultivated on all of the continents except Antarctica; it accounts for about 23% of the world's total land area under cereal production (Wassmann et al. 2009a; Jagadish et al. 2010) and covers more than 161 million ha (with production of about 680 million metric tons). More than 90% of rice is produced and consumed in Asia (Jena and Mackill 2008; Grewal, Manito, and Bartolome 2011; Kumar and Ladha 2011), where this crop's historical importance is significant. Rice cultivation sustained many civilizations in the river deltas of India, China, and Southeast Asia and has become deeply intertwined with the cultures of these regions (Krishnan et al. 2011).

During the thousands of years since its domestication, Asian rice has been cultivated under significantly diverse agro-ecosystems in order to meet different human demands (Zong et al. 2007; Vaughan, Lu, and Tomooka 2008a, 2008b; Xiong et al. 2011). This has resulted in tremendous genetic diversity in rice around the world, as shown by different molecular tools, such as the analysis of restriction fragment length polymorphism (Zhang et al. 1992) and simple sequence repeats. As a consequence, many rice varieties with different characteristics have arisen under natural and human selection (Vaughan, Balazs, and Heslop-Harrison 2007). Yan et al. (2010) studied the genetic diversity of rice collected from all over the world at the United States Department of Agriculture (USDA) and concluded that germplasm accessions obtained from southern Asia, Southeast Asia, and Africa were highly diversified, whereas those from North America and western and eastern Europe had the lowest diversity.

In Latin America, rice is mainly eaten with the common bean (*Phaseolus vulgaris* L.) within all sections of society. In West Africa, rice is developing as a major staple food crop, and its production increased approximately by 170% from the 1970s to the early 2000s (Saito, Azoma, and Sie 2010). However, domestic production caters to only about 60% of current rice consumption (Saito, Sokei, and Wopereis 2012). Rice provides 35–60% of the dietary calories consumed by more than three billion people (Fageria, Slaton and Baligar, 2003). Table 1.1 shows the average protein, lipid, and carbohydrate contents in the seed of principal food crops. Globally, rice is the third most cultivated cereal after wheat and corn. Unlike wheat, 95% of the world's rice is grown in less developed nations located primarily in Asia, Africa, and Latin America. China and India are the world's largest producers and consumers of rice. With the world's population projected to be more than nine billion by the year 2050, the demand for rice will grow faster than for other crops.

TABLE 1.1
Protein, Lipid, and Carbohydrate Contents in the Seed of Principal Cereal and Legume Crop Species

Crop Species	Protein (g kg^{-1})	Lipid (g kg^{-1})	Carbohydrate (g kg^{-1})
Rice	105	22	828
Wheat	113	17	822
Barley	132	35	813
Corn	102	41	842
Sorghum	88	35	849
Oat	125	47	803
Soybean	314	201	444
Dry bean	192	24	755
Lupin	251	41	674
Adzuki bean	216	13	747
Pea	233	15	723
Chickpea	173	51	75
Peanut	193	304	484

Source: Adapted from Shinano, T., M. Osaki, K. Komatsu, and T. Tadano. 1993. Comparison of production efficiency of the harvesting organs among field crops. *Soil Sci. Plant Nutr.* 39:269–280; Fageria, N. K. 2009. *The Use of Nutrients in Crop Plants*. Boca Raton, FL: CRC Press.

The Green Revolution of the 1960s increased world rice production significantly. However, during the past decade, the production potential of modern cultivars has remained stagnant (Jena and Mackill 2008). As the global market for rice is growing quickly, it is imperative to increase production in different rice-growing ecosystems to feed the increasing world population (Khush 2005; Jena and Mackill 2008). Also, it is estimated that it will be necessary to produce about 60% more rice than what is being currently produced in order to meet the food needs of a growing world population by the year 2025 (Fageria 2007). Furthermore, the land available for crop production is decreasing steadily due to urban growth and land degradation. Therefore, an increase in rice production will have to come from the same or even a lesser amount of land. Appropriate rice production practices should be developed and adopted to improve rice yield per unit area. Today, rice yield is stagnated in many rice-producing regions of the world, and many biotic and abiotic stresses are responsible for this dilemma (Prasad 2011). Despite this apparent yield barrier, the quest for higher yield potential continues (Peng et al. 1999; Fageria 2007). Therefore, knowledge of rice plant ecophysiology, the influence of environmental factors on its growth and development and biological processes and functions, is very important. Using this knowledge, it is possible to ascertain the factors that are important in determining growth and development, and yield-forming processes during the growth cycle and also to formulate management strategies to reduce the risk of yield-decreasing factors. The objective of this introductory chapter is to provide information on the impact of climatic factors on the growth and development, and yield formation components during the growth cycle of the rice plant and the association of climate with yield. Most of the discussion is supported by experimental results to make the information presented as practical as possible.

1.2 RICE CULTIVATION ECOSYSTEMS

An *ecosystem* is defined as a crop's growing environment. The description of a crop ecosystem is important for adopting or improving production practices for higher yields. Before discussing the ecophysiology of rice, the different types of cultural systems or ecosystems used to produce rice should be defined because the majority of the discussion here will focus on a specific production system. There is no true consensus on the terminology used to describe the different rice-growing systems and environments. Upland, lowland, dryland, and wetland are often used to describe the different production systems, but these general terms frequently have regional connotations and may share some common characteristics. In 1984, the International Rice Research Institute (IRRI) classified rice into five groups based on seasonal water regime (deficit, excess, or optimal), drainage (poor or good), air temperature (optimum or low), soil (normal or problematic), and topography (flat or undulating). These groups are (1) irrigated lowland, (2) rainfed lowland, (3) deep water, (4) upland, and (5) tidal wetlands. In this discussion, we primarily consider irrigated lowland (or flooded) rice, upland rice, and deep water (or floating) rice.

1.2.1 Irrigated (or Flooded) Rice

Irrigated (or *flooded*) *rice* is defined as rice cultivated on relatively flat lands wherein water is accumulated so that the rice is grown for the entire or part of the growing season in flooded conditions (Fageria, Slaton and Baligar, 2003). Irrigated lowland rice systems account for about 57% of the world's harvested rice area and contribute about 76% of the global rice production (Fageria 2003; Fageria et al. 2003; Zhao et al. 2010; Mohapatra, Panigrahi, and Turner 2011).

The presence of floodwater for part or all of the growing season requires that the rice root system is adapted to largely anaerobic soil conditions. The rice plant has adapted to this environment by transporting oxygen from the aerial portions of the plant to the root system via aerenchyma tissues (Yoshida 1981). A secondary adaptive mechanism is the development of an extensive lateral, fibrous root system located in the surface, 1–2 mm, of oxidized soil at the soil-water interface. Oxygen diffusing through the water layer allows this zone of soil to remain oxidized. For these reasons, flooded rice normally has a shallow, fibrous root system (Wells et al. 1993). The aquatic environment not only influences the development of the root system but also alters the availability of several essential nutrients, affects nutrient uptake and use efficiency as well as fertilization practices, and makes rice especially unique among crop production systems. Rice is probably the world's most diverse and versatile crop. Rice undoubtedly has evolved high levels of adaptability to various ecological habitats, most of which are characterized by their unique hydrological states (Nguyen, Babu, and Blum 1997). Rice is cultivated at an elevation of more than 3000 m in the Himalayas and at sea level in the deltas of the great rivers of Asia (Santos, Fageria, and Prabhu 2003). Floating cultivars grow in water as deep as 5 m in Thailand; in Brazil, rice is grown as a dryland crop much like wheat (*Triticum aestivum* L.) or corn (*Zea mays* L.). In West Africa, it is grown in mangrove swamps (Santos, Fageria, and Prabhu 2003). Elsewhere, it is grown in dryland and wetland conditions and over a wide range of latitudes (Fageria 2001). For example, rice is grown at 53° N latitude in northeastern China; on the equator in central Sumatra; and at 35° S latitude in New South Wales, Australia (Mae 1997). Bouman et al. (2007) reported that the irrigated and rainfed rice ecosystems, which form the major mainstay of food security in Asia, have been highly sustainable because the environment has few adverse impacts.

Rice belongs to the genus *Oryza* in the family Poaceae. It is a small genus of 20–25 species with a pantropical and subtropical distribution (Nayar 2012). There are two species of cultivated rice, *Oryza sativa* and *Oryza glaberrima* (Morishita 1984). *Oryza sativa* is the main species cultivated worldwide. *Oryza glaberrima* is endemic only to West Africa, and *Oryza glaberrima* Steud. is still grown there in extensive, traditional systems in West Africa (Linares 2002; Saito, Sokei, and Wopereis 2012). The usual assumption is that African rice (*Oryza glaberrima* Steud.) originated about 3500 years ago from the annual wide rice, *Oryza barthii* A. Chev.; Asian rice (*Oryza sativa* L.) was introduced in West Africa in the late fifteenth century by Europeans (Nayar 2012). *Oryza glaberrima* is known to be highly weed competitive and resistant to local biotic and abiotic stresses (Jones et al. 1997). However, it has a low yield potential due to its poor resistance to lodging and grain shattering (Jones et al. 1997). These two species (*Oryza sativa* and *Oryza glaberrima*) are classified into the AA genome line, but they are subdivided phyletically into different series. *Oryza sativa* is understood to have originated from the wild species *Oryza perennis*, and *Oryza glaberrima* is thought to have arisen from the wild species *Oryza brevilgulata* (Mae 1997) or *Oryza barthii* (Nayar 2012). The *Oryza sativa* rice species is divided into three groups based on geographical distribution—*japonica, indica*, and *javanica* (Takahashi 1984). Although extensively cultivated under different environmental conditions around the globe, *indica* and *japonica* varieties are adapted to different planting areas where they perform optimally in terms of yield (Xiong et al. 2011). Geographically, *indica* rice is commonly found in tropical and subtropical planting regions, whereas *japonica* varieties are restricted to more temperate regions (Vaughan, Lu, and Tomooka 2008a; Lu, Cai, and Jin 2009). However, progress in breeding has permitted wider adaptation and distribution of *japonica* and *indica* species (Mae 1997). Being adapted to different environments, *indica* and *japonica* varieties have developed diverse morphological, agronomical, physiological, and molecular characteristics (Zhang et al. 1992; Lu, Cai, and Jin 2009; Xiong et al. 2011) that provide valuable genetic resources for breeding high-yielding rice (Khush 2001). *Javanica* species is a subgroup of *japonica* germplasm and thus is often called tropical *japonica* (Glaszmann 1987).

Lowland or flooded rice production is threatened by a growing worldwide water shortage (Tuong and Bouman 2003). In Asia, by 2025, physical water scarcity is projected for more than two million ha of dry season lowland rice and 13 million ha of wet season lowland rice, and an economic water scarcity is expected to hamper most of that continent's 22 million ha of dry season lowland rice (Tuong and Bouman 2003; Zhao et al. 2010). Under these conditions, cultivation of upland or aerobic rice is a promising strategy to save water resources and to maintain rice production (Zhao et al. 2010).

1.2.2 Upland Rice

Rice grown in rainfed, naturally well-drained soils, without surface water accumulation or phreatic water supply, and normally not bunded is called upland or aerobic rice (Fageria 2001). Huke (1982) used the term dryland rice, instead of upland rice, and defined it as field-grown rice that is not bunded, is prepared and seeded under dry conditions, and depends on rainfall for moisture.

Globally, upland rice is planted on about 20 million ha. Of those hectares, about 60% are in Asia, 30% are in Latin America, and 10% are in Africa (Gupta and O'Toole 1986). Upland rice is distinct from lowland rice, which is usually grown in saturated or submerged soil for part or all of the growing season. Upland rice yield is quite low in all rice-growing regions. For example, the average yield of upland rice in Brazil is approximately 2200 kg ha^{-1}, compared with lowland or flooded rice at approximately 5500 kg ha^{-1} (Fageria 2001). In Asia, upland rice yields average only about 1000 kg ha^{-1} versus about 4900 kg ha^{-1} for irrigated lowland rice (George et al. 2002). The low yield of upland rice is associated with many biotic and abiotic stresses. Although upland rice has a low yield, it will continue to be an important component of cropping systems in South America, Africa, and Asia because of its low production cost and the lack of irrigation facilities in these areas. In Brazil, when land is first cleared for crop production, generally upland rice is the first crop planted. This is primarily due to the acidity tolerance of rice; in the second year, the area is used for pasture establishment.

In addition, upland rice has a high degree of drought tolerance and is able to cope with periods of water stress that occur in rainfed production. Breeders in Brazil and China have developed locally adapted upland rice cultivars yielding 5–6 Mg ha^{-1} (Wang et al. 2002; Pinheiro, Castro, and Guimares 2006) with limited water supply (600 mm of rainfall and irrigation water in northern China), resulting in almost double water-use efficiency and 50% water savings relative to conventional lowland rice (Zhao et al. 2010). A study in Laos showed that under rainfed conditions, upland or aerobic rice cultivars were more responsive to N and 86% higher yielding than traditional upland rice varieties across six sites and three N levels (Saito et al. 2007; Zhao et al. 2010). Atlin et al. (2006) reported that in Asia upland rice cultivars produce yield of about 5 Mg ha^{-1} in favorable environmental conditions.

1.2.3 Deep Water and Floating Rice

In addition to lowland and upland rice, deep water rice is a general term used for rice culture, or the variety that is planted, when the standing water for a certain period of time is more than 50 cm. Floating rice is grown where the maximum water depth ranges between 1 and 6 m for more than half of the growth duration (De Datta 1981). Deep water and floating rice are mostly grown in India,

Bangladesh, Thailand, Myanmar, Vietnam, Kampuchea, Mali, Nigeria, and Indonesia (De Datta 1981; Vergara 1985). These deep water rice crops cover more than 10 million ha, and millions of farmers depend on them for survival and security (Vergara 1985).

Although excess water is common to all deep water rice types, growth and development vary according to the crop establishment practice and the onset, depth, speed, and duration of flooding. As flood water rises, plants elongate, produce nodal roots, and may also produce nodal tillers. Water levels may reach 100 cm near the end of the monsoon season, when photoperiod-sensitive rice varieties flower. As the water recedes, grain ripens. Crop duration may be 300 days, and harvest may be on dry or wet fields (Vergara 1985). This type of rice culture requires improvement of varietal and cultural practices to increase yields.

The major distinguishing features of floating rice appear to be a semiprostrate appearance near the base of the plant (even in the early stages of growth under shallow water), the ability to elongate rapidly under rising water conditions (up to 10 cm day^{-1}), the formation of adventitious roots at the higher nodes, and a distinct photoperiodic type of flowering behavior (Jackson et al. 1972). Under deep water conditions, the leaves appear to float on the surface. When the water recedes and flowering occurs, a tangled mass of stems result. However, the upper portion of the stem usually exhibits phototropism, which reduces the damage to the panicle caused by water and mud, and which also facilitates harvesting (Jackson et al. 1972). Figure 1.1 shows the growth stages of the floating rice cultivar from seedling to physiological maturity.

1.3 SOIL USED FOR RICE CULTIVATION

Rice is cultivated on a variety of soils in different rice-growing regions. The soil orders for rice cultivation are Alfisols, Andisols, Aridisols, Entisols, Gelisols, Histosols, Inceptisols, Mollisols, Oxisols, Spodosols, Ultisols, and Vertisols. Characteristics of these soil orders are summarized in Table 1.2.

1.3.1 Lowland Rice

On a global level, lowland rice occupies a larger land area than upland rice. Moorman (1978) summarized that, worldwide, lowland rice is grown on all soil orders identified in the soil classification system (USDA 1975). Hence, the pedogenetic and morphological characteristics of soils used to grow lowland rice also vary considerably. The wide array of soils used to produce rice results in an equally diverse assortment of management practices implemented for successful rice production. Murthy (1978) reported that the soils on which rice grows in India are so extraordinarily varied that there is hardly a type, including salt-affected soils, on which rice cannot be grown without some degree of success. In Brazil, flooded rice is mainly grown on Alfisols, Vertisols, Inceptisols, Histosols, and Entisols (Moraes 1999). In Sri Lanka, rice is grown on Alfisols, Ultisols, Entisols, Inceptisols, and Histosols (Panabokke 1978), and in Indonesia, the main rice soils are Entisols, Inceptisols, Vertisols, Ultisols, and Alfisols (Soepraptohardjo and Suhardjo 1978). Raymundo (1978) reported that

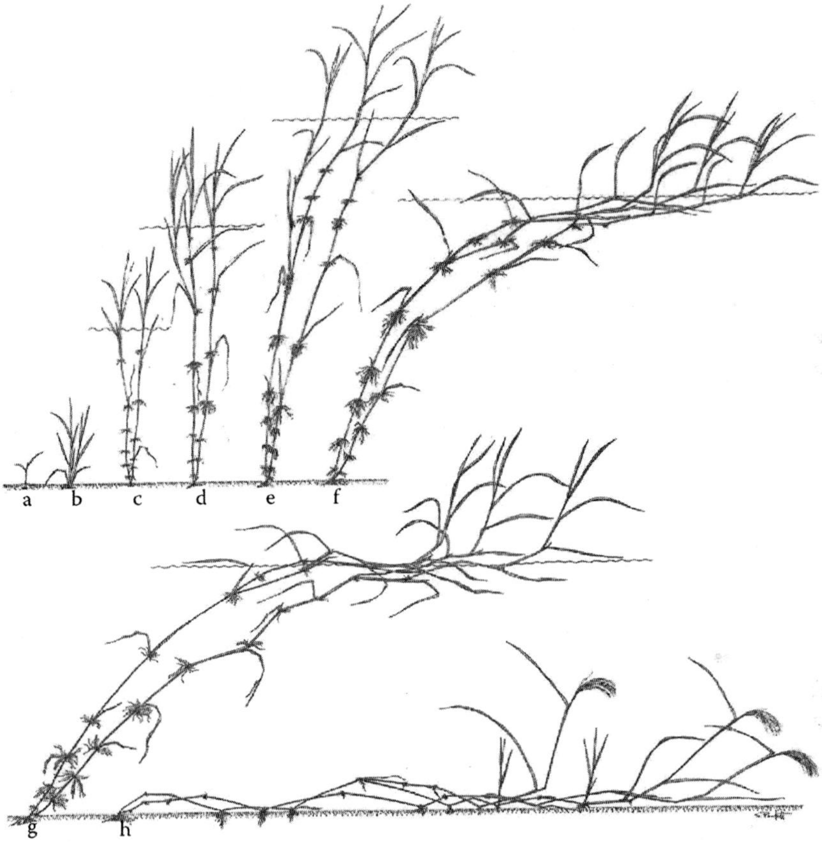

FIGURE 1.1 Floating rice growth stages. (a) Seedling stage, two leaves, plant length about 20 cm; (b) basal tillering stage, preflood period; (c) early elongation stage, nodal tiller emerging, water depth about 75 cm; (d) elongation stage, main culm, basal tiller and nodal tiller, length about 175 cm, water depth about 125 cm; (e) full elongation stage, zigzagging of culms begins near water surface, length about 250 cm, water depth about 175 cm (maximum for season); (f) kneeing begins, prepanicle initiation stage, floating canopy, fully developed nodal roots, culm thickness about 4.5 mm below major site of kneeing, water subsiding, water depth about 135 cm; (g) booting stage, length about 350 cm, water depth about 120 cm; (h) milk stage, all culms prostrate except the terminal sections holding up the ripening panicles above the mass of culms, nodal roots grow down into the soil at some places, and late growth flushes of tillers, which are mostly nonproductive, may develop, and water almost completely receded leaving puddles on heavy soils, length about 350 cm. (Adapted from Catling, H. D., and S. Parfitt. 1981. *An Illustrated Description of a Traditional Deepwater Rice Variety of Bangladesh*. IRRI Research Paper Series Number 60. Los Banõs, Philippines: IRRI.)

in the Philippines, the soils used for wetland rice production are mainly Entisols, Inceptisols, Alfisols, and Vertisols. In Europe, rice is planted on limited areas in Albania, Bulgaria, France, Greece, Hungary, Italy, Portugal, Romania, Spain, and Yugoslavia, where the predominant soil orders are Inceptisols, Entisols, and Vertisols (Matsuo, Pecrot, and Riquier 1978). In the United States, rice is grown

TABLE 1.2
Soil Orders for Rice Cultivation and Their Characteristics

Soil Order	Characteristics
Alfisols	Moderately leached soils with a subsurface zone of clay accumulation and >35% base saturation. These soils result from weathering processes that leach clay minerals and other constituents out of the surface layer and into the subsoil, where they can hold and supply moisture and nutrients to plants. They form primarily under forest or mixed vegetative cover and are productive for most crops. Alfisols make up about 10% of the world's ice-free land surface.
Andisols	Soils formed in volcanic ash. Andisols form from weathering processes that generate minerals with little orderly crystalline structure. These minerals can result in an unusually high water- and nutrient-holding capacities. As a group, Andisols tend to be highly productive soils. They include weakly weathered soils with much volcanic glass as well as more strongly weathered soils. They are commonly found in cool areas with moderate to high precipitation, especially those areas associated with volcanic materials. Andisols make up about 1% of the world's ice-free land surface.
Aridisols	Soils of arid environments with subsurface horizon development. Aridisols are too dry for the growth of mesophytic plants. The lack of moisture greatly restricts the intensity of weathering processes and limits most soil development processes to the upper part of the soils. Aridisols often accumulate gypsum, salt, calcium carbonate, and other materials that are easily leached from soils in more humid environments. Aridisols make up about 12% of the world's ice-free land surface.
Entisols	Soils with little or no morphological development. Entisols show little or no evidence of pedogenic horizon development. Entisols occur in areas of recently deposited parent materials or in areas where erosion or deposition rates are faster than the rate of soil development—such as dunes, steep slopes, and flood plains. They occur in many environments and make up about 16% of the world's ice-free land surface.
Gelisols	Soils with permafrost within two meters of the surface. Gelisols have permafrost near the surface and/or have evidence of cryoturbation (frost churning) and/or ice segregation. Gelisols make up about 9% of the world's ice-free land surface.
Histosols	Organic soils. Histosols have a high content of organic matter and no permafrost. Most are saturated year round, but a few are freely drained. Histosols are commonly called bogs, moors, peats, or mucks. Histosols form in decomposed plant remains that accumulate in water, forest litter, or moss, faster than they decay. If these soils are drained and exposed to air, microbial decomposition is accelerated, and the soils may subside dramatically. Histosols make up about 1% of the world's ice-free land surface.
Inceptisols	Soils with weakly developed subsurface horizons. Inceptisols are found in semiarid to humid environments that generally exhibit only moderate degrees of soil weathering and development. Inceptisols have a wide range of characteristics and occur in a variety of climates. Inceptisols make up about 17% of the world's ice-free land surface.
Mollisols	Grassland soils with high base saturation. Mossisols have a dark colored surface horizon relatively high in organic matter content. These soils are base rich throughout and, therefore, are quite fertile. Mollisols characteristically form under grass in climates that have a moderate to pronounced seasonal moisture deficit. They are extensive on the steppes of Europe, Asia, North America, and South America. Mollisols make up about 7% of the world's ice-free land surface.

TABLE 1.2 (*Continued*)
Soil Orders for Rice Cultivation and Their Characteristics

Soil Order	Characteristics
Oxisols	Intensely weathered soils of tropical and subtropical environments. They are dominated by low activity minerals, such as quartz, kaolinite, and iron oxides, and tend to have indistinct horizons. Oxisols characteristically occur on land surfaces that have been stable for a long time. They have low natural fertility as well as a low capacity to retain additions of lime and fertilizer. Oxisols make up about 8% of the world's ice-free land surface.
Spodosols	Acid forest soils with a subsurface accumulation of metal–humus complexes. Spodosols form from weathering processes that strip organic matter combined with aluminum (with or without iron) from the surface layer and deposit them in the subsoil. In undisturbed areas, a gray eluvial horizon that has the color of uncoated quartz overlies a reddish brown or black subsoil. Spodosols commonly occur in areas of coarse-textured deposits under coniferous forests of humid regions. They tend to be acidic and infertile. Spodosols make up about 4% of the world's ice-free land surface.
Ultisols	Strongly leached soils with a subsurface zone of clay accumulation and <35% base saturation. These soils are found in humid areas and form from fairly intense weathering and leaching processes that result in a clay-enriched subsoil dominated by minerals, such as quartz, kaolinite, and iron oxides. Ultisols make up about 8% of the world's ice-free land surface.
Vertisols	Clayey soils with high shrink/swell capacity. These soils have a high content of expanding clay minerals. They undergo pronounced changes in volume with changes in moisture and have cracks that open and close periodically, showing evidence of soil movement in the profile. Because they swell when wet, Vertisols transmit water very slowly and have undergone little leaching. They tend to be fairly high in natural fertility. Vertisols make up about 2% of the world's ice-free land surface.

Source: Compiled from USDA (United States Department of Agriculture) Natural Resource Conservation Service. 2011. The twelve orders of soil taxonomy. Available at http://soils.usda.gov/technical/soil_orders/. Accessed December 28, 2011.

primarily on Alfisols, Inceptisols, Mollisols, and Vertisols (Flach and Slusher 1978). However, in Florida, a small hectarage of rice is produced on Histosols. Most of the soils used for rice production in the United States (and some other geographic areas) have properties that make them ideally suited for flood-irrigated rice. The soils are relatively young, contain significant amounts of weatherable minerals, and have relatively high base saturations despite the fact that some of these soils are in areas of high precipitation (Flach and Slusher 1978).

Soil parameters for favorable rice yields are optimum soil depth, compact subsoil horizon, good soil moisture retention, good internal drainage, good fertility, and a favorable soil structure. Clayey to loamy clay-textured soils are appropriate for lowland rice production. Permeable, coarse-textured soils are less suitable for flood-irrigated rice production because they have low water- or nutrient-holding capacity.

In Brazil, there are about 35 million ha of poorly drained soils; known locally as "Varzea," they are distributed throughout the country. At present, about two million ha of Varzea are cultivated, primarily with lowland rice, during the rainy season. Generally, Varzea soils have good initial soil fertility, but after two to three years of cultivation, the fertility level is known to decline (Fageria and Baligar 1996). Farming systems need to be developed with improved soil management technologies to bring these areas under successful crop production. A sufficient supply of nutrients is one of the key factors required to improve crop yields and maintain sustainable agricultural production on Varzea soils. Flood-irrigated rice is an important crop that needs to be included in the cropping system of these poorly drained areas during the rainy seasons. During dry periods, other crops can be planted in rotation, provided there is proper drainage. These soils generally have an adequate natural water supply throughout the year, but they are acidic and require routine applications of lime, if legumes are grown in rotation with rice. In the future, these soils may make up the world's largest land area of lowland rice production. The chemical properties of lowland soils from eight Brazilian states where irrigated or flooded rice is planted on Varzea are listed in Table 1.3. These soils are acidic and have medium natural soil fertility. Figure 1.2 shows irrigated or flooded rice growing on Brazilian Inceptisol.

1.3.2 Upland Rice

Upland rice is mostly grown in South America, Africa, and Asia. The soils used for upland rice production in South America and Asia are Oxisols and Ultisols. These soils are acidic and have natural low fertility. With sufficient nutrients and soil

TABLE 1.3
Lowland Soil Chemical Properties, Brazil

State	O.M. (g kg^{-1})	pH in H_2O	P (mg kg^{-1})	K (mg kg^{-1})	Ca ($cmol_c$ kg^{-1})	Mg ($comi_c$ kg^{-1})	Al ($cmol_c$ kg^{-1})
Goiás	42	5.2	15.2	85	4.7	2.6	1.5
Mato Grosso	16	5.1	6.9	68	2.5	1.4	1.3
Mato Grosso do Sul	69	5.3	21.7	75	7.8	3.4	1.1
Paraná	138	4.3	36.4	84	2.6	1.8	4.4
Minas Gerais	25	5.0	17.7	133	3.9	1.6	0.5
Rio Grande do Norte	25	7.1	45.1	168	10.4	6.6	0.1
Piauí	10	5.6	13.6	114.8	10.3	6.7	0.7
Maranhão	8	4.8	1.9	82	6.7	10.7	1.5
Average	42	5.3	19.8	101	6.1	4.4	1.5

Source: Adapted from Fageria, Stone and Santos 2003. Nutrient management for improving lowland rice productivity and sustainability. *Adv. Agron.* 80:63–152.

Note: 0–20 cm soil depth.

FIGURE 1.2 Growth of flooded rice on a Brazilian Inceptisol.

moisture, these soils can produce adequate yields of rice (Fageria 2001). Brazil is the largest upland rice-producing country in the world, and other countries with sizable upland rice production are Colombia, Venezuela, Costa Rica, Panama, Mexico, Bolivia, and Ecuador. In Brazil, upland rice is grown on about two million ha each year (Breseghello et al. 2011); this system is regarded as having a high potential for expansion as the international demand for rice is increasing and water is becoming less readily available (Pinheiro, Castro, and Guimares 2006). Upland rice is generally planted in the central part in Brazil in an area locally known as the Cerrado region. In Cerrado soils, nutrient deficiency is one of the most important yield-limiting factors for crop production. Figure 1.3 shows the response of upland rice to macro- and micronutrients. Because of a deficiency of most macro- and micronutrients in this region, the relative dry weight of upland rice was reduced. In the Cerrado area, the Brazilian state of Mato Grosso is one of the principal regions for upland rice production along with other crops such as corn and soybeans.

A survey was conducted by Fageria and Breseghelo (2004) to evaluate soil fertility and the nutritional status of upland rice plants in 43 sites, covering 33 rural properties, in three municipalities of the Chapada dos Parecis region in the Brazilian state of Mato Grosso (Tables 1.4 and 1.5). Upland rice grain yield was also determined where soil and plant samples were taken for nutritional evaluation. The grain yield of upland rice varied from 975 to 7853 kg ha^{-1} with an average value of 3595 kg ha^{-1}. Soil analysis showed an average value of pH 5.5, O. M. 21.5 g dm^{-3}, P 4.1 mg dm^{-3}, K 31.6 mg dm^{-3}, Ca 17.7 mmol dm^{-3}, Mg 8.4 mmol dm^{-3}, and Al 1.8 mmol dm^{-3}. Average micronutrient

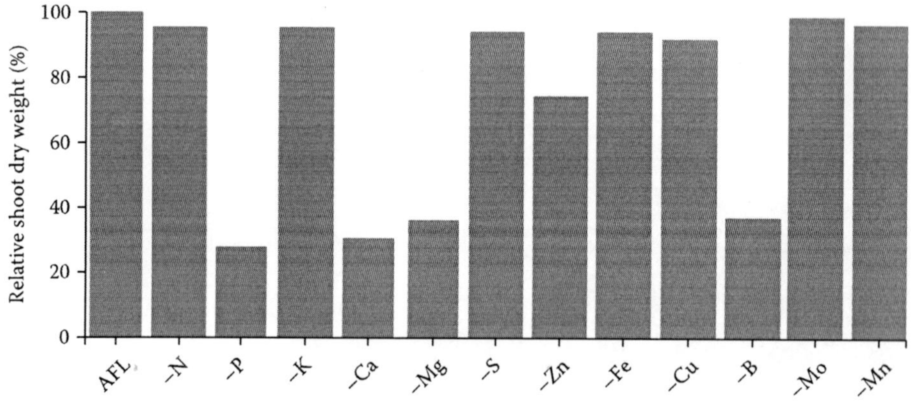

FIGURE 1.3 Relative shoot dry weight of upland rice grown on a Brazilian Oxisol under different nutrient treatments. AFL = adequate fertility level. Minus sign in front of each nutrient indicates that this nutrient was omitted from adequate fertility treatment. (Adapted from Fageria, N. K., and V. C. Baligar. 1997. Response of common bean, upland rice, corn, wheat, and soybean to soil fertility of an Oxisol. *J. Plant Nutr.* 20:1279–1289.)

levels in the soil were 0.51 mg dm^{-3} of Cu, 2.4 mg dm^{-3} of Zn, 61.6 mg dm^{-3} of Fe, and 4.79 mg dm^{-3} of Mn. Similarly, the average base saturation was 28%, Al saturation was 8%, Ca saturation was 18%, Mg saturation was 9%, K saturation was 0.82%, Ca/K ratio was 24, Ca/Mg ratio was 2.23, and Mg/K ratio was 12. Plant analysis showed average values of N 30.3 g kg^{-1}, P 1.68 g kg^{-1}, K 19 g kg^{-1}, Ca 8.7 g kg^{-1}, Mg 2.66 g kg^{-1}, Zn 22 mg kg^{-1}, Cu 8 mg kg^{-1}, Mn 120 mg kg^{-1}, and Fe 106 mg kg^{-1} (Fageria and Bresseghello 2004). Based on soil fertility analysis, there were low levels of P, base saturation, Ca saturation, Mg saturation, and K saturation and deficiencies of Cu, Zn, and Mn in some samples. Plant analysis showed a deficiency of P, followed by deficiencies of K and N.

Figure 1.4 shows upland rice growth on Brazilian Oxisol. A comparison of upland and lowland rice cultures is given in Table 1.6.

1.4 CLIMATIC CONDITIONS

Climatic conditions such as temperature, solar radiation, and soil moisture content are important factors in determining crop productivity. Climatic factors directly affect crop plant growth and development through impacting physiological and biochemical processes, and they indirectly affect plant growth through diseases and insect infestation. Climatic factors such as temperature, solar radiation, and water availability determine growth stage duration in rice. There are optimum temperature, solar radiation, and soil moisture contents for each crop and also for different growth stages. Productivity decreases at lower as well as at higher values of these climatic factors. However, there is always a range of optimum climatic conditions or factors. Some factors, such as temperature and soil moisture content, can be controlled to some extent to ensure higher yields. However, factors such as solar radiation cannot be controlled under field conditions.

TABLE 1.4
Grain Yield of Upland Rice and Soil Chemical Properties of an Oxisol

Soil Sample	Grain Yield ($kg\ ha^{-1}$)	pH in H_2O	O. M. ($g\ dm^{-3}$)	P ($mg\ dm^{-3}$)	K	Ca ($mmol\ dm^{-3}$)	Mg	Al	Cu ($mg\ dm^{-3}$)	Zn	Fe	Mn
1	2660	5.5	21	2.8	56	18.9	6.6	1	0.4	1.3	50	4
2	2180	5.6	14	0.9	16	10.8	8.3	2	0.2	0.4	99	4
3	2360	5.5	16	1.4	14	6.3	4.6	4	0.2	0.4	131	5
4	4472	5.6	26	14.8	28	27.0	6.9	1	0.8	5.6	47	5
5	5355	5.6	26	5.1	34	23.4	6.2	2	0.4	2.5	47	4
6	4000	5.7	20	5.9	36	24.3	7.5	1	0.4	3.6	44	4
7	4806	5.3	20	4.9	37	13.5	4.6	2	0.7	3.6	55	4
8	3981	5.7	20	4.0	32	25.2	13.4	1	0.4	1.8	57	7
9	1730	5.8	20	8.3	50	25.2	8.2	1	0.9	6.3	52	7
10	3287	5.2	28	5.8	73	15.3	5.5	1	0.2	1.4	56	13
11	1216	5.5	18	0.6	17	9.0	6.6	2	0.4	0.3	74	2
12	2013	5.2	15	3.5	37	9.9	5.1	4	0.3	1.4	53	7
13	2544	5.7	20	11.5	28	24.3	10.4	1	0.5	4.1	50	4
14	4824	5.5	22	2.6	34	15.3	8.7	3	0.6	1.5	76	8
15	4529	6.3	24	1.9	20	32.4	24.2	2	0.4	1.3	30	6
16	7853	5.6	22	1.2	20	13.5	9.7	0	1.0	8.7	121	6
17	3386	5.4	22	3.9	23	19.8	7.3	2	0.6	2.6	52	3
18	4330	5.4	20	4.9	25	19.8	8.7	2	0.5	2.7	49	3
19	2750	5.7	22	1.2	34	20.7	8.7	2	0.7	1.5	67	11
20	1477	5.6	23	4.1	51	17.1	8.0	1	0.6	3.3	49	3

(Continued)

TABLE 1.4 (*Continued*)
Grain Yield of Upland Rice and Soil Chemical Properties of an Oxisol

Soil Sample	Grain Yield (kg ha^{-1})	pH in H_2O	O. M. (g dm^{-3})	P	K	Ca	Mg	Al	Cu	Zn	Fe	Mn
				(mg dm^{-3})		(mmol dm^{-3})				(mg dm^{-3})		
21	4302	5.4	20	1.8	30	18.9	8.3	1	0.7	3.0	55	4
22	2793	4.8	23	5.5	33	8.1	2.2	1	0.5	2.4	50	5
23	2510	5.5	17	2.3	25	18.9	10.7	1	1.0	3.0	54	4
24	2546	5.2	24	3.1	33	17.1	7.6	4	0.8	3.5	72	6
25	2200	5.9	24	4.8	33	23.4	10.4	1	0.7	3.4	41	4
26	2895	5.5	23	6.5	16	17.1	5.3	2	0.6	5.7	59	6
27	3074	5.9	16	7.0	28	26.1	9.0	4	0.6	4.5	45	5
28	4608	5.6	15	2.6	33	22.5	11.2	1	0.6	1.3	99	6
29	3882	5.4	19	7.4	33	15.3	6.0	2	0.7	3.7	46	4
30	7471	5.7	16	1.8	25	20.7	12.7	1	0.3	1.3	52	4
31	4237	5.5	20	5.9	25	20.7	7.2	1	0.3	2.1	46	4
32	3850	5.2	20	7.9	30	15.3	5.8	3	0.5	2.2	54	4
33	3550	5.3	12	3.0	12	5.4	4.0	3	0.2	0.8	88	4

34	3410	5.3	26	5.4	55	17.1	7.9	2	0.3	1.4	60	4
35	4960	5.8	31	2.6	66	21.6	17.0	1	0.3	0.7	57	7
36	2700	5.2	27	0.6	20	6.3	5.4	1	0.2	0.3	80	2
37	2640	5.2	23	2.5	27	12.8	5.5	4	0.5	1.4	63	4
38	2978	5.0	30	0.9	14	7.2	4.7	2	0.3	0.4	71	3
39	975	5.4	8	5.4	20	9.0	5.8	2	0.2	0.4	99	2
40	5912	5.3	27	2.6	25	19.8	7.2	2	0.7	1.6	46	3
41	5735	5.9	30	4.9	51	34.2	16.8	0	0.6	1.9	36	4
42	4272	5.3	29	0.9	33	8.1	6.5	4	0.4	0.8	73	4
43	3346	5.8	27	2.6	25	22.5	13.7	1	0.7	3.1	44	3
Minimum	975	4.8	8	0.6	12	5.4	2.2	0	0.2	0.3	30	2
Maximum	7853	6.3	31	14.8	73	34.2	24.2	4	1.0	8.7	131	13
Mean	3595	5.5	21.5	4.12	31.6	17.7	8.4	1.8	0.51	2.40	61.60	4.79
SD	1490	0.28	5.1	2.93	13.7	7.0	4.0	1.1	0.22	181	21.70	2.17

Source: Fageria, N. K., and F. Breseghello. 2004. Nutritional diagnostic in upland rice production in some municipalities of State of Mato Grosso, Brazil. *J. Plant Nutr.* 27:15–28.

TABLE 1.5
Soil Fertility Analysis

Soil Sample	Base Sat. (%)	Al Sat. (%)	Ca Sat. (%)	Mg Sat. (%)	K Sat. (%)	Ratio (Ca/K)	Ratio (Ca/Mg)	Ratio (Mg/K)
1	31	4	22	8	1.64	13	2.86	5
2	24	9	14	10	0.51	26	1.30	20
3	16	26	9	6	0.50	18	1.37	13
4	29	3	23	6	0.60	38	3.91	10
5	27	6	21	6	0.78	27	3.77	7
6	37	3	27	8	1.03	26	3.24	8
7	22	10	16	5	1.11	14	2.93	5
8	39	2	25	13	0.81	31	1.88	16
9	39	3	28	9	1.45	20	3.07	6
10	17	4	11	4	1.40	8	2.78	3
11	19	11	11	8	0.51	21	1.36	15
12	19	20	12	6	1.15	10	1.94	5
13	36	3	25	11	0.73	34	2.34	14
14	26	11	16	9	0.91	18	1.76	10
15	60	3	34	25	0.53	63	1.34	47
16	25	0	14	10	0.53	26	1.39	19
17	28	7	20	7	0.60	34	2.71	12
18	30	6	20	9	0.66	31	2.28	14
19	32	6	22	9	0.93	24	2.38	10
20	27	4	17	8	1.32	13	2.14	6
21	33	3	22	10	0.90	25	2.28	11
22	11	8	8	2	0.85	10	3.68	3
23	35	3	22	12	0.75	29	1.77	17
24	20	14	13	6	0.64	20	2.25	9
25	34	3	23	10	0.84	28	2.25	12
26	22	8	18	5	0.40	42	3.23	13
27	47	10	34	12	0.94	36	2.90	13
28	39	3	26	13	0.96	27	2.01	13
29	27	8	18	7	1.01	18	2.55	7
30	38	3	23	14	0.71	32	1.63	20
31	30	3	21	7	0.66	32	2.88	11
32	22	12	15	6	0.76	20	2.64	8
33	16	24	9	7	0.52	18	1.35	13
34	22	7	14	7	1.17	12	2.16	6
35	31	2	17	13	1.30	13	1.27	10
36	11	8	6	5	0.48	12	1.17	11
37	16	18	11	5	0.59	18	2.29	8
38	9	14	5	3	0.25	20	1.53	13
39	21	12	13	8	0.72	18	1.55	11
40	23	7	16	6	0.53	31	2.75	11
41	50	0	33	16	1.25	26	2.04	13

TABLE 1.5 (*Continued*)
Soil Fertility Analysis

Soil Sample	Base Sat. (%)	Al Sat. (%)	Ca Sat. (%)	Mg Sat. (%)	K Sat. (%)	Ratio (Ca/K)	Ratio (Ca/Mg)	Ratio (Mg/K)
42	12	21	6	5	0.67	10	1.25	8
43	34	3	21	12	0.58	35	1.64	21
Minimum	9	0	5	2	0.25	8	1.17	3
Maximum	60	26	34	25	1.64	63	3.91	47
Mean	28	8	18	9	0.82	24	2.23	12
SD	10.78	6.33	7.36	4.03	0.31	10.62	0.74	7.08

Source: Fageria, N. K., and F. Breseghello. 2004. Nutritional diagnostic in upland rice production in some municipalities of State of Mato Grosso, Brazil. *J. Plant Nutr.* 27:15–28.

FIGURE 1.4 Upland rice growth on a Brazilian Oxisol.

1.4.1 Temperature

Temperature is one of the most important factors affecting the growth of crop plants, including rice. Environmental temperatures affect species distribution, photosynthesis, and many physiological and biochemical processes. In addition, dry matter accumulation and partitioning, and phenological development are also affected by temperature (Jones 1985). Temperature affects the length of the vegetative period,

TABLE 1.6
Comparison of Upland and Lowland Rice Cultures

Lowland	Upland
1. Cultivated on leveled, bunded, undrained soils	Cultivated on undulating or leveled naturally drained soils
2. Water supply through rainfall or irrigation	Water supply through rainfall
3. Water accumulation in the field during major part of crop growth	No water accumulation during crop growth
4. Reduced root zone during major part of crop growth	Oxidized root zone during crop growth
5. Direct seeding or transplanting	Direct seeding
6. Thin and shallow root system	Vigorous and deep root system
7. High tillering	Relatively low tillering
8. Short and thin leaves	Long and thick leaves
9. Environmental conditions are stable and uniform	Environmental conditions are unstable and variable
10. Incidence of diseases and insects low	Incidence of diseases and insects high
11. Weeds are not a serious problem	Weeds are a serious problem
12. Needs high input	Needs low input
13. High cost of production	Low cost of production
14. Stable and high yield	Unstable and low yield

Source: Fageria, N. K., V. C. Baligar, and C. A. Jones. 2011. *Growth and Mineral Nutrition of Field Crops*, 3rd edition. Boca Raton, FL: CRC Press.

thereby influencing the amount of reserves stored in stems, and the length of the grain-filling period, which determines grain yield (Dingkuhn and Kropff 1996). In statistical analyses of environmental factors correlated with growth and yield of crop plants, temperature is often among the most important variables (Jones 1985). Rice originated in tropical or subtropical areas and is a low-temperature-sensitive crop. Krishnan et al. (2011) reported that growth and development of rice is severely damaged below 15°C. Temperature below 15°C causes photo inhibition at high light intensity (Krishnan et al. 2011). Yoshida (1981) reported that the critically low and high temperatures are 20°C and 30°C, respectively. However, Tanaka (1976) and Ishii (1977) noted that temperatures up to 35°C cause no decline in rice net photosynthesis, but temperatures above 39°C do (Murata 1961; Osada 1975; Deng 1984). Leaf temperature can be 3°C higher or lower than air temperature as a result of solar radiation and transpiration (Monteith and Unsworth 1990). Mean annual air temperatures closely approximate mean annual soil temperatures in the tropics, according to Sanchez (1976).

The minimum, maximum, and optimum temperatures for different growth stages of rice are presented in Table 1.7. The growth stage most sensitive to low and high temperatures is about 10 days before and after flowering. At this growth stage, spikelet number and weight are adversely affected. Shimono et al. (2011) reported that spikelet sterility induced by low temperatures during the booting stage (at the young microspore stage, 10–12 days before the heading) has one of

TABLE 1.7
Minimum, Optimum, and High Temperatures for Rice Growth at Different Stages

Growth Stage	Critical Temperature (°C)		
	Low	Optimum	High
Germination	10	20–35	45
Seedling emergence and establishment	12–13	25–30	35
Rooting	16	25–28	35
Leaf elongation	7–12	31	45
Tillering	9–16	25–31	33
Initiation of panicle primordia	15	25–30	35
Panicle differentiation	15–20	25–28	38
Anthesis	22	30–33	35
Ripening	12–18	20–25	30

Source: Compiled from Yoshida, S. 1981. *Fundamentals of Rice Crop Science*. Loa Bãnos, Philippines: IRRI; Fageria, N. K. 1989. *Tropical Soils and Physiological Aspects of Crops*. Brasilia: EMBRAPA.

the strongest impacts on rice yield in areas with cool climate in temperate regions. However, low temperature can also cause problems in tropical areas such as Senegal, Mali, and Brazil (Shimono et al. 2011). Low temperatures disrupt proper pollen development, leading to a shortage of sound pollen at the flowering stage and thereby severe yield losses (Satake 1976). There are differences among rice genotypes in relation to cold tolerance (Shimono et al. 2011).

Shortage of land and water for rice cultivation (Khush 2005), accompanied by an increase in demand, has forced cultivation to extend beyond normal monsoon periods, where temperatures are optimal for growth (Prasad et al. 2006), to warmer summer seasons, where temperature is an important constraint (Jagadish, Craufurd, and Wheeler 2008). Rice crops in Hubei Province in China, in coastal Andhra Pradesh in India, or in the state of Tocantins in the central part of Brazil, for example, may experience daytime temperatures of 50° (Jagadish, Craufurd, and Wheeler 2008). According to climate change models, global warming can adversely affect food production and food security (Lobell et al. 2008). On a decadal timescale, anthropogenic climate change is expected to increase mean surface air temperature by 1.8°C–4.0°C by next decade with an increase in variability around this mean (IPCC 2007). Therefore, rice crops will be cultivated in hotter conditions in current and future climates (Jagadish, Craufurd, and Wheeler 2008). The production of rice may be limited by possible future increases in global temperatures, particularly in rice-growing regions where temperatures for rice cultivation are currently optimum.

Yoshida, Satake, and Mackill (1981) reported that anthesis is the stage when rice is most sensitive to high temperatures. The heat-sensitive processes of anthesis, anther dehiscence, pollination, pollen germination, and—to a lesser extent—pollen tube growth are completed within 45 minutes of the opening of a rice spikelet (Ekanayake, De Datta, and Steponkus 1989), and fertilization is completed within 1.5–4 hours (Cho 1956). A spikelet tissue temperature of ≥33.7°C for an hour at anthesis was sufficient to induce spikelet sterility (Jagadish, Craufurd, and Wheeler 2007). In contrast, spikelet fertility is not affected by high temperatures an hour before or after anthesis, even at 38°C and 41°C (Yoshida, Satake, and Mackill 1981). Rice typically antheses during late morning with peak anthesis occurring between 1000 and 1200 hours. Therefore, it is essential that screening or phenotyping for heat tolerance at anthesis excludes or minimizes the possibility of avoidance or escape by early or late anthesing spikelets (Jagadish, Craufurd, and Wheeler 2008). During flowering, there have been large differences in high-temperature tolerance reported among rice genotypes (Matsui and Omasa 2002). Jagadish, Craufurd, and Wheeler (2008) found that some genotypes of the *indica* and *japonica* species have high-temperature tolerance of up to 41°C and for durations ranging from two hours to the whole crop cycle.

Guimaraes et al. (2010) determined the influence of high leaf temperature on spikelet sterility, grain harvest index (GHI), and grain yield of upland rice genotypes (Figures 1.5 through 1.7). He found that spikelet sterility increased significantly and linearly, whereas GHI and grain yield decreased significantly and linearly, when leaf temperature increased from 30.5°C to 35.5°C. Wassmann et al. (2009b)

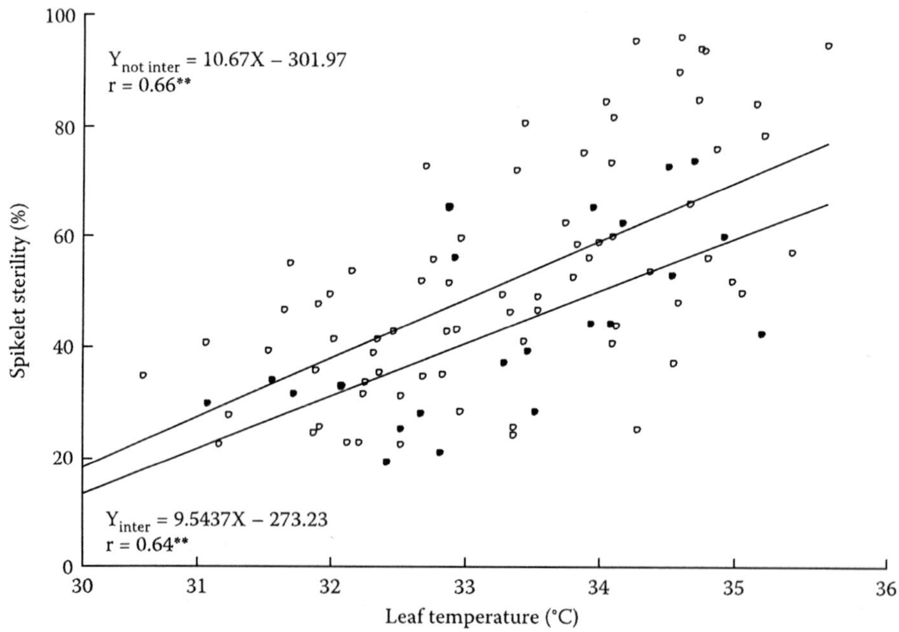

FIGURE 1.5 Relationship between leaf temperature and spikelet sterility in upland rice genotypes. (Adapted from Guimaraes, C. M., L. F. Stone, M. Lorieux, J. P. Oliveira, G. C. O. Alencar, and R. A. A. Dias. 2010. Infrared thermometry for drought phenotyping of inter and intra specific upland rice lines. *Rev. Bras. Eng. Agri. Amb.* 2:148–154.)

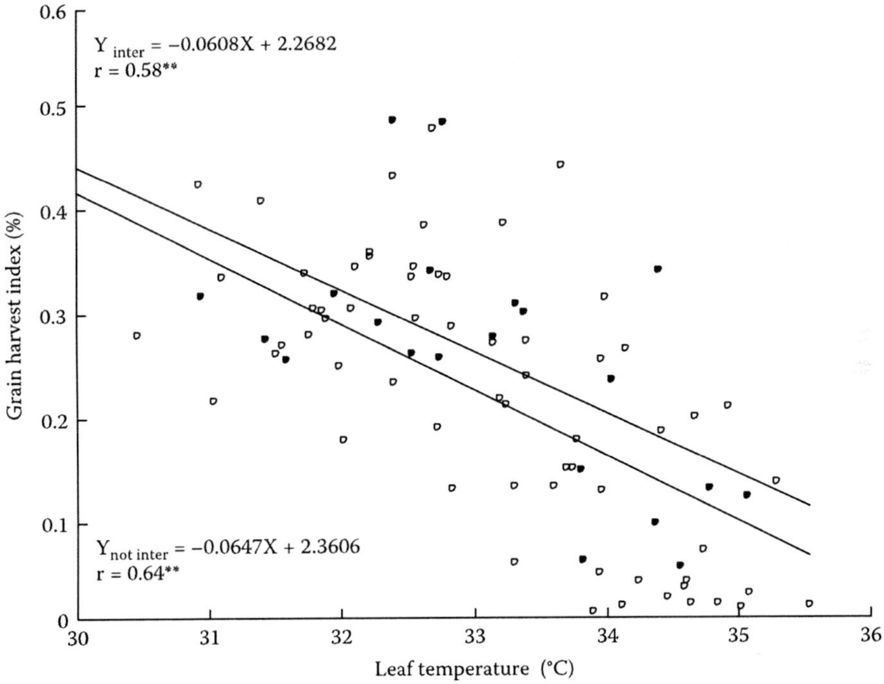

FIGURE 1.6 Relationship between leaf temperature and GHI of upland rice genotypes. (Adapted from Guimaraes, C. M., L. F. Stone, M. Lorieux, J. P. Oliveira, G. C. O. Alencar, and R. A. A. Dias. 2010. Infrared thermometry for drought phenotyping of inter and intra specific upland rice lines. *Rev. Bras. Eng. Agri. Amb.* 2:148–154.)

reported that the growth rate of rice increases linearly in the temperature range of 22°C–31°C, depending on genotype; beyond that range, growth and productivity rapidly decrease. Wassmann et al. (2009b) reviewed the literature and reported that simulation models for rice production indicate a reduction in yield of about 5% per degree rise in mean temperature above 32°C. The critical temperature for spikelet fertility (defined as when fertility exceeds 80%) varies among genotypes but is about 32°C–36°C (Yoshida 1981). Below 20°C and above 32°C, spikelet sterility becomes a major factor, even if growth is sufficient in plant components (Wassmann et al. 2009b). Nishiyama (1995) reviewed literature on temperature effects on rice and reported that optimum air temperature for rice at flowering is about 31°C–32°C, with a minimum of about 25°C. For rice, spikelet sterility increased above and below these values. Satake (1995) reviewed the effects of high temperature on rice and reported that spikelet sterility increased when daily maximum temperatures were higher than 34°C–35°C. However, there was a significant difference among cultivars and spikelet sterility ranging from 10% in a heat-tolerant cultivar to 65% in a heat-susceptible cultivar. Peng et al. (2004) reported that variation in rice grain yield was attributed to changes in nighttime temperatures as a result of possible global warming. Similarly, Mohammed and Tarpley (2009) noted an increase in respiration

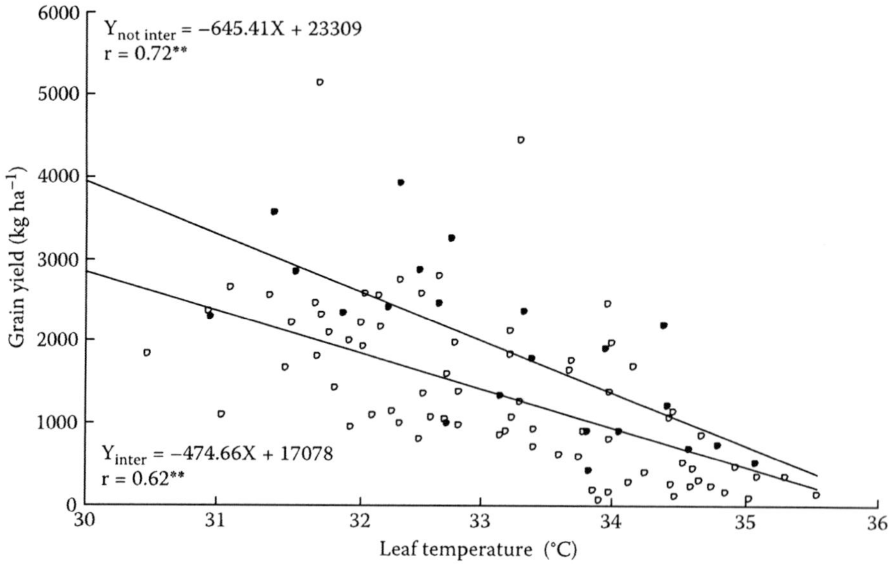

FIGURE 1.7 Relationship between leaf temperature and grain yield of upland rice genotypes. (Adapted from Guimaraes, C. M., L. F. Stone, M. Lorieux, J. P. Oliveira, G. C. O. Alencar, and R. A. A. Dias. 2010. Infrared thermometry for drought phenotyping of inter and intra specific upland rice lines. *Rev. Bras. Eng. Agri. Amb.* 2:148–154.)

rates in rice plants as a result of an increase in night temperature, which was associated with a decrease in yield. High dark respiration increases the proportion of assimilates respired for maintenance and uncoupled respiration (Beevers 1970), thereby affecting plant carbon status (Turnbull, Murthy, and Griffin 2002).

There are also differences among *indica* and *japonica* species of rice in relation to optimum temperature requirements. It is reported that *indica* cultivars are better adapted to high temperatures, whereas *japonica* cultivars require low temperature for better ripening (Yoshida 1981). At lower or higher temperatures than optimum, spikelet sterility (which is responsible for yield) increased or decreased depending on the species type. However, spikelet sterility is also impacted by the duration of lower and high temperatures. Significant effects in spikelet sterility were reported when minimum and maximum temperatures lasted more than six days (Yoshida 1981). One or two days of lower or higher temperatures do not decrease spikelet sterility significantly. Fageria (1984) studied the influence of temperature on upland rice yield and spikelet sterility. Data in Table 1.8 show the effect of minimum temperatures (15°C or 19°C) around flowering (15 days before to 15 days after) for three Brazilian upland rice cultivars in a greenhouse experiment. When the temperature was 15°C for one month around flowering, yield was reduced significantly in all cultivars due to very high grain sterility. Little sterility was observed at 19°C for the same growth stage.

Predictions of global warming effects on crop production vary with current temperatures and the extent of the projected warming (Grant et al. 2011). In temperate

TABLE 1.8
Effect of Temperature on Grain Yield and Spikelet Sterility in Three Brazilian Upland Rice Cultivars

Cultivar	Minimum[a] (°C)	Maximum[a] (°C)	Grain Yield (g pot^{-1})	Grain Sterility (%)
IAC 25	15	30	3.45	96
	19	27	26.00	9
IAC 164	15	30	5.56	97
	19	27	29.54	19
IAC 165	15	30	7.37	97
	19	27	30.73	15

Source: Fageria, N. K. 1984. *Fertilization and Mineral Nutrition of Rice*. Rio de Janeiro: EMBRAPA–CNPAF/Editora Campus.

[a] Minimum and maximum temperature values were observed about one month from flowering.

climates, crop growth may increase due to climate warming. However, in tropical and subtropical regions (where rice is mainly grown), global warming may result in a decrease in crop production (Hatfield et al. 2011). Schlenker and Roberts (2009), Grant et al. (2011), and Hatfield et al. (2011) reported that temperature change affects growth because crops have optimum growth temperatures. When temperatures are extreme (lower or higher than optimum), the physical, physiological, and biochemical processes that are responsible for plant yields are affected. These processes are summarized in Table 1.9. Minimum and maximum temperatures in the Cerrado region of Brazil, where upland rice is mostly grown, are presented in Table 1.10.

1.4.2 Solar Radiation

Light, or solar radiation, is a critical resource because it essentially controls morphogenesis and productivity in crop plants. Colaizzi et al. (2012a) reported that radiation is the primary driver of the energy and water balance of the soil-plant–atmosphere continuum. Radiative transfer in the soil-plant–atmosphere continuum is affected by several interrelated factors, including wavelength (photosynthetically active radiation [PAR] in the 400–700 nm spectrum and near-infrared radiation [NIR] in the 700–3000 nm spectrum] and beam angle, leaf angle distribution, and the spatial distribution of vegetation (Colaizzi et al. 2012b). Its role in photosynthesis is well known. In a sense, crop plants "harvest" solar radiation with their complex biological machinery to produce human food. *Agriculture* could thus be defined as the exploitation of solar radiation with the help of water and nutrients (Venkateswarlu and Visperas 1987). The world's top rice-yielding regions are those that receive the most solar radiation. New South Wales in Australia, Portugal, and the Mediterranean countries of Spain, Morocco, Egypt, Greece, and Italy have long days and sunny skies during the rice-growing season. Northern Japan, the Republic of Korea, and northwestern India also receive high solar radiation because of their higher latitude (Venkateswarlu and Visperas 1987). Modern rice cultivars, although possessing

TABLE 1.9
Effects of Extreme Temperatures on Plants

Process	Effects on Plant Growth and Yield	Reference
Crop water relations	Warming hastens transpiration and soil drying, particularly in warmer climates where vapor pressure deficits increase more with a given rise in temperature. Consequent declines in canopy water potential may induce declines in canopy conductance, CO_2 fixation, and crop growth.	Lavalle et al. 2009; Grant et al. 2011
CO_2 fixation	Warming at lower temperatures, such as those encountered in temperate climates, improves the kinetics of carboxylation and electron transport, thereby increasing CO_2 fixation. Warming in tropical and subtropical regions increases photorespiration relative to carboxylation, thereby slowing CO_2 fixation.	Jordan and Ogren 1984; Bernacchi et al. 2001; Bernacchi, Pimentel, and Long 2003; Grant et al. 2011
Dark respiration	Respiration increases continuously with rising temperature even though CO_2 fixation does not; with warming, increases in respiration increasingly offset those in CO_2 fixation. This offset reduces plant growth rates with warming at higher temperatures, particularly at night.	Aggarwal 2003; Grant et al. 2011
Phenology	Increase in temperature hastens phenological advance, which in annual plants shortens key growth stages, thereby reducing leaf area duration, seasonal CO_2 fixation, and crop growth.	Lavalle et al. 2009; Grant et al. 2011; Hatfield et al. 2011
Grain set	Warming at lower temperatures may reduce the adverse effects of chilling on grain set and raise the grain number and yields. However, warming at higher temperatures can reduce spikelet fertility and reduce grain number and yields. These effects of temperature on grain set occur from a few days before to two weeks after anthesis but can be very large, greatly increasing the sensitivity of crop production to extreme temperatures during this period.	Subedi, Floyd, and Budhathoki 1998; Gibson and Paulsen 1999; Baker 2004; Porter and Semenov 2005; Wheeler et al. 2009; Maqbool, Shafiq, and Lake 2010; Grant et al. 2011

higher yield potentials, have not produced wet season yields equal to those of the dry season when solar radiation is high. There is sufficient evidence that light intensity, particularly low light intensity, acts as a stressor and a determinant of rice productivity in tropical and subtropical climates. Several vital physiological processes, crop growth, and productivity have been examined in Japan, India, the United States, and the Philippines. Lower productivity in tropical and subtropical climates was mainly due to a lower incidence of radiation rather than temperature (Yoshida 1972; Venkateswarlu and Visperas 1987).

TABLE 1.10
Minimum and Maximum Temperatures (°C) in the Cerrado Region, Central Part of Brazil

Month	Year 2010		Year 2011		Year 2012		Average	
	Minimum	Maximum	Minimum	Maximum	Minimum	Maximum	Minimum	Maximum
January	19.6	28.9	19.4	27.5	18.7	26.4	19.2	27.6
February	19.5	30.1	19.1	28.4	18.3	28.2	18.9	28.9
March	19.8	29.0	19.4	27.3	18.8	29.4	19.3	28.6
April	17.7	29.3	18.4	29.2	18.9	29.6	18.3	29.4
May	15.8	29.2	15.9	28.0	16.4	27.2	16.0	28.1
June	14.7	28.8	14.8	27.3	16.7	28.1	15.4	28.0
July	16.2	28.9	15.1	28.6	15.0	28.3	15.4	28.6
August	15.8	30.5	17.2	31.4	16.2	28.4	16.4	30.1
September	19.5	33.1	18.8	32.2	18.9	32.0	19.0	32.4
October	19.3	30.7	18.9	28.0	19.9	32.5	19.4	30.4
November	19.0	27.7	18.6	27.5	19.8	27.6	19.1	27.6
December	19.5	28.5	19.3	27.0	19.7	29.6	19.5	28.4

Note: Data were taken from the record of the meteorological station, National Rice and Bean Research Center, Santo Antônio de Goiás, Brazil.

Solar radiation is also important for photosynthesis in crop plants. Photosynthesis in green leaves uses solar energy in wavelengths from 0.4 to 0.7 μm; this is often referred to as PAR or, simply, light. The ratio of PAR to total solar radiation is close to 0.50 in the tropics and in the temperate regions (Monteith 1972). The average radiation in tropical areas is about twice that of temperate areas (Sanchez 1976). Therefore, according to Sanchez (1976), tropical areas have approximately twice the yield-producing potential per hectare per year in temperate areas, assuming there are no additional limiting factors.

The unit of solar energy is calories per square centimeter per day (cal cm^{-2} day^{-1}). It is sometimes also expressed in kilojoules per square meter per second (kJ m^{-2} s^{-1}). The relationship between these two units is 1 kJ m^{-2} per second = 1.43 cal cm^{-2} per minute (Yoshida 1981). The solar radiation requirements of rice vary according to growth stages. High amounts are needed during reproductive and grain-filling growth stages compared with the vegetative growth stage. The influence of solar radiation on rice yield can be described in the order of vegetative growth stage < grain-filling growth stage < reproductive growth stage. This means that the reproductive growth stage for rice is the most sensitive stage, if solar radiation deficiency occurs. However, the effect of solar radiation is related to grain yield. If grain yield is about 5 Mg ha^{-1}, a solar radiation of 300 cal cm^{-2} per day during the reproductive stage is sufficient. When grain yield is more than 5 Mg ha^{-1}, higher radiation is required. Yoshida (1981) reported that in order to obtain a grain yield of about 7 Mg ha^{-1}, solar radiation of 500–600 cal cm^{-2} per day is required during the reproductive growth stage.

According to Dingkuhn and Kropff (1996), radiation and N resources determine the leaf area index (LAI) and leaf N content and, as a combined result, the crop growth rate (CGR). These authors suggested a simple model for grain yield formation: $Y = S + G \times D$, where Y = grain yield, S = net amount of stem reserves transported to grain, G = the average CGR during the grain-filling period, and D = the length of the grain-filling period. Therefore, it can be concluded that the increase in grain yield of rice was caused by the accumulation and allocation of stem reserves, a prolonged grain-filling period, or an increased CGR during grain filling.

1.4.3 Water Requirements

Water is the main climatic factor that determines yield in rice-growing regions. High crop yield is expected when the soil moisture regime is maintained at an adequate level during the whole crop growth cycle. Water is essential for all physiological and biochemical processes in the plants. It also controls physical, chemical, and biological processes in the soil. The physiological importance of water is reflected in its ecological importance; plant distribution on the earth's surface is controlled by the availability of water wherever temperature permits growth (Kramer 1969). The role of water in crop production is so vast that it is considered one of the most important components of world food security. There is mounting worldwide concern about diminishing water supplies and the need for water conservation to overcome impending deficiencies of food and fiber. Rice consumes about 50% of the total irrigation water used in Asia; it accounts for 24–30% of the withdrawal of total freshwater and 34–43% of the world's irrigation water (Bouman et al. 2007; Kumar and Ladha 2011).

The source of water for plant growth may either fall as natural precipitation or be applied as irrigation. Water requirements of rice vary with soil type, crop management practices, cultivars planted, temperature regime, and solar radiation. Yoshida (1981) reviewed water requirements of irrigated rice in 43 locations in China, Japan, Korea, the Philippines, Vietnam, Thailand, and Bangladesh. Water loss was 1.5–9.8 mm day^{-1} in transpiration, 1.0–6.2 mm day^{-1} in evaporation, and 0.2–15.6 mm day^{-1} in percolation. Total water loss was 5.6–20.4 mm day^{-1}. Yoshida (1981) further reported that on average, some 180–300 mm of water per month is needed to produce a reasonably good crop of rice. In four months, an irrigated rice cultivar will consume about 720–1200 mm of water. The average water requirement of irrigated rice at various locations in Southeast Asia was reported to be about 1240 mm per crop (Kung 1971). Consequently, rice cultivation appears to be limited to areas where rainfall exceeds 1000 mm during the growing season or where other sources of water are available (Yoshida 1983).

Water requirement data differ significantly from region to region and among types of cultivation. Tuong and Bouman (2003) estimated seasonal water input for typical puddled transplanted rice to vary from 660 to 5280 mm depending on growing seasons, climatic conditions, soil types, and hydrological conditions, with 1000–2000 mm as a typical value in most cases. Tripathi (1990) studied seasonal water input in lowland rice in India, which ranged from 1566 mm in a clay loam soil to 2262 mm in a sandy loam soil, with variations due primarily to deep percolation losses. Gupta et al. (2002) estimated water use for lowland rice in the Indo-Gangetic Plains, which varied from 1144 mm in the state of Bihar to 1560 mm in the state of Haryana, India. In the Philippines, water use has been reported at 1300–1500 mm for the dry season and 1400–1900 mm for the wet season for lowland rice (Bouman et al. 2005). In the major rice-growing countries, per capita water availability decreased by 34–76% between 1950 and 2005 and is likely to decline by 18–88% by 2050 (Kumar and Ladha 2011). Hence, improving water use efficiency is fundamental for sustainable production in most rice-growing regions.

The water requirements of upland rice are lower than those of flooded or irrigated rice. Kung (1971) reported that upland rice consumes 380–880 mm water or 2.9–6.3 mm day^{-1}. He further reported that other upland crops, such as soybeans, consume 300–350 mm or 2.3–3.5 mm day^{-1}; corn consumes 350–400 mm or 2.9–3.5 mm day^{-1}; and peanuts consume 400–500 mm or 2.7–3.5 mm day^{-1}. Water consumption by rice varied with age. Fageria (1980) studied the water evapotranspiration and consumption of upland rice (Figure 1.8). The evaporation of upland rice increases with increasing plant age, with a maximum in the reproductive stage, and is then followed by a decrease during the ripening stage.

The response of rice to water stress at the vegetative stage has been reported primarily in terms of reduced height, tillers, and leaf area (IRRI 1975), while, at more sensitive reproductive stages, such as flowering, high spikelet sterility results in the greatest reduction in grain yield (Fageria 1980). The most critical water deficit stage

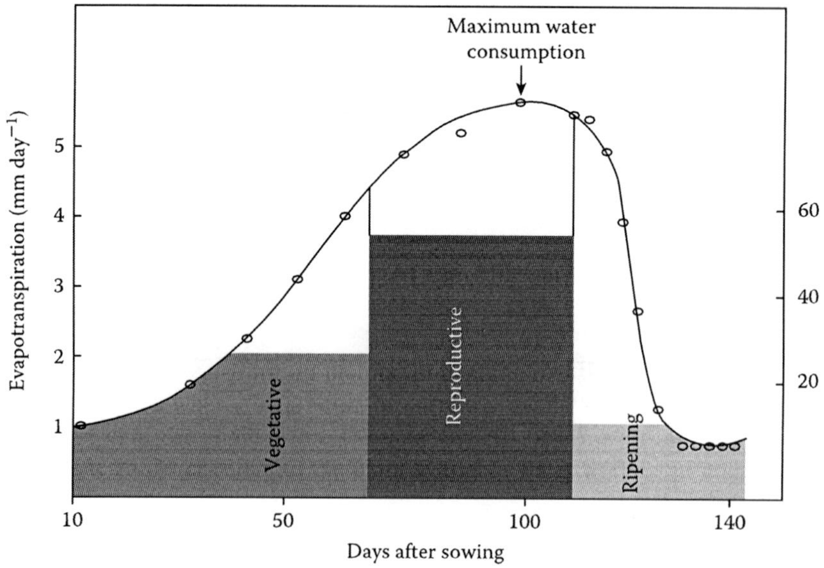

FIGURE 1.8 Evaporation and water consumption during growing season of upland rice in central Brazil. (From Fageria, N. K. 1980. Upland rice response to phosphate fertilization as affected by water deficiency in Cerrado soils. *Pesq. Agropec. Bras.* 15:259–265.)

is the 10-day period leading up to flowering (Matsushima 1962). During this time period, high sterility, brought on by water stress, is irreversible, and adequate water at later stages is totally ineffective in reducing this outcome. However, if water stress occurs during the early vegetative stage plants can recover.

Uneven distribution of rainfall (or drought) during upland rice growth is one of the world's most yield-limiting factors. In general, *drought* can be defined as a temporary condition where the amount of currently available water due to precipitation falls short of a threshold value (Woli et al. 2012). From an agricultural point of view, the term *water* refers to the plant available water in the soil and the threshold value is the atmospheric demand for evaporation. An agricultural drought occurs when the amount of plant available water in the soil resulting from precipitation is not enough to meet the atmospheric demand for evapotranspiration (Woli et al. 2012). Drought is typical in Brazil's Cerrado region where upland rice receives about 1500 mm of rainfall from October to March (Table 1.11). However, due to the uneven distribution of rainfall, a one- to two-week drought period is very common and sometimes reduces rice yield significantly (Fageria, Baligar, and Jones 2011). In experiments in the Philippines, Jana and De Datta (1971) showed that water deficits reduced yields even when annual rainfall was more than 2000 mm. In Asia, approximately 34 million ha of shallow rainfed lowland rice and 8 million ha of upland rice, totaling approximately one-third of the total Asian rice area, are subject to occasional or frequent drought stress (Venuprasad, Lafitte, and Atlin 2007). In the eastern Indian states of Jarkhand, Orissa, and Chhattisgarh alone, total rice production

TABLE 1.11
Precipitation in the Cerrado Region of Brazil

Month	Year 2008	Year 2009	Year 2010	Year 2011	Year 2012	Average
January	347.8	221.4	129.4	208.4	338.0	1245.0
February	292.3	161.9	142.2	260.0	310.8	1167.2
March	264.2	193.2	172.8	339.2	163.0	1132.4
April	248.0	160.2	80.4	49.0	54.0	591.6
May	77.8	12.8	0.2	0.8	15.6	107.2
June	0.0	27.6	3.6	17.8	13.2	62.2
July	0.0	0.0	0.2	0.4	5.2	5.8
August	0.0	52.2	0.0	0.0	0.0	52.2
September	51.1	177.4	25.5	0.4	109.6	364.0
October	122.0	292.8	289.2	290.6	51.6	1046.2
November	219.8	290.8	279.0	195.2	183.4	1168.2
December	230.4	293.2	252.0	200.4	212.6	1188.6
Total	1853.4	1858.5	1374.5	1562.2	1457.0	1621.1

Note: Data were taken from the record of the meteorological station, National Rice and Bean Research Center, Santo Antônio de Goiás, Brazil.

losses during severe droughts (about one every five years) averaged 40%, valued at US$650 million (Venuprasad, Lafitte, and Atlin 2007).

There are relatively few stress-free areas where crops may approach their potential yields. Li, Chen, and Wu (2011) reported that 25% of the world's agricultural land is now affected by drought stress. However, Bot, Nachtergaele, and Young (2000) found that up to 45% of the world's agricultural lands—where 38% of the world's population resides—are subject to continuous or frequent drought. At present, about 18% of global farmland is irrigated (more than 240 million ha); some 40% of the world's food supply is produced on this land (Somerville and Briscoe 2001). Drought is one of the most devastating environmental stressors affecting upland rice yield productivity throughout the world. Areas designated as "drought prone," in the classification of rainfed rice environments (Garrity et al. 1986), are found throughout Asia, Africa, and Latin America. Assessments of rice-breeding priorities target drought tolerance (Garrity and O'Toole 1994) and have found that the most effective method of minimizing the adverse effects of drought is to grow crops during periods of high rainfall and of high soil water availability in order to escape the drought period. Crop duration is important in determining grain yield because early maturing cultivars often escape a terminal stress although late maturing cultivars may be affected by it.

During the growing season, the timing of drought is also important. Rice is particularly sensitive to drought stress during periods of reproductive growth, when even moderate stress can result in drastic reductions in grain yield (Hsiao 1982; O'Toole 1982). It is well known that the period from panicle development to anthesis is the growth stage most susceptible to water stress in rice (O'Toole 1982). Boonjung (1993) has shown that grain yield decreases at the rate of 2% per day when a 15-day stress period (with morning leaf water potential less than −1.0 MPa) occurs later during panicle development. Assuming a reduction of 2% grain yield per day, with the delay in termination of a 15-day stress, a 20-day difference in flowering time between two cultivars of equal yield potential could cause a grain yield difference of about 40%. Cultivars with different phonological development are likely to react differently to a drought, depending on the timing of stress (Maurya and O'Toole 1986). These results suggest that genotypes within the same maturity group should be compared when evaluating germplasm for drought resistance/susceptibility (Garrity and O'Toole 1994). Alternatively, it is possible in some experiments to vary planting dates so that all lines flower at about the same time (Lilley and Fukai 1994).

Drought is the major cause of yield loss in rainfed rice. Therefore, the development of rice cultivars with improved drought tolerance is an important strategy in reducing risk, increasing productivity, and alleviating poverty in areas where the population depends on rainfed production (Venuprasad, Lafitte, and Atlin 2007). Vadell, Cabot, and Medrano (1995) have reported that plant responses to soil drying involve several modifications of morphological and physiological parameters (i.e., a variety of adaptive mechanisms) such as a shift in the allocation of dry matter from shoots to roots (Kramer 1988) and a reduction in leaf expansion (Karamanos, Elston, and Wadsworth 1982), leaflet movements (Satter and Galston 1981), stomata closure, and osmotic adjustment (Morgan 1984). From an agronomic point of view, all these drought-adaptive responses of plants cannot be evaluated without assessing their impact on crop productivity (Karamanos and Travlos 2012).

Several other drought-resistance mechanisms and traits have been identified for rice. For upland conditions, a deep root system with high root length density at depth is useful in extracting soil water, but it does not appear to offer much hope for improving drought resistance in rainfed lowland rice where the development of a hardpan may prevent deep root penetration. In water-limiting environments, genotypes that maintain the highest leaf water potential generally grow best, but it is not known if genotypic variation in leaf water potential is solely caused by root factors. Osmotic adjustment is promising because, potentially, it can counteract the effects of a rapid decline in leaf water potential, and there is large genetic variation for this trait. There is also genotypic variation in the expression of green leaf retention, which appears to be a useful character for prolonged droughts, but it is affected by plant size—which complicates its use as a selection criterion for drought resistance. Rice genotypes having high carbohydrate storage capacity in their stems resist drought better during grain filling than other genotypes because mobilization can partly substitute assimilation (Dingkuhn and Kropff 1996). The conclusion emerging from long-term multiplication drought studies for rice is that rainfed lowland rice is the most successful drought avoider, with the genotypes that produce higher grain yield during drought being those that are able to maintain better plant water status around flowering and grain setting (Serraj et al. 2011).

Several screening methodologies for drought stress in rice have been suggested, and several drought-related traits (root system, leaf water potential, panicle water potential, osmotic adjustment, and leaf temperature) have been identified. However, in this author's opinion, yield is the best parameter or trait for evaluating drought tolerance in crop plants, including rice. In several studies of unselected populations of doubled-haploid lines, broad-sense heritability of grain yield under reproductive stage drought stress was observed to be comparable to that of grain yield estimated in nonstress conditions (Blum et al. 1999; Babu et al. 2003; Lanceras et al. 2004; Venuprasad, Lafitte, and Atlin 2007). This indicates that direct selection for yield under drought stress is likely to be effective. Selection indices that give substantial weight to grain yield under drought stress have been proven highly effective in improving drought tolerance in corn (Edmeades et al. 1999; Monneveux et al. 2006).

Despite our increased understanding of stress physiology, the development of drought-resistant cultivars (i.e., cultivars that produce higher yields than others in drought conditions) has been slow in rice and other crops, mainly because of large genotype by environment (G × E) interactions, which complicate the selection of drought-resistant germplasm (Fukai and Cooper 1995).

Rice is a semiaquatic plant that can grow successfully in standing water. It can transport oxygen or oxidized compounds from the leaves to the roots and into the rhizosphere. The oxygen in the rice leaves and roots comes from atmospheric oxygen absorbed by the leaves and from oxygen released in photosynthesis through the hydrolysis of water. The water requirements of rice, as measured by the transpiration ratio, are similar to those of most major crops and are mostly affected by climate and soil. Tomar and Ghildyal (1975) studied the differences in resistance to water transport between plants grown on upland soils and those grown on flooded soils. They concluded that resistance to water transport in the nonflooded rice was nearly twice as high as in the flooded plants. The nonaerenchymatous roots of nonflooded plants had about 17 times more resistance than aerenchymatous roots of flooded rice.

1.5 GROWTH STAGES

Cultivated rice (*Oryza sativa* L.) is an annual, monocarpic cereal with a C_3 type photosynthesis pathway. Growth of C_3 plants is different from C_4 plants such as corn or millet. Knowledge of a particular crop's growth and development patterns is essential for adopting improved growth cycle management practices in order to produce higher yields. In addition, knowledge of occurrence of growth stages can also be used in many physiological studies to identify the plant growth and development stages that are sensitive to environmental factors (Fageria 2007). *Growth* is defined as the irreversible change in the size of a plant cell or organ (Fageria 1992). On the other hand, *plant development* is defined as the sequence of ontogenetic events, involving growth and differentiation, leading to change in function and morphology (Landsberg 1977). Development is most clearly manifested in the changes that take place in an organism's form, such as crop plants changing from the vegetative to the reproductive stage and from the reproductive stage to maturity. This development can be studied through morphological as well as physiological changes (Fageria 1992).

The rice plant takes three to six months to complete its life cycle from germination to maturity, depending on the variety and environmental conditions (Mae 1997). Rice phenology is generally divided into vegetative (from emergence to panicle primordia initiation), reproductive (from panicle primordia initiation to flowering),

TABLE 1.12
Timing (DAS) of Rice Growth Stages

Growth Stage	DAS	Definition
Germination	5	Coleoptile tip first becomes visible
Tillering initiation	19	First tiller from the main shoot is visible
Active tillering	45	Maximum tillering rate per unit time during crop growth
Panicle primordia initiation	61	Initiation of panicle
Booting	85	Panicle is enclosed by the sheath of the uppermost leaf
Flowering	95	Flowers are visible on the panicles
Physiological maturity	120	Grains are ripened and panicles are ready for harvest

Source: Fageria, N. K., and A. M. Knupp. 2013. Upland rice phenology and nutrient uptake in tropical climate. *J. Plant Nutr.* 36:1–14.
Note: DAS = days after sowing. The cycle of the cultivar was 120 days.

and spikelet filling (from flowering to physiological maturity; Fageria, Baligar, and Jones 2011). Different growth stages of rice are defined in Table 1.12. Figure 1.9 shows growth stages of an upland rice cultivar having a growth cycle of 130 days from sowing to physiological maturity or 125 days from germination to physiological maturity under Brazilian conditions or in the tropics. The vegetative growth stage was 65 days in duration (from germination to initiation of panicle primordia), the reproductive growth stage was 30 days in duration (from panicle primordia initiation to flowering), and the spikelet-filling stage was also 30 days in duration (from flowering to physiological maturity). Similarly, Figure 1.10 displays the growth stages of Brazilian lowland rice, which has a 140-day growth cycle, from sowing to physiological maturity.

The plant growth stages from panicle initiation to flowering and from flowering to physiological maturity or spikelet filling are important for yield determination because during these stages the estimates for seed number and seed weight components of the yield are formed (Fageria, Baligar, and Jones 2011).

1.5.1 Vegetative Growth Stage

The vegetative growth stage extends from germination to panicle primordia initiation. Its main features are increases in root length or weight, plant height, tillering, shoot dry weight (culm plus leaves), and leaf characters or LAI. The tillering number is associated with the panicle number; therefore, the panicle number is an important yield component in rice and is determined during the vegetative growth stage. The differences in a genotype's growth cycle are mainly due to variations in the vegetative growth stage. In the tropics, under favorable environmental conditions, the vegetative growth stage in rice makes up about half of the total growth duration (from germination to physiological maturity). The period between the maximum tiller number stage and the panicle primordia initiation is defined as the vegetative lag period (Murayama 1995).

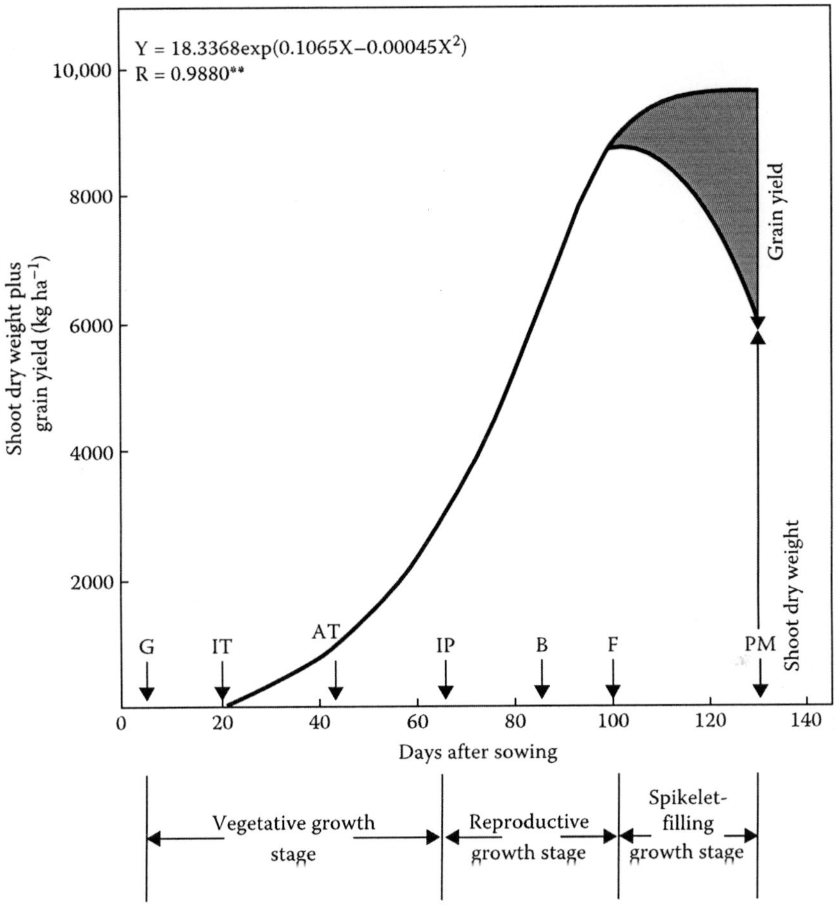

FIGURE 1.9 Shoot dry weight accumulation and grain yield of upland rice during the growth cycle of the crop in the central part of Brazil. G = germination, IT = initiation of tillering, AT = active tillering, IP = initiation of panicle primordia, B = booting, F = flowering, and PM = physiological maturity. (From Fageria, N. K. 2007. Yield physiology of rice. *Plant Nutr.* 30:843–879.)

In order to study their morphology, upland and lowland rice seedlings were photographed at the tillering initiation growth stage; they are depicted in Figures 1.11 and 1.12. The rice seedlings at this stage have four leaves, adventitious and radicle root systems, and an initiation of a tiller from the base of the main culm. The root system and the tops of the upland rice seedling were more vigorous than those of the lowland rice seedling. Both the seedlings were grown under similar environmental conditions. Hence, the difference in morphology (roots and tops) of the two seedlings was a cultivar characteristic.

Initially, cereal seedlings are heterotrophic, depending totally on food mobilized from the endosperm; then they pass through a transition phase, when photosynthesis commences, while endosperm mobilization continues (Salam, Jones, and Jones 1997).

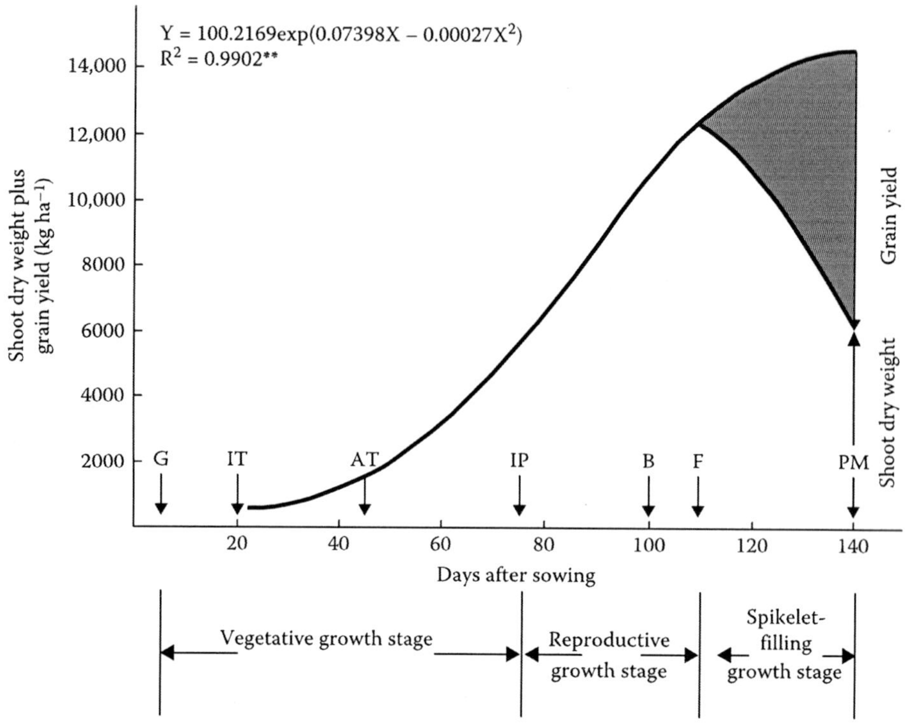

FIGURE 1.10 Shoot dry weight accumulation and grain yield of lowland rice during the growth cycle of the crop in the central part of Brazil. G = germination, IT = initiation of tillering, AT = active tillering, IP = initiation of panicle primordia, B = booting, F = flowering, and PM = physiological maturity. (From Fageria, N. K., V. C. Baligar, and C. A. Jones. 2011. *Growth and Mineral Nutrition of Field Crops*, 3rd edition. Boca Raton, FL: CRC Press.)

Finally, seedlings depend entirely on photosynthesis and become autotrophic. Salam, Jones, and Jones (1997) reported that rice seedlings become autotrophic at the three- to four-leaf stages, somewhere in the range of 14–22 days from emergence. According to rice seedling morphology (see Figure 1.11), in the central part of Brazil, upland rice seedlings become autotrophic 19 days after sowing or 14 days after emergence. At the vegetative growth stage, most of the essential nutrients are actively absorbed. Proteins are vigorously synthesized, leading to the acceleration of tillering and the extension of leaf areas. Photosynthates also increase rapidly (Murayama 1995).

1.5.1.1 Root Growth

The root is an important plant organ that absorbs water and nutrients. It also gives mechanical support to the plants, and it supplies hormones. Furthermore, root materials left in the soil after a crop is harvested improve the soil's organic matter content. A rice seedling starts with a radicle (seminal root), mesocoty root, and nodal roots. However, the rice root system is basically composed of nodal or adventitious roots (Yoshida 1981). Figures 1.11 and 1.12 show radicle and nodal root systems in upland and lowland rice seedlings. Root hairs are mainly responsible for the absorption of

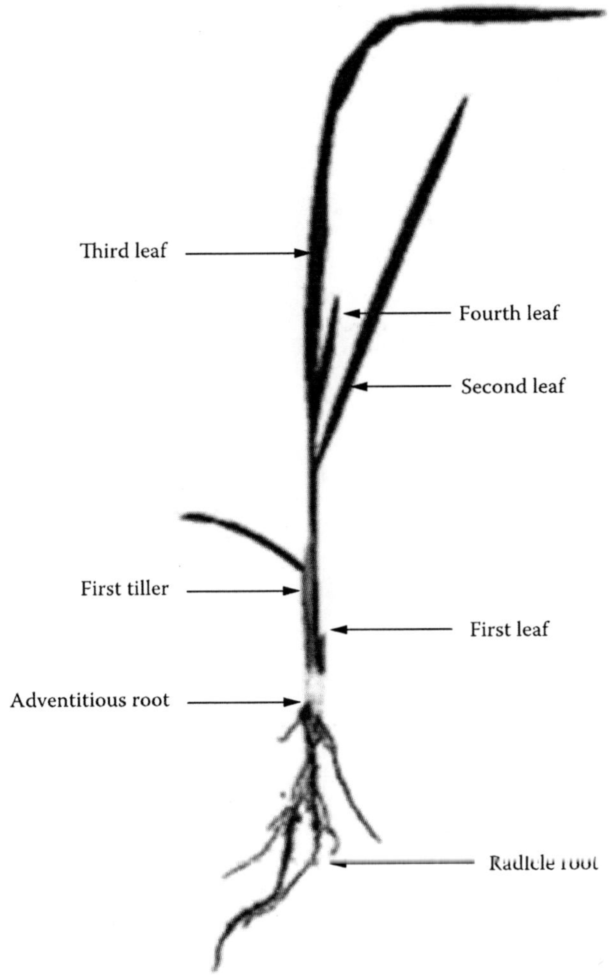

FIGURE 1.11 Morphology of upland rice seedling at the tillering initiation growth stage. (From Fageria, N. K. 2007. Yield physiology of rice. *Plant Nutr.* 30:843–879.)

water and nutrients. Their formation is greatly affected by the root environment. Murata and Matsushima (1975) reported that the emergence of a rice root is closely correlated with the N content of the stem base; active emergence of roots takes place only when the N concentration is more than 10 g kg^{-1} or 1%. The rice root system is also genetically controlled. The root system of aerobic or upland rice is larger and vigorous and has more root hairs than that of lowland or flooded rice. Rice roots possess large air spaces. These air spaces are connected with those in the culms and leaves, providing an efficient air passage system from shoot to root. That is why rice plants can grow well under anaerobic or reduced soil conditions.

Fageria, Baligar, and Jones (2011) reported the results of maximum root length along with other morphological characteristics of lowland rice cultivars having a growth cycle of 120 days (Figure 1.13). Maximum root length followed a significant

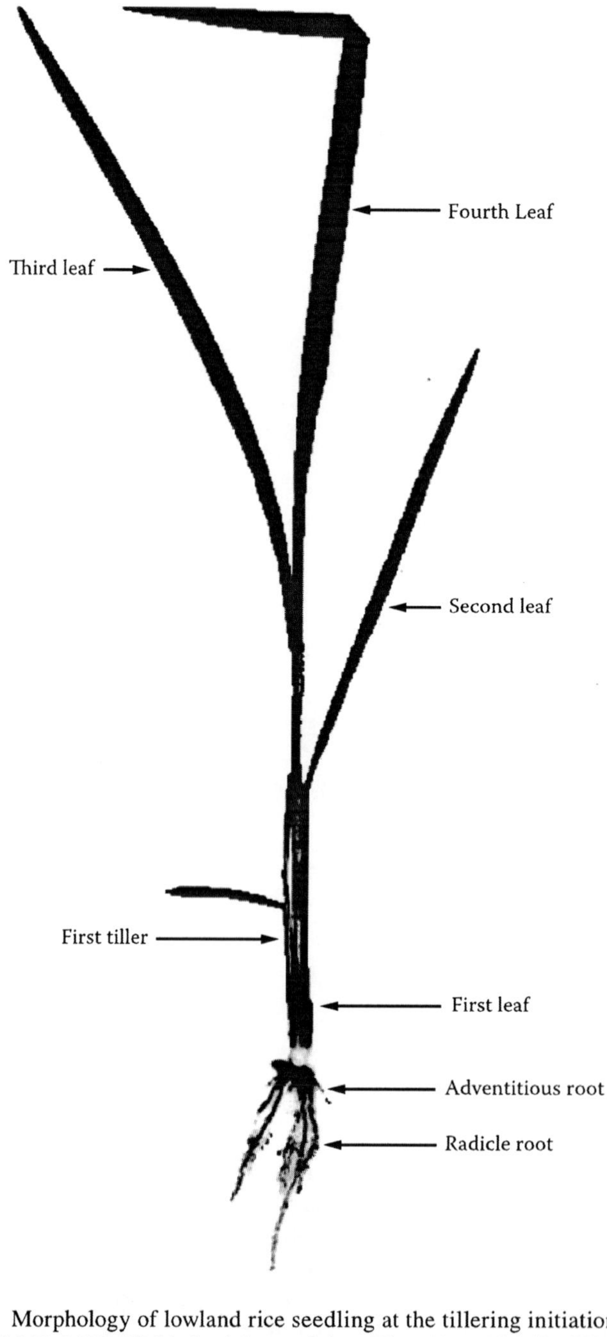

FIGURE 1.12 Morphology of lowland rice seedling at the tillering initiation growth stage. (From Fageria, N. K. 2007. Yield physiology of rice. *Plant Nutr.* 30:843–879.)

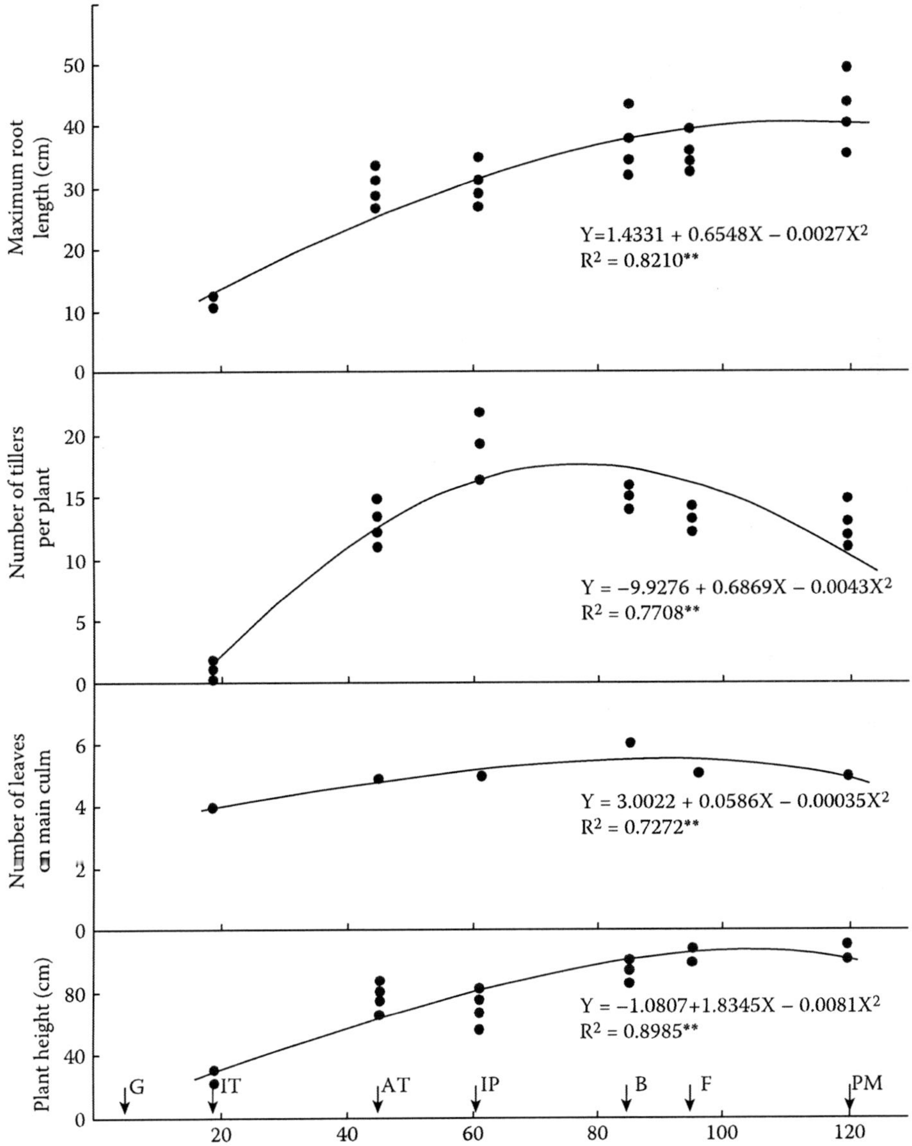

FIGURE 1.13 Relationship between plant age and morphological plant parameters. G = germination, IT = initiation of tillering, AT = active tillering, IP = initiation of panicle, B = booting, F = flowering, and PM = physiological maturity. (From Fageria, N. K., V. C. Baligar, and C. A. Jones. 2011. *Growth and Mineral Nutrition of Field Crops*, 3rd edition. Boca Raton, FL: CRC Press.)

quadratic response with the advancement of plant age from 19 to 120 days after sowing. There was a linear increase in root length from tillering initiation to flowering. Thereafter, root length was more or less constant or reached a plateau. Klepper (1992) reported that the general pattern of root development over the life of the crop shows a shift from a heavy investment in roots during seedling establishment and early vegetative growth in the first part of the growing season to a heavy investment in reproductive structures during the latter part of the season. This may explain why roots reach a plateau during the grain or spikelet-filling growth stage.

The author of this volume also compared the root growth of upland rice during its growth cycle with other annual crops, such as wheat, corn, dry beans, and soybeans (Table 1.13 and Figures 1.14 through 1.18). The root dry weight of five crop species was significantly influenced by plant age, and the root dry weight of rice increased linearly with the advancement of plant age. The root dry weight of the remaining four crop species increased in a quadratic fashion with increasing plant age. Overall, corn produced a higher root weight compared with that of the other crops. This higher root growth of corm may be related to its C_4 photosynthetic pathway.

A root system that extends the root zone in order to more fully extract available soil water and mineral nutrition has the potential to increase the yield under drought conditions (Mambani and Lal 1983; Nguyen, Babu, and Blum 1997). The capacity for water and nutrient uptake may limit rice productivity even in flooded soils (Ingram et al. 1994). The rice plant invests up to 60% of its carbon in the root system (Nguyen, Babu, and Blum 1997). Ingram et al. (1994) and Yu et al. (1995) reported

TABLE 1.13
Root Dry Weight of Five Crop Species during Their Growth Cycle

Plant Age in Days	Upland Rice ($g\ plant^{-1}$)	Corn ($g\ plant^{-1}$)	Wheat ($g\ plant^{-1}$)	Soybean ($g\ plant^{-1}$)	Dry Bean ($g\ plant^{-1}$)
21	0.05	0.22	0.06	0.08	0.27
40	0.53	3.43	0.10	0.40	2.45
61	1.25	7.81	0.17	1.08	3.01
78	2.51	8.04	0.27	1.26	1.49
97	3.62	5.64	0.17	1.40	1.27
118	4.40	4.64			

Regression Analysis

Upland rice age versus root dry wt. $(Y) = -1.2473 + 0.0478X$, $R^2 = 0.9254^*$

Corn plant age versus root dry wt. $(Y) = -6.4108 + 0.3553X - 0.0022X^2$, $R^2 = 0.8652^*$

Wheat plant age versus root dry wt. $(Y) = -0.0955 + 0.0077X - 0.000048X^2$, $R^2 = 0.7002^*$

Soybean plant age versus root dry wt. $(Y) = -0.6792 + 0.0368X - 0.00015X^2$, $R^2 = 0.9434^*$

Dry bean age versus root dry wt. $(Y) = -2.3708 + 0.1655X - 0.0013X^2$, $R^2 = 0.6829^*$

Source: Fageria, N. K. 2012. *The Role of Plant Roots in Crop Production.* Boca Raton, FL: CRC Press.

Note: Crops were harvested at physiological maturity; upland rice and corn were matured at 118 days, and wheat, soybean, and dry bean were matured at 97 days.

*Significant at the 1% probability level.

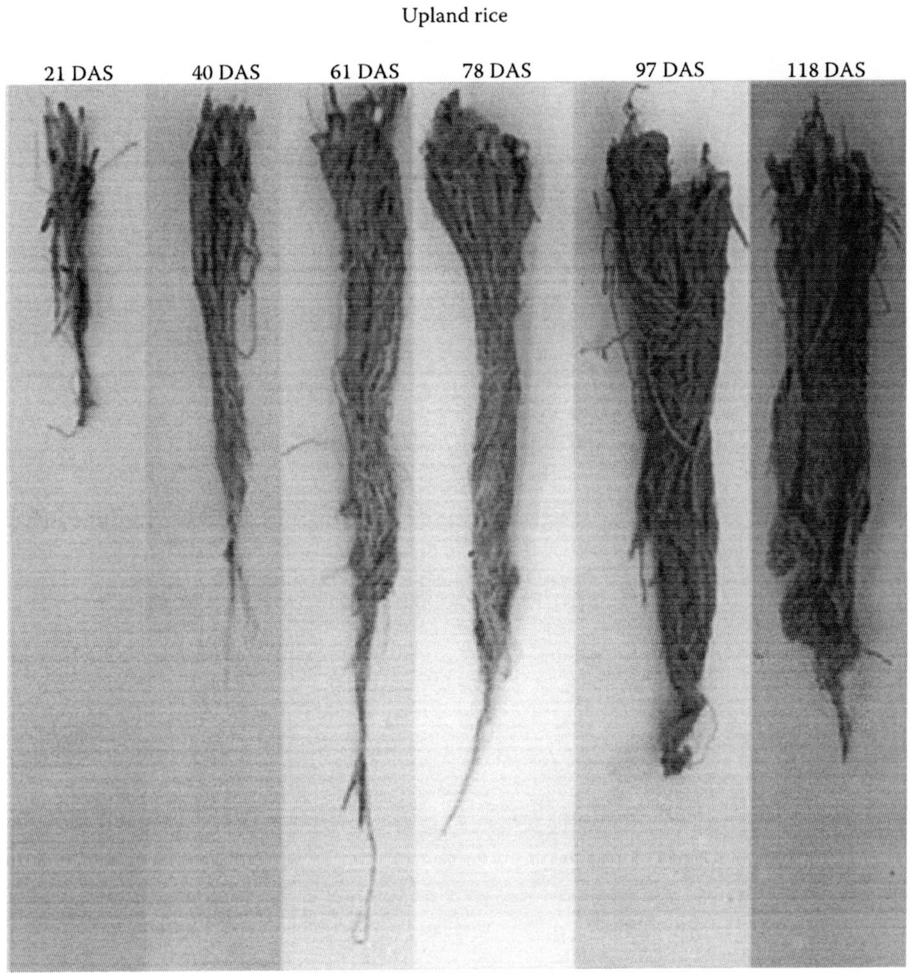

FIGURE 1.14 Root development of upland rice during crop growth cycle. DAS = days after sowing.

that the ability of rice to reach deep soil moisture or to penetrate compacted soil is linked with the capacity to develop a few thick (fibrous) and long root axes. Thick roots persist longer and produce more and larger branch roots, thereby increasing root length density and water uptake capacity (Ingram et al. 1994; Nguyen, Babu, and Blum 1997).

1.5.1.2 Plant Height

For seedlings or juvenile plants, plant height is the distance from ground level to the tip of the tallest leaf. For mature plants, it is the distance from ground level to the tip of the tallest panicle. Plant height is an important trait because it is associated with plant lodging, seedling growth capacity, and weed control (Donald and Hamblin 1976). Shorter plants mean reduced lodging, particularly under

Wheat

21 DAS 40 DAS 61 DAS 78 DAS 97 DAS

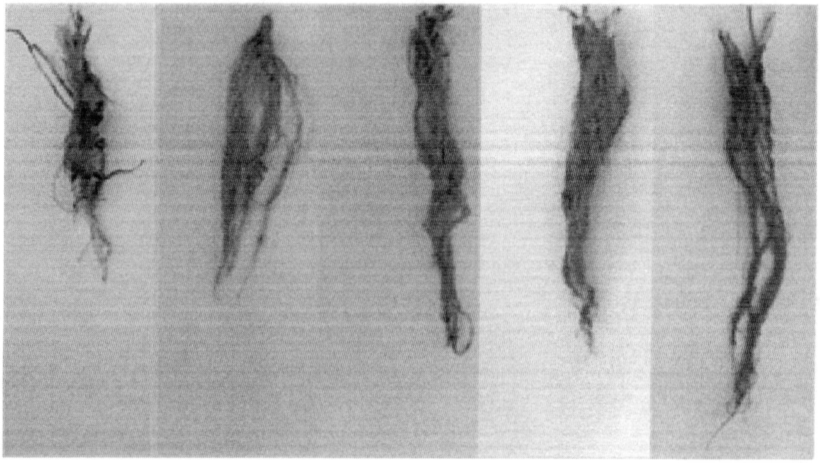

FIGURE 1.15 Root development of wheat during crop growth cycle. DAS = days after sowing.

Soybean

21 DAS 40 DAS 61 DAS 78 DAS 97 DAS

FIGURE 1.16 Root development of soybean during crop growth cycle. DAS = days after sowing.

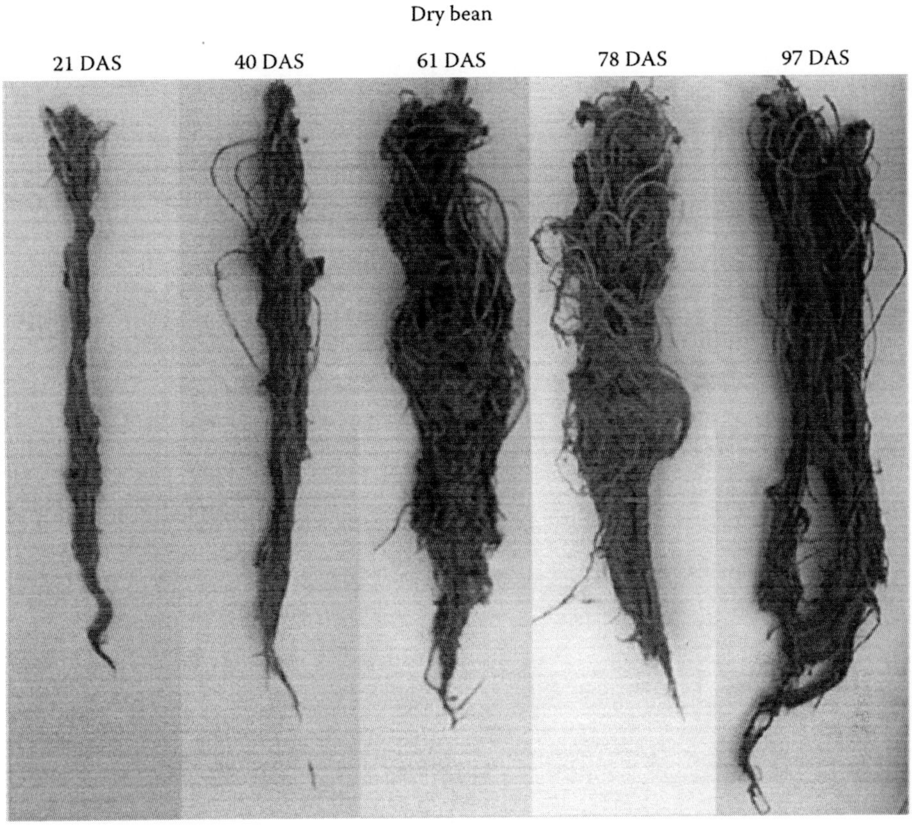

FIGURE 1.17 Root development of dry bean during crop growth cycle. DAS = days after sowing.

high N inputs; if a taller population were planted, a higher harvest index could be obtained, and this would translate to a relatively higher yield (Tabien, Samonte, and McClung 2008).

As early as 1966, scientists at IRRI discovered that plant architecture of traditional cultivars was the main constraint in increasing grain yield. Traditional cultivars usually lodged at high N application levels but produced well if supported to stand (Tabien, Samonte, and McClung 2008). The first decade of breeding at IRRI was focused on dwarfism (Peng and Khush 2003) and the development of IR8, the high-yielding dwarf cultivar. The cultivar IR8 was developed by a cross between the tall tropical cultivar "Peta" from Indonesia and the semidwarf cultivar "Dee-geo-gen" from Taiwan (Fageria, Baligar, and Jones 2011; Fageria, Gheyi, and Moreira 2011). "Deo-geo-woo-gen" and "Taichung Native" tropical rice cultivars from Asia, all carrying the sd-1 gene, were the main sources of semidwarfism in U.S. breeding programs (Tabien, Samonte, and McClung 2008). In Texas, the first semidwarf cultivar with the sd-1 gene from the Taichung Native cultivar was "Bellemont," released in 1981 (Bollich et al. 1983). This was quickly replaced in 1983 by the semidwarf

FIGURE 1.18 Root development of corn during crop growth cycle. DAS = days after sowing.

cultivar "Lemont," which was the first widely grown, semidwarf cultivar in the southern United States (Tabien, Samonte, and McClung 2008). Another semidwarf cultivar known as "Gulfmont," released in 1986, remained a cornerstone of the U.S. rice industry for more than 20 years (Tabien, Samonte, and McClung 2008). Plant heights of recently released U.S. cultivars were more stable across N levels, and the cultivars were less susceptible to lodging (Tabien, Samonte, and McClung 2008).

Old rice cultivars were taller in comparison to modern cultivars and were also more susceptible to lodging with a higher rate of nitrogen application. This scenario changed during the latter half of the twentieth century with a dramatic increase in grain yields of cereals (rice, wheat, and corn); the term "Green Revolution" was used to refer to this change. The green revolution was mainly associated with the development of cultivars with short stature (90–110 cm) that were less susceptible to lodging when heavily fertilized, especially with nitrogen (Yoshida 1981). In addition to lodging resistance, short stature and sturdy culm cultivars give higher yields at close plant spacing compared with taller cultivars. However, taller plants have an advantage in competing with weeds when compared with shorter stature plants. Because grain yields decrease with increasing water depth, under such conditions, intermediate stature (100–130 cm) is more desirable

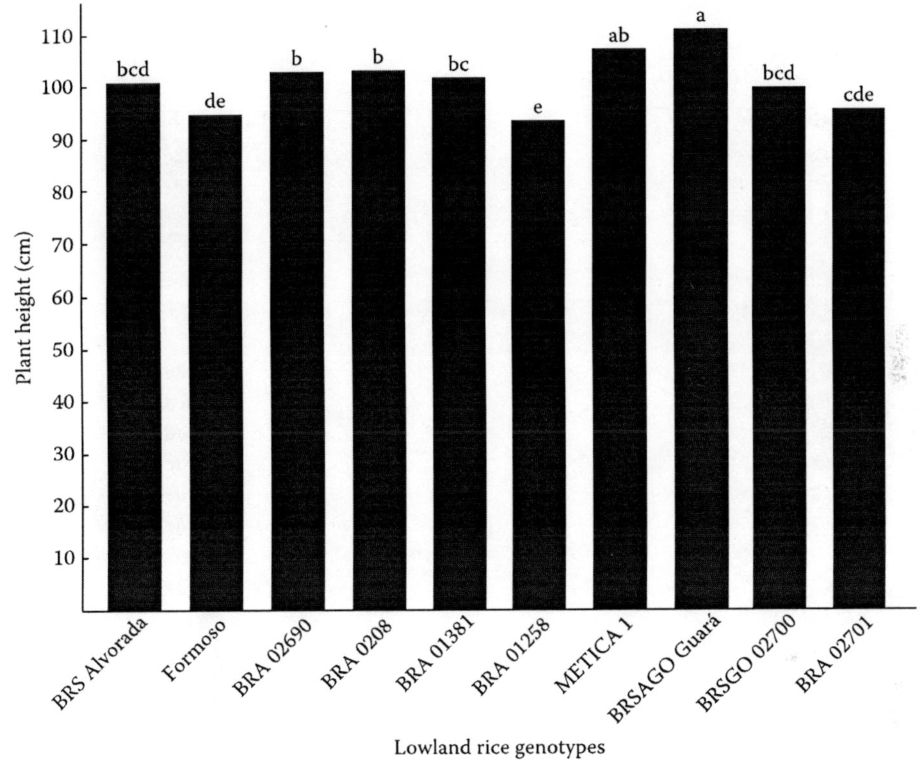

FIGURE 1.19 Plant height of 10 lowland rice genotypes.

than short stature (90–110 cm; Yoshida 1981). An extremely dwarf height is also not good because grain yield increases quadratically with increasing plant height (Fageria et al. 2004).

A marked increase in harvest index and in grain yield per day has been associated with reduced plant height and earlier maturity (Evans, Visperas, and Vergara 1984). Evans, Visperas, and Vergara (1984) reported that despite these changes, there has been no change in photosynthetic rate, CGR, or spikelet weight. The plant height of 10 lowland rice genotypes grown under Brazilian conditions is presented in Figure 1.19. The height of these genotypes varied from 94 to 111 cm, with an average of 102 cm. Similarly, the plant height of 19 upland rice genotypes grown on Brazilian Oxisol is presented Table 1.14. The plant height was significantly ($P < 0.01$) influenced by N and genotype treatments; across two N levels, height varied from 93 to 118 cm among genotypes, with an average value of 103 cm (Table 1.14). There was a significant quadratic relationship between plant height and grain yield (Figures 1.20 and 1.21). Hence, grain yield increases with plant height, but there is a limit to this increase. When plant height is too short, less dry matter is produced, and when it is too high, the plant may lodge and be less responsive to N fertilization (Yoshida 1981). This means intermediate plant height is a better compromise in relation to grain yield (Fageria et al. 2010).

TABLE 1.14
Plant Height of Upland Rice Genotypes[a]

Genotype	Plant Height (cm)
CRO 97505	106cde
CNAs 8993	104cde
CNAs 8812	93g
CNAs 8938	99efg
CNAs 8960	112abc
CNAs 8989	99efg
CNAs 8824	94g
CNAs 8957	105cde
CRO 97422	108bcd
CNAs 8817	109bcd
CNAs 8934	105cde
CNAs 9852	104cde
CNAs 8950	106cde
CNA 8540	93g
CNA 8711	118a
CNA 8170	93g
BRS Primavera	115ab
BRS Canastra	94fg
BRS Carisma	102def
Average	103
F-test	
N level (N)	*
Genotype (G)	*
N × G	NS
CV (%)	4

Source: Fageria, N. K., O. P. Morais, and A. B. Santos. 2010. Nitrogen use efficiency in upland rice. *J. Plant Nutr.* 33:1696–1711.

[a] Across two N rates (0 and 400 mg N kg^{-1}).

*,NS Significant at the 1% probability level and nonsignificant, respectively. Means followed by the same letter in the same column are not significantly different at the 5% probability level by Tukey's test.

Although plant height is influenced by environmental factors, height is a genetically controlled trait; the heritability of dwarfism is high, and it is easy to identify, select, and recombine with other traits (Jennings, Coffman, and Kauffman 1979). Dwarf segregates have a fairly narrow range in height, presumably from minor gene action. Although a few are so short that they are undesirable, the great majority fall within the useful range of 80–100 cm, with some reaching 120 cm under certain conditions (Jennings, Coffman, and Kauffman 1979). During the 1960s, rice breeders made excellent progress in the development of dwarf cultivars that responded to heavy applications of nitrogen (Jennings, Coffman, and Kauffman 1979).

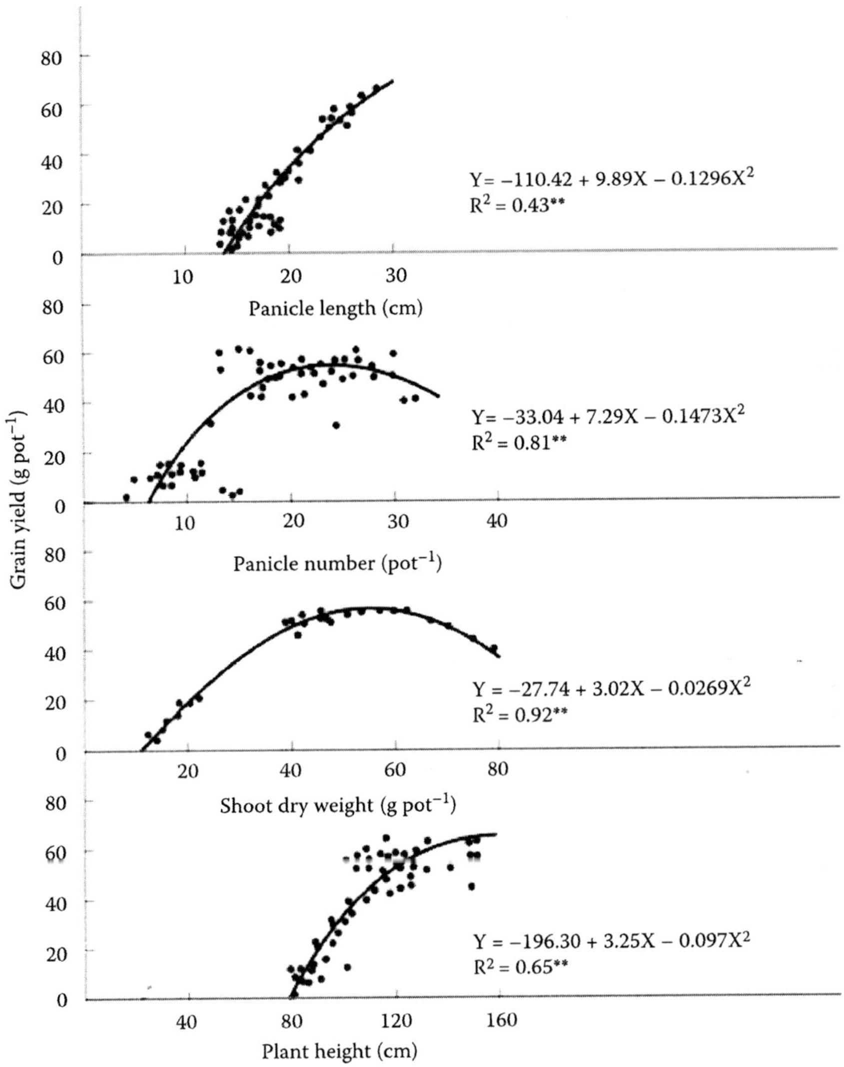

FIGURE 1.20 Relationship among plant height, shoot dry weight, panicle number, panicle length, and grain yield of upland rice. (From Fageria, N. K., O. P. Morais, and A. B. Santos. 2010. Nitrogen use efficiency in upland rice. *J. Plant Nutr.* 33:1696–1711.)

1.5.1.3 Tillering

Rice tillering is a model system for the study of branching in monocotyledons (Liu et al. 2011). It is also an important agronomic tool for population structure and grain production (Ling 2000; Li et al. 2003). Rice tillers develop from tiller buds, and the number of tillers is dynamic and adjustable (Kariali and Mohapatra 2007). Rice tiller buds are axillary buds, which differentiate at the leaf axils. The nutrient

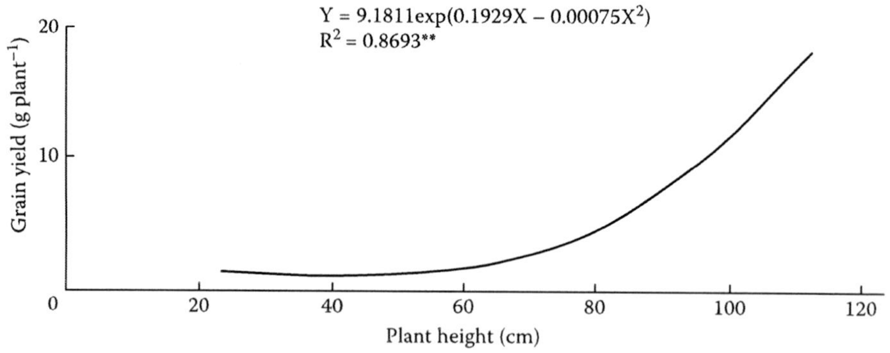

FIGURE 1.21 Relationship between plant height and grain yield of upland rice grown on a Brazilian Oxisol.

supply required for the growth of tiller buds inside the subtending leaf sheaths relies upon the mother stem. After the third leaf has completely emerged, the nutrient supply to tillers shifts from heterotrophy to autotrophy (Hanada 1995). Emergence of tillers is closely linked to the number of leaves, and tillering begins at the four- to five-leaf stages (Murata and Matsushima 1975). In the tropics, under favorable environmental conditions, four- to five-leaf stages in rice generally are attained within 12–17 days after germination. During tillering, protein synthesis is higher than the synthesis of other organic compounds (starch, lignin, and cellulose), and the plant rapidly develops photosynthetic organs (Hayashi 1995).

Tillering is influenced by environmental conditions (light, temperature, soil moisture, and N content of main stem) and also is genetically controlled. In addition to these factors, auxin and cytokinin (CTK) play important roles in the regulation of tiller bud growth (Liu et al. 2011). High tillering capacity basically relates to the maximum use of space and resources. Tillering has special importance in relation to biotic and abiotic stresses due to the compensation process. High tillering compensates for missing plants at low densities, but cultivars with limited tillering capacity lack this plasticity. However, under favorable environmental conditions, heavy-tillering cultivars have no advantage over low-tillering ones with relation to yield. Peng, Khush, and Cassman (1994) reported that moderate tillering contributes greatly to rice yields, but excessive tillering leads to high tiller abortion, poor grain set, small panicle size, and eventually reduced grain yield.

Heavy tillering is not much of an advantage with direct-seeded rice, a common practice in South American mechanized agriculture. Under direct seeding, tillering capacity rarely affects grain yield within conventional seeding rates because total panicle number per square meter depends more on the main culm than on tillers (Yoshida 1981). Low seed rate is required by heavy-tillering cultivars compared with low-tillering cultivars. However, Jennings, Coffman, and Kauffman (1979) reported that a combination of high tillering ability and compact or nonspreading culm arrangement is desirable. Compact culms that are moderately erect allow increased solar radiation to tillers and less mutual shading per unit of land area.

Tillering follows a quadratic increase with the advancement of rice plant age (Fageria, Baligar, and Jones 2011) and is significantly influenced by N and P fertilization (Fageria, Slaton and Baligar 2003; Fageria 2005). Maximum tillering, based on the quadratic function, was attained at 80 days after sowing, and it decreased thereafter (Fageria, Baligar, and Jones 2011). The period in which the increase of tiller number per unit length of time is greatest is defined as the active tillering stage. This stage, during which the tiller number reaches a maximum, is known as the maximum tiller number stage. Tillers that do not produce panicles degenerate, and their number decreases until they become equal to the number of panicles. The growth juncture of this period is called the ineffective tillering stage (Murayama 1995).

Nitrogen and phosphorus fertilization significantly increase tillering in rice (Fageria, Slaton and Baligar 2003). Murata and Matsushima (1975) reported that an N concentration of more than 35 g kg^{-1} (3.5%) is necessary for active tillering. At 25 g N kg^{-1} (2.5%), tillering stops, and below 15 g N kg^{-1} (1.5%), tiller death takes place. Similarly, P concentration is also correlated with tillering, and a P concentration of >2.5 g kg^{-1} (0.25%) in the mother stem is necessary for tillering. About 66–96% of the variation in tillering was apparently due to N fertilization, depending on crop growth stage. Tillering increased with the advancement of the crop growth and, depending on N rate, maximal values were achieved between 35 and 71 days after sowing; values then decreased. Grain yield in cereals is highly dependent on the number of spikelet-bearing tillers produced by each plant (Power and Alessi 1978; Nerson 1980). The number of productive tillers depends on environmental conditions during tiller bud initiation and subsequent developmental stages. Numerous studies have shown that tiller appearance, abortion, or both are affected by environmental conditions, especially nutrient deficiencies (Black and Siddoway 1977; Power and Alessi 1978; Masle 1985).

The decrease in tiller number was attributed to the death of some of the last tillers as a result of their failure to compete for light and nutrients (Fageria, Baligar, and Jones 2011). Another explanation is that during the period of growth beginning with panicle development, competition for assimilates exists between developing panicles and young tillers. Eventually, growth of many young tillers is suppressed, and they may senesce without producing seed (Dofing and Karlsson 1993). A correlation between grain yield and number of tiller m^{-2} at different growth stages is presented in Table 1.15. Tillering was significantly related to grain yield at all the growth stages; however, the highest correlation in the three years of experimentation was obtained at the initiation of the panicle growth stage. This means that for lowland rice, number of tillers determined at this growth stage had more significance than at any other growth stage.

Tiller number is quantitatively inherited. Its heritability is low to intermediate depending on the cultural practices used and the uniformity of the soil. Although often associated with early vigor in short-statured materials, tiller number is inherited independently of all other major characters. In many crosses, tiller erectness or compactness is recessive to a spreading culm arrangement (Jennings, Coffman, and Kauffman 1979). Developing good plant types with high tillering capacity is rather simple. Many sources of heavy tillering are available in traditional tropical

TABLE 1.15
Correlation Coefficients (r) between Grain Yield and Tiller Number

Parameter	First Year	Second Year	Third Year
Tiller number m^{-2} at IT	0.59**	0.41*	0.23^{NS}
Tiller number m^{-2} at AT	0.69**	0.43*	0.34*
Tiller number m^{-2} at IP	0.79**	0.59**	0.68**
Tiller number m^{-2} at B	0.67**	0.52**	0.46**
Tiller number m^{-2} at F	0.70**	0.37*	0.52**
Tiller number m^{-2} at PM	0.77**	0.48**	0.44*

Source: Fageria, N. K., and V. C. Baligar. 2001. Lowland rice response to nitrogen fertilization. *Commun. Soil Sci. Plant Anal.* 32:1405–1429.

Note: IT, initiation of tillering; AT, active tillering; IP, initiation of panicle; B, booting; F, flowering; PM, physiological maturity.

^{*,**,NS}Significant at the 5% and 1% probability levels and nonsignificant, respectively.

rice cultivars. When their culms are shortened, their tillering ability generally does not decrease and may even increase (Jennings, Coffman, and Kauffman 1979).

In addition to tiller number, tiller angle is also an important component of rice plant architecture. Rice plants with relatively vertical tillers are referred to as compact types and those with a wide tiller angle as spreading types (Xu, McCouch, and Shen 1998). Rice of spreading type can escape diseases such as rice sheath blight [*Thanatephorus cucumeris* (A. B. Frank) Donk] (Han et al. 2003), whereas the extremely compact type is inefficient in harvesting light and is more susceptible to pathogen attacks (Xu, McCouch, and Shen 1998; Qian et al. 1994). The perfect ideotype would have a wide, spreading type in the early development stage (because this promotes photosynthetic efficiency) and would become compact once stem elongation has begun (to delay as much as possible the onset of leaf senescence caused by shading; Chen et al. 2008). Genetic analysis has shown that tiller angle in rice is a quantitatively inherited trait (Xu and Shen 1993), with at least four quantitative trait loci (QTL) being responsible for its determination (Xu, McCouch, and Shen 1998; Chen et al. 2008).

1.5.1.4 Shoot Dry Weight

With the advancement of plant age, upland and lowland rice shoot dry weight follows a sigmoid curve (Fageria, Baligar, and Jones 2011). Shoot dry weight increases significantly in the vegetative as well as reproductive growth stages (Table 1.16). The increase in shoot weight is mainly associated with an increase in leaf and culm weights during these growth stages. Rice plant weight mainly consists of organic matter such as proteins and carbohydrates. Carbohydrates are composed of cell wall substances such as cellulose and reserved substances such as starch. The protein metabolism dominates in the vegetative growth stage, and the carbohydrate metabolism dominates in the reproductive growth stage (Murayama 1995). The portion of inorganic matter in rice plant weight is generally small. However, rice straw

TABLE 1.16
Shoot Dry Matter Yield (kg ha^{-1}) of Lowland Rice at Different N Rates

	Days after Sowing					
N Rate	22 (IT)	35 (AT)	71 (IP)	97 (B)	112 (F)	140 (PM)
0	313	815	3065	5650	7694	5278
30	320	860	3709	6913	8953	6764
60	342	1230	3721	8242	11056	7294
90	374	1044	4164	8695	10758	7303
120	380	1229	4313	9570	13378	8215
150	452	1207	4893	10031	12745	8624
180	351	1294	5077	11290	13682	9060
210	351	1130	5841	10384	13490	9423
R^2	0.56*	0.76*	0.97**	0.97**	0.94**	0.96**

Source: Fageria, N. K., and V. C. Baligar. 2001. Lowland rice response to nitrogen fertilization. *Commun. Soil Sci. Plant Anal.* 32:1405–1429.

Note: Values are averages from three years of field trials. IT, initiation of tillering; AT, active tillering; IP, initiation of panicle; B, booting; F, flowering; PM, physiological maturity.

*,**Significant at the 0.05 and 0.01 probability levels, respectively.

generally has a high silicon content. In a mature shoot of rice, silicon content can be as high as 10% (Murayama 1995).

Upland and lowland rice shoot weight decreases from flowering to physiological maturity (Fageria, Baligar, and Jones 2011). Dry matter loss from the vegetative tissues during the interval from flowering to maturity was 35%, suggesting active transport of assimilates to the panicles, which resulted in a grain yield of 3811 kg ha^{-1}. Fageria and colleagues reported a more or less similar reduction in shoot dry weight of upland rice from flowering to physiological maturity (Fageria, Baligar, and Clark 2006; Fageria, Baligar, and Jones 2011).

Increase in shoot weight is important because it is significantly associated with grain yield (Table 1.17). Dry matter production has a higher correlation with grain yield during the booting, flowering, and physiological maturity growth stages than at earlier growth stages (Table 1.17). Shoot weight is a characteristic of genotypes and is also influenced by environmental factors. In Figure 1.22, different genotypes showed significant variability in grain yield and shoot dry weight. Shoot dry weight varied from 6602 kg ha^{-1} for the genotype CNAi 8569 to 4041 kg ha^{-1} for the genotype BRS Biguá, with an average value of 4980 kg ha^{-1}. Similarly, variation in grain yield ranged from 4828 kg ha^{-1} for the genotype BRSGO Guará to 3638 kg ha^{-1} for the genotype BRS Jaburu, with an average value of 4287 kg ha^{-1}. Fageria et al. (1997), and Fageria, Slaton and Baligar (2003) and Fageria and Barbosa Filho (2001) have reported differences in shoot and grain yield for lowland rice genotypes. Variation in shoot dry weight and grain yield among genotypes may be associated with differences in the amount of intercepted PAR by the canopy, the radiation use efficiency, and the GHI (Kiniry et al. 2001; Fageria and Baligar 2005).

TABLE 1.17
Correlation Coefficients (r) between Lowland Rice Grain Yield and Shoot Dry Matter Production during Different Growth Stages

Parameter	First Year	Second Year	Third Year
Dry matter yield at IT	0.36*	0.37*	0.29^{NS}
Dry matter yield at AT	0.71**	0.55**	0.42*
Dry matter yield at IP	0.63**	0.51**	0.63**
Dry matter yield at B	0.72**	0.81**	0.61**
Dry matter yield at F	0.81**	0.80**	0.57**
Dry matter yield at PM	0.78**	0.80**	0.53**

Source: Fageria, N. K., and V. C. Baligar. 2001. Lowland rice response to nitrogen fertilization. *Commun. Soil Sci. Plant Anal.* 32:1405–1429.

Note: IT, initiation of tillering; AT, active tillering; IP, initiation of panicle; B, booting; F, flowering; PM, physiological maturity.

*,**,NS Significant at the 0.05 and 0.01 probability levels and nonsignificant, respectively.

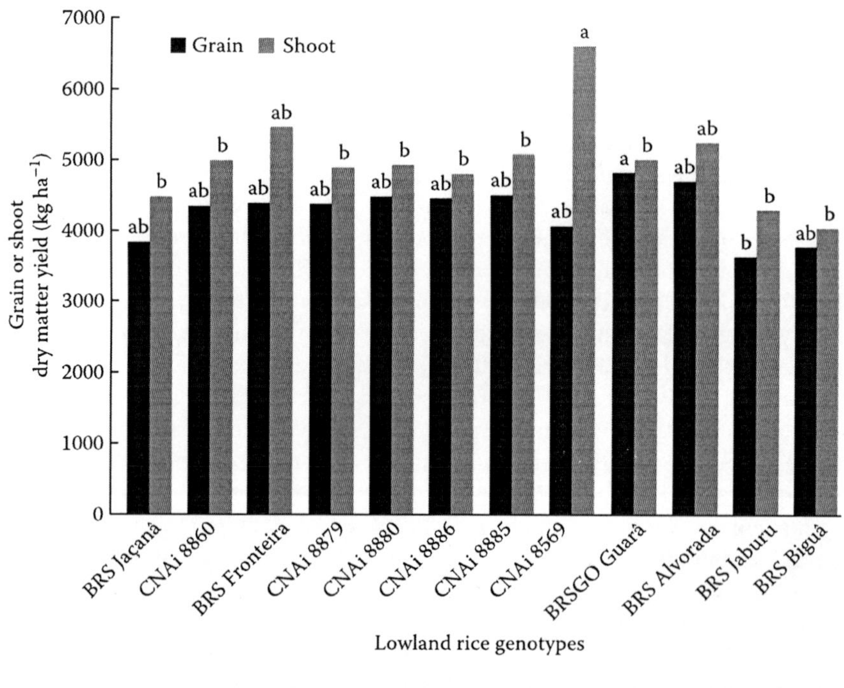

FIGURE 1.22 Grain and shoot dry weight of 12 lowland rice genotypes, a and b on the bars indicates significant differences among the genotypes in grain and shoot yield. (From Fageria, N. K., M. P. Barbosa Filho, and C. M. Guimaraes. 2008. Allelopathy in upland rice in Brazil. *Allelopathy J.* 22:289–298.)

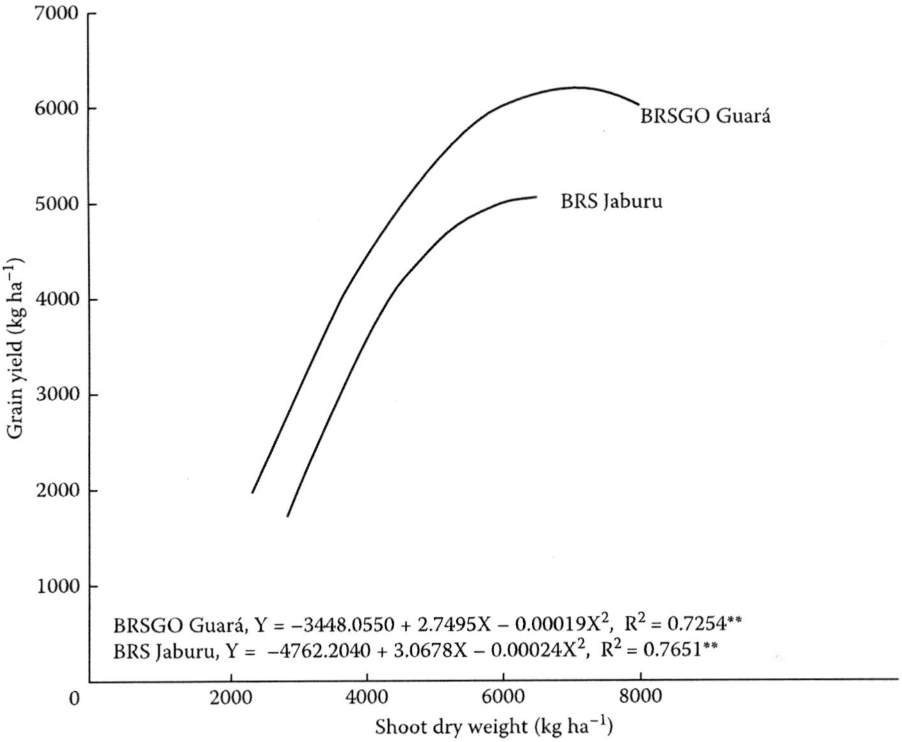

FIGURE 1.23 Relationship between shoot dry weight and grain yield of two lowland rice genotypes. (From Fageria, N. K., A. B. Santos, and V. A. Cutrim. 2008b. Dry matter and yield of lowland rice genotypes as influenced by nitrogen fertilization. *J. Plant Nut.* 31:788–795.)

In Figure 1.23, shoot dry weight was 16% higher compared with grain yield across 12 genotypes. Regression analysis was performed to determine association between shoot dry weight (X) and grain yield (Y) of the highest grain-yield-producing genotype (BRSGO Guará) and the lowest grain-yield-producing genotype (BRS Jaburu). Shoot weights of these two genotypes displayed significant quadratic association with grain yield. Similarly, Fageria, Baligar, and Jones (2011) also reported a significant quadratic relationship between shoot dry weight and grain yield of lowland rice (Figure 1.24). Snyder and Carlson (1984) found that shoot dry matter is positively correlated with grain yield in annual crops. Fageria and Barbosa Filho (2001) reported a highly significant ($r = 0.91^{**}$) positive correlation between shoot dry weight and grain yield of eight lowland rice genotypes. Similarly, Fageria and Baligar (2005) reported that shoot dry weight is an important plant component for determining grain yield in field crops. Fageria and Baligar (2001) and Fageria et al. (2004) also reported a quadratic relationship between shoot dry weight and grain yield in rice. Peng et al. (2000) found that yield improvement figures, released by IRRI in the Philippines for lowland rice cultivars after 1980, were due to increases in biomass production. This means that genotypes producing the highest grain yield should produce reasonably good dry matter yield.

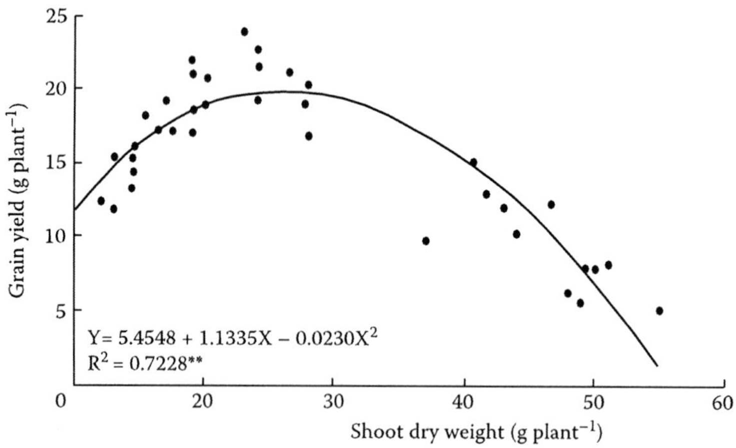

FIGURE 1.24 Relationship between shoot dry weight and grain yield of lowland rice.

Differences have been observed in grain yield among plants or genotypes having the same amount of dry matter due to differences in the utilization of photosynthates (Hayashi 1995). Nitrogen, phosphorous, and potassium fertilization influences shoot dry weight (Fageria, Slaton and Baligar 2003; Fageria and Baligar 2005). Grain yield in cereals is related to biological yield and GHI (Donald and Hamblin 1976). The biological yield of a cereal crop is the total yield of plant tops and is an indication of the yield of the photosynthetic capability of a crop (Yoshida 1981). The biological yield is a function of crop growth duration and CGR at successive growth stages (Tanaka and Osaki 1983). GHI is the ratio of grain to the aboveground biological yield. The GHI is controlled by partition of photosynthates between harvesting and nonharvesting organs during the crop growth cycle. Hence, the economic yield is closely related to the crop growth process. Grain yield or economic yield can be increased either by increasing total dry matter production or by increasing GHI. Some authors have speculated that a further increase in grain yield through breeding in cereals such as rice can only be accomplished with an increase in total biological yield (Rahman 1984) and thus total straw yield. The highest GHI exhibited by California lowland rice cultivars under direct seeding was 0.59 (Roberts et al. 1993).

1.5.1.5 Leaf Morphology

Leaf morphological characters, such as erectness, length, width, thickness, toughness, and senescence, are important in determining the yielding capacity of a cultivar. Erect leaves permit greater penetration and more even distribution of light in the plant canopy and, hence, higher photosynthetic activity. Yoshida (1981) reported that photosynthesis of an erect leafed canopy was 20% higher compared with droopy leafed canopy when the LAI was extremely high (>7). Leaf angle has been closely correlated with N response in rice, barley, and wheat (Yoshida 1972). Erect leaves seem to be the result of a pleiotropic (simultaneous) effect from the dwarf gene. Therefore, this trait follows simple recessive modes of inheritance.

Wallace, Ozbun, and Munger (1972) reported that a single gene controls the short, stiff, upright leaf habit of rice. This unique gene has many pleiotropic effects; the recessive allele gives a reduced number of short internodes; shorter, wider, and more erect leaves; and a larger number of shorter panicles. This gene has been responsible for spectacular yield increases because of better distribution of light over the total leaf canopy surface, a better N response, and a higher harvest index.

The erect leaf trait is highly heritable, easily observed at early flowering, and easy to visually rate in pedigree rows of fixed lines (Jennings, Coffman, and Kauffman 1979). Leaf thickness is also a desirable leaf trait. Thicker leaves usually have higher densities of chlorophyll per unit of leaf area and, therefore, have greater photosynthetic capacities than thinner leaves (Craufurd et al. 1999). Leaf size is directly associated with leaf angle; short leaves tend to be more erect than longer ones. Furthermore, short leaves are usually more evenly distributed throughout the canopy, which permits less mutual shading of leaves and more efficient use of light for photosynthesis (Fageria, Baligar, and Clark 2006). Other leaf characteristics that have been associated with yield include leaf toughness, color, and senescence. Leaf toughness, which is important in preventing breaking during wind and rain, is related to thickness and lignification of leaf tissues.

More leaves remaining green during the grain-filling period is a desirable characteristic of rice cultivars. Slow senescence of the upper two or three leaves is desirable because it allows active photosynthesis and grain filling until the grain is fully mature (Jennings, Coffman, and Kauffman 1979). Slow senescence of the upper leaves is easy to evaluate visually at the physiological maturity of the spikelets. Leaf blades are the major source of remobilized N for panicles. This remobilization of N accelerates leaf senescence and results in a rapid decrease of photosynthetic activity. Preventing or slowing accelerated senescence seems to be a key for high-yielding rice cultivars. Top dressing of N fertilizer before or after heading is one way of preventing a rapid decline of leaf N and photosynthesis. However, its effect on yield was estimated to be only a 10% increase (Wada, Shoji, and Mae 1986; Mae 1997). Fageria and Baligar (1999) also reported that N application during booting or flowering keeps rice leaves greener during the grain-filling growth stage; however, grain yield was not increased. Shoot dry matter production exhibited a greater response to late-season application of N than did grain yield (Fageria and Baligar 1999). Fageria and Baligar (1999) also reported that rice plants need appreciable amounts of N during the vegetative growth stage to form an optimal number of panicles. If less N is applied during the vegetative growth stage, an optimal number of panicles will not form, and N application during booting or flowering will not increase grain yield.

1.5.1.6 Leaf Area Index

LAI is the area of the leaf surface per unit area of land surface. LAI is important in many areas of agronomy and crop production because of its influence on light interception, crop growth, crop water use, and crop weed competition (Sone, Saito, and Futakuchi 2009). Leaf area is measured by a leaf area meter. When this equipment is not available, leaf area can be determined by measuring the length and maximum width of each leaf and computing the area of each leaf based on the length–width method:

$$\text{Leaf area} = K \times L \times W \tag{1.1}$$

where K is the adjustment factor, L is the leaf length, and W is the leaf width. The value of K varies with the shape of the leaf, which in turn is affected by the cultivar, nutritional status, and leaf growth stage. However, under most conditions, the value of 0.75 can be used for all stages of growth except the seedling and maturity growth stages. For these two growth stages, a value of 0.67 should be used (Gomez 1972). The LAI can be calculated as follows:

$$\text{LAI (cm}^2\text{ m}^{-2}) = [\text{A (cm}^2) \times \text{tiller/plants (m}^{-2})]/10{,}000 \tag{1.2}$$

where A is the leaf area multiplied by the number of tillers or plants.

Alternative nondestructive measurement of LAI is possible with plant canopy analyzers such as the LI-COR LAI-2000 (LI-COR, Inc., Lincoln, NE; LI-COR, 1992, 2004) and the Delta-T Devices SunScan (Sone, Saito, and Futakuchi 2009). Sone, Saito, and Futakuchi (2009) reported that by restricting the range of LAI to <4 m^2 m^{-2}, the SunScan can be used to estimate LAI of different upland rice cultivars with various types of canopy development. Sone et al. further reported that the indirect method can reduce labor requirements for monitoring LAI.

Variations in LAI are an important physiological parameter that determines crop yield (Evans and Wardlaw 1976). Light interception by the canopy is strongly influenced by LAI, and it is an important parameter used for estimating yields for many crop growth models that use net photosynthesis, assimilate partitioning, canopy mass, and energy exchange (Fageria, Baligar, and Clark 2006). The LAI of rice increases as crop growth advances, and it reaches a maximum at about heading or flowering (Yoshida 1983). The increase in LAI is caused by an increase in tiller number or leaves on each tiller and in the size of successive leaves. Increasing LAI increases dry matter production, but net canopy photosynthesis cannot increase indefinitely because of increased mutual shading of leaves (Fageria 2007).

Among various environmental factors, N and P nutrients have the most marked effect on LAI by increasing the number of tillers as well as the leaf size (Fageria, Baligar, and Jones 2011). An N topdressing 28–44 days prior to heading markedly increases the leaf area of rice, and an N topdressing 16 days before heading is more effective in maintaining the leaf area or functions of leaves during ripening stage (Oritani 1995). Fageria, Baligar, and Jones (2011) reported that the optimum LAI for upland rice is about 2–3 at 85–100 days after sowing. This can be compared with the optimum LAI of lowland rice, which is about 4–7 (Yoshida 1972). Yoshida (1981) reported that an LAI of 5–6 is necessary to achieve maximum crop photosynthesis during the reproductive growth stage. An LAI of 4 at heading is sufficient to produce about 5 Mg grain ha^{-1} (Yoshida 1981). Oritani (1995) reported that maximum grain yield in lowland rice was achieved with an LAI of 6–7, depending on cultivars.

The low LAI of upland rice may be related to differences in plant type. Rice grown under upland conditions is often subjected to moisture stress and, in general, has fewer tillers and less leaf area than rice grown under lowland conditions (Chang and Vergara 1975; Fageria, Barbosa Filho, and Carvalho 1982). Genetic factors, plant densities, and spacing are other major factors influencing the leaf area of plants grown under field conditions (Fageria, Baligar, and Clark 2006).

1.5.2 Reproductive Growth Stage

The reproductive growth stage starts with the differentiation of panicle primordia and extends up to flowering. This stage includes formation of panicle size or number of spikelets per panicle. The potential size of crop yield is primarily determined in the reproductive growth stage. The reproductive growth stage is characterized by culm elongation, a decrease in tiller number, emergence of flag leaves (the last leaf), booting, heading, and flowering. Some aspects of panicle formation such as panicle size, compactness, and panicle exsertion are important traits of this growth stage. In the tropics, the reproductive growth stage generally is completed in 30 days for rice cultivars having a 130- to 140-day growth cycle (from sowing to physiological maturity). Adverse environmental conditions such as N deficiency, drought, low solar radiation, low or high temperature, and blast disease can reduce panicle size and, hence, grain yield (Fageria 2007).

In the reproductive stages of rice, anthesis and microsporogenesis (Yoshida, Satake, and Mackill 1981; Satake and Yoshida 1978), plants are very sensitive to high (>33°C) temperatures (Yoshida, Satake, and Mackill 1981). Extreme temperatures affect anther dehiscence, pollination, and pollen germination, leading to spikelet sterility (Yoshida, Satake, and Mackill 1981; Gunawardena, Fukai, and Blamey 2003). Water deficit or drought stress has similar effects (Liu et al. 2006) and exacerbates the problem by reducing transpiration cooling, thereby increasing canopy and tissue temperature (Garrity and O'Toole 1995; Cohen et al. 2005). Genotypic variation in heat tolerance during the flowering period (usually five to six days in duration) has been reported in *indica* and *japonica* species (Yoshida, Satake, and Mackill 1981; Matsui and Omasa 2002; Prasad et al. 2006; Jagadish, Craufurd, and Wheeler 2007; Jagadish, Craufurd, and Wheeler 2008; Jagadish et al. 2010). Spikelet fertility and grain yield QTL in rice during drought (Lafitte, Price, and Courtois 2004; Yue et al. 2005) and cold tolerance at microspore genesis (Andaya and Mackill 2003; Jagadish et al. 2010) have also been identified. A detailed description of the reproductive development of the rice plant is given by Counce, Keisling, and Mitchell (2000) and Fageria (2007).

1.5.2.1 Panicle Size

Panicle size is the panicle length or the number of spikelets per panicle. There is an inverse relationship between panicle size and panicle number per unit area because, due to the large number of panicles per unit area, source is a limiting factor to fill large sink size. Panicle size is influenced by nitrogen rate and also by genotype (Table 1.18). The size of each hull, or the length and width of the lemma and palea, reaches its maximum five to eight days before heading (Nakamoto, Inagawa, and Nagato 1987). However, after the flowering stage, the hull increases exclusively in weight. The hull weight continues increasing for about two weeks after anthesis and maintains a constant level thereafter (Matsuzaki 1995).

TABLE 1.18
Influence of Nitrogen Rate on Panicle Length (cm) of 10 Upland Rice Genotypes

Genotype	Low N Rate (0 mg N kg^{-1} Soil)	High N Rate (400 mg N kg^{-1} Soil)
CRO 97505	17.9	23.8
CNAs 8812	16.2	19.1
CNAs 8960	20.8	26.6
CNAs 8989	16.3	18.9
CNAs 8957	19.6	25.4
CNA 8540	13.9	18.4
CNA 8170	18.8	21.2
Primaveira	19.7	25.7
Canastra	17.6	20.9
Carisma	17.7	22.0
Average	17.9	22.2

Source: Fageria, N. K. 2007. Yield physiology of rice. *Plant Nutr.* 30:843–879.

IRRI (1989) suggested the following types of rice plants as being ideal for direct-seeded, irrigated rice and upland rice in order to increase productivity and yield stability:

1. Direct-seeded, irrigated rice: Should have 200–250 spikelets per panicle; three to four panicles per plant; no unproductive tillers; very sturdy stems; 90 cm tall; dark green, erect, thick leaves; vigorous root system; 100–130 days of growth duration; multiple disease and insect resistance; harvest index of 0.6; and 13–15 Mg ha^{-1} yield potential
2. Annual upland rice: Should have 150–200 spikelets per panicle; five to eight panicles per plant; very sturdy stems; erect upper leaves; droopy lower leaves; 130 cm tall; deep, thick roots; 100-day growth duration; multiple disease and insect resistance; 3–4 Mg ha^{-1} yield potential

1.5.2.2 Compact Panicle

Compact panicles are desirable in comparison to the spreading type. Little is known about environmental and genetic influences on panicle compactness. However, selection for the normal compact inflorescence is easy and effective in segregating populations (Jennings, Coffman, and Kauffman 1979).

1.5.2.3 Panicle Exsertion

Panicles should emerge completely from the flag leaf sheath so that part of the internode below the panicle base is exposed. The lower panicle branches often remain enclosed because the upper internode is short. Such enclosed spikelets are sterile or partially filled and are often blackened by secondary pathogens, resulting in yield losses (Jennings, Coffman, and Kauffman 1979).

1.5.3 Spikelet-Filling or Ripening Growth Stage

For rice, the spikelet-filling or ripening stage extends from flowering to physiological maturity. Physiological maturity of grain crops is usually defined as the attainment of maximum seed dry weight (Li, Chen, and Wu 2011). At physiological maturity, the crop has reached the maximum possible grain yield, and grains, which are no longer growing, merely lose water (Calderini, Abeledo, and Slafer 2000). This is sometimes also known as the maturity growth stage. Spikelets filling or weight is determined in this growth stage. In the tropics, the dry weight of the caryopsis increases rapidly up to 15–20 days after flowering and, under temperate conditions, 25–30 days after flowering. Some adverse environmental factors such as drought, low solar radiation, N deficiency, low or high temperature, and panicle blast can increase spikelet sterility and consequently affect grain yield (Yoshida 1981; Wada et al. 2011). Spikelet sterility is also genetically controlled. Like the reproductive growth stage, under Brazilian conditions or in the tropics, the spikelet-filling growth stage has a duration of about 30 days for cultivars having a 130- to 140-day growth cycle from sowing to physiological maturity. In the tropics, the maximum varietal range is about 25–35 days. The period from flowering to maturity often ranges from 45 to 60 days in temperate areas where yields are usually high. The spikelet-filling duration of *japonicas* is often slightly longer than that of *indicas* (Jennings, Coffman, and Kauffman 1979). Spikelet weight and spikelet sterility are the main features of this growth stage.

During the spikelet-filling growth stage, the LAI will decrease due to leaf senescence. This is a normal process in the growth cycle of rice. However, it is important to maintain as many active, green leaves as possible until the linear phase of spikelet growth is completed. The ripening growth stage is subdivided into several stages such as milky, doughy, yellow-ripe, and maturity, which use the size, weight, and color of the rice spikelets or grains as indices (Murayama 1995). During the ripening growth stage, the morphogenesis of the rice plant is already completed, and photosynthates are accumulated in the panicles in the form of starch. Mobile carbohydrates, proteins, and mineral nutrients, which are stored in the leaves, stems, and roots of the plant, also move to the panicles, and the plant becomes gradually senescent (Murayama 1995).

More than 85% of rice grain is composed of carbohydrates, of which the main ingredient is starch (Hayashi 1995). The starch accumulated in rice grains originates from carbohydrates assimilated in laminae after flowering, during the grain-filling growth stage, as well as from the carbohydrates stored in the shoot until flowering (Hayashi 1995). In order to increase grain yield, it is necessary to increase the carbohydrate production during the ripening growth stage and to store carbohydrates. In most cases, the amount of carbohydrates assimilated during the ripening growth stage is much higher, compared with that of carbohydrates stored in the shoot at the flowering stage (Hayashi 1995). Therefore, grain yield is affected mainly by the amount of carbohydrates assimilated during the ripening growth stage, especially in the high-yielding modern cultivars (Hayashi 1995).

In the high-yielding cultivars, N absorption generally remains high during the spikelet-filling growth stage (Osaki et al. 1991b). This means root activity should remain high during maturation, indicating that photosynthates must be translocated to roots. Moreover, when a large amount of N is supplied to the leaves from the roots, photosynthesis should remain high during maturation in order to secure the supply of carbohydrates to the roots. Hence, growth or activity of roots and shoots was assumed to be mutually regulated (Osaki et al. 1991b). Osaki and Tanaka (1978) observed that the photosynthetically fixed carbon in the leaves of rice was rapidly translocated to harvesting organs during ripening. Therefore, a large part of the carbon in the harvesting organs was considered to be derived from concurrently assimilated photosynthate.

1.5.3.1 Spikelet Sterility

Spikelet sterility is an important yield component in rice, and reducing spikelet sterility is one way to improve yield. Overall, the filled spikelet percentage is about 85% in rice, even under favorable conditions (Yoshida 1981). It is possible to increase rice yield by 15% if breeding eliminates spikelet sterility. The increase in photoassimilates during the spikelet-filling growth stage is one way to improve the spikelet-filling rate. However, of the 15% unfilled spikelets, about 5–10% are unfertilized and difficult to eliminate (Yoshida 1981). When the filled spikelet number is more than 85%, yield capacity or sink is yield limiting; when the ripened spikelet number is less than 80%, the assimilate supply or source is yield limiting (Murata and Matsushima 1975).

Tanaka and Matsushima (1963) reported that the amount of carbohydrates stored in the shoot at the flowering stage improved spikelet filling by acting as a buffer substance in a case where a plant faced unfavorable conditions. Hayashi (1995) reported improved spikelet filling by the large amount of carbohydrates accumulated during flowering and that differences exist among cultivars in the amount of carbohydrate accumulation at the flowering growth stage. Furthermore, Hayashi (1995) found that the accumulation of a large amount of carbohydrates in the shoot before flowering reduces spikelet degeneration.

The percentage of ripened spikelets decreases when the number of spikelets per unit area increases (Yoshida 1981). Therefore, there must be an appropriate number of panicles or spikelets per unit area in order to achieve maximum yield. During the spikelet ripening, about 70% of the N absorbed by the shoot will be translocated to the spikelet to maintain N content at a certain level. Hence, less N absorption until flowering may reduce the N level in the spikelet, which induces higher spikelet sterility (Yoshida 1981).

Spikelet sterility is higher with low solar radiation at the spikelet-filling growth stage. When solar radiation is low, the source activity may be insufficient to produce enough carbohydrates to support the growth of all the spikelets. As a result, the number of unfilled spikelets may increase. Low or high temperatures may cause spikelet sterility. Spikelet sterility (which is induced by low temperatures, particularly when those temperatures occur during the reproductive growth stage) is a major constraint on rice production in cool climates (Shimono et al. 2005). The sensitivity of spikelet sterility to low temperatures varies during reproductive growth. The sensitivity is extremely high at the young microspore stage, which is a stage of active cell division, and it decreases as the plant develops beyond this stage (Hayase et al. 1969).

Air temperatures below 20°C may cause a high percentage of sterility if low temperatures persist for a few days at booting or heading (Yoshida 1981). Similarly, high temperatures (>35°C) at anthesis or flowering may cause high spikelet sterility. Incidences of diseases or insects during the reproductive or spikelet-filling growth stages may increase spikelet sterility. Besides breeding cultivars for low-temperature stress, other management practices (such as changes in sowing schedule, water management, and fertilization) have been widely introduced to Japanese farmers in an effort to prevent yield losses resulting from spikelet sterility caused by low temperatures (Wada 1992; Shimono et al. 2005). Satake et al. (1988) reported that proper water management is the most important practice to prevent yield losses that occur due to unusually low temperatures because deep flooding can warm the panicle for longer than can shallower waters.

Yamamoto et al. (1991) reported that, for rice, increasing the number of spikelets per panicle causes overproduction of spikelets on secondary branches; sterility of these spikelets is higher compared with spikelets on primary branches. Kato (1997) found that it is possible to develop genetically a rice panicle type that has a sufficient number of spikelets with a high filled grain percentage by increasing the number of primary branches and by suppressing the number of spikelets on secondary branches.

Data related to spikelet sterility of 12 lowland rice genotypes are presented in Table 1.19. Spikelet sterility was significantly influenced by genotype treatment, and N × G interaction was also significant for this yield component, varying from 5.92 to 28.50%, with an average value of 17.51% at a low N rate. At a higher N rate, spikelet sterility varied from 14.96 to 26.55%, with an average value of 19.95%. Yoshida (1981) reported that under normal conditions, rice spikelet sterility generally is about 15%. Spikelet sterility is genetically controlled and is also influenced by environmental factors such as mineral nutrition, temperature, drought, and diseases (Fageria 2009). As expected, spikelet sterility negatively impacts grain yield (Figure 1.25).

1.5.3.2 Spikelet Weight

Spikelet weight for rice is generally expressed in grams in terms of 1000-grain weight. Spikelet size is rigidly controlled by hull size; under most conditions, the 1000-spikelet weight of rice is a very stable varietal character. Data in Table 1.20 show spikelet weights of 10 lowland rice genotypes, varying from 24.7 to 27.4 g, with an average value of 26.2 g. Therefore, there was a difference of about 11% in spikelet weight between the lowest and the highest weight-producing genotypes.

1.5.3.3 Grain Harvest Index

The contribution of crop improvement to yield increases has been studied for rice grown in Asia (Peng and Khush 2003), North America (Tabien, Samonte, and McClung 2008), and South America (Breseghello et al. 2011). These studies showed increasing yield trends. Often, these increases were attributed to improvement in harvest index, which is associated with traits such as plant height, increase in biomass, or the combination of dry matter accumulation (Tabien, Samonte, and McClung 2008). GHI is the ratio of grain yield to total biological yield. This index is calculated with the help of the equation: GHI = (grain yield/grain + straw yield) (Fageria and Baligar 2005). Values for GHI in cereals and legumes are normally <1.

TABLE 1.19
Spikelet Sterility of 12 Lowland Rice Genotypes as Influenced by N Fertilization

Genotype	Low N Rate (0 mg N kg^{-1} Soil)	High N rate (300 mg N kg^{-1} Soil)
BRS Tropical	22.18abc	22.73ab
BRS Jaçanã	5.92e	19.54bc
BRA 02654	28.50a	24.33ab
BRA 051077	15.06cd	15.15c
BRA 051083	11.71de	14.96c
BRA 051108	22.14abc	22.26ab
BRA 051126	13.01de	16.60c
BRA 051129	13.21de	15.19c
BRA 051130	16.56bcd	19.78bc
BRA 051134	24.26ab	26.55a
BRA 051135	15.14cd	16.56c
BRA 051250	22.44abc	25.80a
Average	17.51a	19.95a
F-test		
N rate (N)	NS	
Genotype (G)	*	
N × G	*	
CV (%) N rate	20.64	
CV (%) genotype	11.93	

Note: Means followed by the same letter in the same column are not significantly different at the 5% probability level by the Tukey's test.

*,NS Significant at the 1% probability level and nonsignificant, respectively.

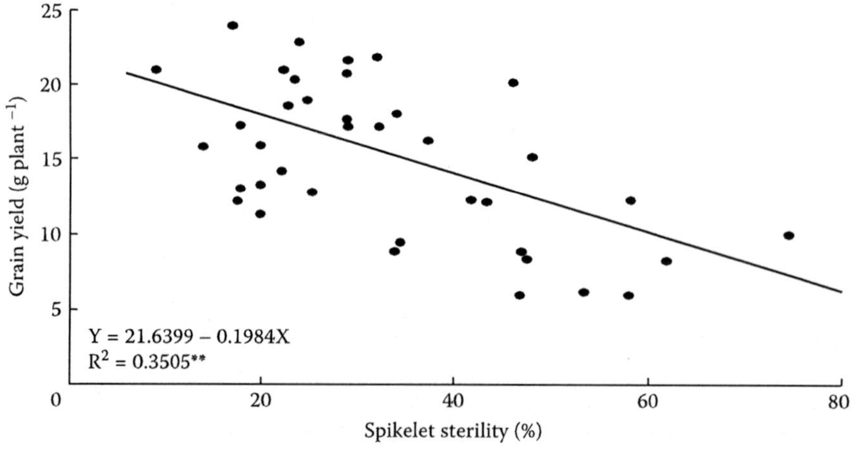

FIGURE 1.25 Relationship between spikelet sterility and grain yield of lowland rice.

TABLE 1.20
Spikelet Weight of 10 Lowland Rice Genotypes Grown in Brazilian Inceptisol

Genotype	1000-Spikelet Weight (g)
BRS Jaçana	25.3bcd
CNAi 8860	26.5abc
BRS Fronteira	27.0ab
CNAi 8879	27.4a
CNAi 8880	26.6abc
CNAi 8886	26.7abc
CNAi 8885	24.7d
CNAi 8569	26.0abcd
BRSGO Guará	25.0cd
BRS Alvorada	26.2abcd
BRS Jaburu	26.5abc
BRS Bigua	26.2abcd
Average	26.2

Source: Fageria, N. K. 2007. Yield physiology of rice. *Plant Nutr.* 30:843–879.

Note: Means in the same column followed by the same letter are not significantly different at the 5% probability level by Tukey's test.

Although GHI is a ratio, and it sometimes is also expressed in percentages. Generally, dry matter has positive associations with grain yield, and N is important for improving GHI (Fageria et al. 2004; Fageria and Baligar 2005). Snyder and Carlson (1984) reviewed GHI for selected annual crops and noted variations from 0.40 to 0.47 for wheat, 0.23 to 0.50 for rice, 0.20 to 0.47 for bunch-type peanuts (*Arachis hypogaea* L.), and 0.39 to 0.58 for dry beans. For major field crops, GHI values of modern crop cultivars are commonly higher than those of older traditional cultivars (Ludlow and Muchow 1990). Mae (1997) reported that GHI for traditional rice cultivars is about 0.30, and about 0.50 for improved semidwarf cultivars.

Rice GHI values varied greatly among cultivars, locations, seasons, and ecosystems and ranged from 0.35 to 0.62, thereby indicating the importance of this variable for yield simulation (Kiniry et al. 2001). Amano et al. (1996) reported a harvest index of 0.67 for *japonica* F_1 hybrid rice in Yunnan Province, south China. In Japan, Osaki et al. (1991a) reported GHI of 0.39 for standard (old) cultivars and 0.47 for modern high-yielding cultivars. Data in Table 1.21 show GHI of 19 upland rice cultivars under two N rates grown on Brazilian Oxisol. GHI varied from 0.36 to 0.52 at the low N rate, with an average value of 0.43. At the higher N rate, GHI varied from 0.33 to 0.57, with an average value of 0.50. The overall increase in GHI with the application of 400 mg N kg^{-1} was 19% compared with 0 mg N kg^{-1} of soil. GHI has a significant quadratic association with grain yield of lowland rice (Figure 1.26).

TABLE 1.21
GHI of 19 Upland Rice Genotypes Grown on Oxisol at Two N Rates

Genotype	0 mg N kg^{-1} Soil	400 mg N kg^{-1} Soil
CRO 97505	0.48a	0.54abcd
CNAs 8993	0.52a	0.57a
CNAs 8812	0.43a	0.49abcd
CNAs 8938	0.43a	0.52abcd
CNAs 8960	0.45a	0.54abc
CNAs 8989	0.49a	0.56abc
CNAs 8824	0.47a	0.50abcd
CNAs 8957	0.40a	0.56ab
CRO 97422	0.40a	0.50abcd
CNAs 8817	0.51a	0.48abcd
CNAs 8934	0.39a	0.46cd
CNAs 9852	0.36a	0.55abc
CNAs 8950	0.43a	0.52abcd
CNA 8540	0.41a	0.48abcd
CNA 8711	0.42a	0.49abcd
CNA 8170	0.41a	0.33e
BRS Primavera	0.49a	0.55abc
BRS Canastra	0.36a	0.44d
BRS Carisma	0.37a	0.47bcd
Average	0.43	0.50
F-test		
N level (N)	*	
Genotype (G)	**	
N × G	**	
CV (%)	9	

Source: Fageria, N. K., V. C. Baligar, and C. A. Jones. 2011. *Growth and Mineral Nutrition of Field Crops*, 3rd edition. Boca Raton, Florida: CRC Press.

Note: Means followed by the same letter in the same column are not significantly different at the 5% probability level by Tukey's test.

*,**Significant at the 5 and 1% probability levels, respectively.

1.6 YIELD AND POTENTIAL YIELD

Because ecophysiology influences growth and yield of rice (the main theme of this chapter), it is important to define yield. Yield is the amount of a specific substance produced (e.g., grain, straw, total dry matter) per unit area (Soil Science Society of America 2008). In the present case, it is grain yield that will be considered for discussion purposes. Grain yield refers to the weight of cleaned and dried grains harvested from a unit area. For rice, grain yield is usually expressed either in kilograms per hectare (kg ha^{-1}) or in metric tons per hectare (Mg ha^{-1}) at 13% or 14% moisture.

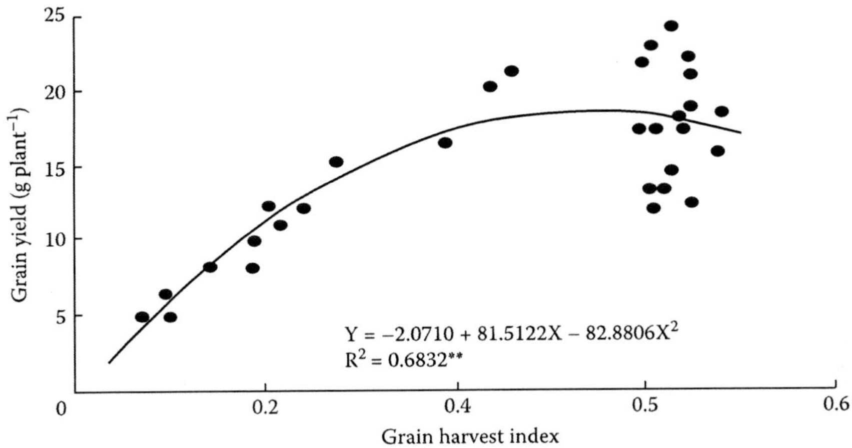

FIGURE 1.26 Relationship between GHI and grain yield of lowland rice.

The yield of a crop is determined by management practices that maintain the productive capacity of a crop ecosystem. These practices include use of crop genotypes, water management, and control of insects, diseases, and weeds.

Potential yield is an estimate of the upper limit of yield increase that can be obtained from a crop plant (Fageria 1992). *Genetic yield potential* is defined as the yield of adapted lines in a favorable environment in the absence of agronomic constraints (Reynolds, Rajaram, and Sayre 1999). Evans and Fischer (1999) defined potential yield as the maximum yield that could be reached by a crop or genotype in a given environment as determined, for example, by simulation models with plausible physiological and agronomic assumptions. These authors further reported that the term potential yield is often used synonymously with yield potential. However, Evans and Fischer (1999) defined yield potential as the yield of a cultivar when grown in an environment to which it is adapted, with abundant nutrients and unlimited water, wherein pests, diseases, weeds, lodging, and other stressors are effectively controlled. They further reported that there is no evidence that a ceiling on yield potential has been reached, but should this occur, average yields could still continue to rise as crop management improves and as plant breeders continue to improve resistance to pests, diseases, and environmental stresses.

There is a large yield gap between yield potential and yield obtained by farmers in most of the rice-growing regions of the world. The *yield gap* is defined as the difference between potential yields and on-farm yields obtained by farmers. Becker et al. (2003) reported that average on-farm yields of irrigated lowland rice in different agro-ecological zones in West Africa ranged from 3.4 to 5.4 Mg ha^{-1} but potential yields ranged from 6.9 to 9.8 Mg ha^{-1}. Yield gaps ranged from 3.2 to 5.9 Mg ha^{-1}, showing considerable scope for increasing yields.

Under tropical conditions, *indica*/*indica* hybrid rice increased yield potential by about 9% (Peng et al. 1999). The higher yield potential of *indica*/*indica* hybrids compared with *indica* inbred cultivars was attributed to the greater biomass production rather than harvest index. Peng et al. (1999) also reported that breeding of new plant types has not yet improved rice yield potential due to poor grain filling and low biomass production. However, in the Philippines, work at IRRI is in progress to remove this yield barrier and increase the yielding potential of rice.

Potential yield is, in a way, the most optimistic estimate of crop yield that is based on present knowledge and available biological material, under ideal management, in an optimum physical environment. Rasmusson and Gengenbach (1984) reported that the genetic potential of a plant or genotype is manifested through the interrelationships among genes, enzymes, and plant growth. These authors further reported that a gene contributes the formation for biosynthesis of an enzyme that functions in a particular metabolic reaction. The combined effect of many genes, through their control of enzymes, results in physiological traits contributing to plant growth, development, and yield (Rasmusson and Gengenbach 1984).

Generally, yield potential is determined by calculating photosynthesis during a spikelet-filling period (Murata and Matsushima 1975). For rice growing in an environment where the daily amount of solar radiation received is 16.7 MJ m^{-2}, assuming an efficiency of 26% in photosynthesis, the net carbohydrate production in a 40-day spikelet-filling period was calculated to be 16.4 Mg ha^{-1} (Austin 1980; Fageria 1992). Yoshida (1981) reported that the theoretical yield potential of lowland rice is about 16 Mg ha^{-1}, but the vagaries of climate scale the yield potential back to 8.5 Mg ha^{-1}. Over the years, rice yields have increased due to advances in breeding and crop management. New rice cultivars have been released that possess a yield potential of >10 Mg ha^{-1} (Ottis and Talbert 2005).

The average world rice yield is <4 Mg ha^{-1}. However, in experiments conducted by this author in the central part of Brazil, a grain yield of 6–7 Mg ha^{-1} for lowland rice is very common (Fageria and Baligar 2001; Fageria and Prabhu 2004). A grain yield of 10 Mg ha^{-1} for lowland rice was even obtained in a field experiment conducted in the central part of Brazil (Fageria and Prabhu 2004). Similarly, in the Brazilian state of Mato Grosso, upland rice on-farm yields varied from 1 to 7.8 Mg ha^{-1} with an average value of 3.6 Mg ha^{-1} (Fageria and Breseghello 2004). Similarly, Peng and Cassman (1998) reported a lowland rice yield of 9 Mg ha^{-1} at IRRI in the Philippines. This indicates that there is large gap between on-farm yields and those obtained at experimental sites. This gap can be reduced significantly if appropriate technology is adopted and the socioeconomic conditions of the rice farmers are improved.

1.7 YIELD COMPONENT ANALYSIS

Grain yield is the product of several components and can be estimated on the basis of the performance of these components. Rice yield is determined by yield components, which are number of panicles, spikelets per panicle, weight of 1000 spikelets, and spikelet sterility or filled spikelet. An analysis of yield components by Yoshida and Parao (1976) indicated that spikelet number m^{-2} alone explains 60% of yield variation, whereas the combination of all the yield components accounts for 81% of variation. Filled grain percentage and grain weight together account for 21% of variation. Therefore, it is very important to understand the management practices that influence yield components and, consequently, grain yield. Fageria and Baligar

(2001) reported that application of N in adequate amounts accounted for about 91% variation in panicles m^{-2}, about 75% variation in spikelet sterility, and about 73% variation in 1000-grain weight. As discussed earlier, the number of panicles is determined during the vegetative growth stage; the spikelet per panicle is determined during the reproductive growth stage; and the weight and spikelet sterility are determined during spikelet-filling or the reproductive growth stage. Hence, adequate N supply throughout the growth cycle of a rice plant is one of the main strategies used to increase grain yield. Rice yield can be expressed in the form of the following equation by taking into account the yield components:

$$\text{Grain yield (Mg ha}^{-1}\text{)} = \text{Number of panicles m}^{-2} \times \text{spikelet per panicle} \times \text{\% filled spikelet} \times \text{1000-spikelet weight (g)} \times 10^{-5} \quad (1.3)$$

Among these yield components, panicles or spikelet per unit area is usually the most variable yield component (Fageria 2007; Fageria, Baliga, and Jones 2011). The number of panicles per unit area is determined during the period up to about 10 days after the maximum tiller number is reached (Murata and Matsushima 1975). Generally, variation in grain yield is on the order of number of panicles or density > spikelet sterility > 1000-grain weight (Fageria 2007). Fageria and Barbosa Filho (2001) reported that correlations between lowland rice grain yield and yield components were number of panicles ($r = 0.96^{**}$), number of grains per panicle ($r = 0.44^{**}$), 1000-grain weight (-28^{NS}), and grain sterility ($r = -0.18^{NS}$). However, in another experiment, Fageria, Baligar, and Jones (2011), and Fageria, Gheyi, and Moreira, (2011) reported significant quadratic association between grain yield and 1000-grain weight (Figure 1.27). Thousand-grain weight is the key component trait for grain yield, and it is composed of three grain shape traits—grain length, grain width, and grain thickness (Zheng et al. 2011).

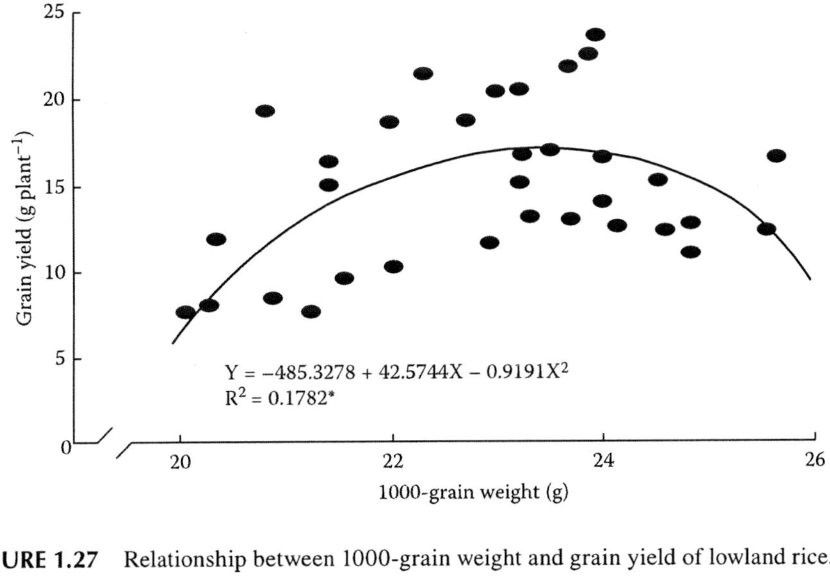

FIGURE 1.27 Relationship between 1000-grain weight and grain yield of lowland rice.

Gravois and Helms (1992) reported that optimum rice yield could not be attained without optimum panicle density of uniform maturity. Similarly, Ottis and Talbert (2005) found a high correlation ($R^2 = 0.85^{**}$) between yield and panicle density. The most important factor for the determination of spikelet number during the reproductive growth stage is the amount of N absorbed, although photosynthesis also contributes to the spikelet number determination (Ishii 1995). Similarly, the specific absorption rate of N per root dry weight during the grain-filling stage is the most important factor for achieving high rice productivity (Osaki et al. 1995).

The grain yield in rice is determined by carbohydrates accumulated in the plant before and after heading. Carbohydrates produced before heading mainly accumulate in the leaf sheath and stem and are known as nonstructural carbohydrates (NSC) and are translocated to the panicles during grain filling. Carbohydrates that form in the plant after heading also are translocated to the grain. Cultivars with a high NSC content in the stem at full heading ripen well even under adverse climatic conditions such as low radiation or extreme temperatures because NSC can compensate for the reduction in the newly assimilated carbohydrates after heading (Okawa, Makino, and Mae 2003; Morita and Nakano 2011). Murata and Matsushima (1975) reported that the contribution of the carbohydrates produced before heading to the final grain yield appeared to be in the range of 20–40%. Hence, about 70% of the grain yield is produced from the carbohydrates produced after heading, and photosynthesis after heading is vital for yield sustainability. The major photosynthetic organs after heading are considered to be the flag leaves. However, the contribution of the second- through fourth-positioned leaves to the grain yield was also fairly large (Ishii 1995).

1.8 RICE RATOONING

Ratooning, the practice of obtaining a second crop from tillers originating from the stubble of a harvested crop, may be one form of potential technology that will increase rice production per unit area and per unit time (Mengel and Wilson 1981; Chauhan, Vergara, and Lopez 1985; Santos, Fageria, and Prabhu 2003). In areas where adequate water is available after the main crop season, rice ratooning could be practiced as an alternative to double cropping. It is particularly suited to areas where monocropping is practiced and where resources go to waste in the off-season because no crops other than rice can be grown under the climate and moisture limitations (Krishnamurthy 1988). Elias (1969) reported that a ratoon crop can be grown with 50–60% less labor because no land preparation or planting is required, and the crop matures with about 60% less water than the main rice crop. Therefore, ratooning rice has a shorter growth duration and requires less labor and inputs than a newly planted crop. Use of this technology reduces cost of production, utilizes resources more efficiently, and lowers environmental pollution.

The importance of ratoon cropping as a way to enhance yield without increasing land area and at a lower per unit production cost has been emphasized (Plucknett, Evenson, and Sanford 1970). Ratooning forage crops, sorghum, sugarcane, and forest trees is successfully practiced in many countries (Mahadevappa and Yogeesha 1988) such as India, Japan, Swaziland, the United States, Colombia, the Philippines, Indonesia, Ethiopia, Puerto Rico, Brazil, and Thailand (Plucknett, Evenson, and

Sanford 1970; Bahar and De Datta 1977; Chauhan, Vergara, and Lopez 1985). However, ratooning is practiced only in limited areas (Bahar and De Datta 1977). In Asia, where most rice is produced and consumed, rice ratooning for large-scale commercial farming has not been accepted, probably because of a lack of varieties with good ratooning ability and a lack of information about management practices (Chauhan, Vergara, and Lopez 1985). Although rice has been ratooned in the Gulf Coast areas of the United States since the early 1960s (Jones 1993), until recently the practice of ratooning had been of minor importance because ratoon grain yields were low and grain quality was generally inferior to that of the main crop (Webb, Bollich, and Scoot 1975). However, rice ratooning has several advantages, such as low production costs (Alvarez 1992), highly efficient water use (Prashar 1970), and reduced growth duration (Haque and Coffman 1980). The introduction of early maturing rice cultivars possessing good ratooning ability initially stimulated early ratoon research efforts in the southern United States, and recently ratooning received renewed interest as a way to lower unit production costs (Jones 1993). Previous studies cite the main disadvantages of rationing as insect, weed, or disease problems (Plucknett, Evenson, and Sanford 1970). However, the situation has changed in the past three decades with research on disease and insect control through host resistance (Khush 1977). All international as well as national rice improvement programs devote major efforts to incorporating genetic resistance to major diseases and insects into their breeding materials. As a result, improved varieties with multiple resistances to several diseases and insects are now available to rice growers (Khush 1981; Tu et al. 1998; Sanchez et al. 2000; Fageria and Scriber 2001). Those varieties and others under development will play crucial roles in increasing the world's rice production. The objective of this section is to summarize rice ratooning management practices. This information may be useful in improving ratoon rice yields. In addition, it is hoped that this information will stimulate interest in ratoon research wherever ratooning is economically feasible.

1.8.1 Ratooning Mechanisms

The word "ratoon" seems to have originated from the Latin *retoñsus*, which means to cut down or mow (Lewis and Short 1958). Winburne (1962) defined ratooning as a basal sucker for propagation, such as in bananas (*Musa acuminata* Colla.), sugarcane (*Saccharum officinarum* L.), and pineapples (*Ananas comosus* L. Merrill). In addition to rice, ratooning is also applied to other crops such as sorghum (*Sorghum bicolor* L. Moench), cotton (*Gossypium hirsutum* L.), and pearl millet (*Pennisetum* glaucum L. Br.). The first harvest of a crop is called the main or principal crop, and each succeeding harvest is designated first ratoon, second ratoon, and so on.

Rice ratooning depends on the ability of dormant buds on the stubble of the first crop to remain viable (Chauhan, Vergara, and Lopez 1985). The buds exist in various stages of development (Nair and Sahadeva 1961). Axillary buds that develop at those nodes grow into ratoon tillers. Ratoon crops will yield well if the main crop stubble is left with two to three nodes; tillers regenerated from higher nodes form more quickly, grow faster, and mature earlier (Chauhan, Vergara, and Lopez 1985). The growth and vigor of ratoon tillers depend on the carbohydrate reserves of the stubble and the root system after the harvest of the main crop. The higher the concentration of carbohydrates in the stubble and the roots, the faster the development of ratoon tillers. In addition, ratoon development is a varietal characteristic (Santos, Fageria, and Prabhu 2003).

1.8.2 Ratoon Crop Management Practices

A good system of crop management requires the use of cultural practices at an adequate level, from sowing to harvesting, in order to produce a fairly good yield from the land at a reasonable cost without degrading the environment. Ratoon crops have been shown to respond to cultural practices applied to the planted (main) crop and the ratoon crop (Mengel and Wilson 1981). Therefore, the management of the ratoon crop begins with the management of the main crop. Naturally, harvest time, cutting height, fertilizer application, water management, plant protection, and weed control for the main crop have a great influence on the growth and yield of the ratoon crop. Therefore, in developing practices for the ratoon crop, special practices required for the main crop need to be worked out as well. Management practices for main crops are discussed by Santos, Fageria, and Prabhu (2003). Cultural practices that promote rapid and uniform new flush are important. Among the cultural practices that affect ratoon are the cutting height of the main crop, N fertilization, irrigation management, and phytosanitary measures. These management practices for a ratoon crop are outlined as follows.

1.8.2.1 Cutting Height

The height for harvest of the main crop, also known as the cutting height, is an important factor in determining the tiller number of a ratoon crop. Stubble height determines the number of buds available for regrowth (Chauhan, Vergara, and Lopez 1985) and ratoon tiller origin (Santos, Fageria, and Prabhu 2003). Different cutting heights result in significantly different grain yields, 1000-grain weights, panicles m^{-2}, and the growth duration of the ratoon crop (Samson 1980). Santos, Fageria, and Prabhu (2003) reported that ratoon characteristics most affected by cutting height are grain yield, tillering, and growth duration. Bahar and De Datta (1977) found that a 15-cm cutting height is better than ground level cutting. However, at harvest, they found similar ratoon crop yields at the ground level and at a 15-cm cutting, if the plots were drained during the main crop harvesting and then irrigated 12 days after harvest. These authors recommended a 15- to 20-cm optimum cutting height for the main crop. Quddus (1981) and Samson (1980) also reported similar results. Zandstra and Sampson (1979) concluded that low cutting heights (<0.5 cm) would produce greater uniformity and higher ratoon yields in cases where water management could be controlled.

Jones (1993) evaluated the effects of main crop harvest cutting height on ratoon crop agronomic performance, yield, and yield components. Higher ratoon yields were obtained at cutting heights of 20–30 cm. Increasing cutting heights had little effect on ratoon panicle number m^{-2} but significantly decreased filled grain number per panicle, which led to decreased ratoon yields at the highest two cutting heights (40 and 50 cm). These findings are in agreement with some earlier studies

(Bahar and De Datta 1977), but they are in conflict with others (Prashar 1970). Ratoon tiller origin varies among cultivars, with some cultivars initiating the majority of ratoons at either basal, near basal, or axillary nodes (Hoff and Bollich 1990). Therefore, cutting heights can have variable effects on ratoon performance, depending on the characteristics of ratoon initiation of the cultivar studied (Jones 1993). Panicle number per meter square and grain number per panicle have been shown to account for up to 85% of the variations in ratoon yield (Jones and Snyder 1987). Jones (1993) found that cutting height had no significant effect on panicles per meter square. However, grain number per panicle was the main yield component affected by cutting height. As cutting height increased, ratoon grain number per panicle decreased by >30%.

1.8.2.2 Nitrogen Fertilization

Among macronutrients, ratoon responds the most to nitrogen application, but adequate quantities of phosphorus and potassium applied to the main crop also significantly increase the growth and yield of ratoon. Rice ratoon crops respond well to N fertilization (Santos, Fageria, and Prabhu 2003); nitrogen stimulates regrowth and tillering. Work in the United States suggests that rates equivalent to 75% of the N applied to the first or planted crop can be utilized by the ratoon (Evatt 1966). When phosphorus and K were adequately supplied to the planted crops, ratoon grain yields did not increase (Mengel and Leonards 1978). Mengel and Wilson (1981) reported that application of 90 kg N ha^{-1} and reflooding immediately after harvesting of the main crop resulted in the highest ratoon yield. Various studies demonstrated that application of nitrogen increases grain yield of ratoon crops, but cultivars differ in their response. In general, the cultivars with greater yield potential in ratoon show greater response to fertilizer application. In order to obtain new flush more rapidly, healthy tillers, and an increase in grain yield, the N should be applied soon after the main crop harvest (Santos, Fageria, and Prabhu 2003). The availability of these elements after harvesting the plants is important for utilizing the carbohydrate reserves accumulated in the culm base and for ratoon growth. Ratoon produced the greatest response with an application of 56 kg ha^{-1} of N soon after the main crop harvest. Santos and Stone demonstrated that the application of 30 kg ha^{-1} of N and 60 kg ha^{-1} of N, after harvesting the main crop, increased grain yield up to 500 kg ha^{-1} (Santos, Fageria, and Prabhu 2003). The grain discoloration caused by *Drechslera oryzae* increased with increasing doses of N.

1.8.2.3 Irrigation Water Management

Adequate water management is essential for ratoon. Compared with the water demand of the main crop, ratoon only requires about 60% of that amount. The low water requirement and its efficient use are considered the principal advantages of a ratoon crop. Irrigation timing for ratoon crops has an effect on the regrowth and yield. Workers in the United States have recommended that water be flushed through the field to supply adequate moisture for regrowth but that flooding be delayed until the new ratoon tillers are 10–15 cm tall (Santos, Fageria, and Prabhu 2003). Bahar and De Datta (1977) found that the cutting height of the first crop influenced when the stubble should be reflooded. When the planted crop was cut at ground level, they obtained yield increases by delaying flooding until 12 days after harvest. However, when the planted crop was cut at 15 cm, timing of the flood during the first 16 days after harvest had no effect on yield. Bollich and Turner (1988) reported that as soon as the main crop is harvested and fertilizer applied, the field should be flooded 8–10 cm deep. If floodwater is not maintained, the main crop grain that falls to the ground during harvest will germinate and compete with the ratoon crop for light and nutrients. Santos, Fageria, and Prabhu (2003) found that the rice ratoon crop performance was affected differently by drainage periods after main crops were harvested depending on environmental conditions. Retardation in the flooding re-initiation decreased grain yield and grain quality when air temperature was unsuitable for rice ratoon crop development. Under suitable air temperature conditions, starting flooding at nine days after the main crop harvest showed the best performance of the ratoon crop and saved 14% of irrigation water. The ratoon has great potential for increasing overall yields where the intensive rice cultivation is limited due to irrigation water shortage. When temperature conditions are less favorable, the commercial value and chemical and physical characteristics of the grain are affected by irrigation management. Irrigation management before the main crop harvest does not affect the ratoon crop (Santos, Fageria, and Prabhu 2003).

1.8.2.4 Phytosanitary Measures

The application of fungicides is necessary for obtaining better yields and grain quality of the ratoon crop. Fungicide foliar applications reduce grain discoloration caused by *Dreschslera oryzae* and increase yield. The grain yield of the main and ratoon crops was negatively correlated with the number of larvae of *Oryzophagus oryzae* in two assessments. Even though a ratoon crop constitutes a guarantee for the survival of the species in the off-season, conditions were not favorable for population buildup. Apparently, adopting specific control measures is not necessary (Mengel and Leonards 1978; Santos, Fageria, and Prabhu 2003).

1.9 ABIOTIC AND BIOTIC STRESSES

Abiotic stresses such as salinity, allelopathy, drought, and soil fertility affect rice growth and development and, consequently, yield. A synthesized discussion about drought is given in the water requirements section of this chapter, and fertility will be discussed in chapters related to mineral nutrition. This section presents a synthesized discussion of salinity, allelopathy, diseases, insects, and weeds.

1.9.1 Soil Salinity

Salt-affected soils are those soils that have been adversely modified for the growth of most crop plants by the presence of soluble salts, with or without high amounts of exchangeable sodium (Soil Science Society of America 2008). Common ions contributing to this problem are Ca^{2+}, Mg^{2+}, Cl^-, Na^+, SO_4^{2-}, HCO_3^-, and, in some cases, K^+ and NO_3^- (Bernstein 1975; Fageria, Gheyi, and Moreira 2011). Salt-affected soils limit crop production worldwide. Civilizations have been destroyed by the encroachment of salinity on soils; as a result, vast areas of land are rendered unfit for agriculture. Salt-affected soils normally occur in arid and semiarid regions where rainfall is insufficient to leach salts from the root zone. However, problems related to salinity are not restricted to arid or semiarid regions. Under appropriate conditions, they can develop even in subhumid and humid regions (Bohn, McNeal, and O'Connor 1979). In addition, salt-affected soils may also be found in coastal areas subject to tides. Generally, salt originates from native soil and irrigation water. Roughly 263 million ha are irrigated worldwide, and in most of that area, salinity is a growing threat (Epstein and Bloom 2005). The irrigated area represents about 20% of the total land used for crop production (Fageria 1992). This represents about 19% of the total area under crop production worldwide. Use of inappropriate levels of fertilizers with inadequate management practices can create saline conditions even in humid conditions.

In the salt-affected environment, there is a preponderance of nonessential elements. Plants must absorb the essential nutrients from a diluted source in the presence of highly concentrated nonessential nutrients (Fageria, Gheyi, and Moreira 2011). This requires extra energy, and plants are not always able to fulfill their nutritional requirements. Salinity imposes two main stresses on plant growth. One is water stress due to the increase in osmotic potential of the rhizosphere as a result of high salt concentration. The other stress is the toxic effect of high concentrations of ions. Hale and Orcutt (1987) reported that if the salt concentration is high enough to lower the water potential by 0.05–0.1 MPa, then a plant is under salt stress. If the salt concentration is not this high, the stress is ion stress and may be caused by one particular species of ion (Hale and Orcutt 1987).

Salt stress reduces plant growth, including leaf area, which reduces the photosynthetic process and nutrient use efficiency. Figure 1.28 shows the effect of soil salinity on the dry weight of rice cultivar tops. Similarly, Figure 1.29 shows the effect of salinity on the growth of roots and tops for lowland rice cultivars. Overall, the top dry weight was constant up to a salinity level of electrical conductivity (EC) of 5 dS m^{-1}. When the salinity level was increased beyond this level, the top dry weight decreased linearly. With increasing salinity levels, the dry weight of lowland rice cultivars' roots was affected less than that of the tops (Figure 1.29).

1.9.1.1 Management Practices to Improve Rice Growth on Salt-Affected Soils

Successful crop production on salt-affected soils depends on soil, water, and plant management practices. Management practices that can improve crop yields and, consequently, nutrient use efficiency of plants grown on salt-affected soils are using soil amendments to reduce the effect of salts, applying farmyard manures to create favorable plant growth environments, leaching salts from the soil profile, and planting salt-tolerant crop species or genotypes within species (Fageria, Gheyi, and Moreira 2011). The addition of fertilizers, especially potassium, may also help in reducing salinity effects and improving nutrient use efficiency. The maintenance of an internal positive turgor potential of plants exposed to saline conditions is an important factor for maintaining growth (Fageria, Gheyi, and Moreira 2011). This is accomplished by the uptake of ions, chiefly K^+, Na^+, and Cl^-, as well as by synthesizing organic metabolites (Yeo 1983). Potassium is the most abundant cation in the

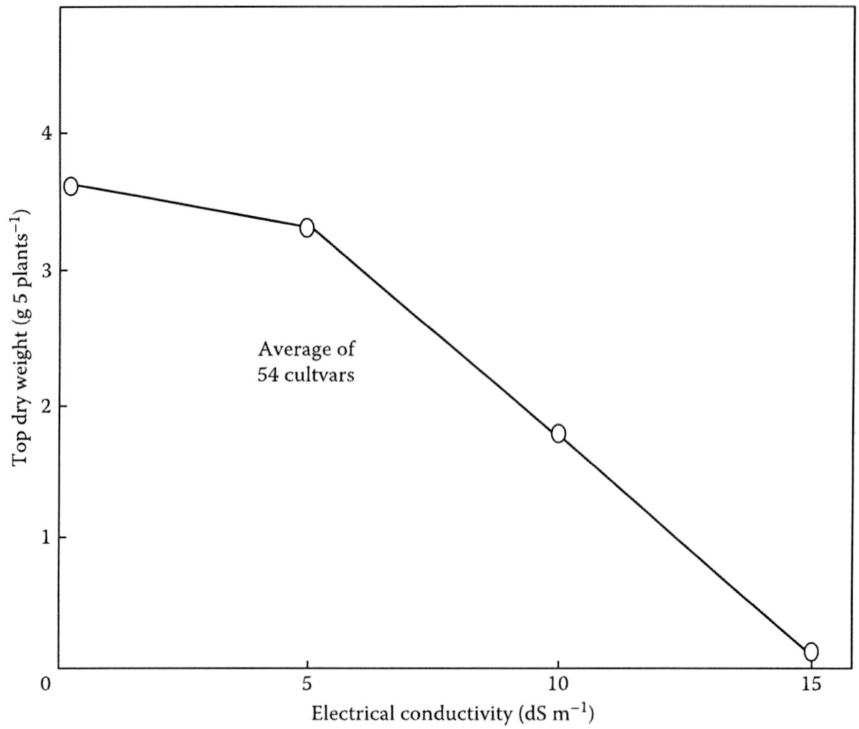

FIGURE 1.28 Influence of soil salinity on top dry weight of 54 lowland rice genotypes. (From Fageria, N. K. 1985. Salt tolerance of rice cultivars. *Plant Soil.* 88:237–243; Fageria, N. K. 1989. *Tropical Soils and Physiological Aspects of Crops.* Brasilia: EMBRAPA.)

cytoplasm and in glycophytes, and it plays an important role in osmotic adjustment (Marschner 1995). Thus, the application of high K^+ fertilization might enhance the osmotic adjustment capacity of plants growing in saline habitats.

Planting salt-tolerant crop species or genotypes within species is an attractive economically and environmentally sound practice that can be used to improve crop growth (including rice) on saline soils (Fageria, Gheyi, and Moreira 2011). Barley, cotton, oats, rye, triticale, sugar beets, guar, and canola or rapseed are all salt-tolerant crop species (Fageria, Gheyi, and Moreira 2011). Hilgard (1906) was among the first to recognize the significance of certain native plants as indicators of the characteristics of soils and to make use of them on salt-affected soil (United States Salinity Laboratory Staff 1973). With the agricultural research advancements of the twentieth century, crop cultivars of limited species have been developed that can grow and produce satisfactorily on salt-affected soils. However, commercial progress in releasing large numbers of crop cultivars has been limited due to a lack of genetic information on salinity and its interaction with environmental factors. But cultivars of corn and alfalfa (United States), rice and tomatoes (Egypt), and watermelons (Israel) are available commercially (Shanon 1990; Nobel and Rogers 1992; Fageria, Gheyi, and Moreira 2011).

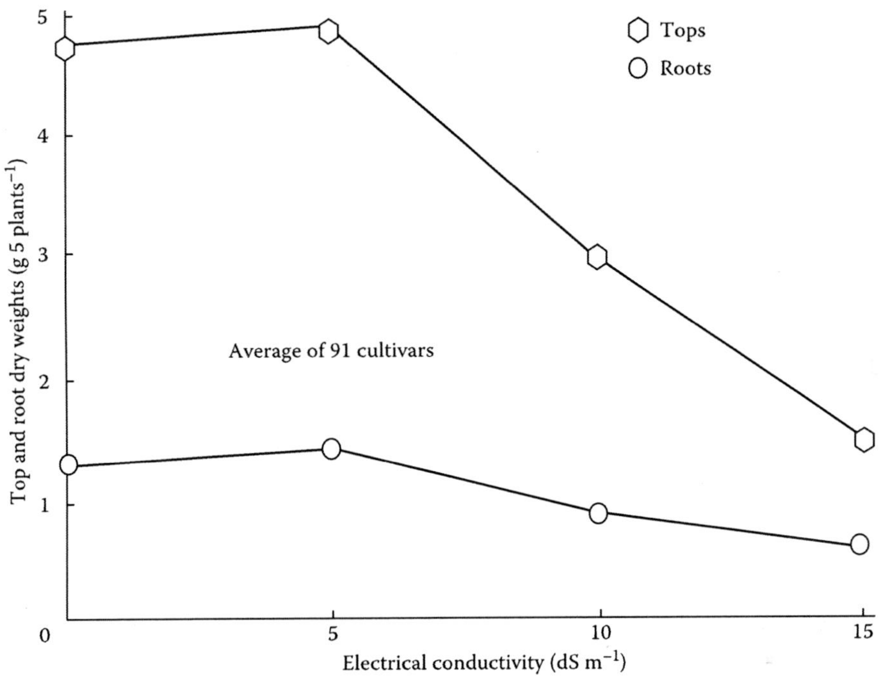

FIGURE 1.29 Influence of soil salinity on top and root dry weights of 91 lowland rice cultivars/genotypes. (Adapted from Fageria, N. K. 1992. *Maximizing Crop Yields*. New York: Marcel Dekker.)

Lowland rice genotypes and their soil salinity classification are displayed in Table 1.22. Similarly, Figure 1.30 shows the growth of two lowland rice genotypes under different salinity levels. Genotype CNA 810137 is highly tolerant to salinity compared with genotype CNA 810162. Similarly, Figures 1.31 through 1.33 show the growth of three lowland rice cultivars at different salinity levels. There were differences in growth among these cultivars at different salinity levels. Data in Table 1.23 show that lowland rice cultivars accumulate Na+ in different amounts; some are highly tolerant of Na+, whereas others are highly susceptible to Na+ toxicity.

A plant species' sensitivity to salinity changes during the ontogeny. It may increase or decrease depending on plant species and on environmental conditions. The sugar beet, for example, is highly tolerant during most of its life cycle, but it is sensitive during germination. In contrast, the salt sensitivity of rice, wheat, and barley usually increases after germination (Maas and Hoffman 1977; Marschner 1995). In corn, salt sensitivity is particularly high at tasseling and low at grain filling (Maas et al. 1983). In plants, tolerance to salinity and associated problems are influenced by a number of physiological, morphological, and ontogenetic characteristics (Chen et al. 2008).

Differences in the tolerance of crop species or genotypes of the same species to salts may be associated with different adaptation or tolerance mechanisms. Tolerance mechanisms used by plants to adapt to salinity can be separated into

TABLE 1.22
Influence of Soil Salinity on Dry Matter Yield (g 5 Plants^{-1}) of Lowland Rice Genotypes and Their Classification Based on Salinity Tolerance

	Electrical Conductivity (dS m^{-1}) and Genotype Classification						
Genotype	**0.29 (control)**	**5**	**10**	**15**	**5**	**10**	**15**
CNA 810098	3.30	3.25	2.76	0.28	T	T	S
CNA 810086	2.90	3.31	1.48	0.42	T	MS	S
CNA 810085	3.50	3.13	0.71	0.11	T	S	S
CNA 810162	3.76	2.85	0.97	0.20	MT	S	S
CNA 810118	5.22	4.99	3.23	0.87	T	MT	S
CNA 810127	3.85	3.07	1.54	0.43	MT	MS	S
CNA 810138	3.76	2.16	1.37	0.12	MS	S	S
CNA 810194	3.97	4.16	2.78	0.10	MT	MT	S

Source: Fageria, N. K. 1985. Salt tolerance of rice cultivars. *Plant Soil.* 88:237–243; Fageria, N. K. 1992. *Maximizing Crop Yields.* New York: Marcel Dekker.

Note: Genotypes were classified on the basis of dry matter yield reduction at each salinity level compared with control treatment. Genotypes were classified as T, 0–20% yield reduction; MT, 21–40% yield reduction; MS, 41–60% yield reduction; and S, >60% yield reduction.

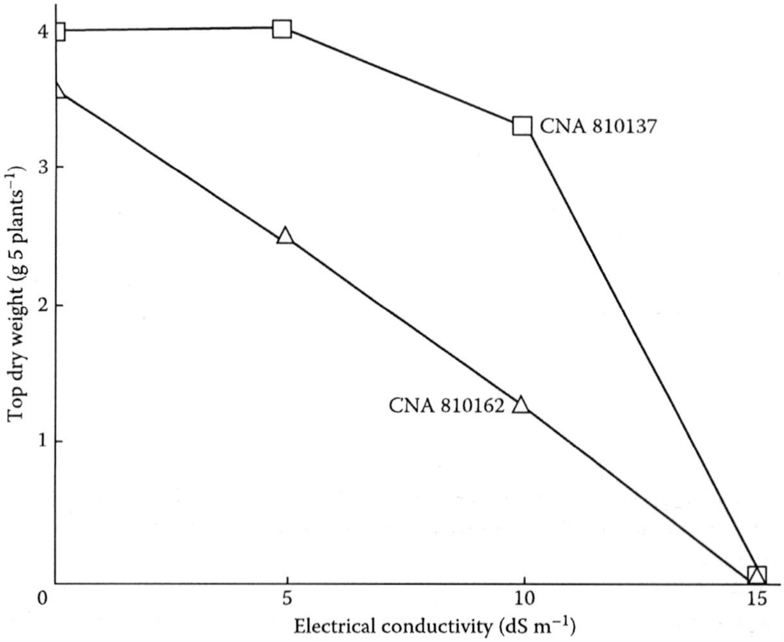

FIGURE 1.30 Influence of soil salinity on top dry weight of two lowland rice genotypes. (From Fageria, N. K. 1985. Salt tolerance of rice cultivars. *Plant Soil.* 88:237–243; Fageria, N. K. 1989. *Tropical Soils and Physiological Aspects of Crops.* Brasilia: EMBRAPA.)

FIGURE 1.31 Growth of lowland rice cultivar Blubelle at different salinity levels. (From Fageria, N. K. 2012. *The Role of Plant Roots in Crop Production*. Boca Raton, FL: CRC Press.)

FIGURE 1.32 Growth of lowland rice cultivar IR22 at different salinity levels. (From Fageria, N. K. 2012. *The Role of Plant Roots in Crop Production*. Boca Raton, FL: CRC Press.)

FIGURE 1.33 Growth of lowland rice cultivar CICA8 at different salinity levels. (From Fageria, N. K. 2012. *The Role of Plant Roots in Crop Production*. Boca Raton, FL: CRC Press.)

TABLE 1.23
Sodium Concentration in the Tops of Lowland Rice Cultivars and Their Classification to Salinity Tolerance

Cultivar	Na^+ Conc. (mol m^{-3})	Classification
Pokkali	39	Tolerant
Nova Bokra	62	Tolerant
IR 2153	50	Tolerant
IR 5	99	Moderately tolerant
IR 58	125	Moderately tolerant
IR 36	150	Susceptible
IR 22	247	Susceptible

Source: Adapted from IRRI. 1994. *Progress Report February 1993*. Los Banõs, Philippines: IRRI.

those that allow the growing cells of the plant to avoid high ion concentrations and those that permit the cells to cope with high ion concentrations upon exposure to salt (Greenway and Munns1980). Salt avoidance mechanisms include exclusion at the root, absorption by xylem parenchyma cells, xylem–phloem exchange systems, distribution of ion gradients between nongrowing and growing portions of the plant, and—for halophytes—sequestration of ions into the salt glands or trichomes (Hasegawa, Bressan, and Handa 1986). In general, exclusion mechanisms are effective at low to moderate levels of salinity, whereas ion accumulation is the primary mechanism used by halophytes at high salt levels, presumably in conjunction with the capacity to compartmentalize ions in the vacuole (Epstein et al. 1980).

Salinity tolerance in glycophytes is mainly associated with the ability to maintain high K^+/Na^+ ratios in shoots (Greenway and Munns 1980; Gorham et al. 1987; Maathuis and Amtmann 1999). A plant's ability to exclude Na^+ from uptake and a cell's ability to retain K (Chen et al. 2005, 2008) may both contribute to this trait. Characters such as yield, survival, vigor, leaf damage, and plant height have been the most commonly used criteria for identifying salinity tolerance (Chen et al. 2008). None of these characters are associated with a plant's ability to maintain an optimal cytosolic K^+/Na^+ ratio. More closely related indices of salinity tolerance have been proposed such as Na^+, K^+, and Cl^- concentrations and K^+/Na^+ tissue content ratio in shoots or roots or the production of specific metabolites in various species (Cuin and Shabala 2005; Chen et al. 2008). In understanding salinity tolerance mechanisms in plants or incorporating tolerance, it is necessary to find salinity tolerance genes in the crops (including rice germplasm) in order to find reliable screening methods and to investigate the genetic behavior of the tolerance (Chen et al. 2008).

1.9.2 Allelopathy

Originally, *allelopathy* was defined as the biochemical interactions among plants of all kinds, including the microorganisms that are typically placed in the plant kingdom (Fageria and Baligar 2003). Since then, the term has undergone several changes, and it is now defined as any direct or indirect harmful or beneficial effect by one plant on another through the production of chemical compounds that escape into the environment (Rice 1984). The International Allelopathy Society (1996) defined allelopathy as any process involving secondary metabolites produced by plants, algae, bacteria, and fungi that influences the growth and development of agricultural and biological systems. This definition considers all biochemical interactions among living systems, including plants, algae, bacteria, and fungi and their environment. Willis (1985) reported that the basic conditions necessary to demonstrate allelopathy in natural systems are (1) a pattern of inhibition of one species or plant by another must be shown; (2) the putatively aggressive plant must produce a toxin; (3) there must be a mode of toxin release from the plant into the environment; (4) there must be toxin transport and/or accumulation in the environment; (5) the afflicted plant must have some means of toxin uptake; and (6) the observed pattern of inhibition cannot be explained solely by physical factors or other biotic factors, especially competition and herbivory. However, Blum, Shafer, and Lehman (1999) reported that no study has ever demonstrated all of these criteria.

In Brazilian Oxisols, upland rice yield is significantly reduced due to auto-allelopathy after two or three years' consecutive monocultural planting in the same area (Tables 1.24 and 1.25). Olofsdotter (2001) reported that the majority of Brazilian upland rice germplasms were allelopathic. Experiments with rice

TABLE 1.24
Upland Rice Grain Yield in Rotation with Soybean and Monoculture Grown on Brazilian Oxisol

Crop Rotation	Grain Yield (kg ha^{-1})
Rice after three years of soybean	4325
Rice after one year of soybean	2577
Rice in monoculture for five years	1160

Source: Guimaraes, C. M., and L. P. Yokoyama. 1998. Upland rice in rotation with soybean. In *Technology for Upland Rice,* eds. F. Breseghello and L. F. Stone, pp. 19–24. Santo Antonio de Goias, Brazil: EMBRAPA Arroz e Feijão.

TABLE 1.25
Grain Yield of Four Crops Grown in Upland Rice–Dry Bean–Corn–Soybean Rotation on Brazilian Oxisol

Crop Rotation[a]	Grain Yield (kg ha^{-1})
First upland rice crop	4887
First dry bean crop after first upland rice crop	1674
First corn crop after first dry bean crop	8424
First soybean crop after first corn crop	1323
Second upland rice crop after first soybean crop	3982
Second dry bean crop after second upland rice crop	2152
Second corn crop after second soybean crop	8578
Second soybean crop after second corn crop	1560

Source: Fageria, N. K. 2001. Nutrient management for improving upland rice productivity and sustainability. *Commun. Soil Sci. Plant Anal.* 32:2603–2629.

[a] Rice and corn were planted during rainy season (November to March), and dry bean and soybean were planted during dry season (June to October) using sprinkler irrigation.

in the Philippines have shown that residual effects of allelochemicals reduced the yields of subsequent rice crops (Olofsdotter 2001). In Taiwan, Chou (1980) reported a 25% reduction in rice yield of a second crop that was attributed to the phytotoxins produced during the decomposition of rice residues left on the soil.

1.9.2.1 Rice Allelochemicals

The organic compounds involved in allelopathy are collectively called allelochemicals (Fageria and Baligar 2003). Rice (1984) and Fageria and Baligar (2003) have reviewed the chemistry of various allelochemical compounds in terrestrial ecosystems. These include simple phenolic acids, aliphatic acids, coumarins, terpenoids, lactones, tannins, flavonoids, alkaloids, cyanogenic glycosides, and glucosinolates.

Phenolic acids have been identified in allelopathic rice germplasm (Fageria et al. 2008) and have previously been described as allelochemicals (Fageria and Baligar 2003). Most are secondary metabolites released into the environment by leaching, volatilization, or exudation from shoots and roots. Many compounds are degradation products released during decomposition of dead tissues. Once these chemicals are released into the immediate environment, they must accumulate in sufficient quantity to affect other plants, persist for some period of time, or be constantly released in order to have lasting effects (Fageria et al. 2008). Abiotic (physical and chemical) and biotic (microbial) factors can influence the phytotoxicity of chemicals in terms of the quality and quantity required to cause injury. After entering the soil, allelochemicals encounter millions of soil microbes. The accumulation of chemicals at phytotoxic levels and their fate and persistence in soil are important determining factors for allelochemical interference. After entry into the soil, all chemicals undergo processes, such as retention, transport, and transformation, which influence their phytotoxic levels (Fageria and Baligar 2003).

Observation of apparent allelopathy in rice has recently drawn great attention, and there is much interest in identification of the allelochemicals (Fageria et al. 2008). Identification of the phytotoxic compounds responsible for allelopathy will allow efficient generation of more allelopathic cultivars through traditional breeding or biologically based genetic alterations. Such cultivars could become important tools in the development of advanced integrated weed management strategies for rice, which would mean less dependence on synthetic herbicides (Fageria et al. 2008). Various researchers have attempted to identify allelochemicals from rice (Fageria and Baligar 2003). Rimando et al. (2001) developed a bioactivity-guided isolation method with the objective of isolating the allelochemicals in rice. Identified compounds were azelaic acid, *p*-coumaric acid, 1H-indole-3-carboxaldehyde, 1H-indole-3-carboxylic acid, 1H-indole-5-carboxylic acid, and 1,2-benzenedicarboxylic acid bis(2-ethylhexyl) ester. Furthermore, Olofsdotter (2001) reported that in genetic mapping of QTL, four QTLs have been identified that correlate with allelopathy expression, measured as barnyardgrass root reduction. This indicates that allelopathy is quantitatively inherited.

Most of the allelochemicals are released into the rhizosphere through root exudates (Fageria and Baligar 2003). They are also released by decomposition of plant tissues. The toxicity of these allelochemicals is primarily controlled by their concentration in the rhizosphere. If they accumulate in high concentration, toxicity may be high and vice versa (Fageria et al. 2008). Abiotic (physical and chemical) and biotic (microbial) factors can influence the phytotoxicity of chemicals in terms of quality and quantity required to cause injury. After entering the soil, allelochemicals are exposed to millions of soil microbes. The accumulation of chemicals to determine phytotoxic levels and their fate and persistence in soil are important factors to determine the allelochemicals interference. After entering the soil, all chemicals undergo various processes (retention, transport and transformation), which influence their phytotoxicity levels (Fageria and Baligar 2003).

1.9.2.2 Ameliorating Upland Rice Allelopathy

Adopting appropriate soil and crop management practices can control the problem of upland rice allelopathy. These practices include (1) using appropriate crop rotation,

(2) improving or stabilizing soil organic matter content, (3) keeping land under pasture, (4) using an adequate rate of fertilizer, (5) adopting a conservation tillage system, (6) using of cover crops, and (7) planting allelochemical-resistant cultivars, if available.

1.9.2.2.1 Use of Appropriate Crop Rotation

Crop rotation is an ancient practice used in agriculture to sustain crop productivity. It has many benefits: improving soil fertility; controlling diseases, insects, and weeds; and improving water use efficiency. In addition, crop rotation brings diversification and reduces a farmer's risk of economic loss (Fageria and Baligar 2003). Generally, because of differences in root systems, a legume crop is planted in rotation with cereals in order to (1) supply N to crops and (2) extract nutrients and water from different soil depths due to differences in root system of legumes compared to cereals.

1.9.2.2.2 Improving Soil Organic Matter

Organic matter is the most important component of soil quality. It influences soil's physical, chemical, and biological properties, which in turn influences crop growth, development, and yields. Soil organic matter neutralizes the toxic chemicals produced during the allelopathic interactions of plants (Fageria and Baligar 2003; Fageria et al. 2008). It may coat mineral surfaces (such as Mn^{2+} and Fe^{2+}), which prevents the direct contact of phenolic acids and mineral ions and thus influences the oxidation of phenolic acids (Fageria and Baligar 2003). Phenolic compounds are polymerized into humic acids by Mn, Fe, Si, and Al oxides (Fageria et al. 2008).

Fageria and Souza (1995) conducted a field experiment in which they grew upland rice and common beans in rotation on central Brazilian Oxisol (Table 1.26). The upland rice yield of the third crop was significantly decreased compared with the first and second crops, even with high fertility treatments. However, unlike with other fertility treatments, medium soil fertility and green manure treatment maintained the upland rice yield of the third crop. Yet, green manure did not help in improving the common bean grain yield. Organic matter supplied by green manure helped in maintaining the yield of upland rice. Adding farmyard manure, green manuring, incorporating crop residues, adopting appropriate crop rotation, and conservation tillage can improve organic matter.

1.9.2.2.3 Keeping Land under Pasture

The use of land for pasture is important in restoring land degraded by upland rice allelochemicals. Experiments conducted at Brazil's National Rice and Bean Research Centre (EMBRAPA) showed that land degraded by upland rice allelochemicals (yield < 1000 kg ha^{-1}) can be restored to normal rice yield (>4000 kg ha^{-1}) by leaving it under pasture (Brachiaria grass) for four years and then planting soybeans during the rainy season and dry beans during the dry period (in irrigated conditions). The soil quality is improved by increasing the organic matter content by pasture soil pasture species crops. In addition, a large proportion of nutrients ingested by animals is returned to the soil through urine and feces (West et al. 1989). Animals retain only a small proportion (about 20% of the nutrients) they ingest, and the rest

TABLE 1.26
Response of Upland Rice and Common Bean to Chemical Fertilization and Green Manuring Grown in Rotation on a Brazilian Oxisol

Fertility Level[1]	First Upland Rice Crop[2]	First Bean Crop	Second Upland Rice Crop[2]	Second Bean Crop	Third Upland rice Crop	Third Bean Crop
	Grain Yield (kg ha^{-1})					
Low	2188a	1935b	2383a	866c	480c	890c
Medium	2428a	2382a	2795a	1831ab	1127b	1242ab
High	2330a	2568a	2657a	2432a	1324b	1486a
Medium + Green Manure	–	2344a	–	1202bc	2403a	1065bc

Source: Fageria, N. K., and N. P. Souza. 1995. Response of rice and common bean crops in succession to fertilization in Cerrado soil. *Pesq. Agropec. Bras.* 30:359–368.

Note: Values followed by the same letter in the same column are statistically not different by the Tukey's test at the 5% probability level.

[1] Soil fertility levels for rice were low (without addition of fertilizers); medium (50 kg N ha^{-1}, 26 kg P ha^{-1}, 33 kg K ha^{-1}, 30 kg ha^{-1} fritted glass material as a source of micronutrients); and high (all the nutrients were applied at the double the medium level). *Cajanus cajan* L. was used as a green manure at the rate of 25.6 t ha^{-1} green matter. For common beans, the fertility levels were low (without addition of fertilizers); medium (35 kg N ha^{-1}, 44 kg P ha^{-1}, 42 kg K ha^{-1}, 30 kg ha^{-1} fritted glass material as a source of micronutrients); and high (all the nutrients were applied at the double the medium level).

[2] In the first and third rice crop, plots with medium + green manure fertility level were planted with green manure and incorporated about 90 days after sowing (at flowering) and hence grain yield was not presented.

is returned to the soil through excreta (Rao et al. 1992). The expected buildup in soil fertility in a grass–legume pasture under grazing could result from a more rapid cycling and greater proportion of nutrients in a plant-available form. Appropriate management of pasture land improves the soil's biological activity and reduces soil erosion. Increased biological activity is beneficial for soil mineralization, humification, texture, porosity, water infiltration, and retention (Fageria and Baligar 2003).

1.9.2.2.4 Using an Adequate Rate of Fertilizers

Using an adequate rate of fertilizers is necessary to improve the soil fertility and consequently improve crop productivity (Fageria 2009). Deficiency of macro- and micronutrients increases the concentration of allelochemicals in crop plants (Rice 1984). This means that nutrient-deficient plant residues that are incorporated in the soil or left in the field after harvest will have higher concentrations of allelochemicals than plant residues supplied with adequate nutrients. It has been also reported that larger amounts of allelochemicals leach from living intact roots, dried roots, and tops of phosphate-deficient plants than from phosphate-sufficient ones (Fageria et al. 2008). Phosphorus deficiency limits upland rice yields in Brazilian Oxisol (Fageria and Baligar 2008) and, therefore, may aggravate the allelopathy problem. In addition, adequate fertilization improves crop production residues, which may also improve the soil organic matter content. Indirectly, this helps reduce the negative effects of allelochemicals.

1.9.2.2.5 Adopting a Conservation Tillage System

Conservation tillage is defined as a system that minimizes or reduces soil and water loss. It consists of reduced tillage, no tillage, or a combination of both, along with planting to keep 30% of crop residues as cover on the soil surface (Soil Science Society of America 2008). Tillage decreases the aggregate stability by increasing the mineralization of organic matter and exposing aggregates to additional raindrop impact energies. Tillage promotes soil organic matter (SOM) loss through crop residue incorporation into soil, physical breakdown of residues, and disruption of macroaggregates (Wright and Hons 2004). In contrast, conservation or no tillage reduces the soil mixing and soil disturbance, which allows SOM accumulation (Fageria et al. 2008). Greater SOM accumulation in conservation tillage has been observed in intensive multiple cropping systems (Wright and Hons 2004). The use of conservation tillage, including no-till, is being considered as part of a strategy to reduce the C loss from agricultural soils (Denef et al. 2004).

1.9.2.2.6 Use of Cover Crops

Cover crops are defined as the close-growing crops that provide soil protection and improvement between the periods of normal crop production, or between trees in orchards and vines in vineyards. When plowed under and incorporated into the soil, cover crops are called green manure crops (Fageria, Baligar, and Jones 2011). The positive role of cover crops in crop production has been known since ancient times, but the importance of this soil ameliorating practice has increased in recent years due to the high cost of chemical fertilizers, the increased risk of environmental pollution, and the need for sustainable cropping systems (Fageria, Baligar, and Bailey 2005). Cover crops can improve the soil's physical, chemical, and biological properties and, consequently, the crop yields. Cover crops can increase cropping system sustainability by (1) reducing soil erosion, (2) increasing SOM and fertility levels, and (3) reducing the global warming potential (Fageria, Baligar, and Bailey 2005).

1.9.2.2.7 Planting Allelochemical-Resistant Cultivars

Planting allelochemical-resistant upland rice cultivars is an important strategy for improving yields on allelochemically degraded lands (Fageria and Baligar 2003). This strategy not only improves the yields but also improves the soil quality and reduces the cost of rice production. Research has shown that there is wide variability in the allelopathic activity of different rice cultivars (Ebana et al. 2001; Olofsdotter 2001).

1.9.3 Diseases

For upland as well as lowland rice, diseases cause substantial losses in yield and quality of grain. Commercial crop loss data are not available and may vary with region, disease severity, and varietal susceptibility. The major rice diseases are rice blast, sheath blight, brown spot, leaf scald, and grain discoloration, and most are caused by fungus. These diseases can be controlled by the use of fungicides, by cultural practices, and by planting disease-resistant cultivars. Integrated management may be a better strategy for controlling these diseases, avoiding crop losses, and reducing environmental degradation.

1.9.3.1 Rice Blast

Rice blast, caused by *Magnaporthe oryzae* B. Couch (anamorph *Pyricularia oryzae* Cavra), is found worldwide and ranks first in order of economic importance for upland and irrigated rice. The pathogen affects all aerial parts of the plant from seedling to maturity. White to gray lens-shaped lesions with dark green borders are the characteristic symptoms on plant leaves. In severe cases, the lesions coalesce, causing partial or total death of the affected leaf. Often called neck blast or rotten neck, the panicle infection at the last node of the culum is most destructive; it produces white panicles similar to drought stress symptoms. Different parts of the panicles such as panicle branches, pedicels, and spikelets are often infected. Yield losses vary depending on the degree of cultivar resistance, cultural practices, and climatic conditions. Leaf blast at the vegetative phase and panicle blast during grain filling can critically affect yield and grain quality. Losses caused by leaf blast are indirect, affecting photosynthesis and respiration (Bastiaans, Rabbinge, and Zadoks 1994), whereas the losses caused by panicle blast are direct because of the impact on yield components (Pinnschmidt, Teng, and Yong 1994). Panicle blast reduces 1000-grain weight and the percentage of spikelets and filled grains (Prabhu and Faria 1982). Rice blast also causes chalky kernels, reducing the grain quality. Yield losses are greater under tropical conditions than under temperate conditions. In Brazil, the average yield loss among four upland rice cultivars was estimated to be 59.6% (Prabhu et al. 2003). The grain weight/panicle decreased with increases in panicle blast severity and resulted in losses of 72 and 31% in widely grown upland rice cultivars Primavera and BRS Bonança, respectively (Araújo et al. 2004).

1.9.3.2 Sheath Blight

Sheath blight, caused by *Rhizoctonia solani* Kuhn [*Thanatephorus cucumeris* (Frank) Donk.], is the second most economically important disease and manifests mainly in irrigated rice, causing severe loss in grain yield and quality worldwide (Lee and Rush 1983; Rush and Lindberg 1996; Sharma et al. 2009). *Rhizoctonia solani* is a semisaprophytic fungus with a broad host range; it affects many crops, including rice, corn, wheat, sorghum, common beans, and soybeans (Zhao et al. 2006). Typical symptoms, found on leaf sheaths near the water line at the late tillering or early internode elongation growth stages, consist of circular to irregular water-soaked spots that are pale green or white in color with a purple brown border. In tropical as well as temperate climates, the disease is widespread throughout the world's rice-growing areas. Yield loses can be as high as 50% for susceptible rice cultivars (Webster and Gunnell 1992). In the United States, yield losses varying from 19 to 41% have been reported (Marchetti and Bollich 1991). Brazilian studies have shown that the disease affects panicle length by 6–8%, empty spikelets by 10–12%, and grain weight/panicle by 30–32% in early- and medium-maturing genotypes of irrigated rice (Araújo, Prabhu, and Silva 2006). At present, the most common control measure for sheath blight is the application of preventive fungicides, which increases production costs and presents an environmental risk. Thus, rice breeders are seeking sheath-blight-resistant semidwarf cultivars with early maturity and high yield potential (Sharma et al. 2009). Rice cultivars with significantly different levels of sheath blight resistance have been identified (Groth and Novick 1992). Rice sheath blight resistance is generally believed to be a typical quantitative trait controlled by several genes. However, a few studies (Xie et al. 1992; Pan et al. 1999) have proposed that sheath blight resistance in some rice cultivars is controlled by only a few major genes. Sheath-blight-resistant QTLs have been identified and reported in several publications (Zou et al. 2000; Pinson, Capdevielle, and Oard 2005; Sharma et al. 2009). Sharma et al. (2009) found that consistent results across several years indicate the stability of the identified QTLs and their potential for improving rice sheath blight resistance using marker-assisted selection.

1.9.3.3 Brown Spot

Brown spot, caused by fungus *Bipolaris oryzae* (Breda de Haan) Shoemaker, is common on the top two leaves soon after panicle emergence and on glumes in upland and irrigated rice. This fungus produces circular to oval spots that are light brown to gray in color and surrounded by a reddish margin. It also causes dark brown to black oval spots on glumes. Brown spot is a major disease in Thailand and is common in West Africa, Brazil, and other Latin American countries. In India, the Bengal famine of 1942 was attributed mainly to yield losses of 40–90% caused by brown spot disease (Padmanabhan 1973). It is often associated with poor soils, and it is difficult to separate brown spot damage and nutritional deficiencies. Losses as high as 50% have been reported in West Africa (Gupta and O'Toole 1986). In a Brazilian experiment conducted with six upland rice cultivars, this leaf infection did not significantly contribute to yield component loss. Reduction in grain weight was mainly due to grain infection where the percentage loss was in the range of 12–30% in 1000-grain weight and 18–82% in filled grains/panicles (Prabhu, Lopes, and Zimmermann 1980).

1.9.3.4 Leaf Scald

Leaf scald, caused by *Monographella albescens* (Thume) has been found in the southern United States, Latin America, West Africa, and throughout Asia. The disease usually manifests itself at the initial booting stage, causing light and dark brown zonate lesions starting from the tip downward and at the leaf edges. Leaf scald also causes pinhead-sized lesions on the culms and spikelets (Webster and Gunell 1992). In Brazilian uplands with continuous rain, the disease affects growth and development in Cerrado soils. Yield losses are indirect, and there are no reliable estimates.

1.9.3.5 Grain Discoloration

Grain discoloration, caused by different fungi and bacteria, occurs in all rice-growing regions of the world. The commonly associated fungi and bacteria are *Bipolaris oryzae, Phoma sorghina, Gerlachia oryzae, Gibberella fujikuroi, Sarocladium oryzae, Alternaria padwickii, Curvularia spp., Pseudomonas glumae, P. fuscovaginae,* and *Erwinia herbicola.* Grain quality is greatly affected, thereby reducing its marketable value (Webster and Gunell 1992). In Brazil, the estimated loss in panicle weight caused by glume blight (*P. sorghina*), under epidemic conditions, ranged from 29 to 45% (Prabhu and Bedendo 1988). Under field conditions, panicle infection of rice cultivars IAC 101 and IAC 103 from *B. oryzae* caused significant reductions in grain weight, grain number, and 1000-grain weight (Malavolta et al. 2007). Empty spikelets in the upland rice cultivar BRS Bonança are positively correlated with grain discoloration (Silva-Lobo et al. 2011).

1.9.4 Insects

Rice is cultivated in different soils and climates, but it grows best in warm and humid regions. Unlike colder regions, these climatic conditions favor insect infestation, which can cause severe damage to lowland as well as upland rice. Crop losses caused by insects are variable and difficult to estimate because severity varies greatly from place to place and from year to year owing to changes in environmental factors. However, Kramer (1967) estimated that average worldwide losses due to insects for the main agricultural crops were about 12.2%. Gupta and O'Toole (1986) reported that yield losses range from a modest 14% to a high of 30% in West Africa, and average about 29% in Brazil. In Brazil, *Elasmopalpus lignosellus* (Zeller) and *Tomaspis* (= Deois) *flavopicata* (Stal) damage rice seedlings and sometimes make it necessary to replant large areas (Gupta and O'Toole 1986). Insects reduce nutrient and water use efficiency, thereby increasing production costs. They also reduce crop quality and, therefore, lower growers' agricultural produce prices (Fageria 1992). The greatest loss (13%) is caused by soil insects, then by vegetative or seedling phase insects (10%), followed by reproductive-phase and ripening-phase insects (7% and 2%, respectively).

At IRRI, in 24 separate experiments with lowland rice during six cropping seasons, De Datta (1981) reported that plots protected from insects produced an average of 5.3 Mg ha^{-1}, whereas unprotected plots averaged 2.9 Mg ha^{-1}. These results showed that the importance of insect control for lowland rice production is highly significant. Primary rice insects include stem borers, leafhoppers and planthoppers, armyworms and cutworms, grain-sucking insects, rice mealy bugs, rice leaffolders, seedling flies, white grubs, termites, rice water weevils, and rice beetles. Insects can be controlled by the use of insecticides, by cultural practices, and by biological control. In addition, the use of insect-resistant cultivars is an important strategy in minimizing insect damage to rice crops. However, an integrated management approach is the best strategy for controlling rice insects in modern agriculture. A detailed discussion of insect control for rice is presented by De Datta (1981) and Gupta and O'Toole (1986).

1.9.5 Weeds

Weeds are a serious problem for upland as well as lowland rice cultivation and are estimated to account for 32% potential and 9% actual yield losses in worldwide rice

TABLE 1.27
Upland Rice Yield Losses from Insects in Philippines

Parameter	First Year	Second Year	Third Year	Mean
Yield of insects, controlled treatment (Mg ha^{-1})	5.1	3.7	3.1	4.0
Yield of insects, uncontrolled treatment (Mg ha^{-1})	3.1	2.5	2.3	2.6
Yield loss (Mg ha^{-1})	2.0	1.2	0.8	1.3
Yield loss (%)	39	32	26	32

Source: Adapted from IRRI. 1988. *Annual Report for 1987*. Los Baños, Philippines: IRRI.

production (Oerke and Dehne 2004). However, the nature and severity of weed problems vary according to the rice ecosystem (Rodenburg and Johnson 2009). After flooding, weeds in lowland rice can be controlled. But in upland rice, weeds continue to be a problem throughout the growing season. They compete with crop plants for light, water, and nutrients. Weeds host insect pests and diseases, require expensive labor and energy to control, reduce harvesting and processing efficiency, and sometimes are poisonous (Gupta and O'Toole 1986). De Datta and Beachell (1972) reported that weeds rank second only to drought stress in reducing upland rice grain yield and quality. Pandey (1996) found that in the uplands of south and Southeast Asia, weed infestation is the major problem for rice crops and may reduce yields by 30–100%. De Datta (1981) reported that weeds reduce upland rice yield from 42 to 100%. A study conducted by IRRI in the Philippines (1988) evaluated upland rice yield losses due to weeds (Table 1.27). Yield losses averaging 32% over three years have been reported.

In areas where new land has been cleared for upland rice cultivation, serious weed problems start in the third year of planting. Upland weeds include grasses, broadleaf weeds, and sedges. Land preparation, manual weeding, interrow cultivation, and herbicides are used to control weeds. A detailed discussion of worldwide upland rice weed species, distribution, ecology, biology, and control measures is presented by Gupta and O'Toole (1986).

In upland rice production systems, the only affordable solution to the weed problem is hand weeding. In more favorable ecosystems, there is a great dependency on herbicides, including pre-emergent varieties that act as insurance against later weed problems (Jensen, Courtois, and Olofsdotter 2008). Weed control methods that reduce herbicide dependency and offer opportunities to spray when only necessary would be of great benefit to most rice-growing areas (Jensen, Courtois, and Olofsdotter 2008). Superior competitive ability of the rice crop itself could be a way to reduce herbicide and labor dependency. But, so far, no crop cultivar has been marketed as having a superior competing ability (Olofsdotter 2001). Breeding for enhanced weed competitiveness, including improved allelopathic expression, could result in superior varieties, thereby reducing the burden of hand weeding, which can represent up to 30% of upland rice labor in traditional cropping systems, or herbicide dependency (Jensen, Courtois, and Olofsdotter 2008).

1.10 HYBRID RICE

A *hybrid* is defined as the first generation progeny that results from a controlled cross fertilization between individuals that differ in one or more genes (Crop Science Society of America 1992). The era of hybrid crop science plant technology started with the successful development and utilization of corn hybrids around 1930 (Virmani 1996). Today, many crop hybrids have been released and utilized in commercial production. Examples of such crops are sorghum, pearl millet, cotton, sunflowers, tomatoes, eggplants, chilies, onions, sugar beets, and rice. Although rice heterosis has been known since 1926, its commercial utilization was not demonstrated until Chinese scientists developed and used commercial rice hybrids that increased varietal yields by about 20%, compared with semi dwarf inbred varieties (Virmani 1996). The hybrid rice yield increase, or advantage over inbred rice, did not require significant additional input (He, Zhu, and Flinn 1987). During the 1980s, the use of hybrid rice in China encouraged IRRI and other institutions to begin research exploring the potential of hybrid rice cultivation (Virmani, Aquino, and Khush 1982; Yuan and Virmani 1988). By 1991, hybrid rice was cultivated on about 17 million ha in China, which was 55% of that country's total rice area, 66% of its total rice production, and 20% of the world's total rice production (Virmani 1996).

As a self-pollinated crop, rice must use an affective male sterility system to develop and produce F_1 hybrids (Virmani 1996). Hybrid rice yields are about 15–20% higher than that of inbred rice. These yield advantages are greater in high-yielding than in low-yielding environments (Virmani 1996). Hybrid rice yield increase is related to larger dry matter production, higher LAI, greater growth rate, and higher GHI (Agata 1990; Virmani 1996). Hybrid rice has also been found to be more tolerant of salinity and has a higher water use efficiency (Virmani 1996). According to Virmani (1996), the use of hybrid rice is one strategy for lifting the rice yield ceiling in order to help meet the world's future demands, which will increase due to rising population and income. For a detailed discussion of hybrid rice production, readers should refer to Virmani (1996).

1.11 CONCLUSIONS

Rice (*Oryza sativa* L.) is a staple food for about half of the world's population, including regions of high population density and rapid growth. Rice is cultivated under diverse agro-ecological systems, but upland and lowland regions are the dominant rice cultivation ecosystems with 76% of the world's rice produced from irrigated lowland rice systems. Major climatic factors that affect rice growth and development are temperature, solar radiation, and water. The optimum temperature for maximum economic yield is about 20–35°C. Solar radiation of 500–600 cal/cm^2/day may produce a yield of about 7 Mg ha^{-1} during the reproductive growth stage. Upland rice requires about 600 mm of water during the approximately 120-day crop growth cycle; the water requirement of irrigated rice is about double that of upland rice. Yield potential of this crop is estimated to be about 16 Mg ha^{-1}; however, the average world yield is less than 4 Mg ha^{-1}. This low yield is associated with biotic and abiotic stresses and the socioeconomic conditions of rice farmers. If these stresses were reduced, the potential for improving rice yield would greatly increase. Rice has a C_3 type metabolism and is adapted to warm, humid climates. Its growth cycle is divided into three stages: vegetative, reproductive, and spikelet filling or ripening. The vegetative growth stage is characterized by seed germination, active tillering, rooting, elongation of plant height, and leaf emergence. The reproductive growth stage begins with the differentiation of panicle primordia. Panicle development and elongation of culms, sometimes called internode elongation, characterize this stage. During the ripening growth stage, most of the photoassimilates go to the panicles and the plant gradually becomes senescent.

The main yield components are panicles per unit area, spikelet per panicle, weight of spikelet, and spikelet sterility or grain filling. The variation in grain yield due to yield components is panicle number > spikelet sterility > spikelet weight. In addition to these components, yield is also positively associated with plant height, shoot dry weight, GHI, and N harvest index. Panicles are developed primarily in the vegetative growth stage; panicle size is determined in the reproductive growth stage; and spikelet weight and sterility are determined in the spikelet-filling growth stage. Any biotic or abiotic stress during the vegetative, reproductive, or spikelet-filling growth stage can reduce the rice yield. However, the reproductive growth stage is more sensitive to environmental stresses than the other two stages. The most sensitive growth stage for rice is about 10 days before and after flowering. The use of higher yield potential genotypes along with adequate crop management practices can reduce the risk of yield-limiting factors so that higher yields are obtained. Higher rice yields achieved by incorporating short, erect, thick, dark green leaves and short-stiff stems clearly demonstrate the merit of including physiological-component traits in plant breeding programs. There still is potential for increasing rice yield by incorporating yield component traits in modern rice cultivars. Rice ratooning is one technology for increasing rice production. Ratooning provides higher resource use efficiency per unit time and per unit land area. However, positive ratoon crop yields are possible by adopting appropriate management practices for the main crop as well as for the ratoon crop. These practices include land preparation; adequate plant density and spacing; appropriate cultivar use; water management; adequate fertilizer application rates; appropriate cutting height; and disease, insect, and weed control. Hybrid rice and genetic engineering are also important strategies to lift the yield ceiling in many of the world's rice-producing regions.

REFERENCES

Agata, W. 1990. Mechanism of high yield achievement in Chinese F_1 rice compared with cultivated rice varieties. *Japn. J. Crop Sci.* 59:270–273.

Aggarwal, P. K. 2003. Impact of climate change on Indian agriculture. *J Plant Biol.* 30:189–198.

Alvarez, J. 1992. Cost and return for rice production on muck soils in Florida. Gainesville, FL: Food and Resource Economics Department, University of Florida.

Amano, T., C. Shi, D. Qin, M. Tsuda, and Y. Matsumoto. 1996. High yielding performance of paddy rice achieved in Yunnan Province of China. I. High yielding ability of Japanese F_1 hybrid rice. Yu-Za 29. *Jpn. J. Crop Sci.* 65:16–21.

Andaya, V. C., and D. J. Mackill. 2003. QTLs conferring cold tolerance at the booting stage of rice using recombinant inbred lines from a *japonica* x *indica* cross. *Theor. Appl. Genet.* 106:1084–1090.

Araújo, L. G., A. S. Prabhu, C. F., Oliveira, and R. F., Berni. 2004. Effect of blast on panicle and yield components of upland rice cultivar Primavera and Bonança. *Summa Phytopathol.* 30:265–270.

Araújo, L. G., A. S. Prabhu, and G. B. Silva. 2006. Unit tiller methods for determining loss of sheath blight on yield components of rice. *Fitopatol. Brás.* 31:119–202.

Atlin, G. N., H. R. Lafitte, D. Tao, M. Laza, M. Amante, and B. Courtois. 2006. Developing rice cultivars for high-fertility upland systems in the Asian tropics. *Field Crops Res.* 97:43–52.

Austin, R. B. 1980. Physiological limitations to cereal yields and ways of reducing them by breeding. In *Opportunities for Increasing Crop Yields*, R. G. Hurd, P. V. Biscoe, and C. Dennis, eds. pp. 3–19. London: Pitman.

Babu, R. C., B, D. Nguyen, V. P. Chamarerk, P. Shanmugasundram, P. Chezhian, S. K. Jeyapraksh, A. Ganesh, et al. 2003. Genetic analysis of drought resistance in rice by molecular markers. *Crop Sci.* 43:1457–1469.

Bahar, F. A., and S. K. De Datta. 1977. Prospects of increasing tropical rice production through ratooning. *Agron. J.* 69:536–540.

Baker, J. T. 2004. Yield responses of southern U.S. rice cultivars to CO_2 and temperature. *Agric. For. Meteorol.* 122:129–137.

Bastiaans, L., R. Rabbinge, and J. C. Zadoks. 1994. Understanding and modeling leaf blast effects on crop physiology and yield. In *Rice Blast Disease*, eds. R. S. Zeigler, S. A. Leong, and P. S. Teng, pp. 357–380. Wallingford: CAB.

Becker, M., D. E. Johnson, M. C. S. Wopereis, and A. Sow. 2003. Rice yield gaps in irrigated systems along an agro-ecological gradient in West Africa. *J. Plant Nutr. Soil Sci.* 166:61–67.

Beevers, H. 1970. Respiration in plants and its regulation. In *Prediction and Measurement of Photosynthetic Productivity*, ed. I. Malek, pp. 209–214. Wageningen, The Netherlands: Center for Agriculture Publishing and Documentation.

Bernacchi, C. J., C. Pimentel, and S. P. Long. 2003. In vivo temperature response functions of parameters required to model RuBP-limited photosynthesis. *Plant Cell Environ.* 26:1419–1430.

Bernacchi, C. J., E. L. Singsaas, C. Pimentel, A. L. Portis, and S. P. Long. 2001. Improved temperature response function for models of Rubisco-limited photosynthesis. *Plant Cell Environ.* 24:253–259.

Bernstein, L. 1975. Effects of salinity and sodicity on plant growth. *Ann. Rev. Phytopatho.* 13:295–312.

Black, A. L., and F. H. Siddoway. 1977. Hard red and durum spring wheat responses to seeding date and NP-fertilization on fallow. *Agro. J.* 69: 885–888.

Blum, A., J. Mayer, G. Golan, and B. Sinmena. 1999. Drought tolerance of a doubled haploid line population of rice in the field. In *Genetic Improvement of Rice for Water-Limited Environments*, ed. O. Ito, pp. 310–330. Los Bãnos, Philippines: IRRI.

Blum, U., S. R. Shafer, and M. E. Lehman. 1999. Evidence for inhibitory allelopathic interactions involving phenolic acids in field soils: Concepts vs. an experimental model. *Crit. Rev. Plant Sci.* 18:673–693.

Bohn, H, B. McNeal, and G. O'Connor. 1979. *Soil Chemistry*. New York: John Wiley.

Bollich, C. N., and F. T. Turner. 1988. Commercial ratoon rice production in Texas, USA. In *Rice Ratooning*, ed. International Rice Research Institute, pp. 209–217. Los Baños, Philippines: IRRI.

Bollich, C. N., B. D. Webb, M. A. Marschner, and J. E. Scott. 1983. Bellmont rice. *Crop Sci.* 23:803–804.

Boonjung, H. 1993. Modelling growth and yield of upland rice under water limitation conditions. PhD thesis, The University of Queensland.

Bot, A. J., F. O. Nachtergaele, and A. Young. 2000. *Land Resource Potential and Constraints at Regional and Country Levels*. World Soil Resource Report 90. Rome: Land and Water Development Division, Food And Agriculture Organization.

Bouman, B. A. M., E. Humphreys, T. P. Tuong, and R. Barker. 2007. Rice and water. *Adv. Agron.* 92:187–237.

Bouman, B. A. M., S. Peng, A. R. Castaneda, and R. M. Visperas. 2005. Yield and water use of irrigated tropical aerobic rice systems. *Agric. Water Manage.* 74:87–105.

Breseghello, F., O. P. Morais, P. V. Pinheiro, A. C. S. Silva, E. M. Castro, E. P. Guimaraes, A. P. Castro, et al. 2011. Results of 25 years of upland rice breeding in Brazil. *Crop Sci.* 51:914–923.

Calderini, D. F., L. G. Abeledo, and G. A. Slafer. 2000. Physiological maturity in wheat based on kernel water and dry matter. *Agron. J.* 92:895–901.

Catling, H. D., and S. Parfitt. 1981. *An Illustrated Description of a Traditional Deepwater Rice Variety of Bangladesh*. IRRI Research Paper Series Number 60. Los Banõs, Philippines: IRRI.

Chang, T. T., and B. S. Vergara. 1975. Varietal diversity and morph agronomic characteristics of upland rice. In *Major Research in Upland Rice*, eds. International Rice Research Institute, pp. 72–90. Los Bãnos, Philippines: IRRI.

Chauhan, J. S., B. S. Vergara, and S. S. Lopez. 1985. *Rice Ratooning*. International Rice Research Institute Research Paper Series Number 102. Los Baños, Philippines: IRRI.

Chen, Z., S. Shabala, N. Mendham, I. Newman, G. Zhang, and M. Zhou. 2008. Combining ability of salinity tolerance on the basis of NaCl-induced K^+ flux from roots of barley. *Crop Sci.* 48:1382–1388.

Chen, Z., M. Zhou, I. Newman, N. Mendham, G. Zhang, and S. Shabala. 2005. Potassium and sodium relations in salinized barley tissues as a basis of differential salt tolerance. *Funct. Plant Biol.* 34:150–162.

Cho, J. 1956. Double fertilization in *Oryza sativa* L. and development of the endosperm with special reference to the aleurone layer. *Bull. Natl. Inst. Agric. Sci.* 6:61–101.

Chou, C. H. 1980. Allelopathic research in the subtropical vegetation in Taiwan. *Comparative Physiol. Ecol.* 5:222–234.

Cohen, Y., V. Alchanatis, M. Meron, Y. Saranga, and J. Tsipris. 2005. Estimation of leaf water potential by thermal imagery and spatial analysis. *J. Exp. Bot.* 56:1843–1852.

Colaizzi, P. D., S. R. Evett, T. A. Howell, F. Li, W. P. Kustas, and M. C. Anderson. 2012a. Radiation model for row crops. I. Geometric view factors and parameter optimization. *Agron. J.* 104:225–240.

Colaizzi, P. D., R. C. Schwartz, S. R. Evett, T. A. Howell, P. H. Gowda, and J. A. Tolk. 2012b. Radiation model for row crops. II. Model evaluation. *Agron. J.* 104:225–240.

Counce, P. A., T. C. Keisling, and A. J. Mitchell. 2000. A uniform, objective, and adaptive system for expressing rice development. *Crop Sci.* 40:436–443.

Craufurd, P. Q., T. R. Wheeler, R. H. Ellis, R. J. Summerfield, and J. H. Williams. 1999. Effect of temperature and water deficit on water-use efficiency, carbon isotope discrimination, and specific leaf area in peanut. *Crop Sci.* 39:136–142.

Crop Science Society of America. 1992. *Glossary of Crop Science Terms*. Madison, WI: Crop Science Society of America.

Cuin, T. A., and S. Shabala. 2005. Exogenously supplied compatible solutes rapidly ameliorate NaCl-induced potassium efflux from barley roots. *Plant Cell Environ.* 46:1924–1933.

De Datta, S. K. 1981. *Principles and Practices of Rice Production*. New York: John Wiley.

De Datta, S. K., and H. M. Beachell. 1972. Varietal response to some factors affecting production of upland rice. In *Rice Breeding*, eds. IRRI, pp. 685–700. Los Baños, Philippines: IRRI.

Denef, K., J. Six, R. Merckx, and K. Paustian. 2004. Carbon sequestration in microaggregates of no-tillage soils with different clay mineralogy. *Soil Sci. Soc. Am. J.* 68:1935–1944.

Deng, N. D. 1984. Effect of light intensity and temperature on the synthesis and photorespiration rate of rice. *J. South China Agric. Coll.* 5:32–38.

Dingkuhn M., and M. Kropff. 1996. Rice. In *Photoassimilate Distribution in Plants and Crops: Source-Sink Relationships*, eds. E. Zamski and A. A. Schaffer, pp. 519–547. New York: Marcel Dekker.

Dofing, S. M., and M. G. Karlsson. 1993. Growth and development of uniculm and conventional tillering barley lines. *Agron J.* 85:58–61.

Donald, C. M., and J. Hamblin. 1976. The biological yields and harvest index of cereals as agronomic and plant breeding criteria. *Adv. Agron.* 28:361–405.

Ebana, K., W. Yan, R. H. Dilday, H. Namai, and K. Okuno. 2001. Variation in the allelopathic effect of rice with water soluble extracts. *Agron. J.* 93:12–16.

Edmeades, G. O., J. Bolanos, S. C. Chapman, H. R. Lafitte, and M. Banziger. 1999. Selection improves drought tolerance in tropical maize populations: I. Gains in biomass, grain yield and harvest index. *Crop Sci.* 39:1306–1315.

Ekanayake, J. S., K. De Datta, and P. L. Steponkus. 1989. Spikelet sterility and flowering response of rice to water stress at anthesis. *Ann. Bot.* 63:257–264.

Elias, R. S. 1969. Rice production and minimum tillage. *Outlook on Agric.* 6:67–70.

Epstein, E., and A. J. Bloom. 2005. *Mineral Nutrition of Plants; Principal and Perspectives*, 2nd edition. Sunderland, MA: Sinauer Associates.

Epstein, E., J. D. Norlyn, D. W. Rush, R. W. Kingsbury, D. B. Kelley, G. A. Coningham, and A. F. Wrona. 1980. Saline culture of crops: A genetic approach. *Science.* 210:399–404.

Evans, L. T., and R. A. Fischer. 1999. Yield potential: Its definition, measurement, and significance. *Crop Sci.* 39:1544–1551.

Evans, L. T., R. M. Visperas, and B. S. Vergara. 1984. Morphological and physiological changes among rice varieties used in the Philippines over the last seventy years. *Field Crops Res.* 8:105–124.

Evans, L. T., and I. F. Wardlaw. 1976. Aspects of the comparative physiology of grain yield in cereals. *Adv. Agron.* 28.301–359.

Evatt, N. S. 1966. High annual yields of rice in Texas through ratoon or double-cropping. *Rice J.* 69:10–12.

Fageria, N. K. 1980. Upland rice response to phosphate fertilization as affected by water deficiency in Cerrado soils. *Pesq. Agropec. Bras.* 15:259–265.

Fageria, N. K. 1984. *Fertilization and Mineral Nutrition of Rice*. Rio de Janeiro: EMBRAPA–CNPAF/Editora Campus.

Fageria, N. K. 1985. Salt tolerance of rice cultivars. *Plant Soil.* 88:237–243.

Fageria, N. K. 1989. *Tropical Soils and Physiological Aspects of Crops*. Brasilia: EMBRAPA.

Fageria, N. K. 1992. *Maximizing Crop Yields*. New York: Marcel Dekker.

Fageria, N. K. 2001. Nutrient management for improving upland rice productivity and sustainability. *Commun. Soil Sci. Plant Anal.* 32:2603–2629.

Fageria, N. K. 2003. Plant tissue test for determination of optimum concentration and uptake of nitrogen at different growth stages in lowland rice. *Commun. Soil Sci. Plant Anal.* 34:259–270.

Fageria, N. K. 2005. Soil fertility and plant nutrition research under controlled conditions: Basic principles and methodology. *J. Plant Nutr.* 28:1975–1999.

Fageria, N. K. 2007. Yield physiology of rice. *Plant Nutr.* 30:843–879.

Fageria, N. K. 2009. *The Use of Nutrients in Crop Plants*. Boca Raton, FL: CRC Press.

Fageria, N. K. 2012. *The Role of Plant Roots in Crop Production*. Boca Raton, FL: CRC Press.

Fageria N. K., and V. C. Baligar. 1996. Response of lowland rice and common bean grown in rotation to soil fertility levels on a Varzea soil. *Fert. Res.* 45:13–20.

Fageria, N. K., and V. C. Baligar. 1999. Yield and yield components of lowland rice as influenced by timing of nitrogen fertilization. *J. Plant Nutr.* 22:23–32.

Fageria, N. K., and V. C. Baligar. 2001. Lowland rice response to nitrogen fertilization. *Commun. Soil Sci. Plant Anal.* 32:1405–1429.

Fageria, N. K., and V. C. Baligar. 2003. Upland rice and allelopathy *Commun. Soil Sci. Plant Anal.* 34:1311–1329.

Fageria, N. K., and V. C. Baligar. 2005. Enhancing nitrogen use efficiency in crop plants. *Adv. Agron.* 88:97–185.

Fageria, N. K., and V. C. Baligar. 2008. Ameliorating soil acidity of tropical Oxisols by liming for sustainable crop production. *Adv. Agron.* 99:345–399.

Fageria, N. K., V. C. Baligar, and B. A. Bailey. 2005. Role of cover crops in improving soil and row crop productivity. *Commun. Soil Sci. Plant Anal.* 36:2733–2757.

Fageria, N. K., V. C. Baligar, and R. B. Clark. 2006. *Physiology of Crop Production.* New York: Haworth Press.

Fageria, N. K., V. C. Baligar, and C. A. Jones. 2011. *Growth and Mineral Nutrition of Field Crops*, 3rd edition. Boca Raton, FL: CRC Press.

Fageria, N. K., and M. P. Barbosa Filho. 2001. Nitrogen use efficiency in lowland rice genotypes. *Commun. Soil Sci. Plant Anal.* 32:2079–2090.

Fageria, N. K., M. P. Barbosa Filho, and J. R. P. Carvalho. 1982. Response of upland rice to phosphorus fertilization on an Oxisol. *Agron. J.* 74:51–56.

Fageria, N. K., M. P. Barbosa Filho, and C. M. Guimaraes. 2008. Allelopathy in upland rice in Brazil. *Allelopathy J.* 22:289–298.

Fageria, N. K., M. P. Barbosa Filho, L. F, Stone, and C. M. Guimãres. 2004. Phosphorus nutrition for upland rice production. In *Phosphorus in Brazilian Agriculture*, eds. T. Yamada and S. R. S. Abdalla, pp. 401–418. Piracicaba, São Paulo: Brazilian Potassium and Phosphate Research Association.

Fageria, N. K., and F. Breseghello. 2004. Nutritional diagnostic in upland rice production in some municipalities of State of Mato Grosso, Brazil. *J. Plant Nutr.* 27:15–28.

Fageria, N. K., H. R. Gheyi, and A. Moreira. 2011. Nutrient bioavailability in salt affected soils. *J. Plant Nutr.* 34:945–962.

Fageria, N. K., and A. M. Knupp. 2013. Upland rice phenology and nutrient uptake in tropical climate. *J. Plant Nutr.* 36:1–14.

Fageria, N. K., O. P. Morais, and A. B. Santos. 2010. Nitrogen use efficiency in upland rice. *J. Plant Nutr.* 33:1696–1711.

Fageria, N. K., and A. S. Prabhu. 2004. Blast control and nitrogen management in lowland rice cultivation. *Pesq. Agropec. Bras.* 39:123–129.

Fageria, N. K., A. B. Santos, and V. C. Baligar. 1997. Phosphorus soil test calibration for lowland rice on an Inceptisol. *Agron. J.* 89:737–742.

Fageria, N. K., and J. M. Scriber. 2001. The role of essential nutrients and minerals in insect resistance in crop plants. In *Insects and Plant Defense Dynamics*, ed. T. N. Ananthakrishnan, pp. 23–54. Plymouth, UK: Science.

Fageria, N. K., N. A. Slaton, and V. C. Baligar. 2003. Nutrient management for improving lowland rice productivity and sustainability. *Adv. Agron.* 80:63–152.

Fageria, N. K., and N. P. Souza. 1995. Response of rice and common bean crops in succession to fertilization in Cerrado soil. *Pesq. Agropec. Bras.* 30:359–368.

Fageria, N. K., L. F. Stone, and A. B. Santos. 2003. *Soil Fertility Management of Irrigated Rice.* Santo Antônio de Goias, Brazil: EMBRAPA Arroz e Feijão.

Flach, K. W., and D. F. Slusher. 1978. Soils used for rice culture in the United States. In *Soils and Rice*, ed. International Rice Research Institute, pp. 199–215. Los Baños, Philippines: IRRI.

Fukai, S., and M. Cooper. 1995. Development of drought-resistant cultivars using physio-morphological traits in rice. *Field Crops Res.* 40:67–86.

Garrity, D. P., L. R. Oldemann, R. A. Morris, and D. Lenka. 1986. Rainfed lowland rice ecosystems: Characterization and distribution. In *Progress in Rainfed Lowland Rice* ed. International Rice Research Institute, pp. 3–23. Manila, Philippines: IRRI.

Garrity, D. P., and J. C. O'Toole. 1994. Screening rice for drought resistance at the reproductive phase. *Field Crops Res.* 39:99–110.

Garrity, D. P., and J. C. O'Toole. 1995. Selection for reproductive stage drought avoidance in rice, using infrared thermometry. *Agron. J.* 87:773–779.

George, T., R. M. Dennis, P. Garrity, B. S. Tubana, and J. Quiton. 2002. Rapid yield loss of rice cropped successively in aerobic soil. *Agron. J.* 94:981–989.

Gibson, L. R., and G. M. Paulsen. 1999. Yield components of wheat grown under high temperature stress during reproductive growth. *Crop Sci.* 39:1841–1846.

Glaszmann, J. C. 1987. Isozymes and classification of Asian rice varieties. *Theor. Appl. Genet.* 74:21–30.

Gomez, K. A. 1972. *Techniques for Field Experiments with Rice*. Los Bãnos, Philippines: IRRI.

Gorham, J., C. Hardy, R. G. Wyn Jones, L. R. Joppa, and C. N. Law. 1987. Chromosomal location of K^+/Na^+ discrimination character in the D genome of wheat. *Theor. Appl. Genet.* 74:584–588.

Grant, R. F., B. A. Kimball, M. M. Conley, J. W. White, G. W. Wall, and M. J. Ottman. 2011. Controlled warming effects on wheat growth and yield: Field measurements and modeling. *Agron. J.* 103:1742–1754.

Gravois, K. A., and R. S. Helms. 1992. Path analysis of rice yield and yield components as affected by seeding rate. *Agron. J.* 88:1–4.

Greenway, H., and R. Munns. 1980. Mechanisms of salt tolerance in non-halophytes. *Ann. Rev. Plant Physiol.* 31:149–190.

Grewal, D., C. Manito, and V. Bartolome. 2011. Doubled haploids generated through another culture from crosses of elite *indica* and *japonica* cultivars and/or lines of rice: Large-scale production, agronomic performance, and molecular characterization. *Crop Sci.* 51:2544–2553.

Groth, D. E., and E. M. Novick. 1992. Selection for resistance to rice sheath blight through the number of infection cushions and lesion type. *Plant Dis.* 76:721–723.

Guimaraes, C. M., L. F. Stone, M. Lorieux, J. P. Oliveira, G. C. O. Alencar, and R. A. A. Dias. 2010. Infrared thermometry for drought phenotyping of inter and intra specific upland rice lines. *Rev. Bras. Eng. Agri. Amb.* 2:148–154.

Guimaraes, C. M., and L. P. Yokoyama. 1998. Upland rice in rotation with soybean. In *Technology for Upland Rice,* eds. F. Breseghello and L. F. Stone, pp. 19–24. Santo Antonio de Goias, Brazil: EMBRAPA Arroz e Feijão.

Gunawardena, T. A., S. Fukai, and F. P. C. Blamey. 2003. Low temperature induced spikelet sterility in rice. I. Nitrogen fertilization and sensitive reproductive period. *Aust. J. Agric. Res.* 54:937–946.

Gupta, P. C., and J. C. O`Toole. 1986. *Upland Rice: A Global Perspective*. Los Baños, Philippines: IRRI.

Gupta, R. K., R. K. Naresh, P. R. Hobbs, and J. K. Ladha. 2002. Adopting conservation agriculture in rice-wheat systems of the Indo-Gangetic Plains: New opportunities for saving on water. In *Proceedings of the International Workshop on Water-Wise Rice Production*, eds. B. A. M. Bouman, H. Hengsdijk, B. Hardy, B. Bindraban, T. P. Toung, and J. K. ladha, pp. 207–222. Los Baños, Philippines; IRRI.

Hale, M. G., and D. M. Orcutt. 1987. *The Physiology of Plants under Stress*. New York: John Wiley.

Han, Y. P., Y. Z. Xing, S. L. Gu, Z. X. Chen, X. B. Pan, and X. L. Chen. 2003. Effect of morphological trait on sheath blight resistance in rice. *Acta. Bot. Sin.* 45:825–831.

Hanada, K. 1995. Differentiation and development of tiller buds. In *Science of the Rice Plant: Physiology*, Vol. 2, eds. T. Matsuo, K. Kumazawa, R. Ishii, K. Ishihara, and H. Hirata, pp. 61–65. Tokyo: Food and Agriculture Policy Research Center.

Haque, M. M., and W. R. Coffman. 1980. Varietal variation and evaluation procedures for ratooning ability in rice. *SABRAO J.* 12:113–120.

Hasegawa, P. M., R. A. Bressan, and A. K. Handa. 1986. Cellular mechanisms of salinity tolerance. *Horticulture Sci.* 21:1317–1324.

Hatfield, J. L., K. J. Boote, B. A. Kimball, L. Ziska, C. Izaurrald, D. Ort, A. Thomson, et al. 2011. Climate impacts on agriculture: Implications for agronomic production. *Agron. J.* 103:351–370.

Hayase, H. T., T. Satake, I. Nishiyama, and N. Ito. 1969. Male sterility caused by cooling treatment at the meiotic stage in rice plants: II. The most sensitive stage to cooling and the fertilizing ability of pistils. *Proc. Crop Sci. Soc. Jpn.* 38:706–711.

Hayashi, H. 1995. Translocation, storage and partitioning of photosynthetic products. In *Science of the Rice Plant: Physiology*, Vol. 2, eds. T. Matsuo, K. Kumazawa, R. Ishii, K. Ishihara, and H. Hirata, pp. 546–565. Tokyo: Food and Agriculture Policy Research Center.

He, G. T., X. Zhu, and J. C. Flinn. 1987. A comparative study of economic efficiency of hybrid and conventional rice production in Jiangsu Province, China. *Oryza.* 24:285–296.

Hilgard, E. W. 1906. *Soils, Their Formation, Properties, Composition, and Relations to Climate and Plant Growth.* New York: MacMillan.

Hoff, B. J., and P. K. Bollich. 1990. Node origin of ratoon tillers of U.S. rice cultivars. In *Proceedings of the 23rd Rice Technology Working Group, Biloxi, Mississippi.* College Station, TX: Texas Agric. Exp. Stn.

Hsiao, T. C. 1982. The soil plant atmosphere continuum in relation to drought and crop production. In *Drought Resistance in Crops with Emphasis on Rice*, ed. IRRI, pp. 39–52. Los Baños, Philippines: IRRI.

Huke, R. E. 1982. *Rice Area by Type of Culture: South, Southeast and East Asia.* Los Bãnos, Philippines: IRRI.

Hussain, S., B. L. Ma, M. F. Saleem, S. A. Anjum, A. Saeed, and J. Iqbal. 2012. Abscisic acid spray on sunflower acts differently under drought and irrigation conditions. *Agron. J.* 104:561–568.

Ingram, K. T., F. D. Bueno, O. S. Namuco, E. B. Yambao, and C. A. Beyrouty. 1994. Rice root traits for drought resistance and their genetic variation. In *Rice Roots: Nutrient and Water Use*, ed. G. J. D. Kirk, pp. 67–77. Los Bãnos, Philippines: IRRI.

International Allelopathy Society. 1996. *First World Congress on Allelopathy: A Science for the Future.* Cadiz, Spain: IAS.

IPCC (Intergovernmental Panel on Climate Change). 2007. Summary for policy makers. In *Climate Change 2007: The Physical Science Basis*, ed. IPPC, p. 9, Geneva, Switzerland: IPCC.

IRRI (International Rice Research Institute). 1975. *Major Research in Upland Rice.* Los Baños, Philippines: IRRI.

IRRI (International Rice Research Institute). 1984. *Terminology for Rice Growing Environments.* Los Bãnos, Philippines: IRRI.

IRRI (International Rice Research Institute). 1988. *Annual Report for 1987.* Los Baños, Philippines: IRRI.

IRRI (International Rice Research Institute). 1989. *IRRI Toward 2000 and Beyond.* Los Bãnos, Philippines: IRRI.

IRRI (International Rice Research Institute). 1994. *Progress Report February 1993.* Los Banõs, Philippines: IRRI.

Ishii, R. 1977. The effect of temperature on the rates of photosynthesis, respiration and the activity of RuBP carboxylase in barley, rice and maize leaves. *Japn. J. Crop Sci.* 46:516–523.

Ishii, R. 1995. Roles of photosynthesis and respiration in the yield-determining process. In *Science of the Rice Plant: Physiology*, Vol. 2, eds. T. Matsuo, K. Kumazawa, R. Ishii, K. Ishihara, and H. Hirata, pp. 691–696. Tokyo: Food and Agriculture Policy Research Center.

Jackson, B. R. A. Yantasast, C. Prechachart, M. A. Choudhary, and S. M. H. Zaman. 1972. Breeding rice for deep water areas. In *Rice Breeding*, ed. IRRI, pp. 515–528. Los Baños, Philippines: IRRI.

Jagadish, S. V. K., J. Cairns, R. Lafitte, T. R. Wheeler, A. H. Price, and P. Q. Craufurd. 2010. Genetic analysis of heat tolerance at anthesis in rice. *Crop Sci.* 50:1633–1641.

Jagadish, S. V. K., P. Q. Craufurd, and T. R. Wheeler. 2007. High temperature stress and spikelet fertility in rice (*Oryza sativa* L.). *J. Exp. Bot.* 58:1627–1635.

Jagadish, S. V. K., P. Q. Craufurd, and T. R. Wheeler. 2008. Phenotyping parents of mapping populations of rice (*Oryza sativa* L.) for heat tolerance during anthesis. *Crop Sci.* 48:1140–1146.

Jana, R. K., and S. K. De Datta. 1971. Effects of solar energy and moisture tension on the nitrogen response of upland rice. In *Proceedings of the International Symposium on Soil Fertility Evaluation*, eds. J. S. Kanwar, W. P. Datta, S. S. Bains, D. R. Bhumbla, and J. D. Biswas, pp. 487–497. New Delhi: Indian Society of Soil Science.

Jena, K. K., and D. J. Mackill. 2008. Molecular markers and their use in marker-assisted selection in rice. *Crop Sci.* 48:1266–1276.

Jennings, P. R., W. R. Coffman, and H. E. Kauffman. 1979. *Rice Improvement*. Los Bãnos, Philippines: IRRI.

Jensen, L. B., B. Courtois, and M. Olofsdotter. 2008. Quantitative trait loci analysis of allelopathy in rice. *Crop Sci.* 48:1459–1469.

Jones, C. A. 1985. *C4 Grasses and Cereals: Growth, Development, and Stress Response*. New York: John Wiley.

Jones, D. B. 1993. Rice ratoon response to main crop harvest cutting height. *Agron. J.* 85:1139–1142.

Jones, D. B., and G. H. Snyder. 1987. Seeding rate and row spacing effects on yield and yield components of ratoon rice. *Agron. J.* 79:627–629.

Jones, M. P., M. Dingkuhn, G. K. Aluko, and M. Semon. 1997. Interspecific *Oryza sativa* L x *O. Glaberrima* Steud. progênies in upland rice improvement. *Euphytica.* 92:237–246.

Jordan, D. B., and W. L. Ogren. 1984. The CO_2/O_2 specificity of ribulose 1,5-bisphosphate carboxylase/oxygenase. *Planta.* 161;308–313.

Karamanos, A. J., J. F. Elston, and R. M. Wadsworth. 1982. Water stress and leaf growth of field bean (*Vicia faba* L.) in the field. Water potentials and laminal expansion. *Ann. Bot.* 49:815–826.

Karamanos, A. J., and I. S. Travlos. 2012. The water relations and some drought tolerance mechanisms of the marama bean. *Agron. J.* 104:65–72.

Kariali, E., and P. K. Mohapatra. 2007. Hormonal regulation of tiller dynamics in differentially tillering rice cultivars. *Plant Growth Regul.* 53:215–223.

Kato, T. 1997. Selection responses for the characters related to yield sink capacity. *Crop Sci.* 37:1472–1475.

Khush, G. S. 1977. Disease and insect resistance in rice. *Adv. Agron.* 29:265–341.

Khush, G. S. 1981. Breeding rice for multiple disease and pest resistance. In *Rice Improvement in China and Other Asian Countries*, ed. International Rice Research Institute, pp. 219–238. Los Baños, Philippines: IRRI.

Khush, G. S. 2001. Green revolution: The way forward. *Nat. Rev. Genet.* 2:815–822.

Khush, G. S. 2005. What it will take to feed 5.0 billion rice consumers in 2030. *Plant Mol. Biol.* 59:1–6.

Kiniry, J. R., G. McCauley, Y. Xie, and J. G. Arnold. 2001. Rice parameters describing crop performance of four U.S. cultivars. *Agron. J.* 93:1354–1361.

Klepper, B. 1992. Development and growth of crop root systems. *Adv. Soil Sci.* 19:1–25.

Kramer, H. H. 1967. Plant protection and world crop production. Bayer Pflanzenschutz, Leverkusen.

Kramer, P. J. 1969. *Plant and Soil Water Relationships: A Modern Synthesis*. New York: McGraw-Hill.

Kramer, P. J. 1988. Water relations of plants. *Plant Cell Environ.* 11:565–568.

Krishnamurthy, K. 1988. Rice ratooning as an alternative to double cropping in tropical Asia. In *Ratooning*, ed. International Rice Research Institute, pp. 3–15. Los Baños, Philippines: IRRI.

Krishnan, P., B. Ramakrishnan, K. R. Reddy, and V. R. Reddy. 2011. High-temperature effects on rice growth, yield, and grain quality. *Adv. Agron.* 111:87–206.

Kumar, V., and J. K. Ladha. 2011. Direct seeding of rice: Recent development and future research needs. *Adv. Agron.* 111:297–413.

Kung, P. 1971. *Irrigation Agronomy in Monsoon Asia.* Rome: Food and Agricultural Organization.

Lafitte, H. R., A. H. Price, and B. Courtois. 2004. Yield responses to water deficit in an upland rice mapping population. Associations among traits and genetic markers. *Theor. Appl. Genet.* 109:1237–1246.

Lanceras, J., G. Pantuwan, B. Jongdee, and T. Toojinda. 2004. Quantative trait loci associated with drought tolerance at reproductive stage in rice. *Plant Physiol.* 135:384–399.

Landsberg, J. J. 1977. Effects of weather on plant development. In *Environmental Effects on Crop Physiology*, eds. J. J. Lndsberg and C. V. Cutting, pp. 289–307. London: Academic Press.

Lavalle, C., F. Micale, T. D. Houston, A. Camia, R. Hiederer, C. Lazar, C. Conte, et al. 2009. Climate change in Europe: 3. Impact on agricultural and forestry. A review. *Agron. Sustainable Dev.* 29:433–446.

Lee, E. N., and M. C. Rush. 1983. Rice sheath blight: A major rice disease. *Plant Dis.* 67:829–832.

Lewis, C. T., and C. A. Short. 1958. *Latin Dictionary.* London: Oxford University Press.

Li, P., J. Chen, and P. Wu. 2011. Agronomic characteristics and grain yield of 30 spring wheat genotypes under drought stress and nonstress conditions. *Agron. J.* 103:1619–1628.

Li, X. Y., Q. Qian, Z. M. Fu, Y. H. Wang, G. S. Xiong, D. L. Zeng, X. Q. Wang, et al. 2003. Control of tillering in rice. *Nature.* 422:618–621.

Lilley, J. M., and Fukai, S. 1994. Effect of timing and severity of water deficit on four diverse rice cultivars. III. Phenological development, crop growth and grain yield. *Field Crops Res.* 37:225–234.

Linares, O. 2002. African rice (*Oryza glaberrima*): History and future potential. *Proc. Natl. Acad. Sci. USA.* 99:16360–16365.

Ling, Q. H. 2000. *The Quality of Crop Population.* Shanghai, China: Shanghai Scientific & Technical Publishers.

Liu, H., H. Mei, X. Yu, G. Zou, G. Liu, and L. Luo. 2006. Towards improving the drought tolerance of rice in China. *Plant Genet. Resour.* 4:47–53.

Liu, Y., Y. Ding, D. Gu, G. Li, Q. Wang, and S. Wang. 2011. The positional effects of auxin and cytokinin on the regulation of rice tiller bud growth. *Crop Sci.* 51:2749–2758.

Lobell, D. B., M. B. Burke, C. Tebaldi, M. D. Mastrandrea, W. P. Falcon, and R. L. Naylor. 2008. Prioritizing climate change adaption needs for food security in 2030. *Science.* 319:607–610.

Lu, B. R., X. X. Cai, and X. Jin. 2009. Efficient *indica* and *japonica* rice identification based on InDel molecular methods: Its implication in rice breeding and evolutionary research. *Prog. Nat. Sci.* 19:1241–1252.

Ludlow, M. M., and R. C. Muchlow. 1990. A critical evaluation of traits for improving crop yields in water-limited environments. *Adv. Agron.* 43:107–153.

Maas, E. V., and G. J. Hofman. 1977. Crop salt tolerance-current assessment. *J. Irrigation Drainage.* 103:115–134.

Maas, E V., G. J. Hoffman, G. D. Chaba, J. A. Poss, and M. C Shannon. 1983. Salt sensitivity of corn at various growth stages. *Irrigation Sci.* 4:45–57.

Maathuis, F. J. M., and A. Amtmann. 1999. K^+ nutrition and Na^+ toxicity: The basis of cellular K^+/Na^+ ratios. *Ann. Bot.* 84:123–133.

Mae, T. 1997. Physiological nitrogen efficiency in rice: Nitrogen utilization, photosynthesis, and yield potential. *Plant Soil.* 196:201–210.

Mahadevappa, M., and H. S. Yogeesha. 1988. Rice ratooning: Breeding, agronomic practices, and seed production potentials. In *Rice Ratooning*, ed. International Rice Research Institute, pp. 177–185. Los Baños, Philippines: IRRI.

Malavolta, V. M. A., E. A. Soligo, D. D., Dias, L. E., Azzini, and C. R. Bastos. 2007. Fungus incidence and quantification of loss caused in seeds of rice genotypes. *Summa Phytopathol.* 33:280–286.

Mambani, B., and R. Lal. 1983. Response of upland rice cultivars to drought stress. III. Screening rice varieties by means of variable moisture along a *Plant Soil.* 73:73–94.

Maqbool, A., S. Shafiq, and L. Lake. 2010. Radiant frost tolerance in pulse crops: A review. *Euphytica.* 172:1–12.

Marchetti, M.A., and C. N. Bollich. 1991. Quantification of the relationship between sheath blight severity and yield loss in rice. *Plant Dis.* 75:773–775.

Marschner, H. 1995. *Mineral Nutrition of Higher Plants*, 2nd edition. New York: Academic Press.

Masle, J. 1985. Competition among tillers in winter wheat: Consequences for growth and development of the crop. In *Wheat Growth and Modeling*, eds. W. Day and R. K. Atkin, pp. 33–54. New York: Plenum Press.

Matsui, T., and K. Omasa. 2002. Rice (*Oryza sativa* L.) cultivars tolerant to high temperature at flowering: Anther characteristics. *Ann. Bot.* 89:683–687.

Matsuo, H., A. J. Pecrot, and J. Riquier. 1978. Rice soils of Europe. In *Soils and Rice*, ed. International Rice Research Institute, pp. 191–198. Los Baños, Philippines: IRRI.

Matsushima, S. 1962. Some experiments on the soil-water-plant relationship in rice cultivation. *Proc. Crop Sci. Soc. Jpn.* 31:115–121.

Matsuzaki, A. 1995. Development and aging of panicles. In *Science of the Rice Plant: Physiology,* Vol. 2, eds. T. Matsuo, K. Kumazawa, R. Ishii, K. Ishihara, and H. Hirata, pp. 156–164. Tokyo: Food and Agriculture Policy Research Center.

Maurya, D. M., and J. C. O'Toole. 1986. Screening upland rice for drought tolerance. In *Progress in Upland Rice Research.* Proceedings of the 1985 Jakarta Conference. Los Baños, Philippines: IRRI.

Mengel, D. B., and W. J. Leonards. 1978. Effects of nitrogen, phosphorus and potassium fertilization on the yield and quality of second crop Labelle rice. *Annu. Progress Report, Louisiana Agric. Exp. Stn.* 70:60–62.

Mengel, D. B., and F. E. Wilson. 1981. Water management and nitrogen fertilization of ratoon crop rice. *Agron. J.* 73:1008–1010.

Mohammed, A. R., and L. Tarpley. 2009. Impact of high nighttime temperature on respiration, membrane stability, antioxidant capacity, and yield of rice plants. *Crop Sci.* 49:313–322.

Mohapatra, P., R. Panigrahi, and N. C. Turner. 2011. Physiology of spikelet development on the rice panicles: Is manipulation of apical dominance crucial for grain yield important? *Adv. Agron.* 110:333–359.

Monneveux, P., C. Sanchez, D. Beck, and G. O. Edmeades. 2006. Drought tolerance improvement in tropical maize source populations: Evidence of progress. *Crop Sci.* 46:180–191.

Monteith, J. L. 1972. Solar radiation and productivity in tropical ecosystem. *J. Appl. Ecol.* 9:747–766.

Monteith, J. L., and M. H. Unsworth. 1990. *Principles of Environmental Physics*, 2nd edition. New York: Edward Arnold.

Moraes, J. F. V. 1999. Soils. In *Rice Culture in Brazil*, eds. N. R. A. Vieira, A. B. Santos, and E. P. Santana, pp. 88–115. Santo Antônio de Goiás, Brazil: EMBRAPA Arroz e Feijão.

Morgan, J. M. 1984. Osmoregulation and water stress in higher plants. *Annu. Rev. Plant Physiol.* 35:299–319.

Moorman, F. R. 1978. Morphology and classification of soils on which rice is grown. In *Soils and Rice*, ed. International Rice Research Institute, pp. 255–272. Los Baños, Philippines: IRRI.

Morishita, K. 1984. Wild plant and domestication. In *Biology of Rice*, eds. S. Tsunoda and N. Takahashi, pp. 3–30. Amsterdam: Elsevier.

Morita, S., and H. Nakano. 2011. Nonstructural carbohydrate content in the stem at full heading contributes to high performance of ripening in heat tolerant rice cultivars. *Crop Sci.* 51:818–828.

Murata, Y. 1961. Studies on the photosynthesis of rice plants and its significance. *Bull. Nat. Inst. Agric. Sci.* 69:1–169.

Murata, Y., and S. Matsushima. 1975. Rice. In *Crop Physiology: Some Case Histories*, ed. L. T. Evans, pp. 73–99. London: Cambridge University Press.

Murayama, N. 1995. Development and senescence of an individual plant. In *Science of the Rice Plant: Physiology*, Vol. 2, ed. T. Matsuo, K. Kumazawa, R. Ishii, K. Ishihara, and H. Hirata, pp. 119–178. Tokyo: Food and Agriculture Policy Research Center.

Murthy, R. S. 1978. Rice soils of India. In *Soils and Rice*, ed. International Rice Research Institute, pp. 3–17. Los Baños, Philippines: IRRI.

Nair, N. R., and P. C. Sahadevan. 1961. A note on vegetative propagation of cultivated rice. *Curr. Sci.* 30:474–476.

Nakamoto, T., S. Inagawa, and Y. Nagato. 1987. An analysis of the growth pattern of glumes in the rice plant. *Jpn. J. Crop Sci.* 56:149–155.

Nayar, N. M. 2012. Evolution of the African rice: A historical and biological perspective. *Crop Sci.* 52:505–516.

Nerson, H. 1980. Effects of population density and number of ears on wheat yield and its components. *Field Crops Res.* 3: 225–234.

Nguyen, H. T., R. C. Babu, and A Blum. 1997. Breeding for drought resistance in rice: Physiology and molecular genetics considerations. *Crop Sci.* 37:1426–1434.

Nishiyama, I. 1995. Factors and mechanisms causing cool weather damage. In *Science of the Rice Plant: Physiology,* Vol. 2, eds. T. Matsuo, K. Kumazawa, R. Ishihara, K. Ishihara, and H. Hirata, pp. 776–793. Tokyo: Food and Agriculture Policy Center.

Nobel, C L., and M. E. Rogers. 1992. Arguments for the use of physiological criteria for improving the salt tolerance in crops. *Plant Soil.* 146:99–107.

Oerke, E. C., and H. W. Dehne. 2004. Safeguarding production-losses in major crops and the role of crop protection. *Crop Prot.* 23:275–285.

Okawa, S., A. Makino, and T. Mae. 2003. Effect of irradiance on the portioning of assimilated carbon during the early phase of grain filling in rice. *Ann. Bot.* 92:357–364.

Olofsdotter, M. 2001. Rice: A step toward use of allelopathy. *Agron. J.* 93:3–8.

Oritani, T. 1995. Mechanisms of aging and senescence. In *Science of the Rice Plant: Physiology*, Vol. 2, eds. T. Matsuo, K. Kumazawa, R. Ishii, K. Ishihara, and H. Hirata, pp. 164–178. Tokyo: Food and Agriculture Policy Research Center.

Osada, A. 1975. Some characteristics in photosynthetic activity of leaves of *indica* rice varieties. In *Rice in Asia*, ed. Assoc. Japn. Agric. Sci. Soc., pp. 210–222. Tokyo: University of Tokyo Press.

Osaki, M., K. Morikawa, T. Shinano, M. Urayama, and T. Tadano. 1991b. Productivity of high-yielding crops. II. Comparison of N, P, K, Ca, and Mg accumulation and distribution among high-yielding crops. *Soil Sci. Plant Nutr.* 37:445–454.

Osaki, M., K. Morikawa, M. Yoshida, T. Shinano, and T. Tadano. 1991a. Productivity of high-yielding crops. I. Comparison of growth and productivity among high-yielding crops. *Soil Sci. Plant Nutr.* 37:331–339.

Osaki, M., T. Shinano, M. Matsumoto, J. Ushiki, M. M. Shinano, M. Urayama, and T. Tadano. 1995. Productivity of high-yielding crops. V. Root growth and specific absorption rate of nitrogen. *Soil Sci. Plant Nutr.* 41:635–647.

Osaki, M. and A. Tanaka. 1978. The ^{14}C retention percentage in the rice plant. *J. Sci. Soil Manure, Jpn.* 49:217–220.
O'Toole, J. C. 1982. Adaptation of rice to drought-prone environments. In *Drought Resistance in Crops with Emphasis on Rice*, ed. International Rice Research Institute, pp. 195–213. Los Baños, Philippines: IRRI.
Ottis, B. V., and R. E. Talbert. 2005. Rice yield components as affected by cultivar and seedling rate. *Agron. J.* 97:1622–1625.
Padmanabhan, S. Y. 1973. The great Bengal famine. *Ann. Rev. Phyopathol.* 11:11–26.
Pan, X. B., M. C. Rush, X. Y. Sha, Q. J. Xie, S. D. Linscombe, S. R. Stetina, and J. H. Oard. 1999. Major gene, nonallelic sheath blight resistance from the rice cultivars Jasmine 85 and Teqing. *Crop Sci.* 39:338–346.
Panabokke, C. R. 1978. Rice soils of Sri Lanka. In *Soils and Rice*, ed. International Rice Research Institute, pp. 19–33. Los Baños, Philippines: IRRI.
Pandey, S. 1996. Socioeconomic context and priorities for strategic research on Asian upland rice ecosystems. In *Upland Rice Research in Partnership*, ed. C. Piggin, pp. 103–124. Los Baños, Philippines: IRRI.
Peng, S., J. Huang, J. E. Sheehy, R. C. Laza, R. M. Visperas, X. Zhong, G. S. Centeno, et al. 2004. Rice yields decline with higher night temperature from global warming. *Proc. Natl. Acad. Sci. USA.* 101:9971–9975.
Peng, S., and G. S. Khush. 2003. Four decades of breeding for varietal improvement of irrigated lowland rice in the International Rice Research Institute. *Plant Prod. Sci.* 6:157–164.
Peng, S. B., and K. G. Cassman. 1998. Upper thresholds of nitrogen uptake rates and associated nitrogen fertilizer efficiencies in irrigated rice. *Agron. J.* 90:178–185.
Peng, S. B., K. G. Cassman, S. S. Virmani, J. Sheehy, and G. S. Khush. 1999. Yield potential trends of tropical rice since the release of IR8 and the challenge of increasing rice yield potential. *Crop Sci.* 39:1552–1559.
Peng, S. B., G. S. Khush, and K. G. Cassman. 1994. Evolution of the new plant ideotype for increased yield potential. In *Breaking the Yield Barrier: Proceedings of a Workshop on Rice Yield Potential in Favorable Environments*, ed. K. G. Kassman, pp. 5–20. Los Baños, Philippines: IRRI.
Peng, S. B., R. C. Laza, R. M. Visperas, A. L. Sanico, K. G. Cassman, and G. S. Khush. 2000. Grain yield of rice cultivars and lines developed in the Philippines since 1966. *Crop Sci.* 40:307–314.
Pinheiro, B. S., E. M. Castro, and C. M. Guimares. 2006. Sustainability and profitability of aerobic rice production in Brazil. *Field Crops Res.* 97:34–42.
Pinnschmidt, H. O., P. S. Teng, and L. Yong. 1994. Methodology for quantifying rice yield effects of blast. In *Rice Blast Disease*, eds. R. S. Zeigler, S. A. Leong, and P. S. Teng, pp. 381–408. Wallingford: CAB.
Pinson, S. R. M., F. M. Capdevielle, and J. H. Oard. 2005. Confirming QTLs and finding additional loci conditioning sheath blight resistance in rice using recombinant inbred lines. *Crop Sci.* 45:503–510.
Plucknett, D. L., J. P. Evenson, and W. G. Sanford. 1970. Ratoon cropping. *Adv. Agron.* 22: 286–330.
Porter, J. R., and M. A. Semenov. 2005. Crop response to climatic variation. *Philos. Trans. R. Soc.* 260:2021–2035.
Power, J. F., and J. Alessi. 1978. Tiller development and yield of standard and semidwarf spring wheat varieties as affected by nitrogen fertilizer. *J. Agric. Sci.* 90: 97–108.
Prabhu, A. S., L. G. Araújo, C. Fuastina, and R. F. Berni. 2003. Estimation of loss caused by blast on yield of upland rice. *Pesq. Agropec. Brás.* 38:1045–1051.
Prabhu, A. S., and I. P. Bedendo. 1988. Glume blight of rice in Brazil: Etilogy, varietal reaction and loss estimates. *Tropical Pest Manage.* 34:85–88.

Prabhu, A. S., and J. C. Faria. 1982. Quantative relationships between leaf blast and neck blast and their effect on grain filling and grain weight in upland rice. *Pesq. Agropec. Brás.* 17: 219–223.

Prabhu, A.S., M. A Lopes, and F. J. P. Zimmermann. 1980. Infection of leaves and grain of rice by *Helminthosporium oryzae* and its effects on yield components. *Pesq. Agropec. Brás.* 15:183–189.

Prasad, P. V. V., K. J. Boote, L. H. Allen, Jr., J. E. Sheehy, and J. M. G. Thomas. 2006. Species, ecotype and cultivar differences in spikelet fertility and harvest index of rice in response to high temperature stress. *Field Crops Res.* 95:398–411.

Prasad, R. 2011. Aerobic rice systems. *Adv. Agron.* 111:207–247.

Prashar, C. R. K. 1970. Paddy ratoons. *World Crops.* 22:145–147.

Qian, Q., P. He, S. Teng, D. L. Zeng, and L. H. Zhu. 1994. QTL analysis of tiller angle in rice (*Oryza sativa* L.). *Acta Genet. Sin.* 28:29–32.

Quddus, M. A. 1981. Effect of several growth regulators, shading and cultural management practices on rice ratooning. MS thesis, University of Philippines.

Rahman, M. S. 1984. Breaking the yield barriers in cereals with special reference to rice. *J. Aust. Inst. Agric. Sci.* 504:228–232.

Rao, I. M., M. A. Ayarza, R. J. Thomas, M. J. Fisher, J. I. Sanz, J. M. Spain, and C. E. Lascano. 1992. Soil plant factors and processes affecting productivity in ley farming. In *Pasture for the Tropical Lowlands*, ed. Centro Internacional de Agriculture Tropical, pp. 145–175. Cali, Colombia: CIAT.

Rasmusson, D. C., and B. G. Gengenbach. 1984. Genetics and use of physiological variability in crop breeding. In *Physiological Basis of Crop Growth and Development*, ed. M. B. Tesar, pp. 291–321. Madison, WI: American Society of Agronomy and the Crop Science Society of America.

Raymundo, M. E. 1978. Rice soils of the Philippines. In *Soils and Rice*, ed. International Rice Research Institute, pp. 115–133. Los Baños, Philippines: IRRI.

Reynolds, M. P., S. Rajaram, and K. D. Sayre. 1999. Physiological and genetic changes of irrigated wheat in the post-green revolution period and approaches for meeting projected global demand. *Crop Sci.* 39:1611–1621.

Rice, E L. 1984. *Allelopathy,* 2nd edition. New York: Academic Press.

Rimando, A. M., M. Olofsdotter, F. E. Dayan, and S. O. Duke. 2001. Searching for rice allelochemicals: An example of bioassay-guided isolation. *Agron. J.* 93:16–20.

Roberts, S. R., J. E. Hills, D. M. Brandon, B. C. Miller, S. C. Scardaci, C. M. Wick, and J. F. Williams. 1993. Biological yield and harvest index in rice: Nitrogen response of tall and semidwarf cultivars. *J. Prod. Agric.* 6:585–588.

Rodenburg, J., and D. E. Johnson. 2009. Weed management in rice-based cropping systems in Africa. *Adv. Agron.* 103:149–218.

Rush, M. C., and G. D. Lindberg. 1996. Rice disease research. *Rice J.* 77:49–52.

Saito, K., G. N. Atlin, B. Linquist, K. Phanthaboon, T. Shiraiwa, and T. Horie. 2007. Performance of traditional and improved upland rice cultivars under nonfertilized and fertilized conditions in northern Laos. *Crop Sci.* 47:2473–2481.

Saito, K., K. Azoma, and M. Sie. 2010. Grain yield performance of selected lowland new rice for Africa and modern Asian rice genotypes in West Africa. *Crop Sci.* 50:281–291.

Saito, K., Y. Sokei, and M. C. S. Wopereis. 2012. Enhancing rice productivity in West Africa through genetic improvement. *Crop Sci.* 52:484–493.

Salam, M. U., J. W. Jones, J. W. G. Jones. 1997. Phasic development of rice seedlings. *Agron. J.* 89:653–658.

Samson, B. T. 1980. Rice ratooning: Effects of varietal type and some cultural management practices. MS thesis. University of the Philippines.

Sanchez, A. C., D. S. Brar, N. Huang, and G. S. Khush. 2000. Sequence tagged site marker-assisted selection for three bacterial blight resistance genes in rice. *Crop Sci.* 40:792–797.

Sanchez, P. A. 1976. *Properties and Management of Soils in the Tropics*. New York: John Wiley.
Santos, A. B., N. K. Fageria, and A. S. Prabhu. 2003. Rice ratooning management practices for higher yields. *Commun. Soil Sci. Plant Anal.* 34:881–918.
Satake, T. 1976. Determination of the most sensitive stage to sterile-type cool injury in rice plants. *Res. Bull. Hokkaido Nat. Agric. Exp. Stn.* 113:1–44.
Satake, T. 1995. High temperature injury. In *Science of the Rice Plant: Physiology,* Vol. 2, eds. T. Matsuo, K. Kumazawa, R. Ishihara, K. Ishihara, and H. Hirata, pp. 805–812. Tokyo: Food and Agriculture Policy Center.
Satake, T., S. Y. Lee, S. Koike, and K. Kariya. 1988. Male sterility caused by cooling treatment at the young microspore stage in rice plants. *Jpn. J. Crop Sci.* 57:234–241.
Satake, T., and S. Yoshida. 1978. High temperature induced sterility in *indica* rice at flowering. *Jpn. J. Crop Sci.* 47:6–17.
Satter, R. L., and A. W. Galston. 1981. Mechanisms of control of leaf movements. *Annu. Rev. Plant Physiol.* 32:83–110.
Schlenker, W., and M. J. Roberts. 2009. Nonlinear temperature effects indicate severe damages to U.S. crop yields under climate change. *Proc. Natl. Acad. Sci.* 106:15594–15598.
Serraj, R., A. Kumar, K. L. McNally, I. Slamet-Loedin, R. Bruskiewich, R. Mauleon, J. Cirns, et al. 2011. Improvement of drought resistance in rice. *Adv. Agron.* 103:41–99.
Shanon, M. C. 1990. Breeding, selection, and the genetics of salt tolerance. In *Salinity Tolerance in Plants*, ed. R. C. Staples and G. H. Toenniessen, pp. 231–254. New York: John Wiley.
Sharma, A., A. M. McClung, S. R. M. Pinson, J. L. Kepiro, A. R. Shank, R. E. Tabien, and R. Fjellstrom. 2009. Genetic mapping of sheath blight resistance QTLs within tropical *japonica* rice cultivars. *Crop Sci.* 49:256–264.
Shimono, H., T. Hasegawa, M. Moriyama, S. Fujimura, and T. Nagata. 2005. Modeling spikelet sterility induced by low temperature in rice. *Agron. J.* 97:1524–1536.
Shimono, H., A. Ishii, E. Kanda, M. Suto, and K. Nagano. 2011. Genotypic variation in rice cold tolerance responses during reproductive growth as a function of water temperature during vegetative growth. *Crop Sci.* 51:290–297.
Shinano, T., M. Osaki, K. Komatsu, and T. Tadano. 1993. Comparison of production efficiency of the harvesting organs among field crops. *Soil Sci. Plant Nutr.* 39:269–280.
Silva-Lobo, V. L., M. G. Lacerda, M. C. Filippi, G. B. Silva, and A.S. Prabhu. 2011. Influence of nitrogen fertilization, sowing time and airspores on severity of grain discoloration in upland rice. *Summa Phytopathol.* 37:110–115.
Snyder, F. W. and G. E. Carlson. 1984. Selecting for partitioning of photosynthetic products in crops. *Adv. Agron.* 37:47–72.
Soepraptohardjo, M., and H. Suhardjo. 1978. Rice soils of Indonesia. In *Soils and Rice*, ed. International Rice Research Institute, pp. 99–113. Los Baños, Philippines: IRRI.
Soil Science Society of America. 2008. *Glossary of Soil Science Terms*. Madison, WI: SSSA.
Somerville, C., and J. Briscoe. 2001. Genetic engineering and water. *Science*. 292:2217.
Sone, C., K. Saito, and K. Futakuchi. 2009. Comparison of three methods for estimating leaf area index of upland rice cultivars. *Crop Sci.* 49:1438–1443.
Subedi, K. D., C. N. Floyd, and C. B. Budhathoki. 1998. Cool temperature induced sterility in spring wheat (*Triticum aestivum* L.) at high altitudes in Nepal: Variation among cultivars in response to sowing date. *Field Crops Res.* 55:141–151.
Tabien, R. E., S. O. P. B. Samonte, and A. M. McClung. 2008. Forty-eight years of rice improvement in Texas since the release of cultivar Bluebonnet in 1944. *Crop Sci.* 48:2097–2106.
Takahashi, N. 1984. Differentiation of ecotypes in *Oryza sativa* L. In *Biology of Rice*, eds. S. Tsunoda and N. Takahashi, pp. 31–67. Amsterdam: Elsevier.
Tanaka, A. 1976. Climate influence on photosynthesis and respiration of rice. In *Climate and Rice*, ed. International Rice Research Institute, pp. 223–247. Los Baños, Philippines: IRRI.

Tanaka, A., and M. Osaki. 1983. Growth and behavior of photosynthesized ^{14}C in various crops in relation to productivity. *Soil Sci. Plant Nutr.* 29:147–158.
Tanaka, T., and S. Matsushima. 1963. Analysis of yield-determining process and the application to yield prediction and culture improvement of lowland rice. *Proc. Crop Sci. Soc. Jpn.* 32:35–38.
Tomar, V. S., and B. P. Ghildyal. 1975. Resistance to water transport in rice plants. *Agron. J.* 67:269–272.
Tripathi, R. P. 1990. Water requirement in rice-wheat system. In *Rice–Wheat Workshop*, Modipuram, India.
Tu, J., Q. Zhang, T. W. Mew, G. S. Khush, and S. K. De Datta. 1998. Transgenic rice cultivar IR72 with Xa21 is resistant to bacterial blight. *Theor. Appl. Genet.* 97:31–36.
Tuong, T. P., and B. A. M. Bouman. 2003. Rice production in water-scarce environments. In *Water Productivity in Agriculture: Limits and Opportunities for Improvement*, eds. J. W. Kijne, R. Barker, and D. Molden, pp. 53–67. Wallingford, UK: CABI.
Turnbull, M. H. R., R. Murthy, and K. L. Griffin. 2002. The relative impacts of daytime and night-time warming on photosynthetic capacity in *Populus deltoides*. *Plant Cell Environ.* 25:1729–1737.
USDA (United States Department of Agriculture). 1975. USDA agriculture handbook 436. In *Soil Taxonomy: A Basic System of Soil Classification for Making and Interpreting Soil Surveys*. Washington, DC: US Government Printing Office.
USDA (United States Department of Agriculture). Natural Resources Conservation Service. 2011. The twelve orders of soil taxonomy. Available at http://soils.usda.gov/technical/soil_orders/.Accessed December 28, 2011
United States Salinity Laboratory Staff. 1973. *Diagnosis and Improvement of Saline and Alkali Soils*. Agriculture Handbook, 60. Washington, DC: USDA-ARS.
Vadell, J., C. Cabot, and H. Medrano. 1995. Diurnal time course of leaf gas exchange rates and related characters in drought acclimated and irrigated *Trifolium subterraneum*. *Aust. J. Plant Physiol.* 22:461–469.
Vaughan, D. A., E. Balazs, and J. S. Heslop-Harrison. 2007. From crop domestication to super-domestication. *Ann. Bot.* 100:893–901.
Vaughan, D. A., B. R. Lu, and N. Tomooka. 2008a. The evolving story of rice evolution. *Plant Sci.* 174:394–408.
Vaughan, D. A., B. R. Lu, and N. Tomooka. 2008b. Was Asia rice (*Oryza sativa*) domesticated more than once? *Rice.* 1:16–24.
Venkateswarlu, B., and R. M. Visperas. 1987. Solar radiation and rice productivity. *IRRI Research Paper Ser. 129*. Los Baños, Philipines: IRRI.
Venuprasad, R., H. R. Lafitte, and G. N. Atlin. 2007. Response to direct selection for grain yield under drought stress in rice. *Crop Sci.* 47:285–293.
Vergara, B. S. 1985. Growth and development of deep water rice plant. *IRRI Paper Ser.* 103:1–38.
Virmani, S. S. 1996. Hybrid rice. *Adv. Agron.* 57:377–462.
Virmani, S. S., R. C. Aquino, and G. S. Khush. 1982. Heterosis breeding in rice. *Oryza sativa* L. *Teor. Appl. Genet.* 63:373–380.
Wada, G., S. Shoji, and T. Mae. 1986. Relation between nitrogen absorption and growth and yield of rice plants. *JARQ.* 20:135–145.
Wada, H., H. Nonami, Y. Yabioshi, A. Maruyama, A. Tanaka, K. Wakamatsu, et al. 2011. Increased ring-shaped chalkiness and osmotic adjustment when growing grains under foehn-induced dry wind condition. *Crop Sci.* 51:1703–1715.
Wada, S. 1992. *Cool Weather Damage in Rice Plants*. Tokyo: Youkendo.
Wallace, D. H., J. L. Ozbun, and H. M. Munger. 1972. Physiological genetics of crop yield. *Adv. Agron.* 24:97–146.
Wang, H. Q., B. A. M. Bouman, D. L. Zhao, C. Wang, and P. F. Moya. 2002. Aerobic rice in northern China: Opportunities and challenges. In *Proceedings of the International*

Workshop on Water-Wise Rice Production, eds. B. A. M. Bouman, H. Hengsdijk, B. Hardy, P. S. Binraban, T. P. Tuong, and J. K. Ladha, pp. 143–154. Los Baños, Philippines: IRRI.
Wassmann, R., S. V. K. Jagadish, S. Heuer, A. Ismail, E. Redona, R. Serraj, R. K. Singh, et al. 2009a. Climate change affecting rice production: The physiological and agronomic basis for possible adaptation strategies. *Adv. Agron.* 101:59–122.
Wassmann, R., S. K. V. Jagadish, K. Sumfleth, H. Pathak, G. Howell, A. Ismail, R. Serraj, et al. 2009b. Regional vulnerability of climate change impacts on Asian rice production and scope for adaptation. *Adv. Agron.* 102:91–133.
Webb, D. B., C. N. Bollich, and J. E. Scoot, J. E. 1975. Comparative quality characteristics of rice from first ratoon crops. *Texas Agric. Exp. Stn. Prog. Rep.*:3324C.
Webster, R. K., and P. S. Gunell. 1992. *Compendium of Rice Diseases*. St. Paul: APS Press.
Wells, B. R., B. A. Huey, R. J. Norman, and R. S. Helms. 1993. Rice. In *Nutrient Deficiencies and Toxicities in Crop Plants*, ed. W. F. Bennett, pp. 15–19. St. Paul, Minnesota: The American Phytopathological Society.
West, C. P., A. P. Malaren, W. F. Wedin, and D. B. Marx. 1989. Spatial variability of soil chemical properties in grazed pastures. *SSSAJ.* 53:784–789.
Wheeler, T., J. Krishana, P. Craufurd, A. Challinor, and M. P. Singh. 2009. Effects of high temperature stress on grain crops in current and future climates. *Aspects Appl. Biol.* 88:23.
Willis, R. J. 1985. The historical bases of the concept of allelopathy. *J. Hist. Biol.* 18:71–102.
Winburne, J. N. 1962. *A Dictionary of Agricultural and Allied Terminology*. East Lansing, MI: Michigan State University Press.
Woli, P., J. W. Jones, K. T. Ingram, and C. W. Fraisse. 2012. Agricultural reference index for drought (ARID). *Agron. J.* 104:287–300.
Wright, A. L. and F. M. Hons. 2004. Soil aggregation and carbon and nitrogen storage under soybean cropping sequences. *SSSAJ.* 68:507–513.
Xie, Q. J., S. D. Linscombe, M. C. Rush, and F. K. Jodari. 1992. Registration of LSBR-33 and LSBR-5 sheath blight resistant germplasm lines of rice. *Crop Sci.* 32:507.
Xiong, Z. Y., S. J. Zhang, B. V. Ford-Lloyd, X. Jin, Y. Wu, H. X. Yan, P. Liu, et al. 2011. Latitudinal distribution and differentiation of rice germplasm: Its implication in breeding. *Crop Sci.* 51:1050–1058.
Xu, Y. B., S. R. McCouch, and Z. T. Shen. 1998. Transgressive segregation of tiller angle in rice caused by complementary gene action. *Crop Sci.* 38:12–19.
Xu, Y. B., and Z. T. Shen. 1993. Genetic analysis of tiller angle in early *indica* rice. *J. Zhe Jiang Agric. Univ.* 5:1–5.
Yamamoto, Y., T. Yoshida, T. Enomoto, and G. Yoshikawa. 1991. Characteristics for the efficiency of spikelet production and the ripening in high-yielding *japonica-indica* hybrid and semidwarf *indica* rice varieties. *Japanese J. Crop Sci.* 60:365–372.
Yan, W. G., H. Agrama, M. Jia, R. Fjellstrom, and A. McClung. 2010. Geographic description of genetic diversity and relationships in the USDA rice world collection. *Crop Sci.* 50:2406–2417.
Yeo, A. R. 1983. Salinity resistance: Physiologies and prices. *Physiol. Plantram.* 58:214–222.
Yoshida, S. 1972. Physiological aspects of grain yield. *Annu. Rev. Plant Physiol.* 23:437–464.
Yoshida, S. 1981. *Fundamentals of Rice Crop Science*. Loa Bãnos, Philippines: IRRI.
Yoshida, S. 1983. Rice. In *Potential Productivity of Field Crops under Different Environments*, ed. International Rice Research Institute, pp. 103–127. Loa Bãnos, Philippines: IRRI.
Yoshida, S., and F. T. Parao. 1976. Climatic influence on yield and yield components of lowland rice in the tropics. In *Climate and Rice*, ed. International Rice Research Institute, pp. 471–494. Los Banõs, Philippines: IRRI.
Yoshida, S., T. Satake, and D. S. Mackill. 1981. High temperature stress in rice. *IRRI Research Paper Ser. 67.* Los Baños, Philipines: IRRI.

Yu, L. X., J. D. Ray, J. C. O'Toole, and H. T. Nguyen. 1995. Use of wax-petrolatum layers for screening rice root penetration. *Crop Sci.* 35:684–687.

Yuan, L. P., and S. S. Virmani. 1988. Organization of a hybrid rice breeding program. In *Hybrid Rice*, ed. International Rice Research Institute, pp. 33–37. Los Baños, Philippines: IRRI.

Yue, B., L. Xiong, W. Xue, Y. Xing, L. Luo, and C. Xu. 2005. Genetic analysis for drought resistance of rice at reproductive stage in field with different types of soil. *Theor. Appl. Genet.* 111:1127–1136.

Zandstra, H. G. and B. T. Sampson. 1979. Rice ratoon management. *Proceedings of International Rice Research Institute*; April 17–21, 1979; IRRI: Los Banos, Philippines.

Zhang, Q. F., M. A. Saghai-Maroof, T. Y. Lu, and B. Z. Shen. 1992. Genetic diversity and differentiation of *indica* and *japonica* rice detected by RFLP analysis. *Theor. Appl. Genet.* 83:495–499.

Zhao, D. L., G. N. Atlin, M. Amante, T. S. Cruz, and A. Kumar. 2010. Developing aerobic rice cultivars for water-short irrigated and drought-prone rainfed areas in the tropics. *Crop Sci.* 50:2268–2276.

Zhao, M., Z. Zhang, S. Zhang, W. Li., D. P. Jeffers, T. Rong, and G. Pan. 2006. Quantative trait loci for resistance to banded leaf and sheath blight in maize. *Crop Sci.* 46:1039–1045.

Zheng, T. Q., Y. Wang, A. J. Ali, L. H. Zhu, Y. Sun, H. Q. Zhai, H. W. Mei, et al. 2011. Genetic effects of background independent loci for grain weight and shape identified using advances reciprocal introgression lines from Lemont × Teqing in rice. *Crop Sci.* 51:2525–2534.

Zong, Y., Z. Chen, J. B. Innes, C. Chen, Z. Wang, and H. Wang. 2007. Fire and flood management of coastal swamp enabled first rice paddy cultivation in east China. *Nature.* 449:459–462.

Zou, J. H., X. B. Pan, Z. X. Chen, J. Y. Xu, J. F. Lu, W. X. Zhai, and L. H. Zhu. 2000. Mapping quantitative trait loci controlling sheath blight resistance in two rice cultivars (*Oryza sativa* L.). *Theor. Appl. Genet.* 101:569–573.

2 Nitrogen

2.1 INTRODUCTION

Nitrogen (N) is one of the most important nutrients in crop production; it is used worldwide to increase and maintain crop production and is considered a key element in maintaining the sustainability and economic viability of cropping systems across the world (Fixon and West 2002; Abbasi and Tahir 2012; Dillon et al. 2012). Before discussing the role of nitrogen in rice production, it is pertinent to define nutrition and discuss essential plant nutrients. *Nutrition* may be defined as the supply and absorption of chemical elements needed for growth and metabolism, and the chemical elements required by an organism are termed *nutrients* (Fageria and Baligar 2005a). Normal growth of higher plants requires 17 essential mineral nutrients (Figure 2.1). On the basis of quantity required by plants, nutrients are divided into macro and micro groups. The essential nutrients such as carbon (C), hydrogen (H), oxygen (O), nitrogen (N), phosphorus (P), potassium (K), calcium (Ca), magnesium (Mg), and sulfur (S) are known as *macronutrients*, whereas iron (Fe), manganese (Mn), zinc (Zn), copper (Cu), boron (B), molybdenum (Mo), chlorine (Cl), and nickel (Ni) are termed *micronutrients*. Further, the essentiality of silicon (Si), sodium (Na), vanadium (V), and cobalt (Co) has been considered but is not proven.

The macro- and micronutrient classification is based simply on the amount required. Macronutrients are needed in larger quantities than are micronutrients. Each essential nutrient performs a biophysical or biochemical function within plant cells. Hence, all nutrients are important for production of higher crop yields. If any nutrient is deficient in the growth medium, plant growth is reduced. The low requirements of plants for micronutrients can be explained by the micronutrients' functioning in enzymatic reactions and their functioning as constituents of growth hormones rather than as components of major plant products such as structural and protoplasmic tissue (Fageria, Baliger, and Jones 2011). To obtain higher yields, essential nutrients should be present in adequate concentrations and in available form, as well as in appropriate balance. An imbalance results when one or more nutrients are either deficient or present in excess supply (Fageria and Baligar 2005a).

Plants obtain C, H, and O from air and water, and the remaining nutrients as inorganic ions or oxides are absorbed from soil solution by growing plant roots. Nutrient absorption by plants is usually referred to as *ion uptake* or *ion absorption* because it is the ionic form in which nutrients are absorbed. Cations and anions may be absorbed independently and may not be absorbed in equal quantities; however, electroneutrality must be maintained in the plant and in the growth medium. Therefore, ionic relationships are extremely important in plant nutrition. Most cations in plant tissues are in the inorganic form, predominantly K^+, Ca^{2+}, and Mg^{2+}, and the majority of anions are in the organic form. The organic ions are synthesized within the

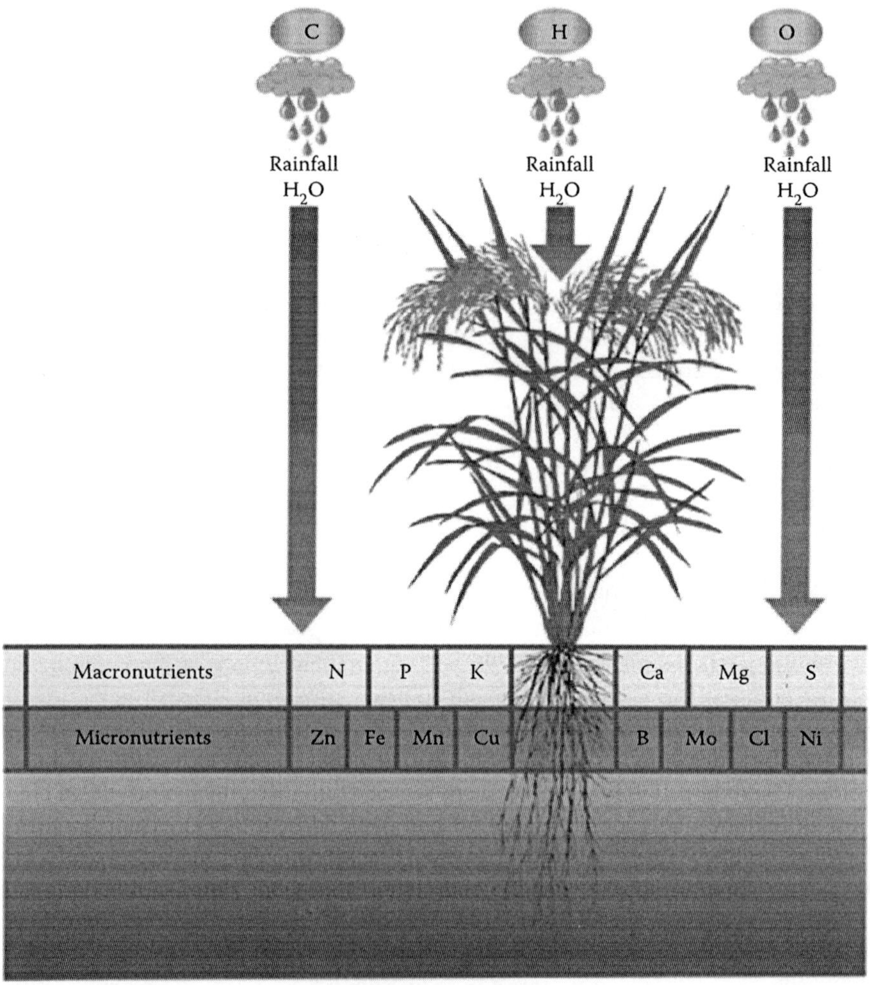

FIGURE 2.1 Essential nutrients for plant growth and development. (Adapted from Fageria, N. K. 1989. Tropical soils and physiological aspects of crops. Brasilia, Brazil: EMBRAPA-CNPAF.)

tissue, whereas inorganic ions are absorbed from the growth medium. Generally, the monovalent cations are absorbed rapidly, whereas the divalent cations, especially Ca^{2+}, are absorbed more slowly. Similarly, the monovalent anions are generally absorbed more rapidly than are polyvalent anions. The so-called physiological acidity or alkalinity of a salt depends upon which ion of the salt, the cation or the anion, is most absorbed rapidly. Thus, a salt like K_2SO_4 would be physiologically acidic because the K would enter the roots more rapidly than would the SO_4^{2-}. By the same token, $CaCl_2$ should be physiologically alkaline because the Ca^{2+} enters slowly and the Cl^- rapidly.

Mineral nutrients can also be absorbed by higher plants when applied as foliar sprays in appropriate concentrations. However, in modern high-yielding cultivars, foliar applications rarely meet nutritional (macronutrients) requirements.

Foliar application of macronutrients requires several sprays, and rain can wash off these sprays. The plant should have sufficient leaf area for absorption. Leaf damage by high nutrient concentrations is a serious practical problem. Despite these drawbacks, under some situations, foliar application is the most effective method of correcting nutritional deficiencies. For example, iron deficiency in calcareous soils can be corrected more efficiently by foliar application of ferrous sulfate solution than by soil application of iron sources.

Nitrogen is one of the nutrients that limits rice yields in all rice-growing soils of the world (Fageria and Baligar 2003a, 2005b; Fageria, Santos, and Cutrim 2007; Dillon et al. 2012). Nitrogen application increased rice yield in Brazil (Fageria and Baligar 2001, 2003b; Fageria, Santos, and Cutrim 2007; Fageria, Baliger, and Jones 2011), China (Yang 1987), India (De Datta 1981), the Philippines (Dobermann et al. 2000), Japan (Yoshida 1981), Africa (Gaudin and Dupuy 1999), and the United States (Wilson et al. 2001; Dillon et al. 2012). Brazil has about 35 million ha of lowland areas, locally known as "Varzea." These areas are distributed throughout the country. At present, less than 2 million ha of these areas are under cultivation (Fageria and Barbosa Filho 2001). Most of these lands are located near natural lakes or river basins and have water supply throughout the year. Climatic conditions are favorable in most parts of Brazil throughout the year; these soils present the most potent agricultural land in the world. In Brazil, lowland rice is the main component of the cropping system on these lands during rainy season, which normally lasts from October to March. During dry periods, other crops such as common bean, corn, soybean, cotton and wheat can be cultivated with subirrigation. These lands can produce two or three crops annually. After adequate drainage system, soil fertility is one of the primary constraints on annual crop production on varzea soils (Fageria and Baligar 1996, 2003b). Among essential plant nutrients, N is one of the more important factors in determining lowland rice yield in the varzea soils of central part of Brazil (Fageria and Baligar 2003a; Fageria, Santos, and Cutrim 2007).

The importance of nitrogen in crop production is widely discussed. Greenwood and Walker (1990) reported high correlations between cereal grain yield and the average rate of N applied in each nation when less than 160 kg N ha^{-1} are applied. Watkins, Hignight, and Wilson (2008) reported that N fertilization for rice in the southern United States can account for approximately 25% of the variable costs associated with commercial production. The recent high cost of fertilizer is another factor that demands judicious use of N fertilizers (Watkins et al. 2010). In addition to economic factors, both N use efficiency (NUE) in rice production and the subsequent impact on environmental quality are under constant scrutiny (Dillon et al. 2012).

Deficiency of N is reported in upland as well as lowland rice (Fageria et al. 2003; Fageria, Baligar, and Jones 2011; Fageria, Moreira, and Coelho 2011; Fageria, Santos, and Coelho 2011). Figure 2.2 shows shoot and root growth of upland rice as influenced by N, P, and K fertilization grown of Brazilian Oxisol. Similarly, Figure 2.3 shows influence of N, P, and K fertilization on grain yield of upland rice grown on Brazilian Oxisol. The shoot and root growth and grain yield were increased with increasing levels of N, P, and K, which shows the importance of these nutrients in upland rice production. In both cases, P was the most yield-limiting nutrient, followed by N and K.

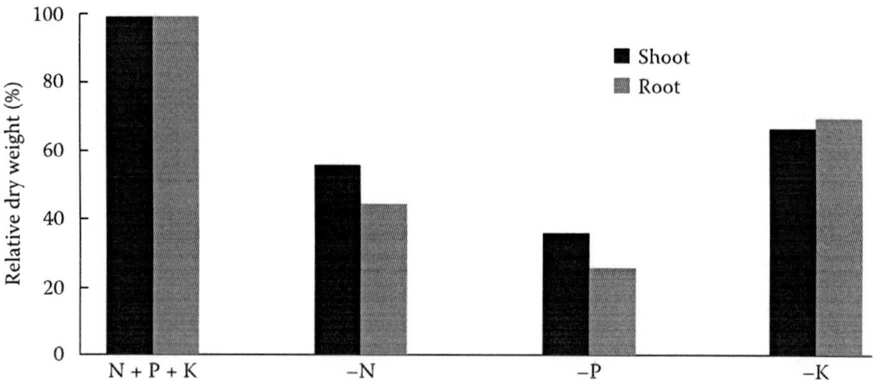

FIGURE 2.2 Influence of N, P, and K fertilization and their omission on relative dry weight of shoot and root of upland rice grown on Brazilian Oxisol. The N rate was 300 mg kg^{-1}, and P and K rates were 200 mg kg^{-1}.

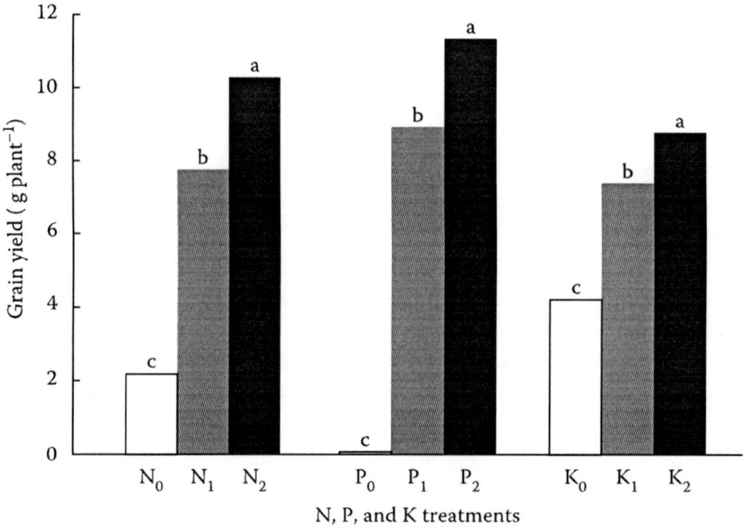

FIGURE 2.3 Influence of N, P, and K treatments on grain yield of upland rice. The levels of N were N_0 = 0 mg N kg^{-1}, N_1 = 150 mg N kg^{-1}, and N_2 = 300 mg N kg^{-1}. The levels of P were P_0 = 0 mg P kg^{-1}, P_1 = 100 mg P kg^{-1}, and P_2 = 200 mg P kg^{-1}. The K levels were K_0 = 0 mg K kg^{-1}, K_1 = 100 mg K kg^{-1}, and K_2 = 200 mg K kg^{-1}.

Rice's uptake of N is greater than its uptake of any other essential nutrient. However, in some modern cultivars, N is absorbed in equal or slightly lower amounts than is K (Fageria, Baligar, and Jones 2011). In addition, the large amount of N required by modern crop cultivars and the limited ability of soils to supply available N cause N to be the most yield-limiting nutrient for crop production on a global basis. One of the greatest boons to the development of a strong and efficient agriculture has been the production and use of commercial nitrogen fertilizers (Tisdale, Nelson, and Beaton 1985). Nitrogen is an important nutrient that should be conserved and carefully regulated. The objective of this chapter is to discuss nitrogen nutrition of upland and lowland rice and to suggest appropriate management practices to improve NUE and consequent yield.

2.2 NITROGEN CYCLE IN SOIL–PLANT SYSTEM

The nitrogen cycle in a soil–plant system is dynamic and is influenced by soil, climate, and plant factors. The *nitrogen cycle* in a soil–plant system is defined as the sequence of biochemical changes of organic and/or inorganic sources of nitrogen. Nitrogen is added to the soil–plant system by chemical fertilizers; biological and nonbiological fixation; and organic manures, including crop residues; and from the atmosphere as precipitation. Depletion of nitrogen in the soil–plant system is due to absorption by plants, absorption on soil colloids, leaching, and loss as a gas through volatilization or denitrification. Nitrogen is also lost by soil erosion in the soil–plant system. The mobility of nitrogen in the soil–plant system is significant, and the cyclic transfer goes on and on. Figures 2.4 and 2.5 depict a simplified nitrogen cycle in soil–plant

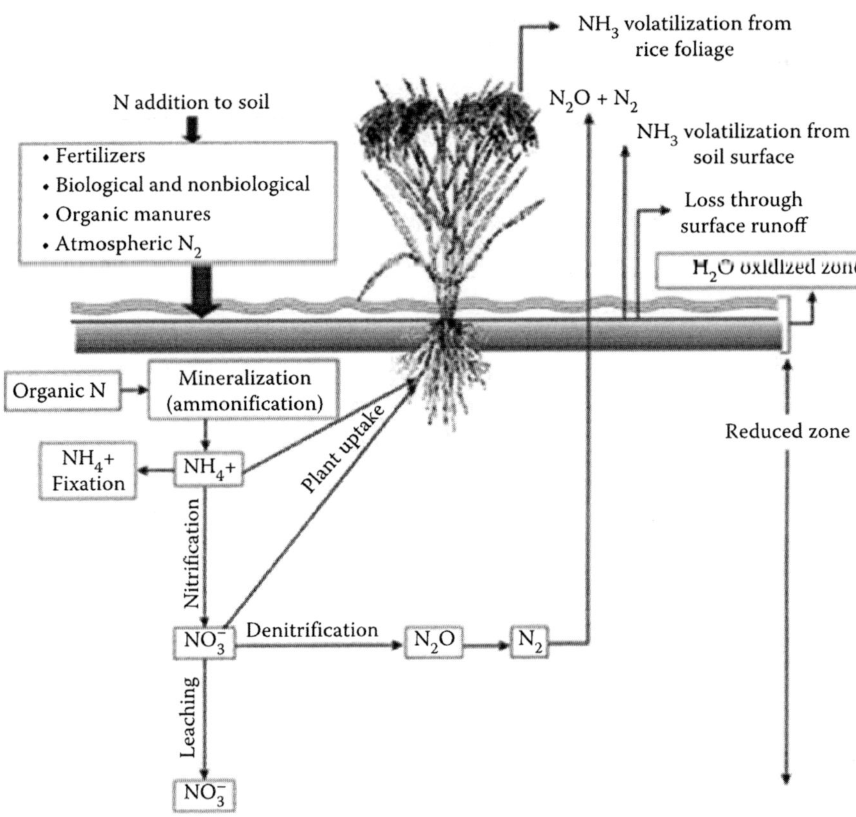

FIGURE 2.4 Nitrogen cycle in soil–plant system of irrigated rice.

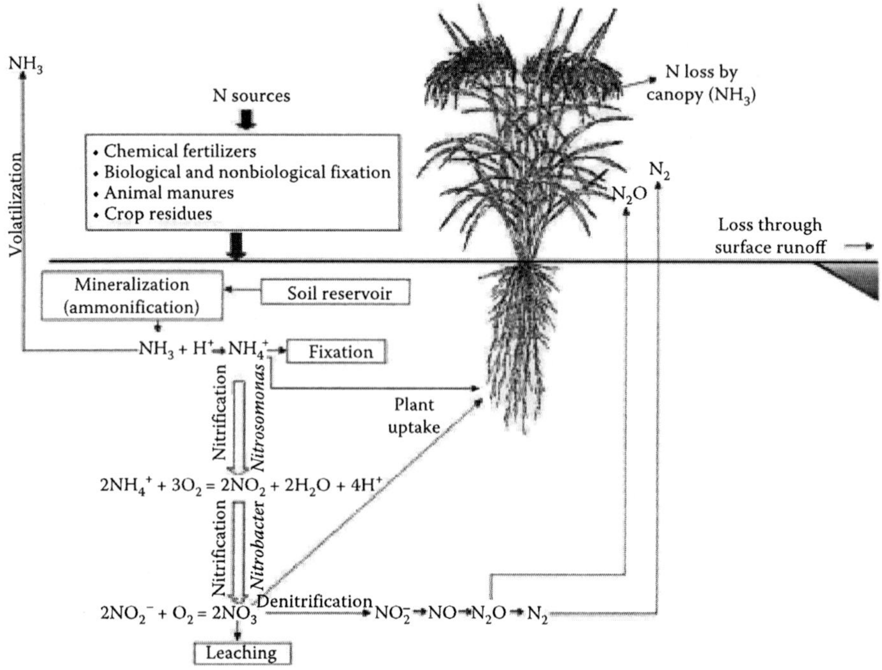

FIGURE 2.5 Nitrogen cycle in soil–plant system of upland rice.

TABLE 2.1
Nutrient Mobility in Soil–Plant Systems

Soil		Plant	
Mobile	**Immobile**	**Mobile**	**Immobile**
Nitrogen (NO_3^-)	Phosphorus (H_3PO_4)	Nitrogen	Calcium
Boron (H_3BO_3)	Potassium (K^+)	Phosphorus	Sulfur
Chlorine (Cl^-)	Calcium (Ca^{2+})	Potassium	Zinc
Sulfur (SO_4^{2-})	Magnesium (Mg^{2+})	Magnesium	Copper
Molybdenum (MoO_4^{2-})	Zinc (Zn^{2+})	Molybdenum	Iron
	Copper (Cu^{2+})		Manganese
	Iron (Fe^{3+} or Fe^{2+})		Boron
	Manganese (Mn^{2+})		Chlorine
	Nickel (Ni^{2+})		Nickel

systems of lowland and upland rice, respectively. In addition, some nutrients are mobile in the soils and plants, whereas others are immobile (Table 2.1).

Soil is the major source of N for plant growth. Most of the nitrogen (≈95%) is present in organic form in the soil. In organic form, it is protected from loss in the soil–plant system. However, it has to be mineralized before it can be absorbed by plants. *Mineralization* is the conversion of an element from an organic form to an inorganic state as a result of microbial activity (Soil Science Society of America 2008). Only 2–3% of N a year is mineralized under normal conditions. The amount of nitrogen in the form of soluble ammonium and nitrate compounds is seldom more than 1–2% of the total present, except where large quantities of inorganic nitrogen fertilizers have been applied (Brady and Weil 2002). Even so, nitrogen released in this forms has long sustained a significant portion of crop needs.

The final products of mineralization are NH_4^+ and NO_3^-. The initial conversion to NH_4^+ is referred to as *ammonification*, and the oxidation of this compound to NO_3^- is termed *nitrification*. Microorganisms that oxidize ammonium to NO_2^-, and finally to NO_3^-, are *Nitrosomonas* and *Nitrobacter*. However, Paul and Clark (1996) and Mengel et al. (2001) reported that *Nitrosolobus* and *Nitrosospira* also oxidize ammonium to NO_2^-. In addition, Bhuija and Walker (1977) reported that in many soils, *Nitrosolobus* plays a much more significant role in nitrification than is generally recognized and that *Nitrosolobus* is more important than *Nitrosomonas*. Collectively, the nitrifying organisms are known as *Nitrobacteria*. The nitrifiers are active over a range of temperatures (2–40°C), and the optimum pH lies between 7 and 9 (Paul and Clark 1996; Mosier, Doran, and Freney 2002). The overall nitrification process is controlled primarily by ammonium and oxygen concentrations. The nitrification in soil–plant systems occurs in two steps, which can be represented by the following equations (Fageria 2009):

$$2NH_4^+ + 3O_2 \leftrightarrow 2NO_2^- + 2H_2O + 4H^+ \tag{2.1}$$

$$2NO_2^- + O_2 \leftrightarrow 2NO_3^- \tag{2.2}$$

Under optimal soil temperature and humidity, nitrification occurs quickly. In addition, nitrification is an oxidation process, and aeration of soil increases nitrification. Plowing and cultivation are recognized means of promoting nitrification. Nitrification results in release of H^+ ions, leading to soil acidification. Furthermore, the enzymatic oxidation of nitrification also releases energy. Utilization of NH_4^+ and NO_3^- by plants and microorganisms constitutes assimilation and immobilization, respectively (Fageria 2009).

The major losses of nitrogen in the soil–plant system are leaching, volatilization, and denitrification. In addition, a small amount of nitrogen is also adsorbed as NH_4^+ on soil colloids and immobilized by microorganisms. Nitrate ion charges negatively and easily leaches under heavy rainfall or irrigation. The amount of N leached depends on soil type, source of N fertilizer, and crops and methods of fertilizer application (Fageria 2009). Also, nitrate is a major factor associated with leaching of such bases as calcium, magnesium, and potassium from the soil. The nitrate and bases move out together. As these bases are removed and replaced by hydrogen, soil becomes more acidic. Nitrogen fertilizers containing such strong acid-forming anions as sulfate increase acidity more than do other carriers, without acidifying anions (Fageria and Gheyi 1999).

Ammonia volatilization is an important process of N loss from soil–plant systems when nitrogen fertilizers are applied to the surface, especially in alkaline soils. In addition, NH_3 volatilization may be high from soils that have low cation exchange capacity and that lack specific adsorption sites for NH_4^+ (Fenn and Hossner 1985). Urea is more susceptible to volatilization than are ammonium-containing fertilizers, such as ammonium sulfate, because the application of urea to soils is alkalinizing, as its amino groups may be protonated (Mengel et al. 2001). Ammonia volatilization can be expressed by the following equation (Bolan and Headley 2003):

$$NH_4^+ + OH^- \leftrightarrow NH_3 + H_2O \tag{2.3}$$

Ammonia volatilization from flooded rice soils is a major mechanism for N loss and a cause of low fertilizer use efficiency by rice. Reviews on NH_3 volatilization from flooded rice soils indicate that losses of ammoniacal-N fertilizer applied directly to flood water may vary from 10 to 50% of the amount applied (Fillery and Vlex 1986; Mikkelsen 1987). However, ammonia volatilization losses have been reported to vary from 20 to 80% in rice production in the United States (Mikkelsen, De Datta, and Obcemea 1978; Beyrouty, Sommers, and Nelson 1988; Griggs et al. 2007; Norman et al. 2009). Dillon et al. (2012) reported that ammonia volatilization from urea occurs in the dry-seeded delayed-flood rice culture when urea is hydrolyzed to ammonium carbonate [$(NH_4)_2CO_3$] by the urease enzyme, and ammonium carbonate decomposes to produce NH_3 and CO_2. However, losses are site specific and soil-management specific; thus, disparities may exist in reported rates of volatilization, depending on rate-controlling factors and methods of measurement. Ammonia volatilization under flooded rice conditions is influenced by many factors: NH_4–N concentration, N source, soil and air temperature, pH, temperature, depth of flooded water, wind speed, and soil cation exchange capacity (Harper et al. 1983; Jayaweera and Mikkelsen 1990; Fageria and Gheyi 1999). Ammonia volatilization from urea fertilizers can be reduced by placing urea deep in the soil and by using polymer-coated urea or urease inhibitors.

Denitrification is a major loss of N from the soil–plant system. It can be both microbial and chemical, but the anaerobic bacterial process dominates (Mosier, Doran, and Freney 2002). Denitrification occurs in the following reductive ways:

$$NO_3^- \text{ (nitrate)} \Rightarrow NO_2^- \text{ (nitrite)} \Rightarrow NO \text{ (nitric oxide)}$$

$$\Rightarrow N_2O \text{ (nitrous oxide)} \Rightarrow N_2 \text{ (dinitrogen)} \tag{2.4}$$

Most denitrifying bacteria exist in the topsoil (0–30 cm), with the number decreasing exponentially down to 120–150 cm. Denitrification is influenced by several factors such as soil pH, temperature, organic C supply, nitrate concentration, aeration, and water status (Aulakh, Doran, and Mosier 1992). High salinity and osmotic stress also affect denitrification (Weier et al. 1993). Soil water tends to moderate oxygen diffusion in soil, and, generally speaking, denitrification occurs only in soil water contents of more than 60% of water-filled pore space (Linn and Doran 1984; Mosier, Doran, and Freney 2002). Due to physical and chemical factors, the exact quantity of N loss due to denitrification is difficult to assess. However, Aulakh, Doran, and Mosier (1992) reported that overall N losses due to denitrification might be about 30% in an agroecosystem. According to Schneider and Haider (1992), denitrification losses in arable soils range from 10 to 20 kg N ha^{-1} per year. Kowalenko and Cameron (1977), using ^{15}N-labeled fertilizers to follow the fate of N, found a total recovery (^{15}N in crop + soil) of 69% in the first year and 54% in the following year. The remaining 31 and 46%, respectively, were assumed to be mainly denitrification losses. In addition, estimation of global denitrification losses ranges from 83 Tg yr^{-1} (Stevenson 1982) to 390 Tg yr^{-1} (Hauck and Tanji 1982; Aulakh, Doran, and Mosier 1992). A large part of N is ultimately returned to the atmosphere through biological denitrification, thereby completing the cycle (Stevenson 1982). Denitrification usually occurs in soil high in organic matter (1) under extended periods of waterlogged conditions and (2) as temperature rises. N_2 is an inert gas that poses no known environmental risk, whereas N_2O is a greenhouse gas that contributes to the destruction of the ozone layer (Fageria and Gheyi 1999). Among different vegetation covers, denitrification was higher in grasslands than in either hardwood or pine forest areas (Lowrance, Vellidis, and Hubbard 1995).

In addition to these losses, a large part of nitrogen is also absorbed by crop plants. The nitrogen that is accumulated in the seeds of crop plants is permanently removed from the soil–plant system. In seeds of cereals, about 50% of the total N uptake accumulates in the seeds and in legumes. The nitrogen accumulation in the seeds varies from 60 to 70%, depending on crop species and genotypes within species. Hence, in a nitrogen management strategy, the removal of N by a crop should be taken into account (Fageria 2009).

2.3 FUNCTIONS

Nitrogen plays a role in many physiological and biochemical processes in plants. It is an important constituent of numerous organic compounds of general importance, such as amino acids, proteins, and nucleic acids, and compounds of secondary plant metabolism, such as alkaloids (Mengel et al. 2001). Nitrogen is a constituent of chlorophyll and many enzymes. Nitrogen also improves root growth, which is especially important in absorption of water and nutrients. Figure 2.6 shows root growth of three upland rice genotypes at 0 and 300 mg N kg^{-1}. Root growth in soil with 300 mg N kg^{-1} was much more voluminous than in soil treated with 0 mg N kg^{-1}. Nitrogen increases plant height, tillering, and consequently panicles and yield. Data in Table 2.2 show an increase in tiller number during the growth cycle of lowland rice. Tillering increased with N fertilization and varied from 66 to 96% due to N application. Figure 2.7 shows growth of upland rice cultivar BRS Primavera with and without N at the vegetative growth stage. Plants with N were green, having higher tillering and plant height compared with plants that did not receive N fertilization. Similarly, Figure 2.8 shows growth of upland rice cultivar BRS Sertaneja at soils with 0, 150, and 300 mg N kg^{-1} at a physiological maturity growth stage. Plant height and panicle number were reduced in soils with 0 mg N kg^{-1} compared to soils with 150 and 300 mg N kg^{-1}. Nitrogen improves plant height (Table 2.3). Nitrogen also improves panicle number and panicle length in upland rice (Table 2.4). Panicle

FIGURE 2.6 Root growth of three upland rice genotypes at two N levels.

TABLE 2.2
Numbers of Tillers in Lowland Rice at Different N Rates during Crop Growth Cycle

	Days after Sowing					
N Rate	**22 (IT)**	**35 (AT)**	**71 (IP)**	**97 (B)**	**112 (F)**	**140 (PM)**
kg ha^{-1}			m^{-2}			
0	506	681	652	541	499	468
30	516	749	715	547	516	495
60	574	880	772	601	571	531
90	599	759	751	597	561	522
120	632	876	812	623	573	569
150	619	862	883	660	580	592
180	557	880	903	662	588	572
210	565	819	934	666	590	581
R^2	0.82*	0.66*	0.96**	0.95**	0.91**	0.92**

Source: Fageria, N. K., and V. C. Baligar. 2001. Lowland rice response to nitrogen fertilization. *Commun. Soil Sci. Plant Anal.* 32:1405–1429.

Note: Values are averages of a three-year field trial.

*,**Significant at the 5 and 1% probability levels, respectively.

number or density in upland rice varied from 3.7 to 10.3 per pot at low N level with an average value across genotypes of 7.9 per pot. Similarly, panicle number at higher N levels varied from 14.3 to 29.0 per pot, with an average value of 21.4 per pot. Overall, the panicle number per pot increased by 171% with the application of N compared with treatment without N fertilization. The number of panicles (X) had a quadratic

FIGURE 2.7 (See color insert.) Upland rice cultivar BRS Primavera growth with adequate rate of P and K but without N (left) and with adequate rate of N + P + K (at right).

FIGURE 2.8 (See color insert.) Growth of upland rice at three N levels.

TABLE 2.3
Plant Height (cm) of Lowland Rice as Influenced by Nitrogen Fertilization

Nitrogen Rate (kg ha^{-1})	First Year	Second Year
0	102.19	95.41
50	104.69	97.80
100	107.97	98.08
150	113.06	99.63
200	112.78	101.89
Average	107.65	98.62

Regression Analysis

N rate vs. plant height (first year) (Y) = 102.23 + 0.59X, $R^2 = 0.94$*
N rate vs. plant height (second year) (Y) = 95.61 + 0.029X, $R^2 = 0.95$*
N rate vs. plant height (average of two years) (Y) = 98.21 + 0.048X, $R^2 = 0.97$*

*Significant at the 1% probability level.

relationship with grain yield ($P < 0.01$; $Y = -32.9349 + 7.2697X - 0.1467X^2$; $R^2 = 0.81$**). This means that panicle number accounted for 81% of the variability in grain yield. Panicle number per unit area is an important component that increases upland rice yield (Fageria 2007a, 2009; Fageria, Baligar, and Jones 2011).

Panicle length varied from 13.9 to 20.8 cm, with an average value of 17.7 cm at low N levels (Table 2.4). Similarly, at high N levels, panicle length varied from 18.4 to 26.6 cm, with an average of 22 cm. Overall, panicle length at higher N level was 24% higher than it was at low N levels. Panicle length (X) showed a significant ($P < 0.01$) quadratic relationship with grain yield ($Y = -110.5796 + 9.9081X - 0.1299X^2$; $R^2 = 0.43$**). Yoshida (1981) reported that the number of spikelets or grains per unit area of rice crop is positively correlated with the amount of N absorbed by the end of spikelet initiation stage or by flowering.

Panicle density also was influenced by N fertilization in lowland rice (Table 2.5). Panicle density varied from 5 per plant produced by genotype BRA 051134 to 8.08 per plant produced by genotype BRA 051083, with an average value of 6.58 at low N rates. At high N rates, panicle density varied from 10 per plant produced by genotypes BRA 051134 and BRA 051135, with an average value of 12.85 per plant. The increase in panicle density was almost twice as great at high N rates as it was at low N rates. Panicle density or panicle number was linearly related to grain yield ($Y = 4.6694 + 1.2607X$, $R^2 = 0.9350$**). Fageria and Barbosa Filho (2001) reported a high correlation ($r = 96$**) between grain yield and panicle density in lowland rice grown in Brazilian Inceptisol.

Nitrogen also reduced spikelet sterility in lowland rice genotypes (Table 2.6). It varied across genotypes, but at higher N rates, the decrease in spikelet sterility was 17% greater than it was at lower N rates. However, Fageria (2007a) reported

** Significant at the 1% probability level.

TABLE 2.4
Influence of Nitrogen on Panicle Density and Panicle Length in Upland Rice Genotypes

	Panicle Density (4 plants^{-1})		Panicle Length (cm)	
Genotype	0 mg N kg^{-1}	400 mg N kg^{-1}	0 mg N kg^{-1}	400 mg N kg^{-1}
CRO 97505	8.0ab	17.7fg	17.9bcd	23.8abcd
CNAs 8993	8.3ab	22.0a–f	16.7de	19.5efgh
CNAs 8812	10.3a	28.0ab	16.2def	19.1fgh
CNAs 8938	7.7abc	20.3c–g	17.5cde	20.6defgh
CNAs 8960	6.0bc	15.0fg	20.8a	26.6a
CNAs 8989	8.0ab	21.7a–g	16.3def	18.9fgh
CNAs 8824	9.3ab	26.3abcd	17.6cde	19.2efgh
CNAs 8957	6.7abc	17.0fg	19.6abc	25.4abc
CRO 97422	7.3abc	18.7efg	16.8de	23.9abcd
CNAs 8817	8.0ab	19.3defg	20.5ab	25.3abc
CNAs 8934	8.7ab	21.3b–g	17.6cde	21.6c–h
CNAs 9852	3.7c	14.3g	18.6abcd	22.6b–g
CNAs 8950	6.7abc	19.7c–g	17.3cde	23.1a–e
CNA 8540	8.7ab	25.3a–e	13.9f	18.4h
CNA 8711	8.7ab	21.0b–g	17.4cde	22.8a–f
CNA 8170	10.0ab	29.0a	18.8abcd	21.2d–h
BRS Primavera	7.7abc	17.0fg	19.7abc	25.7ab
BRS Canastra	8.0ab	25.3a–e	17.6cde	20.9d–h
BRS Carisma	8.3ab	27.0abc	15.3ef	18.7gh
Average	7.9	21.4	17.7	22.0
F-test				
N level (N)	**		**	
Genotype (G)	**		**	
N × G	**		**	
CV (%)	11		5	

Source: Fageria, N. K., O. P. Morais, and A. B. Santos. 2010. Nitrogen use efficiency in upland rice. *J. Plant Nutr.* 33:1696–1711.

Note: Means followed by the same letter in the same column are not significantly different at the 5% probability level by Tukey's test.

**Significant at the 5 and 1% probability levels, respectively.

that spikelet sterility may increase, may decrease, or may have no effect of N rate, depending on genotype. Spikelet sterility increased with increasing N rate in genotype CNAi 8860, had no significant effect of N in genotype CNAi 8569, and decreased in case of genotype BRS Jaburu (Table 2.7).

Nitrogen also improved root growth in upland rice (Table 2.8). Nitrogen × genotype interaction was significant for root length and root dry weight; therefore, data are reported separately for two N rates (Table 2.8). Root length varied from

TABLE 2.5
Panicle Number or Density of 12 Lowland Rice Genotypes as Influenced by N Fertilization

	Panicle Density ($plant^{-1}$)	
Genotype	**0 mg N kg^{-1}**	**300 mg N kg^{-1}**
BRS Tropical	7.25abcd	15.00a
BRS Jaçanã	7.08abcd	13.50ab
BRA 02654	7.75ab	13.91ab
BRA 051077	6.33bcde	14.00ab
BRA 051083	8.08a	14.33a
BRA 051108	7.50abc	14.00ab
BRA 051126	6.42bcde	12.48bc
BRA 051129	5.25e	11.33cd
BRA 051130	5.83de	13.33ab
BRA 051134	5.00e	10.00d
BRA 051135	6.00cde	10.00d
BRA 051250	6.50abcde	12.25bc
Average	6.58b	12.85a
F-test		
N rate (N)	**	
Genotype (G)	**	
N × G	**	
CV (%) N rate	7.28	
CV (%) genotype	5.84	

Source: Fageria, N. K., and A. B. Santos. in press. Lowland rice genotypes evaluation for nitrogen use efficiency. *J. Plant Nutr.*
Note: Means followed by the same letter in the same column are not significant at the 5% probability level by Tukey's test.
**Significant at the 1% probability level.

27 cm in genotype BRA 01600 to 43 cm in genotype BRA 032039, with an average value of 30.49 cm at the low N rate (0 mg N kg^{-1}). Similarly, at high N rates (300 mg N kg^{-1}), root length ranged from 21 cm in genotype BRA 01506 cm to 40.33 cm in genotype BRS Sertaneja, with an average value of 31.99 cm. Overall, root length was 5% higher at the higher N rate than it was at the lower N rate. However, 35% of genotypes produced lower root length at the higher N rate compared with the lower N rate. Fageria (1992) reported a higher root length of rice at low N rates than at high N rates in nutrient solution. Fageria (1992) also reported that at nutrient-deficient levels, root length is higher than it is at high nutrient levels because of the tendency of plants to tap nutrients from deeper soil layers. This is consistent with the findings of Sagi et al. (1997), Pessarakali (2010), and Heshmati and Pessarakli (2011) that under stressful conditions (N deficiency stress in the present study), roots grow more in search of water and/or nutrients.

TABLE 2.6
Spikelet Sterility in Lowland Rice Genotypes at Two N Levels

Genotype	Spikelet Sterility (%)	
	0 mg N kg^{-1}	300 mg N kg^{-1}
Javae	3.57c	4.93b
Rio Formoso	30.30ab	22.20a
CNA 6343	16.90bc	24.53a
CNA 7550	25.22ab	22.87a
CNA 7556	31.90a	21.27a
CNA 7857	30.03ab	23.83a
CNA 8319	30.37ab	19.43ab
CNA 8619	10.0c	13.60ab
Average	22.33	19.08

Source: Fageria, N. K., and M. P. Barbosa Filho. 2001. Nitrogen use efficiency in lowland rice genotypes. *Commun. Soil Sci. Plant Anal.* 32:2079–2089.

Note: Means followed by the same letter in the same column are not significantly different at the 5% probability level by Turkey's test.

TABLE 2.7
Influence of Nitrogen on Spikelet Sterility (%) of Three Lowland Rice Genotypes

N rate (kg ha^{-1})	CNAi 8860	CNAi 8569	BRS Jaburu
0	8.7	24.4	22.1
50	15.7	22.9	21.6
100	17.5	20.3	19.5
150	17.1	24.1	18.7
200	18.1	24.0	17.6
Average	15.4	22.0	19.9

Regression Analysis

N rate (X) vs. spikelet sterility (Y)(CNAi 8860) = $9.3543 + 0.1211X - 0.000403X^2$, $R^2 = 0.7383^{**}$

N rate (X) vs. spikelet sterility (Y)(CNAi 8569) = $24.3312 - 0.0508X - 0.00025X^2$, $R^2 = 0.1641^{NS}$

N rate (X) vs. spikelet sterility (Y)(BRS Jaburu) = $22.3139 - 2.3933X$, $R^2 = 0.4651^{**}$

Source: Fageria, N. K. 2007a. Yield physiology of rice. *J. Plant Nutr.* 30:843–879.

Note: Values are averages of a two-year field trial.

^{**,NS}Significant at the 1% probability level and nonsignificant, respectively.

Root dry weight varied from 0.87 g $plant^{-1}$, produced by genotype BRA 01596, to 1.78 g $plant^{-1}$, produced by genotype BRA 052034, with an average value of 1.38 g $plant^{-1}$ at the lower N rate (0 mg N kg^{-1}). At the higher N rate (300 mg N kg^{-1}), root dry weight ranged from 0.40 g $plant^{-1}$, produced by genotype BRA 01506, to 4.14 g $plant^{-1}$, produced by genotype BRA 052023, with an average value of

TABLE 2.8
Influence of Nitrogen on Root Length and Root Dry Weight of Upland Rice Genotypes

	Root Length (cm)		Root Dry Weight (g plant^{-1})	
Genotype	**0 mg N kg^{-1}**	**300 mg N kg^{-1}**	**0 mg N kg^{-1}**	**300 mg N kg^{-1}**
BRA 01506	34.67ab	21.00cd	0.92a	0.40f
BRA 01596	30.00b	15.67d	0.87a	0.45f
BRA 01600	27.00ab	25.33bcd	1.14a	1.03ef
BRA 02535	31.67ab	30.00abcd	1.33a	3.25abcd
BRA 02601	28.00b	32.67abc	1.11a	3.73ab
BRA 032033	30.00a	28.00abcd	1.12a	2.41cd
BRA 032039	43.00a	34.67abc	1.24a	3.77ab
BRA 032048	28.50b	32.00abc	1.05a	3.62abc
BRA 032051	30.67ab	35.67ab	1.31a	2.82bcd
BRA 042094	30.33b	38.00ab	1.78a	2.31de
BRA 042156	29.33b	33.00abc	1.19a	2.84bcd
BRA 042160	29.67b	32.50abc	1.51a	3.33abcd
BRA 052015	31.00ab	35.00abc	1.67a	2.83bcd
BRA 052023	29.67b	26.50abcd	1.48a	4.14a
BRA 052033	30.33b	31.33abc	1.56a	3.38abcd
BRA 052034	29.00b	37.67ab	1.78a	2.58bcd
BRA 052045	28.67b	37.67ab	1.75a	3.66abc
BRA 052053	29.00b	36.50ab	1.65a	2.72bcd
BRS Primavera	29.33b	36.33ab	1.68a	2.49bcd
BRS Sertaneja	30.00b	40.33a	1.45a	2.68bcd
Average	30.49	31.99	1.38	2.72
F-test				
N rate (N)	NS		*	
Genotype (G)	**		**	
N × G	**		**	
CV (%)	14.25		15.86	

Source: Fageria, N. K. 2010a. Root growth of upland rice genotypes as influenced by nitrogen fertilization. Paper presented at 19th World Congress of Soil Science, Brisbane, Australia, 1–6 August.

Note: Means followed by the same letter in the same column are not significant at the 5% probability level by Tukey's test.

*,**,NS Significant at the 5 and 1% probability levels and nonsignificant, respectively.

2.72 g plant^{-1}. Overall, root dry weight was 97% higher at the higher N rate than at the lower N rate. Fageria and Baligar (2005b) and Fageria (2009) reported that N fertilization improved root dry weight in crop plants, including upland rice. The positive effect of N on root dry matter has been previously documented (Fageria 2009).

Results related to maximum root length and root dry weight of lowland genotypes as influenced by N fertilization are presented in Table 2.9. Maximum root growth was influenced by N rate and genotype treatments. However, N × G interaction was

TABLE 2.9
Maximum Root Length and Root Dry Weight as Influenced by Nitrogen and Genotype Treatments

N Rate/Genotype	Maximum Root Length (cm)	Root Dry Weight (g plant^{-1})	
		0 mg N kg^{-1}	300 mg N kg^{-1}
0 mg N kg^{-1}	26.58a		
300 mg N kg^{-1}	23.61b		
BRS Tropical	29.00a	6.04a	5.59d
BRS Jaçanã	24.00ab	4.85abc	2.67e
BRA 02654	25.83ab	5.88ab	8.56bc
BRA 051077	23.83ab	5.59ab	13.10a
BRA 051083	26.00ab	3.09c	6.29d
BRA 051108	27.66ab	6.76a	9.03b
BRA 051126	24.50ab	5.06abc	6.41d
BRA 051129	21.17b	5.85ab	8.88b
BRA 051130	25.67ab	3.73bc	8.64bc
BRA051134	23.50	5.96ab	5.91d
BRA 051135	23.00ab	4.66abc	5.33d
BRA 051250	27.00ab	4.57abc	6.81cd
Average	24.84	5.17b	7.27a
F-test			
N rate (N)	**	**	
Genotype (G)	*	**	
N × G	NS	**	
CV (%) N rate		8.94	
CV (%) genotype		11.77	

Source: Fageria, N. K., and A. B. Santos. in press. Lowland rice genotypes evaluation for nitrogen use efficiency. *J. Plant Nutr.*

Note: Means followed by the same letter in the same column (separate for N rate and genotypes) are not significantly different at the 5% probability level by Tukey's test.

*,**,NS Significant at the 5 and 1% probability levels and nonsignificant, respectively. Means followed by the same letter in the same column (separate for N rate and genotypes) are not significantly different at the 5% probability level by Tukey's test.

not significant for this growth parameter, indicating that each of the 12 genotypes reacted similarly to changes in N rates. Maximum root length varied from 21.17 to 29 cm, with an average value of 24.84 cm. Root length was higher at low N rates than at high N rates. When there is deficiency of a determined nutrient, roots try to grow longer to take nutrients from lower soil depth (Fageria and Moreira 2011). Visually, roots at lower N rates had fewer and smaller hairs than those at higher N rates.

Root dry weight was influenced by N rate and genotype treatments (Table 2.9). The N × G interaction was also significant for root dry weight, suggesting that genotypes responded differently under two N rates. At the lower N rate, the root dry weight varied from 3.09 g plant^{-1}, produced by genotype BRA 051083, to 6.76 g

FIGURE 2.9 Growth of two lowland rice genotypes at two N levels.

plant^{-1}, produced by genotype BRA 051108, with an average value of 5.17 g plant^{-1}. At the higher N rate, root dry weight varied from 2.67 g plant^{-1}, produced by genotype BRS Jaçanã, to 13.10 g plant^{-1}, produced by genotype BRA 051077, with an average value of 7.27 g plant^{-1}. Application of 300 mg N kg^{-1} produced 41% higher dry weight than did 0 mg N kg^{-1}. Figure 2.9 shows root growth of genotype BRA 02654 and BRA 051077 at low and high N rates, respectively. Roots of both the genotypes were more vigorous at the high N rate than at the lower N rate.

2.4 DEFICIENCY SYMPTOMS

Nutrient deficiency symptoms can be used to assess deficiency versus sufficiency of a determined nutrient in crop plants. It is a cheaper nutrient disorder diagnostic technique than soil testing, tissue analysis, and greenhouse or field experimentation. However, diagnosis of nutrient deficiency in crop plants is confused with other biotic and abiotic stresses, and care should be taken in arriving at conclusions. Nitrogen is a mobile nutrient in plants, and when there is a shortage, it moves from the older part to the younger-growing part of the plants or leaves. Hence, N deficiency first appears in the older leaves in the crop plants, including rice. Nitrogen deficiency reduces plant height and tillering in rice. In the beginning, older leaves turn yellow. When deficiency is advanced, entire leaves on the plant may show yellow coloration. In severe deficiency, leaves become dry at the tips and margins. Plants supplied with adequate N show vigorous vegetative growth and a dark green color. Nitrogen toxicity in crop plants is rare but excess can create nutritional imbalance.

The yellow color of leaves due to N deficiency is related to reduction in chlorophyll content. Chlorophyll is an important pigment in chloroplast, and it plays an indispensable role in light harvesting and energy transfer of photosynthesis (Temnykh et al. 2000; Eckhardt, Grimm, and Hortensteiner 2004; Wu et al. 2011). Consequently, chloroplast development and chlorophyll metabolism are crucial to plants in relation to photosynthesis. So absence or malfunction of key factors (e.g., N deficiency) required for chlorophyll synthesis and/or chloroplast development in plants usually lead to phenotypic changes, such as differences in leaf color (Wu et al. 2011). Figures 2.9 through 2.13 show N deficiency symptoms in upland and lowland rice grown under field and greenhouse conditions.

FIGURE 2.10 Upland rice growth in plots that received nitrogen and that did not receive nitrogen. 0 kg N ha^{-1} indicate plants that did not receive N whereas 100 kg N ha^{-1} indicates those plants that received N.

FIGURE 2.11 Plants with nitrogen deficiency symptoms (0 mg N kg^{-1}) and those without nitrogen deficiency symptoms (300 mg N kg^{-1}).

FIGURE 2.12 Two upland rice genotypes with and without N.

FIGURE 2.13 (See color insert.) Two lowland rice genotypes showing N deficiency symptoms in pots that did not receive N fertilization.

Nitrogen deficiency symptoms normally occur over an area, not on an individual plant. If symptom is found on a single plant, it may be due to disease, insect injury, or genetic variation. Also, earlier symptoms are often more useful than late mature symptoms. Some nutrients are relatively immobile in the plant, whereas others are more mobile (Fageria and Baligar 2005a). In conclusion, the use of visible symptoms has the advantage of direct field application without the need of costly equipment or laboratory support services, as is the case with soil and plant analysis (Fageria and Baligar 2005a). A disadvantage is that sometimes it is too late to correct a nutrient deficiency because the disorder is identified when it is in the advanced stage. For some disorders, considerable yield loss may have already occurred by the time visible symptoms appear.

2.5 UPTAKE IN PLANTS

Uptake of nutrients by crop plants is influenced by climate, soil, and plant factors and their interactions. Most nitrogen fertilizers are water soluble, and nitrogen, after their application, in soil solution can reach plant roots by mass flow, if soil has sufficient humidity. Under optimum soil conditions, the effect of application of nitrogen fertilizers can be observed in five days, as the leaves turn greener than earlier.

2.5.1 Forms of Uptake

Plants absorb N from the soil solution as NO_3^- or NH_4^+. Most plants can absorb both forms of N equally; however, the form of N uptake is mainly determined by its abundance and accessibility. In well-drained soils, NO_3 dominates, whereas under anaerobic conditions and in cold climates, NH_4 is the dominant form. In general, under agricultural conditions, soil NO_3 concentrations range between 0.5 and 10 mM (7–140 mg NO_3 kg^{-1}), whereas NH_4 concentrations are 10–1000 times lower, and they reach the millimolar range only in exceptional cases, such as after fertilization (Wiren, Gazzarrini, and Frommer 1997). Although both forms of N can be absorbed by rice, Fried et al. (1965) showed that excised roots from rice seedlings absorbed NH_4^+ 5–20 times faster than did NO_3^-, with the rate being somewhat dependent on the solution pH. In addition, uptake of NO_3^- requires more energy than does uptake of NH_4^+ ions (Fageria et al. 2011). Most published literature suggests that rice prefers NH_4^+ over NO_3 (Mengel and Viro 1978). When NO_3^- is absorbed, the NO_3^- must be reduced for assimilation and transport within the plant, with the reduction process requiring energy. Takenga (1995) suggested that rice may prefer one N form over another based on plant growth stage. During vegetative growth, NH_4^+ was absorbed more effectively, whereas NO_3 is preferentially absorbed during reproductive growth. Fageria et al. (2003) reported that rice shoot and root dry weights were higher when rice was grown in an NH_4–N culture solution than when it was grown in a NO_3–N culture solution. Rice that is flooded continuously, from early vegetative growth until physiological maturity, does not absorb large amounts of NO_3^- because the NO_3^- is rapidly lost via denitrification after flooding. For this reason, soil or fertilizer NO_3 is of little or no benefit to lowland rice. In most cases, an NH_4^+-forming N fertilizer must be used along with management practices that largely prevent nitrification to optimize N uptake efficiency.

Ammonium sulfate and urea [$CO(NH_2)_2$] are major sources of nitrogen-supplying fertilizers in crop production, including rice (Bufogle et al. 1998; Gaudin and Dupuy 1999; Fageria, Baligar, and Jones 2011). The initial soil pH, due to its influence on potential NH_3 volatilization losses and the length of time required to flood the soil, may influence the efficiency of uptake of these two N sources (Fageria et al. 2003). Ammonium sulfate may be more efficient when applied on alkaline soils or when the N is not incorporated immediately by flooding. Urea, following application to the soil, undergoes hydrolysis to eventually form NH_4. However, under certain soil and environmental conditions, N in the form of NH_3, a gas, may be lost by volatilization. Ammonia volatilization may be a pathway of N loss on alkaline soils.

Marschner (1995) reported that calcifuges, or plants adapted to acidic soils and plants adapted to low soil redox potential (e.g., lowland rice), prefer NH_4. In contrast, calcicoles, or plants with a preference for calcareous soils, prefer utilizing NO_3. De Datta (1981) reported that the N applied at planting should be in the NH_4 form. The N used as topdressing is less critical, as NH_4 and NO_3 forms appear equally effective (Wilson, Wells, and Norman 1994). When the crop is fully established (i.e., panicle initiation), the NO_3^- form of N is rapidly taken up before it can be leached down to the reduced soil layer where it could be lost through denitrification. This may account for the equal performance of NO_3-containing fertilizers (e.g., NH_4NO_3) and other NH_4-containing or NH_4-forming N sources such as ammonium sulfate or urea (De Datta 1981; Fageria et al. 2003).

2.5.2 Uptake in Plant Tissue

Tissue analysis is playing an increasingly important role in the expanding technology of economic crop production. Grain and straw should be analyzed separately to determine nutrient uptake by a crop to get information about fertility depletion. Knowledge of adequate N concentration (content per unit dry weight) and uptake (concentration × dry weight) is an important plant parameter for appropriate N management of a crop. The adequate N concentration and uptake varied with yield level, which is affected with crop management practices. Concentration of nutrients is generally expressed in gram per kilogram for macronutrients and milligram per kilogram for micronutrients. Similarly, uptake is expressed in kilogram per hectare for macronutrients and in gram per hectare for micronutrients for field experiment results. For greenhouse experiments, concentration results are expressed in the same units (g kg^{-1} or mg kg^{-1}) as are those of field experiments, but uptake results are expressed in milligram per milligram (for macronutrients) or microgram per milligram (for micronutrients). When the objective of the experiment is to determine soil fertility depletion, whole tops should be analyzed. When the purpose of the experiment is identifying nutrient deficiency or sufficiency, the top three or four matured leaves should be analyzed. Nutrient uptake varied during crop growth cycle due to change in plant demand, which is related to increases in dry matter production.

Results related to concentration and uptake of nitrogen in the straw of four crop species, including upland rice, are presented in Figure 2.14. Concentration of N in most tissues of crop plants decreased with the advancement of plant age. This was as expected, because with increasing plant age, more dry matter was produced, which diluted the concentration of nutrients accumulated (Fageria and Baligar 2005b; Fageria, Baligar, and Jones 2011). Figure 2.15 presents data related to nitrogen uptake in the shoot of four crop species (excluding grain) during the growth cycle. The N uptake increased in a quadratic fashion with the advancement of plant age. The N uptake in the straw of these crop species followed the pattern of dry matter production (Figure 2.16). Sims and Place (1968), Moore, Gilmour, and Wells (1981),

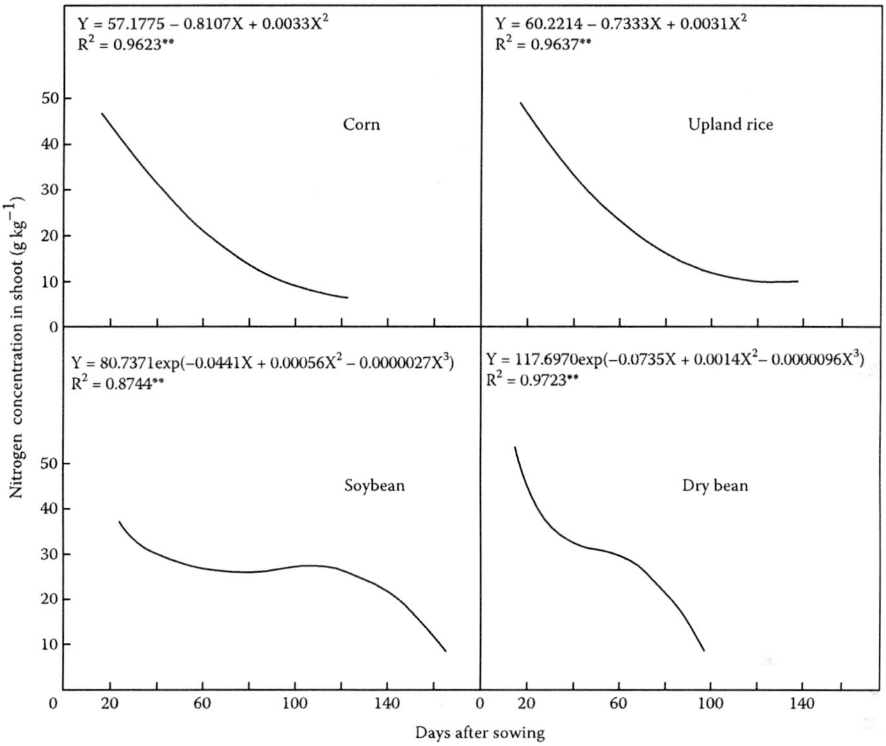

FIGURE 2.14 Nitrogen concentration in four crop species during growth cycle. (From Fageria, N. K. 2004. Dry matter yield and shoot nutrient concentrations of upland rice, common bean, corn and soybean in rotation on an Oxisol. *Commun. Soil Sci. Plant Anal.* 35:961–974.)

and Ntamatungiro et al. (1999) reported that the amount of N accumulated generally paralleled dry matter accumulation and increased with plant age in rice.

Nitrogen concentration in the rice straw increased with the addition of N fertilizer (Table 2.10). In addition, during the growing season, concentration in the straw of lowland rice gradually declines as plant dry matter increases, regardless of N rate (Table 2.9). Although tissue N concentration decreases, total N uptake increases as N fertilizer rate increases (Table 2.11). Approximately one-half of the total dry matter of a mature rice crop is produced by panicle differentiation or the early boot stage (Sims and Place 1968; Guindo, Wells, and Norman 1994; Guindo, Norman, and Wells 1994b). Total N uptake tends to follow the same general pattern as dry matter accumulation (Fageria and Baligar 2001, 2003b). Under optimum fertilization, one-half of the total N is absorbed before one-half of the total dry matter is produced. The remaining one-half to one-third of the N, not in the plant by panicle differentiation, is obtained either from top-dressed N fertilizer, soil N released via N mineralization, or both. The total N content of rice straw and grain at maturity usually contains 20–40% more N than does that supplied by fertilizer and usually represents plant uptake of native soil N (Moore, Gilmour, and Wells 1981; Norman et al. 1992).

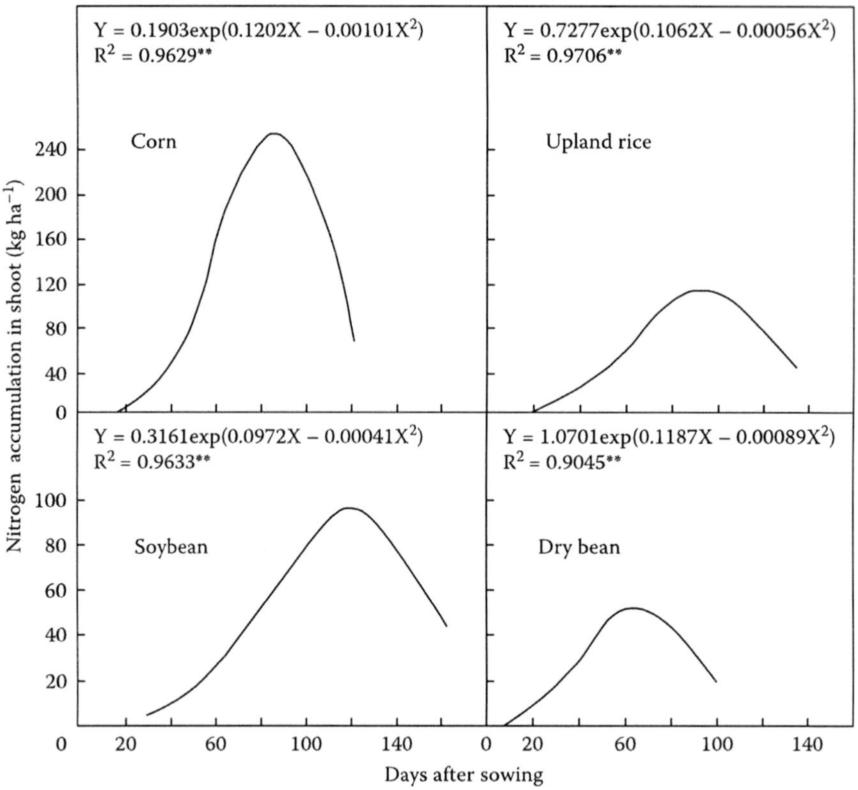

FIGURE 2.15 Nitrogen uptake in four crop species during growth cycle. (From Fageria, N. K. 2004. Dry matter yield and shoot nutrient concentrations of upland rice, common bean, corn and soybean in rotation on an Oxisol. *Commun. Soil Sci. Plant Anal.* 35:961–974.)

Data presented in Tables 2.12 and 2.13 show the relationship between (1) shoot dry weight or grain yield and (2) N concentration and uptake in the shoot or grain of lowland rice grown on Brazilian Inceptisol. Based on regression equations, adequate concentration and uptake for the maximum shoot and grain yield can be calculated (Fageria 2003). The adequate N concentration level in rice shoot varied from 43.4 to 6.5 g kg^{-1}, depending on plant age (Fageria 2003). It decreases with the advancement of plant age, reflecting the dilution effect with the advancement of plant age (Fageria, Baligar and Jones 2011). At physiological maturity, optimum N concentration in the grain was 10.9 g kg^{-1}, which was 68% higher than that of shoot concentration. The N uptake in shoot, as well as in the grain, was related to shoot dry weight and grain yield (Table 2.13). The higher R^2 values of N uptake in shoot after 35 days of growth indicate that N uptake increased in the shoot with increasing plant age. Variability (99%) in N uptake peaked at physiological maturity. This means that N uptake determination during this growth stage is more important for knowing the quantity of N removed from the soil by rice crop and adopting appropriate N management practice to supply the desired N rate.

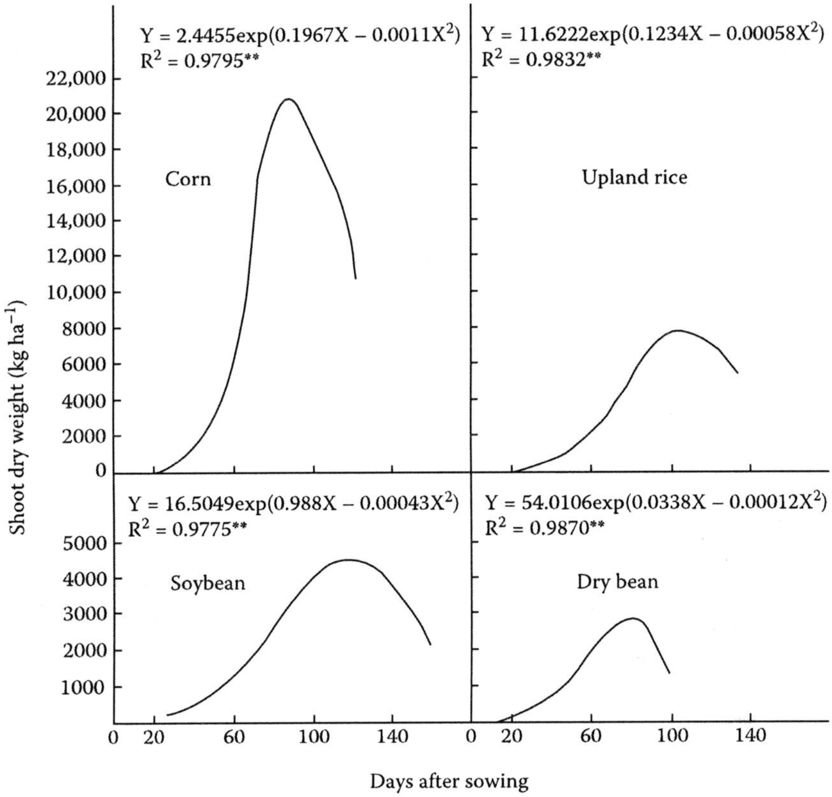

FIGURE 2.16 Shoot dry weight of four crop species during crop growth cycle. (From Fageria, N. K. 2004. Dry matter yield and shoot nutrient concentrations of upland rice, common bean, corn and soybean in rotation on an Oxisol. *Commun. Soil Sci. Plant Anal.* 35:961–974.)

Optimum value of N for maximum shoot yield increased with the advancement of plant age up to flowering and then decreased (Fageria 2003). The decrease may be associated with translocation of major part of N to grain formation after flowering. Total N uptake in shoot and grain was 147 kg ha^{-1}. Accumulated N at harvest or physiological maturity produced 9545 kg ha^{-1} shoot dry weight and 6550 kg ha^{-1} grain weight. Hence, to produce 1 Mg grain yield, rice needs 22 kg N uptake in shoot and grain. Yoshida (1981) reported that to produce one metric ton of rough rice, the N requirement was 20 kg ha^{-1}.

The N uptake in shoot and grain was related to grain yield (Figures 2.17 through 2.20). Accumulation of N in the shoot had a quadratic relationship with grain yield, whereas N accumulation in the grain had a linear relationship with grain yield. Hence, improving N uptake in both shoot and grain can improve grain yield in rice. Fageria (2003) reported a quadratic association with N uptake in shoot and grain yield of lowland rice.

TABLE 2.10
Nitrogen Concentration in Straw of Lowland Rice as Influenced by N Rate and Days after Sowing

N Fertilizer Rate	Days after Seeding (Growth Stage)					
	22 (IT)	35 (AT)	71 (IP)	97 (B)	112 (F)	140 (PM)
kg N ha^{-1}	g N kg^{-1}					
0	40	28	12	9	8	5
30	42	28	11	9	7	6
60	43	30	12	10	9	6
90	44	31	12	11	8	6
120	44	34	13	10	9	6
150	45	32	13	12	9	6
180	46	34	13	13	10	7
210	45	33	15	13	10	7
R^2	0.95**	0.85**	0.84**	0.89**	0.72*	0.75**

Source: Fageria, N. K., and V. C. Baligar. 2001. Lowland rice response to nitrogen fertilization. *Commun. Soil Sci. Plant Anal.* 32:1405–1429.

Note: IT, initiation of tillering; AT, active tillering; IP, initiation of panicle; B, booting; F, flowering; PM, physiological maturity.

*,**Significant at the 5 and 1% probability levels, respectively.

TABLE 2.11
Nitrogen Uptake in Straw and Grain of Lowland Rice under Different N Rates during Crop Growth Cycle across the Three Years

N Fertilizer Rate	Days after Seeding (Growth Stage)						
	22 (IT)	35 (AT)	71 (IP)	97 (B)	112 (F)	140 (PM)	Grain
(kg N ha^{-1})	(kg N ha^{-1})						
0	13	24	35	50	57	28	36
30	13	26	41	62	66	37	41
60	15	37	45	82	95	40	55
90	17	33	52	88	89	41	55
120	17	42	57	91	122	48	66
150	19	40	63	113	113	52	67
180	16	44	67	137	135	59	71
210	16	39	87	130	130	66	74
R^2	0.78*	0.85*	0.96*	0.95*	0.92*	0.98*	0.97*

Source: Fageria, N. K., and V. C. Baligar. 2001. Lowland rice response to nitrogen fertilization. *Commun. Soil Sci. Plant Anal.* 32:1405–1429.

Note: IT, initiation of tillering; AT, active tillering; IP, initiation of panicle; B, booting; F, flowering; PM, physiological maturity.

*Significant at the 1% probability level.

TABLE 2.12
Relationship between Dry Matter Yield of Shoot or Grain (Y) and N Concentration in Shoot or Grain at Different Growth Stages in Lowland Rice

Plant Growth Stage	Regression	R^2
IT (22)[a]	$Y = -439.4654 + 22.5403X - 0.0946X^2$	0.39^{NS}
AT (35)	$Y = -8974.3480 + 586.9736X - 8.4265X^2$	0.74*
IP (71)	$Y = 211.7915 - 34.9390X + 28.1748X^2$	0.88**
B (97)	$Y = -36286.13 + 7325.2430 - 285.4674X^2$	0.77*
F (112)	$Y = -44383.16 + 10690.71X - 485.6974X^2$	0.94**
PM (140)	$Y = -100159.00 + 33792.63X - 2605.362X^2$	0.94**
PM (140)[b]	$Y = 1141085.70 + 27046.20X - 1237.72X^2$	0.78*

Source: Fageria, N. K. 2003. Plant tissue test for determination of optimum concentration and uptake of nitrogen at different growth stages in lowland rice. *Commun. Soil Sci. Plant Anal.* 34:259–270.

Note: IT, initiation of tillering; AT, active tillering; IP, initiation of panicle; B, booting; F, flowering; PM, physiological maturity. Values are averages of a three-year field experiment.

[a] Values in parentheses represent age of the plants in days after sowing.

[b] In this line, values are for grain yield.

*,**,NS Significant at the 5 and 1% probability levels and nonsignificant, respectively.

TABLE 2.13
Relationship between Grain Yield (Y) and N Uptake in the Shoot and Grain of Lowland Rice at Different Growth Stages

Plant Growth Stage	Regression	R^2
IT (22)[a]	$Y = 166.46 + 9.4552X - 0.1565X^2$	0.61^{NS}
AT (35)	$Y = -391.29 + 63.8885X - 0.5898X^2$	0.93*
IP (71)	$Y = 40.32 + 101.2576X - 0.3939X^2$	0.97*
B (97)	$Y = -2069.44 + 185.7829X - 0.6725X^2$	0.94*
F (112)	$Y = -367.39 + 167.8636X - 0.4528X^2$	0.97*
PM (140)	$Y = -2330.74 + 335.1191X - 2.3641X^2$	0.99*
PM (140)[b]	$Y = -3547.09 + 261.4988X - 1.7099X^2$	0.99*

Source: Fageria, N. K. 2003. Plant tissue test for determination of optimum concentration and uptake of nitrogen at different growth stages in lowland rice. *Commun. Soil Sci. Plant Anal.* 34:259–270.

Note: IT, initiation of tillering; AT, active tillering; IP, initiation of panicle; B, booting; F, flowering; PM, physiological maturity.

[a] Values in parentheses represent age of the plants in days after sowing.

[b] In this line, values are for grain yield. Values are averages of a three-year field experiment.

*,NS Significant at the 1% probability level and nonsignificant, respectively.

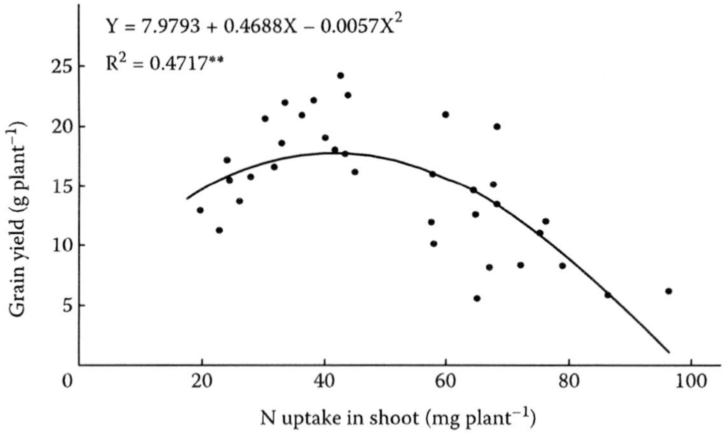

FIGURE 2.17 Relationship between N uptake in shoot and grain yield of lowland rice under greenhouse conditions. (From Fageria, N. K., A. B. Santos, and A. M. Coelho. 2011. Growth, yield and yield components of lowland rice as influenced by ammonium sulfate and urea fertilization. *J. Plant Nutr.* 34:371–386.)

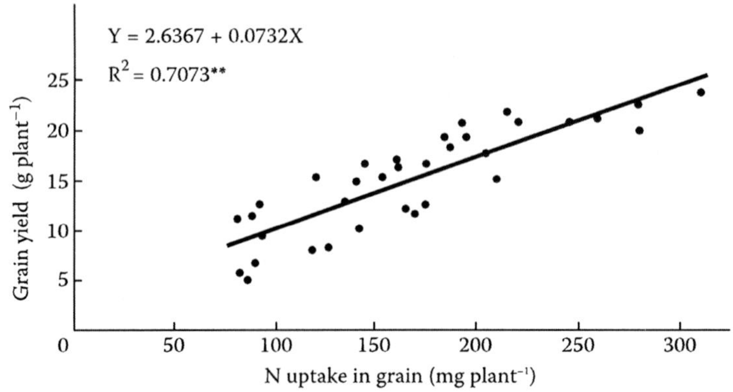

FIGURE 2.18 Relationship between N uptake in grain and grain yield of lowland rice under greenhouse conditions. (From Fageria, N. K., A. B. Santos, and A. M. Coelho. 2011. Growth, yield and yield components of lowland rice as influenced by ammonium sulfate and urea fertilization. *J. Plant Nutr.* 34:371–386.)

2.6 HARVEST INDEX

The proportion of total plant N partitioned to the grain is called the N harvest index (NHI). It is calculated using the following equation (Fageria and Baligar 2005b):

$$\text{NHI} = \frac{\text{N uptake in grain}}{\text{N uptake in grain + straw}} \tag{2.5}$$

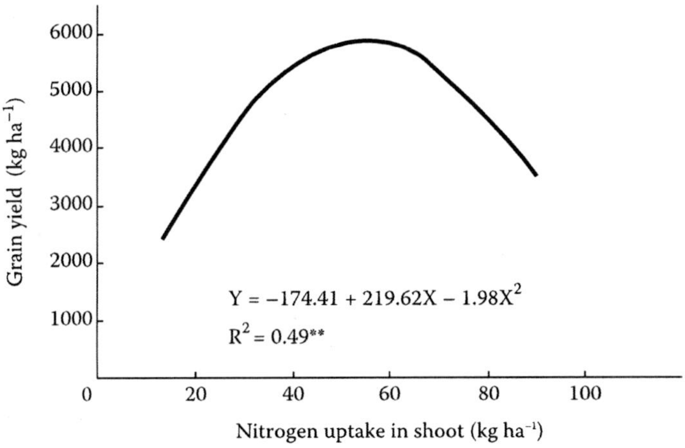

FIGURE 2.19 Relationship between nitrogen uptake in shoot (excluding grain) and grain yield of lowland rice under field conditions. (From Fageria, N. K. 2009. *The Use of Nutrients in Crop Plants.* Boca Raton, FL: CRC Press.)

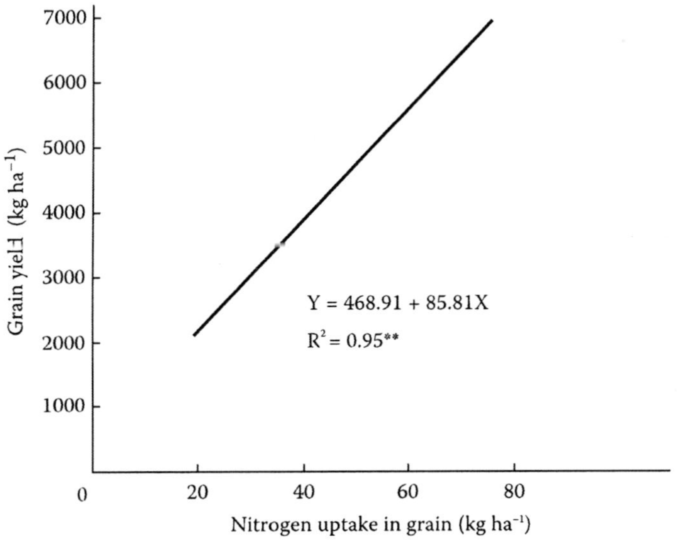

FIGURE 2.20 Relationship between nitrogen uptake in grain and grain yield of lowland rice under field conditions. (From Fageria, N. K. 2009. *The Use of Nutrients in Crop Plants.* Boca Raton, FL: CRC Press.)

Nitrogen in the roots has little influence on the efficiency of N partitioning (Fageria and Baligar 2005b); hence, the NHI ratio refers only to N in the above-ground parts of the plant. This index is useful for measuring N partitioning in crop plants, which provides an indication of how efficiently the plant utilized acquired N for grain production (Fageria and Baligar 2003a). Genetic variability for NHI exists within crop genotypes, and high NHI is associated with efficient utilization of N (Fageria and Baligar 2003b). Thus, variations in NHI are characteristics of genotypes, and this trait may be useful in selecting crop genotypes for higher grain yield (Fageria and Baligar 2003a). Dhugga and Waines (1989) reported that genotypes that accumulate little or no N after anthesis had low grain yields and low NHIs.

Fageria, Moreira, and Coelho (2011) studied the influence of nitrogen rate using two N sources, namely ammonium sulfate and urea, on the NHI (Table 2.14). NHI was significantly and quadratically increased with increasing N rate from 0 to 400 mg kg^{-1} by ammonium and urea source of N fertilizers. For ammonium sulfate, the NHI values varied from 0.51 to 0.87, with an average of 0.79. For urea, it varied from 0.63 to 0.85, with an average of 0.77. Ammonium sulfate accounted for 94% of the variation in NHI, and urea accounted for 93% variation in NHI. Overall, ammonium sulfate produced about 3% higher NHI compared with urea fertilization. Fageria and Baligar (2005b) reported that the NHI values varied across crop species and across genotypes of the same species. Fageria and Barbosa Filho (2001) reported that NHI values varied from 0.44 to 0.66 in lowland rice, depending on genotypes. Similarly, Dingkuhn et al. (1991) reported NHI values ranging from 0.60 to 0.72 for three semidwarf rice cultivars differing in growth duration. Guindo and colleagues (1994) reported NHI values of 0.58 and 0.62 in two lowland rice cultivars. NHI had a linear association with grain yield (Figures 2.21 and 2.22). This means that increasing N uptake in grain can increase grain yield of lowland rice. The NHI varied with genotypes (Figure 2.23), indicating that it is also genetically

TABLE 2.14
Nitrogen Harvest Index of Lowland Rice as Influenced by N Rate and Sources

Nitrogen Rate (mg kg^{-1})	$(NH_4)_2SO_4$	$CO(NH_2)_2$
0	0.86	0.81
50	0.84	0.85
100	0.86	0.85
150	0.87	0.81
300	0.77	0.71
400	0.51	0.63
Average	0.79	0.77
F-test	*	*
CV (%)	10.8	2.4

Regression Analysis

N rate $(NH_4)_2SO_4$ vs. NHI (Y) = 0.83 + 0.00090X – 0.000041X^2, R^2 = 0.94**
N rate $CO(NH_2)_2$ vs. NHI (Y) = 0.83 + 0.000224X – 0 0000018X^2, R^2 = 0.93**

Source: Fageria, N. K., A. Moreira, and A. M. Coelho. 2011. Yield and yield components of upland rice as influenced by nitrogen sources. *J. Plant Nutr.* 34:361–370.
*Significant at the 1% probability level.

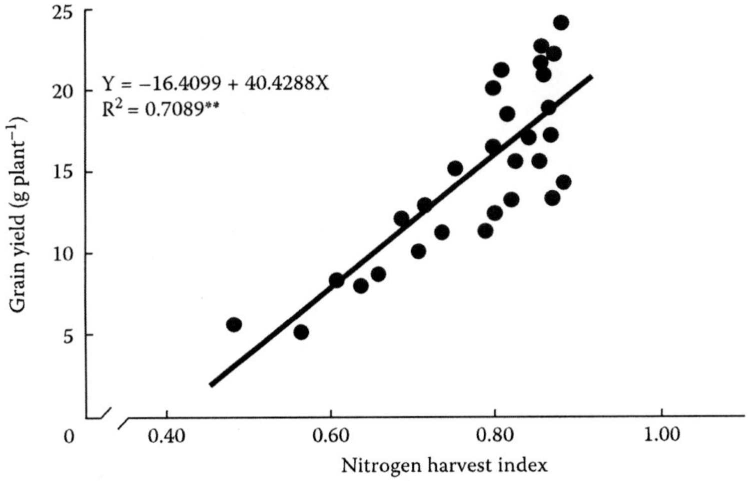

FIGURE 2.21 Relationship between nitrogen harvest index and grain yield of lowland rice under greenhouse conditions. (From Fageria, N. K., A. B. Santos, and A. M. Coelho. 2011. Growth, yield and yield components of lowland rice as influenced by ammonium sulfate and urea fertilization. *J. Plant Nutr.* 34:371–386.)

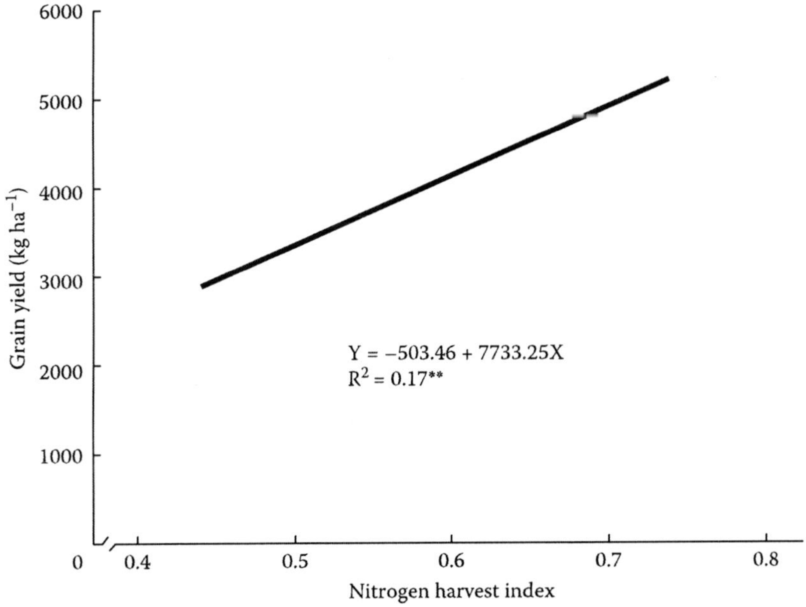

FIGURE 2.22 Relationship between nitrogen harvest index and grain yield of lowland rice under field conditions. (From Fageria, N. K. 2007a. Yield physiology of rice. *J. Plant Nutr.* 30:843–879.)

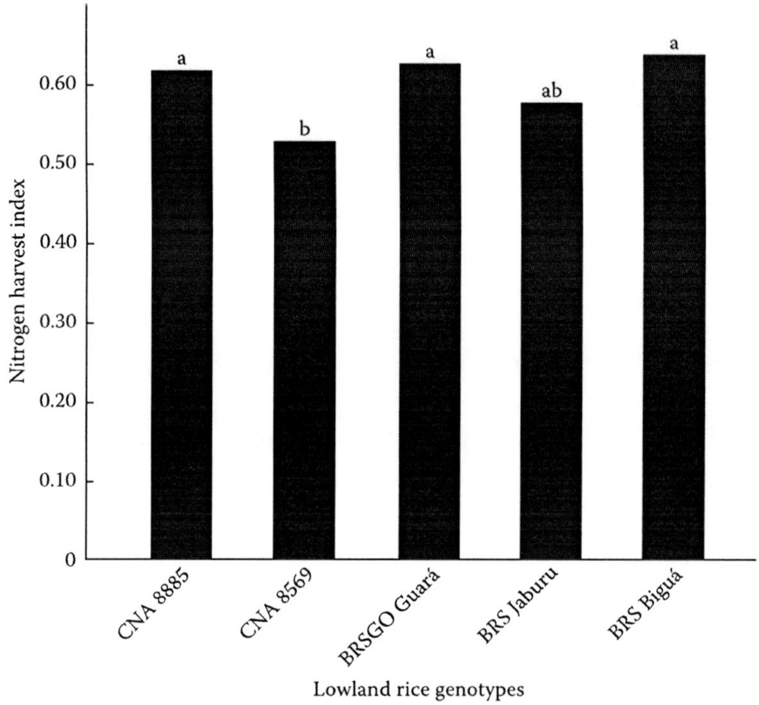

FIGURE 2.23 Nitrogen harvest index of lowland rice genotypes. (From Fageria, N. K. 2007a. Yield physiology of rice. *J. Plant Nutr.* 30:843–879.)

controlled. Hence, use of genotypes with higher NHI can be an important strategy in improving rice yield.

2.7 USE EFFICIENCY

Application of N through chemical fertilizers is the dominant and effective practice in crop production worldwide. However, N applied through chemical fertilizers is not used efficiently, and recovery of N in the soil–plant system seldom exceeds 50% (Lopez-Bellido and Lopez-Bellido 2001; Guarda, Padovan, and Delogu 2004; Fageria and Baligra 2005b; Abbasi and Tahir 2012). Raun and Johnson (1999) reported that only 33% of the total N applied for cereal production in the world is actually removed in the grain, much less than that generally reported. In recent years, fertilizer cost and concern for sustainable soil productivity and ecological stability in relation to chemical fertilizers (especially the N fertilizers) have emerged as important issues (Aulakh et al. 2000). Hence, discussion on NUE is important in crop production from economic and environmental points of view.

NUE is an important index in determining the way the applied N was used by rice crop. Hence, knowledge of this index is fundamental in improving NUE and consequently N management. Most research on N fertilization of rice in Brazil, the United States, and many other rice-growing regions has focused on calibration of N fertilizer rates; plant uptake and utilization; and the magnitude of N loss pathways among various N fertilizer sources, application times, and application methods (Fageria et al. 2003). Information generated from these types of studies help to develop best management practices for the production of high yields while minimizing N losses and costs associated with N fertilization. Nutrient use efficiency can also be expressed in terms of biomass or grain production efficiency per unit of nutrient uptake or application. Regardless of how nutrient use efficiency is expressed, the ultimate goal of this information is to enhance our knowledge on crop growth and production efficiency (Fageria et al. 2003).

In the literature, NUE is defined and calculated in several ways. Fageria and Baligar (2005b) suggested five definitions and methods of calculating NUE in crop plants. These efficiencies are known as *agronomic efficiency* (AE), *physiological efficiency* (PE), *agrophysiological efficiency* (APE), *apparent recovery efficiency* (ARE), and *utilization efficiency* (UE). The determination of NUE in crop plants is an important approach in evaluating the fate of applied chemical fertilizers and their role in improving crop yields. The N use efficiencies are calculated using the following formulae (Fageria and Baligar 2003b; Fageria et al., 2003; Fageria 2009; Fageria, Baligar, and Jones 2011):

$$\text{Agronomic efficiency (AE, kg kg}^{-1}) = \frac{\text{GY}_f \text{ in kg} - \text{GY}_u \text{ in kg}}{\text{N}_a \text{ in kg}} \tag{2.6}$$

where GY_f is the grain yield of the N fertilized plot, GY_u is the grain yield in the unfertilized plot, and N_a is the quantity of N applied.

$$\text{Physiological efficiency (PE, kg kg}^{-1}) = \frac{\text{BY}_f \text{ in kg} - \text{BY}_u \text{ in kg}}{\text{N}_f \text{ in kg} \quad \text{N}_u \text{ in kg}} \tag{2.7}$$

where BY_f is the biological yield (grain plus straw) of the N fertilized plot, BY_u is the biological yield (of grain plus straw) in the unfertilized plot, N_f is the N uptake in the grain and straw of the N fertilized plot, and N_u is the N uptake in the grain and straw of the N unfertilized plot.

$$\text{Agrophysiological efficiency (APE, kg kg}^{-1}) = \frac{\text{GY}_f \text{ in kg} - \text{GY}_u \text{ in kg}}{\text{N}_f \text{ in kg} - \text{N}_u \text{ in kg}} \tag{2.8}$$

where GY_f is the grain yield of the N fertilized plot, GY_u is the grain yield of the N unfertilized plot, N_f is the N uptake in the grain and straw of the fertilized plot, and N_u is the N uptake in the grain and straw of the unfertilized plot.

$$\text{Apparent recovery efficiency (ARE, \%)} = \frac{\text{N}_f \text{ in kg} - \text{N}_u \text{ in kg}}{\text{N}_a \text{ in kg}} \tag{2.9}$$

where N_f is the N uptake in the grain and straw of the N fertilized plot, N_u is the N uptake in the grain and straw of the N unfertilized plot, and N_a is the amount of N applied.

$$\text{Utilization efficiency (UE, kg kg}^{-1}\text{)} = \text{PE} \times \text{ARE} \quad (2.10)$$

The above-mentioned five N use efficiencies for lowland rice were calculated, and the values are presented in Table 2.15. On average, all NUE values were higher at lower N rates and lower at higher N rates. This indicated that rice plants were unable to absorb N when applied in excess because their absorption mechanisms might have been saturated. Under these conditions, there is a possibility for more N being subject to loss by NH_3 volatilization, leaching, and denitrification. It has also been reported by Jarrell and Beverly (1981) that in any experiment with nutritional variables, plants grown at the lowest nutrient concentrations will inevitably have the highest utilization quotient because of dilution effects. Similarly, Fageria, Santos, and Cutrim (2007) calculated these five N use efficiencies in lowland rice genotypes; the results are presented in Table 2.16. There were differences among genotypes in NUE. Overall, 29% of the N applied was recovered, indicating that much N is lost in soil–plant systems and that appropriate management practices are necessary to improve its efficiency.

Decreasing NUE at higher N rates indicates that rice plants could not absorb or utilize N at higher rates or that N loss exceeded the rate of plant uptake. Decreases in N uptake efficiency at higher N rates have been reported by Kurtz et al. (1984). Similarly, Eagle et al. (2000) reported that NUE in rice, which has both physiological and soil N supply components, decreased with increase in soil N supply, indicating that some of the decrease in NUE may have been due to the increased soil N supply.

TABLE 2.15
Nitrogen Use Efficiencies of Lowland Rice as Affected by N Rates

N Rate (kg ha^{-1})	AE (kg kg^{-1})	PE (kg kg^{-1})	APE (kg kg^{-1})	ARE (%)	UE (kg kg^{-1})
30	35	156	72	49	76
60	32	166	73	50	83
90	22	182	75	37	67
120	22	132	66	38	50
150	18	146	57	34	50
180	16	126	51	33	42
210	13	113	46	32	36
Average	23	146	63	39	58
R^2	0.93**	0.62*	0.87**	0.82**	0.90**

Source: Fageria, N. K., and V. C. Baligar. 2001. Lowland rice response to nitrogen fertilization. *Commun. Soil Sci. Plant Anal.* 32:1405–1429; Fageria, N. K., N. A. Slaton, and V. C. Baligar. 2003. Nutrient management for improving lowland rice productivity and sustainability. *Adv. Agron.* 80:63–152; Fageria, N. K., and V. C. Baligar. 2003b. Methodology for evaluation of lowland rice genotypes for nitrogen use efficiency. *J. Plant Nutr.* 26:1315–1333.

Note: AE, agronomic efficiency; PE, physiological efficiency; APE, agrophysiological efficiency; ARE, apparent recovery efficiency; UE, utilization efficiency.

*,**Significant at the 0.05 and 0.01 probability levels, respectively.

TABLE 2.16
Nitrogen Use Efficiency by Five Lowland Rice Genotypes

Genotype	AE (kg kg^{-1})	PE (kg kg^{-1})	APE (kg kg^{-1})	ARE (%)	UE (kg kg^{-1})
CNAi 8886	23	105	56	37	39
CNAi 8569	17	188	69	29	55
BRSGO Guará	21	222	123	29	64
BRS Jaburu	16	114	64	26	30
BRS Biguá	19	145	74	23	33
Average	19	155	77	29	44

Source: Adapted from Fageria, N. K., A. B. Santos, and V. A. Cutrim. 2007. Yield and nitrogen use efficiency of lowland rice genotypes as influenced by nitrogen fertilization. *Pesq. Agropc. Bras.* 42:1029–1034.

Note: AE, agronomic efficiency; PE, physiological efficiency; APE, agrophysiological efficiency; ARE, apparent recovery efficiency; UE, utilization efficiency.

Nitrogen recovery efficiency for flooded rice grown in Asia has been reported to range from 20 to 40% of applied N (De Datta et al. 1987, 1988; Schnier et al. 1990). These values were estimated using ^{15}N labeled fertilizer and by differences in methods for determining N recovery efficiency values as calculated by Cassman et al. (1993). In some of these studies, values ranged from 34 to 64%. Hussain et al. (2000) reported that N recovery efficiency in lowland rice grown in the Philippines was 36%. Hussain et al (2000) also reported that AE of lowland rice produced in the Philippines with the application of 1 kg N was 18 kg. Bronson et al. (2000) reported that recovery efficiency in transplanted rice grown in Asia was higher when the difference method (54%) rather than isotopic dilution method (44%) was used to calculate values.

The equations written above to calculate N use efficiencies are for the data obtained under field conditions. The equations for calculating N use efficiencies under controlled or greenhouse conditions are the same. However, their unit is different than those for field experimental data. Fageria, Morais, and Santos (2010) provided definitions of N use efficiencies and their methods of calculation (Table 2.17). These authors also calculated these efficiencies for upland rice genotypes; results are presented in Table 2.18. AE and UE were different across genotypes. PE, APE, and ARE also varied across genotypes; however, the difference was not statistically significant. The AE varied from 12.8 to 26.7 mg grain produced per mg N applied, with an average value of 21.4 mg grain produced per mg N applied. Average value across PE genotypes was 86.6 mg dry matter production (grain plus straw) per mg N accumulated in grain and straw. Average value of APE was 45.2 mg grain produced per mg N accumulated in grain and straw. Fageria and Barbosa Filho (2001) determined APE in eight lowland rice genotypes and reported an average value of 45.5 mg grain produced per mg N accumulated in grain and straw. Hence, APE of lowland and upland rice is comparable.

The average ARE was 49.2%. The ARE in lowland rice is reported to be in the range of 31–40% in major rice-growing regions of the world (Casman, Dobermann, and Walters 2002). Fageria and Baligar (2001) reported that average ARE in lowland rice in Brazilian Inceptisol was 39%. This means ARE is higher in upland

TABLE 2.17
Definitions and Methods of Calculating Nitrogen Use Efficiency

Nutrient Use Efficiency	Definitions and Formulae for Calculation
Agronomic efficiency (AE)	The *agronomic efficiency* is defined as the economic production obtained per unit of nutrient applied. It can be calculated by: AE (mg mg^{-1}) = $GY_f - GY_u/N_a$, where GY_f is the grain yield of the N fertilized pot (mg), GY_u is the grain yield of the N unfertilized pot (mg), and N_a is the quantity of N applied (mg).
Physiological efficiency (PE)	*Physiological efficiency* is defined as the biological yield obtained per unit of nutrient uptake. It can be calculated by: PE (mg mg^{-1}) = $BY_f - BY_u/N_f - N_u$, where BY_f is the biological yield (grain plus straw) of the N fertilized pot (mg), BY_u is the biological yield of the N unfertilized plot (mg), N_f is the N uptake (grain plus straw) of the N fertilized pot (mg), and N_u is the N uptake (grain plus straw) of the N unfertilized pot (mg).
Agrophysiological efficiency (APE)	*Agrophysiological efficiency* is defined as the economic production (grain yield in case of annual crops) obtained per unit of nutrient uptake. It can be calculated by: APE (mg mg^{-1}) = $GY_f - GY_u/N_f - N_u$, where GY_f is the grain yield of N fertilized pot (mg), GY_u is the grain yield of the N unfertilized pot (mg), N_f is the N uptake (grain plus straw) of the N fertilized pot (mg), and N_u is the N uptake (grain plus straw) of N unfertilized pot (mg).
Apparent recovery efficiency (ARE)	*Apparent recovery efficiency* is defined as the quantity of nutrient uptake per unit of nutrient applied. It can be calculated by: ARE (%) = $(N_f - N_u/N_a) \times 100$, where N_f is the N uptake (grain plus straw) of the N fertilized plot (mg), N_u is the N uptake (grain plus straw) of the N unfertilized pot (mg), and N_a is the quantity of N applied (mg).
Utilization efficiency (UE)	Nutrient *utilization efficiency* is the product of physiological and apparent recovery efficiency. It can be calculated by: UE (mg mg^{-1}) = PE × ARE.

Source: Adapted from Fageria, N. K., O. P. Morais, and A. B. Santos. 2010. Nitrogen use efficiency in upland rice. *J. Plant Nutr.* 33:1696–1711.

than in lowland rice. The highest grain-yield-producing genotype, CNAs 8993, had the highest AE, whereas the lowest-yielding genotype, CNA 8170, had the lowest AE. The genotype CNAs 8992 had the highest PE and APE and reasonably good values of ARE and UE. The AE showed a highly significant ($P < 0.01$) relationship with grain yield (r = 0.77**). The PE was not related to grain yield; however, APE had a highly significant ($P < 0.01$) relationship with grain yield (r = 0.37**).

TABLE 2.18
Nitrogen Use Efficiency of 19 Upland Rice Genotypes

Genotype	Agronomical Efficiency (mg mg^{-1})	Physiological Efficiency (mg mg^{-1})	Agrophysiological Efficiency (mg mg^{-1})	Apparent Recovery Efficiency (%)	Utilization Efficiency (mg mg^{-1})
CRO 97505	23.5ab	100.2a	55.4a	43.5a	42.5ab
CNAs 8993	26.7a	101.4a	59.0a	47.8a	45.5ab
CNAs 8812	24.2ab	80.5a	41.3a	58.2a	47.0ab
CNAs 8938	21.8ab	87.8a	46.3a	48.7a	40.9ab
CNAs 8960	21.7ab	79.3a	46.1a	48.3a	37.3ab
CNAs 8989	24.8aab	72.3a	42.3a	59.3a	42.4ab
CNAs 8824	17.0ab	73.7a	36.7a	47.0a	33.5b
CNAs 8957	22.4ab	72.3a	46.3a	50.7a	35.2ab
CRO 97422	19.8ab	85.7a	47.1a	45.0a	36.6ab
CNAs 8817	17.9ab	72.7a	33.6a	53.2a	38.3ab
CNAs 8934	20.3ab	83.3a	40.5a	50.2a	41.8ab
CNAs 9852	22.9ab	81.7a	48.9a	47.7a	38.4ab
CNAs 8950	21.0ab	88.5a	49.9a	44.7a	37.6ab
CNA 8540	22.2ab	89.7a	45.2a	53.5a	44.1ab
CNA 8711	19.4ab	79.2a	40.0a	48.1a	38.0ab
CNA 8170	12.8b	92.8a	28.0a	45.4a	42.0ab
BRS Primavera	22.6ab	89.8a	51.2a	45.8a	39.9ab
BRS Canastra	21.7ab	125.7a	57.7a	40.6a	47.4ab
BRS Carisma	24.2ab	89.6a	42.8a	56.2a	50.5a
Average	21.4	86.6	45.2	49.2	41.0
F test					
Genotype	*	NS	NS	NS	**
CV (%)	17	23	26	24	12

Source: From Fageria, N. K., O. P. Morais, and A. B. Santos. 2010. Nitrogen use efficiency in upland rice. *J. Plant Nutr.* 33:1696–1711.

Note: Means followed by the same letter in the same column are not significantly different at the 5% probability level by Tukey's test.

*,**,NS Significant at the 5 and 1% probability levels and nonsignificant, respectively.

The ARE (r = 0.31*) and UE (r = 0.40**) were also associated with grain yield (Fageria, Morais, and Santos 2010). The NUE in lowland and upland rice varied across genotypes (Figures 2.24 and 2.25).

The NUE in shoot as well as grain showed a quadratic association with grain yield (Figures 2.26 and 2.27). However, NUE in shoot accounted for only 26% of the variation in grain yield, and 57% of the variability in grain yield was due to NUE in grain (Figures 2.26 and 2.27). This means that NUE in grain is more important than is NUE in shoot in improving grain yield of rice. Fageria and Baligar (2005a) reported significant quadratic association of NUE with grain yield in lowland rice.

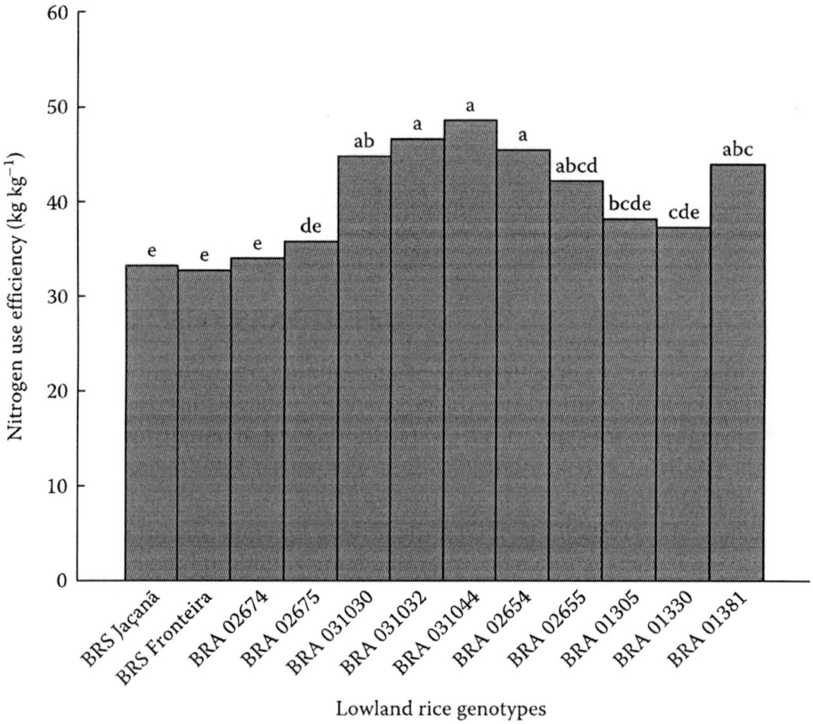

FIGURE 2.24 Agronomic efficiency of 12 lowland rice genotypes.

2.8 MANAGEMENT PRACTICES

Nitrogen management in crop plants has been the subject of considerable research and debate for several decades (Roberts et al. 2012). Inefficient N management practices have contributed to low (30–40%) NUE estimates for cereal crops globally (Raun and Johnson 1999; Cassman et al. 2002). Increased N fertilizer application in excess of crop needs is associated with groundwater contamination by nitrates, N emissions into the air, and soil acidification (Gao et al. 2012). The consequences of these environmental problems are meaningful on a global scale (Gao et al. 2012). In this context, adopting appropriate nitrogen management practices in rice cultivation that improve NUE and reduce losses in the soil–plant system is important for reducing cost of production and environmental pollution. These practices are effective source, adequate rate, appropriate fractional application timing, use of farmyard and green manure to improve soil organic matter content, use of cover crops, use of slow release fertilizers, adopting conservation tillage system, adopting appropriate crop rotation, and planting N efficient genotypes. All these management practices are components of integrated nutrient management. Hence, these practices should be adopted in appropriate combinations and not isolated to obtain good results.

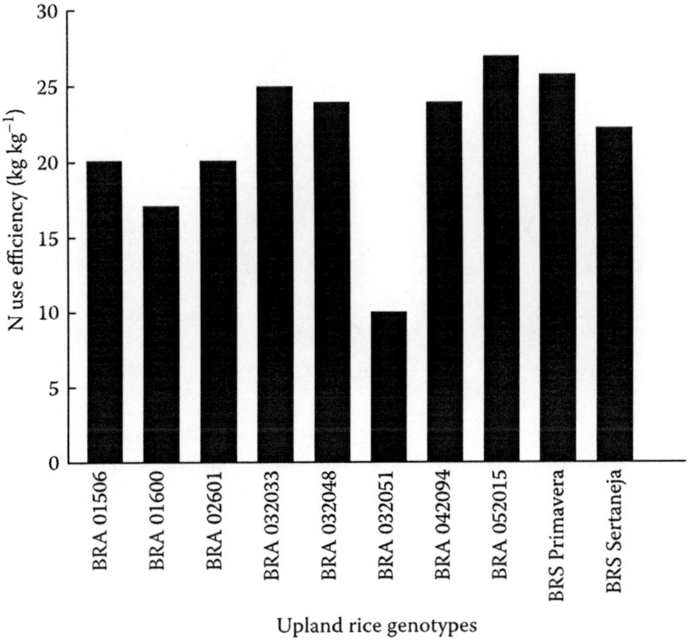

FIGURE 2.25 Agronomic Nitrogen use efficiency of 10 upland rice genotypes.

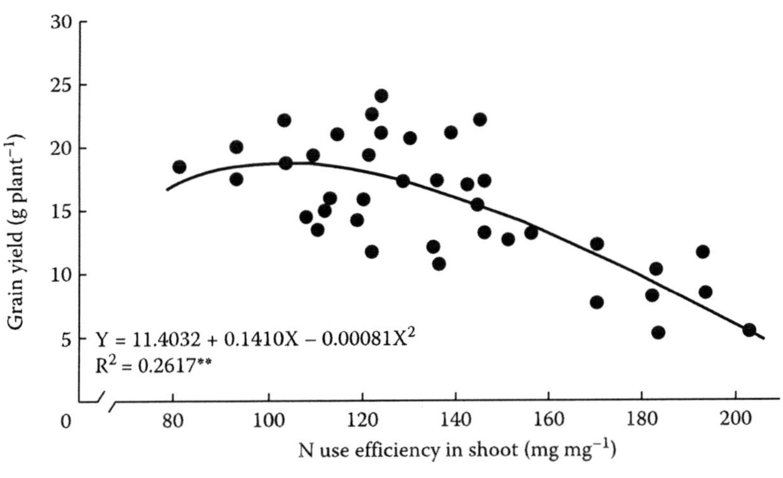

FIGURE 2.26 Relationship between N use efficiency in shoot and grain yield of lowland rice. (From Fageria, N. K., A. B. Santos, and A. M. Coelho. 2011. Growth, yield and yield components of lowland rice as influenced by ammonium sulfate and urea fertilization. *J. Plant Nutr.* 34:371–386.)

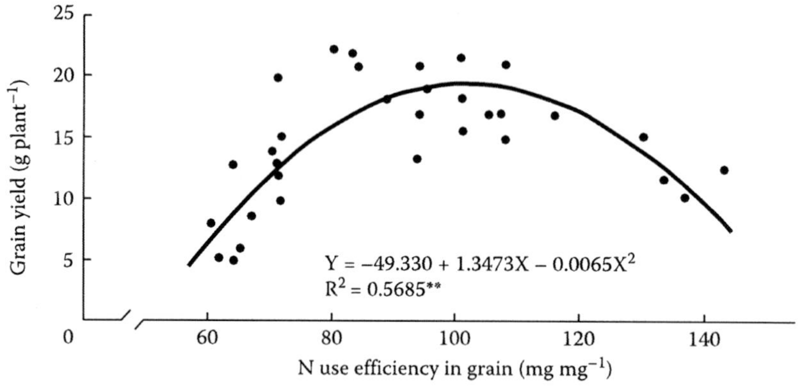

FIGURE 2.27 Relationship between N use efficiency in grain and grain yield of lowland rice. (From Fageria, N. K., A. B. Santos, and A. M. Coelho. 2011. Growth, yield and yield components of lowland rice as influenced by ammonium sulfate and urea fertilization. *J. Plant Nutr.* 34:371–386.)

2.8.1 Sources and Methods of Application

Nitrogen sources and methods of application influence N uptake efficiency in crop plants. Important considerations in selecting the source of N by growers are availability, economics, convenience in storage and handling, and effectiveness of the carrier. The majority of rice production occurs in an anaerobic environment; thus, N fertilizer sources should be ammonium or ammonium forming (Dillon et al. 2012). Generally, urea and ammonium sulfate are the principal sources of N fertilization. Urea is the most widely used N fertilizer in rice production due to its high N content (45%) and relatively low cost (Bufogle et al. 1998; Norman, Wilson, and Slaton 2003; Griggs et al. 2007; Dillon et al. 2012). Ammonium sulfate (21% N) is an excellent source of N, but application costs are greater compared with urea due to its lower N concentration (Dillon et al. 2012). Regardless of which type of N fertilizer is applied before permanent flood established in the delayed-flood system, it is important to establish a flood within a few days following application to maintain N in the ammonium form (Dillon et al. 2012).

Several N-containing fertilizers are available in the market (Table 2.19). In the U.S. agriculture, anhydrous ammonia (NH_3) is an important source of N fertilization. At normal pressure, NH_3 is a gas and is transported and handled as liquid under pressure. It is injected into the soil to prevent loss through volatilization. The NH_3 protonates to form NH_4^+ in the soil and becomes XNH_4^+, which is stable. The major advantages of anhydrous ammonia are its high N analysis (82% N) and low cost of transportation and handling. However, specific equipment is required for storage, handling, and application. Hence, it is not a popular N carrier in developing countries (Fageria and Baligar 2005b). Results obtained by the authors under field conditions show that ammonium sulfate is better than urea as a source of N for lowland rice grown on Brazilian Inceptisols (Figure 2.28). Grain yield expressed in relative yield was significantly increased by urea, as well as ammonium sulfate

TABLE 2.19
Major Nitrogen Fertilizers, Their Chemical Formulae, and N Contents

Common Name	Formula	N (%)
Ammonium sulfate	$(NH_4)_2SO_4$	21
Urea	$CO(NH_2)_2$	45
Anhydrous ammonia	NH_3	82
Ammonium chloride	NH_4Cl	26
Ammonium nitrate	NH_4NO_3	35
Potassium nitrate	KNO_3	14
Sodium nitrate	$NaNO_3$	16
Calcium nitrate	$Ca(NO_3)_2$	16
Calcium cyanamide	$CaCN_2$	21
Ammonium nitrate sulfate	$NH_4NO_3(NH_4)_2SO_4$	26
Nitrochalk	$NH_4NO_3 + CaCO_3$	21
Monoammonium phosphate	$NH_4H_2PO_4$	11
Diammonium phosphate	$(NH_4)_2HPO_4$	18

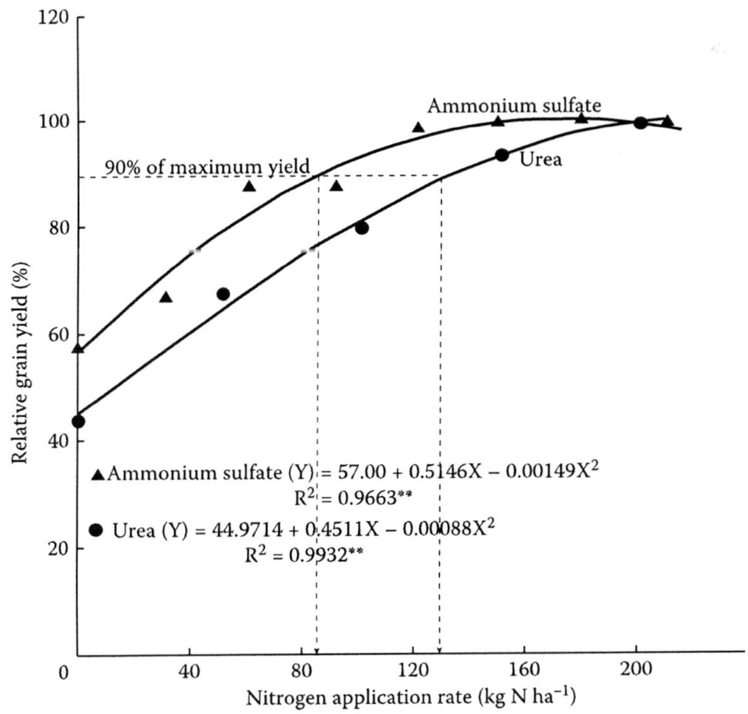

FIGURE 2.28 Relative grain yield of lowland rice as influenced by N applied with ammonium sulfate and urea. (From Fageria, N. K., A. B. Santos, and M. F. Moraes. 2010b. Influence of urea and ammonium sulfate on soil acidity indices in lowland rice production. *Commun. Soil Sci. Plant Anal.* 41:1565–1575.)

fertilization, and the increase was in a quadratic fashion. In fertilizer experiments, 90% of the relative yield is considered an economic index, and this index was used to calculate adequate N rate (Figure 2.28). Ninety percent of the relative yield (corresponding to 5750 kg grain ha^{-1}) was obtained with the application of 84 kg N ha^{-1} in the case of ammonium sulfate. Similarly, in the case of urea, 90% of the relative grain yield (corresponding to 4811 kg grain ha^{-1}) was obtained with the application of 130 kg N ha^{-1}.

Singh et al. (1998) reported that maximum average grain yield of 7700 kg ha^{-1} of 20 lowland rice genotypes was obtained at 150–200 kg N ha^{-1} at the International Rice Research Institute in the Philippines. Aulakh et al. (2000) reported that flooded rice responded to N rates up to 120 kg N ha^{-1} on sandy loam soils in India. In the Philippines, Dobermann et al. (2000) reported that 80–100 kg N ha^{-1} was used for maximal yields in the field experiments during the wet season (rainy season), and 120–150 kg N ha^{-1} was used during the dry season. These authors also reported that N fertilization rates in the Philippines for irrigated-lowland rice, from 1992 onward, increased from 108 to 120 kg N ha^{-1} during the wet season and from 190 to 216 kg N ha^{-1} during the dry season. Fageria and Baligar (1996) also reported increases in grain yields of lowland rice grown on Inceptisol in the central part of Brazil. These authors reported that an average yield of three years (5523 kg ha^{-1}) of lowland rice was achieved with the application of 100 kg N ha^{-1}.

Fageria, Santos, and Coelho (2011) also studied the influence of ammonium sulfate and urea as a source of N on lowland rice yield (Figure 2.29). Grain yield quadratically increased ($P < 0.01$) with the application of ammonium sulfate, as well as urea, from 0 to 400 mg kg^{-1}. Maximum grain yield was obtained with the application of ammonium sulfate to 168 mg N kg^{-1} of soil and urea to 152 mg N kg^{-1} of soil. The variation in grain yield was 5.5–22.8 g $plant^{-1}$ (with an average yield of 15.9 g $plant^{-1}$) by application of ammonium sulfate and 8.0–19.7 g $plant^{-1}$ (with an average value of 14.4 g $plant^{-1}$) by urea fertilization. Ammonium sulfate accounted

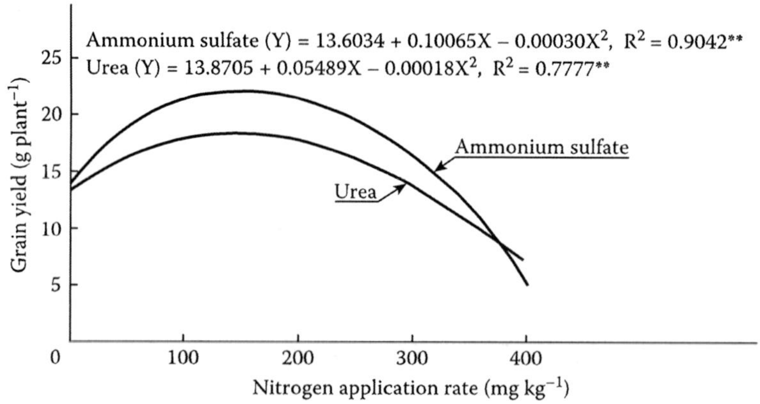

FIGURE 2.29 Influence of nitrogen application rate of two N sources on grain yield of lowland rice. (From Fageria, N. K., A. B. Santos, and A. M. Coelho. 2011. Growth, yield and yield components of lowland rice as influenced by ammonium sulfate and urea fertilization. *J. Plant Nutr.* 34:371–386.)

for 90% variability in grain yield, whereas urea accounted for 78% of the variability in grain yield. Across six N rates, ammonium sulfate produced 10% higher grain yield than did urea. In addition, the two N sources, namely ammonium sulfate and urea (at the rate of 160 mg N kg^{-1}), on average, produce 22.5 and 18.5 g grain yield per plant, respectively. The application of ammonium sulfate at the rate of 160 mg N kg^{-1} produced 22% higher grain yield than did urea at the same rate of N. This means that ammonium sulfate was superior to urea as fertilizer for lowland rice grain.

The higher yield with the application of ammonium sulfate may be related to higher panicle number or density produced when compared with the application of urea (Figure 2.30). The number of panicles increased quadratically with increasing N rates of soil from 0 to 400 mg kg^{-1} by application of both ammonium sulfate and urea as sources of N. Panicle response to N fertilization was similar for both N sources; however, magnitude of response was higher in the case of ammonium sulfate. Ammonium sulfate and urea accounted for 70 and 57%, respectively, of the variability in panicle number. Thus, ammonium sulfate is a superior fertilizer for panicle production in lowland rice compared to urea. Overall, ammonium sulfate produced 8% more panicles than urea. Fageria and Baligar (2001) reported a quadratic increase in panicle number with increasing N rates in lowland rice. Hasegawa (2003) also reported that panicle density was unrelated to the grain yield of high-yielding rice cultivars.

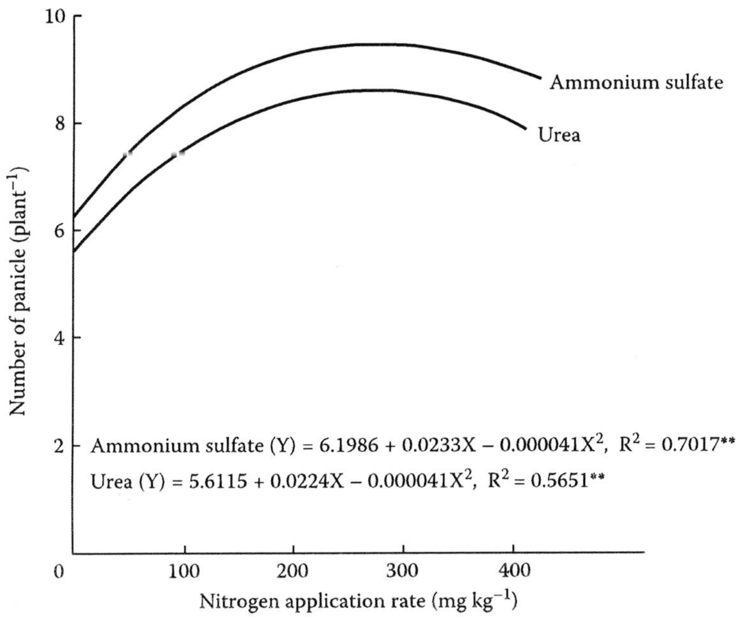

FIGURE 2.30 Influence of N applied with two sources on number of panicles of lowland rice. (From Fageria, N. K., A. B. Santos, and A. M. Coelho. 2011. Growth, yield and yield components of lowland rice as influenced by ammonium sulfate and urea fertilization. *J. Plant Nutr.* 34:371–386.)

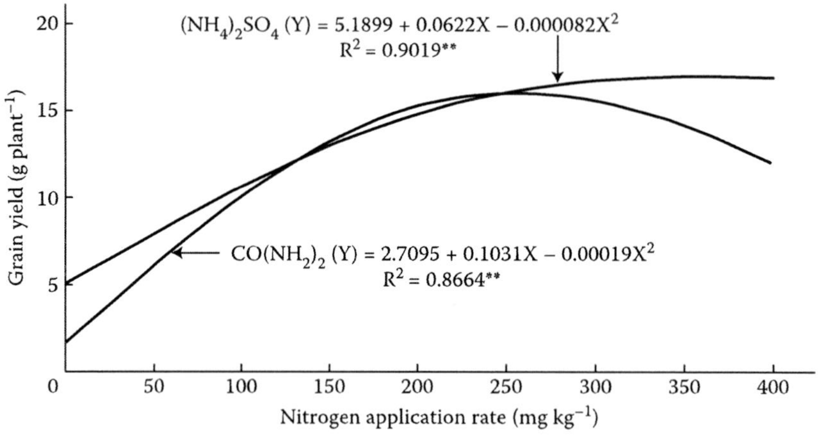

FIGURE 2.31 Relationship between N application rate by two N sources and grain yield of upland rice. (From Fageria, N. K., A. Moreira, and A. M. Coelho. 2011. Yield and yield components of upland rice as influenced by nitrogen sources. *J. Plant Nutr.* 34:361–370.)

Fageria, Baligar, and Jones (2011) also studied the influence of N sources on grain yield of upland rice under controlled conditions (Figure 2.31). Grain yield increased in a quadratic fashion when N rate was increased, using ammonium sulfate and urea sources of N, in the range of 0–400 mg kg^{-1} of soil. Based on regression equation, maximum grain yield was obtained with the application of ammonium sulfate in 380 mg N kg^{-1} of soil. Similarly, maximum grain yield was obtained with the application of urea in 271 mg N kg^{-1} of soil. Fageria, Baligar, and Clark (2006) reported that maximum grain yield of upland rice in Brazilian Oxisol was obtained with the application of ammonium sulfate in 400 mg N kg^{-1} of soil. At the lower as well as the higher N rates, ammonium sulfate produced higher grain yield than did urea. However, at the intermediate N rate (125–275 mg N kg^{-1}), urea produced slightly more grain yield than did ammonium sulfate. Across the six N rates, ammonium sulfate produced 12% higher grain yield than did urea. The superiority of ammonium sulfate at higher N rates compared with urea may be associated with the higher acidity-producing capacity of ammonium sulfate, compared with urea. Rice is an acid-tolerant plant, and its growth was linearly increased when Al saturation in Brazilian Oxisol soil was increased from 0 to 30% (Fageria and Santos 1998). Fageria, Baligar and Jones (2011) also reported that when soil pH of Brazilian Oxisol was raised from 4.6 to 6.8, upland rice plants showed iron deficiency at higher pH level (pH 5.7 in water), and yield was reduced. Uptake of Cu, M, Fe, and Zn were reduced at higher pH (>5.7) (Fageria, Baligar, and Clark 2002). Fageria (2009) also reported that rice plant growth was better at 10 mg Al^{3+} L^{-1} than at zero level of aluminum in the nutrient solution. The superiority of ammonium sulfate over urea may be related to higher shoot dry weight, panicle number, plant height, and root dry weight at higher N rate with application of ammonium sulfate compared with application of urea (Figures 2.32 through 2.35). These traits were associated with grain yield (Table 2.20). The influence of urea and ammonium sulfate at different N rates on growth of upland rice is shown in Figures 2.36 and 2.37. Similarly, the influence of these N sources on root growth is also

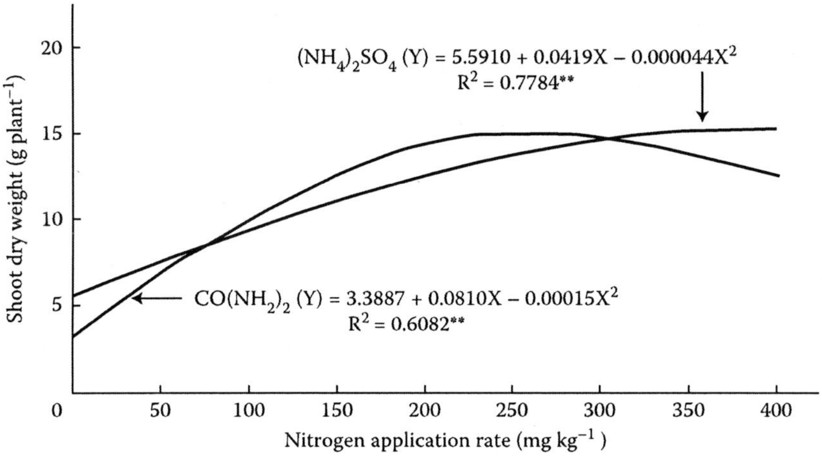

FIGURE 2.32 Relationship between N application rate by two N sources and shoot dry weight (excluding grain yield) of upland rice. (From Fageria, N. K., A. Moreira, and A. M. Coelho. 2011. Yield and yield components of upland rice as influenced by nitrogen sources. *J. Plant Nutr.* 34:361–370.)

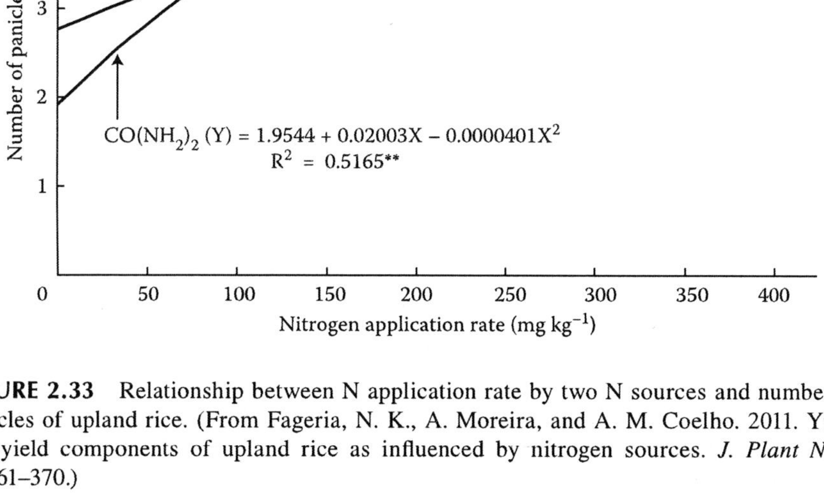

FIGURE 2.33 Relationship between N application rate by two N sources and number of panicles of upland rice. (From Fageria, N. K., A. Moreira, and A. M. Coelho. 2011. Yield and yield components of upland rice as influenced by nitrogen sources. *J. Plant Nutr.* 34:361–370.)

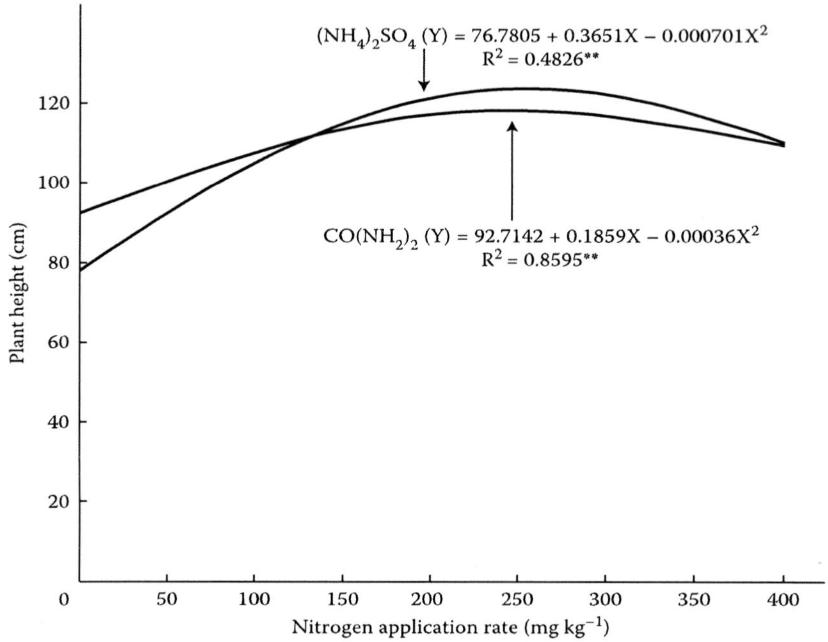

FIGURE 2.34 Relationship between N application rate by two N sources and plant height of upland rice. (Fageria, N. K., A. Moreira, and A. M. Coelho. 2011. Yield and yield components of upland rice as influenced by nitrogen sources. *J. Plant Nutr.* 34:361–370.)

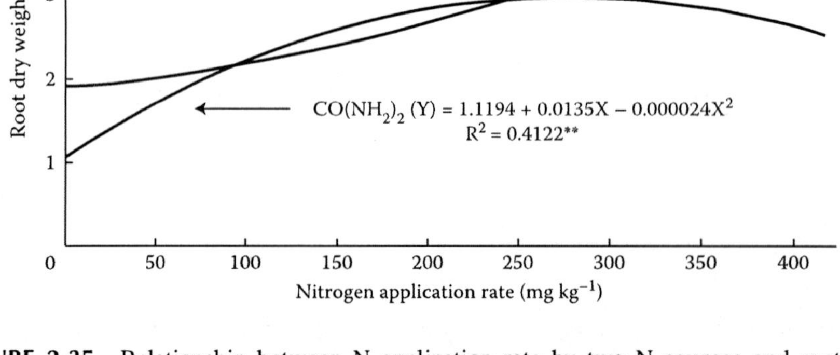

FIGURE 2.35 Relationship between N application rate by two N sources and root dry weight of upland rice. (From Fageria, N. K., A. Moreira, and A. M. Coelho. 2011. Yield and yield components of upland rice as influenced by nitrogen sources. *J. Plant Nutr.* 34:361–370.)

TABLE 2.20
Relationship between Plant Growth and Yield Components (X) and Grain Yield (Y) of Upland Rice

Variable	Regression Equation	R^2
Plant height vs. grain yield	$Y = 0.7296\exp(0.00226X + 0.00021X^2)$	0.8228**
Shoot dry weight vs. grain yield	$Y = -4.8385 + 2.1852X - 0.0553X^2$	0.8932**
Panicle number vs. grain yield	$Y = -6.5792 + 4.9425X$	0.8190**
Root dry weight vs. grain yield	$Y = -8.5675 + 11.8641X - 1.4130X^2$	0.6362**

Source: Fageria, N. K., A. Moreira, and A. M. Coelho. 2011. Yield and yield components of upland rice as influenced by nitrogen sources. *J. Plant Nutr.* 34:361–370.

Note: Values are averages of two N sources.

**Significant at the 1% probability level. Ref. Page 153- Ref. Phongpan, S., S. Vacharotayan, and K. Kumazawa. 1988. Efficiency of urea and ammonium sulfate for wetland rice grown on an acid sulfate soil as affected by rate and time of application. *Fertilizer Research* 15:237–246.

FIGURE 2.36 Influence of urea on upland rice growth.

shown in Figures 2.38 and 2.39. Shoot as well as root growth was better with the application of ammonium sulfate than with the application of urea.

Reddy and Patrick (1978) and Bufogle et al. (1998) reported no differences in straw or grain yield of lowland rice between the two N sources. In a greenhouse study, Phongpan et al. (1988) found no differences in yield of grain and straw between urea and ammonium sulfate in an acid sulfate soil at low N rate (160 mg N kg^{-1}), but at higher N rates (320–480 mg N kg^{-1}), urea consistently produced higher yields

FIGURE 2.37 Influence of ammonium sulfate on upland rice growth.

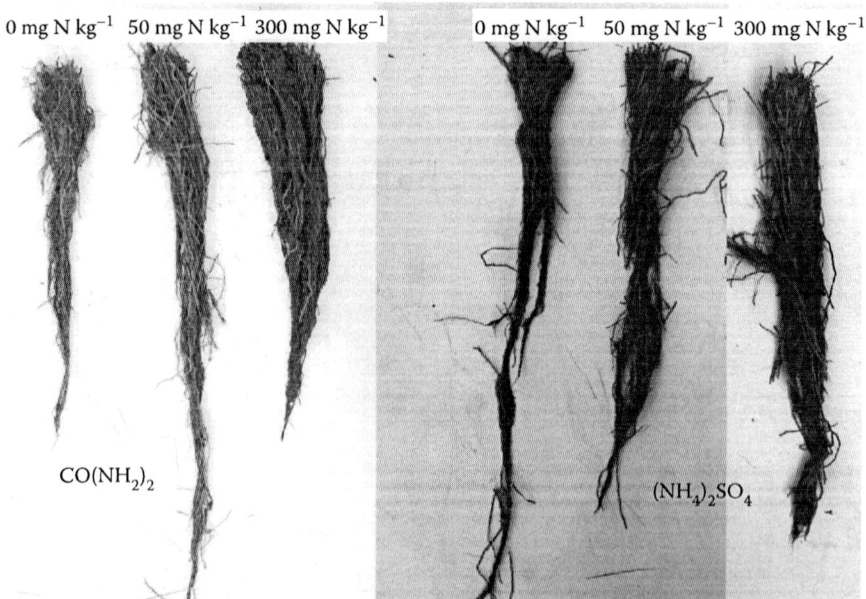

FIGURE 2.38 Root growth of upland rice at two N sources and three rates.

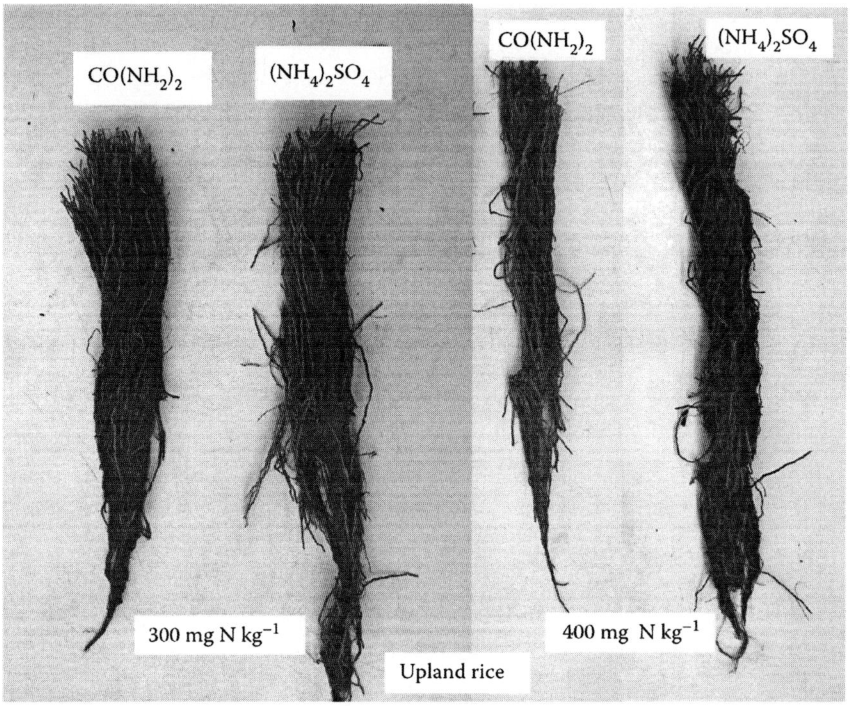

FIGURE 2.39 Influence of two N sources and two N rates on root growth of upland rice.

than did ammonium sulfate. This means that response of lowland rice to ammonium sulfate and urea depends on soil and climatic conditions (Bufogle et al. 1998).

Nitrogen moves by mass flow in the soil–plant system; hence, its application as broadcast or furrow does not make much difference in its absorption by plants. However, broadcast application may reduce its efficiency due to loss by volatilization. Application in furrow or mixing with the soil if applied as broadcast is recommended to avoid loss through volatilization. Nitrogen fertilizers are broadcast and mixed into soil before crop sowing. In addition, it may also be applied in row below seed at sowing and may be banded in rows beside seed at planting or pre-emergence. During postemergence, fertilizers may be side-dressed, injected into subsurface, and top dressed. Mixing fertilizers in soil and injecting into subsurface are more efficient methods of N application compared with broadcast, wherein N is left on the soil surface

Placing urea or ammonium sulfate in the anaerobic layer of flooded rice is an important strategy to avoid N losses by nitrate leaching and denitrification (Fageria and Baligar 2005b).

2.8.2 Adequate Rate

After defining the appropriate N source, adequate rate is important for maximizing crop yield and NUE. Nitrogen is a dynamic and mobile nutrient in soil–plant systems. Most of the N in soil is in the organic form. Mineralization of organic N depends on microbial activity, which is influenced by environmental factors. Further, there are many sources of addition and loss pathways of N in the soil–plant system, which complicate its balance and use by plants. Hence, N concentration changes in the rhizosphere with time and space. Therefore, soil analysis test is not applicable to N, as in the case of immobile nutrients, such as P and K. Hence, a crop response curve showing yield versus N rates is the most efficient and effective method of defining a crop's N requirement (Fageria and Baligar 2005b). Appropriate N application rates in rice production in the mid-southern United States are established by each state based on multiyear and multisite N response trials, which are further refined for specific variety, cultural management, and soil type (Guindo, Wells, and Norman 1994; Snyder and Slaton 2002; Harrell et al. 2011). These established N rates are either applied one time at preflood or in split applications (Harrell et al. 2011). Dobermann et al. (2003) reported that decisions on N fertilizer use require knowledge of the expected crop yield response to N application, which is a function of crop N needs, supply of N from indigenous sources, and the short- and long-term fate of applied fertilizer.

The response of lowland rice to N fertilization in Brazilian Inceptisol was determined (Figure 2.40). There was a quadratic increase in grain yield, with increasing N rate in the range of 0–200 kg ha^{-1}. Fageria and Baligar (2001) also reported a quadratic increase in the lowland rice yield in Brazilian Inceptisol when N was applied in the range of 0–210 kg ha^{-1}. The N rate to obtain 90% of maximum yield, which is considered as an economic rate, was about 135 kg ha^{-1}. Fageria and Baligar (2001) reported that the maximum economic yield of lowland rice cultivated for three consecutive

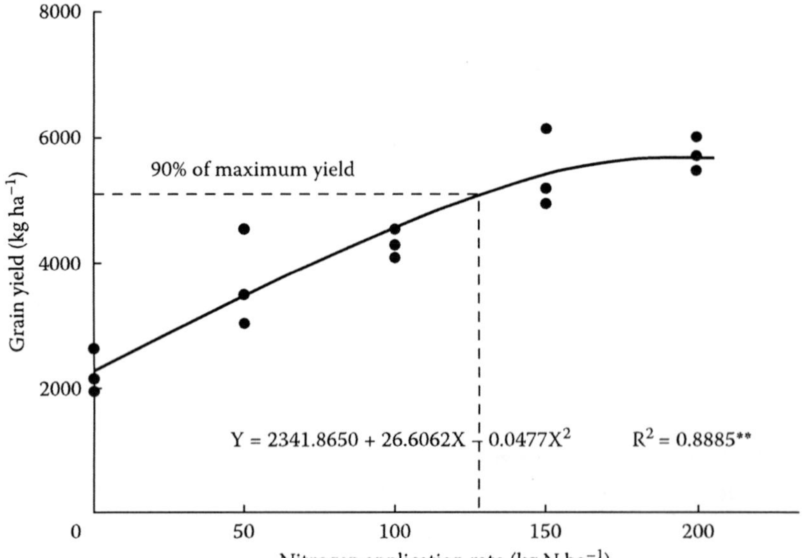

FIGURE 2.40 Relationship between N rate and grain yield of lowland rice. Values are averages of 12 genotypes and two years' field trial. (From Fageria, N. K., A. B. Santos, and V. A. Cutrim. 2008. Dry matter and yield of lowland rice genotypes as influenced by nitrogen fertilization. *J. Plant Nutr.* 31:788–795.)

years in the same area was obtained with the application of 90 kg N ha^{-1}. Dobermann et al. (2000) reported at the International Rice Research Institute in the Philippines that the yield of the irrigated rice cultivar IR 72 increased up to 150 kg N ha^{-1}. Tem Berger and Riethovem (1997) reported that in China, adequate rate of N for lowland rice cultivars of medium growth cycle varied from 100 to 150 kg ha^{-1}. Yield of modern lowland rice cultivars in the State of Rio Grande do Sul of Brazil was obtained with an N rate of 114–126 kg ha^{-1}, depending on the region and type of soil. In the state of Punjab, India, 120 kg N ha^{-1}, applied at key growth stages, is recommended for transplanted rice (Varinderpal-Singh, Yadvinder-Singh, and Gupta 2010). When N is applied at the recommended rate to crops, NUE is higher, and N losses are at a minimum. When N is applied at rates higher than those are necessary for maximum economic yield, N accumulates in the soil profile, and losses are higher.

The response of upland rice to N application in Brazilian Oxisol has also been studied. The response of three upland rice genotypes to N fertilization was quadratic (Figure 2.41). The genotypes' response to N fertilization was similar but different in magnitude. Similarly, the average response of 10 upland rice genotypes to N fertilization was also determined (Figure 2.42). Based on the regression equation, maximum yield was obtained with the application of 129 kg N ha^{-1}. Stone et al. (1999) reported that adequate N rate for upland rice grown on an Oxisol is about 120 kg ha^{-1}. Mahajan et al. (2011) reported that 150 kg N ha^{-1} in four split applications in equal rates (at sowing, 21, 42, and 63 days after sowing) is recommended for direct-seeded aerobic rice in India. Similarly, the study by Mahajan and Timsina (2011) in direct-seeded aerobic rice revealed that the crop responds up to 150 kg N ha^{-1} under weed-free environments. However, under weedy environment, the crop responds

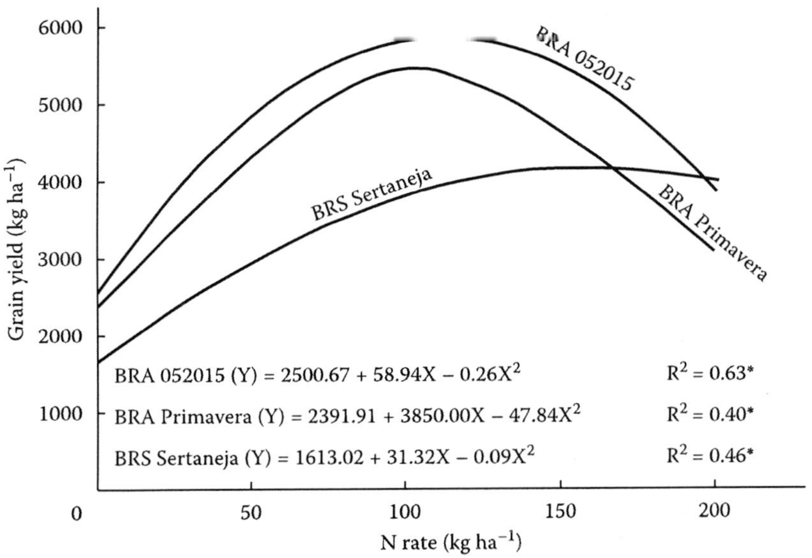

FIGURE 2.41 Relationship between N rate and grain yield of three upland rice genotypes grown on Brazilian Oxisol.

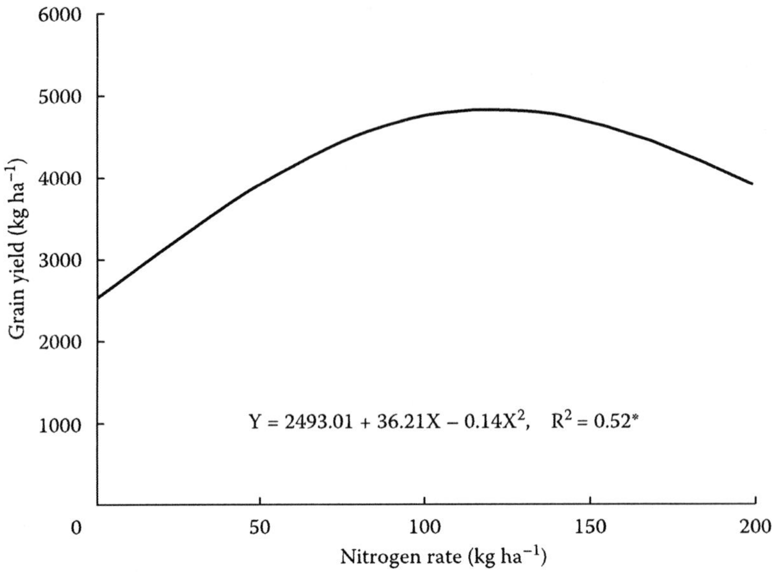

FIGURE 2.42 Relationship between N rate and grain yield of upland rice grown on Brazilian Oxisol. Values are averages of 12 genotypes.

only up to 120 kg N ha^{-1}. In tropical soils, soil organic matter is a major determinant of soil productivity and response to N fertilization. Considering the critical value of 30 g kg^{-1} or 3% of soil organic matter for a probable response to applied N (Foster 1971; Kaizzi et al. 2012), the soil used in the upland rice contained about 2% organic matter. When organic matter in these Brazilian Oxisols was more than 3.2%, response to upland rice was low.

2.8.3 Application Timing during Growth Cycle

Nitrogen is lost from the soil–plant system via volatilization, leaching, denitrification, or runoff (Fageria and Baligar 2005b; Fageria, Baligar, and Clark 2006). This suggests that there is more N available for loss at any time during crop-growing season, if N is applied only once during crop growth. Hence, splitting the N fertilizer applications during crop growth can reduce nitrate leaching and improve NUE. For lowland rice under Brazilian conditions, applying half of the N in a band at sowing, and the remaining six to seven weeks later, should increase both N fertilizer use efficiency and N uptake by minimizing leaching opportunity time and better timing the N application to N uptake (Fageria and Baligar 1999). Fageria and Baligar (1999) reported that AE of N in lowland rice was higher when N was applied in a three-split application (one-third at sowing + one-third at active tillering + one-third at panicle initiation) than when the entire N was applied at sowing. Split application of N in sandy soils and high rainfall areas is most desirable. A study conducted by Fageria and Prabhu (2004) in Brazilian Inceptisol showed that N fractionated into two or three equal doses produced higher grain yield of lowland rice compared with the entire quantity applied at sowing (Figure 2.43).

Fageria and Baligar (1999) studied different timings of N application in a greenhouse experiment in lowland rice (Figure 2.44). Maximum grain yield was obtained with total N (240 mg kg^{-1}) applied at sowing, followed by one-third applied at sowing + one-third at active tillering + one-third at initiation of panicle primordia growth stage. Minimum grain yield was obtained when one-third N was applied at sowing + one-third at initiation of panicle primordia + one-third at flowering, followed by one-third at sowing + one-third at tillering initiation + one-third at flowering. The N applied treatment that received one-third at sowing + one-third at the initiation of panicle primordia + one-third at booting also produced lower yield. Hence, a part of N applied at the late reproductive growth stage or at the initiation of grain filling growth stage produced the lowest grain yield, compared with N applied at the vegetative growth stage or the initiation of reproductive growth stage. The lower yield with late application of N was associated with a lower number of panicles per pot, lower number of filled spikelets, and lower grain harvest index (GHI) (Table 2.21).

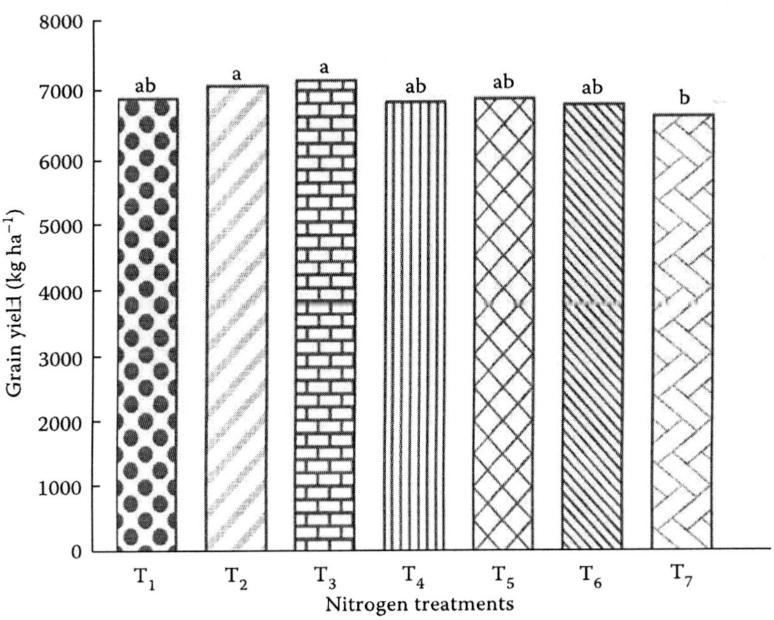

FIGURE 2.43 Grain yield of lowland rice as influenced by N timing treatments. T_1 = all the N applied at sowing, T_2 = 1/3 N applied at sowing + 1/3 N applied at active tillering + 1/3 N applied at the initiation of panicle primordia, T_3 = 1/2 N applied at sowing + 1/2 N applied at active tillering, T_4 = 1/2 N applied at sowing + 1/2 N applied at the initiation of panicle primordia, T_5 = 2/3 N applied at sowing + 1/3 N applied at active tillering, T_6 = 2/3 N applied at sowing + 1/3 N applied at the initiation of panicle primordia, and T_7 = 1/3 N applied at sowing + 2/3 N applied at 20 days after sowing. (From Fageria, N. K., and A. S. Prabhu. 2004. Blast control and nitrogen management in lowland rice cultivation. *Pesq. Agropec. Bras.* 39:123–129.)

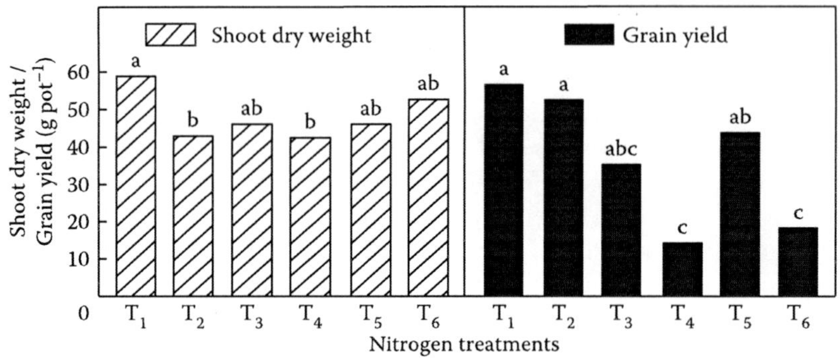

FIGURE 2.44 Shoot dry weight and grain yield of lowland rice as influenced by N timing treatments. T_1 = all the N applied at sowing, T_2 = 1/3 N applied at sowing + 1/3 N applied at active tillering + 1/3 N applied at the initiation of panicle primordia, T_3 = 1/3 N applied at sowing + 1/3 N applied at the initiation of panicle primordia + 1/3 N applied at booting, T_4 = 1/3 N applied at sowing + 1/3 N applied at the initiation of panicle primordia + 1/3 N applied at flowering, T_5 = Zero N applied at sowing + 1/2 N applied at initiation of tillering + 1/2 N applied at the initiation of panicle primordia, and T_6 = 1/3 N applied at sowing + 1/3 N applied at the initiation of tillering + 1/3 N applied at flowering. (From Fageria, N. K., and V. C. Baligar. 1999. Yield and yield components of lowland rice as influenced by timing of nitrogen fertilization. *J. Plant Nutr.* 22:23–32.)

TABLE 2.21
Influence of N Timing Treatments on Panicle Number, Number of Filled Spikelets, and Grain Harvest Index of Lowland Rice

N Timing Treatment	Number of Panicles per Pot	Number of Filled Spikelets per Pot	Grain Harvest Index
Total N applied at sowing	30a	2342a	0.49a
1/3 at sowing + 1/3 at tillering initiation + 1/3 at panicle initiation	22ab	1800ab	0.55a
1/3 at sowing + 1/3 at panicle initiation + 1/3 at booting	15bcd	1428abc	0.43a
1/3 at planting + 1/3 at panicle initiation + 1/3 at flowering	8d	590c	0.25a
1/2 at tillering initiation +1/2 at panicle initiation	19abc	1799ab	0.48a
1/3 at tillering initiation + 1/3 at booting + 1/3 at flowering	12cd	710c	0.32a

Source: Adapted from Fageria, N. K., and V. C. Baligar. 1999. Yield and yield components of lowland rice as influenced by timing of nitrogen fertilization. *J. Plant Nutr.* 22:23–32.

Note: Total N was 240 mg kg^{-1} of soil. Means followed by same letter in the same column are not significantly different at the 5% probability level by Tukey's test. Grain harvest index = grain weight/grain plus straw weight.

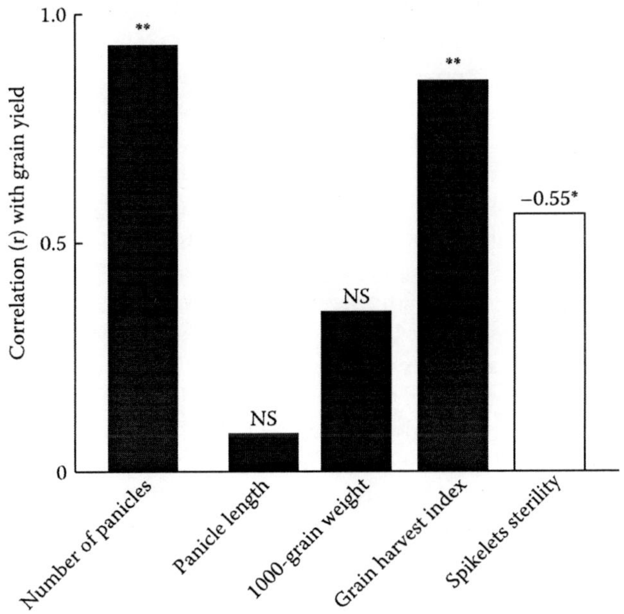

FIGURE 2.45 Correlation (r) of grain yield with yield components of lowland rice. *,**,NSSignificant at the 5 and 1% probability levels and nonsignificant, respectively. (From Fageria, N. K., and V. C. Baligar. 1999. Yield and yield components of lowland rice as influenced by timing of nitrogen fertilization. *J. Plant Nutr.* 22:23–32.)

Number of panicles per pot and GHI had the highest correlation with grain yield (Figure 2.45). NHI and NUE were lower for part of the N applied late or at booting and flowering compared with N applied early, in the vegetative growth stage, or during the initiation of reproductive growth stage (Fageria and Baligar 1999).

In the mid-southern United States, rice famers apply two- or three-split applications of N in drill-seeded, delayed-flood rice production systems (Harrell et al. 2011). The split N application is preferred by rice producers. First, even distribution of fertilizers from a single application is difficult. Second, establishment of a permanent flood in a timely manner and maintaining the permanent flood throughout the season can be problematic (Snyder and Slaton 2002).

2.8.4 Use of Farm Yard and Green Manure

Use of organic animal manure or poultry litter at an adequate rate, in proper form, and the right method (well decomposed and incorporated) brings many benefits, which improve NUE in crop plants. Organic manure has been a valuable resource as a fertilizer and soil amendment in crop production. Studies comparing soils of organically and conventionally managed farming systems have documented higher soil organic matter and total N with the use of organic practices (Clark et al. 1998). Addition of animal manure not only increases the soil inorganic

N pool but also (and perhaps more importantly) increases the seasonal soil N mineralization available to the plant (Ma, Dwyer, and Gregorich 1999). Similarly, Dao and Cavigelli (2003) reported that animal manure has long been used as a source of plant nutrients and for improving soil physical conditions of farmlands. Poultry litter is generated in large quantities in the southeastern United States and has long been recognized as a desirable organic fertilizer because it improves soil fertility by adding essential plant nutrients and organic matter and improves water and nutrient retention (Moore et al. 1995; Adeli, Tewolde, and Jenkins 2012). Improvement in soil physical properties such as porosity, structure, water infiltration rate and available water-holding capacity has been reported by addition of organic manures (Sweeten and Mathers 1985). Manures have also been found to reduce surface crusting, soil compaction, and soil bulk density (Tiarks, Mazurak, and Chesnin 1974). Organic manures have been used effectively in restoring productivity of marginal, less fertile, and eroded soils (Larney and Janzen 1997). Larney and Janzen (1997) reported that more efficient use of N sources produced on farms (e.g., livestock manure, crop residues) may provide an alternative for producers with a desire to restore their eroded soils and, at the same time, reduce their inputs of N fertilizer. These authors also stated that the mechanisms by which the amendments brought about soil restoration was largely explained by their N supplying power.

A complementary use of organic manure and chemical fertilizers has proved to be the best soil fertility management strategy in the tropics (Makinde and Agboola 2002). Well-decomposed farmyard manure contains about 12.9 g kg^{-1} total N, 1.0 g kg^{-1} available P, 4.5 g kg^{-1} exchangeable K, 10.8 g kg^{-1} exchangeable Ca, and 0.7 g kg^{-1} exchangeable Mg (Makinde and Agboola 2002). It has a greater beneficial residual effect on the soil than can be derived from the use of either inorganic fertilizer or organic manure, when applied alone.

In addition to farm yard manure, use of green manuring wherever possible is another practice to improve organic matter and NUE in crop production (Fageria 2007b). There has been a resurgence of interest in green manuring in recent years because of the need to develop sustainable agricultural systems. Green manuring is the practice of turning undecomposed green plant tissue into the soil (Fageria 2007b). Green manuring is an age-old practice for improving soil productivity and sustainability in many parts of the world. Adoption of soil and crop management practice by farmers depends on its agronomic and economic viability. Hence, it is important to state the salient benefits and drawbacks of green manuring. The benefits of green manuring include amelioration of physical, chemical, and biological properties of soil. The addition of symbiotic fixed N_2 in the cropping system and conservation of soil water and organic matter are particularly important. Furthermore, the practice increases biological security by decreasing the problems associated with intensive cropping systems, including problems of weeds, diseases, and insects. The drawbacks of this practice are temporary immobilization of N for succeeding crop (if the green manure crop is not incorporated into soil at proper time and growth stage) and its adjustment within cropping system. Crops with high water requirements (e.g., green manure crop) may be unsuitable under dry conditions.

Traditionally, both legume and nonlegume crops have been used as green manure. However, legumes are considered superior to nonlegumes because they fix atmospheric nitrogen. Major legume species used (e.g., green manure in tropical and temperate climates) are presented in Table 2.22. Legume green manure crops accumulate much N in their tops (Tables 2.23 and 2.24), which can be transferred to the succeeding crops in the rotation. Moreover, they have a lower Carbon/nitrogen ratio (C/N) ratio (Table 2.25) compared with gramineas, which facilitate easy decomposition and lower risk of N immobilization for the succeeding crop. Historically, the C/N ratio is the most widely used index of residue quality and decomposition rate. However, recent studies have shown that besides C/N ratio, ratios of lignin to N, or polyphenol-plus-lignin to N, have also been important in crop residue mineralization and release of nutrients.

2.8.5 Use of Cover Crops

Cover crops can be defined as close-growing crops that provide soil protection and soil improvement between periods of normal crop production, between trees in orchards, or between vines in vineyards (Fageria, Baligar, and Bailey 2005; Baligar and Fageria 2007). Cover crops are grown not for market purposes. However, when plowed under and incorporated into the soil, cover crops may be referred to as *green manure crops.* Cover crops are sometimes called *catch crops.* Cover crops are usually killed on the soil surface before they are mature by using appropriate herbicides. In most studies, cover crops managed as no-till mulches have been killed with glyphosate [N-(phosphono-methyl)glycine], paraquat (1,1-dimethyl-4-bipyridinium ion), or mixture of nonselective, postemergence, and pre-emergence herbicides (Fageria, Baligar, and Bailey 2005). Because many growers often want to reduce the use of chemical inputs, nonchemical methods of killing or suppressing cover crops are desirable. These mechanical methods are mowing, rolling, roll-chopping, undercutting, and partial rototilling. Creamer and Dabney (2002) have reviewed the literature on killing cover crops mechanically.

Cover crops are generally included in cropping systems as nutrient management tools. Cover crops can be leguminous or nonleguminous. Legume cover crops are used as a source of N for the following cash crop, whereas grasses are mainly used to reduce NO_3 leaching and erosion (Thomas et al. 1973). Biological N fixation by leguminous crops potentially reduces the need for N fertilizers for the succeeding crop (Fageria, Baligar, and Bailey 2005). A biculture of a legume and a grass is used with the intention of providing the benefits of both simultaneously (Ranells and Wagger 1996; Fageria, Baligar, and Bailey 2005).

The contribution of N is the most commonly observed primary benefit of leguminous crops (Singh, Singh, and Khind 1992). Both legume and nonlegume cover crops affect N fertilizer management (Bauer and Roof 2004). Legume cover crops fix atmospheric N and reduce N fertilizer needs for succeeding cash crops. The rate of N fixed by cover crops is determined largely by the genetic potential of the legume species and by the amount of plant available N in the soil. Two bacterial species, *Rhizobium* and *Bradyrhizobium*, are responsible for symbotic nitrogen fixation in legumes. The genus *Rhizobium* contains fast-growing, acid-producing bacteria, whereas *Bradyrhizobium* are slow growers that do not produce acid (Brady and

TABLE 2.22
Major Legume Green Manure Crops for Tropical and Temperate Regions

Tropical Region		Temperate Region	
Common Name	**Scientific Name**	**Common Name**	**Scientific Name**
Sunnhemp	*Crotalaria juncea* L.	Hairy vetch	*Vicia villosa* Roth
Sesbania	*Sesbania aculeata* Retz Poir	Barrel medic	*Medicago truncatula* Gaertn
Sesbania	*Sesbania rostrata* Bremek & Oberm	Alfafa	*Medicago sativa* L.
Cowpea	*Vigna unguiculata* L. Walp.	Black lentil	*Lens culinaris* Medikus
Soybean	*Glycine max* L. Merr.	Red clover	*Trifolium ngustif* L.
Clusterbean	*Cyamopsis tetragonoloba*	Soybean	*Glycine max* L. Merr.
Alfalfa	*Medicago sativa* L.	Faba bean	*Vicia faba* L.
Egyptian clover	*Trifolium alexandrinum* L.	Crimson clover	*Trifolium incarnatum* L.
Wild indigo	*Indigofera tinctoria* L.	Ladino clover	*Trifolium repens* L.
Pigeon Pea	*Cajanus cajan* L. Millspaugh	Subterranean clover	*Trifolium subterraneum* L.
Mungbean	*Vigna ngusti* L. Wilczek	Common vetch	*Vicia sativa* L.
Lablab	*Lablab purpureus* L.	Purple vetch	*Vicia benghalensis* L.
Graybean	*Mucuna cinereum* L	Cura clover	*Trifolium ambiguum* Bieb.
Buffalobean	*Mucuna aterrima* L. Piper & Tracy	Sweet clover	*Melilotus officinalis* L.
Crotalaria breviflora	*Crotalaria breviflora*	Winter pea	*Pisum sativum* L.
White lupin	*Lupinus albus* L.	Narrowleaf vetch	*Vicia ngustifólia* L.
Milk vetch	*Astragalus sinicus* L.	Milk vetch	*Astragalus sinicus* L.
Crotalaria	*Crotalaria striata*		
Zornia	*Zornia latifolia*		
Jackbean	*Canavalia ensiformis* L. DC.		
Tropical kudzu	*Pueraria phaseoloides* (Roxb.) Benth.		
Velvetbean	*Mucuna deeringiana* Bort. Merr.		
Adzuki bean	*Vigna angularis*		
Brazilian stylo	*Stylosanthes guianensis*		
Jumbiebean	*Leucaena leucocephala* Lam. De Wit		
Desmodium	*Desmodiumovalifolium* Guillemin & Perrottet		
Pueraria	*Pueraria phaseoloides* Roxb.		

Source: Compiled from various sources by Fageria, N. K. 2007b. Green manuring in crop production. *J. Plant Nutr.* 30:691–719.

TABLE 2.23
Nitrogen Accumulation in Major Legume Green Manure Crops

Crop Species	Growth Duration (days)	N Accumulation (kg ha^{-1})
Glycine max	45	115
Crotalaria juncea	45	169
Cajanus cajan	45	33
Sesbania aculeata	45	225
Vigna radiata	45	75
Dolichos lablab	45	63
Indigofera tinctoria	45	45
Sesbania rostrata	56	176
Sesbania aculeata	56	144
Vigna radiata	45	75
Vigna unguiculata	45	75
Sesbania rostrata	60	219
Sesbania cannabina	60	171
Sesbania aegyptiaca	57	39
Sesbania grandiflora	57	24
Cyamopsistetragonoloba L.	49	91
Astragalussinicus L.	Flowering	65–131
Vicia sativa L.	Flowering	105–210
Melilotusofficinalis L.	Flowering	150–300

Source: Compiled from various sources by Fageria, N. K. 2007b. Green manuring in crop production. *J. Plant Nutr*. 30:691–719.

TABLE 2.24
Dry Matter Yield and Nutrient Accumulation in 91-Day-Old Graybean Green Manure Crop Grown on Inceptisol in the Central Part of Brazil

Dry Matter Yield/Nutrient	Value[a]
Dry matter (kg ha^{-1})	7016
N (kg ha^{-1})	185.2
P (kg ha^{-1})	11.7
K (kg ha^{-1})	161.1
Ca (kg ha^{-1})	72.9
Mg (kg ha^{-1})	18.7
Zn (g ha^{-1})	229.4
Cu (g ha^{-1})	170.0
Mn (g ha^{-1})	1667.6
Fe (g ha^{-1})	2527.8

Source: Adapted from Fageria, N. K., and A. B. Santos. 2007. Response of irrigated rice to green manure and chemical fertilization in the state of Tocantins. *Rev. Bras. Eng. Agri. Ambien*. 11:387–392.

[a] Values are averages of two field experiments.

TABLE 2.25
C/N Ratio of Major Legume Green Manure and Cereal Crops

Crop Species	Growth Stage/Age in Days	C/N Ratio
Corn residues (*Zea mays* L.)	Physiol. maturity	67
Rice straw (*Oryza sativa* L.)	Physiol. maturity	69
Rice straw (*Oryza sativa* L.)	Physiol. maturity	56
Sorghum (*Sorghum bicolor* L. Moench)	Vegetative	22.0
Barley straw (*Hordeum vulgare* L.)	Physiol. maturity	99.1
Ryegrass (*Lolium multiflorum* Lam)	Vegetative	30
Rye (*Secale cereale* L.)	Heading	40
Alfalfa hay (*Medicago sativa* L.)	Not given	15.9
Pea straw (*Pisum sativum* L.)	Physiol. maturity	21
Pea hay (*Pisum sativum* L.)	Not given	15.4
Red clover (*Trifolium pratense* L.)	101 days	13.7
White clover (*Trifolium repens* L.)	101 days	10.7
Yellow trefoil (*Medicago lupulina* L.)	101 days	10.1
Persian clover (*Trifolium resupinatum* L.)	101 days	15.8
Egyptian clover (*Trifolium alexandrinum* L.)	101 days	16.7
Subterranean clover (*Trifolium Subterraneum* L.)	101 days	11.4
Cowpea (*Vigna unguiculata* L. Walp.)	Green pods	13.9
Sunnhemp (*Crotalaria juncea* L.)	Mature pods	20.2
Soybean (*Glycine max* L. Merr.)	Vegetative	17.9
Pigeon pea (*Cajanus cajan* L. Millspaugh)	Not given	25.9
Wild indigo (*Indigofera tinctoria* L.)	Flowering	15.8
Sesbania (*Sesbania rostrata* Bremek & Oberm)	Vegetative	27.8
Sesbania (*Sesbania emerus* Aubl. Urb.)	Vegetative	26.5
Aeschynomene afraspera	Vegetative	23.9
Desmanthus virgatus	Green pods	18.9
Tropical kudzu (*Pueraria phaseoloides*)	Not given	19
Hairy vetch (*Vicia villosa* Roth)	Vegetative	12
Hairy vetch (*Vicia villosa* Roth)	Flowering	18
Hairy vetch (*Vicia villosa* Roth)	Early bloom	17
Crimson clover (*Trifolium incarnatum* L.)	Midbloom	11

Source: Compiled from Fageria, N. K. 2007b. Green manuring in crop production. *J. Plant Nutr.* 30:691–719.

Weil 2002). Soil factors such as pH, moisture content, and temperature also determine N fixation capacity of a legume cover crop. In some cases, the amount of N provided by legume cover crops is adequate to produce optimal yields of subsequent nonleguminous crops; however, higher-N-requiring cereals such as corn (*Zea mays* L.) generally need supplemental N fertilizer. In such crops, N fertilizer rates could be lowered appreciably while maintaining optimal economic yields (Fageria, Baligar, and Bailey 2005). Table 2.26 provides data of nitrogen fertilizer equivalence (NFE) of legume cover crops to succeeding nonlegume crops. The NFE values varied

TABLE 2.26
Nitrogen Fertilizer Equivalence (NFE) of Legume Cover Crops to Succeeding Nonlegume Crops

Legume/Nonlegume Crop	NFE (kg ha^{-1})
Hairy vetch/cotton	67–101
Hairy vetch + rye/corn	56–112
Hairy vetch/corn	78
Hairy vetch/sorghum	89
Hairy vetch/corn	78
Hairy vetch + wheat/corn	56
Crimson clover/cotton	34–67
Crimson clover/corn	50
Crimson clover/sorghum	19–128
Common vetch/sorghum	30–83
Bigflower vetch/corn	50
Subterranean clover/sorghum	12–103
Sesbania/lowland rice	50
Alfalfa/corn	62
Alfalfa/wheat	20–70
Arachis spps/wheat	28
Subterranean clover/wheat	66
White lupin/wheat	22–182
Arachis spps/corn	60
Pigeon pea/corn	38–49
Sesbania/potato	48
Mungbean/potato	34–148
Chickpea/wheat	15–65

Source: Compiled from Fageria, N. K., V. C. Baligar, and B. A. Bailey. 2005. Role of cover crops in improving soil and row crop productivity. *Commun. Soil Sci. Plant Anal.* 36:2733–2757.

from 12 to 182 kg ha^{-1}. Smith, Frye, and Varco (1987) reported that NFE values range from 40 to 200 kg ha^{-1} but more typically are between 75 and 100 kg ha^{-1}. Interseeding red clover (*Trifolium pratense* L.) into small grains is a common practice in the northeastern United States (Singer and Cox 1998), and such practice can provide up to 85 kg N ha^{-1} to the subsequent corn crop (Vyn et al. 1999). Researchers in the southeastern United States have estimated that legumes, such as hairy vetch, can supply well over 100 kg N ha^{-1} to the corn or grain sorghum crops that follow (Wagger 1989; Oyer and Touchton 1990). On prairie soils in Kansas, Sweeney and Moyer (2004) found that grain sorghum following initial kill-down of red clover and hairy vetch yielded up to 131% more than continuous sorghum, with estimated fertilizer N equivalencies exceeding 135 kg ha^{-1}.

Legume cover crops should be inoculated with an appropriate strain of N-fixing bacteria. Perennial legumes fix N during any time of active growth. In annual legumes, N-fixation peaks at flowering. With seed formation, it ceases, and the nodules fall off the roots. Rhizobia return to the soil environment to await their next encounter with legume roots. These bacteria remain viable in the soil for three to five years, but often at too low a level to provide optimal N-fixation capacity when legumes are replanted (Fageria, Baligar, and Bailey 2005).

In addition to fixing N, cover crops have been reported to reduce the potential for NO_3 leaching from farm fields (Fageria, Baligar, and Bailey 2005). In studies reviewed by Fageria, Baligar, and Bailey (2005) and Baligar and Fageria (2007), cover crops reduced both the mass of the N leached and the NO_3 concentration of leachate by 20–80%, compared with no cover crop control. They also determined that grasses and brassicas were two to three times as effective as legumes in reducing NO_3 leaching. Francis, Bartley, and Tabley (1998), Shepherd (1999), and Rasse et al. (2000) reported that incorporating a nonleguminous cover crop in a cropping system has reduced NO_3^- leaching because the cover crop can reduce water percolation and also effectively use NO_3^- that would otherwise leach.

Cover crops accumulate inorganic soil N between main crop seasons, hold it in an organic form, and prevent it from leaching. The N is subsequently released to the next crop as the cover crop residue decomposes. Rye (*Secale cereale* L.), a cereal, is recognized as having great potential as a scavenger of residual inorganic N present after corn harvests (Fageria, Baligar, and Bailey 2005). McCracken et al. (1994) reported that rye was much more effective than vetch (*Vicia villosa* Roth) in reducing NO_3^- leaching. Furthermore, effective erosion control, reduced soil compaction, and suppressed weed emergence by use of rye in field cropping systems have been reported (Blum et al. 1997; Fageria, Baligar, and Bailey 2005). The higher capacity of rye in scavenging residual N compared to vetch was associated with its more rapid establishment of an extensive root system (McCracken et al. 1994).

2.8.6 Use of Controlled-Release Nitrogen Fertilizers and NH_4^+/NO_3^- Inhibitors

The loss of N from soil–plant systems depends on environmental conditions and N sources. Urea is the world's predominant N fertilization source, and it accounts for more than half of all N fertilizers available (Glibert et al. 2006). However, NH_4-N in urea is subject to volatilization loss, after urea hydrolysis and denitrification, and to leaching loss, after nitrification to nitrate (Stumpe, Vlek, and Lindsay 1984; Al-Kanani, Mackenzie, and Barthakur 1991). Use of controlled-release fertilizer is a strategic approach to increasing fertilizer efficiency by synchronizing nutrient release with crop demand and thereby reducing environmental losses (Shoji et al. 2001; Morgan, Cushman, and Sato 2009). Common controlled-release fertilizer types include sulfur-coated urea; polymer-coated, water-soluble fertilizers; and low-solubility and biodegradable fertilizer materials (Fageria and Baligar 2005b; Blackshaw et al. 2011). A few studies comparing polymer-coated urea with urea have indicated that crop yield can be higher, lower, or unchanged depending on the crop and environmental conditions during the growing season (Golden et al. 2009; Noelisch et al. 2009; Blackshaw et al. 2011).

An experiment was conducted under greenhouse conditions, and common urea and polymer-coated urea were compared with regard to lowland rice production. Nitrogen rate treatment affected grain and straw yield (Table 2.27). However, there was no significant effect of N source, and N source × N rate interactions were also not significant for these two traits. Nonsignificant N source × N rate interactions indicate that differences in grain and straw yield were consistent across two N sources. Grain yield and straw yield were increased in a quadratic fashion with the addition of N in the range of 0–400 mg N kg^{-1} soil. Variation in grain yield was 8.55–16.91 g $plant^{-1}$, with an average value of 14.26 g $plant^{-1}$. Similarly, variation in straw yield was 9.51–18.89 g $plant^{-1}$, with an average value of 15.47 g $plant^{-1}$. Maximum grain yield was obtained with the addition of 258 mg N kg^{-1} soil, and maximum straw yield was obtained with the application of 309 mg N kg^{-1} soil. The variation in grain yield and straw yield was 46 and 70%, respectively, with the addition of N. Hence, straw yield showed greater increase with the addition of N compared to grain yield. Fageria and colleagues (2009, 2011) reported significant yield increase with the addition of N in the lowland rice grown on Brazilin lowland or varzea soils. Fageria and Baligar (2003b) reported that N is one of the most important nutrients limiting lowland rice yield in Brazilian lowland soils. Figures 2.46 and 2.47 show the growth of lowland rice with

TABLE 2.27
Grain Yield and Straw Yield of Lowland Rice as Influenced by Sources and Rates of Nitrogen Fertilization

N Rate (mg kg^{-1})	Grain Yield (g $plant^{-1}$)	Straw Yield (g $plant^{-1}$)
0	8.55	9.51
50	14.68	13.56
100	16.91	16.10
150	16.08	17.35
200	13.92	17.44
400	15.43	18.89
Average	14.26	15.47
F-test		
N source (NS)	NS	NS
N rate (NR)	*	*
NS × NR	NS	NS
CV (%)	13.14	20.24

Regression Analysis

N rate vs. grain yield $(Y) = 10.6396 + 0.0493X - 0.000096X^2$, $R^2 = 0.4550^*$

N rate vs. straw yield $(Y) = 10.1981 + 0.0612X - 0.000099X^2$, $R^2 = 0.7009^*$

Source: Fageria, N. K., T. Cobacci, and R. A. Reis, Jr. in press. Comparison of conventional and polymer coated urea as nitrogen sources for lowland rice production. *J. Plant Nutr.*

Note: Values are across two N sources.

^{*,NS}Significant at the 1% probability level and nonsignificant, respectively.

FIGURE 2.46 Lowland rice growth at different N rates applied with conventional urea. (From Fageria, N. K., T. Cobacci, and R. A. Reis, Jr. in press. Comparison of conventional and polymer coated urea as nitrogen sources for lowland rice production. *J. Plant Nutr.*)

FIGURE 2.47 Lowland rice growth at different N rates applied with polymer-coated urea. (From Fageria, N. K., T. Cobacci, and R. A. Reis, Jr. in press. Comparison of conventional and polymer coated urea as nitrogen sources for lowland rice production. *J. Plant Nutr.*)

FIGURE 2.48 Comparison of lowland rice growth at conventional and polymer-coated urea with 400 mg N kg^{-1} soil. (From Fageria, N. K., T. Cobacci, and R. A. Reis, Jr. in press. Comparison of conventional and polymer coated urea as nitrogen sources for lowland rice production. *J. Plant Nutr.*)

the addition of conventional and polymer-coated urea at different N rates. Similarly, Figure 2.48 compares the growth of rice plants at two N sources at 400 mg N kg^{-1} soil.

The influence of common urea and polymer-coated urea on root growth of lowland rice has also been studied. Maximum root length and root dry weight were significantly influenced by N fertilization (Table 2.28). These growth parameters increased in a quadratic fashion with increasing N rates in the range of 0–400 mg N kg^{-1} soil. The root length varied from 26.33 to 35.33 cm, with an average value of 30.61 cm. Similarly, the root dry weight varied from 1.95 to 4.06 g $plant^{-1}$, with an average value of 3.36 g $plant^{-1}$. Maximum root length was achieved with the application of 152 mg N kg^{-1} of soil. Similarly, maximum root weight was achieved with the application of 256 mg N kg^{-1}. Hence, it can be concluded that N had more

TABLE 2.28
Maximum Root Length (MRL) and Root Dry Weight of Lowland Rice as Influenced by N Sources

N Rate (mg kg^{-1})	MRL (cm)	Root Dry Weight (g plant^{-1})
0	26.33	1.95
50	35.33	3.30
100	34.00	4.06
150	30.50	3.78
200	30.83	3.48
400	26.67	3.60
Average	30.61	3.36
F-test		
N source (NS)	NS	NS
N rate (NR)	*	*
NS × NR	NS	NS
CV (%)	16.68	31.94

Regression Analysis

N rate vs. MRL (Y) = 29.4800 + 0.0394X – 0.00013X^2, R^2 = 0.2657*

N rate vs. root dry wt. (Y) = 2.3765 + 0.0136X – 0.000027X^2, R^2 = 0.3054*

Source: Fageria, N. K., T. Cobacci, and R. A. Reis, Jr. in press. Comparison of conventional and polymer coated urea as nitrogen sources for lowland rice production. *J. Plant Nutr.*

*,NS Significant at the 5% probability level and nonsignificant, respectively.

influence on root dry weight than on root length. The variation, due to N rate, in root length and root dry weight was 27 and 31%, respectively, as shown by R^2 values. This also proves that root dry weight is more sensitive to N deficiency compared to root length. Figures 2.49 and 2.50 show root growth of rice at different N rates supplied by conventional and polymer-coated urea. These figures show the importance of N fertilization in improving root growth of rice.

The improvement in root length and root dry weight with the addition of N suggests that use of N at an adequate rate is an important strategy in improving uptake of water and nutrients, and consequently yield. Fageria and Moreira (2011) reported that genetics as well as environmental factors influence root growth. Costa et al. (2002) reported that greater root length and root surface area were obtained at an N fertilizer rate of 128 kg ha^{-1} compared with either the absence of fertilizer N or the higher rate of 255 kg N ha^{-1}. Nitrogen fertilizer improves root growth in soils with low organic matter content (Gregory 1994; Robinson, Linehan, and Gordon 1994). Nitrogen fertilization may increase crop root growth by increasing N availability in soil (Fageria and Moreira 2011). Nitrogen also improves production of lateral roots and root hairs, as well as increasing rooting depth and root length density deep in the profile (Hansson and Andren 1987). Hoad et al. (2001) reported that surface application of nitrogen fertilizer increases root densities in the surface layers of soil.

FIGURE 2.49 Root growth of lowland rice at different N rates applied with conventional urea. (From Fageria, N. K., T. Cobacci, and R. A. Reis, Jr. in press. Comparison of conventional and polymer coated urea as nitrogen sources for lowland rice production. *J. Plant Nutr.*)

In addition to use of slow-release or controlled-release N fertilizers, use of nitrification inhibitors with NH_4^+-N-based fertilizers is recognized as another potential tool to improve NUE and crop yields (Freney, Smith, and Mosier 1992; Ferguson, Lark, and Slater 2003). The greatest benefits from nitrification inhibitors have been reported on coarse-textured soils and soils that remain saturated with water during large parts of the crop-growing season (Fageria and Baligar 2005b). Nitrogen is subject to leaching from coarse-textured soils, and denitrification can be a dominant N loss mechanism in saturated soils. It has also been reported that beneficial effects of nitrification inhibitors are frequently observed at suboptimal N rates (Cerrato and Blackmer 1990).

Inhibition of nitrification retains NH_4^+-N-based N fertilizers in the NH^+ form, which may be retained on cation exchange sites in the soil medium and conserved against leaching in contrast to being readily leached NO_3-N. The nitrification inhibitor [nitrapyrin (2-chloro-6-trichloromethyl) pyridine] has been used to inhibit nitrification of urea and ammoniacal fertilizers (Goring 1962). Furthermore, nitrification inhibitors also reduced N_2O emissions in winter wheat (Bronson and Mosier 1993) and barley (Delgado and Mosier 1996) cropping systems. Controlled-release fertilizers effectively decreased NO_3-N leaching and increased crop yields and NUE in greenhouse (Shoji and Gandeza 1992) and field experiments (Shoji and Kanno 1994). The N_2O emissions were also reported to be reduced with the use of controlled-release fertilizers in lysimeter and field studies (Delgado and Mosier 1996). Urease activity inhibitors such as nBTPT [N-(n-butyl) thiophosphoric triamide] have also been reported to decrease

FIGURE 2.50 Root growth of lowland rice at different N rates applied with polymer-coated urea. (From Fageria, N. K., T. Cobacci, and R. A. Reis, Jr. in press. Comparison of conventional and polymer coated urea as nitrogen sources for lowland rice production. *J. Plant Nutr.*)

the rate of urea hydrolysis, preventing abrupt pH rises around fertilizer granules and consequently diminishing NH_3 volatilization losses (Watson et al. 1994).

2.8.7 Adoption of Conservation Tillage System

Conservation tillage improves soil quality, which enhances N availability and utilization by crop plants (Fageria 2002b). Several beneficial effects of conservation tillage are summarized in this section, which are related to improved NUE by crop plants and consequently higher yields. For example, tillage accelerates the loss of soil organic matter by increasing biological oxidation and often by increasing soil erosion (Schillinger et al. 1999). Because of the decline in organic matter and associated soil quality, most tillage-based farming systems in dry land environments are not sustainable in the long term (Papendick and Parr 1997). One option for maintaining and improving soil quality is reducing or eliminating tillage.

The no-till or minimum tillage crop production system is becoming more common in various parts of the world and is reportedly helpful in improving soil quality (Fageria 2002b). Soil protection from erosion losses; conservation of soil water by increased infiltration and decreased evaporation; increased use of land too steep for conventional production; and reduction in fuel, labor, and machinery costs are among the reasons for increased use of reduced tillage systems (Doran and Linn 1994). A review by Steiner (1974) demonstrated the value of residue management systems for conserving soil water through reduced soil water evaporation. No-tillage production results in changes in chemical and physical properties of soil, including increases in soil organic matter content (Douglas and Gross 1982), aggregate stability (Heard, Kladivko, and Mannering 1988), and macroporosity (Blackwell and Blackwell 1989; Lal, Vleeschauwer, and Ngaje 1990). Collectively and individually, these changes influence plant growth (Dao 1993). The changes can be detrimental, neutral, or beneficial for crop growth and yield, depending on soil texture and structure (Dick and VanDoren 1985), climatic factors such as rainfall (Boyer 1970), and weed control (Kapusta 1979). In general, no-till systems have greater positive effects on crop growth and yield when used on soils characterized by low organic matter levels and well-drained soils than when used on poorly drained soils high in organic matter (Opoku, Vyn, and Swanton 1997).

Sharpley et al. (1991) reported that N in runoff decreased with no-till or reduced tillage system compared with convention tillage. Calvino, Andrade, and Sadras (2003) in the Argentine Pamps reported improvement in NUE of corn planted under no-till system. These authors also reported corn yield increases of 0.9 Mg ha^{-1} from the mid-1990s to 1998 related to no-till and higher plant density. Synchronization of residue N mineralization, fertilizer-N application time, and subsequent crop demand for N can improve NUE of crops planted in conservation tillage systems (Reeves, Wood, and Touchton 1993).

There are conflicting reports in the literature about N balance or use efficiency under no-till compared with conventional tillage systems. Some workers have reported that in conservation tillage, fertilizer N rates have been increased to prevent yield limitations from short-term N immobilization (Wood and Edwards 1992). However, Torbert, Potter, and Morrison (2001) reported that there was no indication of N limitation in the no-till system compared to other tillage systems. It is also reported that nitrate-leaching losses from sandy soils can be greater under no-till than under conventional tillage because of the higher moisture (Thomas et al. 1973). On the contrary, Meisinger et al. (1992) reported that NO_3–N leaching is not a highly efficient process in structural soils due to preferential flow through macropores, which are increased under no-till. However, no-till generally shows a greater infiltration capacity than does conventional tillage because of continuous macropores that are open at the soil surface (Unger and McCalla 1980). Hence, initial leaching losses of surface-applied N could be rapid under no-till if heavy rainfall occurs soon after fertilizer application. Conversely, fertilizer that had time to diffuse into aggregate micropores would be protected from subsequent leaching because of the higher proportion of water flowing through the macropore system under no-till (Cameron and Haynes 1986).

2.8.8 Adopting Appropriate Crop Rotation

Use of appropriate crop rotation is an important strategy in improving NUE in crop plants. Appropriate sequences allow efficient use of soil resources, especially nutrients and water by the crop to increase yield at a systems level (Gan et al. 2003). Rotation of legume and nonlegume crops has been recommended for centuries (Martin and

Leonard 1967). Yield increase associated with crop rotation has been referred to as the *rotation effect* (Pierce and Rice 1988), and yield declines associated with monoculture has been referred to as *monoculture yield declines* (Porter et al. 1997). Crop rotation has generally been thought to reduce risk compared with monoculture (Helmers, Langemeier, and Atwood 1986). *Risk* was defined as the failure to meet an annual pre-hectare net return target (Helmers, Yamoah, and Varvel 2001).

Crop rotation is a planned sequence of crops growing in a regularly recurring succession on the same area of land, as contrasted with continuous culture of one crop or growing a variable sequence of crops (Soil Science Society of America 2008). Crop rotation has been used for thousands of years. Karlen, Varvel, and Bullock (1994) reported that crop rotation was in use in ancient Greece and Rome. MacRae and Mehuys (1985) reported that crop rotation was practiced by the Han Dynasty of China more than 3000 years ago. In an appropriate crop rotation, a legume should be included with cereals. Legumes fix atmospheric nitrogen and hence reduce N requirement of succeeding cereal crops. Crop rotation is effective against diseases, insects, and weeds (Karlen, Varvel, and Bullock 1994).

2.8.9 Use of Chlorophyll Meter

Due to the reported limitations of soil tests based on N recommendations, research has been centered on developing an inseason monitoring approach as a guide to N management decisions (Lofton et al. 2012). Several studies reported that handheld chlorophyll meters can accurately predict N requirement based on a sufficiency index (Wood et al. 1992; Blackmer and Schepers 1995; Waskom et al. 1996):

$$\text{Sufficiency index}\,(\%) = \frac{\text{Fertilizer needed plot}}{\text{Well-fertilized plot}} \times 100 \tag{2.11}$$

According to Varvel, Francis, and Schepers (1997), additional N is recommended when sufficiency index values fall below 95%. A major limitation of using chlorophyll meters to determine N fertilizer recommendations is obtaining a representative sample across a highly variable field (Blackmer and Schepers 1995). In addition to field-scale variability, chlorophyll meters can produce highly variable values within a single plant (Peterson et al. 1993). Therefore, obtaining accurate values in highly variable environments can be costly and time consuming (Lofton et al. 2012).

2.8.10 Use of Efficient Genotypes

Utilization of plant species or genotypes of same species efficient in absorption and utilization of N is an important strategy in improving NUE and sustainable agricultural system. Differences in N uptake and utilization among upland and lowland rice genotypes have been reported by Fageria and Baligar (2005b), Fageria, Santos, and Cutrim (2008), and Fageria, Baligar, and Jones (2011). Figure 2.51 shows response of four lowland rice genotypes to N fertilization. These genotypes differ in yield response to applied N. These genotypes can be grouped into three classes according

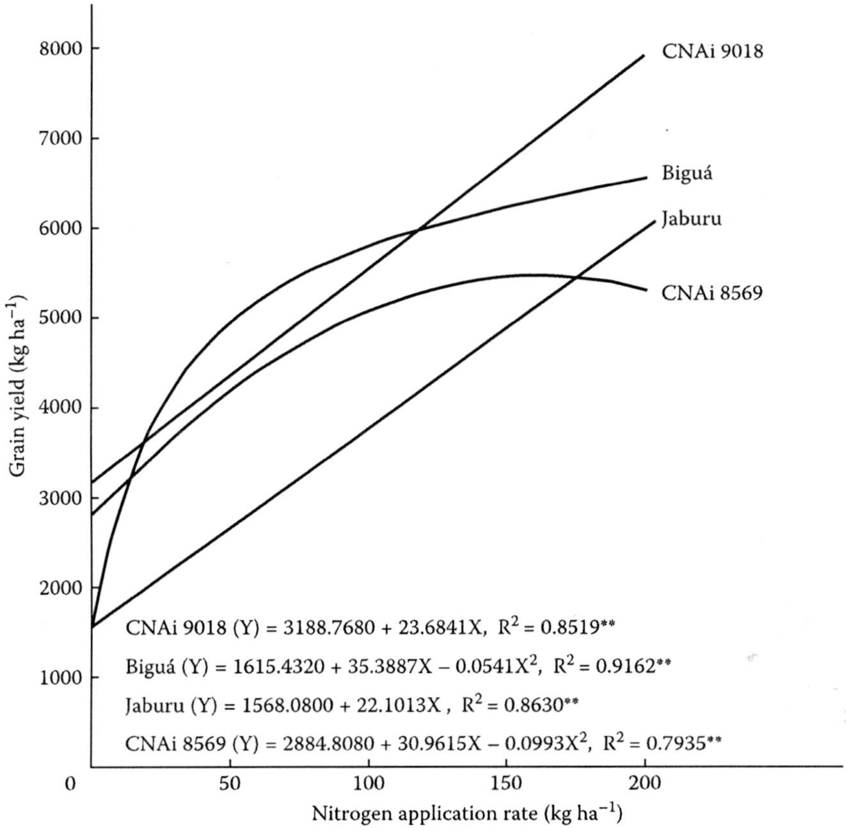

FIGURE 2.51 Response of lowland rice genotypes to nitrogen fertilization. (From Fageria, N. K., and V. C. Baligar. 2003b. Methodology for evaluation of lowland rice genotypes for nitrogen use efficiency. *J. Plant Nutr.* 26:1315–1333.)

to their response to N fertilization. The first group was efficient and responsive. The genotype produced above-average yield compared to the genotypes tested at the low N level and responded well to applied N. The genotype CNAi 9018 falls into this group. The second classification was efficient and nonresponsive. The genotype produced well at low N rates but did not respond well at higher N rates. The genotype CNAi 8569 fall into this group. The third group is a genotype that has low production at low N rates but responds well at higher N rates. These are designated as *inefficient and responsive.* The genotypes Biguá and Jaburu fall into this group. From a practical point of view, the genotypes that are *efficient and responsive* are the most desirable because they can produce well at low soil N levels and also respond well to applied N. Thus, this group can be utilized with both low- and high-input technologies with reasonably good yield. The second most desirable group is *efficient nonresponsive.* Genotypes of this type can be planted under low N levels and still produce more than average yield. The inefficient responsive genotypes can be used in breeding programs for their N-responsive characteristics. Similarly,

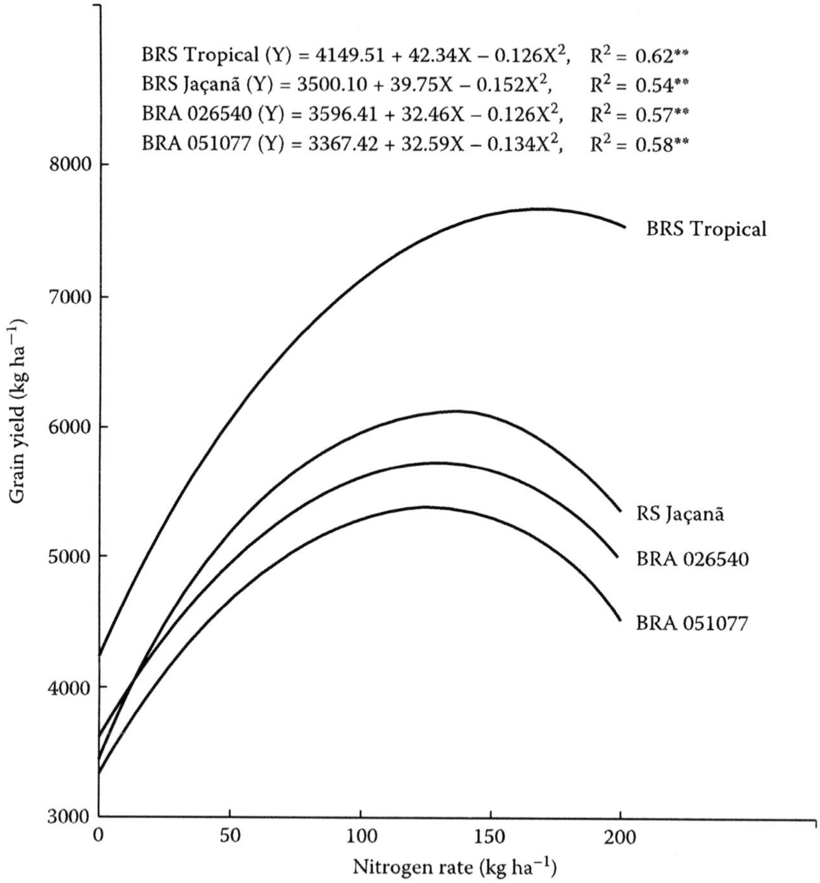

FIGURE 2.52 Response of lowland rice genotypes to nitrogen fertilization.

Figures 2.52 through 2.54 show the response of lowland rice genotypes to N fertilization. All genotypes had a quadratic response when N was applied in the range of 0–200 kg ha^{-1}. However, magnitude of response varied across genotypes.

In the literature, several reasons have been cited as to why some genotypes are more efficient in N utilization than the others (Fageria 2009). Moll, Kamprath, and Jackson (1982) reported that NUE differences among corn hybrids resulted from differing utilization of N already accumulated in the plant prior to anthesis, especially with low N levels. Eghball and Maranville (1991) reported that NUE generally parallels water use efficiency in corn; hence, the two traits can be selected simultaneously where such parallels exist. Kanampiu, Raun, and Johnson (1997) reported that wheat varieties with higher GHI were having higher NUEs. Cox, Qualset, and Rains (1985) reported that wheat varieties that accumulate large amounts of N early in the growing season do not necessarily have high NUE. Plants must convert this accumulated N to grain and must assimilate N after anthesis to produce high NUE. Forms of N uptake (NH_4^+ vs. NO_3^-) may also influence NUE

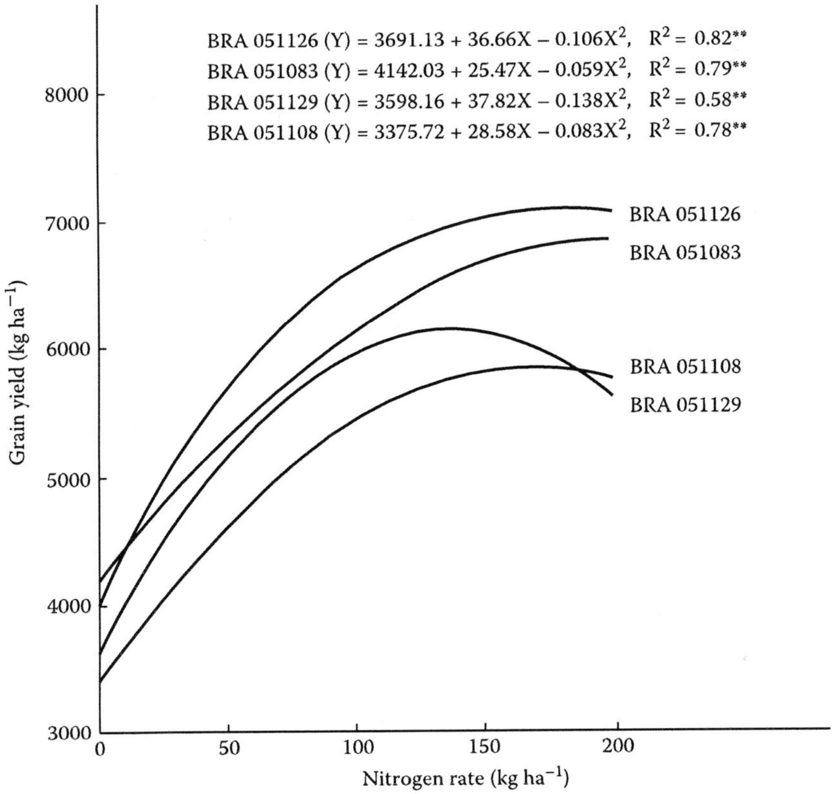

FIGURE 2.53 Relationship between nitrogen rate and grain yield of lowland rice genotypes.

(Thomason et al. 2002). Plants with preferential uptake of NH_4^+ during grain fill may provide greater NUE than plants without this preference (Tsai et al. 1992). Ammonium-N supplied to high-yielding corn genotypes increased yield over plants supplied with NO_3^- during critical ear development (Pan et al., 1984). Salsac et al. (1987) reported that NH_4^+ assimilation required 5 ATP (adenosine triphosphate) mol^{-1} of NH_4^+, whereas NO_3^- assimilation required 20 ATP mol^{-1} of NO_3^-. This energy saving mechanism may be responsible for higher NUE in NH_4^+–N. In addition to the aforementioned reasons, Fageria and Baligar (2003b) summarized soil and plant mechanisms and processes and other factors that influence genotypic differences in plant nutrient efficiency.

Regarding genotypic variability for NUE, Rosielle and Hamblin (1981) reported that heritability for grain yield is usually lower under low N than it is under high N, making potential progress less for low-N than for high-N target environments. Banziger, Betran, and Lafitte (1997) also reported that heritability of grain yield usually decreases under low N. Banziger and Lafitte (1997) reported that secondary traits (e.g., ears per plant, leaf senescence, leaf chlorophyll concentration) are a valuable adjunct in increasing the efficiency of selection for grain yield when broad-sense heritability of grain yield is low under a low-N environment.

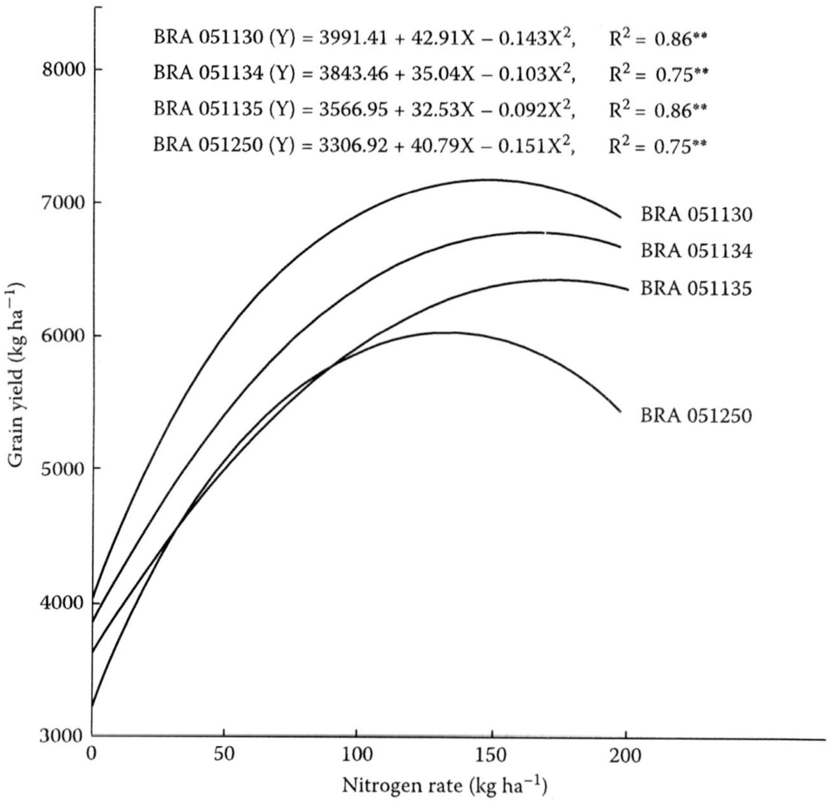

FIGURE 2.54 Relationship between nitrogen rate and grain yield of lowland rice genotypes.

2.9 CONCLUSIONS

Rice has been the staple diet for the majority of the world population, and nitrogen deficiency is one of the major yield-limiting constraints in most rice-producing soils. Nitrogen is absorbed by rice in maximum quantities or sometimes equal to potassium compared to other essential plant nutrients. Its role in rice production is associated with increasing photosynthesis and improving yield components such as panicle and grain weight. Nitrogen is also responsible for decreasing spikelet sterility in rice. In addition, an adequate rate of nitrogen fertilization also improved root growth of rice, which is an important factor considering the importance of roots in absorption of water and nutrients, and consequently yield. Nitrogen is a mobile nutrient in plants; hence, its deficiency symptoms first appear in the older leaves. The yellow color of older leaves is a typical symptom of nitrogen deficiency. In addition, plant height, tillering, and leaf area are reduced in rice when nitrogen is deficient in the soil–plant system.

Crop response to applied N and use efficiency are important criteria for evaluating crop N requirements for maximum economic yield. Based on greenhouse experimental results, the adequate rate of N for growth, yield, and yield components varied across traits. However, on average, 300 mg N kg^{-1} is an adequate rate for most of the growth, yield, and yield components. Similarly, based on field experiment results, N rate for maximum economic rice yield (6500–7000 kg ha^{-1}) varied from 120 to 150 kg ha^{-1} for lowland rice, depending on genotype and soil type. The upland rice N requirement is about 100–130 kg ha^{-1} when grain yield is 5000 kg ha^{-1}.

The timing of application varied from the total required amount of N being applied at sowing to application in two or three installments spread across the crop growth cycle. However, the maximum yield could be expected when half of the required N is applied at sowing and the remaining half is applied at the active or mid-tillering growth stage in lowland and upland rice. Nitrogen applied at booting and flowering did not improve grain yield. Urea and ammonium sulfate are dominant N sources for annual crops, including rice, especially in developing countries. Results obtained under field and greenhouse conditions show that under Brazilian conditions, ammonium sulfate is a better source of N for lowland rice compared to urea. The application of the two sources of N (conventional and polymer-coated urea) did not result in significant differences in the production of rice and yield components. However, NUE of polymer-coated urea was higher than that of conventional urea.

Recovery of N in upland as well as lowland rice is usually less than 50% worldwide. Low recovery of N in rice is associated with its loss by volatilization, leaching, surface runoff, denitrification, and plant canopy. Low recovery of N is responsible for higher cost of crop production and also environmental pollution. Hence, improving NUE is desirable for improving crop yields, reducing production cost, and maintaining environmental quality. To improve NUE in agriculture, integrated N management strategies that take into consideration improved fertilizer, soil, and crop management practices are necessary. Including livestock production with cropping offers one of the best opportunities to improve NUE. Synchrony of N supply with crop demand is essential to ensure adequate quantity of uptake and utilization and optimum yield.

REFERENCES

Abbasi, M. K., and M. M. Tahir. 2012. Economizing nitrogen fertilizer in wheat through combinations with organic manures in Kashmir, Pakistan. *Agron. J.* 104:169–177.

Adeli, A., H. Tewolde, and J. N. Jenkins. 2012. Broiler type and placement effects on corn growth, nitrogen utilization, and residual soil nitrate-nitrogen in a no-till field. *Agron. J.* 104:43–48.

Al-Kanani, T., A. F. Mackenzie, and N. N. Barthakur. 1991. Soil water and ammonia volatilization relationships with surface applied nitrogen fertilizer solutions. *Soil Sci. Soc. Am. J.* 55:1761–1766.

Aulakh, M., T. S. Khera, J. W. Doran, K. Singh, and B. Singh. 2000. Yield and nitrogen dynamics in a rice-wheat system using green manure and inorganic fertilizer. *Soil Sci. Soc. Am. J.* 64:1867–1876.

Aulakh, M. S., J. W. Doran, and A. R. Mosier. 1992. Soil denitrification—Significance, measurement, and effects of management. *Adv. Soil Sci.* 18:1–57.

Baligar, V. C., and N. K. Fageria. 2007. Agronomy and physiology of tropical cover crops. *J. Plant Nutr.* 30:1287–1339.

Banziger, M., F. J. Betran, and H. R. Lafitte. 1997. Efficiency of high-nitrogen selection environments for improving maize low-nitrogen target environment. *Crop Sci.* 37:1103–1109.

Banziger, M., and H. R. Lafitte. 1997. Efficiency of secondary traits for improving maize for low nitrogen target environments. *Crop Sci.* 37:1110–1117.

Bauer, P. J., and M. E. Roof. 2004. Nitrogen, aldicarb, and cover crop effects on cotton yield and fiber properties. *Agron. J.* 96:369–376.

Beyrouty, C. A., L. E. Sommers, and D. W. Nelson. 1988. Ammonium volatilization from surface applied urea as affected by several phosphoroamide compounds. *Soil Sci. Soc. Am. J.* 52:1173–1178.

Bhuija, Z. H., and N. Walker.1977. Autotrophic nitrifying bacteria in acid tea soils from Bangladesh and Sri Lanka. *J Appl. Bact.* 42:253–257.

Blackmer, T. M., and J. S. Schepers. 1995. Use of chlorophyll meter to monitor nitrogen status and schedule fertigation for corn. *J. Prod. Agric.* 8:56–60.

Blackshaw, R. E., X. Hao, R. N. Brandt, G. W. Clayton, K. N. Harker, J. T. O`Donovan, E. N. Johnson, et al. 2011. Canola response to ESN and urea in a four-year no-till cropping system. *Agron. J.* 103:92–99.

Blackwell, P. S., and J. Blackwell. 1989. The introduction of earthworms to an ameliorated, irrigated duplex soil in southeastern Australia and the influence on macro pores. *Aust. J. Soil Res.* 27:807–814.

Blum, U., L. D. King, T. M. Gerig, M. E. Lehman, and A. D. Worsham. 1997. Effects of clover and grain crops and tillage techniques on seedling emergence of some dicotyledonous weed species. *Am. J. Alternate. Agric.* 12:146–161.

Bolan, N. S., and M. J. Hedley. 2003. Role of carbon, nitrogen, and sulfur cycles in soil acidification. In *Handbook of Soil Acidity*, ed. Z. Rengel, pp. 29–56. New York: Marcel Dekker.

Boyer, J. S. 1970. Leaf enlargement and metabolic rates in corn, soybean, and sunflower at various leaf water potentials. *Plant Physiol.* 46:233–235.

Brady, N. C., and R. R. Weil. 2002. *The Nature and Properties of Soils*, 13th edition. Upper Saddle River, NJ: Prentice Hall.

Bronson, K. F., F. Hussain, E. Pasuquin, and J. K. Ladhs. 2000. Use of ^{15}N-labeled soil in measuring nitrogen fertilizer recovery efficiency in transplanted rice. *Soil Sci. Soc. Am. J.* 64:235–239.

Bronson, K. F., and A. R. Mosier. 1993. Nitrous oxide emissions and methane consumption in wheat and corn-cropped systems in northeastern Colorado. In *Agricultural Ecosystems Effects on Trace Gases and Global Climate Change*, eds. L. A. Harper; A. R. Moiser, J. M. Duxbury, and D. E. Rolston, pp. 133–144. Madison, WI: ASA.

Bufogle, A., Jr., P. K. Bollich, J. L. Kovar, C. W. Lindau, and R. R. Macchivelli. 1998. Comparisons of ammonium sulfate and urea as nitrogen sources in rice production. *J. Plant Nutr.* 21:1601–1614.

Calvino, P. A., F. H. Andrade, and V. O. Sadras. 2003. Maize yield as affected by water availability, soil depth, and crop management. *Agron. J.* 95:275–281.

Cameron, K. C., and R. J. Haynes. 1986. Retention and movement of nitrogen in soils. In *Mineral Nitrogen in the Plant–Soil System*, ed. R. J. Haynes, pp. 166–241. Orlando, FL: Academic Press.

Casman, K. G., A. Dobermann, and D. T. Walters. 2002. Agroecosystems, nitrogen use efficiency and nitrogen management. *Ambio.* 31:132–140.

Cassman, K. G., M. J. Kropff, G. Graunt, and S. Peng. 1993. Nitrogen use efficiency of rice reconsidered: What are the key constraints? *Plant Soil.* 155:359–362.

Cerrato, M. E., and A. M. Blackmer. 1990. Effects of nitrapyrin on corn yields and recovery of ammonium-N at 18 site-years in Iowa. *J. Prod. Agric.* 3:513–521.

Clark, M. S., W. R. Horwath, C. Shennan, and K. M. Scow. 1998. Changes in soil chemical properties resulting from organic and low-input farming practices. *Agron. J.* 90:662–671.

Costa, C., L. M. Dwyer, X. Zhou, P. Dutilleul, C. Hamel, L. M. Reid, and D. L. Smith. 2002. Root morphology of contrasting maize genotypes. *Agron. J.* 94:96–101.

Cox, M. C., C. O. Qualset, and D. W. Rains. 1985. Genetic variation for nitrogen assimilation and translocation in wheat. II. Nitrogen assimilation in relation to grain yield and protein. *Crop Sci.* 25:435–440.

Creamer, N. G., and S. M. Dabney. 2002. Killing cover crops mechanically: Review of recent literature and assessment of new research results. *Am. J. Alternative Agric.* 17:32–40.

Dao, T. H. 1993. Tillage and winter wheat residue management effects on water infiltration and storage. *Soil Sci.* 57:1586–1595.

Dao, T. H., and M. A. Cavigelli. 2003. Mineralizable carbon, nitrogen, and water-extractable phosphorus release from stockpiled and composted manure and manure-amended soils. *Agron. J.* 95:405–413.

De Datta, S. K. 1981. *Principles and Practices of Rice Production*. New York: Wiley.

De Datta, S. K., R. J. Buresh, M. I. Samson, and W. Kai-Rong. 1988. Nitrogen use efficiency and nitrogen-15 balances in broadcast seeded flooded and transplanted rice. *Soil Sci. Soc. Am. J.* 52:849–855.

De Datta, S. K., W. N. Obcemea, R. Y. Chen, J. C. Calabio, and R. C. Evangelista. 1987. Effect of water depth on nitrogen use efficiency and nitrogen-15 balance in lowland rice. *Agron. J.* 79:210–216.

Delgado, M. J., and A. R. Mosier. 1996. Mitigation alternatives to decrease nitrous oxides emissions and urea-nitrogen loss and their effect on methane flux. *J Environ. Qual.* 25:1105–1111.

Dhugga, K. S., and J. G. Waines. 1989. Analysis of nitrogen accumulation and use in bread and durum wheat. *Crop Sci.* 29:132–139.

Dick, W. A., and D. M. VanDoren, Jr. 1985. Continuous tillage and rotation combinations effects on corn, soybean, and oat yields. *Agron. J.* 77:459–465.

Dillon, K. A., T. W. Walker, D. L. Harrell, L. J. Krutz, J. J. Varco, C. H. Koger, and M. S. Cox. 2012. Nitrogen sources and timing effects on nitrogen loss and uptake in delayed flood rice. *Agron. J.* 104:466–472.

Dingkuhn, M., H. F. Schnier, S. K. De Datta, K, Dorffling, and C. Javellana. 1991. Relationships between ripening phase productivity and crop duration, canopy photosynthesis and senescence in transplanted and direct seeded lowland rice. *Field Crops Res.* 26:327–345.

Dobermann, A., D. Dawe, R. P. Roetter, and K. G. Cassman. 2000. Reversal of rice yields decline in a long-term continuous cropping experiment. *Agron J.* 67:233–306.

Dobermann, A., C. Witt, S. Abdulrachman, H. C. Gines, R. Nagrajan, T. T. Son, P. S. Tan, et al. 2003. Estimating indigenous nutrient supplies for site-specific nutrient management in irrigated rice. *Agron. J.* 95:924–935.

Doran, J. W., and D. M. Linn. 1994. Microbial ecology of conservation management systems. In *Soil Biology: Effects on Soil Quality*, eds. J. L. Hatfield and B. A. Stewart, pp. 1–27. Boca Raton, FL: Lewis.

Douglas, J. T., and M. J. Goss. 1982. Stability and organic matter content of surface aggregates under different methods of cultivation and in grassland. *Soil Tillage Res.* 2:155–175.

Eagle, A. J., J. A. Bird, W. R. Horwath, B. A. Linquist, S. M. Brouder, J. E. Hill, and C. V. Kessel. 2000. Rice yield and nitrogen utilization efficiency under alternate straw management practices. *Agron. J.* 92:1096–1103.

Eckhardt, U., B. Grimm, and S. Hortensteiner. 2004. Recent advances in chlorophyll biosynthesis and breakdown in higher plants. *Plant Mol. Biol.* 56:1–14.

Eghball, B., and J. W. Maranville. 1991. Interactive effects of water and nitrogen stresses on nitrogen utilization efficiency, leaf water status and yield of corn genotypes. *Commun. Soil Sci. Plant Anal.* 22:1367–1382.

Fageria, N. K. 1989. Tropical soils and physiological aspects of crops. Brasilia, Brazil: EMBRAPA-CNPAF.

Fageria, N. K. 1992. *Maximizing Crop Yields*. New York: Marcel Dekker.

Fageria, N. K. 2002b. Soil quality vs. environmentally-based agriculture. *Commun. Soil Sci. Plant Anal.* 33:2301–2329.
Fageria, N. K. 2003. Plant tissue test for determination of optimum concentration and uptake of nitrogen at different growth stages in lowland rice. *Commun. Soil Sci. Plant Anal.* 34:259–270.
Fageria, N. K. 2004. Dry matter yield and shoot nutrient concentrations of upland rice, common bean, corn and soybean in rotation on an Oxisol. *Commun. Soil Sci. Plant Anal.* 35:961–974.
Fageria, N. K. 2007a. Yield physiology of rice. *J. Plant Nutr.* 30:843–879.
Fageria, N. K. 2007b. Green manuring in crop production. *J. Plant Nutr.* 30:691–719.
Fageria, N. K. 2009. *The Use of Nutrients in Crop Plants*. Boca Raton, FL: CRC Press.
Fageria, N. K. 2010a. Root growth of upland rice genotypes as influenced by nitrogen fertilization. Paper presented at 19th World Congress of Soil Science, Brisbane, Australia, 1–6 August.
Fageria, N. K., and V. C. Baligar. 1996. Response of lowland rice and common bean grown in rotation to soil fertility levels on a varzea soil. *Fert. Res.* 45:13–20.
Fageria, N. K., and V. C. Baligar. 1999. Yield and yield components of lowland rice as influenced by timing of nitrogen fertilization. *J. Plant Nutr.* 22:23–32.
Fageria, N. K., and V. C. Baligar. 2001. Lowland rice response to nitrogen fertilization. *Commun. Soil Sci. Plant Anal.* 32:1405–1429.
Fageria, N. K., and V. C. Baligar. 2003a. Fertility management of tropical acid soils for sustainable crop production. In *Handbook of Soil Acidity*, ed. Z. Rengel, pp. 359–385. New York: Marcel Dekker.
Fageria, N. K., and V. C. Baligar. 2003b. Methodology for evaluation of lowland rice genotypes for nitrogen use efficiency. *J. Plant Nutr.* 26:1315–1333.
Fageria, N. K., and V. C. Baligar. 2005a. Nutrient availability. In *Encyclopedia of Soils in the Environment*, ed. D. Hillel, pp. 63–71. San Diego, CA: Elsevier.
Fageria, N. K., and V. C. Baligar. 2005b. Enhancing nitrogen use efficiency in crop plants. *Adv. Agron.* 88:97–185.
Fageria, N. K., V. C. Baligar, and B. A. Bailey. 2005. Role of cover crops in improving soil and row crop productivity. *Commun. Soil Sci. Plant Anal.* 36:2733–2757.
Fageria, N. K., V. C. Baligar, and R. B. Clark. 2002. Micronutrients in crop production. *Adv. Agron.* 77:185–268.
Fageria, N. K., V. C. Baligar, and R. B. Clark. 2006. *Physiology of Crop Production*. New York: Haworth Press.
Fageria, N. K., V. C. Baligar, and C. A. Jones. 2011. *Growth and Mineral Nutrition of Field Crops*, 3rd edition. Boca Raton, FL: CRC Press.
Fageria, N. K., and M. P. Barbosa Filho. 2001. Nitrogen use efficiency in lowland rice genotypes. *Commun. Soil Sci. Plant Anal.* 32:2079–2089.
Fageria, N. K., and H. R. Gheyi. 1999. *Efficient Crop Production*. Paraiba, Brazil: University of Paraiba.
Fageria, N. K., O. P. Morais, and A. B. Santos. 2010. Nitrogen use efficiency in upland rice. *J. Plant Nutr.* 33:1696–1711.
Fageria, N. K., and A. Moreira. 2011. The role of mineral nutrition on root growth of crop plants. *Adv. Agron.* 110:251–331.
Fageria, N. K., A. Moreira, and A. M. Coelho. 2011. Yield and yield components of upland rice as influenced by nitrogen sources. *J. Plant Nutr.* 34:361–370.
Fageria, N. K., and A. S. Prabhu. 2004. Blast control and nitrogen management in lowland rice cultivation. *Pesq. Agropec. Bras.* 39:123–129.
Fageria, N. K., and A. B. Santos. 1998. Rice and common bean growth and nutrient concentration as influenced by aluminum on an acid lowland soil. *J. Plant Nutr.* 21:903–912.

Fageria, N. K., and A. B. Santos. 2007. Response of irrigated rice to green manure and chemical fertilization in the state of Tocantins. *Rev. Bras. Eng. Agri. Ambien.* 11:387–392.

Fageria, N. K., and A. B. Santos. in press. Lowland rice genotypes evaluation for nitrogen use efficiency. *J. Plant Nutr.*

Fageria, N. K., A. B. Santos, and A. M. Coelho. 2011. Growth, yield and yield components of lowland rice as influenced by ammonium sulfate and urea fertilization. *J. Plant Nutr.* 34:371–386.

Fageria, N. K., A. B. Santos, and V. A. Cutrim. 2007. Yield and nitrogen use efficiency of lowland rice genotypes as influenced by nitrogen fertilization. *Pesq. Agropc. Bras.* 42:1029–1034.

Fageria, N. K., A. B. Santos, and V. A. Cutrim. 2008. Dry matter and yield of lowland rice genotypes as influenced by nitrogen fertilization. *J. Plant Nutr.* 31:788–795.

Fageria, N. K., A. B. Santos, and M. F. Moraes. 2010b. Influence of urea and ammonium sulfate on soil acidity indices in lowland rice production. *Commun. Soil Sci. Plant Anal.* 41:1565–1575.

Fageria, N. K., N. A. Slaton, and V. C. Baligar. 2003. Nutrient management for improving lowland rice productivity and sustainability. *Adv. Agron.* 80:63–152.

Fageria, N. K., T. Cobacci, and R. A. Reis, Jr. in press. Comparison of conventional and polymer coated urea as nitrogen sources for lowland rice production. *J. Plant Nutr.*

Fenn, L. B., and L. R. Hossner. 1985. Ammonium volatilization from ammonium or ammonium forming fertilizers. *Adv. Soil Sci.* 1:123–169.

Ferguson, R. B., R. M. Lark, and G. P. Slater. 2003. Approaches to management zone definition for use of nitrification inhibitors. *Soil Sci. Soc. Am. J.* 67:937–947.

Fillery, I. R. P., and P. L. G. Vlex. 1986. Reappraisal of the significance of ammonia volatilization as an N loss mechanism in flooded rice fields. *Fert. Res.* 9:79–98.

Fixon, P. E., and F. B. West. 2002. Nitrogen fertilizers: Meeting contemporary challenges. *Ambio.* 31:169–176.

Foster, H. L. 1971. Rapid routine soil and plant analysis without automatic equipment: Routine soil analysis. *East. Afr. Agric. For. J.* 37:160–170.

Francis, G. S., K. M. Bartley, and F. J. Tabley. 1998. The effect of winter cover crop management on nitrate leaching losses and crop growth. *J. Agric. Sci.* 131:299–308.

Freney, J. R., C. J. Smith, and A. R. Mosier. 1992. Effect of a new nitrification inhibitor (wax coated calcium carbonate) on transformation and recovery of fertilizer nitrogen by irrigated wheat. *Fert. Res.* 32:1–11.

Fried, M., F. Zsoldos, D. B. Vose, and I. L. Shatokhin. 1965. Characterizing the NO_3 and NH_4 uptake process of rice plants by use of ^{15}N labeled NH_4NO_3. *Physiol. Plant.* 18:313–320.

Gan, Y. T., P. R. Miller, B. G. McConkey, R. P. Zentner, F. C. Stevenson, and C. L. McDonald. 2003. Influence of diverse cropping sequences on durum wheat yield and protein in the semiarid northern Great Plains. *Agron. J.* 95:245–252.

Gao, Q., C. Li, G. Feng, J. Wang, Z. Cui, X. Chen, and F. Zhang. 2012. Understanding yield response to nitrogen to achieve high yield and high nitrogen use efficiency in rainfed corn. *Agron. J.* 104:165–168.

Gaudin, R., and J. Dupuy. 1999. Ammoniacal nutrition of transplanted rice fertilized with large urea granules. *Agron. J.* 91:33–36.

Glibert, P. M., J. Harrison, C. Heil, and S. Seitzinger. 2006. Escalating world-wide use of urea: A global change contributing to coastal eutrophication. *Biogeochemistry.* 77:441–463.

Golden, B. R., N. A. Slaton, R. J. Norman, C. E. Wilson, Jr., and R. E. DeLong. 2009. Evaluation of polymer coated urea for direct seeded, delayed-flood rice production. *Soil Sci. Soc. Am. J.* 73:375–383.

Goring, C. A. I. 1962. Control of nitrification of ammonium fertilizers and urea by 2-chloro-6-(trichloromethyl) pyridine. *Soil Sci.* 93:431–439.

Greenwood, D. J., and A. Walker. 1990. Modeling soil productivity and pollution. *Phil. Trans. R. Soc. London*. 42:103–128.
Gregory, P. J. 1994. Root growth and activity. In *Physiology and Determination of Crop Yield*, ed. G. A. Peterson, pp. 65–93. Madison, WI: ASA, CSSA, and SSSA.
Griggs, B. R., R. J. Norman, C. E. Wilson, Jr., and N. A. Slaton. 2007. Seasonal accumulation and partitioning of nitrogen-15 in rice. *Soil Sci. Soc. Am. J*. 71:745–751.
Guarda, G., S. Padovan, and G. Delogu. 2004. Grain yield, nitrogen use efficiency and baking quality of old and modern Italian bread-wheat cultivars grown at different nitrogen levels. *Eur. J. Agron*. 21:181–192.
Guindo, D., R. J. Norman, and B. R. Wells. 1994. accumulation of fertilizer nitrogen-15 by rice at different stages of development. *Soil Sci. Soc. Am. J*. 58:840–845.
Guindo, D., B. R. Wells, and R. J. Norman. 1994. Cultivar and nitrogen rate influence on nitrogen uptake and partitioning in rice. *Soil Sci. Soc. Am. J*. 58:840–845.
Hansson, A. C., and O. Andren. 1987. Root dynamics in barley, Lucerne, and meadow fescue investigated with a minirhizotron technique. *Plant Soil*. 103:33–38.
Harper, L. A., V. R. Catchpoole, R. Davis, and K. L. Weir. 1983. Ammonium volatilization: Soil, plant, and microclimate effects on diurnal and seasonal fluctuations. *Agron. J*. 75:212–218.
Harrell, D. L., B. S. Tubana, T. W. Walker, and S. B. Phillips. 2011. Estimating rice grain yield potential using normalized difference vegetation index. *Agron. J*. 103:1717–1723.
Hasegawa, H. 2003. High yielding rice cultivars perform best even at reduced nitrogen fertilizer rate. *Crop Sci*. 43:921–926.
Hauck, R. D., and K. K. Tanji. 1982. Nitrogen transfers and mass balances. In *Nitrogen in Agricultural Soils*, ed. F. J. Stevenson, pp. 891–925. Madison, WI: American Society of Agronomy.
Heard, J. R., E. J. Kladivko, and J. R. Mannering. 1988. Soil macro porosity, hydraulic conductivity and air permeability of silt soils under long-term conservation tillage in Indiana. *Soil Tillage Res*. 11:1–18.
Helmers, G. A., M. R. Langemeier, and J. A. Atwood. 1986. An economic analysis of alternative cropping systems for east-central Nebraska. *Am. J. Altern. Agric*. 1:153–158.
Helmers, G. A., C. F. Yamoah, and G. E. Varvel. 2001. Separating the impacts of crop diversification and rotations on risk. *Agron. J*. 93:1337–1340.
Heshmati, G. A., and M. Pessarakali. 2011. Threshold model in studies of ecological recovery in bermudagrass (*Cynodon dactlyon* L.) under nutrient stress conditions. *J. Plant Nutr*. 34:2183–2192.
Hoad, S. P., G. Russell, M. E. Lucas, and I. J. Bingham. 2001. The management of wheat, barley, and oat root systems. *Adv. Agron*. 74:193–246.
Hussain, F., K. F. Bronson, Y. Singh, B. Singh, and S. Peng. 2000. Use of chlorophyll meter sufficiency indices for nitrogen management of irrigated rice in Asia. *Agron. J*. 92:875–879.
Jarrell, W. M., and R. B. Beverly. 1981. The dilution effect in plant nutrition studies. *Adv. Agron*. 34:197–224.
Jayaweera, G. R., and D. S. Mikkelsen. 1990. Ammonia volatilization from flooded soil systems: A computer model I. Theoretical aspects. *Soil Sci. Soc. Am. J*. 54:1447–1455.
Kaizzi, K. C., J. Byalebeka, O. Semalulu, I. Alou, W. Zimwanguyizza, A. Nansamba, P. Musinguzi, et al. 2012. Sorghum response to fertilizer and nitrogen use efficiency in Uganda. *Agron. J*. 104:83–90.
Kanampiu, F. K., W. R. Raun, and G. V. Johnson. 1997. Effect of nitrogen rate on plant nitrogen loss in winter wheat varieties. *J. Plant Nutr*. 20:389–404.
Kapusta, G. 1979. Seedbed tillage and herbicide influence on soybean (*Glycine max*) weed control and yield. *Weed Sci*. 27:520–526.
Karlen, D. L., G. E. Varvel, and D. G. Bullock. 1994. Crop rotations for the 21st century. *Adv. Agron*. 53:1–45.

Kowalenko, C. G., and D. R. Cameron. 1977. Nitrogen transformations in soil–plant systems in three years of field experiments using tracer and non-tracer methods on an ammonium-fixing soil. *Can. J. Soil Sci.* 58:195–208.

Kurtz, L. T., L. V. Boone, T. R. Peck, and R. G. Hoeft. 1984. Crop rotations for efficient nitrogen use. In *Nitrogen in Crop Production*, ed. R. D. Hauck, pp. 295–306. Madison, WI: ASA, CSSA, and SSSA.

Lal, R., D. Vleeschauwer, and R. M. Ngaje. 1990. Changes in properties of a newly cleared tropical Alfisol as affected by mulching. *Soil Sci. Soc. Am. J.* 44:823–827.

Larney, F. J., and H. H. Janzen. 1997. A stimulated erosion approach to assess rates of cattle manure and phosphorus fertilizer for restoring productivity to eroded soils. *Agric. Ecosys. Environ.* 65:113–126.

Linn, D. M., and J. W. Doran. 1984. Effect of water-filled pore space on carbon dioxide and nitrous oxide production in tilled and non-tilled soils. *Soil Sci. Soc. Am. J.* 48:1267–1272.

Lofton, J., B. S. Tubana, Y. Kanke, J. Teboh, and H. Viator. 2012. Predicting sugarcane response to nitrogen using a canopy reflectance based response index value. *Agron. J.* 104:106–113.

Lopez-Bellido, R. J., and L. Lopez-Bellido. 2001. Efficiency of nitrogen in wheat under Mediterranean conditions; Effect of tillage, crop rotation and N fertilization. *Field Crops Res.* 71:31–46.

Lowrance, R., G. Vellidis, and R. K. Hubbard. 1995. Denitrification in a restored riparion forest wetland. *J. Environ. Qual.* 24:808–815.

Ma, B. L., L. M. Dwyer, and E. G. Gregorich. 1999. Soil nitrogen amendment effects on nitrogen uptake and grain yield of maize. *Agron. J.* 91:650–656.

MacRae, R. J., and G. R. Mehuys. 1985. The effect of green manuring on the physical properties of temperate area soils. *Adv. Soil Sci.* 3:71–94.

Mahajan, G., B. S. Chauhan, and M. S. Gill. 2011. Optimal nitrogen fertilization timing and rate in dry seeded rice in northwest India. *Agron. J.* 103:1676–1682.

Mahajan, G., and J. Timsina. 2011. Effect of nitrogen rates and weed control methods on weed abundance and yield of direct-seeded rice. *Arch. Agron. Soil Sci.* 57:239–250.

Makinde, E. A., and A. A. Agboola. 2002. Soil nutrient changes with fertilizer type in cassava-based cropping system. *J. Plant Nutr.* 25:2303–2313.

Marschner, H. 1995. *Mineral Nutrition of Higher Plants*. New York: Academic Press.

Martin, J. H., and W. H. Leonard. 1967. *Principles of Field Crop Production*. New York: Macmillan.

McCracken, D. V., S. M. Smith, J. H. Grove, C. T. MacKown, and R. L. Blevins. 1994. Nitrate leaching as influenced by cover cropping and nitrogen source. *Soil Sci. Soc. Am. J.* 58: 1476–1483.

Meisinger, J. J., V. A. Bandel, J. S. Angle, B. E. Okeefe, and C. M. Reynolds. 1992. Presidedress soil nitrate test evaluation in Maryland. *Soil Sci. Soc. Am. J.* 56:1527–1532.

Mengel, K., A. Kirkby, H. Kosegarten, and T. Appel. 2001. *Principles of Plant Nutrition*, 5th edition. Dordrecht, The Netherlands: Kluwer Academic.

Mengel, K., and M. Viro. 1978. The significance of plant energy status for the uptake and incorporation of NH_4-N by young rice plants. *Soil Sci. Plant Nutr.* 24:407–416.

Mikkelsen, D. S. 1987. Nitrogen budgets in flooded soils used for rice production. *Plant Soil.* 100:71–97.

Mikkelsen, D. S., S. K. De Datta, and W. N. Obcemea. 1978. Ammonium volatilization losses from flooded rice soils. *Soil Sci. Soc. Am. J.* 42:725–730.

Moll, R. H., E. J. Kamprath, and W. A. Jackson. 1982. Analysis and interpretation of factors which contribute to efficiency of nitrogen utilization. *Agron. J.* 74:562–564.

Moore, P. A., T. C. Daniel, D. R. Edwards, and D. M. Miller. 1995. Effect of chemical amendments on ammonia volatilization from poultry litter. *J. Environ. Qual.* 24:293–300.

Moore, P. A., J. T. Gilmour, and B. R. Wells. 1981. Seasonal patterns of growth and soil nitrogen uptake by rice. *Soil Sci. Soc. Am. J.* 45:875–879.

Morgan, K. T., K. E. Cushman, and S. Sato. 2009. Release mechanisms for slow and controlled release fertilizers and strategies for their use in vegetable production. *Horttechnology*. 19:10–12.

Mosier, A. R., J. W. Doran, and J. R. Freney. 2002. Managing soil denitrification. *J. Soil Water Conserv.* 57:50–513.

Noelisch, A. J., P. P. Motavalli, K. A. Nelson, and N. R. Kitchen. 2009. Corn response to conventional and slow release nitrogen fertilizer across a claypan landscape. *Agron. J.* 101:607–614.

Norman, R. J., D. Guindo, B. R. Wells, and C. E. Wilson, Jr. 1992. Seasonal accumulation and partitioning of N-15 in rice. *Soil Sci. Soc. Am. J.* 56:1521–1527.

Norman, R. J., C. E. Wilson, Jr., and N. A. Slaton. 2003. Soil fertilization and mineral nutrition in U.S. mechanized rice culture. In *Rice: Origin, History, Technology, and Production*, eds. C. W. Smith, and R. H. Dilday, pp. 331–411. Hoboken, NJ: John Wiley.

Norman, R. J., C. E. Wilson, Jr., N. A. Slaton, B. R. Griggs, J. T. Bushong, and E. E. Gbur. 2009. Nitrogen fertilizer sources and timing before flooding dry seeded, delayed flood rice. *Soil Sci. Soc. Am. J.* 73:2184–2190.

Ntamatungiro, S., R. J. Norman, R. W. McNew, and B. R. Wells. 1999. Comparison of plant measurements for estimating nitrogen accumulation and grain yield by flooded rice. *Agron. J.* 91:676–685.

Opoku, J., T. J. Vyn, and C. J. Swanton. 1997. Modified no-till systems for corn following wheat on clay soils. *Agron. J.* 89:549–556.

Oyer, L. J., and J. T. Touchton. 1990. Utilization legume cropping systems to reduce nitrogen fertilizer requirements for conservation-tilled corn. *Agron. J.* 82:1123–1127.

Pan, W. L., E. J. Kamprath, R. H. Moll, and W. A. Jackson. 1984. Prolificacy in corn: Its effects on nitrate and ammonium uptake and utilization. *Soil Sci. Soc. Am. J.* 48:1101–1106.

Papendick, R. I., and J. F. Parr. 1997. No-till farming: The way of the future for sustainable dry land agriculture. *Ann. Arid Zone.* 36:193–208.

Paul, E. A., and F. E. Clark. 1996. *Microbiology and Biochemistry*. London: Academic Press.

Pessarakali, M. 2010. *Handbook of Plant and Crop Stress*. Boca Raton, FL: CRC Press.

Peterson, T. A., T. M. Blackmer, D. D. Francis, and J. S. Schepers. 1993. *Using a Chlorophyll Meter to Improve N Management*. Coop. Ext. Serv. Neb-Guide G93-1171A. Lincoln, NE: University of Nebraska.

Pierce, F. J., and C. W. Rice. 1988. Crop rotation and its impact on efficiency of water and nitrogen use. In *Cropping Strategies for Efficient Use of Water and Nitrogen*, ed. W. L. Hargrove, pp. 21–42. Madison, WI: ASA, CSSA, and SSSA.

Porter, P. M., J. G. Lauer, W. E. Lueschen, J. H. Ford, T. R. Hoverstad, E. S. Oplinger, and R. K. Crookston. 1997. Environment affects the corn and soybean rotation effect. *Agron. J.* 89:441–448.

Ranells, N. N., and M. G. Wagger. 1996. Nitrogen release from grass and legume cover crop monocultures and bicultures. *Agron. J.* 88:777–782.

Rasse, D. P., J. T. Ritchie, W. R. Peterson, J. Wei, and A. J. M. Smucker. 2000. Rye cover crop and nitrogen fertilization effects on nitrate leaching in inbred maize fields. *J. Environ. Qual.* 29:298–304.

Raun, W. R., and G. V. Johnson. 1999. Improving nitrogen use efficiency for cereal production. *Agron. J.* 91:357–363.

Reddy, K. R., and W. H. Patrick, Jr. 1978. Utilization of labeled urea and ammonium sulfate by lowland rice. *Agron. J.* 70:465–467.

Reeves, D. W., C. W. Wood, and J. T. Touchton. 1993. Timing nitrogen applications for corn in a winter legume conservation-tillage system. *Agron. J.* 85:98–106.

Roberts, D. F., R. B. Ferguson, N. R. Kitchen, V. I. Adamchuk, and J. F. Shanahan. 2012. Relationships between soil based management zones and canopy sensing for corn nitrogen management. *Agron. J.* 104:119–129.

Robinson, D., D. J. Linehan, and D. C. Gordon. 1994. Capture of nitrate from soil by wheat in relation to root length, nitrogen inflow, and availability. *New Phytolo.* 128:297–305.

Rosielle, A. A., and J. Hamblin. 1981. Theoretical aspects of selection for yield in stress and non-stress environments. *Crop Sci.* 21:943–946.

Sagi, M., N. A. Savidov, N. P. Lvov, and S. H. Lips. 1997. Nitrate reductase and molybdenum cofactor in annual ryegrass as affected by salinity and nitrogen source. *Physiol. Plantarum.* 99:546–553.

Salsac, L., S. Chaillou, J. F. Morot-Gaudry, C. Lesaint, and E. Jolivoe. 1987. Nitrate and ammonium nutrition in plants. *Plant Physiol. Biochem.* 25:805–812.

Schillinger, W. F., R. J. Cook, and R. I. Papendick. 1999. Increased dryland cropping intensity with no-till barley. *Agron. J.* 1:744–752.

Schneider, U., and K. Haider. 1992. Denitrification and nitrate leaching losses in intensively cropped water shed. *Z. Pflanzenernahr. Bodenk.* 155:135–141.

Schnier, H. F., M. Dingkuhn, S. K. De Datta, E. P. Marquesse, and J. E. Faronilo. 1990. Nitrogen-15 balance in transplanted and direct seeded flooded rice as affected by different methods of urea application. *Biol. Fertl. Soils.* 10:89–96.

Sharpley, A. N., S. J. Smith, J. R. Williams, O. R. Jones, and G. A. Coleman. 1991. Water quality impacts associated with sorghum culture in the southern plains. *J. Environ. Q.* 20:239–244.

Shepherd, M. A. 1999. The effectiveness of cover crops during eight years of a UK sandland rotation. *Soil Use Manage.* 15:41–48.

Shoji, S., J. Delegado, A. Mosier, and Y. Miura. 2001. Use of controlled release fertilizers and nitrification inhibitors to increase nitrogen use efficiency and to conserve air and water quality. *Commun. Soil Sci. Plant Anal.* 32:1051–1070.

Sims, T. R., and G. A. Place. 1968. Growth and nutrient uptake of rice at different growth stages and nitrogen levels. *Agron. J.* 60:692–696.

Singer, J. W., and W. J. Cox. 1998. Agronomics of corn production under different crop rotations. *J. Prod. Agric.* 11:462–468.

Singh, U., J. K. Ladha, E. G. Castillo, G. Punzalan, A. Tirol-Padre, and M. Duqueza. 1998. Genotype variation in nitrogen use efficiency in medium and long duration rice. *Field Crops Res.* 58:35–53.

Singh, Y., B. Singh, and C. S. Khind. 1992. Nutrient transformations in soils amended with green manures. *Adv. Soil Sci.* 20:237–309.

Smail, V. 2001. *Enriching the Earth.* Cambridge, MA: MIT Press.

Smith, M. S., W. W. Frye, and J. J. Varco. 1987. Legume winter cover crops. *Adv. Soil Sci.* 7:95–139.

Snyder, C. S., and N. A. Slaton. 2002. Rice production in the United States: An overview. *Better Crops.* 16:30–35.

Soil Science Society of America. 2008. *Glossary of Soil Science Terms.* Madison, WI: Soil Science Society of America.

Shoji, S., and A. T. Gandeza. 1992. *Controlled Release Fertilizers with Polyolefin Resin Coating.* Sendai, Japan: Konno Printing Co.

Shoji, S., and H. Kanno. 1994. Use of polyolefin-coated fertilizers for increasing fertilizers efficiency and reducing nitrate leaching and nitrous oxide emissions. *Fer. Res.* 39:147–152.

Steiner, R. 1974. *Agriculture: A Course of Eight Lectures.* London: Bio-Dynamic Agriculture Assoc.

Stevenson, F. J. 1982. Origin and distribution of nitrogen in soil. In *Nitrogen in Agricultural Soils*, ed. F. J. Stevenson, pp. 1–42. Madison, WI: ASA, CSSA, and SSSA.

Stone, L. F., P. M. Silveira, J. A. A. Moreira, and L. P. Yokoyama. 1999. Rice nitrogen fertilization under supplemental sprinkler irrigation. *Pesq Agropec. Bras.* 34:927–932.

Stumpe, J. M., P. L. G. Vlek, and W. L. Lindsay. 1984. Ammoniacal volatilization from urea and urea phosphates in calcareous soils. *Soil Sci. Soc. Am. J.* 48:921–927.

Sweeney, D. W., and J. L. Moyer. 2004. In-season nitrogen uptake by grain sorghum following legume green manures in conservation tillage systems. *Agron. J.* 96:510–515.

Sweeten, J. M., and A. C. Mathers. 1985. Improving soils with livestock manure. *J. Soil Water Conserv.* 40:206–210.

Takenga, H. 1995. Internal factors in relation to nutrient absorption. In *Science of the Rice Plant: Physiology*, Vol. 2, eds. T. Matsuo, K. Kumazawa, R. Ishii, K. Ishihara, and H. Hirata, pp. 294–309. Tokyo: Food and Agriculture Policy Research Center.

Tem Berger, H. F. M., and J. J. M. Riethoven. 1997. Application of a simple rice nitrogen model. In *Plant Nutrition for Sustainable Food Production and Environment*, eds. T. Ando, K. Fuzita, T. Mae, H. Matsumoto, S. Mori, and J. Sekiya, pp. 793–798. Dordrecht, The Netherlands: Kluwer Academic.

Temnykh, S., W. D. Park, N. Ayers, S. Cartinhour, N. Hauck, L. Lipovich, Y. G. Cho, et al. 2000. Mapping and genome organization of microsatellite sequences in rice (*Oryza sativa* L.). *Theor. Appl. Genet.* 100:697–712.

Thomas, G. W., R. L. Blevins, R. E. Phillips, and M. A. Mcmahon. 1973. Effect of a killed sod mulch on nitrate movement and maize yield. *Agron. J.* 65:736–739.

Tiarks, A. E., A. P. Mazurak, and L. Chesnin. 1974. Physical and chemical properties of soil associated with heavy applications of manure from cattle feedlots. *Soil Sci. Soc. Am. Proc.* 38:826–830.

Tisdale, S. L., W. L. Nelson, and J. D. Beaton. 1985. *Soil Fertility and Fertilizers*, 4th edition. New York: Macmillan.

Thomason, W. E., W. R. Raun, G. V. Johnson, K. W. Freeman, K. J. Wynn, and R. W. Mullen. 2002. Production system techniques to increase nitrogen use efficiency in winter wheat. *J. Plant Nutr.* 25:2261–2283.

Torbert, H. A., K. N. Potter, and J. E. Morrison, Jr. 2001. Tillage system, fertilizer nitrogen rate, and timing effect on corn yields in the Texas Blackland Prairie. *Agron. J.* 93:1119–1124.

Tsai, C. Y., I. Dweikat, D. M. Huber, and H. L. Warren. 1992. Interrelationship of nitrogen nutrition with maize (*Zea mays* L.) grain yield, nitrogen use efficiency and grain quality. *J. Sci. Food Agric.* 58:1–8.

Unger, P. W., and T. M. McCalla. 1980. Conservation tillage systems. *Adv. Agron.* 33:1–58.

Varinderpal-Singh, B. S., H. S. T. Yadvinder-Singh, and R. K. Gupta. 2010. Need based nitrogen management using the chlorophyll meter and leaf color chart in rice and wheat in South Asia: A review. *Nutr. Cycling Agroecosyst.* 88:361–380.

Varvel, G. E., D. D. Francis, and J. S. Schepers. 1997. Ability for in season correction of nitrogen deficiency in corn using chlorophyll meter. *Soil Sci. Soc. Am. J.* 61:1233–1239.

Vyn, T. J., K. L. Janovicek, M. H. Miller, and E. G. Beauchamp. 1999. Soil nitrate accumulation and corn response to preceding small-grain fertilization and cover crops. *Agron. J.* 91:17–24.

Wagger, M. G. 1989. Cover crop management and nitrogen rate in relation to growth and yield of no-till corn. *Agron. J.* 81:533–538.

Waskom, R. M., D. G. Westfall, D. E. Spellman, and P. N. Soltanpur. 1996. Monitoring nitrogen status of corn with a portable chlorophyll meter. *Commun. Soil Sci. Plant Anal.* 27:545–560.

Watkins, B., J. Hignight, and C. E. Wilson, Jr. 2008. Estimating 2009 cost of production, rice, clay soils, eastern Arkansas. AG 1210-11-08. Univ. of Arkansas, Div. Agric. Coop. Ext. Serv. Little Rock, AR.

Watkins, K. B., J. A. Hignight, R. J. Norman, T. L. Roberts, N. A. Slaton, C. E. Wilson, Jr., and D. L. Frizzell. 2010. Comparison of economic optimum nitrogen rates for rice in Arkansas. *Agron. J.* 102:1099–1108.

Watson, C. J., H. Miller, P. Poland, D. J. Kilpatrick, M. D. B. Allen, M. K. Garrett, and C. B. Christianson. 1994. Soil properties and the ability of the urease inhibitor N-(n-butyl) thiphosphoric triamide (nBTPT) to reduce ammonia volatilization from surface applied urea. *Soil Biol. Biochem.* 9:1165–1169.

Weier, K. K., J. D. Doran, J. F. Power, and D. T. Walters. 1993. Denitrification and the dinitrogen/nitrous oxide ratio as affected by soil water, available carbon, and nitrate. *Soil Sci. Soc. Am. J.* 57:66–72.

Wilson, C. E., Jr., B. R. Wells, and R. J. Norman. 1994. Fertilizer nitrogen uptake by rice from urea-ammonium nitrate solution vs. granular urea. *Soil Sci. Soc. Am. J.* 58:1825–1828.

Wilson, C. R., Jr., N. A. Slaton, R. J. Norman, and D. M. Miller. 2001. Efficient use of fertilizer. In *Rice Production Handbook*, ed. N. A. Slaton, pp. 51–74. Little Rock, AR: University of Arkansas.

Wiren, N. V., S. Gazzarrini, and W. B. Frommer. 1997. Regulation of mineral nitrogen uptake in plants. In *Plant Nutrition for Sustainable Food Production and Environment*, eds. T. Ando, K. Fujita, T. Mae, H. Tsumoto, S. Mori, and J. Sekiya, pp. 41–49. Dordrecht, The Netherlands: Kluwer Academic.

Wood, C. W., and J. H. Edwards. 1992. Agroecosystem management effects on soil carbon and nitrogen. *Agric. Ecosyst. Environ.* 39:123–138.

Wood, C. W., D. W. Reeves, R. R. Duffied, and K. L. Edmistern. 1992. Field chlorophyll measurements for evaluation of corn nitrogen status. *J. Plant Nutr.* 15:487–500.

Wu, L., R. Li, Q. Shu, H. Zhao, D. Wu, J. Li, and R. Wang. 2011. Characterization of a new green-revertible albino mutant in rice. *Crop Sci.* 51:2706–2715.

Yang, X. 1987. Physiological mechanisms of nitrogen efficiency in hybrid rice. PhD Dissertation. Hangzhou, China: Zhejiang Agricultural University.

Yoshida, S. 1981. *Fundamentals of Rice Crop Science*. Los Baños, Philippines: IRRI.

3 Phosphorus

3.1 INTRODUCTION

Phosphorus (P) deficiency is one of the major limitations for crop production, particularly in low-input agricultural systems around the world. The problem is widespread for upland and lowland rice grown on acidic soils. In the tropics, the soils used for rice production are mostly Oxisols and Ultisols, which have naturally low levels of P and high immobilization capacity due to Al and Fe oxides (Fageria, Slaton, and Baligar 2003). Because of low natural phosphorus and high immobilization capacity, a heavy dose of P is needed for these soils to achieve high production (Yost et al. 1979; Fageria, Barbosa Filho, and Carvalho 1982). Phosphorus deficiency has been identified as one of the major limiting factors for rice production in highly weathered Oxisols and Ultisols worldwide (Sanchez and Salinas 1981; Fageria, Barbosa Filho, and Carvalho 1982; Haynes 1984; Fageria 2001). For example, in Brazil, upland rice is grown mostly on Oxisols and Ultisols in the centrally located Cerrado region, which makes up about 22% of Brazil's total land area. These soils are acidic and have low fertility and low P (Lopes and Cox 1977; Fageria, Barbosa Filho, and Carvalho 1982). Brazil also has some 35 million ha of lowlands, known locally as *Varzea*, that represent one of the world's largest agricultural lowland regions. However, at present, less than 2% of these lowlands are used for crop production. Lowland rice is the main crop grown during the rainy season, with other crops being grown during the dry season. Varzea soils are acidic, and P fixation is a major problem (Fageria et al. 1991, 1997; Fageria and Baligar 1996). Adequate P application rate is an important factor for rice production on these soils. Table 3.1 shows the influence of applied P on straw yield, grain yield, and panicle density for upland rice grown on central Brazilian Oxisol. All were significantly increased with the addition of P. Figures 3.1 and 3.2 show the dry matter yield of shoots and grain for 20 upland rice genotypes at 25 and 200 mg P kg^{-1} soil. There were significant differences among genotypes for shoot dry weight and grain yield. However, compared with a low P level, a higher P rate significantly increased shoot and grain yield, thereby indicating the importance of P fertilization for upland rice production.

Because of P fixation, plants rarely absorb more than 20% of the total P fertilizer applied (Friesen et al. 1997). Uptake of P by plants is governed by the soil's ability to supply P to plant roots and by the desorption characteristics of the soil (Roy and De Datta 1985). The supply of P to plant roots depends on the concentration of inorganic P in the soil solution and on the capacity of the soil to maintain this concentration (Fageria, Slaton, and Baligar 2003).

Application of P is often essential for profitable agricultural production. However, accumulation of soil P in excess of crop needs has the potential to enrich surface runoff

TABLE 3.1
Response of Upland Rice to Phosphorus Application on Brazilian Oxisol

P Level (mg kg^{-1})	Straw Yield (g 4 $plants^{-1}$)	Grain Yield (g 4 $plants^{-1}$)	Panicle Number (4 $plants^{-1}$)
0	48.7	16.7	21
50	54.2	53.9	23
100	57.6	7.9	24
150	58.2	56.2	24
200	62.0	57.1	25
400	83.5	58.9	34
R^2	0.99*	0.90*	0.99*

Source: Fageria, N. K. 2001. Nutrient management for improving upland rice productivity and sustainability. *Commun. Soil Sci. Plant Anal.* 32:2603–2629.
*Significant at the 1% probability level.

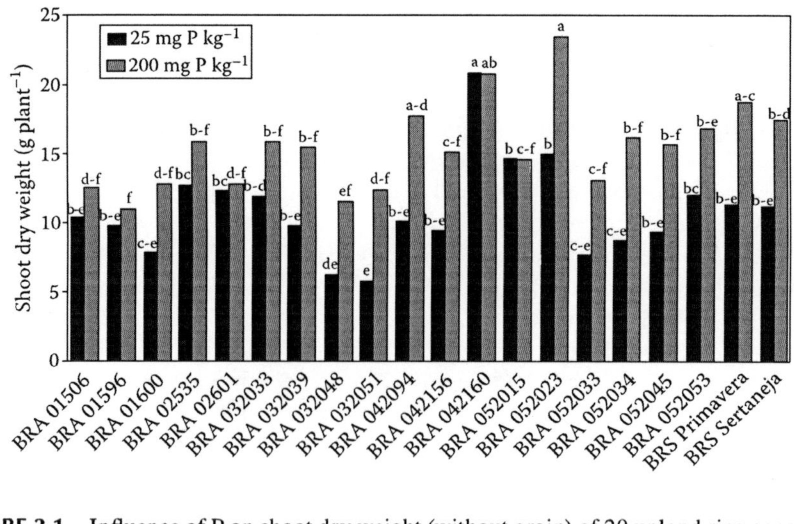

FIGURE 3.1 Influence of P on shoot dry weight (without grain) of 20 upland rice genotypes grown on Brazilian Oxisol. Means followed by the same letter on the same on the same bar are not significantly different at the 5% probability level by Tukey's test.

with P causing eutrophication. Eutrophication has been identified as the main cause of impaired surface water quality (Fageria, Slaton, and Baligar 2003). Because of the increased growth of undesirable algae and aquatic weeds and the oxygen shortages caused by their death and decomposition, eutrophication restricts water use for fisheries, recreation, industry, and drinking (Abrams and Jarrell 1995). Therefore, appropriate P management is important not only for higher rice yields but also for environmental protection. The objective of this chapter is to discuss the importance of P in upland and

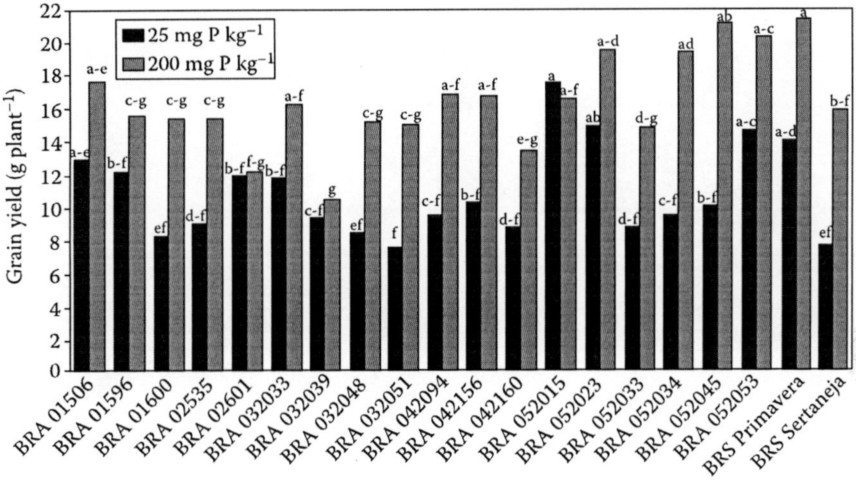

FIGURE 3.2 Influence of P on grain yield of 20 upland rice genotypes grown on Brazilian Oxisol. Means followed by the same letter on the same on the same bar are not significantly different at the 5% probability level by Tukey's test.

lowland rice production and the management strategies necessary to achieve maximum economic yield and to maintain sustainability of rice-based cropping systems.

3.2 CYCLE IN SOIL–PLANT SYSTEM

Knowledge of the P cycle with regard to a soil–plant system is important to understand its transformation, immobilization, losses, and availability to plants. A simplified version of the P cycle in a soil–plant system is shown in Figure 3.3. A major part of soluble P is immobilized in the root zone, and a small part is absorbed by the plants (Fageria 2009). The higher degree of P fixation occurs at very low and very high pH. In acidic soils, immobilization of P occurs due to high levels of aluminum, iron, and manganese ions that form insoluble hydroxyl phosphate precipitates, as shown in the following equations:

$$Al^{3+} + H_2PO_4^- \text{ (soluble P)} + 2H_2O \leftrightarrow Al(OH)_2H_2PO_4 \text{ (insoluble P)} + 2H^+ \quad (3.1)$$

$$Fe^{3+} + H_2PO_4^- \text{ (soluble P)} + 2H_2O \leftrightarrow Fe(OH)_2H_2PO_4 \text{ (insoluble P)} + 2H^+ \quad (3.2)$$

$$Mn^{3+} + H_2PO_4^- \text{ (soluble P)} + 2H_2O \leftrightarrow Mn(OH)_2H_2PO_4 \text{ (insoluble P)} + 2H^+ \quad (3.3)$$

Similarly, when lime is added to the acidic soils to correct the soil acidity, P fixation can occur, as shown in the following equation (Brady and Weil 2002):

$$\underset{\substack{\text{Monocalcium phosphate}\\\text{(soluble)}}}{Ca(H_2PO_4)_2 \cdot H_2O} + 2H_2O \xrightarrow{CaCO_3} \underset{\substack{\text{Dicalcium phosphate}\\\text{(slightly soluble)}}}{2(CaHPO_4 \cdot 2H_2O)} + CO_2\uparrow \xrightarrow{CaCO_3} \underset{\substack{\text{Tricalcium phosphate}\\\text{(very low solubility)}}}{Ca_3(PO_4)_2} + CO_2\uparrow + 5H_2O \quad (3.4)$$

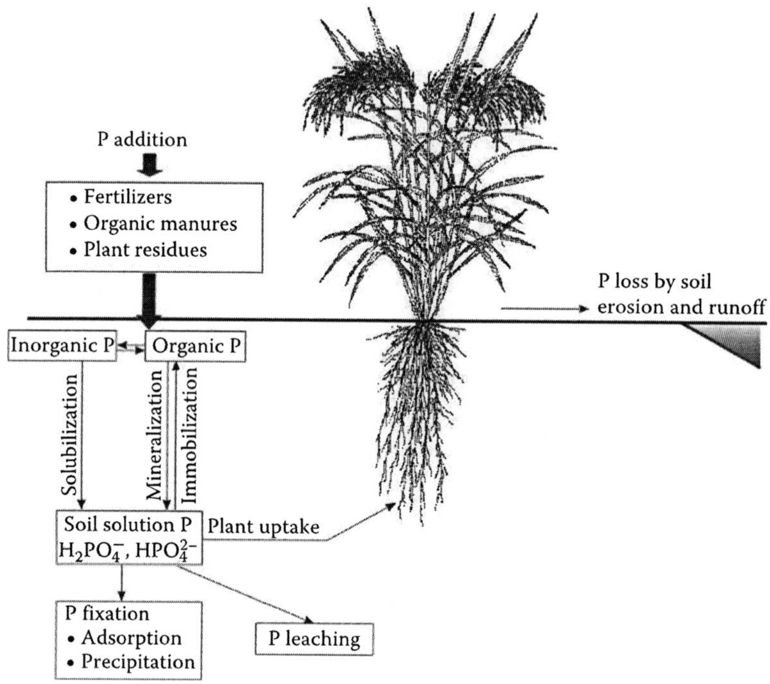

FIGURE 3.3 Phosphorus cycle in soil–plant system. (From Fageria, N. K. 2009. *The Use of Nutrients in Crop Plants*. Boca Raton, FL: CRC Press.)

Maximum P availability for plants generally is in the pH range of 6.0–7.0 (Brady and Weil 2002). Figure 3.4 shows P fixation as a function of reaction time of soluble P in Oxisol. The extractable P was reduced when reaction time was increased from zero to 28 days. However, it was more or less constant when reaction time was increased from 28 days to 43 days. In addition, some P is also lost by erosion and runoff from the soil surface. Soil P can be divided into organic and inorganic fractions with the inorganic P fraction being recognized as the pool that controls P availability to plants. The inorganic P fraction is made up of at least five basic categories, including aluminum phosphates (Al–P), iron phosphates (Fe–P), calcium phosphates (Ca–P), reductant soluble or the occluded phosphates (RS–P), and the readily available orthophosphate forms (PO_4^{-3}, HPO_4^{-2}, or $H_2PO_4^-$) of P (Chang and Jackson 1957; Fageria, Slaton, and Baligar 2003). The predominant form of orthophosphate present in the soil solution changes with soil pH. Orthophosphoric acid dissociates into three forms, as shown in the following equations (Yoshida 1981):

$$H_3PO_4 \leftrightarrow H_2PO_4^- + H^+ \tag{3.5}$$

$$H_2PO_4^- \leftrightarrow HPO_4^{2-} + H^+ \tag{3.6}$$

$$HPO_4^{2-} \leftrightarrow PO_4^{3-} + H^+ \tag{3.7}$$

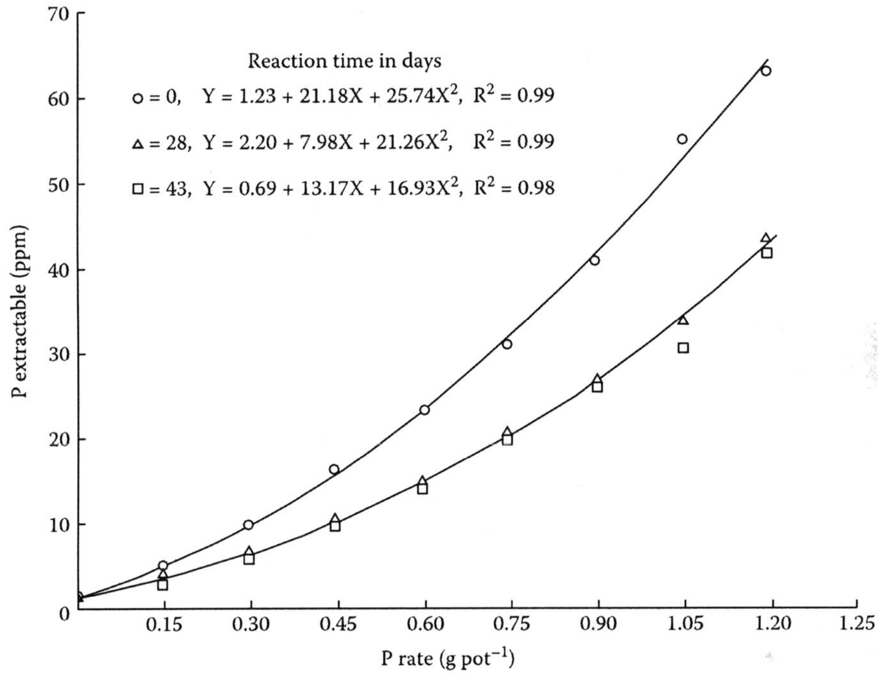

FIGURE 3.4 Relationship between P rate and extractable soil P at three reaction timings.

At pH 7.2, $H_2PO_4^-$ and HPO_4^{2-} ions are present in equal amounts; at pH 5, $H_2PO_4^-$ is the dominant species and $H_2PO_4^{2-}$ is practically absent (Yoshida 1981).

The reduction process that occurs in the soil following flooding normally increases the P availability to rice. Therefore, on many soils, P availability is not a yield-limiting factor. Rice yields may not respond to P fertilization, but upland crops growing on the same soil, such as corn and soybeans, might show dramatic responses to P fertilization. However, the response depends on the soil type. For example, in Brazilian Inceptisols, flooded rice responds to P fertilization (Fageria and Barbosa Filho 2007; Fageria and Santos 2008; Fageria, Baligar, and Jones 2011). Phosphorus deficiency has been reported in several upland crops that follow rice in planting rotation (Willett 1982; Wells et al. 1995). The alternate anaerobic–aerobic soil conditions reduce the availability of P to upland crops that follow rice in the rotation. The soils' P sorption capacity and the bonding energy of P increase under alternate anaerobic–aerobic conditions (Sanyal and De Datta 1991). Flooding decreases the crystallinity of ferrous hydroxides; this increases their sorption capacity and the insoluble Fe–P fraction and reduces P desorption.

Iron phosphates are the primary source of P for lowland rice because their availability is quickly affected by the anaerobic conditions created by flooding (Patrick and Mahapatra 1968; Ponnamperuma 1972; Goswami and Banerjee 1978). Willett (1986, 1989) reported that P availability increases after flooding from the (1) reductive dissolution of ferric oxides; (2) the liberation of sorbed- and RS–P; (3) changes in soil pH that increase the solubility of Fe-, Al-, and Ca-P; and (4) the desorption of surface P.

The relatively insoluble ferric phosphates are reduced to the more soluble ferrous phosphates, resulting in hydrolysis of P compounds. For acidic soils, soil pH generally increases following soil submergence, and this increases the solution concentrations of Fe and Al phosphates. Flooding the soil usually reduces hydrated ferric oxides to ferrous hydroxides, which releases part of the RS–P (Patrick and Mikkelsen 1971).

Turner and Gilliam (1976) illustrated how flooding the soil significantly influenced P availability on calcareous soils where Ca–P predominates over Fe–P. They found that increased P availability in flooded or saturated soils occurred, at least partially, from the decreased tortuosity that increased the P diffusion rate by 10- to 100-fold. However, Sah and Mikkelson (1986) observed that the Ca–P fraction of some, but not all, clay soils in California increased for several weeks after flooding from the formation and precipitation of insoluble Ca–P, which would then decrease the availability of P after flooding.

An interesting aspect of the P chemistry of flooded soils is that more P is released from the soil into the solution under reduced conditions than under oxidized conditions, if the soil solution is initially low in phosphorus, but reduced soils also have a greater sorption capacity for P (Patrick and Reddy 1978). Roy and De Datta (1985) and Khalid, Patrick, and Peterson (1979) suggested that rice required 0.12–0.20 mg P L^{-1} in the soil solution for optimum growth. Hossner, Freeouf, and Folsom (1973) concluded that the minimum soil solution P concentration required to produce 90% rice yield was >0.10 mg P L^{-1}.

Several forms of soil Fe are highly correlated with P sorption (Willett and Higgins 1978; Khalid, Patrick, and Peterson 1979). Soil P released under reduced conditions has been related to oxalate-extractable Fe (Khalid, Patrick, and Peterson 1979; Shahandeh, Hossner, and Turner 1994). Evans and Smillie (1976) and Fox and Kamprath (1970) reported that soil clay and available Fe content strongly influenced P sorption under aerobic soil conditions. Ammonium oxalate-extractable Fe was highly correlated with P sorption under anaerobic conditions (Willett and Higgins 1978; Khalid, Patrick, and Peterson 1979; Shahandeh, Hossner, and Turner 1994). Shahandeh, Hossner, and Turner (1994) reported that 84% of the added P was sorbed under anaerobic conditions when ammonium oxalate-extractable Fe was ~3000 mg Fe kg^{-1}. Oxalate extracts poorly crystalline Fe oxides (Campbell and Schwertmann 1984), which are the most reactive Fe oxides in the soil because of their small size and high surface area (Shahandeh, Hossner, and Turner 1994).

Although our understanding of P availability for rice grown under flooded soil conditions has increased during the past 40 years, this remains an area that requires additional research. The general relationships of P availability are well characterized but need to be better defined with regard to specific physical and chemical properties of soil and then correlated to plant uptake of P (Fageria, Slaton, and Baligar 2003).

3.3 FUNCTIONS

Phosphorus is one of the major essential nutrients needed for the growth and reproduction of higher plants. It is required for the synthesis of phospholipids, nucleotides, adenosine triphosphate (ATP), glycophosphates, and other phosphate esters. Plant growth and yield are dramatically reduced by P deficiency because P is a component of high-energy

FIGURE 3.5 (See color insert.) Upland rice growth on Brazilian Oxisol at three P levels.

compounds like ATP and is an essential component of the genetic material required for seed production (Fageria, Slaton, and Baligar 2003). Phosphorus increases the number of tillers and, consequently, the number of panicles in upland as well as lowland rice (Fageria, Zimmermann, and Lopes 1977; Fageria, Barbosa Filho, and Carvalho 1982). Phosphorus also increases the leaf area index (LAI), which is responsible for solar radiation interception and photosynthesis (Fageria, Barbosa Filho, and Carvalho 1982). Figure 3.5 shows growth of upland rice at three P levels. Similarly, Figure 3.6 shows growth of two upland rice genotypes at two P levels. In both the genotypes, growth was higher at the higher P level compared with the lower P level. Figure 3.7 shows root growth of upland rice genotypes at two P levels. Root growth of two genotypes was more vigorous in volume compared with lower P level. Hence, it can be concluded that P improves the growth of tops and roots of upland rice grown on Brazilian Oxisols.

3.4 DEFICIENCY SYMPTOMS

Phosphorus deficiency symptoms occur primarily in seedling rice at the onset of tillering when the rice begins to rapidly accumulate dry matter (Fageria, Slaton, and Baligar 2003). Symptoms include severe stunting with plants having erect and dark

FIGURE 3.6 Growth of two upland rice genotypes at two P levels.

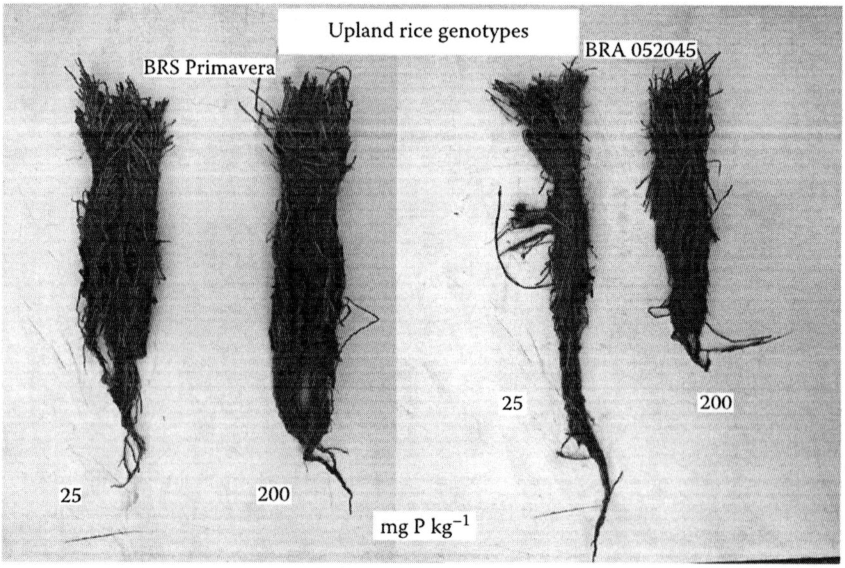

FIGURE 3.7 Root growth of two upland rice genotypes at two P levels.

green leaves. Phosphorus deficiency reduces seedling height, tiller number, stem diameter, leaf size, and leaf duration. When P is deficient, cell and leaf expansion are retarded more than chlorophyll formation. Thus, the chlorophyll content per unit leaf area increases, but the photosynthetic efficiency per unit of chlorophyll decreases (Marschner 1995). Phosphorus is not a component of chlorophyll, so the concentration of chlorophyll in P-deficient rice becomes comparatively high, and the leaf color changes from green to dark green. If P deficiency persists, the older leaves may turn orangish and desiccate from the leaf tip back toward the base. Rice maturity can be delayed by as much as 10–12 days by P deficiency (Fageria 1980). For many crops, Phosphorus deficiency symptoms include a reddish or purple tint on leaves due to the accumulation of anthocyanins (Hewitt 1963). However, the leaf purpling symptom has not been observed in P-deficient rice (Fageria and Barbosa Filho 1994; Fageria and Gheyi 1999). Phosphorus-deficient plants are more susceptible to some rice diseases. In Arkansas, excessive brown spot (*Bipolaris oryzae*) is commonly observed on nutritionally stressed rice (Fageria, Slaton, and Baligar 2003). Phosphorus fertilization has also significantly increased rice root growth, the number of panicles, and the grain weight of rice grown on P-deficient soils (Fageria and Gheyi 1999). When P is deficient, rice does not respond to the application of N, K, or other fertilizers. Color photographs depicting typical P deficiency symptoms of rice are available in various publications (Wallace 1961; Mueller 1974; Cheaney and Jennings 1975; Ishizuka 1978; Yoshida 1981; Fageria 1984; Bennett 1993; Fageria and Barbosa Filho 1994; Fageria, Baligar, and Jones 2011). Figure 3.8 through 3.10 show P deficiency in upland rice grown under field and greenhouse conditions.

3.5 UPTAKE IN PLANT TISSUE

Nutrient concentration (content per unit of dry weight) and uptake (concentration × dry matter yield) are important plant tissue analysis parameters in relation to the mineral status of plants. Generally, concentration results are used to identify the nutrient deficiency or sufficiency level in plants at a given growth stage. Nutrient uptake data are used to determine the amount of nutrients removed by a crop in order to replenish the removed nutrients from the soil and maintain soil fertility for sustainable crop production. The concentration of a given nutrient in plants is mostly affected by the growth stage and is determined by analyzing plant part. Yield level is the most important factor affecting nutrient uptake by crop plants.

Because phosphorus is a mobile element inside the plant, the P concentration of individual leaves generally declines as leaf age increases (Fageria, Slaton, and Baligar 2003). The top leaves have the highest P concentration, and the bottom leaves have the lowest, especially when plant available P is limited (Westfall, Flinchum, and Stansel 1973). Sims and Place (1968) reported that tissue P concentration varied less across plant development stages than N concentrations, which is generally true when P is not a growth-limiting factor (Fageria, Slaton, and Baligar 2003). This author studied P concentration in the shoot (straw) of upland rice, corn, soybeans, and dry beans during their growth cycle (Figure 3.11). In these four crop species, there was a significant quadratic decrease in the concentration of P with the advancement of

FIGURE 3.8 Upland rice plot on Brazilian Oxisol with P deficiency in the front (0 kg P_2O_5 ha^{-1}) and without P deficiency in the back (100 kg P_2O_5 ha^{-1}).

plant age. At the first sampling or at the beginning of growth stage (20 days after sowing), the P concentration in the shoot was almost equal in two cereals (rice and corn) and also almost equal in two legumes (soybeans and dry beans). Overall, it was lower in legumes compared with cereals. At physiological maturity, the P concentration in two cereals was almost equal (1 g kg^{-1} or 0.1%). However, among legumes, it was higher in soybeans (about 2 g kg^{-1} or 0.2%) compared with dry beans (about 1.2 g kg^{-1} or 0.12%).

The decrease in P concentration during growth stages indicates that plant tissue sampling for food crops should be taken at different growth stages. Furthermore,

FIGURE 3.9 (See color insert.) Upland rice plants without P and with P.

FIGURE 3.10 Growth of upland rice genotypes on Brazilian Oxisol at low and high P levels.

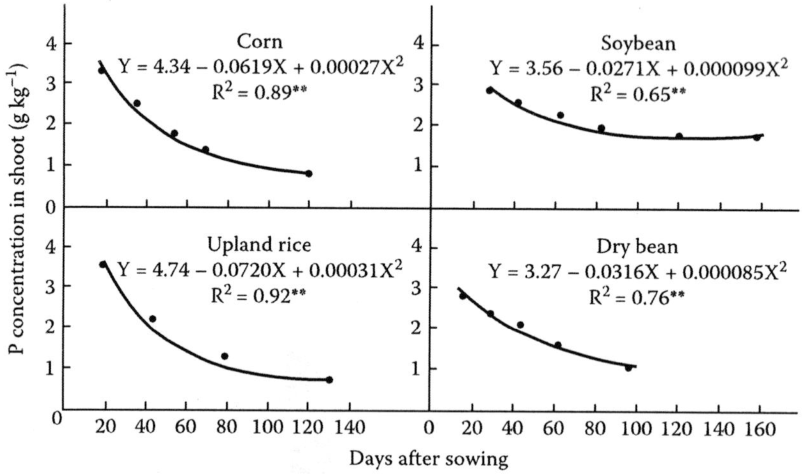

FIGURE 3.11 Relationship between plant age and P concentration in the shoot (straw) of four crop species.

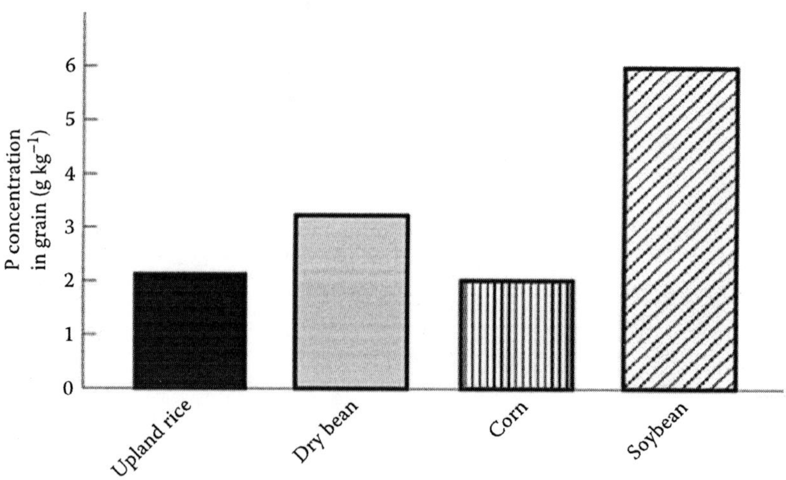

FIGURE 3.12 Phosphorus concentration in the grain of four crop species.

the decrease in P concentration with the advancement of plant age was associated with an increase in dry matter yield, which is known as the dilution effect in the mineral nutrition of plants (Fageria, Baligar, and Jones 2011). The decrease in plant nutrient concentration (including P) with the advancement of plant age is widely reported (Fageria 1992; Osaki 1995; Fageria, Baligar, and Clark 2006; Fageria, Baligar, and Jones 2011). The P concentration in the grain of these four crop species was also determined (Figure 3.12); the order was soybean > dry bean > rice = corn.

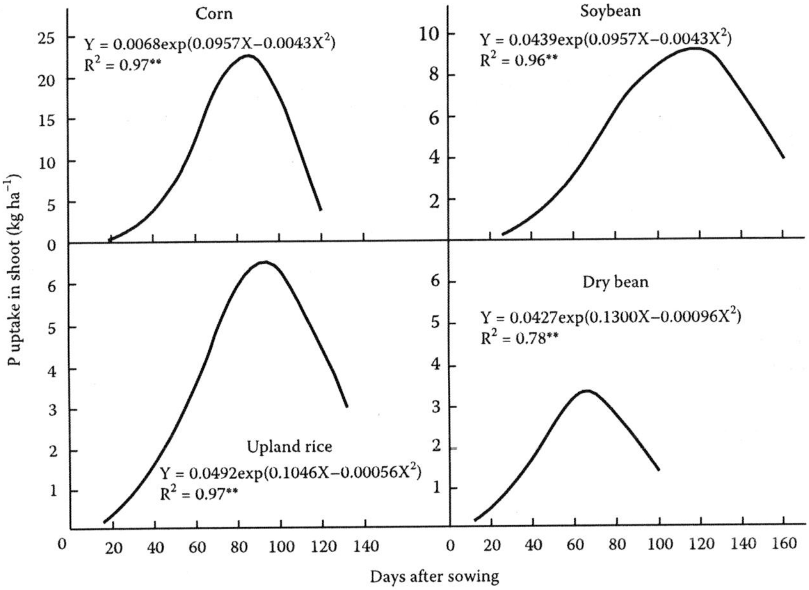

FIGURE 3.13 Relationship between plant age and P uptake in the shoot (straw) of four crop species.

Hence, legumes require a higher amount of P in the grain compared with cereals. Osaki et al. (1991) reported higher P harvest index (PHI) and P concentration in the seeds of soybeans compared with rice and corn. Data related to P uptake in the shoot (straw) of four crop species during the growth cycle are presented in Figure 3.13. The increase in P uptake was quadratic with the advancement of plant age. Maximum P uptake was at 80 days after sowing in corn, 90 days after sowing in upland rice, 120 days after sowing in soybeans, and 65 days after sowing in dry beans. The amount of P uptake was corn > soybean > upland rice > dry bean. The uptake of P in the straw of the four crop species followed the pattern of dry matter accumulation. The decrease in P uptake during the grain-filling growth stages in four crops was related to the export of P in the grain (Figure 3.14). Higher uptake of P in the grain of corn and upland rice was related to a higher grain yield for these crops compared with soybeans and dry beans (Figure 3.15).

Fageria, Slaton, and Baligar (2003) also determined P uptake in the straws of lowland rice during growth stages and in the grain at maturity. Table 3.2 shows the P accumulation in the shoots and grain of the lowland rice cultivar "Javaé" as influenced by plant age and P application rates on Brazilian Inceptisol. Total P uptake by lowland rice during the growing season increases with plant development. This increase is quadratic for all the growth stages. Fageria and Santos (2008) also reported that P uptake in the straws and grain followed quadratic responses in lowland rice. The total aboveground P uptake by a high-yielding rice crop may approach 60 kg P ha^{-1}, but it more commonly ranges from 25 to 50 kg P ha^{-1} with 60–75% of the plant's total P contained in the panicles at maturity.

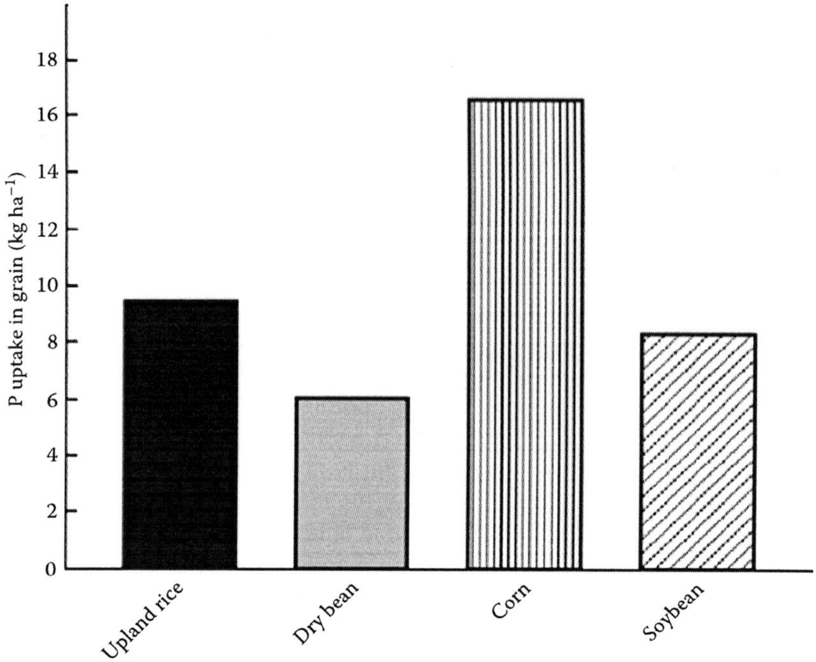

FIGURE 3.14 Phosphorus uptake in the grain of four crop species.

Seasonal P uptake and dry matter accumulation tend to follow similar patterns. The accumulation of P is closely related to plant age. Fageria, Santos, and Baligar (1997) showed that plant age accounted for 70% of the variability in seasonal P accumulation. For P-deficient soils, total P uptake also increases as P fertilizer rate increases (Table 3.2). In straw, P accumulates with increasing plant age until flowering. After flowering, during the ripening period, straw P content decreases as P is translocated from the straw to the developing grain. A significant portion of P is either translocated from the root system or absorbed between flowering and maturity to satisfy grain P requirements. At maturity, about two-thirds of the total plant P can be accounted for in the aboveground biomass at flowering (Table 3.2). The straw P content at maturity is only about one-third of the total P content at flowering and the grain content at maturity.

3.6 HARVEST INDEX

The PHI is important in determining the portioning of P in the rice plant. The average PHI (grain P/straw plus grain P) generally ranges from 0.60 to 0.75 (Fageria, Slaton, and Baligar 2003). Fageria (2009) determined the grain harvest index (GHI) (grain yield/grain plus straw yield) and the PHI of four crop species (Table 3.3). GHI values varied from 0.42 to 0.64. The values of PHIs were 0.77 for upland rice, 0.79 for corn, 0.86 for dry beans, and 0.89 for soybeans (Table 3.6). The GHI and

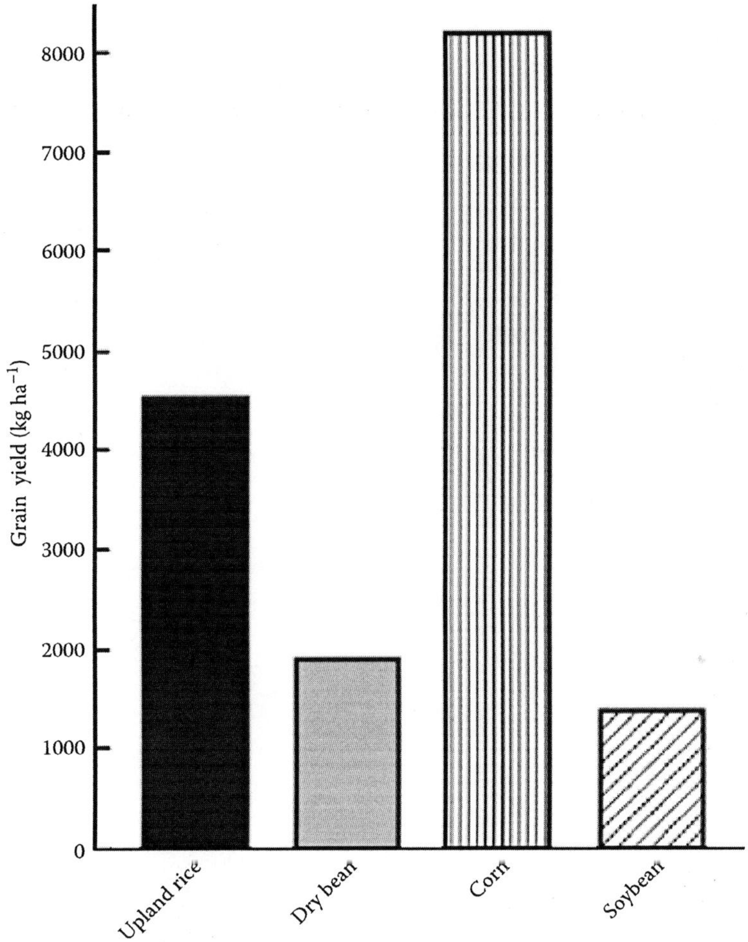

FIGURE 3.15 Grain yield of four crop species.

the PHI were higher for legumes than for cereals, indicating that P requirements are higher for legumes than for cereals. The harvest of cereal and legume grains removes a significant portion of the total P taken up during the growing season. Thus, a major portion of P accumulated by rice and other crops cannot be recycled for use by those crops that follow in the rotation, even though the straw of the rice and the other crops is incorporated back into the soil (Fageria, Slaton, and Baligar 2003).

3.7 USE EFFICIENCY

Knowledge of P use efficiency is very important to determine what happened to the applied P fertilizers in the rice rhizosphere and how much P is utilized by the plants in dry matter or grain yield production. This information is useful for applied

TABLE 3.2
Phosphorus Accumulation in the Shoot and Grain (kg ha^{-1}) of the Lowland Rice Cultivar "Javaé" as Influenced by Plant Age and P Application Rates on Brazilian Inceptisol[a]

P Fertilizer	Days after Sowing (Growth Stage)							
Rate	18 (IT)	35 (AT)	63 (IP)	81 (B)	91 (F)	119 (PM)	119 (grain)	Total
0	0.20	0.6	4.6	9.9	12.8	4.5	16.4	20.1
87	0.33	1.1	6.4	13.8	16.5	4.6	19.0	23.6
175	0.43	1.5	7.4	13.5	18.1	6.5	20.1	26.6
262	0.53	2.1	7.8	15.3	22.4	8.7	25.1	33.7
350	0.60	2.3	8.0	14.4	18.4	6.4	25.0	31.4
437	0.61	1.7	7.0	13.0	19.0	7.5	23.1	30.6
Average	0.45	1.5	6.9	13.3	17.9	6.3	21.4	27.8
R^2	0.99**	0.91**	0.95**	0.88**	0.83**	0.67*	0.90**	
Regression	Q	Q	Q	Q	Q	Q		

Source: Adapted from Fageria, N. K., N. A. Slaton, and V. C. Baligar. 2003. Nutrient management for improving lowland rice productivity and sustainability. *Adv. Agron.* 80:63–152.

Note: IT, initiation of tillering; AT, active tillering; IP, initiation of panicle; B, booting; F, flowering; PM, physiological maturity; Q, quadratic. Phosphorus rates were broadcast.

[a] Values are measured in kg P ha^{-1}.

*,**Significant at the 5 and 1% probability levels, respectively.

TABLE 3.3
Grain Harvest Index and Phosphorus Harvest Index of Four Crop Species

Crop Species	GHI	PHI
Upland rice	0.42	0.77
Corn	0.43	0.79
Dry bean	0.64	0.86
Soybean	0.53	0.89

Source: Fageria, N. K. 2009. *The Use of Nutrients in Crop Plants*. Boca Raton, FL: CRC Press.

Note: GHI = grain yield/grain plus straw yield. PHI = P uptake in grain/P uptake in grain plus straw.

chemical fertilizer management to reduce crop production costs and environmental pollution. The definitions of P use efficiency and their methods of calculation are given in Table 3.4.

With the exception of physiological and utilization efficiencies, P use efficiency calculated as agronomic efficiency (AE), physiological efficiency (PE),

TABLE 3.4
Definitions and Methods of Calculating P Use Efficiency

P Use Efficiency	Definitions and Formulae for Calculation
Agronomic efficiency (AE)	*Agronomic efficiency* is defined as the economic production obtained per unit of nutrient applied. It can be calculated by: AE (mg mg^{-1}) = $GY_f - GY_u/P_a$, where GY_f is the grain yield of the P fertilized pot (mg), GY_u is the grain yield of the unfertilized pot (mg), and P_a is the quantity of P applied (mg)
Physiological efficiency (PE)	*Physiological efficiency* is defined as the biological yield obtained per unit of nutrient uptake. It can be calculated by: PE (mg mg^{-1}) = $BY_f - BY_u/P_f - P_u$, where BY_f is the biological yield (grain plus straw) of the fertilized pot (mg), BY_u is the biological yield of the unfertilized pot (mg), P_f is the P uptake (grain plus straw) of the fertilized pot, and P_u is the P uptake (grain plus straw) of the unfertilized pot (mg)
Agrophysiological efficiency (APE)	*Agrophysiological efficiency* is defined as the economic production (grain yield in case of annual crops) obtained per unit of nutrient uptake. It can be calculated by: APE (mg mg^{-1}) = $GY_f - GY_u/P_f - P_u$, where GY_f is the grain yield of P fertilized pot (mg), GY_u is the grain yield of the unfertilized pot (mg), P_f is the P uptake (grain plus straw) of the fertilized pot (mg), and P_u is the P uptake (grain plus straw) of unfertilized pot (mg)
Apparent recovery efficiency (ARE)	*Apparent recovery efficiency* is defined as the quantity of nutrient uptake per unit of nutrient applied. It can be calculated by: ARE (%) = $(P_f - P_u/P_a) \times 100$, where P_f is the P uptake (grain plus straw) of the P fertilized pot (mg), P_u is the P uptake (grain plus straw) of the unfertilized pot (mg), and P_a is the quantity of P applied (mg)
Utilization efficiency (UE)	Nutrient *utilization efficiency* is the product of physiological and apparent recovery efficiency. It can be calculated by: UE (mg mg^{-1}) = PE × ARE

agrophysiological efficiency (APE), apparent recovery efficiency (ARE), and utilization efficiency (UE) varied significantly among upland rice genotypes (Table 3.5). Across the 18 genotypes, AE was 63 mg grain produced per mg P applied, and PE was 927 dry matter produced (grain + straw) per mg P accumulated in grain plus straw. Similarly, APE was 358 mg grain produced per mg P accumulated in grain plus straw and ARE was 18%. The UE was 155 mg grain plus straw produced per mg P applied. Fageria et al. (2004) reported that the efficiency of P use varies among rice genotypes. Phosphorus use efficiencies are generally higher than use efficiencies for N and K (Fageria et al. 2004) with the exception of recovery efficiency. The low recovery efficiency of P is associated with the high P fixation capacity of Brazilian Oxisols (Goedert 1989). All P use efficiencies have a significant positive association with the grain yield of rice (Table 3.6). Hence, improving P use efficiency is fundamental to improve grain yield of rice.

Fageria and Barbosa Filho (2007) also determined the P use efficiency of lowland rice under field conditions, using thermophosphate (water insoluble) as a source of

TABLE 3.5
Phosphorus Use Efficiency by Upland Rice Genotypes

Genotype	AE (mg mg^{-1})	PE (mg mg^{-1})	APE (mg mg^{-1})	ARE (%)	UE (mg mg^{-1})
CRO 97505	73.4ab	945	424ab	17.4abc	159.9
BRS Liderança	81.6a	817	382abc	21.3ab	173.7
BRS Curinga	67.1abc	823	347abc	19.3abc	158.8
CNAs 8938	63.0abcd	923	365abc	17.3abc	159.4
CNAs 8960	72.5ab	827	381abc	19.1abc	156.8
BRS Colosso	79.9a	737	378abc	21.2ab	156.1
CNAs 8824	61.8abcd	811	350abc	17.7abc	143.8
CNAs 8957	72.6ab	793	352abc	20.7abc	162.3
CRO 97442	68.8abc	732	375abc	18.5abc	137.4
CNAs 8817	68.6abc	847	370abc	18.5abc	155.7
BRS Aroma	53.8abcd	917	314abc	17.3abc	154.0
CNAs 8950	64.0abc	834	350abc	18.2abc	151.2
BRS Talento	41.8bcd	801	240c	16.8abc	131.8
BRS Caripuna	26.7d	1187	237c	11.0c	128.7
Primavera	76.7ab	808	348abc	22.3a	178.6
Canastra	59.3abcd	1211	450ab	13.2abc	156.9
Maravilha	34.6cd	1441	294bc	11.6bc	164.4
Carisma	62.1abcd	1237	483a	13.8abc	155.7
Average	62.7	927.3	357.8	17.5	154.7

Note: AE, agronomic efficiency; PE, physiological efficiency; APE, agrophysiological efficiency; ARE, apparent recovery efficiency; UE, utilization efficiency. Definitions and methods of calculating these efficiencies are given in Table 3.1. Means followed by the same letter in the same column are not significantly different at the 5% probability level by Tukey's test. The P source was triple superphosphate.

TABLE 3.6
Relationship between Phosphorus Use Efficiencies (X) and Grain Yield (Y) across 18 Upland Rice Genotypes

Plant Variable	Regression Equation	R^2
AE vs. grain yield	$Y = -0.015 + 1.20X$	0.99*
PE vs. grain yield	$Y = 181.82 - 0.17X + 0.000051X^2$	0.34*
APE vs. grain yield	$Y = -120.00 + 0.93X - 0.00103X^2$	0.60*
ARE vs. grain yield	$Y = 1.43 + 4.21X$	0.67*
UE vs. grain yield	$Y = -28.47 + 0.67X$	0.41*

Note: AE, agronomic efficiency; PE, physiological efficiency; APE, agrophysiological efficiency; ARE, apparent recovery efficiency; and UE, utilization efficiency.
*Significant at the 1% probability level.

TABLE 3.7
Phosphorus Use Efficiency in Lowland Rice under Different P Rates

P Rate (kg ha^{-1})	AE (kg kg^{-1})	PE (kg kg^{-1})	APE (kg kg^{-1})	RE (%)	UE (kg kg^{-1})
131	15.5	604.4	300.9	6.3	39.6
262	12.7	536.8	269.5	5.1	27.1
393	10.5	521.8	477.4	3.8	19.7
524	6.8	443.6	277.8	3.4	14.8
655	6.2	439.3	296.6	2.7	10.9
Average	10.3	509.2	324.4	4.3	22.4

Regression Analysis

P rate (X) vs. agronomic efficiency (Y) = 17.66 – 0.0178X, $R^2 = 0.52^{**}$

P rate (X) vs. physiological efficiency (Y) = 636.22 – 0.3232X, $R^2 = 0.23^{NS}$

P rate (X) vs. agrophysiological efficiency (Y) = 324.70 – 0.00025X, $R^2 = 0.02^{NS}$

P rate (X) vs. recovery efficiency (Y) = 6.93 – 0.0067X, $R^2 = 0.43^{**}$

P rate (X) vs. utilization efficiency (Y) = 43.27 – 0.0529X, $R^2 = 0.50^{**}$

Source: Fageria, N. K., and M. P. Barbosa Filho. 2007. Dry matter and grain yield, nutrient uptake, and phosphorus use efficiency of lowland rice as influenced by phosphorus fertilization. *Commun. Soil Sci. Plant Anal.* 38:1289–1297.

Note: AE, agronomic efficiency; PE, physiological efficiency; APE, agrophysiological efficiency; RE, recovery efficiency; and UE, utilization efficiency. Methods of calculating these efficiencies are the same as presented in Table 3.4. The grain yield and P application unit in this case will be in kilogram rather than milligram. Values are averaged across two years. The P source was thermophosphate.

^{**,NS}Significant at the 1% probability levels and nonsignificant, respectively.

P fertilizer (Table 3.7). Agronomic, physiological, agrophysiological recovery and utilization efficiencies decreased with increasing P rates (Table 3.7). Across P rates, 10.3 kg rice grain yield was produced with the application of 1 kg P. Similarly, 509.2 kg dry matter (straw plus grain) was produced with the accumulation of 1 kg P in grain plus straw. In the case of APE, across P rates, 324.4 kg grain yield was produced with the accumulation of 1 kg P in grain plus straw. Average recovery efficiency was 4.3% and UE was 22.4 kg grain yield with the utilization of 1 kg P. The highest efficiency is usually obtained with the first increment of nutrient, with additional increments providing smaller increases (Fageria, Baligar, and Jones 2011). Singh et al. (2000) reported that APEs in lowland rice varied from 235 to 316 kg grain per kg P. Similarly, Witt et al. (1999) reported an APE value of 385 kg grain per kg P when all production factors were at normal levels. Sahrawat and Sika (2002) reported apparent recovery of applied P in the range of 4.8–11% by rice in an Ultisol. The low recovery efficiency may be associated with a high rate of P fixation in this soil because of iron and aluminum oxides (Abekoe and Sahrawat 2001).

Fageria, Baligar, and Jones (2011) reported that phosphorus use efficiency expressed as AE decreased in a quadratic fashion with increasing P rate in the range of 22–88 kg ha^{-1} (Figure 3.16). At 22 kg P ha^{-1}, AE was 186 kg grain

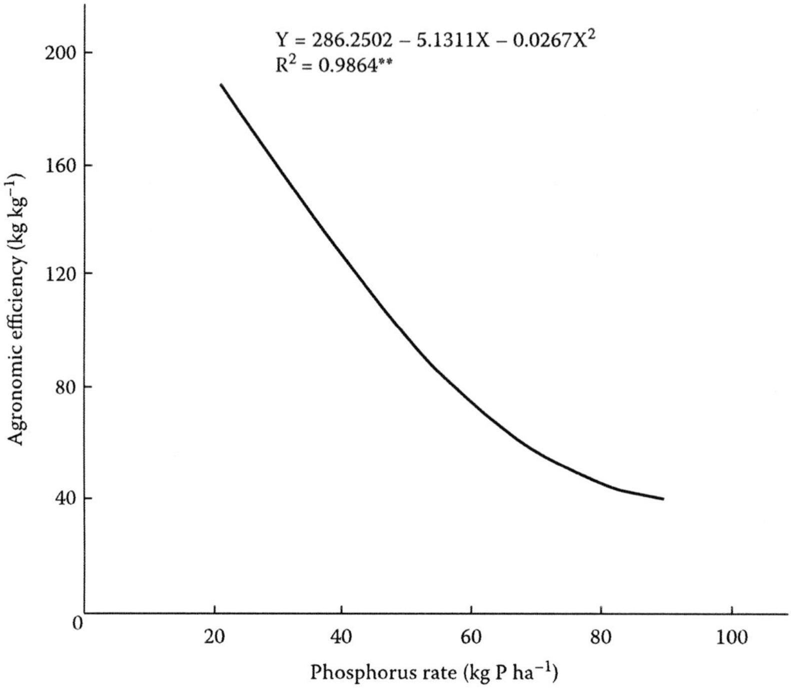

FIGURE 3.16 Relationship between P rate and agronomic efficiency of lowland rice genotypes. Values are averages of 12 genotypes and two years of field trials. (From Fageria, N. K., A. B. Santos, and A. B. Heineman. 2011. Lowland rice genotypes evaluation for phosphorus use efficiency in tropical lowland. *J. Plant Nutr.* 34:1087–1095.)

produced per kg P applied and 40 kg grain produced per kg P applied at 88 kg P ha^{-1} (Figure 3.16). With the increasing P rates, grain yield increased, whereas P use efficiency decreased due to low plant capacity in the absorption and utilization of P (Fageria, Slaton, and Baligar 2003). The decrease in P use efficiency is also associated with a relative decrease in grain yield with successive increments in P rates (Fageria 1992).

Fageria, Baligar, and Jones (2011) also studied the AE of 12 lowland rice genotypes grown on Brazilian Inceptisol (Figure 3.17). There were significant differences among genotypes in relation to P use efficiency. Genotype BRS Jaçana produced maximum P use efficiency, and genotype CNAi 8569 produced minimum P use efficiency. The author also studied the agronomic P use efficiency of upland rice genotypes (Figure 3.18). Upland rice varieties also differ significantly in P use efficiency. Genotype BRA 032046 produced maximum P use efficiency (about 73 kg grain per kg P applied), and genotype BRA 0322051 produced minimum P use efficiency (about 22 kg grain per kg P applied). Variations in P use efficiency in lowland as well as upland rice have been reported by Fageria and Santos (2002) and Fageria, Baligar, and Jones (2011).

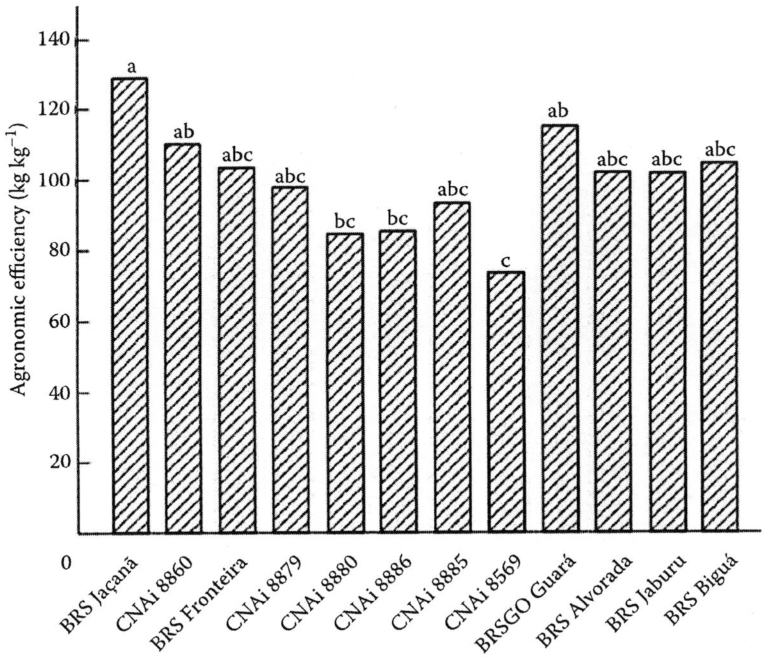

FIGURE 3.17 Agronomic P use efficiency of 12 lowland rice genotypes.

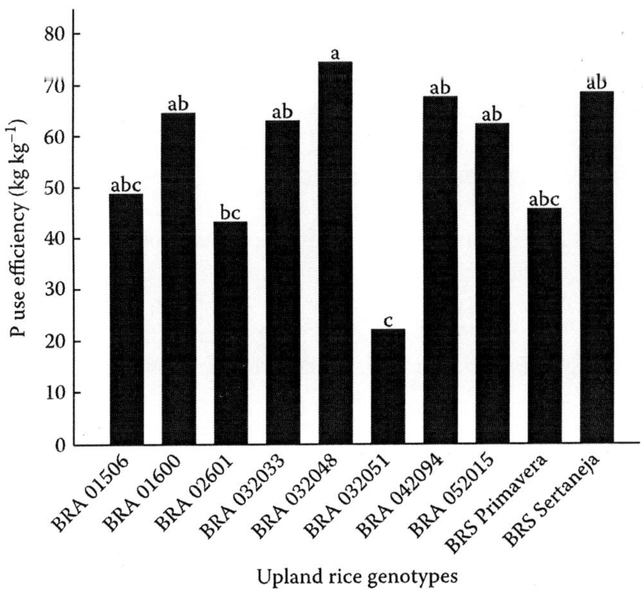

FIGURE 3.18 Agronomic P use efficiency of 10 upland rice genotypes.

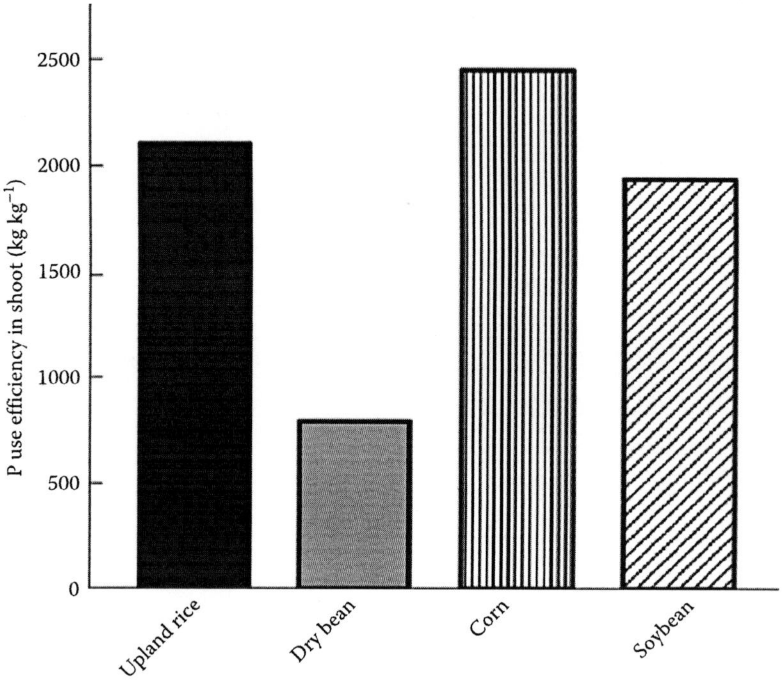

FIGURE 3.19 P use efficiency in the shoot (straw) of four crop species.

The P use efficiency in shoot (straw) and grain of upland rice, dry beans, corn, and soybeans was compared (Figures 3.19 and 3.20). Phosphorus use efficiency in shoot as well as grain followed the pattern of corn > upland rice > soybeans > dry beans. The higher P use efficiency in cereals compared with legumes may be related to the higher yield of corn and upland rice compared with dry beans and soybeans. The P use efficiency in crop plants expressed in terms of dry matter production per unit of P uptake is at a maximum when compared with N and K (Fageria, Baligar, and Clark 2006). This is in contrast to the P recovery efficiency of applied fertilizer in soils. The P recovery efficiency of applied fertilizers in soil by crop plants is less than 20%, whereas N recovery efficiency is 50% or less, and K recovery efficiency is close to 40% (Baligar, Fageria, and He 2001).

3.8 MANAGEMENT PRACTICES

Adopting appropriate soil and crop management practices can improve P use efficiency, reduce the cost of rice production, increase yield, and also avoid environmental pollution. Important beneficial management practices are liming acidic soils, using an appropriate P source, using effective application methods, and adopting the optimal timing of application. In addition, the use of adequate rate is fundamental to achieving maximum economic yield and improving P use efficiency.

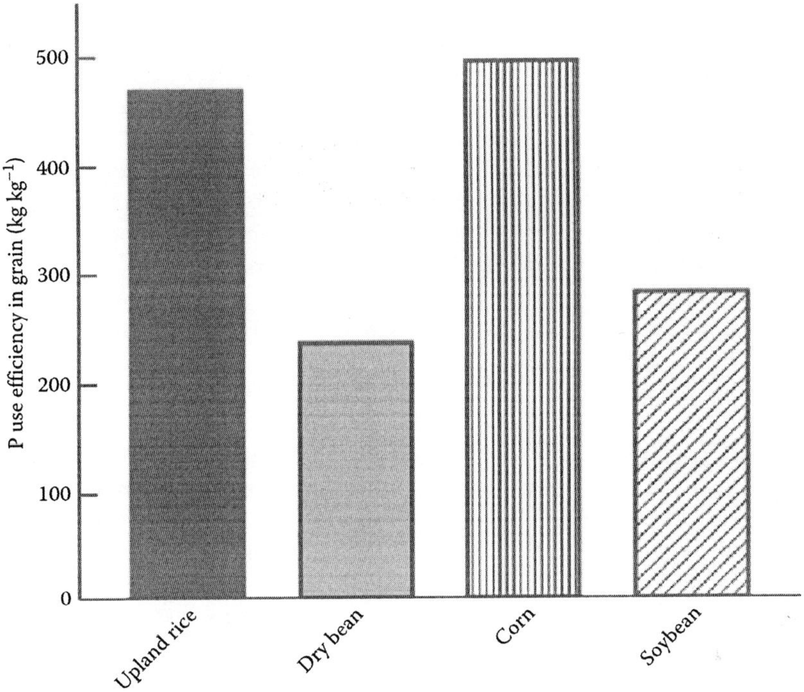

FIGURE 3.20 P Use efficiency in grain of four crop species.

Furthermore, the use of P-efficient genotypes is also a very attractive strategy for efficient rice production. These practices are discussed in the following sections.

3.8.1 Liming Acidic Soils

Soil acidity is one of the most yield-limiting factors for crop production. Land area affected by acidity is estimated at 4 billion ha, which represents approximately 30% of the world's total ice-free land area (Sumner and Noble 2003). About 16.7% of Africa, 6.1% of Australia and New Zealand, 9.9% of Europe, 26.4% of Asia, and 40.9% of the United States have acidic soils (Von Uexkull and Mutert 1995). These soils cover a significant part of at least 48 developing countries located mainly in tropical areas and are found more frequently in Oxisols and Ultisols in South America and in Oxisols in Africa (Narro et al. 2001). In tropical South America, 85% of the soils are acidic, and approximately 850 million ha of this area are underutilized (Fageria and Baligar 2001). Acidic and low-fertility Oxisols and Ultisols cover about 43% of the tropics (Sanchez and Logan 1992). Most of central Brazil is tropical savanna (known as the Cerrado), which covers about 205 million ha or 23% of the country. Most of the soils in this region are Oxisols (46%), Ultisols (15%), and Entisols (15%), with low natural soil fertility, high aluminum saturation, and high P fixation capacity (Fageria and Baligar 2008).

Liming is effective for overcoming these factors and improving crop production on acidic soils. Called the *workhorse*, lime is the foundation of acidic soil crop production. The amount of lime necessary for crops grown on acidic soils is determined by quality of liming material, status of soil fertility, crop species and cultivar within species, crop management practices, and economic considerations (Fageria and Baligar 2008). Soil pH, base saturation, and aluminum saturation are important acidity indices that are used to determine liming rates for reducing plant constraints on acidic soils. Crop responses to liming rates also impact crop liming recommendations for production on acidic soils (Fageria and Baligar 2008).

Liming reduces P immobilization when applied to an acidic soil. Acidic soils (Oxisols and Ultisols) are naturally deficient in total P as well as plant-available P; significant portions of applied P are immobilized due to either P precipitation as insoluble Fe/Al phosphates or chemisorption to Fe/Al-oxides and clay minerals (Nurlaeny, Marschner, and George 1996). Smyth and Cravo (1992) reported that Oxisols are notorious for P immobilization because they have a higher iron oxide content in their surface horizons than any other kind of soil. The P fixation capacity in Oxisols is directly related to the surface area and clay contents of the soil material and inversely related to SiO_2/R_2O_3 ratios (Curi and Camargo 1988).

Bolan et al. (1999) reported that in variable charge soils, a decrease in pH increases the anion exchange capacity, thereby increasing the retention of P. Hence, improving crop yields on these soils requires high rates of P application (Sanchez and Salinas 1981; Fageria 1989). There are conflicting reports regarding the effects of liming on P availability in highly weathered acidic soils (Haynes 1984). Liming has been shown to increase, decrease, or have no effect on P availability (Haynes 1982; Fageria 1984; Mahler and McDole 1985; Anjos and Rowell 1987; Curtin and Syers 2001). However, in a recent study of Brazilian Oxisols, Fageria and Santos (2005) reported a linear increase in Mehlich 1 extractable P with increasing soil pH in the range of 5.3–6.9 (at an average of 0–10 cm and 10–20 cm soil depth; Figure 3.21). Mansell et al. (1984) and Edmeades and Perrott (2004) reported that in acidic soils of New Zealand, the primary benefit of liming occurs through an increase in the availability of P by decreasing P adsorption and stimulating the mineralization of organic P. Fageria (1984) also reported that in Brazilian Oxisols, there was a quadratic increase in the Mehlich 1 extractable P in the pH range of 5–6.5, and thereafter it was decreased. An increase in the availability of P in the pH range of 5–6.5 was associated with a release of P ions from Al and Fe oxides, which were responsible for P fixation (Fageria 1989). At higher pH (>6.5), the reduction of extractable P was associated with the precipitation of P as Ca phosphate. These increases in extractable P or a release of this element in the pH range of 5–6.5 and a reduction in the higher pH range (>6.5) can be explained by the following equations:

$$AlPO_4(\text{P fixed}) + 3OH^- \Leftrightarrow Al(OH)_3 + PO_4^{3-}(\text{P released}) \tag{3.8}$$

$$Ca(H_2PO_4)_2(\text{soluble P}) + 2Ca^{2+} \Leftrightarrow Ca_3(PO_4)_2(\text{insoluble P}) + 4H^+ \tag{3.9}$$

The liming of acidic soils results in the release of P for plant uptake; this is often referred to as the *P spring effect* of lime (Bolan, Adriano, and Curtin 2003). Bolan, Adriano, and Curtin (2003) reported that in soils high in exchangeable and soluble

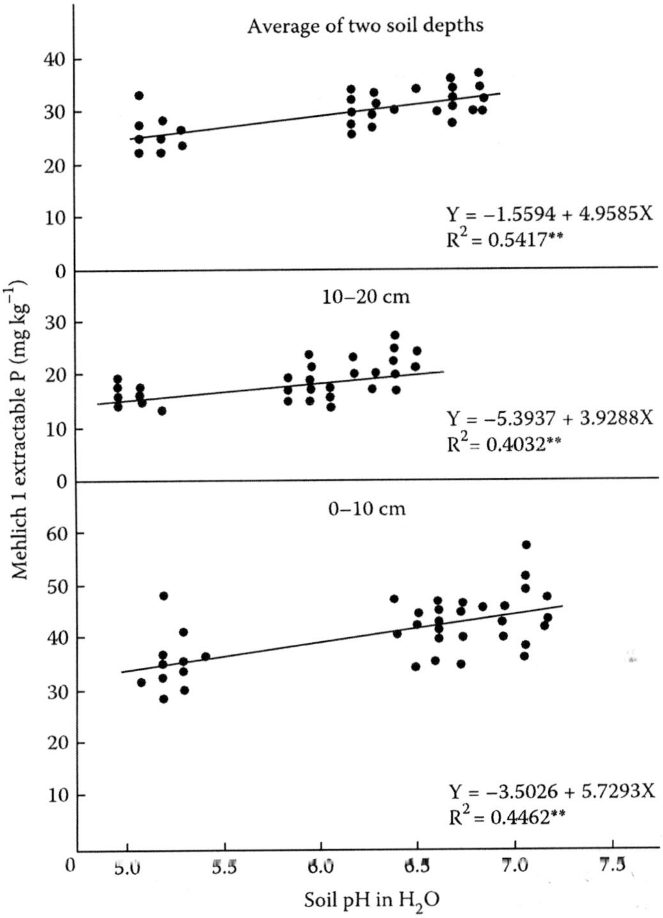

FIGURE 3.21 Relationship between soil pH and Mehlich 1 extractable soil phosphorus in Brazilian Oxisol. (From Fageria, N. K., and A. B. Santos. 2005. Influence of base saturation and micronutrient rates on their concentration in the soil and bean productivity in Cerrado soil in no-tillage system. Paper presented at the *VIII National Bean Congress*, Goiânia, Brazil, October.)

Al, liming might increase plant P uptake by decreasing Al, rather than by increasing P availability. This may be due to improved root growth where Al toxicity is alleviated, allowing a greater volume of soil to be explored (Fageria and Baligar 2008).

3.8.2 Sources, Methods, and Timing of Application

There are a large number of P fertilizers that can be applied to supply the necessary P requirements of crops on P-deficient soils. The major P carriers, their P content, and their solubility is given in Table 3.8. Among these, water-soluble sources are

TABLE 3.8
Major Phosphorus Fertilizers, Their P Content, and Their Solubility

Common Name	Chemical Composition	P_2O_5 Content (%)	Solubility
Simple superphosphate	$Ca(H_2PO_4)_2 + CaSO_4$	18–22	Water soluble
Triple superphosphate	$Ca(H_2PO_4)_2$	46–47	Water soluble
Monoammonium phosphate	$NH_4H_2PO_4$	48–50	Water soluble
Diammonium phosphate	$(NH_4)_2HPO_4$	54	Water soluble
Phosphoric acid	H_3PO_4	55	Water soluble
Thermophosphate (yoorin)	$[3MgO{\cdot}CaO{\cdot}P_2O_5 + 3(CaO{\cdot}SiO_2)]$	17–18	Citric acid soluble
Rock phosphates	Apatites	24–40	Citric acid soluble
Basic slag	$Ca_3P_2O_8{\cdot}Cao+CaO{\cdot}SiO_2$	10–22	Citric acid soluble

more effective for annual crops. Water-insoluble or citric acid-soluble sources of P are effective for acidic soils. Because liming is an essential and effective practice for improving annual crop yield on acidic soils (Fageria 2009), rock phosphates are not effective on limed acidic soils or on soils having a higher pH (>6.0; Foth and Ellis 1988). Dissolution of phosphate rocks in soil requires an adequate supply of acid (H^+) as shown in the following equation (Bolan et al. 1997):

$$Ca_{10}(PO_4)_6F_2 + 12H^+ \leftrightarrow 10Ca^{2+} + 6H_2PO_4^- + 2F^- \quad (3.10)$$

Bolland and Gilkes (1990) reported that the efficiency of rock phosphates was much lower than that of superphosphates tested under arid conditions in Australia. Tisdale, Nelson, and Beaton (1985) reported that water-insoluble phosphates and those with low water solubility may give equally good results under some conditions, but they are not as suitable as the water-soluble materials. Tisdale, Nelson, and Beaton (1985) also reported that the thermal process phosphates, when finely ground, are satisfactory sources of phosphorus for most crops on acidic soils. Generally, however, they have failed to produce favorable results on neutral and alkaline soils.

The immobile nature of P nutrients further demands that the P source should be water soluble to be effective for crop production. Several different views exist regarding the proportion of water-soluble P necessary for commercial phosphate fertilizers. The European Economic Community (EEC) requires 93% of the available P in commercial normal and triple superphosphate fertilizers to be in water-soluble form (Fageria and Gheyi 1999). Research has concluded that there would probably be no significant yield increase from commercial phosphate fertilizers containing >80% of their available P in water-soluble form (Mullins 1988; Mullins and Sikora 1990). Bartos et al. (1992) tested five monoammonium phosphate (MAP) fertilizers, representing the major U.S. sources of phosphate rock (PR) (Florida, North Carolina, and Idaho), using sorghum as a test crop. They concluded that 57–68% water-soluble P is necessary to obtain 90% of the maximum yield.

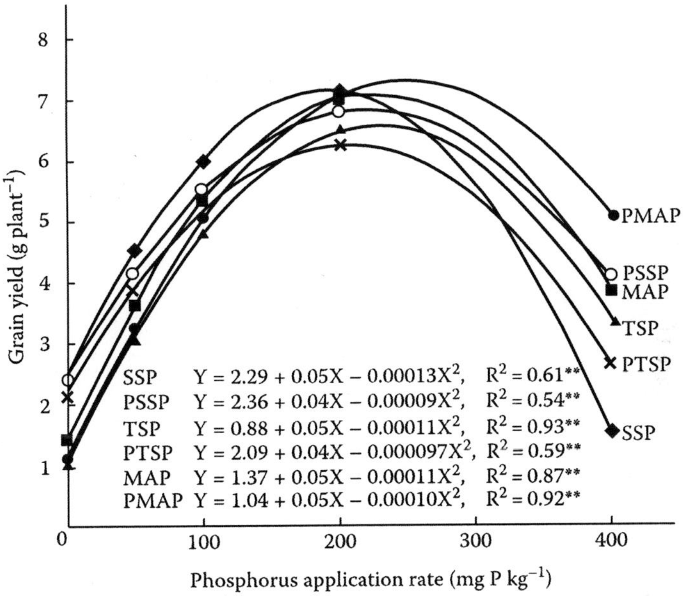

FIGURE 3.22 Response of upland rice to different sources of phosphorus fertilization. SSP, simple superphosphate; PSSP, polymer-coated simple superphosphate; TSP, triple superphosphate; PTSP, polymer-coated triple superphosphate; MAP, monoammonium phosphate; PMAP, polymer-coated monoammonium phosphate.

Also, the response of upland rice to different P sources applied to a Brazilian Oxisol was studied. To classify the P source efficiency in grain production, grain yield was plotted against P rate (Figure 3.22). Upland rice response to different sources was significant, quadratic in definition, and varied in magnitude. Maximum grain yield was achieved with a P source order of polymer-coated monoammonium phosphate (PMAP) > simple superphosphate (SSP) = monoammonium phosphate (MAP) > polymer-coated simple superphosphate (PSSP) > triple superphosphate (TSP) > polymer-coated triple superphosphate (PTSP). There is a difference of about 13% between the higher grain-yield-producing P source and the lower grain-yield-producing P source. The PMAP source can be considered the best among these six P sources due to higher yield at adequate P values as well as at higher P values.

Also, the response of lowland rice to eight P sources applied to Brazilian Inceptisol was studied (Figure 3.23). Data in Figure 3.23 show that the response curves were quadratic for all the P sources evaluated. However, the magnitude of response differed. For example, ammoniated simple superphosphate (ASSP) produced the lowest grain yield, and PSSP, TSP, and PTSP produced the maximum yield at average P rate (273 mg P kg^{-1}). Based on the grain yield and average P rate of maximum grain yield, which is 273 mg kg^{-1}, P sources were classified for P use efficiency in the order of PSSP = TSP > PTSP > polymer-coated, ammoniated simple superphosphate (PASSP) > SSP > MAP > ASSP. Engelstad and Terman (1980) reviewed the literature on AE of different P sources and reported that there was a quadratic increase in

Variable	Regression equation	R^2	PRMV (mg kg^{-1})
	Grain yield		
PR of SSP vs. GY	$Y = -0.041 + 0.07X - 0.00012X^2$	0.90**	292
PR of PSSP vs. GY	$Y = 0.83 + 0.08X - 0.00015X^2$	0.96**	267
PR of ASSP vs. GY	$Y = 0.38 + 0.03X - 0.000084X^2$	0.88**	179
PR of PASSP vs. GY	$Y = 1.48 + 0.05X - 0.000077X^2$	0.88**	325
PR of TSP vs. GY	$Y = 1.06 + 0.08X - 0.00015X^2$	0.94**	267
PR of PTSP vs. GY	$Y = 1.75 + 0.06X - 0.000098X^2$	0.81**	306
PR of MAP vs. GY	$Y = 0.34 + 0.04X - 0.000074X^2$	0.90**	270
PR of PMAP vs. GY	$Y = 1.21 + 0.05X - 0.00012X^2$	0.82**	208
Average	$Y = 0.83 + 0.06X - 0.00011X^2$	0.96**	273

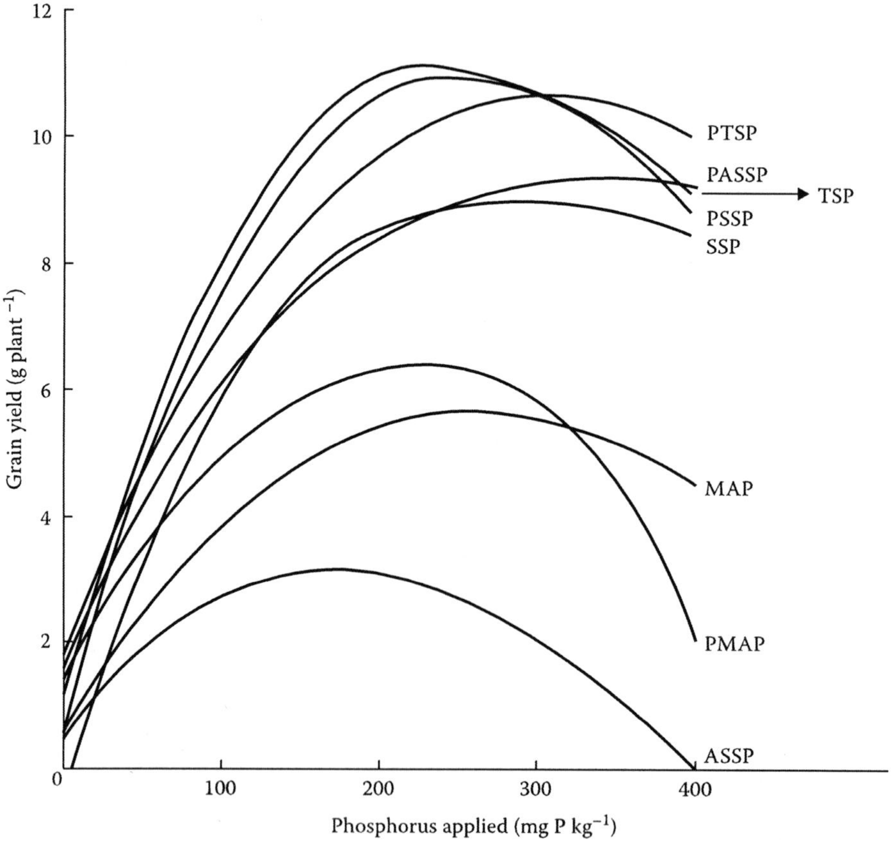

FIGURE 3.23 Relationship between P rate of different P sources and grain yield of lowland rice. PRMV = Phosphorus rate for maximum value (grain yield).

rice yield with the addition of P from 0 to 88 kg ha^{-1}, and response also varied with the P source. Figures 3.24 through 3.31 show the growth of lowland rice at different P sources and rates. Lowland rice growth increased with increasing P rate in a quadratic fashion.

In low-P-content soils, phosphorus should be applied in bands when annual crops are sown. This reduces the fertilizer P fixation to a minimum because it allows

FIGURE 3.24 Response of lowland rice to P fertilization with simple superphosphate.

FIGURE 3.25 Response of lowland rice to P fertilization with polymer-coated simple superphosphate.

FIGURE 3.26 Response of lowland rice to P fertilization with ammoniated simple superphosphate.

FIGURE 3.27 Response of lowland rice to P fertilization with polymer-coated, ammoniated simple superphosphate.

FIGURE 3.28 Response of lowland rice to P fertilization with triple superphosphate.

FIGURE 3.29 Response of lowland rice to P fertilization with polymer-coated triple superphosphate.

FIGURE 3.30 Response of lowland rice to P fertilization with monoammonium phosphate.

FIGURE 3.31 Response of lowland rice to P fertilization with polymer-coated monoammonium phosphate.

the crop the best opportunity to compete with the soil P utilization (Mengel et al. 2001). If soil P content is higher or in the adequate range for crop growth, band and broadcast application will be equally effective. However, in P fixing acidic soils or in low-P-content soils, band application is more effective; in such soils, broadcast application may require three to four times more P than that required for band application. Band application may concentrate the P in a limited soil volume, thereby saturating the soil phosphate adsorption capacity and increasing the soluble phosphate in the soil (Mengel et al. 2001). Early season growth may also be possible with band placement of fertilizer, even in soils that have considerable available P (Foth and Ellis 1988). Banded P usually increases early crop growth more than broadcast placement because of increased uptake (Eghball, Sander, and Skopp 1990; Barber 1995).

Smyth and Cravo (1990) reported that banded P provided maximum yields of corn and cowpea crops at lower rates than broadcast P in a Brazilian Amazon Oxisol. Banded P rates of 22 kg ha^{-1} for corn and 44 kg ha^{-1} for cowpeas would sustain near maximum yields during five years of continuous crop production. Fiedler, Sander, and Peterson (1989) reported that in winter wheat, medium and low levels of soil P seed application resulted in two to four times more profit from fertilizer P application than broadcast P application. Banding P fertilizer increases fertilizer effectiveness for two primary reasons. First, banding reduces soil fertilizer contact, which affects the degree of fertilizer P adsorption or precipitation and reduces what is often referred to as immobilization or fixation. Second, banding relatively close to the plant results in a greater chance of root contact with the fertilizer.

3.8.3 Adequate Rate

Use of an adequate rate of P is very important for crop production (including rice crops); this can produce higher yields, reduce production costs, and lessen environmental pollution. Phosphorus is an immobile nutrient in the soil; soil test calibration data are the best criteria for deciding the adequate rate for a given crop. However, some scientists are also using plant analysis/nutrient accumulation data for determining the adequate nutrient application rate for a given crop. For soil test calibration data, several field experiments should be conducted to provide average values for each agro-ecological region. The amount of nutrient extracted in a soil is of little use until it has been calibrated to crop response in field experiments. Figure 3.32 shows lowland rice grain yield as a function of phosphorus applied in an Inceptisol. Grain yield increased with broadcast P fertilization, with a maximum yield obtained at about 290 kg P ha^{-1}. Relative grain yield of this experiment was plotted against soil extractable P for calibration of soil test P (Figure 3.33).

Four categories were established for the soil P test: very low (VL), low (L), medium (M), and high (H). The 0–70% relative yield zone is called VL, the 70–90% relative yield zone is called L, the 90–100% relative yield zone is called M, and more than 100% relative yield is called the high soil P test. These zones, as suggested by Raij (1991), were selected for a soil P calibration study under Brazilian conditions. The soil P test availability indices and P fertilizer recommendations for lowland rice in Inceptisol, calculated on the basis of Figures 3.32 and 3.33, are presented in Table 3.9.

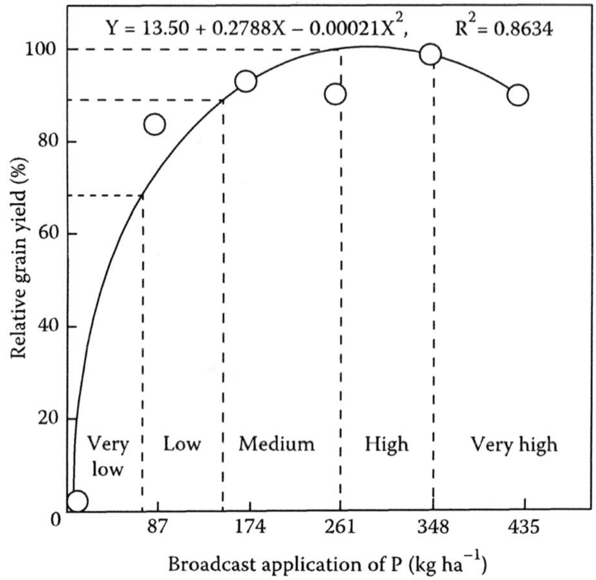

FIGURE 3.32 Relationship between P applied as broadcast and relative grain yield of lowland rice. (From Fageria, N. K. 1996. *The Study of Liming and Fertilization for Rice and Common Bean in Cerrado Region.* Goiania, Brazil: National Rice and Bean Research Center.)

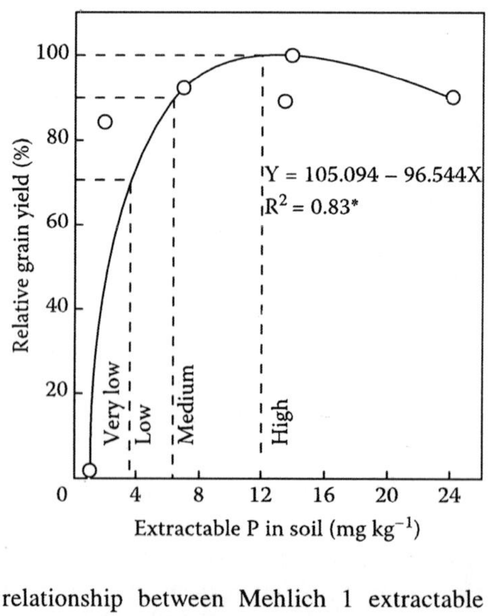

FIGURE 3.33 The relationship between Mehlich 1 extractable P and relative grain yield of lowland rice. (From Fageria, N. K. 1996. *The Study of Liming and Fertilization for Rice and Common Bean in Cerrado Region.* Goiania, Brazil: National Rice and Bean Research Center.)

TABLE 3.9
Soil P Test Availability Indices and P Fertilizer Recommendations for Lowland Rice in Inceptisol

Soil P Test (mg kg^{-1})	Interpretation	Relative Yield (%)	Broadcast P Application (kg ha^{-1})	Band P Application (kg ha^{-1})
0–3.6	Very low	0–70	100	66
3.6–6.4	Low	70–90	170	66
6.4–12.0	Medium	90–100	275	44
>12.0	High	>100	>275	22

Source: Fageria, N. K., V. C. Baligar, and C. A. Jones. 2011. *Growth and Mineral Nutrition of Field Crops*, 3rd edition. Boca Raton, FL: CRC Press.

In addition to a soil test calibration study, P fertilizer recommendations can also be made based on crop response to P fertilization data. Experiments relating P rate with grain yield for lowland as well as upland rice were conducted. The results showing lowland rice genotypes' response to P fertilization are presented in Table 3.10; data related to regression analysis to calculate adequate P rates are presented in Table 3.11. Year and P rate significantly influenced grain yield. However, year × P rate interaction for grain yield was not significant; therefore, the average values of two years were pooled (Table 3.10). Grain yield for 12 genotypes increased significantly in a quadratic fashion with increasing P rate in the range of 0–88 kg P ha^{-1}. Adequate P rates for maximum grain yield varied from genotype to genotype. However, across 12 genotypes, maximum grain yield was obtained with the application of 54 kg P ha^{-1} (124 kg PO_5 ha^{-1}). Genotype BRS Jaçanã was the most efficient, and genotype CNAi 8569 was most inefficient in terms of P use efficiency. The increase in grain yield of genotypes was associated with an increase in panicle number and an increase in shoot dry weight with increasing P rate (Table 3.12). The increase in panicle number (Y) with increasing P rate (X) was significant and quadratic. The variability in panicle number due to P fertilization was 83% (Table 3.12). Similarly, shoot dry weight also increased in a quadratic fashion with increasing P rate in the range of 0–88 kg P ha^{-1}. These two plant parameters were significantly and positively associated with grain yield. The regression equations were panicle number (X) versus grain yield (Y) = $-2845.48 + 32.61X - 0.031X^2$, $R^2 = 0.92^{**}$, and shoot dry weight (X) versus grain yield (Y) = $-4953.37 + 1.69X - 0.000055X^2$, $R^2 = 0.82^{**}$. These two regression equations show that grain yield significantly increased in a quadratic fashion with increasing panicle number and shoot dry weight. The variability in grain yield was 92% due to panicle number and 82% due to shoot dry weight. Fageria (2007) reported that among yield components, panicle number is the most effective component in increasing rice yield. Similarly, Peng et al. (2000) also reported that shoot dry weight is responsible for improving grain yield of those lowland rice cultivars released by the International Rice Research Institute (IRRI)

TABLE 3.10
Grain Yield of 12 Lowland Rice Genotypes as Influenced by Phosphorus Fertilization

	Phosphorus Rate (kg P ha^{-1})				
Genotype	**0**	**22**	**44**	**66**	**88**
BRS Jaçnã	932.9	5903.2	7373.3	6750.0	5681.2
CNAi 8860	1153.4	588.8	6361.2	5470.7	4874.7
CNAi 8879	1174.9	5517.8	6500.3	4736.2	5038.8
BRS Fronteira	1259.0	5185.2	6190.5	4979.3	4182.0
CNAi 8880	1738.0	5346.7	4794.7	5477.8	4994.7
CNAi 8886	1698.3	5377.8	6017.0	5377.1	4318.0
CNAi 8885	1289.0	5139.8	5853.0	5046.3	4854.2
CNAi 8569	58.0	3482.8	3684.0	3402.7	1565.0
BRS Guará	248.9	5100.2	6163.3	5303.3	4826.2
BRS Alvorada	64.2	4435.7	4802.8	4255.2	3764.7
BRS Jaburu	41.6	3779.8	5384.7	4704.3	4441.5
BRS Biguá	128.6	3844.3	5717.5	5026.7	5005.5
F-test					
Year (Y)			*		
P rate (P)			*		
Y × P			NS		
Genotype (G)			*		
Y × G			*		
CV (%)			24.3		

Source: Fageria, N. K., A. B. Santos, and A. B. Heineman. 2011. Lowland rice genotypes evaluation for phosphorus use efficiency in tropical lowland. *J. Plant Nutr.* 34:1087–1095.

Note: Values are averages of two years of field data. Initial soil P level of the experimental area was 2.3 mg kg^{-1} by Mehlich 1 extracting solution.

*,NS Significant at the 1% probability level and nonsignificant, respectively.

in the Philippines after 1980. These authors also reported that before 1980, GHI was responsible for improving rice yields at IRRI.

Although the response to P fertilization of the 12 genotypes was similar, the magnitude in grain yield was different (see Table 3.10). At the lowest P rate (control treatment), genotype BRS Jaburu produced the lowest grain yield (42 kg ha^{-1}) and genotype CNAi 8880 produced the maximum grain yield (1738 kg ha^{-1}). Similarly, at the highest P rate (88 kg ha^{-1}), genotype CNAi 8569 produced a minimum grain yield of 1565 kg ha^{-1} and genotype BRA Jaçana produced a maximum grain yield of 5681 kg ha^{-1}. At an intermediate P rate (44 kg ha^{-1}), genotype CNAi 8569 produced a minimum grain yield (3684 kg ha^{-1}) and genotype BRS Jaçanã produced a maximum grain yield (7373 kg ha^{-1}). Therefore, rice genotypes varied in response magnitude at low, medium, and high P rates. Data for three genotypes and an average of the 12 genotypes are also presented in graphic form (Figures 3.34 and 3.35).

TABLE 3.11
Regression Equations Showing Relationship between Phosphorus Rate (X) and Grain Yield (Y) of 12 Lowland Rice Genotypes

Genotype	Regression Equation	R^2	P Rate for Maximum Grain Yield (kg ha^{-1})
BRS Jaçnã	$Y = 1234.94 + 231.06X - 2.09X^2$	0.87*	55
CNAi 8860	$Y = 1655.52 + 182.48X - 1.68X^2$	0.82*	54
CNAi 8879	$Y = 1657.66 + 172.19X - 1.60X^2$	0.73*	54
BRS Fronteira	$Y = 1564.98 + 177.11X - 1.72X^2$	0.80*	51
CNAi 8880	$Y = 2148.82 + 120.44X - 1.03X^2$	0.67*	58
CNAi 8886	$Y = 1973.31 + 163.50X - 1.16X^2$	0.74*	70
CNAi 8885	$Y = 1656.87 + 156.74X - 1.42X^2$	0.73*	55
CNAi 8569	$Y = 279.20 + 156.29X - 1.62X^2$	0.88*	48
BRS Guará	$Y = 659.75 + 205.91X - 1.86X^2$	0.89*	55
BRS Alvorada	$Y = 500.62 + 170.98X - 1.57X^2$	0.82*	54
BRS Jaburu	$Y = 255.94 + 177.80X - 1.48X^2$	0.88*	60
BRS Biguá	$Y = 323.35 + 180.07X - 1.48X^2$	0.86*	61
Average of 12 genotypes	$Y = 1156.88 + 175.02X - 1.61X^2$	0.90*	54

Source: Fageria, N. K., A. B. Santos, and A. B. Heineman. 2011. Lowland rice genotypes evaluation for phosphorus use efficiency in tropical lowland. *J. Plant Nutr.* 34:1087–1095.

Note: Values are averages of two years' field trials.

*Significant at the 1% probability level.

TABLE 3.12
Influence of Phosphorus on Number of Panicles and Shoot Dry Weight

P Rate (kg ha^{-1})	Panicle Number (m^{-2})	Shoot Dry Weight (kg ha^{-1})
0	69	4331
22	353	8410
44	375	7770
66	376	7593
88	377	7516

Regression Analysis

P rate (X) vs. panicle number (Y) $= 97.82 + 10.55X - 0.087X^2$, $R^2 = 0.83$*

P rate (X) vs. shoot dry weight (Y) $= 4892.14 + 127.17X - 1.16X^2$, $R^2 = 0.61$*

Note: Values are averages of 12 genotypes and two years of field trials.

*Significant at the 1% probability level.

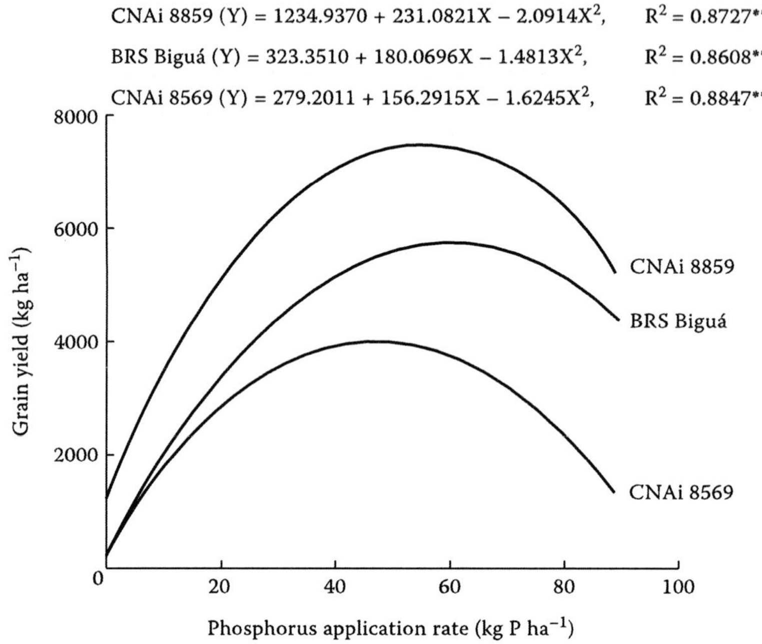

FIGURE 3.34 Response of three lowland rice genotypes to P fertilization. Values are averages of two years of field experiments.

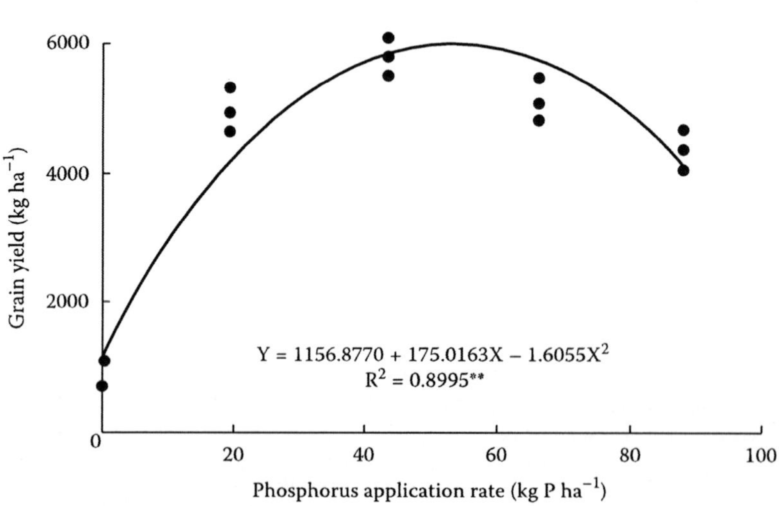

FIGURE 3.35 Response of lowland rice to P fertilization. Values are averages of 12 lowland rice genotypes.

Data related to the response of upland rice genotypes to P fertilization are presented in Figures 3.36 through 3.38. The response curves were quadratic for all the genotypes evaluated. The initial soil P level was about 2 mg kg^{-1} by Mehlich 1 extracting solution. Upland rice in Brazilian Oxisols and Ultisols generally respond to P fertilization when the P level in the soil is less than 5 mg kg^{-1} by the Mehlich 1

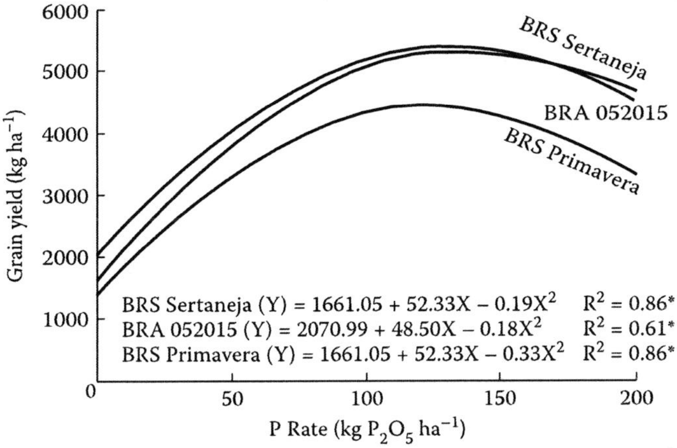

FIGURE 3.36 Response of upland rice genotypes to phosphorus fertilization.

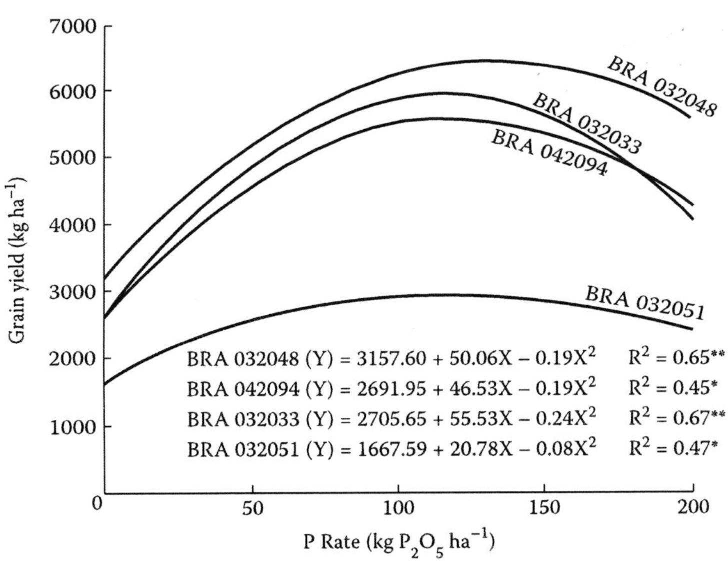

FIGURE 3.37 Response of four upland rice genotypes grown on Brazilian Oxisol to P fertilization. Initial soil P level was about 2 mg kg^{-1} soil by Mehlich 1 extracting solution.

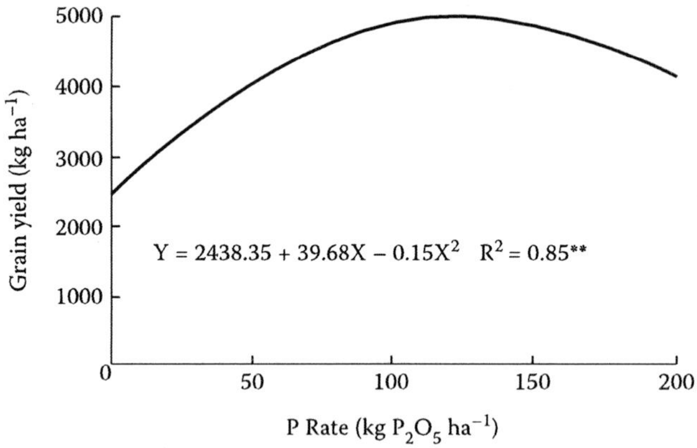

FIGURE 3.38 Response of upland rice to P fertilization. Values are averages of 10 upland rice genotypes grown on Brazilian Oxisol. Initial soil P level was about 2 mg kg^{-1} soil by Mehlich 1 extracting solution.

extracting solution. A phosphorous fertilizer rate of 100–120 kg ha^{-1} (P_2O_5) is recommended when the P level is below 3 mg kg^{-1} of soil. When the P level is higher than 3 mg kg^{-1}, the P fertilizer rate can be lowered to 60–80 kg P_2O_5 ha^{-1}, preferably applied in the furrow at sowing.

3.8.4 Use of Efficient Genotypes

The use of P-efficient genotypes is an important strategy for improving crop yield in P-deficient soils and for reducing production cost. This strategy not only lowers the cost of production but also avoids environmental pollution. Fageria, Baligar, and Li (2008) reported that in the twenty-first century, nutrient efficient plants will play a major role in increasing crop yields; this is mainly due to the limited land and water resources available for crop production, the higher cost of inorganic fertilizer inputs, declining trends in global crop yields, and increasing environmental concerns. Furthermore, at least 60% of the world's arable lands have mineral deficiencies or elemental toxicity problems, and on such soils, fertilizers and lime amendments are essential for achieving improved crop yields. Fertilizer inputs are increasing farmers' production costs, and there is a major concern for environmental pollution due to excess fertilizer inputs. The higher demands for food and fiber by increasing world populations further enhance the importance of nutrient-efficient cultivars that are also higher producers. Nutrient-efficient plants are defined as those plants that produce higher yields per unit of nutrient, applied or absorbed, than other plants (standards) do under similar agroecological conditions (Fageria, Baligar, and Li 2008).

Genetic variability among rice genotypes has been reported for upland as well as lowland rice by various authors (Fageria et al. 1988; Fageria, Wright, and Baligar 1988; Fageria, Slaton, and Baligar 2003; Fageria, Baligar, and Jones 2011; Fageria,

TABLE 3.13
Grain Yield of 10 Upland Rice Genotypes and Their Classification for P Use Efficiency

Genotype	Grain Yield (kg ha^{-1})	GYEI at 100 (kg P ha^{-1})
BRA 01506	4371abc	1.01E
BRA 01600	4320bc	1.11E
BRA 02601	3518cd	0.92ME
BRA 032033	4622ab	1.51E
BRA 032048	5320a	1.57E
BRA 032051	2485e	0.79ME
BRA 042094	4450abc	1.41E
BRA 052015	4215bcd	1.35E
BRS Primavera	3334de	1.14E
BRS Sertaneja	4062bcd	1.38E
Average	4094	1.22

Note: Means followed by the same letter in the same column are not significant at the 5% probability level by the Tukey's test. E, efficient; ME, moderately efficient.

Santos, and Heineman 2011). Data in Table 3.13 show that grain yield of upland rice genotypes was significantly different among genotypes. The grain yield efficiency index (GYEI) was calculated by using the following equation:

$$\text{GYEI} = \frac{\text{GY}_1}{\text{AGY}_1} \times \frac{\text{GY}_2}{\text{AGY}_2} \tag{3.11}$$

where GY_1 is grain yield at zero P rate, AGY_1 is the average grain yield of 10 genotypes at zero P rate, GY_2 is grain yield at high P rate, and AGY_2 is average grain yield of 10 genotypes at high P rate. Genotypes that produced GYEI more than 1.0 were classified as efficient, GYEI between 0.50 and 1.0 as moderately efficient, and GYEI less than 0.50 as inefficient. This index is commonly used to separate crop genotypes that are nutrient efficient and nutrient inefficient (Fageria 2009).

Based on GYEI, genotypes were classified as efficient and inefficient groups for P use efficiency (see Table 3.13). Two genotypes (BRA 02601 and BRA 032051) were classified as moderately efficient in P use efficiency, and the remaining eight genotypes were classified as efficient in P use efficiency. The P use efficiency of genotypes was in the order of BRA 032048 > BRA 032033 > BRA 042093 > BRS Sertaneja > BRA 052015 > BRA Primavera > BRA 01600 > BRA 01506 at 100 kg. The difference in P use efficiency of rice genotypes may be related to variations in P uptake and UE (Fageria 2009). Variation in growth of upland rice genotypes at two P levels is shown in Figure 3.39. Similarly, variation in root growth of upland rice genotypes is shown in Figures 3.40 and 3.41.

FIGURE 3.39 Response of two upland rice genotypes to P fertilization grown on Brazilian Oxisol.

FIGURE 3.40 Root growth of two upland rice genotypes at two P levels.

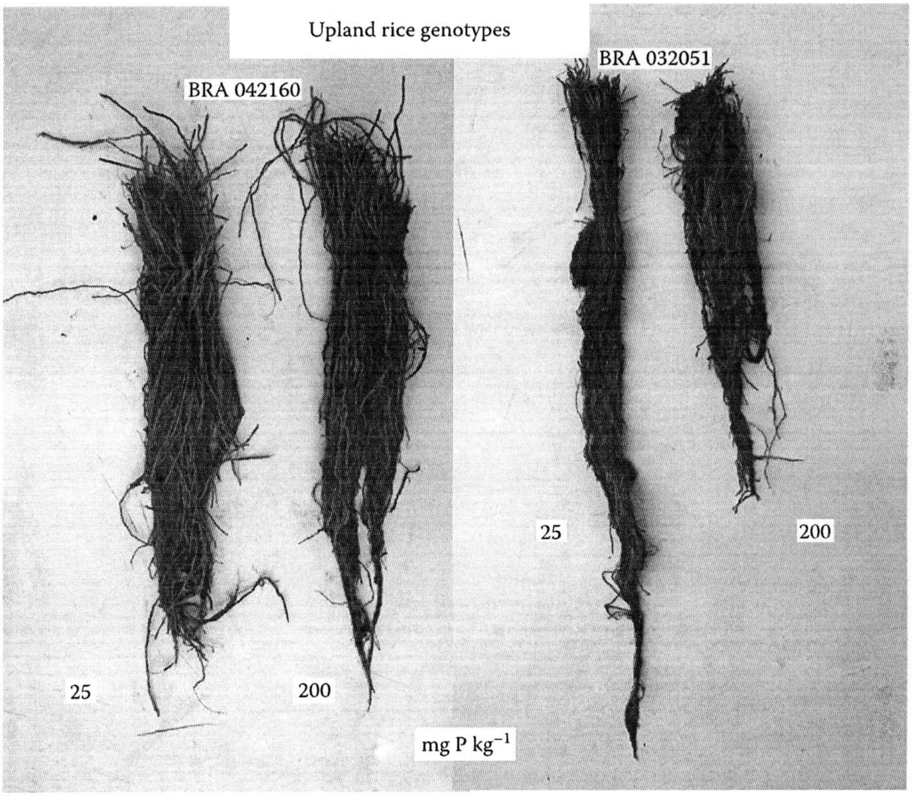

FIGURE 3.41 Root growth of two upland rice genotypes at two P levels.

3.9 CONCLUSIONS

Phosphorus is one of the nutrients that limit rice yield in most of the world's rice-growing regions. Its deficiency is more serious in tropical acidic soils due to low natural levels and high immobilization capacities of these soils. In acidic soils, most of the P is immobilized by iron, manganese, and aluminum oxides. Phosphorus participates in many biochemical and physiological plant functions. Using adequate rates of P improved root growth, which has special importance for upland rice in the absorption of water and nutrients. Phosphorus deficiency decreased tillering, LAI, grain weight, and—consequently—yield. Phosphorus deficiency was also responsible for low rice grain quality. Because phosphorus is a mobile nutrient in plants, a deficiency first appears in the older leaves. The P-deficient rice plants turn dark green in color, and in severe cases, leaves turn orange. The concentration of P in plant tissue is higher during the vegetative growth stage and decreases with the plant's age. Phosphorus uptake (concentration × dry matter) increases with advanced plant age and follows the pattern of dry matter accumulation (grain plus straw). Phosphorus accumulation varies with yield level. However, rice plants accumulate about 28 kg P ha^{-1} in straw and grain when grain yield is more than

6 Mg ha^{-1}. Overall, in order to produce 1 Mg grain, rice accumulates about 4 kg P in grain and straw. Internal P use efficiency in rice is much higher than that of N and K use efficiency. However, P recovery efficiency of rice from soil is much lower (<20%) compared with N (≈40% lowland rice and ≈45% upland rice) and K (>40%) calculated by difference method. PHI varies with genotypes and environmental conditions, but overall it is about 72%. Therefore, more P accumulates in grain than in straw. For rice, higher economic yield is obtained when appropriate P management practices are adopted. These practices include adequate rate, appropriate source, and methods and timing of application. In addition, liming acidic soils; controlling insects, diseases, and weeds; and using P-efficient genotypes can improve yield and P use efficiency. The adequate rate of P should be determined by soil test calibration studies. As P is an immobile nutrient in soil, in acidic and alkaline soils, it should be applied in furrows near the seeds in order to have higher uptake and lesser immobilization. Water-soluble P sources are superior compared with water-insoluble P sources (such as rock phosphates).

REFERENCES

Abekoe, M. K., and K. L. Saharawat. 2001. Phosphate retention and extractability in soils of the humid zone in West Africa. *Geoderma.* 102:175–187.

Abrams, M. M., and W. M. Jarrell. 1995. Soil phosphorus as a potential nonpoint source for elevated stream phosphorus levels. *J. Environ. Qual.* 24:132–138.

Anjos, J. T., and D. L. Rowell. 1987. The effect of lime on phosphorus adsorption and barley growth in three acid soils. *Plant Soil.* 103:75–82.

Baligar, V. C., N. K. Fageria, and Z. L. He. 2001. Nutrient use efficiency in plants. *Commun. Soil Sci. Plant Anal.* 32:921–950.

Barber, S. A. 1995. *Soil Nutrient Bioavailability: A Mechanistic Approach*, 2nd edition. New York: John Wiley.

Bartos, J. M., G. L. Mullins, J. C. Williams, F. J. Sikora, and J. P. Copeland. 1992. Water-insoluble impurity effects on phosphorus availability in monoammonium phosphate fertilizers. *Soil Sci. Soc. Am. J.* 56:972–976.

Bennett, W. F. 1993. Plant nutrient utilization and diagnostic plant symptoms. In *Nutrient Deficiencies and Toxicities in Crop Plants*, ed. W. F. Bennett, pp. 1–7. St. Paul, MN: American Phytopathological Society.

Bolan, N. S., D. C. Adriano, and D. Curtin. 2003. Soil acidification and liming interactions with nutrient and heavy metal transformation and bioavailability. *Adv. Agron.* 78:215–272.

Bolan, N. S., J. Elliott, P. E. H. Gregg, and S. Weil. 1997. Enhanced dissolution of phosphate rocks in the rhizosphere. *Biol. Fert. Soils.* 24:169–174.

Bolan, N. S., R. Naidu, J. K. Syers, and R. W. Tillman. 1999. Surface charge and solute interactions in soils. *Adv. Agron.* 67:88–141.

Bolland, M. D. A., and R. J. Gilkes. 1990. Rock phosphates are not effective fertilizers in Western Australia soils: A review. *Fert. Res.* 22:79–85.

Brady, N. C., and R. R. Well. 2002. *The Nature and Properties of Soils*, 13th edition. Upper Saddle River, NJ: Prentice Hall.

Campbell, J. M., and U. Schwertmann. 1984. Iron oxides mineralogy of placic horizons. *J. Soil Sci.* 35:569–582.

Chang, S. C., and M. L. Jackson. 1957. Fractionation of soil phosphorus. *Soil Sci.* 84:133–144.

Cheaney, R. L., and P. R. Jennings. 1975. *Field Problems of Rice in Latin America.* Cali, Colombia: Centro Internacional de Agricultura Tropical.

Curi, N., and O. A. Camargo. 1988. Phosphorus adsorption characteristics of Brazilian Oxisols. In *Proceedings of the 8th International Soil Classification Workshop*. Part 1. eds. F. H. Beinroth, M. N. Camargo, and H. Eswarn, pp. 56–63. Washington, DC: Soil Management Support Services, U.S. Department of Agriculture.

Curtin, D., and J. K. Syers. 2001. Lime-induced changes in indices of phosphate availability. *Soil Sci. Soc. Am. J.* 65:147–152.

Edmeades, D. C., and K. W. Perrott. 2004. The calcium requirements of pastures in New Zealand: A review. *N. Z. J. Agric. Res.* 47:11–21.

Eghball, B., D. H. Sander, and J. Skopp. 1990. Diffusion, adsorption and predicted longevity of banded phosphorus fertilizer in three soils. *Soil Sci. Soc. Am. J.* 54:1161–1165.

Engelstad, O. P., and G. L. Terman. 1980. Agronomic effectiveness of phosphate fertilizers. In *The Role of Phosphorus in Agriculture*, ed. R. C. Dinauer, pp. 311–332. Madison, WI: ASA, CSSA, and SSSA.

Evans, L. K., and G. W. Smillie. 1976. Extractable iron and aluminum and their relationship to phosphate retention in Irish soils. *Ir. J. Agric. Res.* 15:65–73.

Fageria, N. K. 1980. Influence of phosphorus application on growth, yield and nutrient uptake by irrigated rice. *R. Bras. Ci. Solo.* 4:26–31.

Fageria, N. K. 1984. *Fertilization and Mineral Nutrition of Rice*. Rio de Janeiro/Brasilia: EMBRAPA-CNPAF/Editora Campus.

Fageria, N. K. 1989. *Tropical Soils and Physiological Aspects of Crop Production*. Brasilia: EMBRAPA-DPU, Brasilia, EMBRAPA-CNPAF. Document, 18.

Fageria, N. K. 1992. *Maximizing Crop Yields*. New York: Marcel Dekker.

Fageria, N. K. 1996. *The Study of Liming and Fertilization for Rice and Common Bean in Cerrado Region*. Goiania, Brazil: National Rice and Bean Research Center.

Fageria, N. K. 2001. Nutrient management for improving upland rice productivity and sustainability. *Commun. Soil Sci. Plant Anal.* 32:2603–2629.

Fageria, N. K. 2007. Yield physiology of rice. *J. Plant Nutr.* 30:843–879.

Fageria, N. K. 2009. *The Use of Nutrients in Crop Plants*. Boca Raton, FL: CRC Press.

Fageria, N. K., and V. C. Baligar. 1996. Response of lowland rice and common bean grown in rotation to soil fertility levels on a Varzea soil. *Fert. Res.* 45:13–20.

Fageria, N. K., and V. C. Baligar. 2001. Improving nutrient use efficiency of annual crops in Brazilian acid soils for sustainable crop production. *Commun. Soil Sci. Plant Anal.* 32:1303–1319.

Fageria, N. K., and V. C. Baligra. 2008. Ameliorating soil acidity of tropical Oxisols by liming for sustainable crop production. *Adv. Agron.* 99:345–399.

Fageria, N. K., V. C. Baligar, and R. B. Clark. 2006. *Physiology of Crop Production*. New York: Haworth.

Fageria, N. K., V. C. Baligar, and C. A. Jones. 2011. *Growth and Mineral Nutrition of Field Crops*, 3rd edition. Boca Raton, FL: CRC Press.

Fageria, N. K., V. C. Baligar, and Y. C. Li. 2008. The role of nutrient efficient plants in improving crop yields in the twenty first century. *J. Plant Nutr.* 31:1121–1157.

Fageria, N. K., and M. P. Barbosa Filho. 1994. *Nutritional Deficiency in Rice: Identification and Correction*. Goiania, Brazil: EMBRAPA Arroz e Feijão.

Fageria, N. K., and M. P. Barbosa Filho. 2007. Dry matter and grain yield, nutrient uptake, and phosphorus use efficiency of lowland rice as influenced by phosphorus fertilization. *Commun. Soil Sci. Plant Anal.* 38:1289–1297.

Fageria, N. K., M. P. Barbosa Filho, and J. R. P. Carvalho. 1982. Response of upland rice to phosphorus fertilization on an Oxisol of central Brazil. *Agron. J.* 74:51–56.

Fageria, N. K., M. P. Barbosa Filho, L. F. Stone, and C. M. Guimaraes. 2004. Phosphorus nutrition in upland rice production. In *Phosphorus in Brazilian Agriculture*, eds. T. Yamada and S. R. S. Abdalla, pp. 401–418. Piracicaba: Brazilian Phosphorus and Potassium Institute.

Fageria, N. K., and H. R. Gheyi. 1999. *Efficient Crop Production*. Campina Grande, Brazil: Federal University of Paraiba.
Fageria, N. K., O. P. Morais, V. C. Baligar, and R. J. Wright. 1988. Response of rice cultivars to phosphorus supply on an Oxisol. *Fert. Res.* 16:195–206.
Fageria, N. K., and A. B. Santos. 2002. Lowland rice genotypes evaluation for phosphorus use efficiency. *J. Plant Nutr.* 25:2793–2802.
Fageria, N. K., and A. B. Santos. 2005. Influence of base saturation and micronutrient rates on their concentration in the soil and bean productivity in Cerrado soil in no-tillage system. Paper presented at the *VIII National Bean Congress*, Goiânia, Brazil, October.
Fageria, N. K., and A. B. Santos. 2008. Lowland rice response to thermophosphate fertilization. *Commun. Soil Sci. Plant Anal.* 39:873–889.
Fageria, N. K., A. B. Santos, and V. C. Baligar. 1997. Phosphorus soil test calibration for lowland rice on an Inceptisol. *Agron. J.* 89:737–742.
Fageria, N. K., A. B. Santos, and A. B. Heineman. 2011. Lowland rice genotypes evaluation for phosphorus use efficiency in tropical lowland. *J. Plant Nutr.* 34:1087–1095.
Fageria, N. K., A. B. Santos, I. D. G. Lins, and S. L. Camargo. 1997. Characterization of fertility and particle size of Várzea soils of Mato Grosso and Mato Grosso do Sul states of Brazil. *Commun. Soil Sci. Plant Anal.* 28:37–47.
Fageria, N. K., N. A. Slaton, and V. C. Baligar. 2003. Nutrient management for improving lowland rice productivity and sustainability. *Adv. Agron.* 80:63–152.
Fageria, N. K., R. J. Wright, and V. C. Baligar. 1988. Rice cultivar evaluation for phosphorus use efficiency. *Plant Soil.* 11:105–109.
Fageria, N. K., R. J. Wright, V. C. Baligar, and C. M. R. Sousa. 1991. Characterization of physical and chemical properties of Varzea soils of Goias state of Brazil. *Commun. Soil Sci. Plant Anal.* 22:1631–1646.
Fageria, N. K., F. J. P. Zimmermann, and A. M. Lopes. 1977. Irrigated rice response to phosphorus, zinc and lime application. *Rev. Brs. Ci. Solo.* 1:72–76.
Fiedler, R. J., D. H. Sander, and G. A. Peterson. 1989. Fertilizer phosphorus recommendations for winter wheat in terms of method of phosphorus application, soil pH, and yield goal. *Soil Sci. Soc. Am. J.* 53:1282–1287.
Foth, H. D., and B. G. Ellis. 1988. *Soil Fertility.* New York: John Wiley.
Fox, R. L., and E. J. Kamprath. 1970. Phosphate sorption isotherms for evaluating the phosphate requirements of soils. *Soil Sci. Soc. Am. Proc.* 34:902–907.
Friesen, D. K., I. M. Rao, R. J. Thomas, A. Obserson, and J. I. Sanz. 1997. Phosphorus acquisition and cycling in crop and pasture system in low fertility tropical soils. In *Plant Nutrition for Sustainable Food Production and Environment*, eds. T. Ando, K. Fujita, T. Mae, H. Matsumoto, S. Mori, and J. Sekiya, pp. 493–498. Dordrecht, The Netherlands: Kluwer Academic.
Goedert, W. J. 1989. Cerrado region: Agricultural potential and politics for its development. *Pesq. Agropec. Bras.* 24:1–17.
Goswami, N. N., and N. K. Banerjee. 1978. Phosphorus, potassium, and other macroelements. In *Soils and Rice*, ed. IRRI, pp. 561–580. Los Banos, Philippines: IRRI.
Haynes, R. J. 1982. Effects of liming on phosphate availability in acid soils. A critical review. *Plant Soil.* 68:289–308.
Haynes, R. J. 1984. Lime and phosphate in the soil-plant system. *Adv. Agron.* 37:249–315.
Hewitt, E. J. 1963. The essential nutrients: Requirements and interactions in plants. In *Plant Physiology*, ed. F. C. Steward, pp. 137–360. New York: Academic Press.
Hossner, L. R., J. A. Freeouf, and B. L. Folsom. 1973. Solution phosphorus concentration and growth of rice (*Oryza sativa L.*) in flooded soils. *Soil Sci. Soc. Am. Proc.* 37:405–408.
Ishizuka, S. 1978. *Nutrient Deficiencies of Crops*. Taipei, Taiwan: Food & Fertilizer Technology Center.

Khalid, R. A., W. H. Patrick Jr., and F. J. Peterson. 1979. Relationship between rice yield and soil phosphorus evaluated under aerobic and anaerobic conditions. *Soil Sci. Plant Nutr.* 25:155–164.

Lopes, A. S., and F. R. Cox. 1977. A survey of the fertility status of surface soils under Cerrado vegetation in Brazil. *Soil Sci. Soc. Am. J.* 41:742–747.

Mahler, R. L., and R. E. McDole. 1985. The influence of lime and phosphorus on crop production in northern Idaho. *Commun. Soil Sci. Plant Anal.* 16:485–499.

Mansell, G. P., R. M. Pringle, D. C. Edmeades, and P. W. Shannon. 1984. Effects of lime on pasture production on soils in the North Island of New Zealand. III. Interaction of lime with phosphorus. *N. Z. J. Agric. Res.* 27:363–369.

Marschner, H. 1995. *Mineral Nutrition of Higher Plants*, 2nd edition. New York: Academic Press.

Mengel, K., A. Kirkby, H. Kosegarten, and T. Appel. 2001. *Principles of Plant Nutrition*, 5th edition. Dordrecht, The Netherlands: Kluwer Academic.

Mueller, K. E. 1974. *Field Problems of Tropical Rice*. Los Banos, Philippines: IRRI.

Mullins, G. L. 1988. Plant availability of P in commercial superphosphate fertilizers. *Commun. Soil Sci. Plant Anal.* 19:1509–1525.

Mullins, G. L., and F. J. Sikora. 1990. Field evolution of commercial monoammonium phosphate fertilizers. *Soil Sci. Soc. Am. J.* 54:1469–1472.

Narro, L., S. Pandey, C. D. Leon, F. Salazar, and M. P. Arias. 2001. Implication of soil-acidity tolerant maize cultivars to increase production in developing countries. In *Plant Nutrient Acquisition: New Perspectives*, eds. N. Ae, J. Arihara, K. Okada, and A. Srinivasan, pp. 447–463. Tokyo: Springer.

Nurlaeny, N., H. Marschner, and E. George. 1996. Effects of liming and mycorrhizal colonization on soil phosphate depletion and phosphate uptake by maize (*Zea mays* L.) and soybean (*Glycine max* L.) grown in two tropical acid soils. *Plant Soil.* 181:275–285.

Osaki, M. 1995. Ontogenetic changes of N, P, and K contents in individual leaves of field crops. *Soil Sci. Plant Nutr.* 41:429–438.

Osaki, M., K. Morikawa, T. Shinano, M. Urayama, and T. Tadano. 1991. Productivity of high-yielding crops. II. Comparison of N, P, K, Ca, and Mg accumulation and distribution among high-yielding crops. *Soil Sci. Plant Nutr.* 37:445–454.

Patrick, W. H., Jr., and I. C. Mahapatra. 1968. Transformation and availability to rice of nitrogen and phosphorus in waterlogged soils. *Adv. Agron.* 20:323–359.

Patrick, W. H., Jr., and D. S. Mikkelsen. 1971. Plant nutrient behavior in flooded soil. In *Fertilizer Technology and Use*, 2nd edition, ed. R. A. Olson, pp. 187–215. Madison, WI: SSSA.

Patrick, W. H., Jr., and C. N. Reddy. 1978. Chemical changes in rice soils. In *Soils and Rice*, ed. IRRI, pp. 361–379. Los Banos, Philippines: IRRI.

Peng, S., R. C. Laza, R. M. Vesparas, A. L. Sanico, K. G. Cassman, and G. S. Khush. 2000. Grain yield of rice cultivars and lines developed in the Philippines since 1966. *Crop Sci.* 40:307–314.

Ponnamperuma, F. N. 1972. The chemistry of submerged soils. *Adv. Agron.* 24:29–96.

Raij, B. V. 1991. *Soil Fertility and Fertilization*. São Paulo, Brazil: Editora Agronomica Ceres.

Roy, A. C., and S. K. De Datta. 1985. Phosphate sorption isotherm for evaluating phosphorus requirements of wetland rice soils. *Plant Soil.* 86:185–196.

Sah, R. N., and D. S. Mikkelsen. 1986. Transformations of inorganic phosphorus during the flooding and draining cycles of soil. *Soil Sci. Soc. Am. J.* 50:62–67.

Sanchez, P. A., and T. J. Logan. 1992. Myth and science about the chemistry and fertility of soils in the tropics. In *Myths and Science of the Soils of the Tropics*, eds. R. Lal and P. A. Sanchez, pp. 35–46. Madison, WI: SSSA.

Sanchez, P. A., and J. G. Salinas. 1981. Low input technology for managing Oxisols and Ultisols in tropical America. *Adv. Agron.* 34:279–398.

Sanyal, S. K., and S. K. De Datta. 1991. Chemistry of phosphorus transformations in soil. *Adv. Soil Sci.* 16:1–120.

Sahrawat, K. L., and M. Sika. 2002. Direct and residual phosphorus effects on soil test values and their relationships with grain yield and phosphorus uptake of upland rice on an Ultisol. *Commun. Soil Sci. Plant Anal.* 33:321–332.

Shahandeh, H., L. R. Hossner, and F. T. Turner. 1994. Phosphorus relationships in flooded rice soils with low extractable phosphorus. *Soil Sci. Soc. Am. J.* 58:1184–1189.

Sims, J. L., and G. A. Place. 1968. Growth and nutrient uptake of rice at different growth stages and nutrient levels. *Agron. J.* 60:692–696.

Singh, Y., A. Dobermann, B. Singh, K. F. Bronson, and C. S. Khind. 2000. Optimal phosphorus management strategies for wheat-rice cropping on a loamy sand. *Soil Sci. Soc. Am. J.* 64:1413–1422.

Smyth, T. J., and M. S. Cravo. 1990. Phosphorus management for continuous corn-cowpea production in a Brazilian Amazon Oxisol. *Agron. J.* 82:305–309.

Smyth, T. J., and M. S. Cravo. 1992. Aluminum and calcium constraints to continuous crop production in a Brazilian Oxisol. *Agron. J.* 84:843–850.

Sumner, M. E., and A. D. Noble. 2003. Soil acidification: The world story. In *Handbook of Soil Acidity*, ed. Z. Rengel, pp. 1–28. New York: Marcel Dekker.

Tisdale, S. L., W. L. Nelson, and J. D. Beaton. 1985. *Soil Fertility and Fertilizers*, 4th edition. New York: Macmillan.

Turner, F. T., and J. W. Gilliam. 1976. Increased P processes in flooded soils. *Plant Soil.* 45:365–377.

Von Uexkull, H., and Mutert, E. 1995. Global extent, development and economic impact of acid soils. *Plant Soil.* 171:1–15.

Wallace, T. 1961. *The Diagnosis of Mineral Deficiencies in Plants by Visual Symptoms*, 2nd edition. New York: Chemical Publishing.

Wells, B. R., R. K. Bacon, R. Dilday, J. T. Kelly, and P. A. Dickson. 1995. Response of wheat following rice to fall fertilization. In *Arkansas Soil Fertility Studies 1994*, ed. W. E. Sabbe, pp. 16–20. Fayetteville, AR: Arkansas Agricultural Experiment Station. Research Series 443.

Westfall, D. G., W. T. Flinchum, and J. W. Stansel. 1973. Distribution of nutrients in the rice plant and effect of two nitrogen levels. *Agron. J.* 65:236–238.

Willett, I. R. 1982. Phosphorus availability in soils subjected to short periods of flooding and drying. *Aust. J. Soil Res.* 20:131–138.

Willett, I. R. 1986. Phosphorus dynamics in relation to redox processes in flooded soils. *Trans. 13th Hamb. Int.Cong. Soil Sci. Soc.* 6:748–755.

Willett, I. R. 1989. Causes and predictions of changes in extractable phosphorus during flooding. *Aust. J. Soil Res.* 27:45–54.

Willett, I. R., and M. L. Higgins. 1978. Phosphorus sorption by reduced and reoxidized rice soils. *Aust. J. Soil Res.* 16:319–326.

Witt, C., A. Dobermann, S. Abulrachman, H. C. Gines, G. H. Wang, R. Nagrajan, S. Satawathananont, et al. 1999. Internal nutrient efficiencies of irrigated lowland rice in tropical and subtropical Asia. *Field Crops Res.* 63:113–138.

Yoshida, S. 1981. *Fundamentals of Rice Crop Science*. Los Banos, Philippines: IRRI.

Yost, R. S., E. J. Kamparth, E. Lobato, and G. C. Naderman. 1979. Phosphorus response of corn on an Oxisol as influenced by rates and placement. *Soil Sci. Soc. Am. J.* 43:338–343.

4 Potassium

4.1 INTRODUCTION

Potassium (K), which limits plant growth and development, is the third most important nutrient after nitrogen and phosphorus. In most of the rice-growing soils, the response of rice to potassium fertilization is not as high as it is to nitrogen and phosphorus (Fageria, Slaton, and Baligar 2003). Figure 4.1 shows the response of upland rice grown on Brazilian Oxisol to nitrogen, phosphorus, and potassium fertilization. Rice growth was most affected by phosphorus, followed by nitrogen and potassium. Therefore, the response of rice to N, P, and K was in the order of phosphorus > nitrogen > potassium. The reason for the low response may be related to the availability of nonexchangeable K when the K level falls below the critical limit. Many soils used for the continuous production of rice or for rice–wheat rotations can be cropped for extended periods without needing supplemental K to maintain production (Dobermann et al. 1996). Rice grown in rotation with legumes, such as soybeans, may require annual K input due to the greater K requirement of the legume crop (Fageria, Stone, and Santos 2003).

Lowland and upland rice, especially high-yielding cultivars, absorb potassium in greater quantities than any other essential nutrient (Tables 4.1 through 4.3). De Datta and Mikkelsen (1985) reported that a single lowland rice crop producing 9.8 Mg ha^{-1} of grain in about 115 days took up 218 kg N ha^{-1}, 31 kg P ha^{-1}, 258 kg K ha^{-1}, and 9 kg S ha^{-1}. As grain yield increases, the demand for plant nutrients (particularly K) increases (Fageria 1984). Therefore, adequate annual application of K to rice crops is essential for maintaining soil productivity and the sustainability of rice-based cropping systems.

The response of upland and lowland rice to potassium is not well-documented. Data related to long-term K experiments with rice are lacking, and short-term experiments do not provide a real picture of the response of rice to K fertilization (Fageria et al. 1990a, 1990b). In South America, the savannas of Colombia, Venezuela, Bolivia, and Guyana, as well as the Cerrado region of Brazil and other areas, are important in the production of crops such as upland rice. In these areas, most soils are Oxisols and Ultisols, which are generally characterized by extreme acidity and low levels of available nutrients. Deficiency of N, P, K, Ca, Mg, S, and Zn are very common in the Oxisol–Ultisol regions of South America (Sanchez and Salinas 1981; Fageria et al. 1990b). Portela (1993) also reported that old, highly weathered soils are often low in potassium. Similarly, lowland areas in South America are mostly dominated by Inceptisols; these soils are acidic in reaction and deficient in K (Fageria et al. 2010). In addition, the introduction of high-yielding rice cultivars has increased the response of upland and lowland rice to K fertilization (Fageria, Slaton, and Baligar 2003; Fageria, Baligar, and Jones 2011). Figures 4.2 and 4.3 show the

FIGURE 4.1 (See color insert.) Response of upland rice, grown on Brazilian Oxisol, to N, P, and K fertilization.

TABLE 4.1
Nutrient Accumulation in the Straw of Lowland Rice as Influenced by Phosphorus Treatments at Harvest

P Rate	N	P	K	Ca	Mg
		kg ha^{-1}			
0	30.5	2.7	57.5	9.2	4.6
131	61.0	7.4	117.5	19.5	11.9
262	56.0	8.1	120.8	18.4	12.1
393	59.5	7.7	133.5	17.9	13.7
524	69.8	11.3	138.0	19.5	14.8
655	46.3	10.0	132.1	18.4	11.2
F-test					
Year (Y)	**	**	**	NS	NS
P rate (P)	**	**	**	*	**
Y × P	NS	NS	NS	NS	NS
CV (%)	38	38	29	33	36
R^2	0.31*	0.54**	0.62**	0.43**	0.59**

Source: Adapted from Fageria, N. K., and A. B. Santos. 2008. Lowland rice response to thermophosphate fertilization. *Commun. Soil Sci. Plant Anal.* 39:873–889.

Note: Values are averaged across two years.

^{*,**,NS}Significant at the 5 and 1% probability levels and nonsignificant respectively.

TABLE 4.2
Nutrient Accumulation in the Grain of Lowland Rice as Influenced by Phosphorus Treatments at Harvest

P Rate	N	P	K	Ca	Mg
		kg ha^{-1}			
0	13.8	1.4	2.9	0.8	0.9
131	39.4	5.1	8.9	2.4	3.7
262	52.2	9.5	10.8	3.4	5.3
393	63.5	11.7	14.6	4.4	6.2
524	56.2	10.9	13.6	4.2	5.9
655	59.3	11.3	12.9	4.0	6.2
F-test					
Year (Y)	**	**	*	**	NS
P rate (P)	**	**	**	**	**
Y × P	NS	NS	NS	NS	NS
CV (%)	21	18	30	35	30
R^2	0.90**	0.95**	0.80**	0.78**	0.83**

Source: Adapted from Fageria, N. K., and A. B. Santos. 2008. Lowland rice response to thermophosphate fertilization. *Commun. Soil Sci. Plant Anal.* 39:873–889.

Note: Values are averaged across two years.

*,**,NS Significant at the 5 and 1% probability levels and nonsignificant respectively.

TABLE 4.3
Uptake of Macro- and Micronutrients by Upland Rice Grown on Brazilian Oxisol

Nutrient	Straw	Grain	Total	Required to Produce 1 Mg Grain[1]
N (kg ha^{-1})	56	70	126	28
P (kg ha^{-1})	3	9	12	3
K (kg ha^{-1})	150	56	206	45
Ca (kg ha^{-1})	23	4	27	6
Mg (kg ha^{-1})	12	5	19	4
Zn (g ha^{-1})	161	138	299	65
Cu (g ha^{-1})	35	57	92	20
Fe (g ha^{-1})	654	117	771	169
Mn (g ha^{-1})	1319	284	1603	351
B (g ha^{-1})	53	30	83	18

Source: Adapted from Fageria, N. K., and L. F. Stone. 2008. Micronutrient deficiency problems in South America. In *Micronutrient Deficiencies in Global Crop Production*, ed. B. J. Alloway, pp. 245–266. New York: Springer.

[1] Accumulation of nutrients in the straw and grain to produce 1 Mg grain.

FIGURE 4.2 (See color insert.) Response of upland rice, grown on Brazilian Oxisol, to potassium fertilization.

response of upland and lowland rice to K fertilization applied to Oxisols (upland rice) and Inceptisols (lowland rice).

In the last few decades, crop yields in developed as well as developing countries have significantly increased through the introduction of modern production technologies, but supplies of K^+ in the soils have been rapidly depleted. Pretty and Stangel (1985) reported that 17% of Africa's total land area, 21% of Asia's total land area, and 29% of Latin America's total land area are K^+ deficient. Most of the K^+-deficient soils on these three continents are acidic savanna. Buol et al. (1975) estimated that one-quarter of the soils in the tropics and subtropics have a low K^+ status. Fageria et al. (1990a) reported that 50% of Amazon Basin soil has low K^+ reserves. Fageria (1989) also reported that many soils of the tropical and temperate regions are unable to supply sufficient K^+ to field crops. Therefore, application of this element in adequate amounts is essential for obtaining optimal crop yields and maintaining soil fertility for sustainable crop production.

Foth and Ellis (1988) found that most of the soils in the western half of the United States are minimally or moderately weathered and contain 1.7–2.5% K. By contrast, many soils of the southern coastal plains contain only 0.3% K. Therefore, in the United States, crops grown on high K soils do not respond to

FIGURE 4.3 Response of lowland rice, grown on Brazilian Inceptisol, to potassium fertilization.

K, but crops grown on low K soils do respond to K fertilization. Rice is highly efficient in scavenging plant available soil K due to its fibrous root system and the increased availability of K after flooding. However, on some soils, K deficiency may occur if rice and rotation crops are grown without regular applications of K fertilizer to replace the K removed by the harvested crops. Prior to the early 1990s, K deficiencies were rare in U.S. rice-producing areas. However, K deficiency is now recognized as an annual problem in many soils because rice and rotation crop yields have increased, soils have been mined of K, and production practices have changed (Slaton, Cartwright, and Wilson 1995; Williams and Smith 2001; Fageria, Slaton, and Baligar 2003).

More than 95% of the world's KCl is consumed by the agricultural sector (Stone 2008). World consumption of muriate of potash (fertilizer grade KCl) increased from an estimated 38 Tg in 2001 to 49 Tg in 2007 (Stone 2008; Nelson et al. 2010). Global K removal from soils has far exceeded KCl use (Nelson et al. 2010). From 1997 to 1999, KCl applications replaced approximately 73% of K removed from soils in developed countries; however, in developing countries, KCl applications replaced only 20% of K removed (Nelson et al. 2010). Because potassium deficiency in annual crops is widely reported in many countries (Fageria 2009; Fageria, Santos, and Moraes 2010; Nelson et al. 2010), the objective of this chapter is to discuss the worldwide importance of K in rice production, the K cycle in soil–plant systems, K deficiency symptoms, K uptake and use efficiency in rice, and management practices to maximize upland as well as lowland rice yields.

4.2 CYCLE IN SOIL–PLANT SYTEM

The potassium cycle involves addition, mineralization, transformation, adsorption, leaching, and uptake by crop plants (Figure 4.4). Major additional K sources are organic manures, inorganic chemical fertilizers, and weathering of soil parent materials (mainly feldspar and mica). Feldspar is composed of microcline ($KAlSi_3O_8$) and orthoclase ($KAlSi_3O_8$). Mica is composed of muscovite [$K(Si_3Al)Al_2O_{10}(OH)_2$] and biotite [$K(Si_3Al)(Mg,Fe^{2+})_3O_{10}(OH)_2$]. Feldspars have a higher K content than micas. Potassium is also found in secondary clay minerals, such as illite, vermiculite, chlorite, and montmorillonite, in small concentrations.

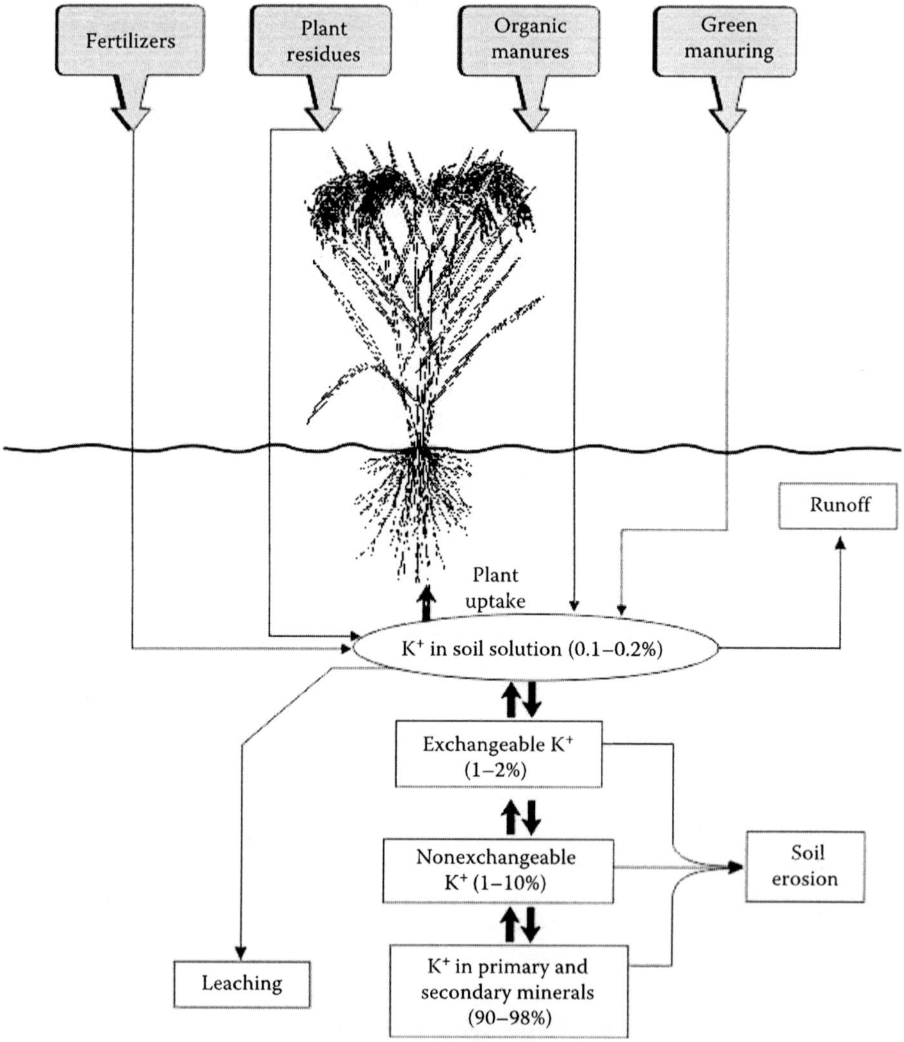

FIGURE 4.4 Potassium cycle in soil–plant system. (From Fageria, N. K. 2009. *The Use of Nutrients in Crop Plants*. Boca Raton, FL: CRC Press.)

Soil K can be divided into four fundamental categories: (1) 0.1–0.2% as soil solution K, (2) 1–2% as exchangeable K, (3) 1–10% as nonexchangeable K, and (4) 90–98% as mineral K (Barber 1995; Brady and Weil 2002; see Figure 4.4). Potassium ions move from one category to another whenever the removal or addition of K disturbs the equilibrium between these soil K pools. Equilibration between the soil solution and exchangeable K pools is rapid and usually complete within hours. However, the equilibration time between nonexchangeable and exchangeable K is much slower, requiring days or even months. The conversion of K from the mineral form via weathering is extremely slow, varies among the soil K minerals, and has little significance in supplying plants with K during a single season (Barber 1995). Most of the K used by plants during a single season comes from the soil solution and exchangeable K pools and is supplemented by K released from the nonexchangeable pools. The quantity of nonexchangeable K^+ in soils depends on the clay content and the type of clay minerals (Sparks and Huang 1985).

Soil solution K can be adsorbed on soil colloids, can be taken up by plants, or may be lost by leaching in humid climates or by flooded rice. Equilibrium may thus be established between adsorbed K^+ and free K^+ in soil solution. The K^+ level in the soil solution resulting from this equilibrium depends a great deal on the selectivity of the adsorption sites. If there are many adsorption sites for K^+, the concentration of K^+ in the soil solution tends to be low and vice versa (Mengel et al. 2001).

Potassium is considered relatively immobile in soil and moves primarily by diffusion in the soil–plant system, especially in upland crops. Teo et al. (1995) found that diffusion, mass flow, and contact exchange accounted for ~57.8, 42, and <0.3%, respectively, of K uptake by flooded rice. Although considered immobile, a significant amount of K can be lost via leaching on some soils following displacement from the exchange complex after flooding. Leaching is a significant problem in the humid tropical regions having acidic soils with low cation-exchange capacities. Liming an acidic soil to raise its pH can reduce leaching losses of K because of the complementary ion effect and increasing the soil cation exchange capacity (CEC) (Brady and Weil 2002).

Leaching is high in sandy soils and soils having low CEC. Data in Table 4.4 show soil extractable K in upland and lowland rice at different soil depths after several cultivations. There was a significant effect of K level and depths in upland rice, but in lowland rice, statistical significance was observed only for soil depth. The leaching of K in both the cultures was evident, especially at higher K rates. Potassium may also be lost by soil erosion. The fixation may occur spontaneously with some minerals in aqueous suspensions or as a result of heating to remove interlayer water in others. Fixed K ions are exchangeable only after expansion of the interlayer space (Soil Science Society of America 2008). In soils where upland and lowland rice are grown, the leaching and fixation of K are not as great as those of N and P.

Like N, potassium present in soil solution as a K^+ cation does not form any gas that can be lost to the atmosphere. Its transformation or solubilization in the soil is related to CEC and mineral weathering rather than to microorganisms (Brady and Weil 2002). Further, unlike N and P, it does not cause environmental problems such as eutrophication or toxic gases when removed from the soil–plant system.

TABLE 4.4
Effect of K Fertilizer Application Rates on Soil Extractable K (mg kg^{-1}) at Different Soil Depths in Upland and Lowland Rice

K Rate (kg ha^{-1})	Soil Depth (cm)	Upland Rice	Lowland Rice
0	0–20	35	27
	20–40	19	22
	40–60	14	26
	60–80	10	23
42	0–20	45	28
	20–40	25	24
	40–60	15	24
	60–80	12	20
84	0–20	53	32
	20–40	34	29
	40–60	18	28
	60–80	14	23
126	0–20	58	31
	20–40	42	31
	40–60	22	28
	60–80	17	24
168	0–20	63	33
	20–40	50	29
	40–60	28	27
	60–80	19	22
F-test			
K level (K)		**	NS
Soil depth (SD)		**	*
K × SD		**	NS

Source: Adapted from Fageria, N. K., V. C. Baligar, R. J. Wright, and J. R. P. Carvalho. 1990a. Lowland rice response to potassium fertilization and its effect on N and P uptake. *Fertilizer Res.* 21:157–162; Fageria, N. K., R. J. Wright, V. C. Baligar, and J. R. P. Carvalho. 1990b. Upland rice response to potassium fertilization on a Brazilian Oxisol. *Fertilizer Res.* 21:141–147.

Note: The soil K was determined after the fourth cultivation in upland rice and after the fifth cultivation in lowland rice. Potassium was applied each year as broadcast using potassium chloride fertilizer.

*,**,NS Significant at the 5 and 1% probability levels and nonsignificant respectively.

4.3 FUNCTIONS

The importance of potassium as an essential nutrient for crop plant growth and development is well known. An adequate supply or balance of K, along with N and P, determines the response of crop plants to these nutrients. Potassium regulates osmotic potential in plants and reduces the incidence of diseases (Epstein and Bloom 2005; Fageria 2009). It also increases root development in rice (Figures 4.5 and 4.6), thereby improving water and nutrient uptake (Fageria 1984; Xiaoe et al. 1997).

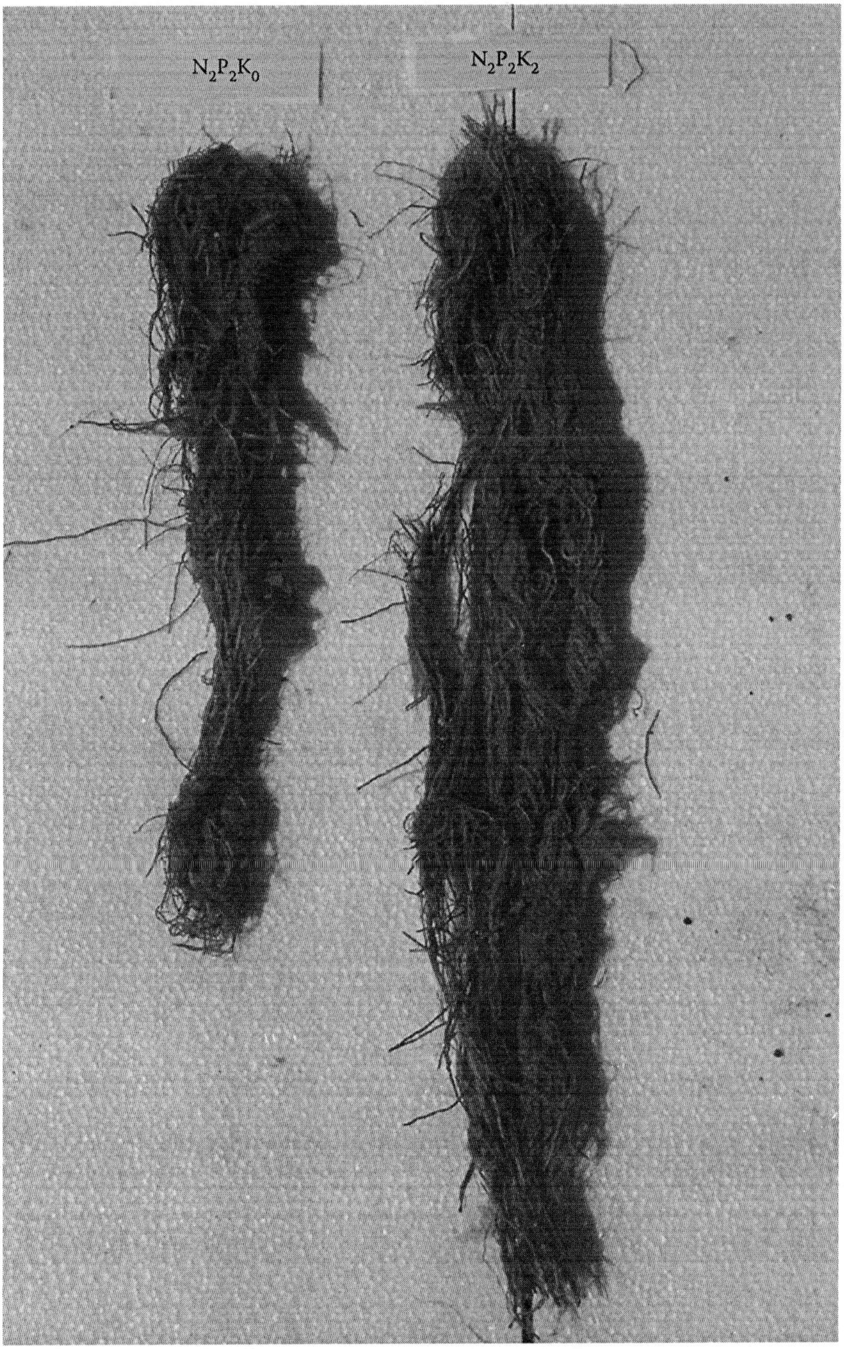

FIGURE 4.5 Root growth of upland rice grown on Brazilian Oxisol. The $N_2P_2K_0$ = 300 mg N + 200 mg P + 0 mg K kg^{-1} soil, and $N_2P_2K_2$ = 300 mg N + 200 mg P + 200 mg K kg^{-1} soil.

FIGURE 4.6 Root growth of two lowland rice genotypes, grown on Brazilian Inceptisol, at two K levels.

Along with nitrogen, potassium improves the photosynthetic activity of rice leaves (Xiaoe et al. 1997). Xiaoe et al. (1997) reported that topdressing K at the late growth stage of rice could prevent early leaf senescence to a certain degree, especially in hybrids. Figure 4.7 shows the growth of lowland rice with and without K fertilization. Growth of plants with K added was much higher than that of plants that did not receive K fertilization. In addition, K-deficient plants had a lower panicle number and panicles were small in size. Figure 4.7 shows the importance of K addition in Brazilian Inceptisols, where lowland rice is mainly grown.

Potassium-deficient rice plants commonly have high instances of several diseases that infest the leaves, stems, and panicles (Slaton, Cartwright, and Wilson 1995). As K deficiency progresses, rice plants usually develop severe disease infestation due to the plants' reduced ability to resist infection. Diseases that are normally insignificant problems, such as brown leaf spot and stem rot (*Sclerotium oryzae*), may become severe, in addition to the occurrence of common diseases such as rice blast (*Pyricularia grisea*). The lodging associated with K-deficient rice may be related to the increased incidence and severity of stem diseases. Potassium is known to play an important role in the lignification of vascular bundles, a factor that contributes to the higher susceptibility of K-deficient plants to lodging and disease. The yield loss of K-deficient rice may actually be a combination of losses from insufficient K nutrition

FIGURE 4.7 (See color insert.) Growth of lowland rice, grown on Brazilian Inceptisol with and without K fertilization.

and losses caused directly by diseases that infest K-deficient rice. Limited research suggests that yields of rice low or deficient in K do not respond to K fertilizer applications made after the panicle differentiation growth stage (Fageria, Slaton, and Baligar 2003). Thus, K fertilizer must be applied during vegetative growth for maximum yield production. Research has not been conducted to determine if a portion of the yield loss attributed to K deficiency can be prevented or recovered by controlling or suppressing these diseases with timely fungicide applications. Additional research efforts are needed to verify whether application of K fertilizer during the reproductive growth stage is beneficial to K-deficient rice (Fageria, Slaton, and Baligar 2003).

4.4 DEFICIENCY SYMPTOMS

Potassium deficiency symptoms initially appear on the lowest, oldest rice leaves because K is highly mobile in the plant. The onset of K deficiency is difficult to diagnose because the only initial difference between K-sufficient and -deficient rice is the color of the lower leaves. This deficiency symptom can easily be confused with N deficiency. Potassium deficiency symptoms include stunted plants with little or no reduction in tillering, droopy and dark green upper leaves, and chlorosis of the interveinal areas and margins of the lower leaves starting at the leaf tip. Leaf tips will eventually die and turn brown with the progression of severe K deficiency. The droopy leaves associated with K deficiency are not always noted because Na may substitute for K in osmotic regulation functions. Potassium deficiency reduces grain size and weight resulting in a direct yield loss. In the United States, K deficiency symptoms are seldom observed during vegetative growth. More commonly, K deficiency symptoms are observed during early reproductive growth beginning at panicle initiation (Fageria, Slaton, and Baligar 2003).

As shown in Table 4.5, potassium deficiency increases spikelet sterility. Spikelet sterility varied from 26–61.5%, with an average value of 38.1% at the lower K level. At the higher K level, spikelet sterility varied from 20.3–35.6%, with an average value of 26.6%. Overall, spikelet sterility decreased 30% at the higher K level compared to the lower K level. However, there was also a difference in spikelet sterility among lowland rice genotypes at lower as well as higher K levels. Therefore, spikelet sterility is also controlled genetically in rice. Figures 4.8 and 4.9 show K deficiency symptoms in upland and lowland rice plant leaves, respectively.

4.5 UPTAKE IN PLANT TISSUE

Determination of nutrient uptake (concentration and accumulation) in plant tissue is an important parameter in the diagnosis of nutrient deficiency or sufficiency. In addition, nutrient uptake also serves as a guideline for maintaining soil fertility to know the amount of nutrients that have been removed from the soil in straw and grain. The removed amount should be added to the soil to maintain the sustainability of the cropping systems. The K concentration (content per unit dry matter) in straw and grain of lowland rice during the growth cycle of a crop was determined (Table 4.6), and it was found that there was a significant quadratic decrease in the K concentration from 22 days after sowing to 140 days after sowing or physiological maturity. The K concentration

TABLE 4.5
Influence of K Fertilization on Spikelet Sterility (%) of 10 Lowland Rice Genotypes

Genotype	Low K (0 mg kg^{-1} soil)	High K (200 mg kg^{-1} soil)
CNAi 8859	29.3b	20.3
CNAi 8860	41.0ab	20.7
BRS Fronteira	40.3ab	25.8
BRA 01435	61.5a	24.0
BRA 01436	39.1ab	24.9
BRA 01258	45.7ab	29.1
BRA 01322	26.0b	28.1
CNAi 9018	42.6ab	35.6
CNAi 9025	26.2b	26.5
BRS Alvorada	29.1b	30.9
Average	38.1	26.6
F-test		
K level (K)	*	*
Genotype (G)	*	*
K × G	*	*

Note: Means in the same column followed by the same letter are not significantly different at the 5% probability level by Tukey's test.
*Significant at the 5% probability level.

FIGURE 4.8 (See color insert.) Potassium deficiency symptoms in the leaves of upland rice plants grown on Brazilian Oxisol.

FIGURE 4.9 (See color insert.) Potassium deficiency symptoms in the leaves of lowland rice plants grown on Brazilian Inceptisol.

TABLE 4.6
Yield, Concentration, and Uptake of K in Straw and Grain of Lowland Rice at Different Growth Stages

Plant Age in Days	Straw/Grain Yield[a] (kg ha^{-1})	Conc.[a] (g kg^{-1})	Uptake[a] (kg ha^{-1})
22	351	32.2	11.2
35	1130	32.9	36.9
71	5841	22.0	128.1
97	10,384	19.3	194.6
112	13,490	16.5	221.4
140	9423	15.9	156.2
140 (grain)	6389	3.1	19.7
Total (straw + grain)	15,812	19.0	179.9

Regression Analysis[b] (n = 6)

Plant age vs. straw yield (Y) = $-6362.26 + 277.43X - 1.11X^2$, $R^2 = 0.89$*
Plant age vs. K conc. in straw (Y) = $40.45 - 0.32X + 0.001X^2$, $R^2 = 0.96$**
Plant age vs. K uptake in straw (Y) = $-112.44 + 5.47X - 0.025X^2$, $R^2 = 0.95$**

[a] Values are averages of three crop years (field experimentation).
[b] Regression analysis includes data of straw only.
*,**Significant at the 5 and 1% probability levels, respectively.

in the tops of 22 rice plants was 32.2 g kg^{-1} or 3.22% and decreased to 15.9 g kg^{-1} or 1.59% at harvest. The R^2 showed that there was a 96% variation in K concentration due to plant age. The decrease in K concentration with the advancement of plant age was related to an increase in dry matter yield. This decrease in mineral nutrition is known as the dilution effect. For example, dry matter yield was 351 kg ha^{-1} at 22 days after sowing and reached 6389 kg ha^{-1} at physiological maturity (Table 4.6).

When compared at different growth stages, K concentration in the grain was much lower (3.1 g kg^{-1}) than that of straw. Uptake of K also increased in a quadratic fashion with the advancement of plant age. It was 11.2 kg ha^{-1} at 22 days after sowing and reached 156.2 kg ha^{-1} at physiological maturity, a 14-fold increase. Uptake of K in grain was also very small compared to that of straw. This was associated with a lower concentration of K in the grain. When values were transformed into percentage of distribution in the plant (straw and grain) at harvest, it was 87% in straw and the remaining 13% was exported to grain. Therefore, it can be concluded that a large part of K remains in the rice straw at harvest, and incorporation of straw into soil can be an important K management practice for K recycling in a rice crop.

Fageria et al. (2010) also studied K uptake and distribution in straw and grain of lowland rice as influenced by the K application rate. As shown in Table 4.7, potassium uptake increased in the shoot as well as in the grain up to 250 kg K ha^{-1} and then decreased. The K uptake pattern in the shoot as well as in the grain followed the grain and shoot dry matter yield pattern with increasing K rates. Across the K rate,

TABLE 4.7
Potassium Uptake and Distribution in Shoot and Grain of Rice under Different K Rates

K Rate (kg ha^{-1})	K Uptake in Shoot (kg ha^{-1})	K Uptake in Grain (kg ha^{-1})	Total K Uptake (kg ha^{-1})	K Distribution in Shoot (%)	K Distribution in Grain (%)
0	87.5	13.0	100.5	87	13
125	91.3	16.7	108.0	85	15
250	108.0	21.8	129.8	83	17
375	87.3	13.6	100.9	87	13
500	103.4	18.5	121.9	85	15
625	86.0	19.5	105.5	82	18
Average	93.9	17.2	111.1	85	15

Source: Fageria, N. K., A. B. Santos, A. Moreira, and M. F. Moraes. 2010. Potassium soil test calibration for lowland rice on an Inceptisol. *Commun. Soil Sci. Plant Anal.* 41:2595–2601.

K uptake was 94 kg ha^{-1} in shoot and 17 kg ha^{-1} in grain. This means that shoot's K uptake was 5.5 times higher than that of grain. Similarly, overall distribution of K in shoot was 85%, and in grain, translocation of K was only 15% of the total K uptake. The higher concentration of K in shoot compared to grain is widely reported for rice (Fageria et al. 1990a, 1990b; Fageria, Baligar, and Jones 2011). Therefore, incorporation of rice straw in soil after crop harvest may contribute significantly to maintaining the K status of rice soils.

The concentration and uptake of K in upland rice as affected by applied K levels was also studied (Table 4.8). Potassium concentration and uptake increased with increasing levels of K fertilizer. As expected, the increase in K concentration was associated with an increase in K levels in soil. The increase in K uptake was related to an increase in dry matter yield. Fageria, Santos, and Moraes (2010) also studied K concentration in upland rice genotypes at low (0 mg kg^{-1}) and high (200 mg kg^{-1}) K levels (Table 4.9). Potassium concentration (content per unit of shoot or grain dry weight) in shoot and grain was influenced by K rate, genotype (G), and K × G interaction. Values of K concentration at two K levels are presented in Table 4.9. Potassium concentration in shoot at low K level (0 mg kg^{-1}) varied from 5.5 to 9.0 g kg^{-1} depending on genotype, with an average value of 7.3 g kg^{-1}. Similarly, K concentration in shoot at high K level (200 mg kg^{-1}) varied from 10.5 to 13.0 g kg^{-1}, with an average value of 11.67 g kg^{-1}. Overall, K concentration in shoot at 200 mg K kg^{-1} soil was 60% higher when compared to 0 mg K kg^{-1} of soil. However, K concentration was slightly higher at low K concentration compared to high K concentration in grain overall. This may be due to the high grain yield at high K level. As noted earlier, this is called the dilution effect (Fageria, Baligar, and Jones 2011). The K concentration in shoot was positively associated with grain yield (Table 4.10). However, K concentration in grain was not significantly associated with grain yield.

TABLE 4.8
Straw Yield, Concentration, and Uptake of K in Straw of Upland Rice

K Rate (kg ha^{-1})	Straw Yield (kg ha^{-1})	Conc. (g kg^{-1})	Uptake (kg ha^{-1})
0	2401	22.2	53
42	2893	21.5	56
84	2828	22.7	65
126	3115	21.7	68
168	2905	29.8	89

Regression Analysis

K rate vs. straw yield (Y) = $2432.11 + 10.08X - 0.042X^2$, $R^2 = 0.84^*$

K rate vs. K conc. in straw (Y) = $22.61\exp(2.65X + 0.000024X^2)$, $R^2 = 0.82^*$

K rate vs. K uptake in straw (Y) = $49.40 + 0.20X$, $R^2 = 0.88^*$

Source: Adapted from Fageria, N. K., R. J. Wright, V. C. Baligar, and J. R. P. Carvalho. 1990b. Upland rice response to potassium fertilization on a Brazilian Oxisol. *Fertilizer Res.* 21:141–147.

*Significant at the 1% probability level.

TABLE 4.9
Potassium Concentration in Shoot and Grain of Upland Rice as Influenced by K and Genotype Treatments

	K Concentration in Shoot (g kg^{-1})		K Concentration in Grain (g kg^{-1})	
Genotype	K_0	K_{200}	K_0	K_{200}
BRS Bonança	6.0c	12.0	1.5	2.0
BRS Primavera	7.5b	12.5	1.5	2.0
BRSMG Curinga	9.0a	10.5	2.5	1.0
BRA 032033	8.0ab	13.0	1.5	1.0
BRA 01596	5.5c	11.5	1.5	1.0
BRA 02582	8.0ab	12.5	1.5	1.5
Average	7.3	11.67	1.67	1.42
F-test				
K rate (K)	*		*	
Genotype (G)	*		*	
K × G	*		*	
CV (%)	14.1		31.1	

Source: Fageria, N. K., A. B. Santos, and M. F. Moraes. 2010. Yield, potassium uptake, and use efficiency in upland rice genotypes. *Commun. Soil Sci. Plant Anal.* 41:2676–2684.

*Significant at the 5% probability level.

TABLE 4.10
Potassium Uptake Traits (X) in Upland Rice Association with Grain Yield (Y)

K Uptake Trait	Regression Equation	R^2
K conc. in shoot	$Y = 9.4715 + 0.4339X$	0.1576*
K conc. in grain	$Y = 21.6379 - 9.8800X + 2.6465X^2$	0.1262^{NS}
K uptake in shoot	$Y = 12.1350 + 0.0058X$	0.1568*
K uptake in grain	$Y = 12.9373 + 0.0319X$	0.1714*
KUER in shoot	$Y = 17.1310 - 0.0309X$	0.1303*
KUER in grain	$Y = 11.9189 + 0.0023X$	0.0372^{NS}

Source: Fageria, N. K., A. B. Santos, and M. F. Moraes. 2010. Yield, potassium uptake, and use efficiency in upland rice genotypes. *Commun. Soil Sci. Plant Anal.* 41:2676–2684.

Note: X, potassium uptake traits; Y, grain yield; KUER, potassium use efficiency ratio, which was calculated with the following equation: Grain or straw yield/K uptake in grain or straw.

^{*,NS}Significant at the 5% probability level and nonsignificant, respectively.

Potassium uptake (concentration × dry weight of shoot or grain) in the shoot was significantly influenced by K rate and genotype; however, the interaction between K and genotype was not significant (Table 4.11). The K rate, genotype, and K × G interaction significantly influenced potassium uptake in grain. Potassium uptake in shoot varied from 118.56 to 166.89 mg plant^{-1}, with an average value of 148.58 mg plant^{-1} under genotype treatment. Under K rate treatment, K uptake value in shoot varied from 107.80 to 189.35 mg plant^{-1}. The increase in K uptake at 200 mg kg^{-1} level was about 76% higher when compared to zero K level. This higher value was associated with higher shoot dry weight at a higher K level (Fageria et al. 2010). In grain, the K uptake value varied from 13.63 mg plant^{-1} to 31.65 mg plant^{-1}, with an average value of 20.51 mg plant^{-1} at zero K level, and 12.4 to 37.0 mg plant^{-1}, with an average value of 21.03 mg plant^{-1} at 200 mg K kg^{-1} level. An F-test showed K × G interaction significant for grain K uptake (Table 4.11). However, Tukey's test could not separate genotype means significantly at two K rates (Table 4.11). Potassium uptake in the shoot as well as in the grain was significantly associated with grain yield (see Table 4.10). This means that increasing K uptake in shoot as well as in grain can improve grain yield of upland rice genotypes.

4.6 HARVEST INDEX

The potassium harvest index (K uptake in grain/K uptake in grain plus straw) was significantly influenced by K rate and genotype treatments (Table 4.12). It varied from 0.09 to 0.17, with an average value of 0.13. This means that a maximum amount of absorbed K was retained in shoot, and overall, 13% was accumulated in grain. Fageria, Barbosa Filho, and Carvalho (1982) reported that in upland rice, large quantities of K remain in shoot and a small amount is translocated to grain. Similarly, Fageria (1991) reported that 76–86% of the absorbed K was retained in shoot and 11–21% was translocated to grain depending on the cultivar of upland rice. Fageria, Barbosa Filho, and Carvalho (1982) recommended that incorporation of rice straw

TABLE 4.11
Uptake of K in the Shoot of Upland Rice across Two K Rates and in Grain as Influenced by K Rate and Genotypes

		K Uptake in Grain (mg plant^{-1})	
K Rate (mg kg^{-1})	K Uptake in Shoot (mg plant^{-1})	K_0	K_{200}
0	107.80b		
200	189.35a		
Genotype			
BRS Bonança	130.08bc	15.6a	24.1a
BRS Primavera	166.81a	22.35a	37.0a
BRSMG Curinga	160.44a	31.65a	15.5a
BRA 032033	150.69ab	18.3a	12.4a
BRA 01596	118.56c	21.55a	17.25a
BRA 02582	164.89a	13.63a	16.9a
Average	148.58	20.51	21.03
F-test			
K rate (K)	**	*	
Genotype (G)	**	*	
K × G	NS	*	
CV (%)	14.6	38.1	

Source: Fageria, N. K., A. B. Santos, and M. F. Moraes. 2010. Yield, potassium uptake, and use efficiency in upland rice genotypes. *Commun. Soil Sci. Plant Anal.* 41:2676–2684.

Note: Means followed by the same letter in the same column are not significantly different by Tukey's test.

*,**,NS Significant at the 5 and 1% probability levels and nonsignificant, respectively. Means followed by the same letter in the same column are not significantly different by Tukey's test.

as a cultural practice after harvesting be implemented to supply large amounts of K to succeeding crops.

4.7 USE EFFICIENCY

Nutrient use efficiency is an important parameter for determining how the applied nutrient has been utilized in the production of straw or grain or both. K use efficiency in rice differs from genotype to genotype and also with environmental conditions. Baligar and Fageria (1997) calculated K use efficiency of six lowland rice genotypes (Table 4.13). Agronomic efficiency (AE) varied from 44 to 80 kg grain produced per kg K applied, with an average value of 66 kg grain produced per kg K applied. Physiological efficiency varied from 89 to 119 kg grain plus straw produced per kg of K uptake in grain plus straw, with an average value of 98 kg grain plus straw produced per kg K accumulated in grain plus straw. Agrophysiological efficiency varied from 26 to 58 kg grain produced per kg of K uptake in grain plus straw, with an average value of 43 kg grain produced per kg K uptake in grain plus straw. The K recovery efficiency varied from 51 to 81%, with an average value of 67%. Fageria,

TABLE 4.12
Potassium Harvest Index as Influenced by Potassium Rate and Upland Rice Genotype Treatment

K Rate (mg kg^{-1})	K Harvest Index
0	0.16a
200	0.10b
Genotype	
BRS Bonança	0.14ab
BRS Primavera	0.15ab
BRSMG Curinga	0.13abc
BRA 032033	0.11bc
BRA 01596	0.17a
BRA 02582	0.09c
Average	0.13
F-test	
K rate (K)	*
Genotype (G)	*
K × G	NS
CV (%)	30.8

Source: Fageria, N. K., A. B. Santos, and M. F. Moraes. 2010. Yield, potassium uptake, and use efficiency in upland rice genotypes. *Commun. Soil Sci. Plant Anal.* 41:2676–2684.

Note: Means followed by the same letter in the same column do not differ significantly at the 5% probability level by Tukey's test.

*,NS Significant at the 1% probability level and nonsignificant, respectively.

Slaton, and Baligar (2003) reported that K use efficiencies varied from genotype to genotype and were also influenced by environmental conditions.

Fageria, Santos, and Moraes (2010) also calculated K use efficiency in upland rice under greenhouse conditions, and the efficiency expressed in different ways (agronomic, physiological, agrophysiological, recovery, and utilization) is presented in Table 4.14. AE varied from 4.11 to 11.33 mg mg^{-1}, with an average value of 8.90 mg mg^{-1}. Physiological efficiency values varied from 17.63 to 90.07 mg mg^{-1}, with an average value of 52.09 mg mg^{-1}. Agrophysiological efficiency values varied from 6.57 to 36.62 mg mg^{-1}, with an average value of 27.36 mg mg^{-1}. Potassium recovery efficiency varied from 4.87 to 38.55%, with an average value of 27.36%. The K utilization efficiency varied from 8.39 to 19.56 mg mg^{-1}, with an average value of 15.01 mg mg^{-1}. Fageria and colleagues (2000, 2011) reported more or less similar trends in K use efficiency in upland rice genotypes grown on Brazilian Oxisol.

Fageria et al. (2010) determined AE of lowland rice grown on Brazilian Inceptisol. AE (kg grain produced per kg K applied) decreased significantly in a quadratic fashion with an increasing K rate in the range of 125–625 kg ha^{-1} ($Y = 46.3669 - 0.1701X + 0.00016 X^2$, $R^2 = 92^{**}$). At the 125 kg K ha^{-1} rate, AE was 30 kg grain kg^{-1}

TABLE 4.13
Potassium Use Efficiency of Lowland Rice Genotypes

	Agronomic Efficiency	Physiological Efficiency	Agrophysiological Efficiency	Recovery Efficiency
Genotype	Δ kg Grain kg^{-1} K Fertilizer Added	Δ kg Grain Δ kg^{-1} K Uptake	Δ kg Grain Δ kg^{-1} Total K Uptake	% Fertilizer Uptake
Aliança	76	89	39	81
CAN 5751	64	119	51	58
CAN 6804	54	89	58	51
CAN 7238	80	100	48	75
CAN 7268	44	101	26	73
Metica 1	78	89	37	61
Average	66	98	43	67

Source: Baligar, V. C., and N. K. Fageria. 1997. Nutrient use efficiency in acid soils: Nutrient management and plant use efficiency. In *Plant-Soil Interactions at Low pH: Sustainable Agriculture and Forestry Production*, eds. A. C. Moniz, A. M. C. Furlani, N. K. Fageria, C. A. Rosolem, and H. Cantarells, pp. 75–95. Campinas, Brazil: Brazilian Soil Science Society.

Note: Formulae used to calculate the listed efficiencies are given in Chapter 3.

TABLE 4.14
Potassium Use Efficiency in Six Upland Rice Genotypes

Genotype	AE (mg mg^{-1})	PE (mg mg^{-1})	APE (mg mg^{-1})	ER (%)	EU (mg mg^{-1})
BRS Bonança	4.11	17.63	6.57	35.50	8.39
BRS Primavera	11.33	45.61	29.47	38.55	17.53
BRSMG Curinga[a]	9.78	–	–	4.87	11.89
BRA 032033	10.00	57.35	36.62	28.70	16.08
BRA 01596	10.44	49.81	33.12	31.13	16.58
BRA 02582	7.75	90.07	34.20	25.41	19.56
Average	8.90	52.09	27.99	27.36	15.01

Source: Fageria, N. K., A. B. Santos, and M. F. Moraes. 2010. Yield, potassium uptake, and use efficiency in upland rice genotypes. *Commun. Soil Sci. Plant Anal.* 41:2676–2684.

Note: AE, agronomic efficiency; PE, physiological efficiency; APE, agrophysiological efficiency; ER, recovery efficiency; EU, utilization efficiency.

[a] Due to negative values, PE and APE values were not presented. Formulae to calculate these efficiencies are given in Chapter 3.

K applied and dropped to 5 kg grain kg^{-1} K applied at 625 kg K ha^{-1}. The variability in AE due to K rate was 92%, indicating the importance of K rate in determining this efficiency. Fageria and colleagues (1992, 2011) reported the decrease in the AE of K with increasing K rate. Potassium use efficiency in rice is low compared to N and P efficiency. To produce one metric ton of grain, lowland rice requires 24 kg N,

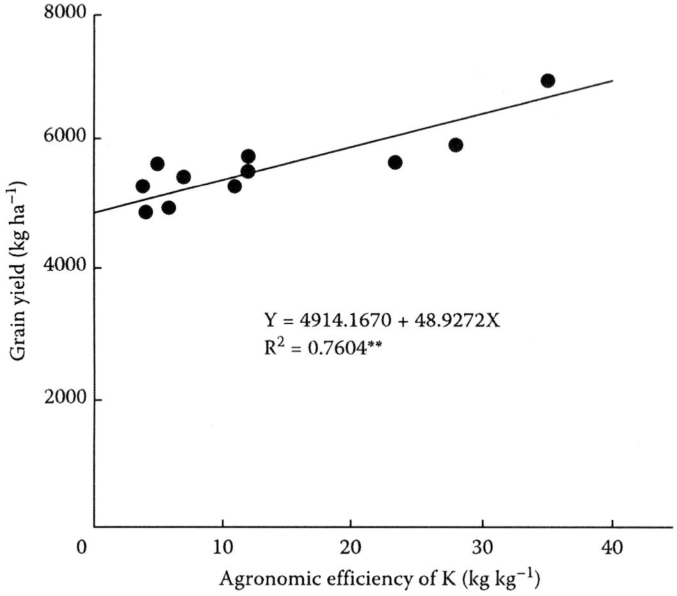

FIGURE 4.10 Relationship between the agronomic efficiency of potassium and the grain yield of lowland rice. (From Fageria, N. K., A. B. Santos, A. Moreira, and M. F. Moraes. 2010. Potassium soil test calibration for lowland rice on an Inceptisol. *Commun. Soil Sci. Plant Anal.* 41:2595–2601.)

5 kg P, and 41 kg K in shoot and grain (Fageria, Baligar, and Clark 2006). Low K use efficiency is associated with a higher quantity of K retained in shoot and a small part translocated to grain (Fageria 1992). The translocation of N and P in rice grain is about 60% (Fageria, Baligar, and Clark 2006). Improvement in AE is important for increasing grain yield of rice (Figure 4.10). The variability in grain yield due to AE was 76%.

4.8 MANAGEMENT PRACTICES

Fertilizer management practices are important strategies for increasing crop yields and reducing fertilizer costs and environmental pollution. The principal K management practices for a rice crop are to use appropriate source, effective methods, and optimal timing of K fertilizer application. These practices are discussed in the following section.

4.8.1 Sources, Methods, and Timing of Application

Potassium chloride is the major source of single K^+ fertilization; potassium sulfate and potassium–magnesium sulfate are other minor sources of K^+ for field crops. The reasons for widespread use of potassium chloride are lower production costs and higher analysis. Potassium sulfate may be useful in areas where S deficiency is reported. However, it should not be used in flooded rice where sulfate reduction and hydrogen sulfide toxicity are a problem (De Datta and Mikkelsen 1985). Similarly, the use of KCl may be discouraged in saline soils because KCl can increase salt concentrations in the rhizosphere. Potassium is also used in formulated fertilizer mixtures as $N–P_2O–K_2O$. In Brazil, fertilizer mixtures of 4–30–16 or 4–20–20 are very common and are used for fertilization of annual crops. Nutrient contents of these fertilizer mixtures are expressed in percentages. Principal K^+ fertilizers are presented in Table 4.15. All the potassium fertilizers are highly water soluble, with the exception of potassium metaphosphate (Fageria 1989). In general, crop yields are comparable whether one or the other source is used, although this may depend on the soil, the crop, or the manner of application (Stewart 1985). The fertilizer source selected by farmers mainly depends on cost of transportation and handling convenience in application.

Most potassium reaches plant roots by the diffusion process; this is considered one of the most limiting steps in K uptake during the crop cycle. Therefore, potassium should be applied near the plant roots or in a furrow to facilitate its uptake by roots. Adequate soil moisture is one of the factors that control the uptake of K by plants. Soils with high cation exchange capacity or fine texture have lower K diffusion rates compared with coarsely textured soils with the same amount of water content. Other factors affecting K uptake by upland and lowland rice plants are pH, soil aeration, crop species, and genotypes within species, and these must be regulated to maximize K uptake.

Potassium fertilization should take place at sowing in the band or furrow. However, in coarsely textured soil, K fertilizers can be applied as topdressing along with nitrogen fertilizers. Fageria (1991) reported that in upland and lowland rice, split application of K produced higher yields compared to all K applied at sowing in Brazilian Oxisol and Inceptisol. Similarly, Santos et al. (1999) reported that fractional or split application of K increased its use efficiency by lowland or flooded rice. In the case of coarsely textured and low CEC, the split application of K can have better results than total K applied at sowing.

TABLE 4.15
Principal Potassium Fertilizers and Their Potassium Content and Solubility

Common Name	Formula	K_2O (%)	Solubility
Potassium chloride	KCl	60	Water soluble
Potassium sulfate	K_2SO_4	50	Water soluble
Potassium–magnesium sulfate	$K_2SO_4.MgSO_4$	23	Water soluble
Potassium nitrate	KNO_3	44	Water soluble
Kainit	$MgSO_4+KCl+NaCl$	12	Water soluble
Potassium metaphosphate	KPO_3	40	Low water solubility

Source: Fageria, N. K. 2009. *The Use of Nutrients in Crop Plants*. Boca Raton, FL: CRC Press.

4.8.2 Adequate Rate

Using an adequate rate of chemical fertilizers in crop production is essential not only to improve yield but also to reduce cost of crop production and environmental pollution. The most important criterion for supplying an adequate rate of an immobile nutrient, such as K, in the planting soil is the use of soil test data. Soil test data should be generated by conducting multisite field experiments in different agroecological regions. A field experiment for lowland or flooded rice in the central part of Brazil using different rates of K was conducted (Table 4.16). The year × K interaction was significant, indicating a variation in grain yield from year to year. In the first year, grain yield varied from 2961 kg ha^{-1} at zero rate of K to 5880 kg ha^{-1} at 125 kg K ha^{-1}, about a twofold variation. In the second year, grain yield varied from 2631 kg ha^{-1} at zero rate of K to 6244 kg ha^{-1} at 125 kg K ha^{-1}, about a 2.4-fold variation. This change in grain yield from year to year was expected because of the different climatic conditions from one year to the next (Fageria 1992). These results also suggested that field experiments should be repeated to make meaningful fertilizer recommendations.

Results of K soil test calibration to applied K are presented in Figure 4.11. There was a quadratic response of increasing soil extractable K in the range of 14–86 mg K kg^{-1} Mehlich 1 extractable soil K in the first year. In the second year, rice yield also

TABLE 4.16
Grain Yield of Lowland Rice as Influenced by Potassium Fertilization

	Grain Yield (kg ha^{-1})		
K Rate (kg ha^{-1})	**First Year**	**Second Year**	**Average**
0	2961	2631	2796
125	5880	6244	6062
250	5526	5590	5558
375	4853	5092	4973
500	4962	5076	5019
625	5091	5329	5210
F-test			
Year (Y)	*		
K rate (K)	*		
Y × K	*		
CV (%)	6.8		

Regression Analysis

K rate (X) vs. first year grain yield (Y) = $3641.72 + 10.3063X - 0.0138X^2$, $R^2 = 0.4466$**
K rate (X) vs. second year grain yield (Y) = $3467.83 + 12.3452X - 0.0163X^2$, $R^2 = 0.4567$**
K rate (X) vs. average grain yield (Y) = $3555.04 + 11.3227X - 0.0151X^2$, $R^2 = 0.4674$**

Source: Fageria, N. K., A. B. Santos, A. Moreira, and M. F. Moraes. 2010. Potassium soil test calibration for lowland rice on an Inceptisol. *Commun. Soil Sci. Plant Anal.* 41:2595–2601.
*Significant at the 1% probability level.

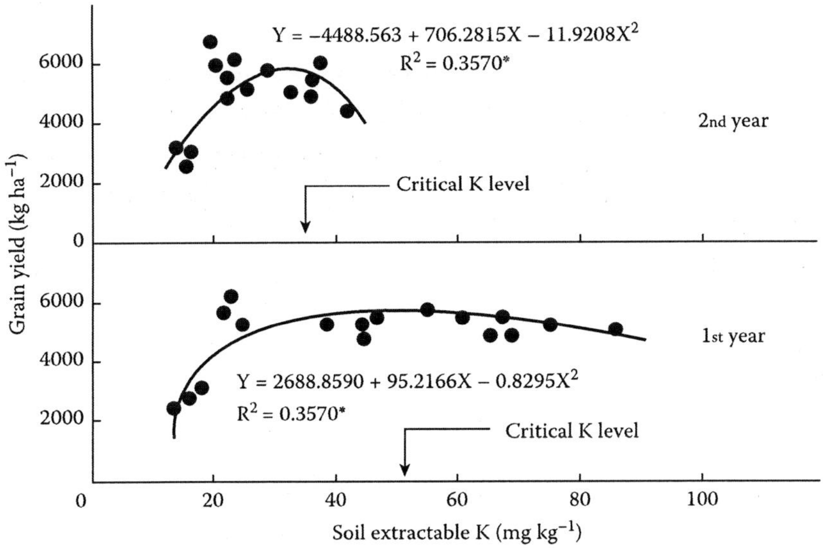

FIGURE 4.11 Relationship between Mehlich 1 extractable soil K and the grain yield of lowland rice. (From Fageria, N. K., A. B. Santos, A. Moreira, and M. F. Moraes. 2010. Potassium soil test calibration for lowland rice on an Inceptisol. *Commun. Soil Sci. Plant Anal.* 41:2595–2601.)

increased in a quadratic fashion with the K soil test in the range of 15–42 mg kg^{-1}. In the first year, based on a regression equation, the maximum grain yield was obtained at 57 mg K kg^{-1} of soil. In the second year, the maximum grain yield was obtained at 30 mg kg^{-1} extractable soil K. Fageria et al. (1990a) reported that the Mehlich 1 extractable soil K level associated with maximum rice yield was 59 mg kg^{-1} for the first crop and 34 mg kg^{-1} for the second crop in Brazilian Inceptisol. Results of a recent study fall in this K level range. The maximum grain yield obtained at the low K level (30 mg kg^{-1}) in the second year indicates that there was a significant reduction in soil K content during two years of rice cultivation. It also suggests that in the second year, nonexchangeable K might have played an important role in the supply of K to the rice crop. Sparks (1987) and Hinsinger (1998) reported that availability of nonexchangeable K to plants is associated with a very low level of K in soil, which may be due to chemical changes in the rhizosphere. Potassium recommendations based on soil test K for upland and lowland rice grown on Brazilian Oxisols and Inceptisols are presented in Table 4.17. These recommendations are a result of the author's personal experience.

4.8.3 Use of Efficient Genotypes

Use of K-efficient rice genotypes is an important and very attractive strategy from an economic and environmental point of view. In the twenty-first century, nutrient-efficient plants will play a greater role in increasing crop yields (Fageria, Baligar, and Li 2008).

TABLE 4.17
Mehlich 1 Soil Test K Availability Indices and K Fertilizer Recommendations for Upland and Lowland Rice

Soil K Test ($mg\ kg^{-1}$)	K Test Interpretation	Band K Application Recommendations ($kg\ K_2O\ ha^{-1}$)
0–25	Very low	125
25–50	Low	90
50–80	Adequate	75
>80	High	60

Limited land and water resources, higher costs of inorganic fertilizer inputs, and declining future crop yields will create greater demand for cultivars with a higher nutrient efficiency. Furthermore, at least 60% of the world's arable land has mineral deficiencies or elemental toxicity problems. A growing demand for food and fiber by an ever-increasing world population further enhances the importance of nutrient-efficient plants. Epstein and Bloom (2005) and Fageria, Baligar, and Li (2008) reported that higher nutrient use efficiency in plants must be fully explored to increase food production to feed the growing human population, and this has to be achieved without accelerating environmental degradation from excessive fertilizer use.

During the last three decades, extensive research has been conducted to identify nutrient-efficient plant species, or genotypes within species, and to understand mechanisms of nutrient efficiency in crop plants. So far, the success of nutrient-efficient cultivars is limited. The main reason for this limited success is a lack of understanding regarding plant genetics and the impact of environmental variables on nutrient use efficiency in plants (Fageria, Baligar, and Li 2008). There is a need for multidisciplinary efforts on the part of plant breeders, soil scientists, physiologists, and agronomists to develop nutrient-efficient cultivars that are suitable for the world's agroecological regions. During the twenty-first century, agricultural scientists have tremendous challenges as well as opportunities for developing nutrient-efficient crop plants. In the previous century, breeding for nutritional traits had been proposed as a strategy for improving efficient fertilizer use or obtaining higher yields in low-input agricultural systems. This strategy for developing nutrient-efficient crop cultivars should continue as a top priority during the twenty-first century.

Previously published studies have outlined the differences among upland and lowland rice genotypes with regard to K use efficiency (Fageria, Baligar, and Li 2008; Fageria, Baligar, and Jones 2011). Twenty upland rice genotypes at low (0 mg K kg^{-1}) and high (200 mg K kg^{-1}) K levels were evaluated. Based on the grain yield efficiency index, genotypes were classified as efficient, moderately efficient, and inefficient in K use (Figure 4.12). Genotypes differed significantly in K use efficiency. The genotypes BRS Primavera and BRA 1600 were the most efficient, and the genotypes BRSMG Curinga and BRA 02582 were the least efficient in K use.

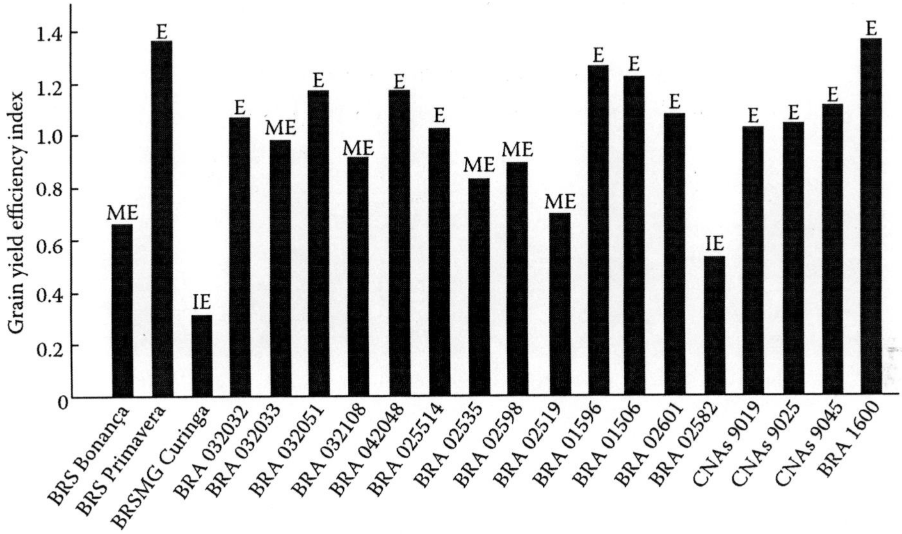

FIGURE 4.12 Classification of upland rice genotypes for K use efficiency.

In addition, the grain harvest index (GHI), 1000-grain weight, panicle length, and spikelet sterility of these genotypes at two K levels were determined (Tables 4.18 and 4.19). The GHI varied from 0.36 to 0.53, with an average value of 0.46 at the low K level, and from 0.37 to 0.53, with an average value of 0.47 at the high K level. Fageria, Castro, and Baligar (2004) reported GHI values from 0.38 to 0.55, with an average value of 0.45 in 20 upland rice genotypes. The highest-yielding genotype, BRS Primavera, had a near maximum GHI at the low as well as at the high K level. Similarly, the lowest-yielding genotype, BRA 02582, had the lowest GHI at the low as well as at the high K level. This demonstrates that GHI is an important plant trait in determining grain yield in rice. The GHI has a linear association with grain yield (Figure 4.13). Fageria (2009) reported a significant ($P < 0.01$) positive correlation between GHI and grain yield in rice.

The 1000-grain weight varied from 18.6 to 26.0 g, with an average value of 23.3 g at the low K level (see Table 4.18). Similarly, 1000-grain weight at the high K level varied from 21.1 g to 28.4 g, with an average value of 25.4 g. Overall, the 1000-grain weight was 9% higher at the high K level compared to the low K level. The 1000-grain weight had a linear correlation with grain yield (Figure 4.14). Hasegawa (2003) reported a positive correlation between 1000-grain weight and grain yield of rice. However, the association was not significant. Chau and Bhargava (1993) also reported a nonsignificant correlation between 1000-grain weight and grain yield in rice.

The panicle length varied from 17.7 to 23.9 cm, with an average value of 21.1 cm at the low K level (0 mg kg^{-1} soil; see Table 4.19). Similarly, at the high K level (200 mg kg^{-1} soil), the panicle length varied from 18.2 to 25.3 cm, with an average value of 21.8 cm. Overall, potassium application resulted in 3% increase in panicle length compared to control treatment; however, the effect was nonsignificant. The panicle length had a significant positive quadratic association with grain yield

TABLE 4.18
Grain Harvest Index (GHI) and 1000-Grain Weight of 20 Upland Rice Genotypes as Influenced by K Levels

Genotype	Grain Harvest Index		1000-Grain Weight (g)	
	K_0	K_{200}	K_0	K_{200}
BRS Bonança	0.44abcde	0.45def	22.8abcd	25.2abcde
BRS Primavera	0.49abcd	0.51abc	23.0abcd	25.1abcde
BRSMG Curinga	0.44abcde	0.48bcdef	23.2abcd	27.5abcd
BRA 032032	0.51ab	0.53ab	23.7abcd	27.9ab
BRA 032033	0.48abcd	0.51abcd	24.2abcd	26.8abcde
BRA 032051	0.49abcd	0.50abcde	25.3ab	27.8abc
BRA 032108	0.46abcd	0.46cdef	18.6e	21.1f
BRA 042048	0.44bcde	0.43fg	25.5ab	27.5abc
BRA 025514	0.41de	0.44ef	22.7abcd	24.3bcdef
BRA 02535	0.42bcde	0.47cdef	20.9de	23.8def
BRA 02598	0.49abcd	0.51abcd	21.4cde	24.6bcdef
BRA 02519	0.41cde	0.37gh	26.0a	24.5bcdef
BRA 01596	0.53a	0.54a	22.2bcd	23.4ef
BRA 01506	0.50abc	0.51abcd	24.9abc	25.5abcde
BRA 02601	0.41de	0.50abcd	22.8abcd	26.3abcde
BRA 02582	0.36e	0.37h	22.5abcd	23.4ef
CNAs 9019	0.48abcd	0.50abcde	24.2abcd	24.2sdef
CNAs 9025	0.41de	0.47cdef	25.4ab	28.4a
CNAs 9045	0.47abcd	0.46cdef	25.7ab	27.0abcde
BRA 1600	0.49abcd	0.50abcd	22.6abcd	24.4bcdef
Average	0.46	0.47	23.3	25.4
F-test				
K level (K)	NS		**	
Genotypes (G)	**		**	
K × G	**		*	
K × G	4.5		4.9	

Note: Means in the same column followed by the same letter are not significantly different at 5% probability level by Tukey's test.

*,**,NS Significant at the 5 and 1% probability levels and nonsignificant, respectively.

across genotypes (Figure 4.15). Fageria and colleagues (2000, 2006) reported a significant correlation between panicle length and grain yield of upland rice genotypes. Spikelet sterility varied from 3.7 to 21.5%, with an average value of 12.8% at the lower K level. Similarly, at the high K level, spikelet sterility varied from 5.6 to 24.8%, with an average value of 12.0%. Overall, application of 200 mg K kg^{-1} soil reduced spikelet sterility by about 7% compared to control treatment. Spikelet sterility, as expected, had a significant negative association with grain yield (Figure 4.16). Fageria, Castro, and Baligar (2004) reported a significant negative

TABLE 4.19
Panicle Length and Spikelet Sterility of 20 Upland Rice Genotypes as Influenced by Genotype and K Levels

Genotype	Panicle Length (cm)		Spikelet Sterility (%)	
	K_0	K_{200}	K_0	K_{200}
BRS Bonança	17.7g	18.2e	13.3abc	11.3ab
BRS Primavera	23.4ab	22.7abc	19.8ab	16.7ab
BRSMG Curinga	19.1efg	19.9de	11.2abc	6.0b
BRA 032032	20.8cdef	21.4bcd	14.5abc	14.7ab
BRA 032033	19.4efg	20.0de	15.9abc	11.4ab
BRA 032051	23.0abcd	22.7abc	14.0abc	6.0b
BRA 032108	23.4ab	23.6ab	9.4ab	17.5ab
BRA 042048	23.9a	23.2ab	16.7abc	9.6ab
BRA 025514	17.7g	18.5e	6.3abc	3.6b
BRA 02535	22.4abcd	25.3a	15.4abc	14.7ab
BRA 02598	22.9abcd	23.3ab	15.6abc	12.1ab
BRA 02519	20.5def	21.9bcd	5.3bc	18.6ab
BRA 01596	22.3abcd	22.5bcd	9.6abc	6.5b
BRA 01506	21.3bcde	22.2bcd	5.9abc	6.9b
BRA 02601	18.7efg	20.1de	12.6abc	6.7b
BRA 02582	18.6fg	20.0de	21.5a	24.8a
CNAs 9019	20.4def	22.3bcd	20.9ab	23.1a
CNAs 9025	23.2abc	23.5ab	17.4abc	13.5ab
CNAs 9045	22.5abcd	23.4ab	6.7abc	11.2ab
BRA 1600	20.9bcdef	20.5cde	3.7c	5.6b
Average	21.1	21.8	12.8	12.0
F-test				
K level (K)	NS		*	
Genotypes (G)	**		**	
K × G	*		*	
K × G	3.7		39.3	

Note: Means in the same column followed by the same letter are not significantly different at 5% probability level by Tukey's test.

*,**,NS Significant at the 5 and 1% probability levels and nonsignificant, respectively.

association between grain yield and spikelet sterility in upland rice genotypes grown on Brazilian Oxisol.

Also, lowland rice genotypes were evaluated for K use efficiency. Grain yield and GHI were significantly influenced by K, genotype, and K × G interaction (Table 4.20). The significant K × G interactions for these two traits indicated that some genotypes were highly responsive to K fertilization although others were not. Thus, genotype selection is an important strategy for improving lowland rice yield in Brazilian soils.

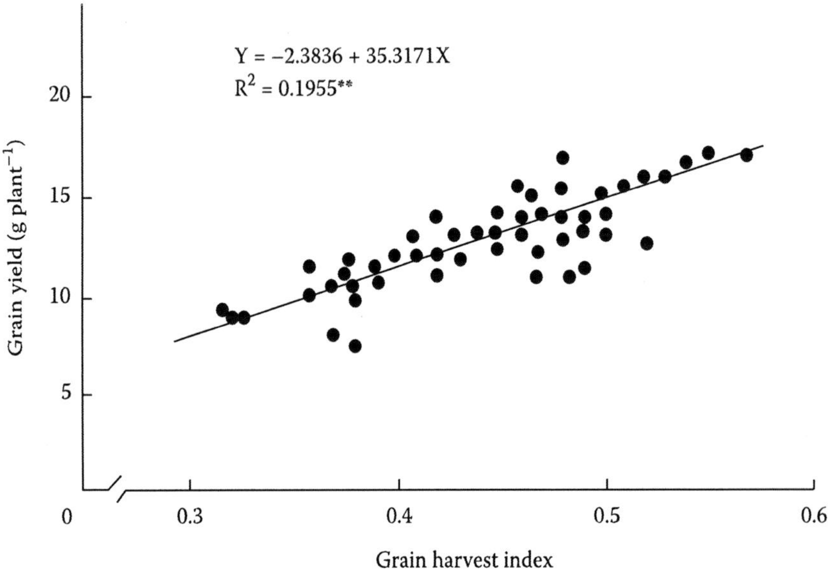

FIGURE 4.13 Relationship between grain harvest index and grain yield of upland rice.

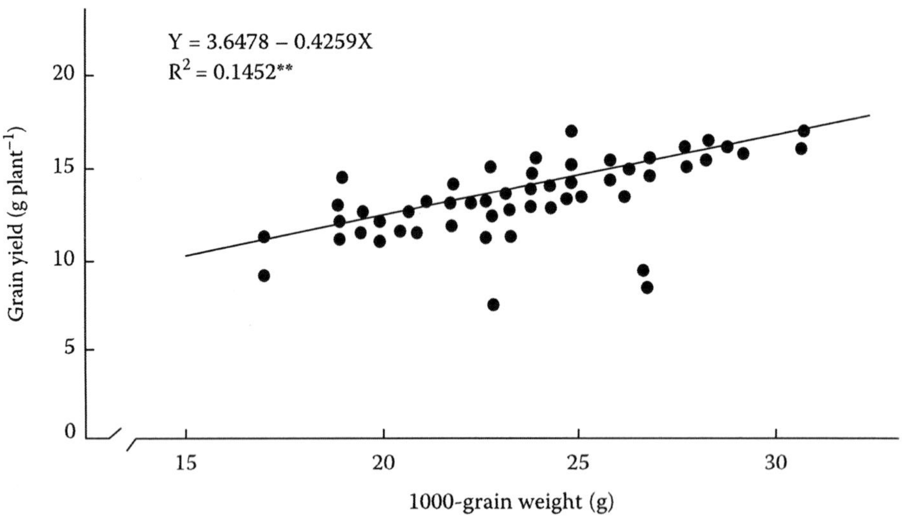

FIGURE 4.14 Relationship between 1000-grain weight and grain yield of upland rice.

Grain yield varied from 4.95 to 11.74 g plant^{-1} at the low K level, with an average value of 8.10 g plant^{-1}. Similarly, at the high K level, grain yield varied from 9.12 to 21.11 g plant^{-1}, with an average value of 15.42 g plant^{-1}. The increase in grain yield with the addition of 300 mg K kg^{-1} soil was 90% compared with 0 mg K kg^{-1} soil. Maschmann et al. (2010) and Fageria et al. (2010) reported a significant increase in

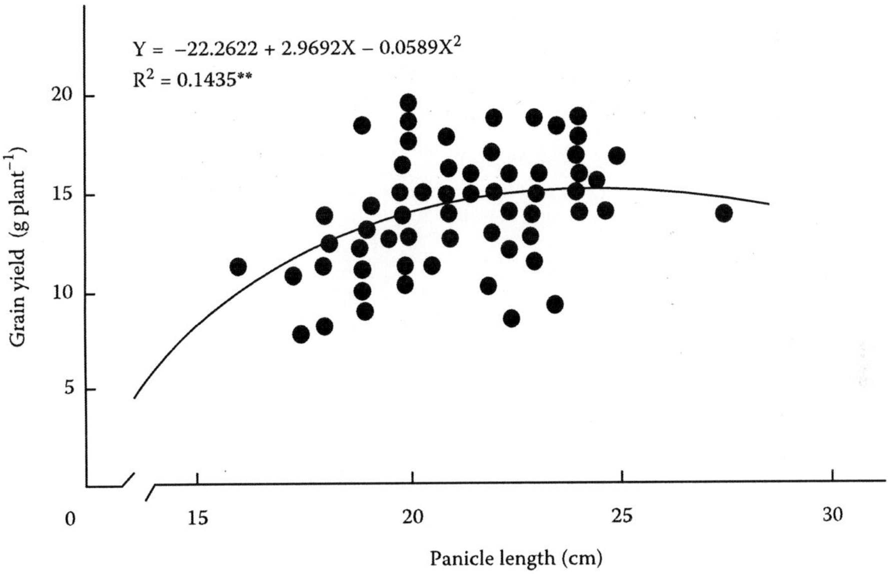

FIGURE 4.15 Relationship between panicle length and grain yield of upland rice.

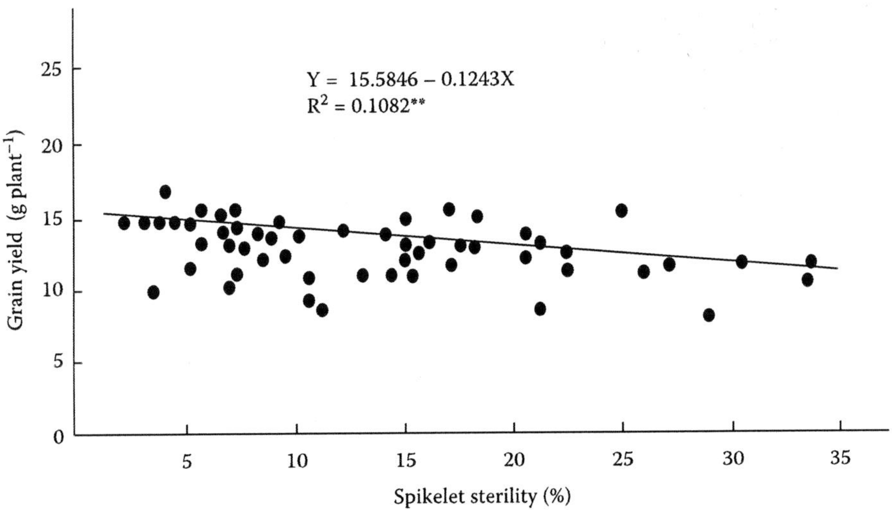

FIGURE 4.16 Relationship between spikelet sterility and grain yield of upland rice.

TABLE 4.20
Grain Yield and Grain Harvest Index (GHI) of 12 Lowland Rice Genotypes as Influenced by K Levels

	Grain Yield (kg plant^{-1})		GHI	
Genotype	0 mg K kg^{-1}	300 mg K kg^{-1}	0 mg K kg^{-1}	300 mg K kg^{-1}
BRS Tropical	11.74a	20.67a	0.45abcd	0.48abc
BRS Jaçanã	9.35ab	17.57abc	0.48ab	0.44cd
BRA 02654	9.48ab	21.11a	0.38de	0.53ab
BRA 051077	7.63bc	16.62bcd	0.49ab	0.41cd
BRA 051083	9.57ab	19.67ab	0.41bcde	0.54a
BRA 051108	5.46cd	14.41cde	0.36e	0.38d
BRA 051126	9.25ab	15.60cd	0.48ab	0.49abc
BRA 051129	10.88a	16.14bcd	0.52a	0.55a
BRA 051130	4.95d	9.12f	0.36e	0.39d
BRA 051134	5.51cd	11.04ef	0.35e	0.44bcd
BRA 051135	6.12cd	9.27f	0.46abc	0.37d
BRA 051250	7.28bcd	13.86de	0.40cde	0.50abc
Average	8.10b	15.42a	0.43b	0.46a
F-test				
K level (K)	*		*	
Genotype (G)	*		*	
K × G	*		*	
CV (%)(K)	13.16		3.75	
CV (%)(G)	8.40		6.41	

Note: Means followed by same letter in the same column are not significantly different at the 5% probability level by Tukey's test. Values for average are compared in the same line for statistically significant differences.

*Significant at the 5% probability level.

lowland rice yield with the addition of K in lowland soils of Brazil and the United States. The GHI varied from 0.35 to 0.52, with an average value of 0.43 at the low K level. At the high K level, GHI varied from 0.37 to 0.55, with an average value of 0.46. Overall, GHI increased 7% with the addition of K fertilizer compared to a control treatment. Snyder and Carlson (1984) reviewed the GHI of annual crops and noted variations from 0.23 to 0.50 for rice. Rice GHI values varied greatly among cultivars, locations, seasons, and ecosystems and ranged from 0.35 to 0.62, indicating the importance of this variable for yield stimulation (Kiniry et al. 2001). Mae (1997) reported that the GHI of traditional rice cultivars is about 0.30 and 0.50 for improved, semidwarf varieties. Figures 4.17 through 4.20 show the growth of lowland rice genotype tops at low and high K levels. Similarly, Figures 4.21 and 4.22 show root growth of lowland rice genotypes at low and high K levels. Shoot and root growth increased with K fertilization. However, there was a difference in the top and root growth among the genotypes.

FIGURE 4.17 Growth of tops of two lowland rice genotypes.

FIGURE 4.18 Growth of tops of two lowland rice genotypes.

FIGURE 4.19 Growth of tops of two lowland rice genotypes.

FIGURE 4.20 Growth of tops of two lowland rice genotypes.

FIGURE 4.21 Growth of roots of two lowland rice genotypes.

FIGURE 4.22 Growth of roots of two lowland rice genotypes.

4.9 CONCLUSIONS

Generally, rice does not respond to K fertilization to the degree noted for either nitrogen or phosphorus. However, K deficiency in rice has been reported in many of the world's rice-growing regions. In South America, rice is planted on many soil orders. However, upland rice is mainly grown on Oxisols and Ultisols, and lowland rice is planted on Inceptisols. These soils are acidic in reaction and are deficient in many macro- and micronutrients, including K.

Potassium plays an important role in many physiological and biochemical processes in plants. It is also responsible for reducing disease and infestation and improving root growth. Potassium is highly mobile in plants; therefore, deficiency symptoms initially appear on the older leaves. Potassium deficiency symptoms may be related to reduction in growth and chlorosis of the interveinal areas and margins of the lower leaves starting at the leaf tip. Leaf tips will eventually die and turn brown with the progression of severe K deficiency.

Rice accumulates maximum amounts of K in comparison to other essential nutrients. However, a major part of K (about 85%) is found in straw, and a small part (about 15%) translocates to grain. On the contrary, a higher part of N and P is exported to grain. The internal efficiency of K (grain yield per unit accumulation of K in grain plus straw) is also lower than that of N and P. Concentration (content per unit dry matter) decreases with plant age due to an increase in dry matter, known as the dilution effect. Important management practices that can be adopted to increase yield and reduce cost of production and environmental pollution are adequate rate use, efficient method, and planting K-efficient genotypes.

Adequate rate can be determined by soil testing. When Mehlich 1 extracting K is lower than 50 mg kg^{-1}, application of about 90 kg K_2O ha^{-1} is recommended. If the K level is higher than 80 mg kg^{-1}, about 60 kg K_2O ha^{-1} can produce a reasonably good yield of upland and lowland rice. Because K mobility in the soil is mainly by diffusion, band application is recommended for higher absorption in K-deficient soils. Potassium uptake by plants roots is influenced by soil moisture content, pH, aeration, the presence of Ca^{2+} and Mg^{2+}, CEC, and soil temperature. Therefore, adequate regulation of these factors may improve K uptake and utilization by rice plants.

REFERENCES

Baligar, V. C., and N. K. Fageria. 1997. Nutrient use efficiency in acid soils: Nutrient management and plant use efficiency. In *Plant-Soil Interactions at Low pH: Sustainable Agriculture and Forestry Production*, eds. A. C. Moniz, A. M. C. Furlani, N. K. Fageria, C. A. Rosolem, and H. Cantarells, pp. 75–95. Campinas, Brazil: Brazilian Soil Science Society.

Barber, S. A. 1995. *Soil Nutrient Bioavailability: A Mechanistic Approach*, 2nd edition. New York: Wiley.

Brady, N. C., and R. R. Weil. 2002. *The Nature and Properties of Soils,* 13th edition. Upper Saddle River, NJ: Prentice Hall.

Buol, S. W., P. A. Sanchez, R. B. Cote, Jr., and M. A. Granger. 1975. Soil fertility capability classification. In *Soil Management in Tropical America.* Proceedings of CIAT Seminar, February 1974, pp. 126–141. Raleigh, NC: North Carolina State University.

Chau, N. M., and S. C. Bhargava. 1993. Physiological basis of higher productivity in rice. *Indian J. Plant Physiol.* 36:215–219.
De Datta, S. K., and D. S. Mikkelsen. 1985. Potassium nutrition of rice. In *Potassium in Agriculture*, ed. R. D. Munson, pp. 665–699. Madison, WI: American Society of Agronomy.
Dobermann, A., K. G. Cassman, P. C. S. Cruz, M. A. Adviento, and M. F. Pampolino. 1996. Fertilizer inputs, nutrient balance, and soil nutrient supplying power in intensive, irrigated rice systems. I. Effective soil K-supplying capacity. *Nutr. Cycling Agroecosyst.* 46:11–21.
Epstein, E., and A. J. Bloom. 2005. *Mineral Nutrition of Plants: Principles and Perspectives*, 2nd edition. Sunderland, MA: Sinauer.
Fageria, N. K. 1984. *Fertilizer and Mineral Nutrition of Rice.* Rio de Janeiro, Brazil: EMBRAPA-CNPAF/Editora Campus.
Fageria, N. K. 1989. *Tropical Soils and Physiological Aspects of Crops.* Goiânia, Brazil: EMBRAPA-CNPAF.
Fageria, N. K. 1991. Response of rice to fractional applied potassium in Brazil. *Better Crops* 7:19.
Fageria, N. K. 1992. *Maximizing Crop Yields.* New York: Marcel Dekker.
Fageria, N. K. 2000. Potassium use efficiency of upland rice genotypes. *Pesq. Agropec. Bras.* 35:2155–2120.
Fageria, N. K. 2009. *The Use of Nutrients in Crop Plants.* Boca Raton, FL: CRC Press.
Fageria, N. K., V. C. Baligar, and R. B. Clark. 2006. *Physiology of Crop Production.* New York: The Haworth Press.
Fageria, N. K., V. C. Baligar, and C. A. Jones. 2011. *Growth and Mineral Nutrition of Field Crops*, 3rd edition. Boca Raton, FL: CRC Press.
Fageria, N. K., V. C. Baligar, and Y. C. Li. 2008. The role of nutrient efficient plants in improving crop yields in the twenty-first century. *J. Plant Nutr.* 31:1121–1157.
Fageria, N. K., V. C. Baligar, R. J. Wright, and J. R. P. Carvalho. 1990a. Lowland rice response to potassium fertilization and its effect on N and P uptake. *Fertilizer Res.* 21:157–162.
Fageria, N. K., M. P. Barbosa Filho, and J. R. P. Carvalho. 1982. Response of upland rice to phosphorus fertilization on an Oxisol of central Brazil. *Agron. J.* 74:51–56.
Fageria, N. K., E. M. Castro, and V. C. Baligar. 2004. Response of upland rice genotypes to soil acidity. In *The Red Soils of China: Their Nature, Management and Utilization,* eds. M. J. Wilson, Z. He, and X. Yang, pp. 219–237. Dordrecht, The Netherlands: Kluwer Academic.
Fageria, N. K., and A. B. Santos. 2008. Lowland rice response to thermophosphate fertilization. *Commun. Soil Sci. Plant Anal.* 39:873–889.
Fageria, N. K., A. B. Santos, and M. F. Moraes. 2010. Yield, potassium uptake, and use efficiency in upland rice genotypes. *Commun. Soil Sci. Plant Anal.* 41:2676–2684.
Fageria, N. K., A. B. Santos, A. Moreira, and M. F. Moraes. 2010. Potassium soil test calibration for lowland rice on an Inceptisol. *Commun. Soil Sci. Plant Anal.* 41:2595–2601.
Fageria, N. K., N. A. Slaton, and V. C. Baligar. 2003. Nutrient management for improving lowland rice productivity and sustainability. *Adv. Agron.* 80:63–152.
Fageria, N. K., and L. F. Stone. 2008. Micronutrient deficiency problems in South America. In *Micronutrient Deficiencies in Global Crop Production*, ed. B. J. Alloway, pp. 245–266. New York: Springer.
Fageria, N. K., L. F. Stone, and A. B. Santos. 2003. *Soil Fertility Management of Irrigated Rice.* Santo Antonio de Goiás, Brazil: EMBRAPA-Rice and Bean Research Center.
Fageria, N. K., R. J. Wright, V. C. Baligar, and J. R. P. Carvalho. 1990b. Upland rice response to potassium fertilization on a Brazilian Oxisol. *Fertilizer Res.* 21:141–147.
Foth, H. D., and B. G. Ellis. 1988. *Soil Fertility.* New York: John Wiley.

Hasegawa, H. 2003. High yielding rice cultivars perform best even at reduced nitrogen fertilizer rate. *Crop Sci.* 43:921–926.
Hinsinger, P. 1998. How do plant roots acquire mineral nutrients? Chemical processes involved in the rhizosphere. *Adv. Agron.* 64:225–265.
Kiniry, J. R., G. McCauley, Y. Xie, and J. G. Arnold. 2001. Rice parameters describing crop performance of four U. S. cultivars. *Agron. J.* 93:1354–1361.
Mae, T. 1997. Physiological nitrogen efficiency in rice: Nitrogen utilization, photosynthesis, and yield potential. *Plant Soil.* 196:201–210.
Maschmann, E. T., N. A. Slaton, R. D. Cartwright, and R. J. Norman. 2010. Rate and timing of potassium fertilization and fungicide influence rice yield and stem rot. *Agron. J.* 102:163–170.
Mengel, K., E. A. Kirkby, H. Kosegarten, and T. Apple. 2001. *Principles of Plant Nutrition,* 5th edition. Dordrecht, The Netherlands: Kluwer Academic.
Nelson, K. A., P. P. Motavalli, W. E. Stevens, D. Dunn, and C. G. Meinhardt. 2010. Soybean response to preplant and foliar-applied potassium chloride with strobilurin fungicides. *Agron. J.* 102:1657–1663.
Portela, E. A. C. 1993. Potassium supplying capacity of northeastern Portuguese soils. *Plant Soil.* 154:13–20.
Pretty, K. M., and P. J. Stangel. 1985. Current and future use of world potassium. In *Potassium in Agriculture*, ed. R. D. Munson, pp. 99–128. Madison, WI: ASA, CSSA, and SSSA.
Sanchez, P. A., and J. G. Salinas. 1981. Low-input technology for managing Oxisols and Ultisols in tropical America. *Adv. Agron.* 34:279–405.
Santos, A. B., N. K. Fageria, L. F. Stone, and C. Santos. 1999. Water and potassium fertilization management for irrigated rice cultivation. *Pesq. Agropec. Bras.* 34:565–573.
Slaton, N. A., C. D. Cartwright, and C. E. Wilson, Jr. 1995. Potassium deficiency and plant diseases observed in rice fields. *Better Crops.* 79:12–14.
Snyder, F. W., and G. E. Carlson. 1984. Selecting for partitioning of photosynthetic products in crops. *Adv. Agron.* 37:47–72.
Soil Science Society of America. 2008. *Glossary of Soil Science Terms*. Madison, WI: SSSA.
Sparks, D. L. 1987. Potassium dynamics in soils. *Adv. Soil Sci.* 6:1063.
Sparks, D. L., and P. M. Huang. 1985. Physical chemistry of soil potassium. In *Potassium in Agriculture*, ed. R. D. Munson, pp. 201–276. Madison, WI: American Society of Agronomy.
Stewart, J. A. 1985. Potassium sources, use, and potential. In *Potassium in Agriculture*, ed. R. D. Munson, pp. 83–98. Madison, WI: ASA, CSSA, and SSSA.
Stone, K. 2008. Potash. In *Canadian Minerals Yearbook*. pp. 36.1–36.12. Ottawa, Canada: Natural Resources Canada.
Teo, Y. H., C. A. Beyrouty, R. J. Norman, and E. E. Gbur. 1995. Nutrient uptake relationship to root characteristics of rice. *Plant Soil.* 171:297–302.
Williams, J., and S. G. Smith. 2001. Correcting potassium deficiency can reduce rice stem diseases. *Better Crops.* 85:7–9.
Xiaoe, Y., V. Romheld, H. Marschner, V. C. Baligar, and D. C. Martens. 1997. Shoot photosynthesis and root growth of hybrid and conventional rice cultivars as affected by N and K levels in the root zone. *Pedosphere.* 7:35–42.

5 Calcium and Magnesium

5.1 INTRODUCTION

Calcium (Ca) and magnesium (Mg) are essential plant nutrients. Along with sulfur, they are sometimes referred to as secondary elements, which are required by plants in large amounts for normal growth and development. Calcium and magnesium deficiencies are very common in Oxisols, Ultisols, Alfisols, and Inceptisols on which upland rice is grown (Clark 1982; Fageria and Souza 1991; Fageria 2000; Fageria and Baligar 2008). Data in Table 5.1 show the response of upland rice to soil pH. The dry matter yield of straw, grain yield, panicle density, and 1000-grain weight were significantly increased with increase in soil pH or addition of lime (Ca^{2+} and Mg^{2+}). With a change in pH or the addition of lime, the variability was 99% in straw yield, 91% in grain yield, and 94% in panicle density. Calcium is usually the predominant soil cation and is present in relatively large amounts, especially on alkaline soils. Ca^{2+} is a nontoxic mineral nutrient, even in high concentrations, and is very effective in detoxifying high concentrations of other mineral elements in plants (Marschner 1995).

Calcium and magnesium are much more stable in soil than sulfur. Highly weathered Oxisols and Ultisols may be low in Ca^{2+} and Mg^{2+} due to excessive leaching. Sandy soils may also contain low levels of Ca^{2+} and Mg^{2+}, thereby increasing the likelihood of crops being deficient in these nutrients (Fageria and Baligar 2008).

In acidic soils, liming is one of the most effective and commonly used practices for supplying Ca^{2+} and Mg^{2+} and eliminating Al toxicity (Fageria, Baligar, and Edwards 1990; Fageria and Baligar 2001, 2008). Data in Table 5.2 show that liming significantly increased pH and exchangeable Ca^{2+} and Mg^{2+} in Oxisol at two soil depths. The values of these chemical properties were higher at the 0–20 soil depth compared with the 20–40 soil depth. In addition, liming also increased base saturation (Figure 5.1). Figures 5.2 and 5.3 show the relationship between lime rate, soil pH, and soil extractable Ca + Mg. Similarly, liming also decreased Al toxicity. Figure 5.4 shows a change in aluminum species with the change in soil pH. The toxic species of Al^{3+} decreased significantly when soil pH was increased from 3.5 to 6.

Known as the foundation of crop production, lime is a "workhorse" in acidic soils. The amount of lime needed for crops grown on acidic soils is determined by the quality of liming material, the status of soil fertility, the crop species and cultivar within the species, crop management practices, and economic considerations (Fageria and Baligar 2008).

Changes in a soil's chemical properties with the use of dolomitic lime [$CaMg(CO_3)_2$] can be explained with the following equations (Fageria and Baligar 2005):

$$CaMg(CO_3)_2 + 2H^+ \Leftrightarrow 2HCO_3^- + Ca^{2+} + Mg^{2+} \tag{5.1}$$

TABLE 5.1
Response of Upland Rice to Soil pH Levels

Soil pH in H_2O	Straw Yield (g plant^{-1})	Grain Yield (g plant^{-1})	Panicle Density (plant^{-1})
4.6	15.91	11.00	5.00
5.7	14.96	11.53	4.58
6.2	12.24	11.73	4.50
6.4	10.62	9.49	3.92
6.6	7.78	6.83	3.42
6.8	6.03	5.15	3.25
R^2	0.99**	0.91*	0.94*

Source: Fageria, N. K. 2000. Upland rice response to soil acidity in Cerrado soil. *Pesq. Agropec. Bras.* 35:2303–2307.

*,**Significant at the 5 and 1% probability levels, respectively.

TABLE 5.2
Soil pH and Exchangeable Ca and Mg after Harvest of Eight Crops in Rotation

Lime Rate (Mg ha^{-1})	pH in H_2O	Ca (cmol$_c$ kg^{-1})	Mg (cmol$_c$ kg^{-1})
		0–20 cm soil depth	
0	5.6	1.92	1.09
4	6.0	2.33	1.14
8	6.2	3.00	1.21
12	6.4	3.10	1.25
16	6.5	3.30	1.25
20	6.8	3.77	1.39
R^2	0.83**	0.72**	0.23*
		20–40 cm soil depth	
0	5.5	1.76	0.98
4	5.9	1.92	1.09
8	6.1	2.30	1.11
12	6.2	2.42	1.14
16	6.3	2.57	1.16
20	6.7	3.25	1.32
R^2	0.78**	0.78**	0.32**

Source: Adapted from Fageria, N. K. 2001a. Effect of lime on upland rice, common bean, corn, and soybean production in Cerrado soil. *Pesq. Agropec. Bras.* 36:1419–1424.

Note: The crops were upland rice, dry beans, corn, and soybeans grown in rotation for four years.

*,**Significant at the 5 and 1% probability levels, respectively.

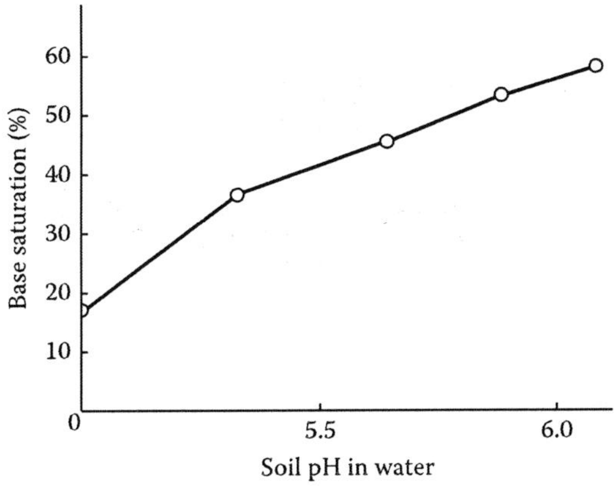

FIGURE 5.1 Influence of soil pH on base saturation. (From Fageria, N. K. 1989. *Tropical Soils and Physiological Aspects of Crops*. Goiania, Brazil: EMBRAPA-CNPAF.)

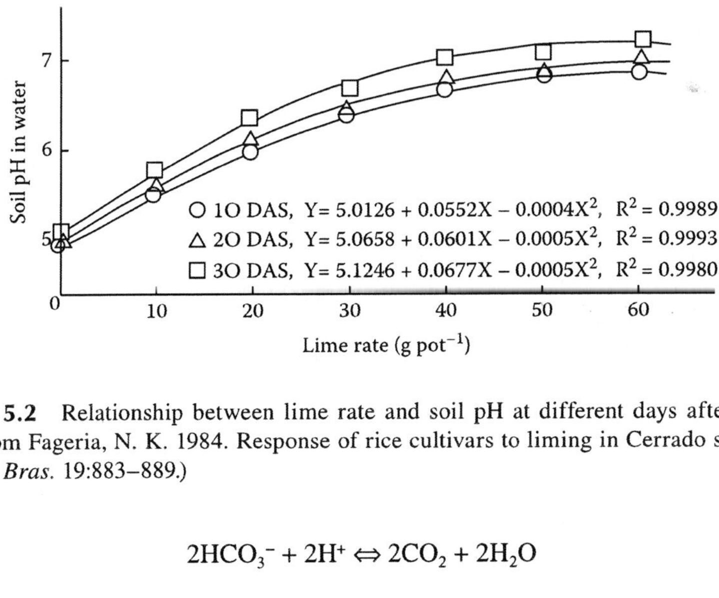

FIGURE 5.2 Relationship between lime rate and soil pH at different days after incubation. (From Fageria, N. K. 1984. Response of rice cultivars to liming in Cerrado soil. *Pesq. Agropec. Bras.* 19:883–889.)

$$2HCO_3^- + 2H^+ \Leftrightarrow 2CO_2 + 2H_2O \tag{5.2}$$

$$CaMg(CO_3)_2 + 4H^+ \Leftrightarrow Ca^{2+} + Mg^{2+} + 2CO_2 + 2H_2O \tag{5.3}$$

The above equations show that the acidity neutralizing reactions of lime occur in two steps. In the first step, Ca^{2+} and Mg^{2+} react with H on the exchange complex, and H is replaced by Ca^{2+} and Mg^{2+} on the exchange sites (negatively charged particles of clay or organic matter), forming HCO_3^-. In the second step, HCO_3^- reacts with H^+ to form CO_2 and H_2O to increase pH. Soil moisture and temperature as well as quantity and quality of liming material mainly determine the reaction rate of lime. To get maximum benefits from liming in order to improve crop yields,

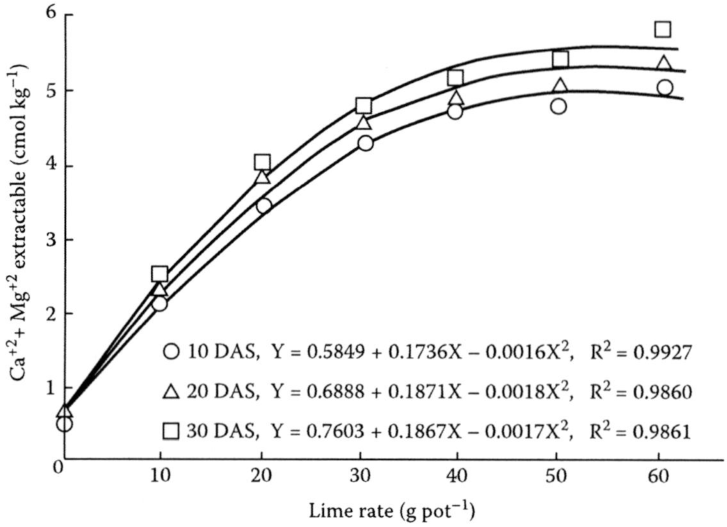

FIGURE 5.3 Relationship between lime rate and exchangeable Ca + Mg in Cerrado soil of Brazil. (From Fageria, N. K. 1984. Response of rice cultivars to liming in Cerrado soil. *Pesq. Agropec. Bras.* 19:883–889.)

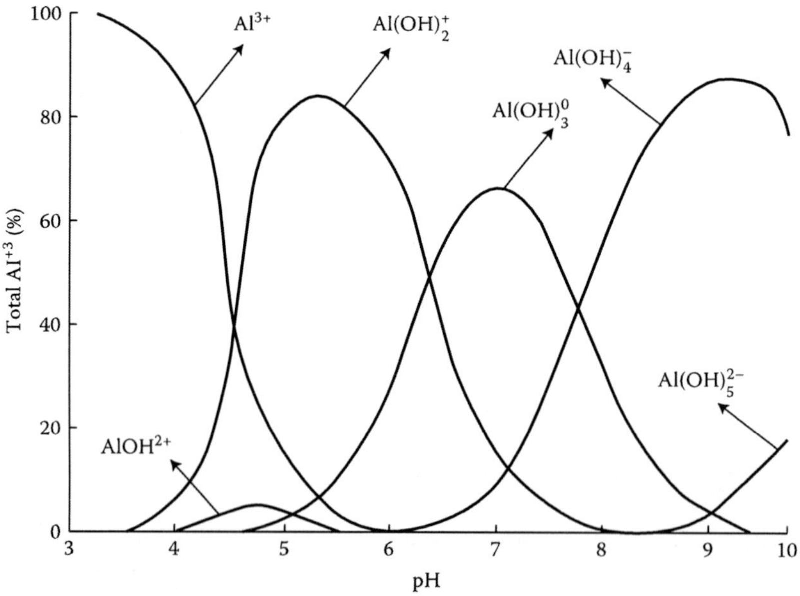

FIGURE 5.4 Relationship between aluminum species and soil pH. (Adapted from Kinraide, T. B. 1991. Identity of rhizotoxic aluminum species. In *Plant Soil Interactions at Low pH*, eds. R. J. Wright, V. C. Baligar, and R. P. Murrmann, pp. 717–728. Dordrecht, The Netherlands: Kluwer Academic.)

liming materials should be applied in advance of crop sowing and thoroughly mixed into the soil to enhance the lime's reaction with soil exchange acidity. The objective of this chapter is to discuss the calcium and magnesium nutrition of upland and lowland rice.

5.2 CYCLE IN SOIL–PLANT SYSTEM

The calcium and magnesium cycle includes addition, transformation, availability to plants, and loss in the soil–plant system. Mineralization of soil parent materials also contributes to soil solution calcium and magnesium. The calcium cycle in the soil–plant system is shown in Figure 5.5. The calcium and magnesium content in soil depends on the type of soil and the extent of weathering. Highly weathered and leached soils under humid conditions (soils such as Spodosols of the temperate zone and Oxisols and Ultisols of the humid tropics) are generally low in calcium and magnesium content. Leaching of Mg is higher compared with Ca^{2+} ions because the thicker water mantle surrounding Mg^{2+} causes it to be less tightly adsorbed to

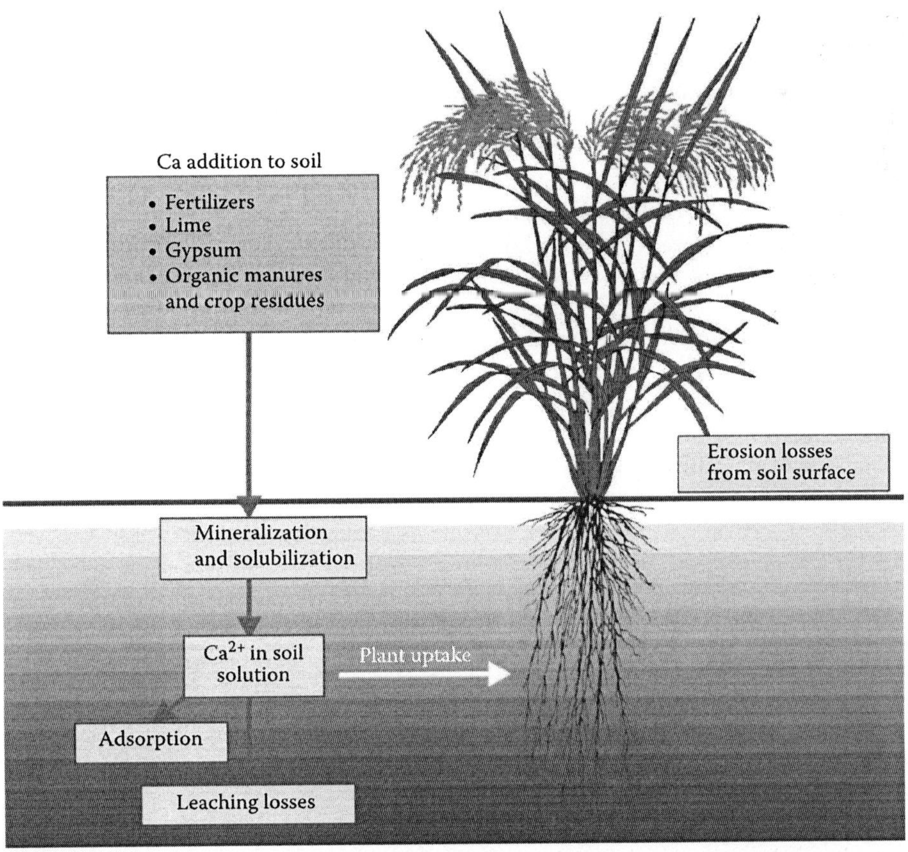

FIGURE 5.5 Calcium cycle in soil–plant system.

soil colloids (Mengel et al. 2001). A major source for the addition of Ca^{2+} and Mg^{2+} is dolomitic lime. Calcium can also be supplied by gypsum ($CaSO_4.2H_2O$) (23% Ca), simple superphosphate (20% Ca), triple superphosphate (13% Ca), and thermophosphate (21% Ca). Mg can be supplied by magnesium oxide (60% Mg), magnesium sulfate (9.5% Mg), and thermophosphate (11% Mg). Similarly, Ca and Mg can be supplied by organic manures and crop residues.

Calcium does not fix in the soil or make complexation in soil-like P. However, clay minerals and other soil components may fix Mg. Most of the Ca and Mg required by plants can be transported to the root surface by mass flow. It can also be absorbed by root interception. Tisdale, Nelson, and Beaton (1985) reported that Ca and Mg availability to plants is determined by soil pH, soil texture, type of soil colloid, cation exchange capacity (CEC), total Ca supply, and ratio of Ca^{2+} to other cations in the soil solution.

The effect of pH on the uptake of Ca^{+} and Mg^{2+} is not as great as on the uptake of N, P, and other micronutrients. Under Ca^{2+} deficiency field conditions, soil pH is usually increased by adding dolomitic lime, which contains Ca^{2+} and Mg^{2+}. Therefore, increasing soil pH increases Ca^{2+} and Mg^{2+} levels and uptake. This increase occurs because more Ca^{2+} and Mg^{2+} are added to the plant growth medium; the lime removes Al and its negative impact on Ca^{2+} and Mg^{2+} uptake (Clark 1984).

5.3 FUNCTIONS

Calcium is involved in cell division and cell elongation. It influences the pH of cells and the structural stability and permeability of cell membranes. Calcium acts as a regulator ion in the translocation of carbohydrates through its effects on cells and cell walls (Bennett 1993). It plays a role in mitosis and is cited for its beneficial impact on plant vigor, straw stiffness, and grain and seed formation (Follett, Murphy, and Donahue 1981). Calcium acts as an activator for only a few enzymes and is a part of certain structural compounds such as calcium oxalate and calcium pectate (Bould, Hewitt, and Needham 1984). Clarkson and Hanson (1980) describe the function of calcium as follows: "simply put, Ca^{2+} can tie diverse macromolecules together." Clark (1984) summarized Ca^{2+} functions in plants as stabilizing membranes, increasing cell wall rigidity and elongation, stimulating enzyme activities and metabolism, and producing phytohormones. Foy (1974) reported that the primary role of Ca^{2+} in plant nutrition has been to detoxify the effects of heavy metals and other cation excesses.

Similarly, magnesium is a component of chlorophyll and a cofactor for many enzymatic reactions (Fageria, Baligar, and Jones 2011). Enzyme reactions that require Mg^{2+} include PO_4^{3-} or nucleotide transfer (phosphatases, kinases, ATPases, synthases, nucleotide transferases), carboxyl transfer (carboxylases, dicarboxylases), and interaction with enzymes as dehydrogenases, mutases, and lyases (Clarkson and Hanson 1980; Clark 1984). Evidence indicates that Mg^{2+} may regulate enzyme activity by binding to enzymes apart from their substrate sites (Clark 1984). Magnesium helps in the translocation of P in plants and in the formation of sugars, oils, and fats. It also activates the formation of polypeptide chains from amino acids (Tisdale, Nelson, and Beaton 1985).

5.4 DEFICIENCY SYMPTOMS

Calcium does not move in the plants from older to newer parts. Therefore, calcium deficiency symptoms first appear in the younger plant tissues or growing points. Calcium deficiency is characterized by malformation and disintegration of a plant's terminal portions. Although calcium deficiency symptoms in rice have been studied in controlled conditions, they rarely appear under field conditions. Figure 5.6 shows calcium-deficient rice plants in a solution culture.

Magnesium is easily translocated from older to newer plant or leaf parts. Therefore, magnesium deficiency symptoms first appear in the older leaves of a plant. Because magnesium is related to chlorophyll, a deficiency of Mg results in a characteristic discoloration of the leaves. Rice plant leaves become purplish red with

FIGURE 5.6 (See color insert.) Rice plants with calcium deficiency symptoms. (From Fageria, N. K., and M. P. Barbosa Filho. 1994. *Nutrient Deficiency in Rice: Identification and Correction*. Goiania, Brazil: EMBRAPA-SPI/EMBRAPA-CNPAF.)

FIGURE 5.7 (See color insert.) Magnesium deficiency symptoms in rice plants. (From Fageria, N. K., and M. P. Barbosa Filho. 1994. *Nutrient Deficiency in Rice: Identification and Correction*. Goiania, Brazil: EMBRAPA-SPI/EMBRAPA-CNPAF.)

green veins. Figure 5.7 shows magnesium deficiency symptoms in rice plants grown in a solution culture. Figure 5.8 also shows Mg-deficient rice leaves. Magnesium deficiency symptoms in rice plants rarely appear under field conditions.

5.5 UPTAKE IN PLANT TISSUE

The first step in a plant's uptake of a nutrient is an adequate concentration in the rhizosphere or at the root surface. Most Ca^{2+} and Mg^{2+} are transported to the roots by mass flow. After reaching the root surface, Ca^{2+} and Mg^{2+} may be absorbed by active or passive processes. Uptake of calcium and Mg ions is genetically controlled and also is influenced by environmental factors, especially concentrations of these ions in the soil solution. Calcium and Mg concentrations (content per unit of dry matter) are generally lower in cereals than in legumes. In rice plants, calcium and Mg uptake is generally lower than K uptake. This may be related to monovalent (K^+) versus divalent (Ca^{2+} or Mg^{2+}) ions.

Calcium concentration is fairly high in legumes and may reach as high as 20–30 g kg^{-1} in the leaves (Clark 1984). However, in cereals (including rice), the Ca requirement for most metabolic and physiological processes is quite low. Brady and Weil (2002) reported that plants generally use calcium in amounts second only to that of N and K. These authors also noted that the Ca amounts taken up are greater for dicots (0.5 to >2% of dry matter) than for monocots (0.15–0.5%). Wallace and Soufi (1975) reported that in many plants, Ca^{2+} may be required at concentrations similar to those of micronutrients. Wallace (1966, 1972) noted that 0.8–2 g kg^{-1} of Ca^{2+} in the foliar parts of plants is sufficient for growth and development. One reason this amount might be so low is because of the reduced concentrations of heavy metals

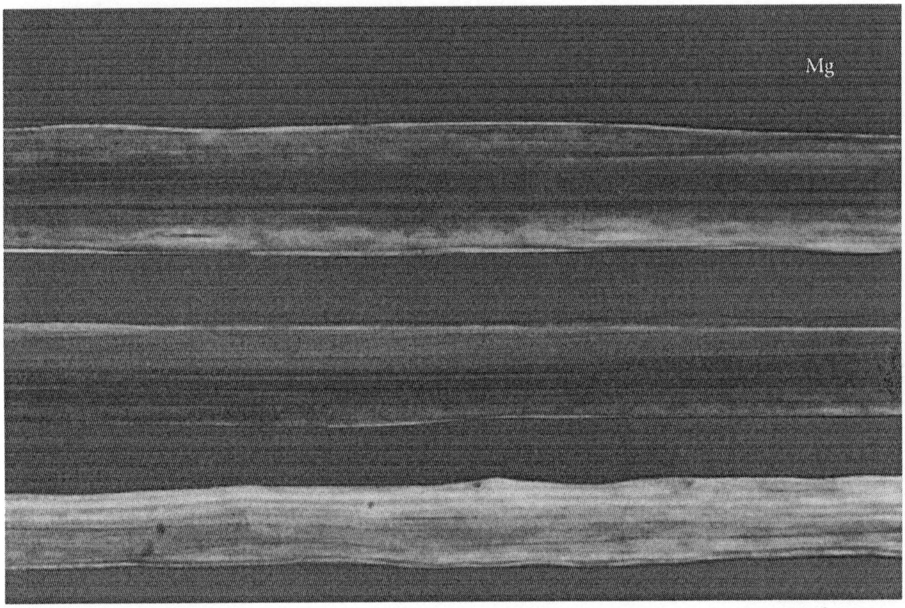

FIGURE 5.8 (See color insert.) Magnesium deficiency symptoms in rice leaves. (From Fageria, N. K., and M. P. Barbosa Filho. 1994. *Nutrient Deficiency in Rice: Identification and Correction*. Goiania, Brazil: EMBRAPA-SPI/EMBRAPA-CNPAF.)

available to plants (Clark 1984). One of the main functions of Ca^{2+} is to detoxify the toxicity of heavy metals. The Mg uptake is lower than Ca and may range from 0.15 to 0.75% (Brady and Weil 2002).

The Ca and Mg concentrations and their uptake in lowland rice straw and grain during the growth cycle were studied (Table 5.3). In straw, calcium as well as magnesium concentrations decreased quadratically with advanced plant age (Table 5.4). The calcium concentration was 4 g kg^{-1} at the beginning of tillering and decreased to 2.6 g kg^{-1} at physiological maturity. Similarly, the Mg concentration in straw was 2.6 g kg^{-1} at the start of the tillering growth stage and decreased to 1.2 g kg^{-1} at maturity. In straw, uptake of Ca as well as magnesium increased significantly in a quadratic fashion with advanced plant age. The Ca uptake increased from 1.5 kg ha^{-1} at the start of the tillering growth stage to 25.8 kg ha^{-1} at physiological maturity (Table 5.3). Similarly, straw's Mg uptake increased from 0.9 kg ha^{-1} at the start of tilling to 15.4 kg ha^{-1} at physiological maturity. The decrease in Ca and Mg concentrations in straw with advanced plant age was related to an increase in straw weight at advanced plant age (Table 5.5). Straw yield was 351 kg ha^{-1} at the beginning of the tillering growth stage and increased to 9423 kg ha^{-1} at physiological maturity. In the field of mineral nutrition, the decrease in Ca and Mg concentration with advanced plant age is known as the dilution effect (Fageria, Baligar, and Jones 2011). Calcium concentration was higher in rice straw than Mg concentration. However, Ca concentration in grain was lower than Mg because Ca is more mobile (Fageria, Slaton, and Baligar 2003). In rice straw of 100 days' growth (about the

TABLE 5.3
Concentration and Uptake of Ca and Mg in the Lowland Rice Straw and Grain at Different Growth Stages

	Concentration (g kg^{-1})		Uptake (kg ha^{-1})	
Growth Stage/Days after Sowing	**Ca**	**Mg**	**Ca**	**Mg**
IT (22)	4.0	2.6	1.5	0.9
AT (35)	4.0	2.3	4.6	2.6
IP (71)	3.2	1.8	18.4	10.3
B (97)	2.9	1.9	30.6	19.6
F (112)	2.9	1.8	37.8	24.4
PM (140)	2.6	1.6	25.8	15.4
PM (140) (grain)	0.80	1.2	5.0	7.0
F-test	*	*	*	*
CV (%)	7	6	14	14

Source: Adapted from Fageria, N. K., and V. C. Baligar. 2003. Fertility management of tropical acid soils for sustainable crop production. In *Handbook of Soil Acidity*, ed. Z. Rengel, 359–385. New York: Marcel Dekker.

Note: Values are averages of three years of field experiments and those in parentheses represent plant age in days after sowing. IT, initiation of tillering; AT, active tillering; IP, initiation of panicle primordia; B, booting; F, flowering; PM, physiological maturity.

*Significant at the 1% probability.

TABLE 5.4
Relationship between Plant Age and Concentration and Uptake of Ca and Mg in the Straw of Lowland Rice Plants

Variable	Regression Equation	R^2
Plant age vs. Ca conc. in straw	$Y = 4.59 - 0.024X + 0.000074X^2$	0.86**
Plant age vs. Mg conc. in straw	$Y = 2.83 - 0.016X + 0.000058X^2$	0.79**
Plant age vs. Ca uptake in straw	$Y = -18.40 + 0.85X - 0.0036X^2$	0.88**
Plant age vs. Mg uptake in straw	$Y = -12.01 + 0.53X - 0.0023X^2$	0.84**

Note: Values are averages of three years' field experiment.

**Significant at the 1% probability level.

booting stage), Fageria, Baligar, and Jones (2011) reported that the adequate concentration range of Ca was 2.5–4.0 g kg^{-1} and of Mg was 1.7–3.0 g kg^{-1}. De Datta (1981) reported the critical concentrations of Ca as 1.5 g kg^{-1} and of Mg as 1.0 g kg^{-1} in rice straw at maturity. In order to produce 1 Mg grain yield of lowland rice, it is necessary to have 4.8 kg Ca and 3.5 kg Mg in straw and grain at maturity.

The concentration and uptake of Ca and Mg in upland rice during the growth cycle was also studied (Table 5.6). Concentration of these elements was significantly decreased with advanced plant age, whereas the uptake of both nutrients was

TABLE 5.5
Straw and Grain Yield of Lowland Rice at Different Growth Stages

Growth Stage/Plant Age	Straw and Grain Yield (kg ha^{-1})
IT (22)	351
AT (35)	1130
IP (71)	5841
B (97)	10384
F (112)	13490
PM (140)	9423
PM (140) (grain)	6389
F-test	*
CV (%)	14

Regression Analysis

Plant age vs. straw yield $(Y) = -6362.42 + 277.42X - 1.11X^2$, $R^2 = 0.86$**

Plant age vs. straw plus grain yield $(Y) = -4310.17 + 185.61X - 0.37X^2$, $R^2 = 0.87$**

Note: Values are averages of three years of field experiment, and those in parentheses represent plant age in days after sowing. IT, initiation of tillering; AT, active tillering; IP, initiation of panicle primordia; B, booting; F, flowering; PM, physiological maturity.

*Significant at the 1% probability level.

significantly increased with advanced plant age (Table 5.7). The decrease in concentration of both elements was related to an increase in straw dry matter with advanced plant age (Table 5.8). Straw yield increased in a quadratic fashion with advanced plant age; when grain yield was added to straw, the increase was linear (Table 5.8). As is the case with lowland rice, the major portion of the Ca and Mg remained in straw, and 16% Ca and 27% Mg were translocated to grain. An accumulation of 6 kg Ca and 4 kg Mg is necessary in grain and straw in order to produce 1 Mg grain upland rice. These values are slightly higher than lowland rice. Clark (1984) reviewed the concentrations of Ca and Mg in principal food crops, including rice (Table 5.9). The range of Ca and Mg varied among crop species and growth stages.

5.6 HARVEST INDEX

The harvest indices of calcium and magnesium are important for determining the distribution of these elements in grain and straw. Calcium harvest index values of different crops species are presented in Table 5.10. In upland rice, only 7% of the total Ca accumulated (grain plus straw) was translocated to grain. For lowland rice, Ca translocation to grain was about 17%, and for corn it was 19%. In dry beans, the calcium harvest index was 19–29% and in soybeans 23%. This means that legume seeds have a higher Ca content than cereals; this has special significance for human nutrition.

TABLE 5.6
Concentration and Uptake of Ca and Mg in Straw and Grain of Upland Rice during Growth Cycle

Growth Stage/Days after Sowing	Concentration ($g\ kg^{-1}$)		Uptake ($kg\ ha^{-1}$)	
	Ca	Mg	Ca	Mg
IT (19)	3.3	2.7	0.29	0.23
AT (43)	3.8	2.7	4.06	2.88
IP (69)	3.1	2.4	11.21	8.61
B (90)	2.6	2.1	15.19	12.23
F (102)	2.8	2.2	20.07	16.17
PM (130)	2.6	2.2	23.45	14.47
PM (130) (grain)			4.33	5.38
F-test	*	*	*	*
CV (%)	14	16	31	24

Note: Values are averages of two years of field experimentation, and those in parentheses represent plant age in days after sowing. IT, initiation of tillering; AT, active tillering; IP, initiation of panicle primordia; B, booting; F, flowering; PM, physiological maturity.
*Significant at the 1% probability.

TABLE 5.7
Relationship between Plant Age and Concentration and Uptake of Ca and Mg in Straw of Upland Rice Plants

Variable	Regression Equation	R^2
Plant age vs. Ca conc. in straw	$Y = 3.66 - 0.0066X + 0.000018X^2$	0.66*
Plant age vs. Mg conc. in straw	$Y = 3.00 - 0.012X + 0.000045X^2$	0.84*
Plant age vs. Ca uptake in straw	$Y = -3.06 + 0.15X - 0.000067X^2$	0.95**
Plant age vs. Mg uptake in straw	$Y = -3.55 + 0.17X - 0.00013X^2$	0.97**

Note: Values are averages of two years of field experiment.
*,**Significant at the 5 and 1% probability levels, respectively.

The magnesium harvest index (defined as the quantity of Mg^{2+} accumulated in grain as a percentage of total Mg accumulated in grain plus shoot) for cereals and legumes is presented in Table 5.11. Overall, the Mg^{2+} harvest index was higher for legumes than for cereals. This means that the Mg^{2+} requirement of legumes is higher than that of cereals. A relation between Mg^{2+} harvest index (X) and grain yield (Y) of cereals and legumes cited in Table 5.11 was calculated as $Y = 15370.18 - 454.4832X + 3.6653X^2$, $R^2 = 0.3187^{NS}$. This means that Mg^{2+} accumulation in cereals and legumes has no effect on grain yield.

TABLE 5.8
Straw and Grain Yield of Upland Rice at Different Growth Stages

Growth Stage/Plant Age	Straw and Grain Yield (kg ha^{-1})
IT (19)	88
AT (43)	1075
IP (69)	3466
B (90)	5987
F (102)	7613
PM (130)	6343
PM (130) (grain)	4568
F-test	*
CV (%)	14

Regression Analysis

Plant age vs. straw yield (Y) = $-3021.30 + 136.66X - 0.45X^2$, $R^2 = 0.0.89^{**}$

Plant age vs. straw plus grain yield (Y) = $-3061.65 + 107.08X$, $R^2 = 0.96^{**}$

Source: Fageria, N. K. 2001b. Response of upland rice, dry bean, corn and soybean to base saturation in Cerrado soil. *Rev. Bras. Eng. Agricola e Amb.* 5:416–424.

Note: Values are averages of two years of field experiments, and those in parentheses represent plant age in days after sowing. IT, initiation of tillering; AT, active tillering; IP, initiation of panicle primordia; B, booting; F, flowering; PM, physiological maturity.

*Significant at the 1% probability level.

TABLE 5.9
Ranges of Ca and Mg Concentrations Considered to Be Sufficient for Principal Food Crops

Crop	Plant Part	Stage of Growth	Ca Conc. (g kg^{-1})	Mg Conc. (g kg^{-1})
Rice	Leaves	MT to PD	1.6–3.9	1.6–3.9
Corn	Leaves	3rd–4th leaf stage	9.0–16.0	3.0–8.0
Corn	Ear leaf	Silk	2.1–10.0	2.0–4.0
Corn	Stalk	Silk	1.0–3.0	1.0–3.0
Wheat and oats	Leaves	Head emergence	2.0–5.0	1.5–5.0
Barley	Leaves	Not given	3.0–12.0	1.2–2.1
Soybeans	Leaves	Not given	3.6–20.0	2.6–10.0
Peanuts	Upper stem and leaves	Various	7.5–20.0	3.0–8.0
Dry beans	Leaves	YM	8.0–30.0	2.5–7.0

Source: Adapted from Clark, R. B. 1984. Physiological aspects of calcium, magnesium, and molybdenum deficiencies in plants. In *Soil Acidity and Liming*, 2nd edition, ed. F. Adams, pp. 99–170. Madison, WI: ASA, CSSA, and SSSA.

Note: MT to PD, Midtillering to panicle differentiation; YM, Younger mature leaves.

TABLE 5.10
Calcium Harvest Index in Principal Field Crops

Crop Species	Ca Harvest Index (%)
Upland rice	7
Lowland rice	17
Lowland rice	16
Corn	19
Dry beans	29
Dry beans	19
Soybeans	23

Source: Fageria, N. K. 2009. *The Use of Nutrients in Crop Plants*. Boca Raton, FL: CRC Press.
Note: Calcium harvest index (%) = (Ca uptake in grain/Ca uptake in grain plus straw) × 100.

TABLE 5.11
Grain Yield and Magnesium Harvest Index in Principal Field Crops

Crop Species	Grain Yield (kg ha^{-1})	Mg Harvest Index (%)
Upland rice	4559	25
Lowland rice	4797	32
Lowland rice	6389	29
Corn	8501	31
Dry beans	1912	36
Dry beans	3409	46
Soybeans	1441	42
Soybeans	3038	33

Source: Fageria, N. K. 2009. *The Use of Nutrients in Crop Plants*. Boca Raton, FL: CRC Press.
Note: Magnesium harvest index (%) = (Mg uptake in grain/Mg uptake in grain plus straw) × 100.

5.7 USE EFFICIENCY

Calcium use efficiency, which is defined as straw or grain yield per unit of Ca uptake, varies with crop species (Table 5.12). It was higher for the production of cereal and legume grains than that of cereal or legume shoots. The higher Ca efficiency for cereal and legume grain production when compared with shoots was associated with a lower amount of Ca accumulation in the grain. Similar to that of Ca, magnesium use efficiency at different growth stages in shoot or grain of upland rice, corn, dry beans, and soybeans is presented in Table 5.13. In upland rice, Mg use efficiency (Y) has a quadratic association with plant age (X) (Y = 279.2875 +

TABLE 5.12
Calcium Use Efficiency in Upland Rice, Corn, Dry Bean, and Soybean during Growth Cycle Grown on Brazilian Oxisols

Age in Days after Sowing	Upland Rice ($kg\ kg^{-1}$)	Age in Days after Sowing	Corn ($kg\ kg^{-1}$)	Age in Days after Sowing	Dry Beans ($kg\ kg^{-1}$)	Age in Days after Sowing	Soybeans ($kg\ kg^{-1}$)
19	299	18	176	15	107	27	86
43	266	35	191	29	49	41	59
69	326	53	226	43	47	62	60
90	406	69	272	62	59	82	71
102	392	84	297	84	72	120	85
130	281	119	379	96	99	158	74
130 (grain)	1324	119 (grain)	1001	96 (grain)	345	158 (grain)	245
Regression	Quadratic		Linear		Quadratic		Quadratic
R^2	0.42^{NS}		0.99^{**}		0.82^{*}		0.11^{NS}

Source: Fageria, N. K. 2009. *The Use of Nutrients in Crop Plants*. Boca Raton, FL: CRC Press.
Note: Calcium use efficiency = (Shoot or grain yield/Ca accumulated in shoot or grain). Values are averages of two years of field trials. Crops grown on Brazilian Oxisols.
^{*,**,NS}Significant at the 5 and 1% probability levels and nonsignificant, respectively.

$3.1627X - 0.0122X^2$, $R^2 = 0.9231^*$). Based on a regression equation, maximum Mg use efficiency was achieved at 130 days of plant growth. In the case of corn, Mg use efficiency was linear with advanced plant age from 18 to 119 days in the growth cycle.

The regression equation relating plant age (X) and Mg use efficiency in shoot (Y) was $Y = 202.4203 + 3.4986X$, $R^2 = 0.9761^{**}$. Therefore, variability in Mg use efficiency was 92% for upland rice and 98% for corn with advanced plant age. In the case of dry beans and soybeans, Mg^{2+} use efficiency had a quadratic association with plant age, but the effect was not significant. The regression equation for dry beans was $Y = 221.8043 + 0.1319X - 0.00218X^2$, $R^2 = 0.1036$, and for soybeans, the equation was $Y = 184.8034 + 0.8373X - 0.0056X^2$, $R^2 = 0.2255$. In the grain, Mg^{2+} use efficiency followed the order of corn > upland rice > dry beans > soybeans. Overall, Mg^{2+} use efficiency in shoot as well as grain was higher in cereals than in legumes. The higher Mg use efficiency in cereals compared with legumes was associated with the low amount of Mg accumulated in upland rice and corn compared with dry beans and soybeans. It is well known that dicotyledons generally have higher contents of bivalent cations than monocotyledons, and the reverse holds true for nonovalent cations (Loneragan and Snowball 1969; Fageria 2009).

TABLE 5.13
Magnesium Use Efficiency in Upland Rice, Corn, Dry Bean, and Soybean during Crop Growth Cycles

Upland Rice		Corn		Dry Beans		Soybeans	
Age in Days after Sowing	Mg Use Efficiency (kg kg^{-1})	Age in Days after Sowing	Mg Use Efficiency (kg kg^{-1})	Age in Days after Sowing	Mg Use Efficiency (kg kg^{-1})	Age in Days after Sowing	Mg Use Efficiency (kg kg^{-1})
19	345	18	279	15	224	27	227
43	379	35	323	29	230	41	195
69	422	53	393	43	212	62	195
90	492	69	436	62	219	82	202
102	477	84	463	84	236	120	247
130	476	119	643	96	203	158	160
130 (grain)	856	119 (grain)	955	96 (grain)	434	158 (grain)	247
Regression	Quadratic		Linear		Quadratic		Quadratic
R^2	0.9231*		0.9761**		0.1036^{NS}		0.2255^{NS}

Source: Fageria, N. K. 2009. *The Use of Nutrients in Crop Plants*. Boca Raton, FL: CRC Press.

Note: Magnesium use efficiency = (Shoot or grain yield/Mg accumulated in shoot or grain). Values are averages of two years of field trials.

^{*,**,NS}Significant at the 5 and 1% probability levels and nonsignificant, respectively.

5.8 MANAGEMENT PRACTICES

The adoption of appropriate management practices is necessary for improving Ca and Mg use and for efficient rice production. These management practices include using an effective source of these nutrients, using an appropriate source and timing of application, adequate rate, and planting Ca- and Mg-efficient genotypes.

5.8.1 Effective Sources

Calcium is not normally formulated into mixed fertilizers but rather is present as a component of the materials supplying other nutrients, particularly phosphorus (Tisdale, Nelson, and Beaton 1985). The primary sources of Ca are liming materials; their use can correct acidity as well as supply Ca. Some liming materials contain both Ca and Mg and can be used to correct soil acidity as well as to supply these elements. In addition, gypsum ($CaSO_4{\cdot}H_2O$) is a very good source of Ca for lengthy time periods. Gypsum is used in crop production to amend acidic soils, and it improves soil and water quality. Over time, gypsum neutralizes heavy metals and loosens soil, thereby improving air and water circulation. Nearly all heavy soils containing a high level of clay can benefit from gypsum.

Magnesium sources such as potassium–magnesium sulfate, magnesium sulfate, and magnesium oxides are the most widely used materials in dry fertilizer formulations. Important Ca and Mg sources are listed in Table 5.14. Among these sources, dolomitic lime (which contains Ca^{2+} and Mg^{2+}) is the most effective and the least costly source of Ca^{2+} and Mg^{2+} for acidic soils.

5.8.2 Appropriate Methods and Timing of Application

When Ca^{2+} and Mg^{2+} are supplied in acidic soils through liming, special attention should be given to method, frequency, depth, and timing of application. Liming

TABLE 5.14
Principal Carriers of Calcium and Magnesium

Commercial Name	Chemical Formula	Ca or Mg (%)
Dolomitic lime	$CaMg(CO_3)_2$	18–21 Ca and 7.8–12 Mg
Calcitic lime	$CaCO_3$	28.4–32.0 Ca
Dolomite lime	$MgCO_3$	3.6–7.2 Mg
Gypsum	$CaSO_4{\cdot}2H_2O$	23 Ca
Simple superphosphate	$Ca(H_2PO_4){\cdot}H_2O$, $CaSO_4{\cdot}2H_2O$	18–21 Ca
Triple superphosphate	$Ca(H_2PO_4){\cdot}H_2O$	12–14 Ca
Thermophosphate	Variable	21 Ca
Magnesium oxide	MgO	55–60 Mg
Magnesium chloride	$MgCl_2{\cdot}10H_2O$	8–9 Mg
Magnesium nitrate	$Mg(NO_3)_2$	16 Mg
Magnesium sulfate	$MgSO_4$	9.5 Mg
Thermophosphate	Variable	11 Mg

material is applied in large quantities in order to bring about the desired chemical changes in acidic soils. Therefore, the best method of application is to broadcast it as uniformly as possible and thoroughly mix it throughout the soil profile. Broadcasting machines are available for uniform application of liming materials (Fageria and Baligra 2008).

Liming frequency is mainly determined by cropping intensity, the crop species planted, and by a soil's Ca^{2+}, Mg^{2+}, Al, and pH levels after each harvest. Lime's effects are long lasting but not permanent. After several crops, Ca^{2+} and Mg^{2+} move downward and beyond the reach of roots. These elements are taken up by crops and, to some extent, are lost through soil erosion. Acid-forming fertilizers and decomposing organic matter lower the soil pH and release more aluminum to the soil solution and cation exchange sites on soil particles. When values of exchangeable Ca^{2+} and Mg^{2+}, and pH fall below optimum levels for a given crop species, liming should be repeated. This means that soil samples should be taken periodically to determine changes in soil chemical properties and to decide liming frequency. Because the effects of lime last longer than those of most other amendments, it is rarely necessary to lime more frequently than every three years (Caudle 1991). Lime studies of Brazilian Oxisols in the Cerrado region of Brazil showed that after four years and eight crops grown (two crops each year) in rotation (rice, common beans, corn, and soybeans), the soil maintained pH levels and enough Ca^{2+} and Mg^{2+} to produce optimal crop yields (Fageria 2001a). The residual effect of coarse lime material is greater than that of finer lime because larger particles react more slowly with soil acidity and tend to remain in the soil longer.

Liming material should be mixed thoroughly and as deeply as possible to improve crop-rooting systems in acidic soils. With the machinery currently available, it is generally mixed to a depth of 20–30 cm. A depth greater than 30 cm requires more power and costs more in terms of labor and energy. The timing of lime application is important for achieving desirable results. Lime should be applied as far in advance of crop planting as possible in order to allow it to react with soil colloids and to bring about significant changes in soil chemical properties. Soil moisture and temperature determine how lime reacts with soil colloids. In Brazil, most studies indicate that liming should be done three months before sowing a crop. However, other studies carried out at the National Rice and Bean Research Center have shown that significant chemical changes in Brazilian Oxisols can take place four to six weeks after applying liming materials, if the soil has sufficient moisture (Fageria 1984; 2001a), and it is not necessary to wait any longer after applying lime.

5.8.3 Adequate Rate

Using an adequate rate of Ca and Mg for maximum crop species yield is an important consideration for economic and ecological reasons. If liming is used as a source of Ca and Mg, the quantity of liming material required is determined on the basis of appropriate levels of soil pH, base saturation, and aluminum saturation adjustment. Appropriate levels of these acidity indices vary with soil type, soil fertility, plant species, and crop genotypes within species (Fageria and Baligar 2003). In addition, crop response curves related to lime rate and yield are another criterion that can be used to define lime requirements for any given crop species. Crop response curves for lime levels should be determined for each crop species in different agroecological regions to make liming recommendations effective and economical. If experimental data relating lime rate and crop yield are not available, empirical equations can be used to determine the rate for a given crop. These equations are as follows (Fageria and Baligra 2008):

$$\text{Lime rate (Mg ha}^{-1}) = (2 \times Al^{3+}) + [2 - (Ca^{2+} + Mg^{+})] \quad (5.4)$$

where values of Al^{3+}, Ca^{2+}, and Mg^{2+} are expressed in $cmol_c\ kg^{-1}$. If the values of Ca^{2+} and Mg^{2+} cations are more than 2 $cmol_c\ kg^{-1}$, only Al multiplied by a factor of 2 is considered. This criterion was originally suggested by Kamprath (1970) for tropical soils and is still primarily used for liming recommendations for Brazilian acidic soils (Paula et al. 1987; Raij 1991; Raij and Quaggio 1997). Alvarez and Ribeiro (1999) recommended that the factor used to multiply Al should be varied according to soil texture. These authors suggested that in sandy soil with clay content of 0–15%, a factor of 0–1 should be used; for medium-textured soils with clay content of 15–35%, a factor of 1–2 should be used; for clayey soil with clay content of 35–60%, a factor of 2–3 should be used; and for heavy clayey soil with clay content of 60–100%, a factor of 3–4 should be used.

In addition, base saturation is another important chemical property of soils that is used as a criterion for liming recommendations. Base saturation is defined as the proportion of the CEC occupied by exchangeable bases. It is calculated as (Fageria, Baligar, and Zobel 2007)

$$\text{Base saturation (\%)} = \Sigma(\text{Ca, Mg, K, Na})/\text{CEC} \times 100 \quad (5.5)$$

where CEC is the sum of Ca, Mg, K, Na, H, and Al expressed in $cmol_c\ kg^{-1}$.

In Brazil, Na^{+} is generally not determined because of the very low levels of this element in Oxisols there (Raij 1991). Hence, Na is not considered for calculating CEC or base saturation. For crop production, base saturation levels in soil may be grouped into very low (lower than 25%), low (25–50%), medium (50–75%), and high (>75%) (Fageria and Gheyi 1999). Very low and low base saturation means a predominance of adsorbed hydrogen and aluminum on the exchange complex. Deficiencies of calcium, magnesium, and potassium are likely to occur in soils with low CEC and very low to low percent base saturation. The quantity of lime required by the base saturation method is calculated using the following formula (Fageria, Baligar, and Edwards 1990; Fageria and Baligar 2008):

$$\text{Lime rate (Mg ha}^{-1}) = [\text{CEC } (B_2 - B_1)/\text{TRNP}] \times \text{df} \quad (5.6)$$

where CEC = total exchangeable cations (Ca^{2+}, Mg^{2+}, K^{+}, H^{+} + Al^{3+}) in $cmol_c\ kg^{-1}$, B_2 = desired optimum base saturation, B_1 = existing base saturation, TRNP = total relative neutralizing power of liming material, and df = depth factor, 1 for a 20-cm depth and 1.5 for a 30-cm depth. For Brazilian Oxisols, the desired optimum base saturation for most cereals is in the range of 50–60%; for legumes, it is in the range of 60–70% (Fageria, Baligar, and Edwards 1990). However, there may be exceptions, such as for upland rice, which is very tolerant to soil acidity and can produce a good yield at base saturations lower than 50%.

TABLE 5.15
Adequate Acidity Indices for Upland Rice

Acidity Indices	Value
pH in H_2O	5.6
Base saturation (%)	40.0
Ca ($cmol_c\ kg^{-1}$)	1.9
Mg ($cmol_c\ kg^{-1}$)	1.2
Ca saturation (%)	24.0
Mg saturation (%)	15.0
K saturation	3.0
Ca/Mg ratio	1.8
Ca/K ratio	9.0
Mg/K ratio	6

Source: Adapted from Fageria, N. K. 2001b. Response of upland rice, dry bean, corn and soybean to base saturation in Cerrado soil. *Rev. Bras. Eng. Agricola e Amb.* 5:416–424.

The adequate acidity indices for upland rice production in Brazilian Oxisols were determined (Table 5.15). These indices can be used to determine the lime rate for upland rice production. In a greenhouse experiment, Fageria, Cobucci, and Knupp (in press) also studied the influence of lime on upland rice growth and yield components. Plant height, straw yield, grain yield, and panicle density were significantly influenced by lime treatments (Table 5.16). Lime increased these growth and yield components in a quadratic fashion with increasing lime rates in the range of 0–4.28 g kg^{-1} soil (Table 5.17). Based on a quadratic regression equation, maximum plant height was achieved with the addition of 1.97 g lime kg^{-1} soil. Similarly, maximum straw yield was achieved with the addition of 1.37 g lime kg^{-1} soil. Maximum grain yield and panicle density were obtained with the application of 1.11 and 1.16 g lime kg^{-1} soil, respectively. The variation in these growth and yield parameters due to liming was 46% for plant height, 70% for straw yield, 54% for grain yield, and 57% for panicle density. With the addition of lime, improvement in growth and yield components of upland rice grown in Brazilian Oxisols has been reported by Fageria and Zimmermann (1998) and Fageria (1984, 2000). Fageria (2000) and Fageria, Castro, and Baligar (2004) also noted that the response of upland rice to liming varied from cultivar to cultivar. Some cultivars were more responsive to liming than others (Fageria, Castro, and Baligar 2004). In a recent experiment, Fageria, Cobucci, and Knupp (in press) also determined optimum acidity indices, such as pH, exchangeable Ca, base saturation (BS), calcium saturation (CaS), and acidity saturation (AS), for upland rice grain yield (Table 5.18). Maximum grain yield was obtained at a pH of about 6.0. For maximum yield, exchangeable Ca was about 1.7. Maximum grain yield was obtained with a BS of about 60% and CaS of 47%. Rice plants can tolerate AS up to 42%. Figure 5.9 shows the response of upland rice to lime applied to Brazilian Oxisol.

TABLE 5.16
Influence of Lime Rates on Plant Height, Straw Yield, Grain Yield, and Panicle Density of Upland Rice

Lime Rate ($g\ kg^{-1}$)	Plant Height (cm)	Straw Yield ($g\ plant^{-1}$)	Grain Yield ($g\ plant^{-1}$)	Panicle Density ($plant^{-1}$)
0	123c	14.88c	13.11bc	3.81
0.71	135ab	16.89b	15.55ab	5.71a
1.42	136a	18.37a	16.22a	5.56a
2.14	137a	18.99a	14.29ab	5.06ab
2.85	129abc	11.67d	11.08c	3.18c
4.28	124bc	9.77e	10.14c	2.56c
F-test	*	*	*	*
CV(%)	3.95	3.91	10.31	14.90

Source: Fageria, N. K., T. Cobucci, and A. M. Knupp. in press. Influence of lime and gypsum on growth and yield of upland rice and changes in soil chemical properties. *J. Plant Nutr.* 36.

Note: abc means followed by the same letter in the same column are statistically not different at 5% by Tukeys test.

*Significant at the 1% probability level.

TABLE 5.17
Association between Lime Rate and Plant Height, Straw Yield, Grain Yield, and Panicle Density of Upland Rice

Variable	Regression Equation	R^2	LR ($g\ kg^{-1}$) for Maximum PH, SY, GY, PD
LR vs. PH	$Y = 125.87 + 9.96X - 2.53X^2$	0.46*	1.97
LR vs. SY	$Y = 15.48 + 2.74X - 1.00X^2$	0.70*	1.37
LR vs. GY	$Y = 14.14 + 1.18X - 0.53X^2$	0.54*	1.11
LR vs. PD	$Y = 4.46 + 0.93X - 0.34X^2$	0.57*	1.16

Source: Fageria, N. K., T. Cobucci, and A. M. Knupp. in press. Influence of lime and gypsum on growth and yield of upland rice and changes in soil chemical properties. *J. Plant Nutr.* 36.

Note: LR, lime rate; PH, plant height; SY, straw yield; GY, grain yield; PD, panicle density.

*Significant at the 1% probability level.

In addition to lime, gypsum is also used to correct Ca deficiency in highly weathered soils. The impact of gypsum on plant height, grain yield, and panicle density of upland rice grown on Brazilian Oxisol under field conditions was studied (Table 5.19). Plant height, grain yield, and panicle density were significantly increased with the addition of gypsum in the range of 0–6 Mg ha^{-1}. Based on regression equations, maximum plant height and grain yield were obtained with the addition of

TABLE 5.18
Relationship between Soil Acidity Indices and Grain Yield of Upland Rice

Variable	Regression Equation	R^2	AVMGY
pH in H_2O vs. GY	Y = −100.72 + 38.83 − 3.24X²	0.58**	5.99
Ca ($cmol_c$ kg^{-1}) vs. GY	Y = 11.32 + 3.42X − 1.01X²	0.27*	1.69
BS (%) vs. GY	Y = 3.59 + 0.37X − 0.0031X²	0.43**	59.67
CaS (%) vs. GY	Y = 3.34 + 0.49X − 0.0052X²	0.39*	47.11
AS (H+Al) (%) vs. GY	Y = 9.93 + 0.26X − 0.0031X²	0.43*	41.94

Source: Fageria, N. K., T. Cobucci, and A. M. Knupp. in press. Influence of lime and gypsum on growth and yield of upland rice and changes in soil chemical properties. *J. Plant Nutr.* 36.
Note: AVMGY, adequate value for maximum grain yield; GY, grain yield; BS, base saturation; CaS, calcium saturation; and AS, acidity saturation.
*,**Significant at the 5 and 1% probability levels, respectively.

FIGURE 5.9 Response of upland rice to lime grown on Brazilian Oxisol.

3.9 Mg gypsum ha^{-1}. Maximum panicle density was obtained with the addition of 4.3 Mg gypsum ha^{-1}. Figures 5.10 through 5.14 show the response to gypsum of upland rice grown on Brazilian Oxisol at 40 days' plant growth, or age. Similarly, the response of upland rice to gypsum application is shown at physiological maturity (Figures 5.15 through 5.19).

TABLE 5.19
Influence of Gypsum on Plant Height, Grain Yield, and Panicle Density of Upland Rice Grown on Brazilian Oxisol under Field Conditions

Gypsum Rate ($Mg\ ha^{-1}$)	Plant Height (cm)	Grain Yield ($kg\ ha^{-1}$)	Panicle Density ($plant^{-1}$)
0	81.5	2473.42	204.58
1	84.8	3630.83	246.67
2	92.8	4446.94	263.33
3	93.8	4395.42	285.00
4	90.3	4117.08	277.50
6	90.0	4157.22	276.67
F-test	*	**	**
CV (%)	5.48	16.32	10.28

Regression Analysis

Gypsum vs. plant height $(Y) = 81.31 + 6.08X - 0.79X^2$, $R^2 = 0.43$**

Gypsum rate vs. grain yield $(Y) = 2673.41 + 945.75X - 120.48X^2$, $R^2 = 0.52$**

Gypsum vs. panicle density $(Y) = 208.60 + 36.32X - 4.22X^2$, $R^2 = 0.56$**

*,**Significant at the 5 and 1% probability levels, respectively.

FIGURE 5.10 Response of upland rice at of 0 and 1 Mg gypsum ha^{-1} grown on Brazilian Oxisol.

Fageria, Cobucci, and Knupp (in press) also studied the influence of gypsum on growth and yield components of upland rice under greenhouse conditions. Application of gypsum significantly increased plant height, straw yield, grain yield, and panicle density in a quadratic fashion (Tables 5.20 and 5.21). Gypsum rate was 1.03 g kg^{-1} for maximum plant height, 1.53 g kg^{-1} for straw yield, 1.13 g kg^{-1} for grain yield, and 1.0 g kg^{-1} for panicle density (Table 5.21). The variation in growth and yield components due to gypsum application was in the order of grain yield > plant height > straw yield > panicle density. The increase in grain yield was associated with the improvement of these plant traits (plant height, straw yield, and panicle density). These traits were significantly and positively related to grain yield. The regression equations showing these relationships were plant

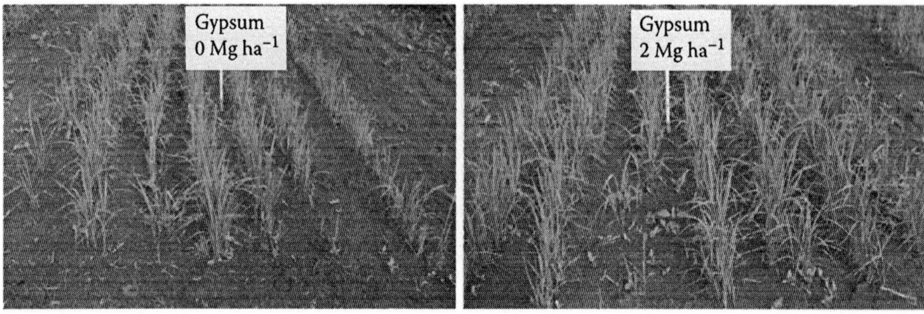

FIGURE 5.11 Response of upland rice at of 0 and 2 Mg gypsum ha^{-1} grown on Brazilian Oxisol.

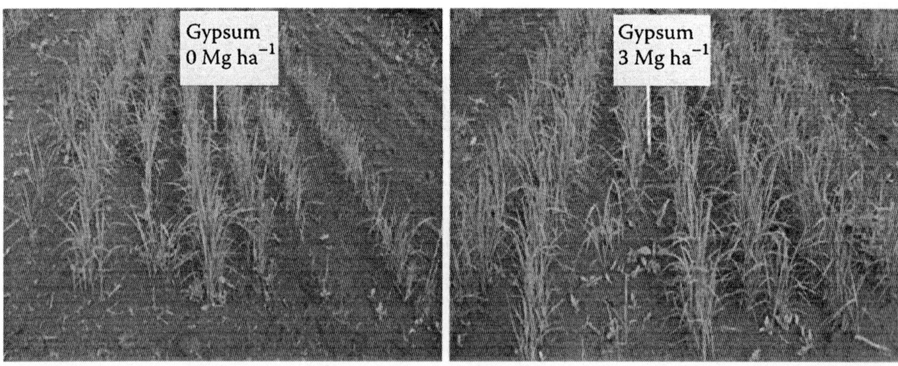

FIGURE 5.12 Response of upland rice at of 0 and 3 Mg gypsum ha^{-1} grown on Brazilian Oxisol.

FIGURE 5.13 Response of upland rice at of 0 and 4 Mg gypsum ha^{-1} grown on Brazilian Oxisol.

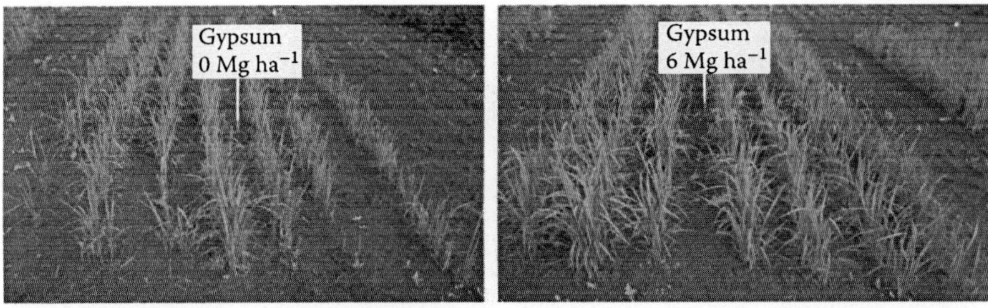

FIGURE 5.14 Response of upland rice at of 0 and 6 Mg gypsum ha^{-1} grown on Brazilian Oxisol.

FIGURE 5.15 Response of upland rice at of 0 and 1 Mg gypsum ha^{-1} grown on Brazilian Oxisol.

height versus grain yield ($Y = -12.92 + 0.26X$, $R^2 = 0.35^{**}$), straw yield versus grain yield ($Y = -44.19 + 6.23X - 0.15X^2$, $R^2 = 0.25^{*}$), and panicle density ($Y = -8.76 + 6.72 - 0.37X^2$, $R^2 = 0.66^{**}$). Rice response to gypsum has been reported under field as well as greenhouse conditions (Momuat et al. 1983; Mazid 1986; Chien, Hellums, and Henao 1987). Figure 5.20 shows upland rice growth at two gypsum levels. Similarly, upland rice root growth was also influenced by gypsum application (Figure 5.21).

The improvement in rice growth may be associated with the favorable chemical changes in the soil from the application of gypsum (Table 5.22). Gypsum significantly increased exchangeable Ca and Mg, effective cation exchange capacity (ECEC), and BS (Table 5.22). However, extractable H + Al were significantly decreased with increasing gypsum levels. The increase in Ca, Mg ECEC, and

FIGURE 5.16 Response of upland rice at of 0 and 2 Mg gypsum ha^{-1} grown on Brazilian Oxisol.

FIGURE 5.17 Response of upland rice at of 0 and 3 Mg gypsum ha^{-1} grown on Brazilian Oxisol.

BS was in a quadratic fashion with an increasing gypsum rate in the range of 0–2.28 g kg^{-1} (Table 5.23). There was a slight increase in soil pH with the addition of gypsum, but it was not significant. The improvement in upland rice growth and yield might have been due to improvement in extractable Ca, Mg, BS, and decreasing soil acidity because of the gypsum addition. Shainberg et al. (1989) reported that gypsum ameliorates subsoil acidity and improves crop yields. Similarly, Alcordo and Rechcigl (1993) reported that gypsum use can be helpful

FIGURE 5.18 Response of upland rice at of 0 and 4 Mg gypsum ha^{-1} grown on Brazilian Oxisol.

FIGURE 5.19 Response of upland rice at of 0 and 6 Mg gypsum ha^{-1} grown on Brazilian Oxisol.

in improving Ca and S in reducing soil acidity and consequently improving crop yields. In addition, Figure 5.21 shows improvement of root growth with the addition of gypsum. The improvement of root growth is important for absorption of water and nutrients from greater soil depths in low fertility soils such as Oxisols of Brazil's Cerrado region.

TABLE 5.20
Influence of Gypsum on Plant Height, Straw Yield, Grain Yield, and Panicle Density of Upland Rice Grown under Greenhouse Conditions

Gypsum Rate ($g\ kg^{-1}$)	Plant Height (cm)	Straw Yield ($g\ plant^{-1}$)	Grain Yield ($g\ plant^{-1}$)	Panicle Density ($plant^{-1}$)
0	120.75	16.12	15.49	5.50
0.28	125.00	19.19	18.69	6.50
0.57	129.50	20.13	19.42	6.75
1.14	128.50	20.22	21.91	7.25
1.71	126.75	20.20	18.09	5.75
2.28	116.50	20.43	16.30	5.25
F-test	*	*	*	*
CV(%)	2.84	7.33	6.47	11.46

Source: Fageria, N. K., T. Cobucci, and A. M. Knupp. in press. Influence of lime and gypsum on growth and yield of upland rice and changes in soil chemical properties. *J. Plant Nutr.* 36.
*Significant at the 1% probability level.

TABLE 5.21
Association between Gypsum Rate and Plant Height, Straw Yield, Grain Yield, and Panicle Density of Upland Rice

Variable	Regression Equation	R^2	GR ($g\ kg^{-1}$) for Maximum PH, SY, GY, PD
GR vs. PH	$Y = 120.99 + 17.54X - 8.53X^2$	0.66*	1.03
GR vs. SY	$Y = 17.06 + 5.03X - 1.63X^2$	0.47*	1.53
GR vs. GY	$Y = 15.97 + 8.71X - 3.87X^2$	0.68*	1.13
GR vs. PD	$Y = 5.72 + 2.41X - 1.20X^2$	0.46*	1.00

Source: Fageria, N. K., T. Cobucci, and A. M. Knupp. in press. Influence of lime and gypsum on growth and yield of upland rice and changes in soil chemical properties. *J. Plant Nutr.* 36.
Note: GR, gypsum rate; PH, plant height; SY, straw yield; GY, grain yield; PD, panicle density.
*Significant at the 1% probability level.

5.8.4 Use of Tolerant/Efficient Genotypes

Crop species and genotypes within species differ significantly with respect to their tolerance of soil acidity and their Ca and Mg use efficiency (Sanchez and Salinas 1981; Foy 1984; Yang et al. 2000; Garvin and Carver 2003; Fageria and Baligar 2008; Fageria, Baligar, and Jones 2011). Hence, lime requirements also vary from species to species and among cultivars within species. Many plant species tolerant of acidity originate in regions with acidic soil, suggesting that adaptation to soil constraints is part of the evolution process (Sanchez and Salinas 1981; Foy 1984). One example of this evolution is the tolerance of acidic soil by Brazilian upland rice cultivars. In Brazilian Oxisols,

FIGURE 5.20 Influence of gypsum on upland rice growth.

upland rice grows very well without liming, as long as other essential nutrients are supplied in adequate amounts and water is not a limiting factor (Fageria 2000, 2001a).

Experimental results obtained with upland rice on Brazilian Oxisols are good examples of evaluation of crop acidity tolerance. Fageria, Castro, and Baligar (2004) reported that grain yield and yield components of 20 upland rice genotypes were

FIGURE 5.21 Root growth of upland rice at different gypsum rates or pH levels.

TABLE 5.22
Influence of Gypsum Rate on pH, Ca, Mg, H+Al, Effective Cation Exchange Capacity, and Base Saturation after Harvest of Upland Rice

GR ($g\ kg^{-1}$)	pH in H_2O	Ca ($coml_c\ kg^{-1}$)	Mg ($coml_c\ kg^{-1}$)	H+Al ($coml_c\ kg^{-1}$)	ECEC ($cmol_c\ kg^{-1}$)	BS (%)
0	4.90	0.90	0.26	3.30	1.32	27.11
0.28	5.40	1.26	0.30	2.40	1.68	40.10
0.57	5.70	1.30	0.40	1.60	1.74	51.87
1.14	5.53	2.10	0.30	1.90	2.43	56.23
1.71	5.06	3.80	0.23	3.13	4.13	56.56
2.28	5.3	6.10	0.20	2.80	6.39	69.43
F-test	NS	**	*	**	**	**
CV (%)	4.58	17.00	23.52	16.36	16.27	6.88

Source: Fageria, N. K., T. Cobucci, and A. M. Knupp. in press. Influence of lime and gypsum on growth and yield of upland rice and changes in soil chemical properties. *J. Plant Nutr.* 36.

Note: GR, gypsum rate; ECEC, effective cation exchange capacity; BS, base saturation.

*,**Significant at the 5 and 1% probability levels, respectively.

TABLE 5.23
Relationship between Gypsum Rate and Soil pH, Extractable Ca, Mg and H+Al, Effective Cation Exchange Capacity, and Base Saturation

Variable	Regression Equation	R^2
GR vs. pH	$Y = 5.11 + 0.71X - 0.31X^2$	0.21^{NS}
GR vs. Ca	$Y = 0.93\exp(0.65X + 0.078X^2)$	0.96**
GR vs. Mg	$Y = 0.28 + 0.10X - 0.066X^2$	0.37*
GR vs. H+Al	$Y = 2.92 - 1.71X + 0.80X^2$	0.33*
GR vs. ECEC	$Y = 1.36\exp(0.41X + 0.12X^2)$	0.95**
GR vs. BS	$Y = 31.11 + 29.07X - 6.03X^2$	0.86**

Source: Fageria, N. K., T. Cobucci, and A. M. Knupp. in press. Influence of lime and gypsum on growth and yield of upland rice and changes in soil chemical properties. *J. Plant Nutr.* 36.
Note: GR, gypsum rate; ECEC, effective cation exchange capacity; BS, base saturation.
^{*,**,NS}Significant at the 5 and 1% probability levels and nonsignificant, respectively.

TABLE 5.24
Grain Yield and Panicle Number of Five Upland Rice Genotypes at Two Soil Acidity Levels in Brazilian Oxisols

	Grain Yield (g pot^{-1})		Panicle Number (pot^{-1})	
Genotype	High Acidity (pH 4.5)	Low Acidity (pH 6.4)	High Acidity (pH 4.5)	Low Acidity (pH 6.4)
CRO97505	74.3	52.0	38.0	28.3
CNAs8983	55.2	42.9	29.0	25.7
Primavera	53.0	47.2	25.0	21.7
Canastra	51.6	38.9	32.0	26.3
Bonança	48.8	36.5	26.3	20.7
Carisma	50.8	17.5	43.3	17.7
Average	66.7	47.0	38.7	28.1

Source: Compiled from Fageria, N. K., E. M. Castro, and V. C. Baligar. 2004. Response of upland rice genotypes to soil acidity. In *The Red Soils of China: Their Nature, Management and Utilization*, eds. M. J. Wilson, Z. He, and X. Yang, pp. 219–237. Dordrecht, The Netherlands: Kluwer Academic.

significantly decreased at low soil acidity (limed to pH 6.4) as compared with high soil acidity (without lime, pH 4.5), demonstrating the tolerance of upland rice genotypes to soil acidity. In Table 5.24, data are presented showing grain yield and panicle number for six upland rice genotypes at two acidity levels. These authors also found that grain yield had a significant negative correlation with soil pH, Ca saturation, and BS. Furthermore, grain yield had a significant positive correlation with soil Al and

TABLE 5.25
Critical Soil Al Saturation for Important Field Crops at 90–95% of Maximum Yield

Crop	Type of Soil	Critical Al Saturation (%)
Cassava	Oxisols/Ultisols	80
Upland rice	Oxisols/Ultisols	70
Cowpea	Oxisols/Ultisols	55
Cowpea	Oxisols	42
Peanut	Oxisols/Ultisols	65
Peanut	Oxisols	54
Soybean	Oxisols	19
Soybean	Oxisols	27
Soybean	Oxisols/Ultisols	15
Soybean	Not given	<20
Corn	Oxisols	19
Corn	Oxisols/Ultisols	29
Corn	Oxisols/Ultisols	25
Corn	Oxisols	28
Mungbean	Oxisols/Ultisols	15
Mungbean	Oxisols/Ultisols	5
Coffee	Oxisols/Ultisols	60
Sorghum	Oxisols/Ultisols	20
Common bean	Oxisols/Ultisols	10
Common bean	Oxisols/Ultisols	8–10
Common bean	Oxisols/Ultisols	23
Cotton	Not given	<10

Source: Compiled from various sources by Fageria, N. K., V. C. Baligar, and C. A. Jones. 2011. *Growth and Mineral Nutrition of Field Crops*, 3rd edition. Boca Raton, FL: CRC Press.

H + Al, confirming that upland rice genotypes were tolerant of soil acidity. Fageria (1989) reported growth stimulation of Brazilian rice cultivars at 10 mg Al^{3+} L^{-1} in nutrient solution compared with a control (no Al) treatment. Okada and Fischer (2001) suggested that tolerance of upland rice to soil acidity is due to the relationship between the regulation of cell elongation and legend-bound Ca at the root apoplast.

In a greenhouse experiment using Oxisol, Fageria and Morais (1987) evaluated 48 upland rice genotypes for their response to lime. Rice genotypes responded to lime, but the response varied from genotype to genotype. Among tested genotypes, IAC 47 and CNA104-B-68-B-2 were more productive at low as well as at higher levels of Ca and Mg. Clark (1984) reviewed the literature on the response of crop genotypes to Ca and Mg and concluded that genotypes differ significantly in Ca and Mg use efficiency. Table 5.25 shows the tolerance of rice to Al saturation or acidity in comparison to other annual crops.

5.9 CONCLUSIONS

Calcium and magnesium deficiencies are more commonly found in rice grown on Oxisols and Ultisols in the humid tropics, such as Brazil's Cerrado region. Ca^{2+} and Mg^{2+} are part of many biochemical and physiological processes in plants. Ca^{2+} is immobile in plant tissues; therefore, if its availability is low in the growth medium, deficiency symptoms appear first in the younger leaves or growing plant parts. On the other hand, Mg^{2+} is mobile in plant tissues, and its deficiency first appears in the older leaves if its concentration is lower than necessary for plant growth. In most of the growing regions of the world, the visual symptoms of Ca^{2+} and Mg^{2+} rarely appear in rice plants under field conditions. The amount of Ca^{2+} and Mg^{2+} required by rice plants is quite low for most of the metabolic and physiological functions. If the concentration of Ca^{2+} and Mg^{2+} in the upper well-developed rice is around 0.2%, 2 g kg^{-1} is sufficient for healthy growth and development.

Concentrations of Ca and Mg decrease with advanced plant age. In order to define sufficiency or deficiency levels, plant samples should be taken at different growth stages. To produce 1 Mg grain yield, lowland rice requires about 4.8 kg Ca and 3.5 kg Mg, and upland rice requires about 6 kg Ca and 4 kg Mg. The slightly higher Ca and Mg requirements of upland rice, in comparison to lowland rice, may be related to higher grain yields of lowland rice and higher nutrient use efficiency. Lowland rice accumulates around 31 kg Ca ha^{-1} and 22 kg Mg ha^{-1} with a grain yield of about 6500 kg ha^{-1}. Similarly, upland rice accumulates about 28 kg Ca and 20 kg Mg to produce a grain yield of about 4500 kg ha^{-1}. Compared to Ca, concentration of Mg is higher in grain, and this may be related to its higher mobility.

Dolomitic lime is the most effective and least costly source of Ca and Mg for crops in acidic soils. The quantity of liming material should be determined with crop response curves in order to determine lime rates for different agroecological regions. Rice is quite tolerant of soil acidity, but it has to be grown in rotation with legumes such as soybeans and dry beans. Experimental findings have shown that maximum economic yield for most annual crops grown on Oxisols and Ultisols can be obtained with the application of 4–6 Mg ha^{-1} of lime (Fageria and Baligar 2008). Liming acidic soils not only improves crop yields but also helps to maintain environmental quality and ultimately animal and human health. However, overliming can significantly reduce the bioavailability of micronutrients (Zn, Cu, Fe, Mn, B), which decreases with increasing pH (Fageria, Baligar, and Clark 2002). This can result in plant nutrient deficiencies, particularly that of Fe in upland rice. In addition to lime, amending acidic soil with gypsum is another way to supply Ca to crop plants. For upland rice production, application of about 4 Mg gypsum ha^{-1} can sustain Ca supply and crop yield for three to four years. Root growth of upland rice increases with the addition of gypsum; this practice is important for improving water and nutrient uptake in Cerrado soils, which are deficient in most macro- and micronutrients. Water use efficiency will also improve due to a vigorous root system.

Planting acid tolerant crop species and genotypes within species is another cost-effective and environmentally sound approach for improving crop production on acidic soils. Genetic variation in acid tolerance of crop plants, including rice, has been widely reported in the literature.

REFERENCES

Alcordo, I. S., and J. E. Rechcigl. 1993. Phosphogypsum in agriculture: A review. *Adv. Agron.* 49:55–118.

Alvarez, V. H., and A. C. Ribeiro. 1999. Liming. In *Recommendations for Using Amendments and Fertilizers, 5th Approximation*, eds. A. C. Ribeiro, P. T. G. Guimarães, and V. H. Alvarez, pp. 43–60. Viçosa, Brazil: Soil Fertility Commission for State of Minas Gerais.

Bennett, W. F. 1993. Plant nutrient utilization and diagnostic plant symptoms. In *Nutrient Deficiencies and Toxicities in Crop Plants*, ed. W. F. Bennett, pp. 1–7. St. Paul, MN: American Phytopathology Society.

Bould, C., E. J. Hewitt, and P. Needham. 1984. *Diagnosis of Mineral Disorders in Plants*, Vol. 1. New York: Chemical Publishing.

Brady, N. C., and R. R. Weil. 2002. *The Nature and Properties of Soils*, 13th edition. Upper Saddle River, NJ: Prentice Hall.

Caudle, N. 1991. *Groundworks 1, Managing Soil Acidity*. Raleigh, NC: North Carolina State University, TropSoils Publications.

Chien, S. H., D. T. Hellums, and J. Henao. 1987. Greenhouse evaluation of elemental sulfur and gypsum for flooded rice. *Soil Sci. Soc. Am. J.* 51:120–123.

Clark, R. B. 1982. Plant response to mineral element toxicity and deficiency. In *Breeding Plants for Less Favorable Environments*, eds. M. N. Christiansen and C. F. Lewis, pp. 71–142. New York: Wiley.

Clark, R. B. 1984. Physiological aspects of calcium, magnesium, and molybdenum deficiencies in plants. In *Soil Acidity and Liming*, 2nd edition, ed. F. Adams, pp. 99–170. Madison, WI: ASA, CSSA, and SSSA.

Clarkson, D. T., and J. B. Hanson. 1980. The mineral nutrition of higher plants. *Annu. Rev. Plant Physiol.* 31:239–298.

De Datta, S. K. 1981. *Principles and Practices of Rice Production*. New York: Wiley.

Fageria, N. K. 1984. Response of rice cultivars to liming in Cerrado soil. *Pesq. Agropec. Bras.* 19:883–889.

Fageria, N. K. 1989. *Tropical Soils and Physiological Aspects of Crops*. Goiania, Brazil: EMBRAPA-CNPAF.

Fageria, N. K. 2000. Upland rice response to soil acidity in Cerrado soil. *Pesq. Agropec. Bras.* 35:2303–2307.

Fageria, N. K. 2001a. Effect of lime on upland rice, common bean, corn, and soybean production in Cerrado soil. *Pesq. Agropec. Bras.* 36:1419–1424.

Fageria, N. K. 2001b. Response of upland rice, dry bean, corn and soybean to base saturation in Cerrado soil. *Rev. Bras. Eng. Agricola e Amb.* 5:416–424.

Fageria, N. K. 2009. *The Use of Nutrients in Crop Plants*. Boca Raton, FL: CRC Press.

Fageria, N. K., and V. C. Baligar. 2001. Improving nutrient use efficiency of annual crops in Brazilian acid soils for sustainable crop production. *Commun. Soil Sci. Plant Anal.* 32:1303–1319.

Fageria, N. K., and V. C. Baligar. 2003. Fertility management of tropical acid soils for sustainable crop production. In *Handbook of Soil Acidity*, ed. Z. Rengel, 359–385. New York: Marcel Dekker.

Fageria, N. K., and V. C. Baligar. 2005. Enhancing nitrogen use efficiency in crop plants. *Adv. Agron.* 88:97–185.

Fageria, N. K., and V. C. Baligar. 2008. Ameliorating soil acidity of tropical Oxisols by liming for sustainable crop production. *Adv. Agron.* 99:346–399.

Fageria, N. K., V. C. Baligar, and R. B. Clark. 2002. Micronutrients in crop production. *Adv. Agron.* 77:185–268.

Fageria, N. K., V. C. Baligar, and D. G. Edwards. 1990. Soil-plant nutrient relationships at low pH stress. In *Crops as Enhancers of Nutrient Use*, eds. V. C. Baligar and R. R. Duncan, pp. 475–507. San Diego, CA: Academic Press.

Fageria, N. K., V. C. Baligar, and C. A. Jones. 2011. *Growth and Mineral Nutrition of Field Crops*, 3rd edition. Boca Raton, FL: CRC Press.

Fageria, N. K., V. C. Baligar, and R. W. Zobel. 2007. Yield, nutrient uptake and soil chemical properties as influenced by liming and boron application in common bean in a no-tillage system. *Commun. Soil Sci. Plant Anal.* 38:1637–1653.

Fageria, N. K., and M. P. Barbosa Filho. 1994. *Nutrient Deficiency in Rice: Identification and Correction.* Goiania, Brazil: EMBRAPA-SPI/EMBRAPA-CNPAF.

Fageria, N. K., E. M. Castro, and V. C. Baligar. 2004. Response of upland rice genotypes to soil acidity. In *The Red Soils of China: Their Nature, Management and Utilization*, eds. M. J. Wilson, Z. He, and X. Yang, pp. 219–237. Dordrecht, The Netherlands: Kluwer Academic.

Fageria, N. K., T. Cobucci, and A. M. Knupp. in press. Influence of lime and gypsum on growth and yield of upland rice and changes in soil chemical properties. *J. Plant Nutr.* 36.

Fageria, N. K., and H. R. Gheyi. 1999. *Efficient Crop Production.* Campina Grande, Brazil: Federal University of Paraiba.

Fageria, N. K., and O. P. Morais. 1987. Evaluation of rice cultivars for utilization of calcium and magnesium in the Cerrado soil. *Pesq. Agropec. Bras.* 22:667–672.

Fageria, N. K., N. A. Slaton, and V. C. Baligar. 2003. Nutrient management for improving lowland rice productivity and sustainability. *Adv. Agron.* 80:63–152.

Fageria, N. K., and C. M. R. Souza. 1991. Upland rice, common bean, and cowpea response to magnesium application on an Oxisol. *Commun. Soil Sci. Plant Anal.* 22:1805–1816.

Fageria, N. K., and F. J. P. Zimmermann. 1998. Influence of pH on growth and nutrient uptake by crop species in an Oxisol. *Commun. Soil Sci. Plant Anal.* 29:2675–2682.

Follett, R. H., L. S. Murphy, and R. L. Donahue. 1981. *Fertilizers and Soil Amendments.* Englewood Cliffs, NJ: Prentice Hall.

Foy, C. D. 1974. Effects of soil calcium availability on plant growth. In *The Plant Root and Its Environment*, ed. E. W. Carlson, pp. 565–600. Charlottesville, VA: University Press of Virginia.

Foy, C. D. 1984. Physiological effects of hydrogen, aluminum and manganese toxicity in acid soils. In *Soil Acidity and Liming*, 2nd edition, ed. F. Adams, 57–97. Madison, WI: ASA, CSSA, and SSSA.

Garvin, D. F., and B. F. Carver. 2003. Role of genotype in tolerance to acidity and aluminum toxicity. In *Handbook of Soil Acidity*, ed. Z. Rengel, pp. 387–406. New York: Marcel Dekker.

Kamprath, E. J. 1970. Exchangeable aluminum as a criterion for liming leached mineral soils. *Soil Sci. Soc. Am. Proc.* 34:252–254.

Kinraide, T. B. 1991. Identity of rhizotoxic aluminum species. In *Plant Soil Interactions at Low pH*, eds. R. J. Wright, V. C. Baligar, and R. P. Murrmann, pp. 717–728. Dordrecht, The Netherlands: Kluwer Academic.

Loneragan, J. F., and K. Snowball. 1969. Calcium requirements of plants. *Aust. J. Agric. Res.* 20:465–478.

Marschner, H. 1995. *Mineral Nutrition of Higher Plants*, 2nd edition. New York: Academic Press.

Mazid, S. A. 1986. The response of major crops to sulphur in agricultural soils of Bangladesh: An overview. In *Proceedings of the International Symposium on Sulphur in Agricultural Soils*, eds. S. Portch and S. G. Hussain, pp. 18–28. Dhaka: The Bangladesh Research Council and Sulphur Institute.

Mengel, K., E. A. Kirkby, H. Kosegarten, and T. Appel. 2001. *Principles of Plant Nutrition*, 5th edition. Dordrecht, The Netherlands: Kluwer Academic.
Momuat, E. O., C. P. Mamaril, A. P. Umar, and C. J. S. Momuat. 1983. Sulfur deficiency and how to cure it in south Sulawest. In *The Role of Research Results of Rice and Upland Food Crops in Agricultural Development*, ed. Central Research Institute for Food Crops, pp. 295–307. Bogor, Indonesia: Central Research Institute for Food Crops.
Okada, K., and A. J. Fischer. 2001. Adaptation mechanisms of upland rice genotypes to highly weathered acid soils of South American savannas. In *Plant Nutrient Acquisition: New Perspectives*, eds. N. Ae, J. Arihara, K. Okada, and A. Srinivasan, pp. 185–200. Tokyo: Springer.
Paula, M. B., F. D. Nogueira, H. Andrade, and J. E. Pitts. 1987. Effect of liming on dry matter yield of wheat in pots of low humic gley soil. *Plant Soil.* 97:85–91.
Raij, B. V. 1991. *Soil Fertility and Fertilization.* São Paulo, Brazil: Agronomy Editor Ceres.
Raij, B. V., and J. A. Quaggio. 1997. Methods used for diagnosis and correction of soil acidity in Brazil: An overview. In *Plant Soil Interactions at Low pH: Sustainable Agriculture and Forestry Production*, eds. A. C. Moniz, A. M. C. Furlani, R. E. Schaffert, N. K. Fageria, C. A. Rosolem, and H. Cantarella, pp. 205–214. Campinas, Brazil: Brazilian Soil Science Society.
Sanchez, P. A., and J. G. Salinas. 1981. Low-input technology for managing Oxisols and Ultisols in tropical America. *Adv. Agron.* 34:280–240.
Shainberg, I., M. E. Sumner, W. P. Miller, M. P. W Farina, M. A. Pavan, and M. V. Fey. 1989. Use of gypsum on soils: A review. *Adv. Soil Sci.* 9:1–111.
Tisdale, S. L., W. L. Nelson, and J. D. Beaton. 1985. *Soil Fertility and Fertilizers*, 4th edition. New York: MacMillan.
Wallace, A. 1966. *Current Topics in Plant Nutrition*. Anna Arbor, MI: Edwards Brothers.
Wallace, A. 1972. *Regulation of the Micronutrient Status of Plants by Chelating Agents and Other Factors*. Anna Arbor, MI: Edwards Brothers.
Wallace, A., and S. M. Soufi. 1975. Low and variable critical concentrations of calcium in plant tissues. *Commun. Soil Sci. Plant Anal.* 6:331–337.
Yang, Z. M., M. Sivaguru, W. J. Horst, H. Matsumoto, and Z. M. Yang. 2000. Aluminum tolerance is achieved by exudation of citric acid from roots of soybean (*Glycine max*). *Physiol. Plantraum.* 110:72–77.

6 Sulfur

6.1 INTRODUCTION

Sulfur (S) is one of the most abundant elements in the earth's crust and is essential for all biological systems (Tisdale, Nelson, and Beaton 1985). It may occur in several forms, such as native elements, sulfates, sulfides, and organic combinations with carbon and nitrogen. Sulfur is designated as one of the secondary nutrients in a system that classifies nitrogen, phosphorus, and potassium as primary fertilizer nutrients. It is as important as nitrogen, phosphorus, and potassium, but its deficiency and the quantity required are relatively low. In addition, fertilizers such as ammonium sulfate contain about 24% S, simple superphosphate contains about 12% S, magnesium sulfate contains 13% S, potassium sulfate contains 17% S, and gypsum contains 19% S. Sulfur is added to soils along with N and P fertilizers and also through rainfall (especially in industrialized regions), from the atmosphere, and through fungicides. However, modern technology has developed higher grade N and P fertilizers that do not contain S. Furthermore, industries release less and less S into the atmosphere in order to comply with environmental regulations. Under these conditions, S deficiency is expected in modern high-yielding crop cultivars, including rice.

Highly weathered Oxisols and Ultisols may be low in plant-available S due to excessive leaching. Sandy soils may also contain low levels of S, increasing the likelihood of crop deficiencies. Sulfur deficiency has been reported in nearly all rice-producing regions of the world, including Indonesia, Brazil, India, Bangladesh, Thailand, and the United States (De Datta 1981; Wells et al. 1993; Fageria, Slaton, and Baligar 2003). Yamaguchi (1997) reported that symptoms associated with S deficiency in rice often occur in irrigated Vertisols of the lower Volta in Ghana, Africa. Sulfur deficiency symptoms may have resulted from farmers applying urea and high-analysis NPK fertilizers that contained little S. Blair, Mamaril, and Momuat (1978) suggested that the low S content of most tropical soils was the primary cause of S deficiency. Numerous papers have been published on the responses to S by crops growing on highly weathered or intensely leached soils (Fox 1980; Fox and Blair 1986). Sulfur deficiency of rice has increased because of numerous reasons, including (1) increased crop removal of S via higher yields, (2) use of fertilizers lacking S, (3) reduced industrial emissions of S lowering the input of atmospheric S, (4) reduction in soil organic matter, (5) leaching and weathering processes, (6) erosion, and (7) crop management practices (De Datta 1981; Fageria, Slaton, and Baligar 2003). The objective of this chapter is to discuss sulfur nutrition of rice, including management practices for improving S use efficiency and consequently rice crop yield.

6.2 CYCLE IN SOIL–PLANT SYSTEM

A simplified S cycle in the soil–plant system is shown in Figure 6.1. In addition to the S content of soil organic matter, sulfur can be added through fertilizers, organic manures and crop residues, pesticides, and atmospheric pollution. The major path of S loss from the soil–plant system is as gases, volatilization loss from the soil surface, and loss by soil erosion and leaching. Plant uptake of S can also reduce the S level in the soil–plant system. Fox and Blair (1986) reported that compared with S accretions, the amount of S removed by harvested products and by leaching, volatilization, or erosion is frequently large. Such disequilibrium will eventually lead to S deficiency unless corrective action is taken. Sulfur from the atmosphere enters the soil portion of the S cycle either through dry or wet deposition. Sulfate is a source of acidity during wet deposition because it commonly associates with H ions during that process (Foth and Ellis 1988).

Throughout the world, the S content of soils ranges from 0.01 to 0.5% (Tabatabai and Bremner 1972). The mean concentration for the surface meter of soil has been estimated at 0.085% S and the total pedospheric pool at 2.6 × 10^5 Tg S (Freney and Williams 1983). About 1.1 × 10^4 Tg S is organic S (Trudinger 1986). In well-drained aerated soils, most inorganic S occurs as water-soluble SO_4^{2-}; SO_4^{2-} adsorbed to clays and Fe and Al oxides; and, particularly, in calcareous soils, a coprecipitated or cocrystallized $CaSO_4$ impurity in $CaCO_3$. Highly weathered soils, which normally

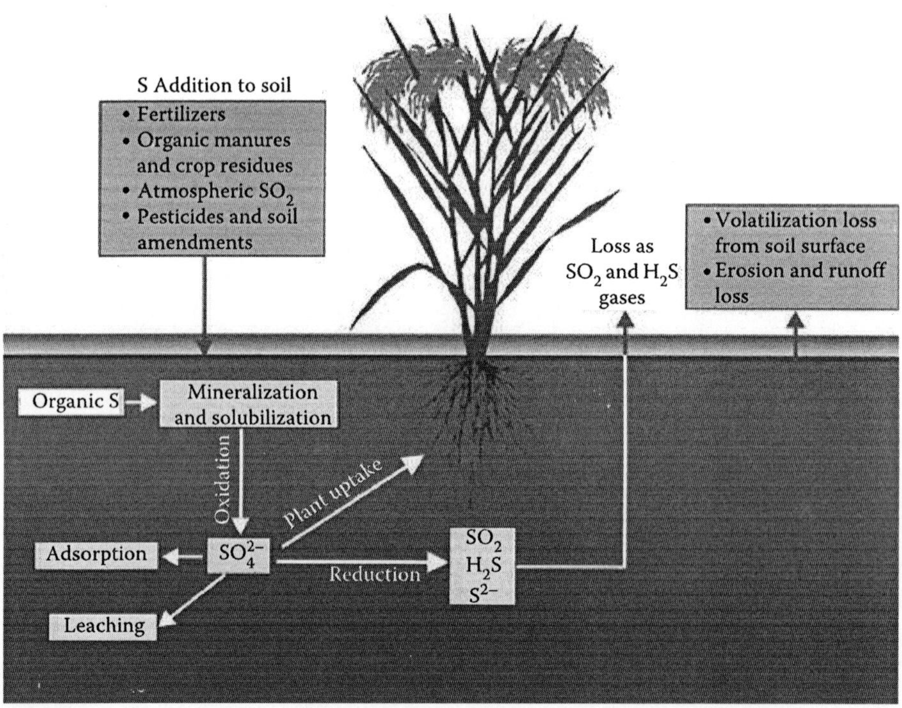

FIGURE 6.1 Sulfur cycle in soil–plant system.

contain considerable quantities of 1:1 clays and Fe and Al hydrous oxides, can accumulate large amounts of adsorbed SO_4^{2-} (Bohn et al. 1986). In anaerobic conditions, such as flooded rice, a significant quantity of sulfide can accumulate, which is rapidly reoxidized to SO_4^{2-} once aerobic conditions are restored (Trudinger 1986).

The main sulfur-bearing minerals in rocks and soils are gypsum ($CaSO_4.2H_2O$), anhydrite ($CaSO_4$), epsomite ($MgSO_4.7H_2O$), mirabilite ($Na_2SO_4.10H_2O$), pyrite and marcasite (FeS_2), sphalerite (ZnS), chalcopyrite ($CuFeS_2$), and cobaltite (CoAsS) (Tisdale, Nelson, and Beaton 1985). More than 90% of the total S in the A horizon of soils exists in organic form. Hence, many of the reactions of sulfur in soils are closely associated with organic matter and the activity of microorganisms (Lindsay 1979). The most important group of sulfur-oxidizing organisms is the autotrophic bacteria belonging to the genus *Thiobacillus*. The N:S ratio in surface soils is relatively constant and averages about 10:1.3 (Foth and Ellis 1988). The SO_4^{2-} ion is relatively mobile in the soil solution, and sulfur, like N, is subject to biological and chemical oxidation–reduction reactions (Tisdale, Nelson, and Beaton 1985). Sulfate ions reach the root surface by mass flow and diffusion. Plant-available S in the rice root zone is quickly depleted in many soils by plant uptake, SO_4^{2-} leaching, and the reduction of SO_4^{2-} to sulfide (S^{2-}). Consequently, the soil SO_4^{2-} concentration declines after flooding but may be accompanied by the accumulation of S^{2-}, which can be toxic to plants and may also be lost from the soil as H_2S gas. Thus, the availability of soil S decreases as soil reduction proceeds. The rate of SO_4 reduction in submerged soils depends on a number of soil properties. In neutral and alkaline soils, SO_4 concentrations as high as 1500 mg SO_4^{2-} kg^{-1} may be reduced to zero within six weeks of submergence (Ponnamperuma 1972). The reduction of SO_4 begins at an oxidation–reduction potential of −0.15 to −0.2 V at pH 6.5 to 7.0 (Takai 1978).

In lowland or flooded rice, the ionic forms of S undergo marked changes following flooding (Patrick and Mikkelsen 1971). Flooded soils frequently become sufficiently reduced due to restricted oxygen supply and microbial activity to reduce SO_4^{2-} to S^{2-}. Because Fe^{3+} reduction to Fe^{2+} precedes SO_4^{2-} reduction, Fe^{2+} is present in the soil solution by the time S^{2-} is produced. The formation of insoluble FeS may prevent the formation of H_2S and protect microorganisms and aquatic plants from the toxic effects of H_2S gas (Patrick and Reddy 1978).

In oxidized soils, such as upland rice or aerobic rice, sulfur is strongly bound with hydrous oxides of aluminum and iron, such as Oxisols and Ultisols of the tropics. Similarly, adsorption of SO_4^{2-} ions in strongly acidic soils is common; it is almost negligible when pH is raised more than 6.5. Addition of $CaCO_3$ to soils has been shown to lead to an increase in soluble SO_4^{2-} on incubation (Freney 1986). In a number of soils studied by Williams (1967), the amount of S mineralized was directly proportional to pH up to a value of 7.5. Above pH 7.5, mineralization increased more rapidly suggesting that chemical hydrolysis may be affecting the process (Freney 1986). In other soils, the amount of SO_4^{2-} mineralized was related to the amount of $CaCO_3$ added and not to the pH of the final mixture (Williams 1967). Bohn et al. (1986) reviewed the effect of pH on sulfate adsorption and reported that sulfate adsorption in soils is strongly pH dependent. The amount adsorbed decreases with increasing pH from 4.0 to 7.0, due to a decrease in the electrostatic potential of the adsorption plane.

6.3 FUNCTIONS

Sulfur is an essential nutrient for all plants and animals because it is a cofactor of essential amino acids (e.g., cysteine and methionine), several coenzymes (e.g., biotin, coenzyme A, thiamin pyrophosphate, and lipoic acid), thioredoxins, and sulpholipids (Zhao et al. 1997). There are many other S compounds in plants that are not essential but that may be involved in defense mechanisms against pests and pathogens or that may contribute to the special taste and odor of some plants (Bennett and Wallsgrove 1994). Apart from effects on yield, a crop's S nutrition often has a strong influence on food quality because of its essential role in the synthesis of amino acids, proteins, and some secondary metabolites (Zhao et al. 1997). Sufficient crop S supply is also important for nutritional quality of legumes and for processing cereal grains (Randall and Wrigley 1986; Fageria, Slaton, and Baligar 2003).

Sulfur deficiency inhibits protein synthesis in plants. In the green leaves, the major part of the protein is located in the chloroplasts where the chlorophyll molecules comprise prosthetic groups of the chromoproteid complex (Marschner 1995). Sulfur-deficient plants have decreased chlorophyll content and their leaves show yellowing. Sulfur in its nonreduced form is a component of sulfolipids and is thus a structural component of all biological membranes (Marschner 1995). In addition to direct nutrient deficiency response, S has a beneficial impact on soil physical condition (Tisdale 1970), on crop quality (Allaway and Thompson 1966), and on disease and insect damage (Haneklaus, Bloem, and Schnug 2007). Sulfur adequately improves plant height, shoot dry weight, panicle density, and grain yield of upland rice (Figure 6.2). Figures 6.3 and 6.4 show improvements (compared with control treatments) in root growth of upland and lowland rice with the addition of 80 mg S kg^{-1}.

FIGURE 6.2 (See color insert.) Upland rice growth on Brazilian Oxisol at different S levels.

FIGURE 6.3 Root growth of upland rice at two S levels.

FIGURE 6.4 Root growth of lowland rice at two S levels.

6.4 DEFICIENCY SYMPTOMS

Visual symptoms can be useful in both diagnosis and correction of nutrient deficiencies. This is particularly so in the case of S because applications of SO_4^{2-} to the growing crop will correct a deficiency and increase production and quality, all in the same growing season (Duke and Reisenauer 1986). Sulfur deficiency symptoms are similar to those of nitrogen. Plants are stunted and light green to yellow in color. The only difference is that S is immobile in plants. If S is deficient in a plant's growth medium, deficiency symptoms first appear in the younger leaves. If the deficiency persists, the whole plant turns yellow. Figure 6.5 shows S deficiency symptoms in lowland rice plants. Leaves are yellow in color, there are few panicles, and the panicles do not emerge properly from the boot. Sulfur deficiency is responsible for decreasing plant height, shoot dry weight, panicle density, and grain yield in upland rice grown on Brazilian Oxisol (Table 6.1).

Plant height varied from 99.50 cm produced at 0 mg S kg^{-1} treatment to 114.25 cm produced at 30 mg S kg^{-1} treatment (see Table 6.1). At 40 mg S kg^{-1} and 80 mg S kg^{-1} soil treatments, there was a small decrease in plant height. The increase in plant height was significant and quadratic when S was applied in the range of 0–80 mg kg^{-1} (Table 6.2). The increase in plant height with the addition of 30 mg S kg^{-1} was 13% compared with the control treatment (0 mg S kg^{-1}). The variation in plant height was 30% with the addition of sulfur fertilization. Improvement in plant height with the addition of S may be related to improvement in the photosynthesis process of the plants because S is necessary in chlorophyll formation, although it is not a component of chlorophyll (Fageria 2009). Chlorophyll is important in improving plant photosynthesis and, consequently, growth and development (Fageria and Gheyi 1999).

Plant height was significantly and quadratically related to grain yield (Figure 6.6). Based on a quadratic regression equation, maximum grain yield was achieved at 111 cm plant height. Plant height is an important trait in determining plant response to N fertilization or soils having high fertility. The green revolution was mainly associated with the development of cultivars with short stature (90–110 cm) that were less susceptible to lodging when heavily fertilized, especially with nitrogen (Fageria 2007a, 2009). In addition to lodging resistance, short stature and sturdy culm cultivars give higher yields at close plant spacing compared with taller cultivars. Taller cultivars may lodge due to higher N application or soils having higher fertility because nitrogen is one of the most important nutrients in determining upland rice yield in Brazilian Oxisols (Fageria 2007a).

Shoot dry weight significantly increased with the addition of S fertilizer in the soil under investigation (see Tables 6.1 and 6.2). This increase was quadratic when S was applied from 0 to 80 mg kg^{-1} soil. Based on a regression equation, maximum shoot dry weight was achieved with the application of 46 mg S kg^{-1} soil (see Table 6.2). The increase in shoot weight with the addition of S was related to increases in plant height, tillering, and number of leaves (Fageria 2009). Shoot dry weight also had a significant quadratic relationship to grain yield (Figure 6.7). The variation in grain yield due to shoot dry weight was 57%. Maximum grain yield was achieved at a shoot dry weight of 12 g $plant^{-1}$. Fageria et al. (2004) reported a significant quadratic relationship between shoot dry weight and grain yield in upland rice grown on Brazilian Oxisol.

Grain yield in cereals is related to biological yield and grain harvest index (GHI; Donald and Hamblin 1976). The biological yield of a cereal crop is the total yield of plant tops and is an indication of a crop's photosynthetic capability (Yoshida 1981). Dry matter production in rice has been reported to be significantly related to intercept photosynthetically active radiation (Kiniry et al. 2001). Crop growth rate depends on the amount of radiation intercepted by the crop and on the efficiency of intercepted radiation conversion into dry matter (Sinclair and Horie 1989). The production of sufficient shoot dry matter is important for improving the grain yield of rice (Fageria 2007a).

FIGURE 6.5 (See color insert.) Sulfur deficiency symptoms in lowland rice plants.

TABLE 6.1
Plant Height, Shoot Dry Weight, Grain Yield, and Panicle Density of Upland Rice as Influenced by Sulfur Fertilization

Sulfur Rate (mg kg^{-1})	Plant Height (cm)	Shoot Dry Weight (g plant^{-1})	Panicle Density (plant^{-1})	Grain Yield (g plant^{-1})
0	99.50c	9.79b	2.37b	7.57b
10	112.50a	12.88a	4.50a	11.83a
20	112.75a	13.01a	4.25a	10.78a
30	114.25a	13.28a	4.62a	12.05a
40	106.75b	12.65a	4.18a	12.00a
80	109.75ab	12.44a	4.31a	9.96ab
F-test	*	*	*	*
CV (%)	2.21	9.11	18.06	12.30

Note: Means followed by the same letter in the same column are statistically not significant at the 5% probability level by the Tukey's test.
*Significant at the 5% probability level.

TABLE 6.2
Relationship between Sulfur Rate and Plant Height, Shoot Dry Weight, Panicle Density, and Grain Yield of Upland Rice

Variable	Regression Equation	R^2	S Rate (mg kg^{-1}) for MV
S rate vs. PH	$Y = 104.08 + 0.39X - 0.0041X^2$	0.30*	48
S rate vs. SDW	$Y = 10.68 + 0.12X - 0.0013X^2$	0.38**	46
S rae vs. PD	$Y = 2.99 + 0.072X - 0.00071X^2$	0.34**	51
S rate vs. GY	$Y = 8.53 + 0.18X - 0.0020X^2$	0.47**	45

Note: MV, maximum value; PH, plant height; SDW, shoot dry weight; PD, panicle density; and GY, grain yield.
*,**Significant at the 5 and 1% probability levels, respectively.

Panicle density was significantly increased with the addition of S fertilizer in the range of 0–80 mg kg^{-1}. Maximum panicle density was achieved at 51 mg S kg^{-1} soil (see Table 6.2). Increase in panicle number with the addition of S was expected in S-deficient soils because S behaves like N in the plants, and N increases panicle number or density in rice (Fageria 2009). Panicle density has a significant linear relationship with grain yield (Figure 6.8). The variability in upland rice grain yield was 50% due to panicle density. Several studies show that panicle density is an important trait for rice yield determination (Yoshida 1981; Fageria 2007a, 2009).

Grain yield of upland rice in the Brazilian Oxisol was significantly increased with the addition of S fertilization (see Tables 6.1 and 6.2). Maximum grain yield was obtained with the addition of 45 mg S kg^{-1} soil. The variation in grain yield was 47% in the soil under investigation. The improvement in upland rice grain yield was

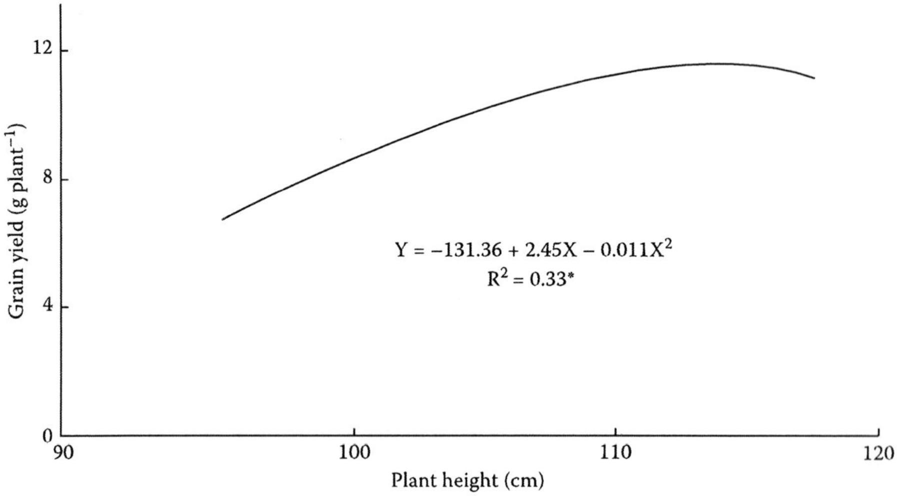

FIGURE 6.6 Relationship between plant height and grain yield of upland rice.

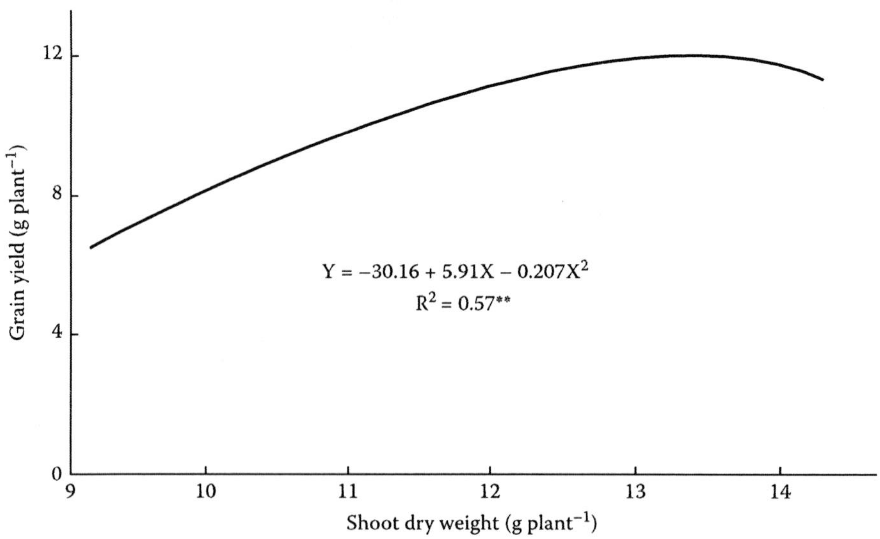

FIGURE 6.7 Relationship between shoot dry weight and grain yield of upland rice.

associated with increases in plant height, shoot dry weight, and panicle density. These are important rice growth parameters and yield components (Fageria 2007a). As shown in Table 6.3, sulfur and genotype treatments significantly increased the panicle density of 12 upland rice genotypes. The S × genotypes interaction was significant for plant height; panicle density of genotypes varied with the S levels. Similarly, sulfur also significantly increased the plant height, panicle density, and grain yield of lowland rice grown on Brazilian Inceptisol (Table 6.4).

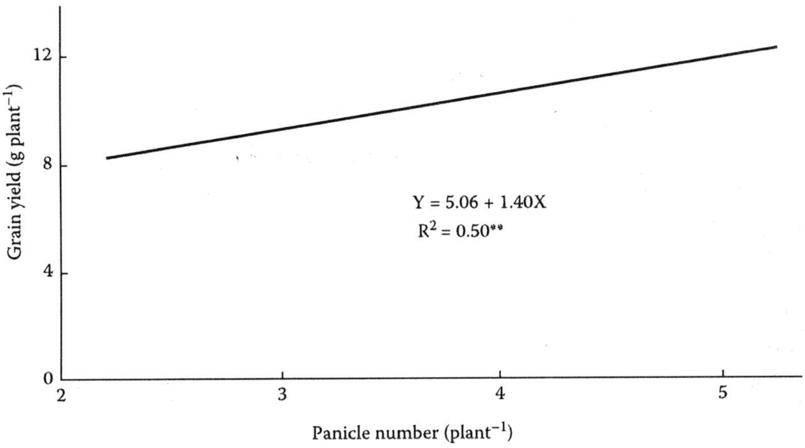

FIGURE 6.8 Relationship between panicle number or density and grain yield of upland rice.

TABLE 6.3
Plant Height (cm) of 12 Upland Rice Genotypes as Influenced by Sulfur and Genotype Treatments

	Sulfur Level (kg ha^{-1})				
Genotype	**0**	**10**	**20**	**40**	**80**
BRA 01506	86.00bc	94.66a	102.66ab	90.33a	94.00a
BRA 01600	84.66bcd	85.33a	95.66abc	88.66a	92.00a
BRA 02601	75.33de	91.66a	81.66c	86.33a	81.66a
BRA 032033	72.00e	83.66a	92.33abc	88.66a	83.66a
BRA 032048	80.00cde	88.33a	88.00bc	103.33a	83.00a
BRA 032051	81.00cde	97.33a	97.00abc	106.33a	96.00a
BRA 042094	96.33a	103.33a	109.66a	106.00a	103.00a
BRA 052015	81.33cde	98.33a	91.00abc	92.33a	96.66a
BRS Primavera	91.33ab	102.00a	96.00abc	92.00a	104.00a
BRS Sertaneja	85.66bc	93.66a	87.33bc	91.66a	96.33a
BR Monica	89.33abc	93.33a	91.00abc	97.33a	95.33a
BR Pepita	85.33bc	93.66a	92.33abc	82.00a	90.66a
Average	84.02	93.77	93.72	93.75	93.02
F-test					
Sulfur (S)	*				
Genotype (G)	**				
S × G	*				
CVS (%)	12.52				
CVG (%)	7.02				

Note: Means followed by the same letter in the same column are statistically not significant at the 5% probability level by the Tukey's test.

*,**Significant at the 5 and 1% probability levels, respectively.

TABLE 6.4
Influence of Sulfur Fertilization on Plant Height, Shoot Dry Weight, and Grain Yield of Lowland Rice

Sulfur Rate ($mg\ kg^{-1}$)	Plant Height (cm)	Shoot Dry Weight ($g\ plant^{-1}$)	Grain Yield ($g\ plant^{-1}$)
0	92.37	13.78	5.19
10	97.25	34.35	19.85
20	98.62	34.08	18.41
30	100.12	27.71	21.31
40	100.00	26.67	20.46
80	97.50	28.30	18.45
120	94.12	29.46	23.64
F-test	*	*	*
CV (%)	2.58	13.69	21.43

*Significant at the 1% probability level.

Sulfur and genotype treatments also significantly increased the panicle density of 12 upland rice genotypes grown on Brazilian Oxisol (Table 6.5). Overall, increase in panicle density was 14% with the addition of 10 kg S ha^{-1} (compared with a control treatment). Similarly, the overall increase in panicle density was 16% with the addition of 20 kg S ha^{-1}. Fageria, Baligar, and Jones (2011) reported that S, like N, improved panicle density in rice.

6.5 UPTAKE IN PLANT TISSUE

Knowledge of nutrient uptake, including concentration (content per unit dry matter) and accumulation (concentration × dry matter), is fundamental for diagnosing nutrient deficiency or sufficiency in crop plants. In addition, this information about nutrient removal from the soil–plant system is important for maintaining soil fertility for a sustainable cropping system. Nutrients are removed differently by straw and grain. Straw can be removed for animal consumption or energy needs and can also be incorporated in the soil after harvest of a grain crop. However, grains are used for human or animal consumption; their nutrients are definitely removed from the soil under cultivation.

Generally, nutrient concentration is used to define critical concentration, which serves as an index for determining nutrient sufficiency or deficiency in crop plants. It varies during a crop's growth cycle, and plant analysis should be done at different growth stages (Fageria, Slaton, and Baligar 2003; Fageria and Baligar 2005). The critical S concentration in rice tissue, like that of N, varies with the stage of plant development and the plant part that is sampled. Wells et al. (1993) reported that the critical concentration of S varies from approximately 2.5 g S kg^{-1} at tillering to 1.0 g S kg^{-1} at heading. Yoshida (1981) found that the critical S concentration in straw needed for maximum dry weight production varies from 1.6 g S kg^{-1} at tillering to 0.7 g S kg^{-1} at maturity. In straw, the critical N:S ratio for maximum biomass production varies from

TABLE 6.5
Influence of Sulfur on Panicle Density (m^{-2}) of 12 Upland Rice Genotypes

	Sulfur Level (kg ha^{-1})				
Genotype	**0**	**10**	**20**	**40**	**80**
BRA 01506	301.11a	315.55ab	344.44ab	291.11bc	313.33a
BRA 01600	279.99a	285.55ab	345.55ab	292.22bc	306.66a
BRA 02601	240.00a	305.55ab	301.11bc	327.77abc	272.22a
BRA 032033	240.00a	280.00b	302.22bc	330.00abc	278.89a
BRA 032048	255.55a	304.44ab	301.11bc	347.78ab	276.66a
BRA 032051	270.00a	324.44ab	338.88abc	352.22ab	320.00a
BRA 042094	314.44a	343.33a	354.44a	366.66a	343.33a
BRA 052015	271.11a	327.77ab	292.22c	310.00abc	322.22a
BRS Primavera	303.33a	345.55a	333.33abc	291.11bc	346.66a
BRS Sertaneja	286.66a	312.22ab	292.22c	281.11c	321.11a
BR Monica	1153.00a	311.11ab	311.11abc	324.44abc	317.77a
BR Pepita	273.33a	312.22ab	313.33abc	280.00c	301.11a
Average	276.29	313.98	319.17	316.20	309.99
F-test					
Sulfur (S)	*				
Genotype (G)	*				
S × G	*				
CVS (%)	10.97				
CVG (%)	5.96				

Note: Means followed by the same letter in the same column are statistically not significant at the 5% probability level by the Tukey's test.
*Significant at the 1% probability level.

23 at active tillering to 13 at maturity. Fageria, Baligar, and Jones (2011) reported that at tillering, adequate concentrations of S in the uppermost mature leaves were 2.0–6.0 g S kg^{-1}. Suzuki (1995) reported 1.0 g S kg^{-1} as a critical level in the rice shoot at tillering and 0.55 g S kg^{-1} in rough rice grains. Wang (1976) concluded that the critical concentration of S in rice straw should be 0.5 g S kg^{-1} for optimum grain yield. Slaton et al. (2001) observed late-season S deficiency symptoms when rice flag leaves, immediately before panicle emergence from the boot, contained <1.5 g S kg^{-1}. Rice grain S concentrations vary between 0.34 g S kg^{-1} for S-deficient plants and 1.6 g S kg^{-1} for plants that had no response to S application (De Datta 1981).

Wang, Liem, and Mikkelsen (1976) determined S uptake in the straw and grain of lowland rice grown in the Amazon Basin in the Brazilian state of Para (Table 6.6). Grain and straw S contents increased as S and N fertilizer rates increased. In addition, at low rates of S fertilization, grain S content was greater than straw S content, but straw and grain S contents were nearly equal at high rates of S fertilization. Wang (1976) found that lowland rice grain yields of 5–7 t ha^{-1} removed S between 5 kg ha^{-1} and 9 kg ha^{-1}. The rate of S removal by lowland rice was affected by the cultivar, S application rate, and N fertilization. In Arkansas, total rice S uptake at

TABLE 6.6
Sulfur Uptake (kg ha^{-1}) in Shoot and Grain of Lowland Rice (Average of Two Cultivars) under Five Different S Rates and Two N Rates

	Aboveground Plant Content, kg S ha^{-1}					
	Shoot		Grain		Total	
S Rate (kg S ha^{-1})	60 kg N ha^{-1}	120 kg N ha^{-1}	60 kg N ha^{-1}	120 kg N ha^{-1}	60 kg N ha^{-1}	120 kg N ha^{-1}
0	0.89	0.99	1.57	1.92	2.46	2.91
25	2.93	3.82	3.30	4.37	6.23	8.19
50	3.41	4.45	3.37	3.81	6.78	8.26
100	4.12	4.72	3.50	3.80	7.62	8.52
Mean	2.84	3.50	2.94	3.48	5.78	6.98

Source: Adapted from Wang, C. H., T. H. Liem, and D. S. Mikkelsen. 1976. *Sulfur Deficiency—A Limiting Factor in Rice Production in the Lower Amazon Basin. II. Sulfur Requirement for Rice Production*. New York: IRI Research Institute.

maturity generally averages about 25 kg ha^{-1} with crop removal by harvested grain representing about 30% of total plant uptake (Wilson et al. 2001; Fageria, Slaton, and Baligar 2003).

6.6 USE EFFICIENCY

Data related to sulfur use efficiency (defined as quantity of grain produced per kg S applied) for upland and lowland rice is presented in Tables 6.7 and 6.8. In upland and lowland rice, sulfur use efficiency varied from genotype to genotype and also with S rate applied. In upland rice, S use efficiency varied from 135 to 465 kg grain produced per kg S applied at the 10 kg S ha^{-1}, with an average value of 239 kg grain produced per kg S applied. At 20 kg S ha^{-1}, S use efficiency varied from 62 to 166 kg grain produced with one 1 kg S applied, with an average value of 133 kg grain produced per kg S applied. The variation in S use efficiency among crop species or genotypes within species has been reported by Tisdale et al. (1985).

Values of S use efficiency were higher at the lower S rate and decreased with increased S rate. Nutrient use efficiency (defined as grain yield per unit nutrient applied) decreased with the increase in nutrient levels (Fageria, Baligar, and Jones 2011). Fageria, Slaton, and Baligar (2003) reported that the decrease in nutrient use efficiency at higher nutrient levels is related to saturation of the root capacity mechanism for nutrient uptake when nutrient concentration is higher in the root zone. Sulfur use efficiency of upland rice genotypes was higher compared with lowland rice genotypes. This may be related to higher loss of S in flooded soils due to oxidation–reduction processes and leaching compared with oxidized soils or aerobic rice. In flooded rice, inorganic S is reduced to FeS, FeS_2, and H_2S. The H_2S formation in the flooded soils occurs with microbial mineralization of organic S, resulting in loss in the soil–plant system (Fageria 2009).

TABLE 6.7
Sulfur Use Efficiency (SUE) in Upland Rice Genotypes at Different S Levels

	SUE (kg grain per kg S applied)			
Genotype	**10 kg S ha^{-1}**	**20 kg S ha^{-1}**	**40 kg S ha^{-1}**	**80 kg S ha^{-1}**
BRA 01600	190	102	81	48
BRA 01600	295	153	86	25
BRA 02601	135	166	44	17
BRA 032033	465	211	83	41
BR Monica	186	62	72	37
BR Pepita	161	101	17	19
Average	239	133	64	31

Note: $\text{SUE (kg kg}^{-1}) = \dfrac{\text{GY at S levels in kg} - \text{GY at zero S level in kg}}{\text{S rate applied in kg}}$,
where GY = grain yield.

TABLE 6.8
Sulfur Use Efficiency (SUE) in Lowland Rice Genotypes at Different S Levels

	SUE (kg grain per kg S applied)			
Genotype	**10 kg S ha^{-1}**	**20 kg S ha^{-1}**	**40 kg S ha^{-1}**	**80 kg S ha^{-1}**
BRS Tropical	95	35	21	5
BRS Jaçanã	43	15	13	–
BRA 02654	54	13	–	–
BRA 051077	64	31	17	–
BRA 051083	150	60	–	12
BRA 051108	166	64	32	–
BRA 051126	154	71	39	23
BRA 051129	31	2	9	3
BRA 051130	78	53	36	11
BRA 051134	62	50	9	10
BRA 051135	70	53	14	8
BRA 051250	38	32	13	21
Average	84	40	20	11

Note: $\text{SUE (kg kg}^{-1}) = \dfrac{\text{GY at S levels in kg} - \text{GY at zero S level in kg}}{\text{S rate applied in kg}}$,
where GY = grain yield. Values of one genotype at 40 kg S ha^{-1} and four genotypes at 80 kg S ha^{-1} were negative; hence, SUE values were not presented.

6.7 MANAGEMENT PRACTICES

Management practices that can be adopted to maximize S use efficiency and rice yield are effective source, appropriate method and timing of application, adequate rate, and planting S-efficient genotypes. These management practices are discussed in the following sections.

6.7.1 Effective Sources

Several sources or fertilizers containing S available in the market are listed in Table 6.9. Some S sources can supply other essential nutrients that can be used to save transport and application costs. For example, ammonium sulfate contains 24% S and 21% N. This fertilizer can supply N and S when applied in adequate rates. Similarly, gypsum contains about 17–19% S and 23% Ca. In addition, some fertilizers contain more than three nutrients. For example, triple superphosphate contains 45% P_2O_5, 13% Ca, and 1.4% S, and simple superphosphate contains 20% P_2O_5, 20% Ca, and 12–14% P_2O_5. There are several other fertilizer sources containing two or more

TABLE 6.9
Principal Sulfur Carriers

Fertilizer/Amendment	Formula	S (%)
Ammonium sulfate	$(NH_4)_2SO_4$	24.0
Ammonium polysulfide	NH_4S_x	45
Ammonium sulfate-nitrate	$(NH_4)_2SO_4.NH_4NO_3$	12.0
Ammonium thiosulfate solution	$(NH_4)_2S_2O_3+H_2O$	26.0
Magnesium sulfate	$MgSO_{4.}7H_2)$	13
Gypsum (by-product)	$CaSO_4.2H_2O$	17.0
Single superphosphate	$Ca(H_2PO_4)_2 + CaSO_4.2H_2O$	12.0
Triple superphosphate	$Ca(H_2PO_4)_2.H_2O$	1.4
Copper sulfate	$CuSO_4.5H_2O$	13.0
Zinc sulfate	$ZnSO_4.H_2O$	18.0
Elemental sulfur	S	100
Sodium sulfate	Na_2SO_4	23
Potassium sulfate	K_2SO_4	18.0
Manganese sulfate	$MnSO_4.4H_2O$	14.5
Iron sulfate	$FeSO_4.7H_2O$	11.5
Sulfur dioxide	SO_2	50.0
Sulfuric acid	H_2SO_4	32.7
Ferrous ammonium sulfate	$Fe(NH_4)_2SO_4$	16

Source: Adapted from Follett, R. H., L. S. Murphy, and R. L. Donahue. 1981. *Fertilizers and Soil Amendments*. Englewood Cliffs, NJ: Prentice-Hall; Tisdale, S. L., W. L. Nelson, and J. D. Beaton. 1985. *Soil Fertility and Fertilizers*, 4th edition. New York: Macmillan; Fageria, N. K. 2009. *The Use of Nutrients in Crop Plants*. Boca Raton, FL: CRC Press; Fageria (1989).

nutrients that can be used to supply S as well as other nutrients. The important criteria used to select the effective source are cost and market availability of the fertilizers.

6.7.2 Appropriate Methods and Timing of Application

As far as method of sulfur fertilizer application is concerned, efficiency of fertilizers containing S is higher when band applied in S-deficient soils. Band application can have higher uptake efficiency due to its availability near the root surface. In addition, S is required for tillering, which starts about 15 days after sowing (depending on rice cultivars and environmental conditions). Therefore, a major portion of the S should be applied at sowing. Fertilizers, such as ammonium sulfate, which contains S, can be top-dressed to supply N and S to a rice crop during the crop growth cycle. If S is applied as phosphogypsum, a large amount is needed; it should be applied as broadcast and mixed well with the soil before the rice crop is sown (Fageria 2009).

6.7.3 Adequate Rate

Adequate rate for a nutrient should be determined through field experiments in different agro-climatic regions. The selected areas for S experiments should be low in organic matter content as well as in available S. Several rates of sulfur (low to high) should be used to get accurate results. The minimum rate used should be five, and the treatments should be repeated at least four times. The S levels selected should give a quadratic response for the rice crop to determine the adequate rate for maximum grain yield. Fageria (2005, 2007b) discussed basic principles and methodology for conducting controlled and field experiments related to mineral nutrition. These articles should be consulted for detailed information about selecting treatments and analysis and for interpretation of research data in the field of mineral nutrition.

Field experiments were conducted to determine adequate S rates for upland and lowland rice, and data are presented in Tables 6.10 through 6.13. Data in Table 6.10 show responses of 12 upland rice genotypes to S fertilization. Sulfur and genotype treatments significantly affected grain yield. The S × genotype interaction for grain yield was also significant, indicating different responses of genotypes to different S levels. The increase in grain yield was 53 and 54% with the addition of 10 and 20 kg S ha^{-1}, respectively. Regression equations were calculated relating S rate versus grain yield to determine adequate S rate (Table 6.11). Among 12 genotypes, four had significant quadratic regression equations. Regression equations (average of 12 genotypes), relating S rate and grain yield were also quadratic and significant. Based on the quadratic regression equations, the S rate for maximum grain yield varied from 44 to 66 kg ha^{-1}, with an average value of 49 kg S ha^{-1}.

Results related to responses of lowland or flooded rice genotypes to S fertilization are presented in Table 6.12. Grain yield was not influenced by S treatment, but S × genotype interaction was significant. However, the increase in grain yield was 21% with the addition of 10 kg S ha^{-1} compared with the control treatment. Regression equations were also calculated relating S rate and grain yield (see Table 6.13). Among 12 genotypes, six displayed a significant quadratic relationship with S rate and grain yield. The S rate for maximum grain yield varied from 29 to 57 kg ha^{-1}, with an

TABLE 6.10
Influence of Sulfur and Genotypes on Grain Yield (kg ha^{-1}) of 12 Upland Rice Genotypes

	Sulfur Level (kg ha^{-1})				
Genotype	**0**	**10**	**20**	**40**	**80**
BRA 01506	2832.04def	4731.85ab	4879.44a	6079.26a	6634.44a
BRA 01600	1962.03fg	4841.66ab	5025.18a	5415.74ab	3941.85ab
BRA 02601	2267.96efg	3618.33ab	5597.59a	4041.11abc	3621.29ab
BRA 032033	1887.78g	6320.55a	5880.18a	4988.88ab	4961.48ab
BRA 032048	3676.67cd	6147.96a	6330.74a	3654.62bc	4032.78ab
BRA 032051	4847.59ab	3433.70ab	4179.25ab	4154.44abc	3515.00ab
BRA 042094	4060.00bc	4619.44ab	3900.74ab	4585.74ab	3663.15ab
BRA 052015	1836.85g	5243.52a	5022.22a	4230.18abc	4431.66ab
BRS Primavera	1962.03fg	1800.37b	1799.25b	1809.26c	2088.14b
BRS Sertaneja	5100.18a	5404.07a	3842.59ab	5366.11ab	5720.74ab
BR Monica	2990.18de	4847.96ab	4322.22a	5882.04ab	5965.18ab
BR Pepita	3239.44cd	4850.92ab	5264.07a	3938.89abc	4773.33ab
Average	3036.89b	4661.11a	4670.29a	4512.46a	4445.75a
F-test					
Sulfur (S)	*				
Genotype (G)	*				
S × G	*				
CVS (%)	14.76				
CVG (%)	23.67				

Note: Means followed by the same letter in the same column are statistically not significant at the 5% probability level by the Tukey's test.

*Significant at the 1% probability level.

TABLE 6.11
Relationship between Sulfur Rate and Grain Yield of Upland Rice Genotypes

Variable	Regression Equation	R^2	VMY
S rate vs. GY (BRA 01506)	$Y = 3148.63 + 110.42X - 0.84X^2$	0.63**	66
S rate vs. GY (BRA 01600	$Y = 2629.46 + 148.66X - 1.67X^2$	0.51*	45
S rate vs. GY (BRA 02601)	$Y = 2727.24 + 103.55X - 1.17X^2$	0.37*	44
S rate vs. GY (BR Monica)	$Y = 3269.22 + 93.39X - 0.74X^2$	0.72**	63
S rate vs. GY (average of 12 genotypes)	$Y = 3511.97 + 60.25X - 0.62X^2$	0.54**	49

Note: VMY, value for maximum yield in kg S ha^{-1}; GY, grain yield. Among 12 genotypes cited in Table 6.2, regression equations were significant only for four genotypes and the average of 12 genotypes.

*,**Significant at the 5 and 1% probability levels, respectively.

TABLE 6.12
Influence of Sulfur and Genotypes on Grain Yield (kg ha^{-1}) of 12 Lowland Rice Genotypes

	Sulfur Level (kg ha^{-1})				
Genotype	**0**	**10**	**20**	**40**	**80**
BRS Tropical	4208.67bc	5160.27a	4910.83a	5046.38a	4622.77ab
BRS Jaçanã	4278.33bc	4711.11a	4569.72a	4790.55a	3229.16ab
BRA 02654	4976.39a	5512.78a	5235.55a	4634.72a	3697.50ab
BRA 051077	3920.28cd	4557.22a	4542.77a	4605.28a	2779.72b
BRA 051083	3461.11e	4965.00a	4662.77a	3298.33a	4446.66ab
BRA 051108	3583.61de	5242.50a	4866.39a	4850.00a	4219.44ab
BRA 051126	2737.50f	4272.50a	4150.55a	4290.27a	4565.55ab
BRA 051129	3983.89bc	4291.94a	4024.44a	4344.16a	4220.00ab
BRA 051130	3916.66cd	4693.89a	4986.11a	5361.11a	4826.39ab
BRA 051134	4078.33bc	4701.66a	5086.38a	4437.78a	4881.94ab
BRA 051135	4088.89bc	4789.44a	5150.55a	4651.38a	4729.16ab
BRA 051250	4359.16b	4734.44a	5008.33a	4865.27a	6071.66a
Average	3966.06	4802.73	4766.22	4595.16	4232.50
F-test					
Sulfur (S)	NS				
Genotype (G)	**				
S × G	*				
CVS (%)	30.92				
CVG (%)	16.59				

Note: Means followed by the same letter in the same column are statistically not significant at the 5% probability level by the Tukey's test.

*,**,NS Significant at the 5 and 1% probability levels and nonsignificant, respectively.

TABLE 6.13
Relationship between Sulfur Rate and Grain Yield of Lowland Rice Genotypes

Variable	Regression Equation	R^2	VMY
S rate vs. GY (BRS Jaçanã)	$Y = 4298.33 + 34.15X - 0.59X^2$	0.55**	29
S rate vs. GY (BRA 051077)	$Y = 4002.27 + 46.30X - 0.77X^2$	0.31*	30
S rate vs. GY (BRA 051108)	$Y = 4051.61 + 54.15X - 0.66X^2$	0.33*	41
S rate vs. GY (BRA 051126)	$Y = 3160.85 + 55.64X - 0.49X^2$	0.42*	57
S rate vs. GY (BRA 051130)	$Y = 4010.90 + 60.92X - 0.64X^2$	0.34*	48
S rate vs. GY (BRA 051250)	$Y = 4520.10 + 9.87X - 0.11X^2$	0.45*	45
S rate vs. average 12 genotypes	$Y = 4215.21 + 30.23X - 0.38X^2$	0.31*	40

Note: VMY, value for maximum yield in kg S ha^{-1}; GY, grain yield. Among 12 genotypes cited in Table 6.4, regression equations were significant only for six genotypes and the average of 12 genotypes.

*,** Significant at the 5 and 1% probability levels, respectively.

average value of 40 kg S ha^{-1}. Figures 6.9 through 6.12 compared upland rice growth at 0 and 10 kg S ha^{-1} to 0 and 80 kg S ha^{-1} at 40 days after sowing, under field conditions. Similarly, Figures 6.13 through 6.16 show the growth of upland rice at different S levels at physiological maturity.

Soil analysis can be used as a guide for applying sulfur to a crop when sufficient field trial data are not available. Jones (1986) interpreted S soil analysis data as low, medium, and high. These classification groups may give an idea whether a crop

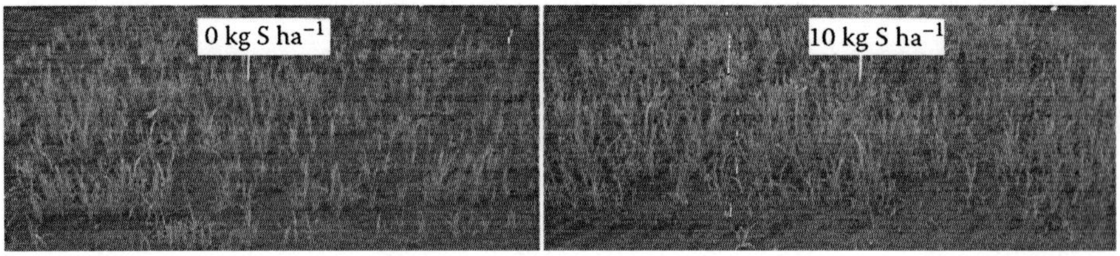

FIGURE 6.9 Growth of upland rice at 0 and 10 kg S ha^{-1} at 40 days after sowing.

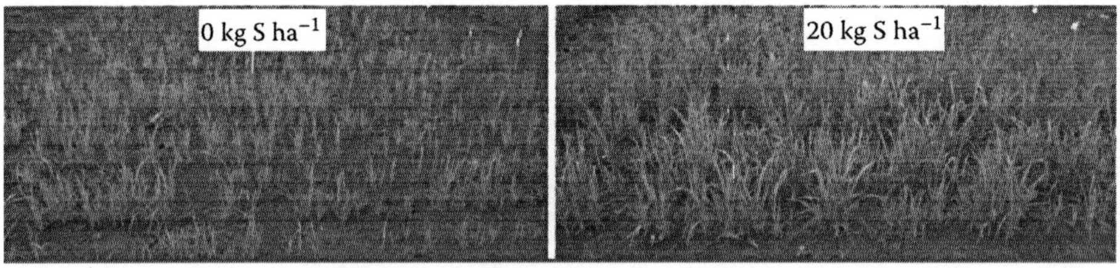

FIGURE 6.10 Growth of upland rice at 0 and 20 kg S ha^{-1} at 40 days after sowing.

FIGURE 6.11 Growth of upland rice at 0 and 40 kg S ha^{-1} at 40 days after sowing.

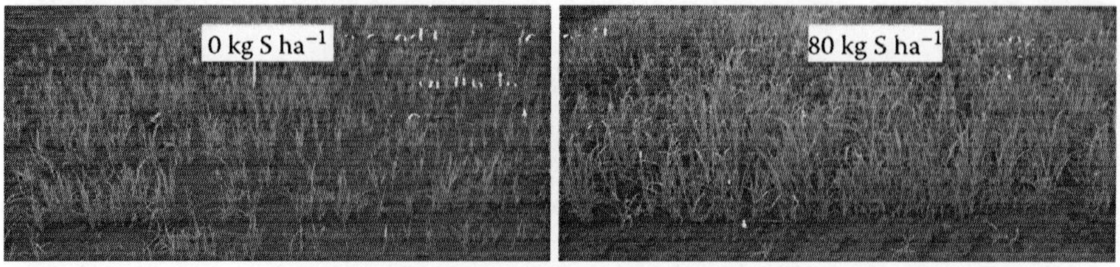

FIGURE 6.12 Growth of upland rice at 0 and 80 kg S ha^{-1} at 40 days after sowing.

FIGURE 6.13 Growth of upland rice at 0 and 10 kg S ha^{-1} at physiological maturity.

FIGURE 6.14 Growth of upland rice at 0 and 20 kg S ha^{-1} at physiological maturity.

will respond to the applied S as a fertilizer or not. These classifications are given in Table 6.14. Foth and Ellis (1988) reported that although variable with location and crop, the critical level of S is expected to be from 3 to 8 mg kg^{-1}.

6.7.4 Use of Efficient Genotypes

Using S-efficient genotypes is an important strategy for improving S use efficiency and for reducing the cost of rice production. Nutrient-efficient genotypes not only

FIGURE 6.15 Growth of upland rice at 0 and 40 kg S ha^{-1} at physiological maturity.

FIGURE 6.16 Growth of upland rice at 0 and 80 kg S ha^{-1} at physiological maturity.

reduce the cost of crop production but also are important in the reduction of environmental pollution. Results obtained show that there are significant differences among upland rice genotypes in grain yield and panicle density at two S levels (Table 6.15). Grain yield was noticeably influenced by sulfur and genotype treatments, and sulfur × genotype interaction was also significant. The sulfur × genotype interactions suggested that the response of upland rice varied with the changes in S levels, and selection for S use efficiency should be done at more than one S level.

Grain yield varied from 2.05 g $plant^{-1}$ produced by genotype BRS CIRAD 302 to 10.67 g $plant^{-1}$ produced by genotype AB072041, with an average value of

TABLE 6.14
Classification of Soil Analysis Data for Sulfur Fertilization in Crop Plants

Extractable S (mg kg^{-1})	Interpretation	Predicted Response
0–3	Low	High response is expected
4–10	Medium	Variable response is expected
>10	High	No response is expected

Source: Adapted from Jones, M. B. 1986. Sulfur availability indexes. In *Sulfur in Agriculture*, ed. M. A. Tabatabai, pp. 549–566. Madison, WI: ASA, CSSA, and SSSA.

7.28 g $plant^{-1}$ at the lower S level. The difference in grain yield between the lowest and the highest grain-yield-producing genotypes was about 5% at the lower S level. At the higher S level, grain yield varied from 12.50 g $plant^{-1}$ produced by genotype AB072083, with an average value of 8.81 g $plant^{-1}$. The difference in grain yield between the lowest and the highest grain-yielding genotypes was about 1.4 times. This means variation among genotypes was higher at the lower S level than the higher S level. Overall, the increase in grain yield was 20% with the addition of 50 mg S kg^{-1} compared with 0 mg S kg^{-1} treatment. Published results about the response of upland rice genotypes to S fertilization are limited. However, Malavolta et al. (1987) reported a significant yield increase in upland rice grown on Brazilian Oxisol with the addition of S fertilization. These authors also found that the yield increase in upland rice grown on Oxisol with S content of 2.3 mg kg^{-1} was 20%.

Panicle density varied from 2.08 to 5.83 $plant^{-1}$, with an average value of 3.21 panicles $plant^{-1}$ at the lower S level (see Table 6.15). At the higher S level, panicle density varied from 2.08 to 6.83 panicles $plant^{-1}$. Overall, the increase in panicle density at the higher S level was about 20% compared with the lower S level. Panicle density is genetically controlled and is also influenced by environmental factors, especially mineral nutrition (Fageria, Baligar, and Jones 2011). The genotype AB072083, which produced the highest panicle density at the high S level, also produced the highest grain yield. Therefore, panicle density is an important trait for increasing grain yield. Fageria (2007a, 2009) and Fageria, Morais, and Santos (2010) reported a significant quadratic association between rice grain yield and panicle density.

In the same experiment, the influence of S on maximum root length (MRL) and root dry weight (RDW) was also studied (Table 6.16). The MRL was significantly influenced by the S level and genotype treatments, and sulfur × genotype interaction was also significant for MRL (Table 6.16). The RDW was not influenced by S treatment, but genotype and S × G interaction were significant for this trait. The MRL varied from 20.66 to 28.33 cm, with an average value of 23.96 cm at the lower S level. At the higher S level, the MRL varied from 21.33 to 36.66 cm, with an average value of 28.26 cm. The increase in MRL was 18% at the higher S level compared with the lower S level. The RDW varied from 0.33 to 2.53 g $plant^{-1}$, with an average value of 1.41 g $plant^{-1}$ at the lower S level. At the higher S level,

TABLE 6.15
Influence of Sulfur and Genotype Treatments on Grain Yield and Panicle Density of 30 Upland Rice Genotypes

	Grain Yield (g plant^{-1})		Panicle Density (plant^{-1})	
Genotype	**0 mg S kg^{-1}**	**50 mg S kg^{-1}**	**0 mg S kg^{-1}**	**50 mg S kg^{-1}**
AB072083	8.52bcd	12.50a	5.83a	6.83a
AB072017	8.89abc	12.40ab	4.00b–e	5.25a–d
AB072041	10.67a	12.36ab	3.50b–i	6.41ab
AB072001	10.00ab	10.83abc	3.66b–g	4.41c–g
AB072063	8.67bc	9.14c–i	2.58f–j	3.08fgh
AB072078	7.83c–h	8.77e–j	2.17j	4.41e–h
AB072050	6.68d–j	9.57c–h	4.08bcd	4.91b–e
AB072047	7.06c–i	9.74c–g	4.08bcd	4.75b–f
AB072084	9.97ab	10.60b–e	4.75ab	5.25a–d
AB072085	8.16b–f	8.89d–j	4.50bc	5.33abc
AB072007	6.00hij	8.43f–j	3.75b–f	4.75b–f
AB072044	7.96c–g	8.92d–j	3.33c–j	3.58d–h
AB072048	8.88abc	9.66c–g	3.00d–j	3.33e–h
AB072035	7.56c–i	9.17c–i	3.08d–j	3.25e–h
BRA032048	7.48c–i	8.18g–j	2.91d–j	2.83gh
AB052033	7.57c–i	9.19c–i	2.25ij	2.83gh
AB062008	8.42b–e	9.80c–g	2.75e–j	3.25e–h
AB062037	10.00ab	10.64a–d	2.50f–j	3.00gh
AB062041	6.46f–j	7.36ijl	2.08j	2.75gh
AB062045	4.75j	5.37mn	2.41g0j	3.25e–h
AB062138	8.76abc	9.47c–h	3.16d–j	3.58d–h
BRSGO Serra Dourada	5.67ij	10.07c–f	2.75e–j	3.25e–h
BRS Pepita	6.47e–j	7.14jlm	2.33hij	2.08h
BRS Sertaneja	5.96hij	7.99g–j	2.25ij	3.00gh
BRS Monarca	6.12g–j	8.04g–j	3.17d0j	3.25e–h
BRS Primavera	4.94j	6.09lm	2.41g–j	3.25e–h
BRSMG Caçula (CMG 1152)	4.77j	5.37mn	2.91d–j	2.75gh
BRS CIRAD 302	2.05l	3.52n	3.58b–h	4.17c–g
Primavera CL (07SEQCL 441)	4.97j	7.52ijl	3.58b–h	4.08c–g
Curinga CL (07SEQCL 563)	7.07c–i	7.73h–l	2.83d–j	3.41e–h
Average	7.28b	8.81a	3.21b	3.84a
F-test				
S level (S)	*		*	
Genotype (G)	*		*	
S × G	*		*	
CVS (%)	12.82		17.87	
CVG (%)	7.13		13.03	

Note: Means followed by the same letter in the same column are statistically not different at the 5% probability level by Tukey's test. Average values were compared in the same line.

*Significant at the 1% probability level.

TABLE 6.16
Influence of Sulfur and Genotype Treatments on Maximum Root Length and Root Dry Weight of 30 Upland Rice Genotypes

	MRL (cm)		RDW (g plant^{-1})	
Genotype	0 mg S kg^{-1}	50 mg S kg^{-1}	0 mg S kg^{-1}	50 mg S kg^{-1}
AB072083	25.33abc	29.33a–e	2.21abc	2.68ab
AB072017	27.66ab	31.33a–d	2.11a–e	2.83a
AB072041	21.66bc	24.00b–e	1.73a–g	2.43abc
AB072001	25.66abc	26.66b–e	2.04a–e	2.28a–d
AB072063	25.00abc	30.00a–d	1.67a–g	1.61b–g
AB072078	23.00abc	28.00b–e	2.25ab	1.58b–g
AB072050	20.66c	27.66b–e	2.53a	2.06a–f
AB072047	21.66bc	28.66a–e	2.16a–d	2.17a–e
AB072084	28.33a	31.66abc	1.78a–g	1.17d–g
AB072085	23.66abc	28.66a–e	1.17e–j	1.52b–g
AB072007	26.00abc	27.66b–e	1.07f–j	1.64a–g
AB072044	21.33c	25.00b–e	1.22d–j	1.88a–g
AB072048	25.00abc	21.33e	1.25d–j	1.56b–g
AB072035	21.66bc	23.00de	0.84g–j	1.12d–g
BRA 032048	22.33abc	24.66b–e	1.08f–j	1.30c–g
AB052033	22.00bc	26.66b–e	1.30b–i	1.36c–g
AB062008	23.33abc	31.33a–d	1.29c–i	1.73a–g
AB062037	24.00abc	29.66a–e	1.46b–h	1.71a–g
AB062041	25.00abc	26.00b–e	1.87a–f	1.77a–g
AB062045	23.33abc	23.66cde	0.55hij	0.99efg
AB062138	27.66ab	31.00a–d	1.58a–g	2.13a–e
BRSGO Serra Dourada	24.33abc	28.66a–e	1.83a–f	1.91a–g
BRS Pepita	24.00abc	31.33a–d	1.34b–i	2.07a–f
BRS Sertaneja	24.66abc	29.00a–e	0.47ij	1.47c–g
BRS Monarca	23.33abc	29.33a–e	1.21d–j	2.09a–f
BRS Primavera	21.66bc	32.33ab	0.33j	0.91fg
BRSMG Caçula (CMG 1152)	25.00abc	28.66a–e	0.62hij	0.82g
BRS CIRAD 302	24.00abc	36.66a	1.27c–j	1.51b–g
Primavera CL (07SEQCL 441)	23.00abc	27.66b–e	0.55hij	1.57b–g
Curinga CL (07SEQCL 563)	24.66abc	28.33a–e	1.39b–i	2.47abc
Average	23.96b	28.26a	1.41a	1.74a
F-test				
S level (S)	*		NS	
Genotype (G)	*		*	
S × G	*		*	
CVS (%)	10.33		56.09	
CVG (%)	8.78		19.00	

Note: Means followed by the same letter in the same column are statistically not different at the 5% probability level by Tukey's test. Average values were compared in the same line. MRL, maximum root length; RDW, root dry weight.

*,NS Significant at the 1% probability level and nonsignificant, respectively.

FIGURE 6.17 Root growth of two upland rice genotypes at two S levels at harvest.

the RDW varied from 0.82 to 2.83 g plant^{-1}, with an average value of 1.74 g plant^{-1}. Few studies have assessed the influence of S on the root growth of crop plants, including rice. However, Fageria and Moreira (2011) reported that the influence of S on root growth is similar to that of Nitrogen. Fageria (2013) noted an increase in root length and RDW of upland rice in Brazilian Oxisol with the addition of S. Figures 6.17 through 6.20 show the influence of sulfur on root growth of upland rice genotypes grown on Brazilian Oxisol. It is very clear from these figures that S improves root growth of upland rice genotypes. This has special significance relating to the absorption of water and nutrients in upland rice, which is subject to abiotic stress in the Brazilian Cerrado region where it is mostly grown.

Based on the grain yield efficiency index (GYEI), upland rice genotypes were classified into efficient, moderately efficient, and inefficient groups with regard to S use efficiency. The GYEI was calculated using the following formula (Fageria 2009):

$$\text{GYEI} = \frac{\text{GY at low S level}}{\text{AGY of 30 genotypes at low S level}} \times \frac{\text{GY at high S level}}{\text{AGY of 30 genotypes at high S level}} \tag{6.1}$$

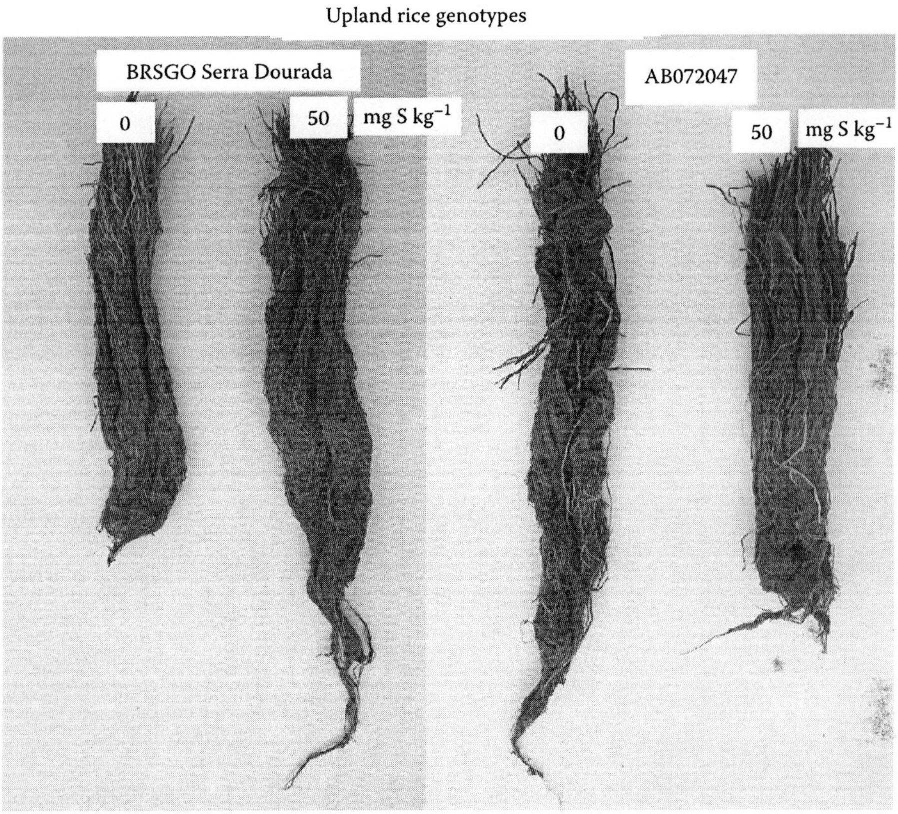

FIGURE 6.18 Root growth of two upland rice genotypes at two S levels at harvest.

where GY is grain yield and AGY is average grain yield. With regard to S use efficiency, genotypes having a GYEI higher than 1.0 were considered S efficient; inefficient genotypes were in the range of 0 to 0.5 GYEI; and genotypes in between these two limits were classified as moderately efficient. Although this is an arbitratory classification system, it is useful in separating high-yielding, stable, nutrient-efficient genotypes from low-yielding, unstable, nutrient-inefficient genotypes (Fageria 2009).

The GYEI was significantly different among upland rice genotypes (Table 6.17). It varied from 0.11 to 2.06, with an average value of 1.05. Based on GYEI, genotypes were classified as efficient, moderately efficient, and inefficient in S use efficiency. Genotypes that were classified as efficient were AB072083, AB072017, AB072041, AB072001, AB072063, AB072078, AB072085, AB072084, AB072085, AB072044, AB072048, AB072035, AB052033, AB062008, AB062037, and AB062138. Moderately efficient genotypes were AB072050, AB072007, BRA 032048, AB062041, BRSGO Serra Dourada, BRS Pepita, BRS Sertaneja, BRS Monarca, Primavera CL (07SEQCL 441), and Curinga CL (07SEQCL 563). The inefficient genotypes were AB062045, BRS Primavera, BRSMG Caçula (CNG 1152), and BRS CIRAD 302.

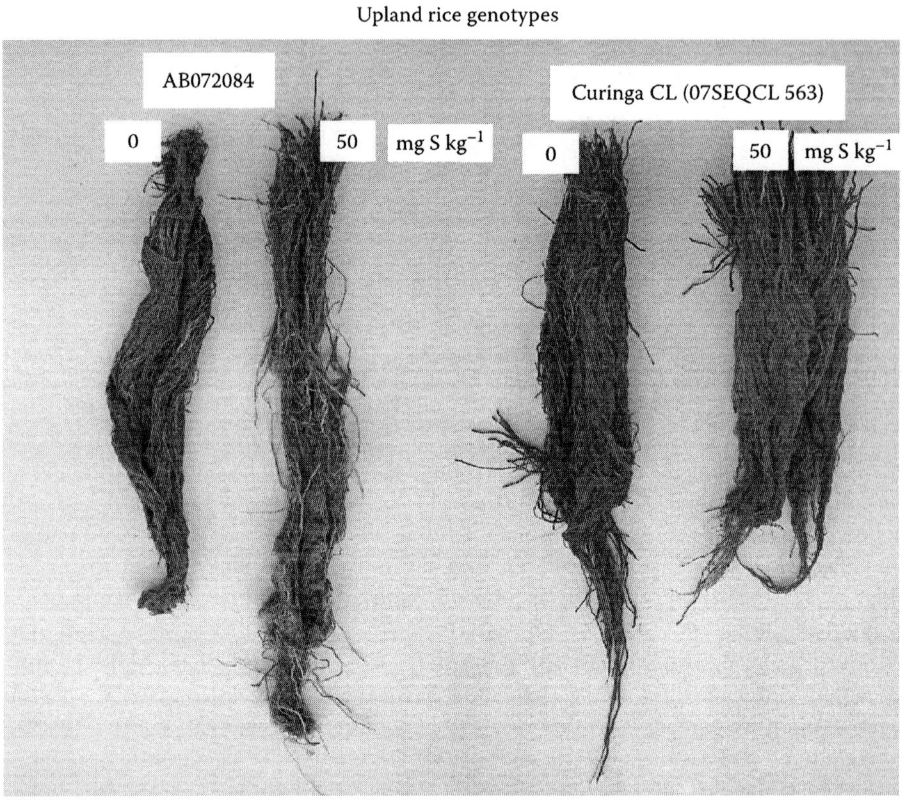

FIGURE 6.19 Root growth of two upland rice genotypes at two S levels at harvest.

Therefore, among 30 genotypes, 53.33% were classified as efficient, 33.33% were classified as moderately efficient, and the remaining 13.33% were classified as inefficient with regard to S use efficiency.

The influence of S on the grain yield and GHI of lowland rice genotypes grown on Brazilian Inceptisol was also studied. The sulfur × genotype interaction was found for grain yield and GHI, which indicated that genotypes responded differently under two sulfur rates (Table 6.18). Grain yield varied from 5.18 g plant^{-1} produced by genotype BRS Tropical to 18.13 g plant^{-1} produced by genotype BRA 051129, with an average value of 13.14 g plant^{-1}. The variation between the lowest and the highest grain-yield-producing genotypes was almost 3.5 times at the lower S rate. Similarly, at the higher S rate, the grain yield varied from 10.41 g plant^{-1} produced by genotype BRA 051134 to 24.83 g plant^{-1} produced by genotype BRS Jaçanã, with an average yield of 18.99 g plant^{-1}. The genotype BRS Tropical also produced as good a grain yield as BRS Jaçanã at the high S level. These two cultivars have been released by the National Rice and Bean Research Center for planting in the lowland soil in the central part of Brazil. Results of this study indicate that the cultivar BRS Tropical is highly sensitive to S deficiency compared with other genotypes. The increase

FIGURE 6.20 Root growth of two upland rice genotypes at two S levels at harvest.

in grain yield was 45% with the addition of S compared with a control treatment. Sulfur deficiency has been reported in lowland rice from various countries, such as Indonesia, India, Bangladesh, and Thailand (De Datta 1981).

The GHI varied from 0.16 to 0.51, with an average value of 0.39 at the lower S rate. Similarly, at the higher S rate, the GHI varied from 0.30 to 0.52, with an average value of 0.44. The increase in GHI was 13% with the addition of S fertilizer compared with a control treatment. The cultivar BRS Tropical produced the lowest grain yield at the low S level and had the lowest GHI, whereas the cultivar BRS Jaçanã produced the highest grain yield at the higher S rate and had almost the maximum GHI (0.51). Grain yield was significantly and positively related to GHI (r = 0.86**). The significant association between grain yield and GHI in rice is reported by Fageria and Baligar (2005), Fageria (2007a), and Fageria, Morais, and Santos (2010). Therefore, it can be concluded that GHI is an important trait in determining grain yield in lowland rice, and S fertilization can improve this trait in this crop.

In this experiment, the influence of S on root growth was also studied. Root length and RDW were significantly influenced by sulfur and genotype treatments and sulfur × genotype interactions (Table 6.19). Root length varied from 18 to 24 cm, with an average value of 21.38 cm at the lower S rate. At the higher S rate, root length

TABLE 6.17
Grain Yield Efficiency Index and Classification of Upland Rice Genotypes for Sulfur Use Efficiency

Genotype	Grain Yield Efficiency Index	Classification
AB072083	1.66abc	E
AB072017	1.70ab	E
AB072041	2.06a	E
AB072001	1.69ab	E
AB072063	1.24de	E
AB072078	1.07d–g	E
AB072050	1.00d–g	ME
AB072047	1.07d–g	E
AB072084	1.65bc	E
AB072085	1.13def	E
AB072007	0.79f–j	ME
AB072044	1.11d–g	E
AB072048	1.34bcd	E
AB072035	1.08d–g	E
BRA 032048	0.96d–h	ME
AB052033	1.09d–g	E
AB062008	1.29cd	E
AB062037	1.66bc	E
AB062041	0.74f–j	ME
AB062045	0.40jl	IE
AB062138	1.29cd	E
BRSGO Serra Dourada	0.89e–h	ME
BRS Pepita	0.73g–j	ME
BRS Sertaneja	0.74f–j	ME
BRS Monarca	0.77f–j	ME
BRS Primavera	0.47ijl	IE
BRSMG Caçula (CMG 1152)	0.40jl	IE
BRS CIRAD 302	0.11l	IE
Primavera CL (07SEQCL 441)	0.58hij	ME
Curinga CL (07SEQCL 563)	0.85e–i	ME
Average	1.05	
F-test		
Genotype	*	
CV (%)	11.71	

Note: Means followed by the same letter in the same column are statistically not different at the 5% probability level by Tukey's test. E, efficient; ME, moderately efficient; IE, inefficient.

*Significant at the 1% probability level.

TABLE 6.18
Grain Yield and Grain Harvest Index as Influenced by Sulfur and Genotype Treatments

	Grain Yield (g $plant^{-1}$)		Grain Harvest Index	
Genotype	0 mg S kg^{-1}	80 mg S kg^{-1}	0 mg S kg^{-1}	80 mg S kg^{-1}
BRS Tropical	5.18c	24.45a	0.16b	0.45ab
BRS Jaçanã	14.54ab	24.83a	0.41a	0.51a
BRA 026540	14.83ab	17.02ab	0.42a	0.41ab
BRA 051077	11.62abc	22.37ab	0.34a	0.48a
BRA 051083	13.35ab	20.82ab	0.41a	0.52a
BRA 051108	13.67ab	22.66ab	0.46a	0.48a
BRA 051126	10.03bc	18.85ab	0.34a	0.47a
BRA 051129	18.13a	17.61ab	0.51a	0.47a
BRA 051130	16.59ab	14.47ab	0.44a	0.38ab
BRA 051134	11.96abc	10.41b	0.36a	0.30b
BRA 051135	13.59ab	13.59ab	0.39a	0.37ab
BRA 051250	14.17ab	20.76ab	0.41a	0.46ab
Average	13.14b	18.99a	0.39b	0.44a
F-test				
S level (S)	**		*	
Genotype (G)	**		**	
S × G	**		**	
CV (%) (S)	28.29		14.64	
CV (%) (G)	21.64		13.23	

Note: Means followed by the same letter in the same column do not differ significantly at the 5% probability level by Tukey's test. For average values, means were compared across the same line.

*,**,NS Significant at the 5 and 1% probability levels and nonsignificant, respectively.

varied from 20.33 cm, with an average value of 25.22 cm. The increase in RDW was 18% at the higher S rate compared with the lower S rate. RDW varied from 2.18 to 4.67 g $plant^{-1}$ at the lower S level, with an average value of 3.38 g $plant^{-1}$. At the higher S rate, RDW varied from 2.51 to 4.49 g $plant^{-1}$, with an average value of 3.58 g $plant^{-1}$. Figures 6.21 through 6.23 show root growth of lowland rice genotypes at low (control) and high S fertilization (80 mg kg^{-1}). Root growth of all six genotypes was more vigorous at the higher S level compared with the low S level. The improvement in RDW with the addition of S was 6% compared with a control treatment. Significant differences in root length and RDW among lowland rice genotypes at lower as well as at higher S rates are important genetic variations. These differences are also influenced by environmental conditions. Therefore, it can be concluded that the use of S-efficient genotypes along with adequate S fertilizers can improve root length, RDW, and the uptake of water and nutrients in favor of higher yields. Root length also displayed a significant positive quadratic correlation (r = 0.43**)

TABLE 6.19
Maximum Root Length and Root Dry Weight of 12 Lowland Rice Genotypes as Influenced by Sulfur and Genotype Treatments

	Root Length (cm)		Root Dry Weight (g plant^{-1})	
Genotype	**0 mg S kg^{-1}**	**80 mg S kg^{-1}**	**0 mg S kg^{-1}**	**80 mg S kg^{-1}**
BRS Tropical	18.00e	20.33d	4.10abc	4.42ab
BRS Jaçanã	22.66ab	24.00bc	3.36abcd	4.49a
BRA 026540	24.00a	25.66bc	4.17ab	4.07abc
BRA 051077	22.33abc	25.33	4.67a	3.68abc
BRA 051083	20.00cde	22.66cd	3.54abcd	3.17abc
BRA 051108	22.33abc	25.00bc	3.73abcd	3.85abc
BRA 051126	19.66de	30.33a	3.30abcd	4.01abc
BRA 051129	20.00cde	23.66bc	3.47abcd	2.73bc
BRA 051130	22.00abcd	25.66bc	2.98bcd	3.74abc
BRA 051134	21.33bcd	30.66a	2.18d	2.51c
BRA 051135	21.66abcd	26.00b	2.52cd	3.32abc
BRA 051250	22.67ab	23.33bcd	2.59bcd	2.98abc
Average	21.38b	25.22a	3.38a	3.58a
F-test				
S level (S)	**		NS	
Genotype (G)	**		**	
S × G	**		*	
CV (%) (S)	3.57		22.75	
CV (%) (G)	4.38		15.62	

Note: Means followed by the same letter in the same column do not differ significantly at the 5% probability level by Tukey's test. For average values, means were compared across the same line.

*,**,NS Significant at the 5 and 1% probability levels and nonsignificant, respectively.

with grain yield. Root growth variation among crop genotypes and improvement with the addition of nutrients is reported by Fageria, Baligar, and Clark (2006) and Fageria (2009).

Based on the GYEI, as described earlier for upland rice, lowland rice genotypes were classified as efficient, moderately efficient, and inefficient (Table 6.20). Efficient genotypes were BRS Jaçanã, BRA 026540, BRA 051077, BRA 051083, BRA 051108, BRA 051129, and BRA 051250. Moderately efficient genotypes were BRS Tropical, BRA 051126, BRA 051130, and BRA 051135. Genotype BRA 051134 was inefficient in S use. This genotype was also the lowest yield producing across S levels. Uptake and the requirements for S differ greatly among species, among cultivars within species, and with the crop's stages of development (Gerloff 1963). Variations in S use efficiency among lowland rice genotypes have been reported by Fageria, Slaton, and Baligar (2003).

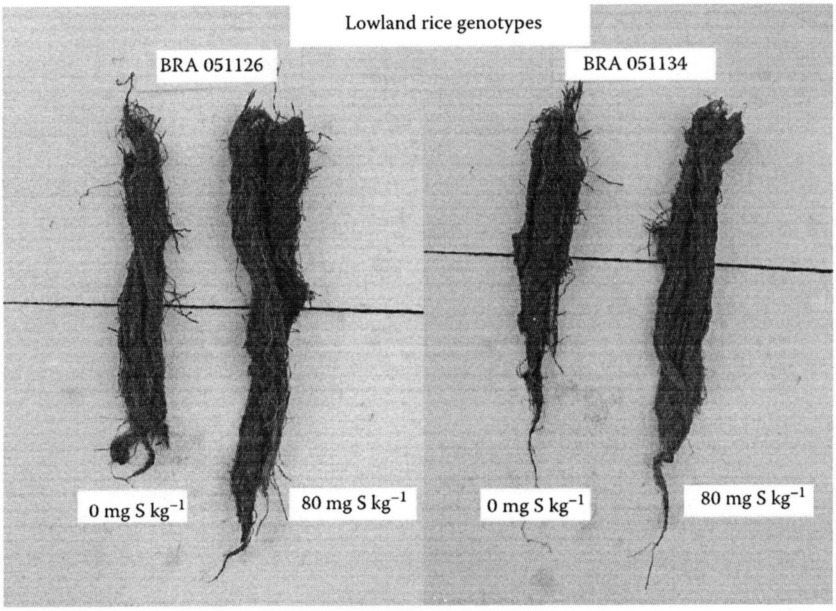

FIGURE 6.21 Root growth of two lowland rice genotypes at two S levels at harvest.

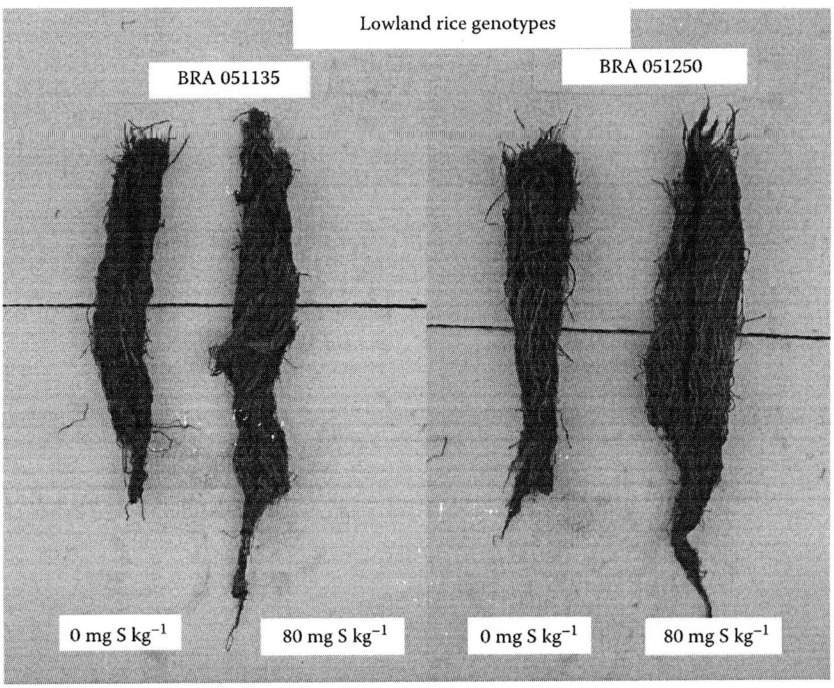

FIGURE 6.22 Root growth of two lowland rice genotypes at two S levels at harvest.

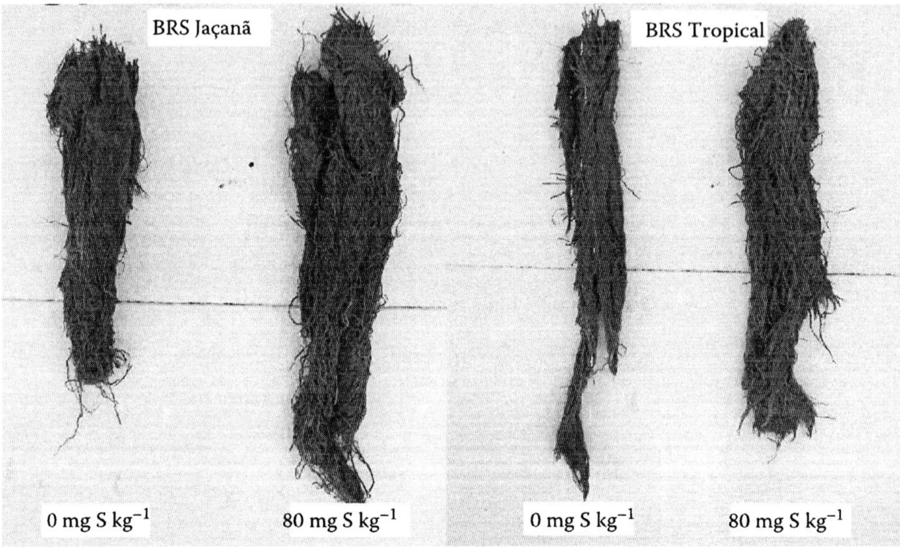

FIGURE 6.23 Root growth of two lowland rice genotypes at two S levels at harvest.

TABLE 6.20
Classification of Genotypes for S Use Efficiency Based on Grain Yield Efficiency Index

Genotype	Grain Yield Efficiency Index
BRS Tropical	0.52ME
BRS Jaçanã	1.44E
BRA 026540	1.02E
BRA 051077	1.03E
BRA 051083	1.12E
BRA 051108	1.28E
BRA 051126	0.75ME
BRA 051129	1.31E
BRA 051130	0.96ME
BRA 051134	0.49IE
BRA 051135	0.71ME
BRA 051250	1.17E
Average	0.98

Note: E, efficient; ME, moderately efficient; IE, inefficient.

6.8 CONCLUSIONS

Sulfur is required in an adequate level for the growth and development of rice, and its deficiency has been reported in many rice-growing regions worldwide. Much of the S required by plants is derived from soils where most of the S remains in organic form; its mineralization is required for plant uptake. However, fertilizers containing S can be used in sulfur-deficient soils. Sulfur improves tillering in rice and, consequently, panicle density. Sulfur fertilization also improves shoot dry weight and GHI. All these growth- and yield-contributing traits are positively associated with rice yield. Sulfur also improves root growth, which has special significance in the absorption of water and nutrients. Sulfur is not a mobile nutrient in plants; its deficiency symptoms initially appear in the younger leaves and are similar to those of N, including the yellowing of leaves. Sulfur-deficient plants produce fewer panicles, and panicle length is also less than that of plants supplied with adequate amounts of S. In addition, S-deficient plant panicles remain in the boot and do not emerge completely. Sulfur use efficiency (defined as grain yield per kg S applied as calculated by the difference method) decreases with increasing S rate. It is higher in upland rice genotypes than lowland rice genotypes, suggesting a higher loss of S in flooded rice than in aerobic or upland rice.

In Brazilian Oxisols and Inceptisols, the response of upland and lowland rice to S fertilization is obtained under controlled and field conditions. In upland and lowland rice, the response varied from genotype to genotype. Overall, maximum yield of upland rice was obtained with the application of 49 kg S ha^{-1}, and application of 40 kg S ha^{-1} along with ammonium sulfate obtained maximum yield for lowland rice. One of the reasons for variability in upland and lowland rice genotypes in relation to yield was associated with different root geometry (as evaluated under controlled conditions). In upland as well as lowland rice, root growth was better in the pots supplied with S, compared with those subjected to a control treatment.

Modern production agriculture requires efficient, sustainable, and environmentally sound management practices. Under these conditions, increasing rice yield per unit area through appropriate S management practices has become an essential component of modern production technology. Adoption of proper S management strategies that include the use of appropriate sources, rates, and application times used in conjunction with high-yielding rice cultivars bred for high-nutrient efficiency may reduce production costs and improve rice yields.

REFERENCES

Allaway, W. H., and J. F. Thompson. 1966. Sulfur in the nutrition of plants and animals. *Soil Sci.* 101:240–247.

Bennett, R. N., and R. M. Wallsgrove. 1994. Secondary metabolites in plant defense mechanisms. *New Phytol.* 127:617–633.

Blair, G. J., C. P. Mamaril, and E. Momuat. 1978. *Sulfur Nutrition of Wetland Rice.* IRRI Res. Paper Ser. 21. Los Baños, Philippines: IRRI.

Bohn, H. L., N. J. Barrow, S. S. S. Rajan, and R. L. Parftt. 1986. Reactions of inorganic sulfur in soils. In *Sulfur in Agriculture*, eds. M. A. Tabatabai, pp. 233–249. Madison, WI: ASA, CSSA, and SSSA.

De Datta, S. K. 1981. *Principles and Practices of Rice Production.* New York: John Wiley.

Donald, C. M., and J. Hamblin. 1976. The biological yield and harvest index of cereals as agronomic and plant breeding criteria. *Adv. Agron.* 28:361–405.

Duke, S. H., and H. M. Reisenauer. 1986. Roles and requirements of sulfur in plant nutrition. In *Sulfur in Agriculture*, ed. M. A. Tabatabai, pp. 123–168. Madison, WI: ASA, CSSA, and SSSA.

Fageria, N. K. 1989. *Tropical Soils and Physiological Aspects of Crop Production.* Brasilia: EMBRAPA-DPU and EMBRAPA-CNPAF.

Fageria, N. K. 2005. Soil fertility and plant nutrition research under controlled conditions: Basic principles and methodology. *J. Plant Nutr.* 28:1–25.

Fageria, N. K. 2007a. Yield physiology of rice. *J. Plant Nutr.* 30:843–879.

Fageria, N. K. 2007b. Soil fertility and plant nutrition research under field conditions: Basic principles and methodology. *J. Plant Nutr.* 28:1–25.

Fageria, N. K. 2009. *The Use of Nutrients in Crop Plants.* Boca Raton, FL: CRC Press.

Fageria, N. K. 2013. *The Role of Plant Roots in Crop Production.* Boca Raton, FL: CRC Press.

Fageria, N. K., and V. C. Baligar. 2005. Nutrient availability. In *Encyclopedia of Soils in the Environment*, ed. D. Hillel, pp. 63–71. San Diego, CA: Elsevier.

Fageria, N. K., V. C. Baligar, and R. B. Clark. 2006. *Physiology of Crop Production.* New York: Howarth Press.

Fageria, N. K., V. C. Baligar, and C. A. Jones. 2011. *Growth and Mineral Nutrition of Field Crops*, 3rd edition. Boca Raton, FL: CRC Press.

Fageria, N. K., M. P. Barbosa Filho, L. F, Stone, and C. M. Guimares. 2004. Phosphorus nutrition for upland rice production. In *Phosphorus in Brazilian Agriculture*, eds. T. Yamada and S. R. S. Abdalla, pp. 401–418. Piracicaba, São Paulo: Brazilian Potassium and Phosphate Research Association.

Fageria, N. K., and H. R. Gheyi. 1999. *Efficient Crop Production.* Campina Grande, Brazil: Federal University of Paraiba.

Fageria, N. K., P. P. Morais, and A. B. Santos. 2010. Nitrogen use efficiency in upland rice genotypes. *J. Plant Nutr.* 33:1696–1711.

Fageria, N. K., and A. Moreira. 2011. The role of mineral nutrition on root growth of crop plants. *Adv. Agron.* 110:251–330.

Fageria, N. K., N. A. Slaton, and V. C. Baligar. 2003. Nutrient management for improving lowland rice productivity and sustainability. *Adv. Agron.* 80:63–152.

Foth, H. D., and B. G. Ellis. 1988. *Soil Fertility.* New York: John Wiley.

Fox, R. L. 1980. Responses to sulfur by crops growing in highly weathered soils. *Sulphur Agric.* 4:16–22.

Fox, R. L., and G. J. Blair. 1986. Plant response to sulfur in tropical soils. In *Sulfur in Agriculture*, ed. M. A. Tabatabai, pp. 405–434. Madison, WI: ASA, CSSA, and SSSA.

Freney, J. R. 1986. Forms and reactions of organic sulfur compounds in soils. In *Sulfur in Agriculture*, ed. M. A. Tabatabai, pp. 207–232. Madison, WI: ASA, CSSA, and SSSA.

Freney, J. R., and C. H. Williams. 1983. The sulfur cycle in soil. In *The Global Biogeochemical Sulfur Cycle*, eds. M. V. Ivanov and J. R. Freney, pp. 129–201. New York: John Wiley.

Gerloff, G. C. 1963. Comparative mineral nutrition of plants. *Annu. Rev. Plant Physiol.* 14:107–124.

Haneklaus, S., E. Bloem, and E. Schnug. 2007. Sulfur and plant disease. In *Mineral nutrition and plant disease*, eds. L. E. Datnoff, W. H. Elmer, and D. M. Huber, pp.101–118. St. Paul, Minnesota: The American Phytopathology Society.

Jones, M. B. 1986. Sulfur availability indexes. In *Sulfur in Agriculture*, ed. M. A. Tabatabai, pp. 549–566. Madison, WI: ASA, CSSA, and SSSA.

Kiniry, J. R., G. McCauley, Y. Xie, and J. G. Arnold. 2001. Rice parameters describing crop performance of four U. S. cultivars. *Agron. J.* 93:1354–1361.

Lindsay, W. L. 1979. *Chemical Equilibrium in Soils.* New York: John Wiley.

Malavolta, E., G. C. Vitti, C. A. Rosolem, N. K. Fageria, and P. T. G. Guimaraes. 1987. Sulphur responses of Brazilian crops. *J. Plant Nutr.* 10:2153–2158.

Marschner, H. 1995. *Mineral Nutrition of Higher Plants*, 2nd edition. New York: Academic Press.

Patrick, W. H. Jr., and D. S. Mikkelsen. 1971. Plant nutrient behavior in flooded soil. In *Fertilizer Technology and Use*, 2nd edition, ed. R. A. Olson, pp. 187–215. Madison, WI: SSSA.

Patrick, W. H. Jr., and C. N. Reddy. 1978. Chemical changes in rice soils. In *Soils and Rice*, ed. IRRI, pp. 361–379. Los Baños, Philippines, IRRI.

Ponnamperuma, F. N. 1972. The chemistry of submerged soils. *Adv. Agron.* 24:29–96.

Randall P. J., and C. W. Wrigley. 1986. Effects of sulfur supply on the yield, composition, and quality of grain from cereals, oilseeds, and legumes. *Adv. Cereals Sci. Tech.* 8:171–206.

Sinclair, T. R., and T. Horie. 1989. Leaf nitrogen, photosynthesis, and crop radiation use efficiency. A review. *Crop Sci.* 29:90–98.

Slaton, N. A., R. D. Cartwright, C. E. Wilson, Jr., and R. J. Norman. 2001. Symptoms and diagnosis of late-season sulfur deficiency of rice in Arkansas. Ark. Agric. Exp. Stn. Res. Ser. 485. In *B. R. Wells Rice Research Studies 2000*, eds. R. J. Norman and J. F. Meullenet, pp. 388–394. Fayetteville, AR: University of Arkansas.

Suzuki, A. 1995. Metabolism and physiology of sulfur. In *Science of Rice Plant: Physiology*, Vol. 2, eds. T. Matsuo, K. Kumazawa, R. Ishii, K. Ishihara, and H. Hirata, pp. 395–401. Tokyo: Food and Agricultural Policy Research Center.

Tabatabai, M. A., and J. M. Bremner. 1972. Forms of sulfur, and carbon, nitrogen and sulfur relationships in Iowa soils. *Soil Sci.* 114:380–386.

Takai, Y. 1978. An oxidation-reduction process in the soil under submerged conditions. In *Pedology on Paddy Soils*, ed. K. Kawaguchi, pp. 23–55. Tokyo: Kodansha.

Tisdale, S. L. 1970. The use of sulphur compounds in irrigated aridland agriculture. *Sulphur Inst. J.* 6:2–7.

Tisdale, S. L., W. L. Nelson, and J. D. Beaton. 1985. *Soil Fertility and Fertilizers*, 4th edition. New York: Macmillan.

Trudinger, P. A. 1986. Chemistry of the sulfur cycle. In *Sulfur in Agriculture*, eds. M. A. Tabatabai, pp. 1–22. Madison, WI: ASA, CSSA, and SSSA.

Wang, C. H. 1976. Sulphur fertilization of rice. In *The Fertility of Paddy Soils and Fertilizer Application for Rice*, ed. Food and Fertilizer Technology Center. Taipei, Taiwan: Food and Fertilizer Technology Center.

Wang, C. H., T. H. Liem, and D. S. Mikkelsen. 1976. *Sulfur Deficiency—A Limiting Factor in Rice Production in the Lower Amazon Basin. II. Sulfur Requirement for Rice Production.* New York: IRI Research Institute.

Wells, B. R., B. A. Huey, R. J. Norman, and R. S. Helms. 1993. Rice. In *Nutrient Deficiencies and Toxicities in Crop Plants*, ed. W. F. Bennett, pp. 15–19. St. Paul, MN: American Phytopathological Society.

Williams, C. H. 1967. Some factors affecting the mineralization of organic sulphur in soils. *Plant Soil.* 26:205–223.

Wilson, C. E., Jr., N. A. Slaton, R. J. Norman, and D. M. Miller. 2001. Efficient use of fertilizer. In *Rice Production Handbook*, ed. N. A. Slaton, pp. 51–74. Litttle Rock, AR: Arkansas Cooperative Extension Service.

Yamaguchi, J. 1997. Sulfur status of rice and lowland soils in West Africa. In *Plant Nutrition for Sustainable Food Production and Environment*, eds. T. Ando, K. Fujita, T. Mae, H. Matsumoto, S. Mori, and J. Sekiya, pp. 813–814. Dordrecht, The Netherlands: Kluwer.

Yoshida, S. 1981. *Fundamentals of Rice Crop Science*. Los Bãnos, Philippines: IRRI.

Zhao, F. J., P. J. A. Withers, E. J. Evans, J. Monagham, S. E. Salmin, P. R. Shewry, and S. P. McGrath. 1997. Sulphur nutrition: An important factor for the quality of wheat and rapeseed. In *Plant Nutrition for Sustainable Food Production and Environment*, eds. T. Ando, K. Fujita, T. Mae, H. Matsumoto, S. Mori, and J. Sekiya, pp. 917–922. Dordrecht, The Netherlands: Kluwer.

7 Zinc

7.1 INTRODUCTION

Zinc (Zn) is classified as a heavy metal among those micronutrients that are essential for plant growth and development. Zinc deficiency is more common than deficiency of other micronutrients, and it occurs mainly on highly weathered tropical soils, on soils low in organic matter content, and on sandy soils. Zinc deficiency is widespread in the world's rice-growing regions. Graham (2008) reported that half of the world's soils are intrinsically deficient in Zn. Similarly, Alloway (2008) also noted that Zn deficiency in crop production is found worldwide. Zinc deficiency in rice has been reported in South America (Fageria and Stone 2008; Fageria, Baligar, and Jones 2011), Asia (De Datta 1981; Singh 2008), Australia (Graham 2008), China (Zou et al. 2008), Turkey (Cakmak 2008), Europe (Sinclair and Edwards 2008), the United States (Brown 2008), and Africa (Waals and Laker 2008).

A global study by the Food and Agricultural Organization (FAO) of the United Nations showed that about 30% of the world's cultivated soils are Zn deficient (Sillanpaa 1982). In addition, about 50% of the world's soils that are used for cereal production contain low levels of plant-available Zn (Graham, Ascher, and Hynes 1992; Welch 1993). De Datta (1981) reported that Zn deficiency is the second most serious nutritional disorder limiting lowland rice yields in the Philippines. Zinc deficiency in crop plants reduces not only the grain yield but also the nutritional quality of the grain. Consumption of large quantities of cereal-based foods with low Zn concentrations, poor bioavailability of Zn, or both are thought to be major factors in the widespread occurrence of Zn deficiency in humans (Welch 1993).

Fageria and colleagues (2002, 2012) reported that the micronutrient deficiency in food crops, very common in various parts of the world, is due to low natural levels of micronutrients in the soils, use of high-yielding cultivars, liming of acidic soils, interactions among macro- and micronutrients, sandy and calcareous soils, increased use of high analysis fertilizers with low amounts of micronutrients, and decreased use of animal manures, composts, and crop residues.

Zinc deficiency is widespread in crops grown in soils of the world's arid and semi-arid regions, resulting in severe reductions in plant growth and yield (Graham 1984; Torun et al. 2003). Plants grown in soils exhibiting Zn deficiency are generally low in organic matter (<2%) and high in $CaCO_3$ (Sillanpaa and Vlek 1985). Low organic matter in soils has been proposed as a major reason for the widespread occurrence of Zn deficiency in plants (Cakmak et al. 1996). Several reports have noted the beneficial effects of soil application of organic materials on plants' Zn nutrition (Torun et al. 2003). Addition of organic materials to soil can increase concentration of soluble organic ligands that form readily soluble Zn complexes and enhance Zn uptake by plants (Torun et al. 2003). Other mechanisms responsible for increasing Zn uptake by plants are decreasing Zn's adsorption rate onto soil clay minerals (Shuman 1995), reducing soil pH during organic matter decomposition (Arnesen and Singh 1998), increasing Zn diffusion in soil (Alvarez, Rico, and Obrador 1997), and releasing Zn during organic matter degradation (Qiao and Ho 1997).

Zinc deficiency is most commonly found in a wide range of soils in tropical and temperate climates (Graham, Ascher, and Hynes 1992; Cakmak et al. 1996; Grewal and Graham 1999; Fageria, Baligar, and Clark 2002). Zinc deficiency is very common in upland rice grown on Brazilian Oxisols (Fageria 2000a, 2001). Figure 7.1 shows the response of two upland rice genotypes grown on Brazilian Oxisol to Zn application. Similarly, Figure 7.2 shows the response of lowland or flooded rice to Zn application. Soils with low available Zn are also widespread in Australia and constitute one of the limiting factors for sustainable crop production there (Graham 2008). Rashid and Fox (1992) studied Zn fertilizer requirements for wheat, sorghum, rice, millet, soybeans, cowpeas, and corn in Zn-deficient soils of Hawaii and reported that fertilizer Zn requirements for near-maximum grain yield were highest for cowpeas (7.5 mg kg^{-1} soil) and lowest for wheat (0.5 mg kg^{-1} soil). Data on Zn uptake and use efficiency are limited for food crops grown on Brazilian Oxisols. The objective of this chapter is to discuss zinc nutrition of rice in order to improve its efficiency and, consequently, rice yield.

FIGURE 7.1 Response of two upland rice genotypes on Brazilian Oxisol to an application of zinc.

FIGURE 7.2 Response of lowland rice to zinc fertilization.

7.2 CYCLE IN SOIL–PLANT SYSTEM

Availability of Zn to plants, or its concentration in the soil solution, is regulated by sorption–desorption reactions at the surface of soil colloids (Fageria, Slaton, and Baligar 2003). Desorption controls the amount and release rate of Zn into the soil solution for plant uptake. Desorption of Zn into the soil solution is controlled by how strongly the Zn is adsorbed onto the surface of soil colloids. Other forms of Zn are also associated with organic matter, carbonates, and oxide minerals, and Zn is found in primary and secondary minerals. Thus, the availability of Zn is influenced by a number of soil characteristics, including soil pH, organic matter content, $CaCO_3$ content, cation exchange capacity, clay content and mineralogy, and the quantity and types of Fe, Al, and Mn oxides (Harter 1991; Hazra and Mandal 1996; Singh, McLaren, and Cameron 1997). After flooded soils are drained, they contain relatively large amounts of amorphous Fe and Mn oxides that have larger surface area and greater adsorption capacity than their crystalline forms (Sah and Mikkelsen 1986; Quang and Dufey 1995). Zinc uptake by rice depends not only on its concentration in the soil solution but also on other factors, particularly the concentrations of Fe^{2+} and Mn^{2+} present. High concentrations of Fe^{2+} and Mn^{2+} in the soil solution negatively impact Zn absorption (Sajwan and Lindsay 1986; Mandal, Hazra, and Mandal 2000). High soil pH and the presence of free $CaCO_3$ decrease the availability of Zn in soils. The solubility of soil Zn is highly pH dependent and decreases 100-fold for each unit increase in pH (Tisdale, Nelson, and Beaton 1985). The uptake, translocation, metabolism, and plant use of Zn are inhibited by high P availability or high rates of P fertilizer applications (Lindsay 1979).

Unlike that of the redox elements (Fe and Mn), the concentration of Zn in the soil solution after flooding generally decreases with time. However, immediately after flooding, Zn concentrations may increase briefly (Mikkelsen and Kuo 1976; Gilmour 1977). A decrease in soil solution Zn concentration may be due to precipitation of $ZnFe_2O_4$ from the increased Fe solubility after flooding (Sajwan and Lindsay 1986) or precipitation of ZnS under highly reduced soil conditions (Kittrick 1976).

Plant uptake of Zn depends not only on the plant species, cultivar within species, and plant age but also on the predominant forms of Zn in the soil (i.e., amount of Zn associated with water-soluble and exchangeable Zn fractions). Major factors affecting the availability of Zn include the soil pH, total soil Zn, Zn fertilizer source, soil organic matter content, and soil texture (Adriano 1986; Chlopecka and Adriano 1996). Of these, soil pH has the greatest impact on Zn availability in most soils. Zinc deficiency is most likely to occur on coarse-textured soils with high pH and low soil Zn, soils disturbed by land leveling, and highly eroded soils (Fageria, Slaton, and Baligar 2003).

The availability of Zn to plants decreases with increases in soil pH. Most pH-induced Zn deficiency occurs within the range 6.0–8.0, and calcareous soils are particularly prone to this nutritional problem (Tisdale, Nelson, and Beaton 1985). Liming acid soils, especially ones low in Zn, will reduce uptake of Zn^{2+}. This depressive action usually is attributed to the effect that increasing pH has on lowering Zn^{2+} solubility. However, it is possible that some Zn^{2+} could be adsorbed on the surface of freely added particles of liming agents such as $CaCO_3$ (Tisdale, Nelson, and Beaton 1985). Barber (1995) noted that soil pH had strong effects on Zn adsorption. This occurred because concentrations of Zn in soil solutions decreased 30-fold for every unit pH increase between the ranges of 5.0 and 7.0. Reduction of Zn concentrations in soil solution was believed to be due to adsorption on hydrous oxide surfaces (Fageria, Baligar, and Clark 2006). Table 7.1 provides information on decreases in Zn uptake of four crop species grown on Brazilian Oxisol with increased levels of $CaCO_3$. Similarly, results in Table 7.2 show a significant decrease in Zn uptake by upland rice with increasing soil pH in Brazilian Oxisol. Figure 7.3 shows the micronutrient cycle, including zinc, in an upland rice soil–plant system.

7.3 FUNCTIONS

Zinc is an essential micronutrient for plant growth and is an important catalytic component of several enzymes (Fageria and Baligar 2005; Fageria, Baligar, and Jones 2011). Zinc is closely involved in the N metabolism of plants, and in Zn-deficient plants, protein synthesis is significantly reduced (Mengel et al. 2001). Zinc deficiency reportedly reduces root system development in crop plants (Fageria 2004, 2009). This may adversely affect absorption of water and nutrients and, consequently, growth and yield. Figure 7.4 shows root growth of upland rice at two Zn levels. Root growth was much better at the 20 mg Zn kg^{-1} treatment compared with the control treatment. Epstein and Bloom (2005) reported that plant flowering and fruiting are significantly reduced under conditions of severe Zn deficiency.

TABLE 7.1
Zinc Uptake by Four Crop Species Grown with Different Levels of Lime and P

Lime Level (g kg^{-1} soil)	P Level (mg kg^{-1} soil)	Upland Rice (g pot^{-1})	Wheat (g pot^{-1})	Common Bean (g pot^{-1})	Maize (g pot^{-1})
0	0	201	69	458	408
0	50	3324	758	982	346
0	175	3910	1090	683	1280
2	0	142	28	158	270
2	50	2740	410	373	796
2	175	1871	496	393	1522
4	0	39	12	130	161
4	50	902	232	324	345
4	175	1226	292	379	739
F-test					
Lime		*	*	*	*
Phosphorus		*	*	*	*
Lime × P		*	*	*	NS

Source: Fageria, N. K., F. J. P. Zimmermann, and V. C. Baligar. 1995. Lime and phosphorus interactions on growth and nutrient uptake by upland rice, wheat, common bean, and corn in an Oxisol. *J. Plant Nutr.* 18:2519–2532.

*,NS Significant at 1% probability level and not significant, respectively.

TABLE 7.2
Zinc Uptake by Upland Rice as Influenced by Soil pH in Brazilian Oxisol

Soil pH in H_2O	Zn Uptake (μg plant^{-1})
4.6	1090
5.7	300
6.2	242
6.4	262
6.6	163
6.8	142
R^2	0.98*

Source: Fageria, N. K. 2000b. Upland rice response to soil acidity in Cerrado soil. *Pesq. Agropec. Bras.* 35:2303–2307.

*Significant at the 1% probability level.

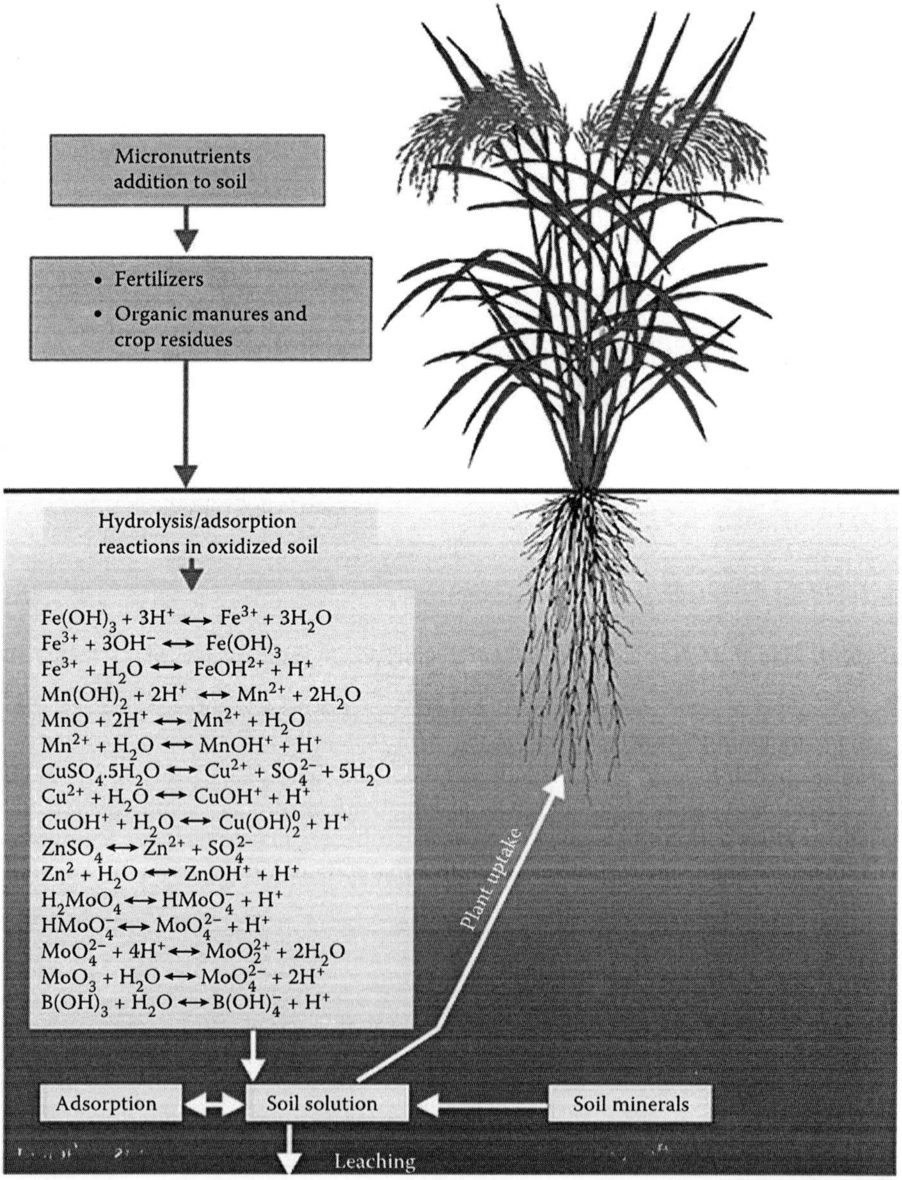

FIGURE 7.3 Micronutrient cycle, including zinc, in soil–plant system of upland rice or oxidized soil.

Zinc is a cofactor for several enzymes that are involved with N metabolism (e.g., glutamate dehydrogenase) and anaerobic metabolism (e.g., alcohol dehydrogenase). The reduction of acetaldehyde to ethanol in anaerobic metabolism requires alcohol dehydrogenase. The alcohol dehydrogenase activity of seedling rice roots increases dramatically after flooding and remains high for several weeks compared with rice

FIGURE 7.4 Upland rice genotype BRA 042160 growth at 0 and 20 mg Zn kg^{-1} soil.

seedlings that are not flooded (Pedrazzini and McKee 1984). When Zn is deficient, the activity of alcohol dehydrogenase is depressed, anaerobic root metabolism decreases, and the ability of the seedling rice to withstand anaerobic soil conditions is reduced (Moore and Patrick 1988). This is one reason why Zn deficiency symptoms are more dramatic after flooding rather than before flooding. When Zn deficiency is diagnosed, draining the flood is commonly recommended to aid in plant recovery (Wilson et al. 2001). Removal of the flood allows seedlings to resume aerobic respiration as oxygen is reintroduced into the soil. The activity of glutamate dehydrogenase was not affected by Zn fertilization in studies conducted by Moore and Patrick (1988).

7.4 DEFICIENCY SYMPTOMS

Rice is considered susceptible to Zn deficiency. The symptoms of Zn deficiency in rice are well documented. Zinc, like the other micronutrients, is not very mobile within the plant; thus, deficiency symptoms are first observed in the youngest leaves. Zinc deficiency most commonly affects seedling rice plants, but if the deficiency is mild and not corrected, symptoms can also affect plants in the reproductive growth phase (Fageria, Slaton, and Baligar 2003). In the early stages of Zn deficiency, the youngest leaves usually become chlorotic, especially at the leaf base. As Zn deficiency progresses, the midribs and base of older leaves may also turn yellow or pale green with brown blotches and streaks appearing on the lower leaves (Yoshida 1981). Brown spots usually develop near the tip of the leaf blade as yellowing begins. Leaf collars may also be stacked as internode elongation is inhibited (Wilson et al. 2001). Zinc deficiency tends to be more severe where high rates of N and P are applied (Fageria, Baligar, and Jones 2011). Zinc-deficient rice plants do not respond to N fertilization (Cheaney and Jennings 1975). Adequate Zn levels in the soil increase tillering and consequently the number of panicles per unit area of lowland rice (Fageria 2001). The application of high rates of P fertilizer is also known to aggravate Zn deficiency. The major reasons for P-induced Zn deficiency are believed to be the formation of Zn phosphate in soil solutions and/or an inhibitory effect of the excessive P on the metabolic functions of the Zn within the plant (Shimada 1995). Color photographs of rice plant zinc deficiency symptoms appear in reports by Cheaney and Jennings (1975), Yoshida (1981), Fageria (1984), Wells et al. (1993), and Fageria and Barbosa Filho (1994). In addition, Figures 7.5 and 7.6 show Zn deficiency symptoms in upland and lowland rice grown on Brazilian Oxisols and Inceptisols, respectively. Figure 7.7 shows the growth of upland rice genotypes at two Zn levels on Brazilian Oxisol.

7.5 UPTAKE IN PLANT TISSUE

Nutrient uptake by crop plants is influenced by climatic, soil, and plant factors. Climatic factors that influence nutrient uptake are temperature, precipitation, and solar radiation. Soil factors that determine nutrient uptake are moisture, bulk density, texture, organic matter content, pH, cation exchange capacity, nutrient concentration in the soil solution, and soil biological properties (Fageria et al. 2012). Plant factors that determine nutrient uptake are root system, genetic makeup of each plant species, and variety within species; all of these affect a plant's capacity to absorb and store nutrients. Plant breeding and crop nutrition are therefore inextricably linked in order to produce the most nutritious crops possible (Fageria et al. 2012).

Zinc concentration in upland rice at different growth stages was studied, and the results were compared with other crops such as corn, dry beans, and soybeans. Zinc concentration (zinc content per unit of shoot dry weight) in the shoot of corn and rice is significantly decreased ($P < 0.01$) with advanced plant age (Figure 7.8). In dry beans and soybeans, Zn concentration increased quadratically with advanced plant age; however, this increase was not significant. At 18 days after sowing, zinc concentration in the corn shoot was 36 mg kg^{-1}, and it decreased to 15 mg kg^{-1} at harvest.

FIGURE 7.5 (See color insert.) Zinc deficiency symptoms in upland rice grown on Brazilian Oxisol.

At 19 days after sowing, Zn concentration in the rice shoot was 40 mg kg^{-1}, and it decreased to 26 mg kg^{-1} at harvest. Similarly, Zn concentration in the shoot of dry beans decreased from 43 mg kg^{-1} at 15 days after sowing to 19 mg kg^{-1} at harvest. At 27 days after sowing, Zn concentration in the soybean shoot was 31 mg kg^{-1}, and it decreased to 16 mg kg^{-1} at harvest. The decrease in Zn concentration in corn and rice with advanced plant age is associated with an increase in the shoot dry weight up to a certain growth stage. Such decreases in nutrient concentrations have been reported in annual crops and are known as the dilution effect (Fageria, Baligar, and Jones 2011). The values of Zn concentration reported in Figure 7.8 can be used in the interpretation of plant tissue analysis results for Zn at different growth stages for rice, corn, dry beans, and soybeans. Fageria (2000a) reported more or less similar concentrations of Zn in rice, corn, dry beans, and soybeans in the early stage of plant growth. The significant variation in Zn concentration suggests that plant tissue analysis should be done during different growth stages for better interpretation of such results.

Fageria, Slaton, and Baligar (2003) reviewed the literature on Zn concentration in rice straw. Zinc concentration of plants typically ranges from 30 to 100 mg Zn kg^{-1}, depending upon species (Shimada 1995). Zinc deficiency of rice occurs primarily on seedling and tillering stages of rice; hence, most studies evaluating critical tissue Zn concentrations have emphasized these growth stages. Zinc deficiency of rice seedling is likely when leaf and/or whole plant concentrations are <15 mg Zn kg^{-1} (Adriano 1986).

FIGURE 7.6 (See color insert.) Zinc deficiency symptoms in upland and lowland rice grown on Brazilian Oxisol and Inceptisol, respectively.

During the vegetative growth stages, the plant part sampled is not critical for rice. Although Zn is considered immobile in the plant, whole seedlings or individual leaves have similar Zn concentrations (Fageria, Slaton, and Baligar 2003). Fageria, Baligar, and Jones (2011) reported that the Zn sufficiency range in rice shoots at tillering was 20–150 mg kg^{-1}. Yoshida (1981) reported plant tissue analysis criteria for classifying the Zn nutritional status of rice. In his system, whole seedling Zn concentrations <10 mg kg^{-1}, 10–15 mg kg^{-1}, 15–20 mg kg^{-1}, and >20 mg kg^{-1} are considered deficient, probably deficient, low, and sufficient, respectively. Research conducted all over

FIGURE 7.7 Growth of upland rice genotypes at 0 and 20 mg Zn kg^{-1} soil in a greenhouse experiment (top) and growth of two genotypes at two Zn levels (bottom). The soil used in these studies was Oxisol.

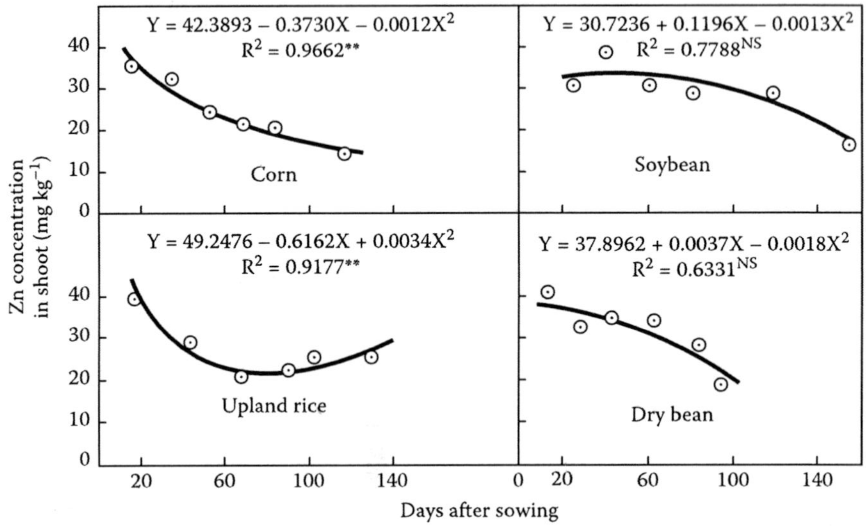

FIGURE 7.8 Relationship between plant age and zinc concentration in the shoots of four food crops. (From Fageria, N. K., M. P. Barbosa Filho, and A. B. Santos. 2008. Growth and zinc uptake and use efficiency in food crops. *Commun. Soil Sci. Plant Anal.* 39:2258–2269.)

the world agrees that seedling Zn concentrations <15–20 mg kg^{-1} are low or deficient and require Zn fertilization for optimum rice growth.

Zinc uptake (concentration × shoot dry weight) in the shoots of corn, rice, soybeans, and dry beans significantly increased with advanced plant age as follows: up to 89 days after sowing in corn, 108 days after sowing in soybeans, 104 days after sowing in rice, and 72 days after sowing in dry beans (Figure 7.9). After these ages, Zn uptake in the shoots of the four crop species decreased. Fageria, Baligar, and Jones (2011) and Karlen, Flannery, and Sadler (1988) reported similar patterns of Zn uptake in upland rice and corn, respectively. Zinc uptake more or less followed the shoot dry matter accumulation pattern in the four crops (Figure 7.10). With advanced plant age, shoot dry weight of the four crop species significantly increased ($P < 0.01$) in an exponential quadratic fashion up to a certain age, and then it decreased (Figure 7.10). The shoot dry weight of corn increased up to 89 days after sowing and then decreased. In the case of soybeans, the shoot dry weight increased up to 115 days after sowing and then decreased. Rice and dry bean shoot dry weight increased up to 107 and 78 days after sowing, respectively, and then decreased. The increase in the shoot dry weight of corn was linear from 35 to 89 days after sowing. Similarly, the increase in the shoot dry weight of soybeans was linear from 27 to 115 days after sowing. The increase of the shoot dry weight of rice was linear from 30 to 107 days after sowing, and in case of dry beans, the linear increase in the shoot dry weight occurred between 20 and 78 days after sowing.

During these growth stages, the increase in the shoot dry weight of these crops was associated with an increase in the number of tillers and leaves in case of rice (Fageria, Barbosa Filho, and Carvalho 1982), an increase in the weight of culm and leaves of corn, and an increase in trifoliate and branches of dry beans and soybeans

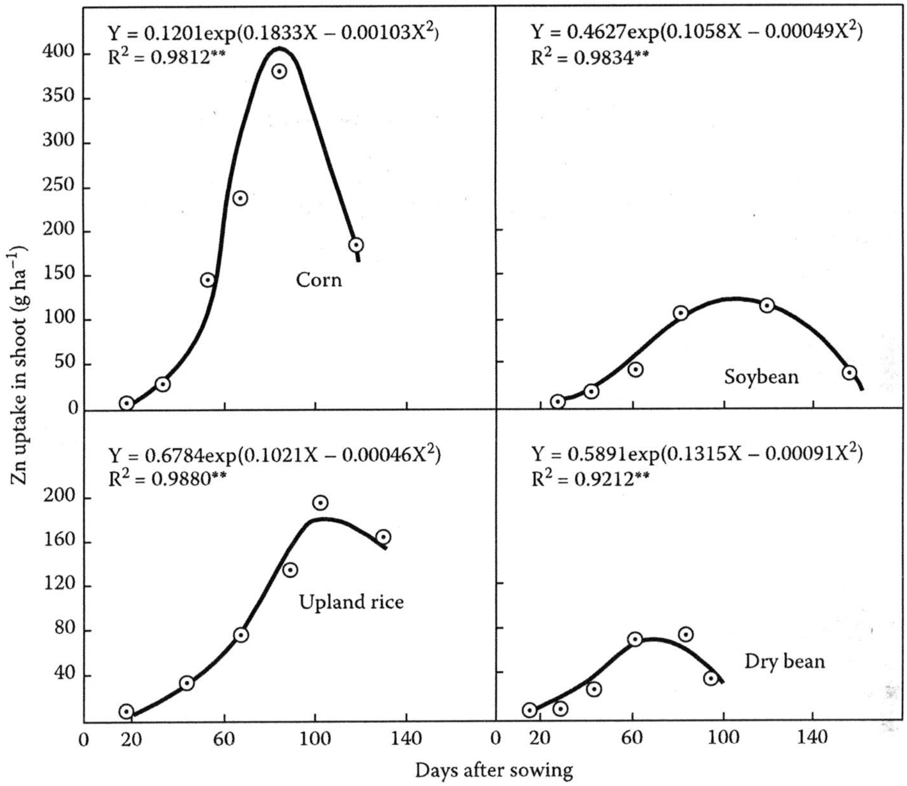

FIGURE 7.9 Relationship between plant age and zinc uptake in the shoots of four food crops. (From Fageria, N. K., M. P. Barbosa Filho, and A. B. Santos. 2008. Growth and zinc uptake and use efficiency in food crops. *Commun. Soil Sci. Plant Anal.* 39:2258–2269.)

(Fageria, Baligar, and Clark 2006). The practical application of quantifying linear growth stages of these crops reveals that more nutrients will be absorbed or required during this growth duration.

The decrease in the shoot dry weight 89 days after sowing in corn and 107 days after sowing in rice was associated with the translocation of photosynthetic products to the grain. Fageria and colleagues (2006, 2011) reported a more or less similar reduction in the shoot dry weight of upland rice and corn during the reproductive growth stage. The decrease in the shoot dry weight of dry beans (115 days after sowing) and soybeans (78 days after sowing) was also associated with the translocation of photoassimilates to the grain as well as the fall of some of the leaves during maturity. Fageria, Baligar, and Clark (2006) reported a similar decrease in the shoot dry weight during the pod and grain formation growth stages in these crops.

Corn had the maximum shoot dry weight, followed by that of rice. Dry beans had a minimum shoot dry weight, and soybeans had an intermediate level of shoot dry weight production. The maximum shoot dry weight of corn was associated with the C_4 photosynthetic pathway of this crop. Rice, dry beans, and soybeans have C_3

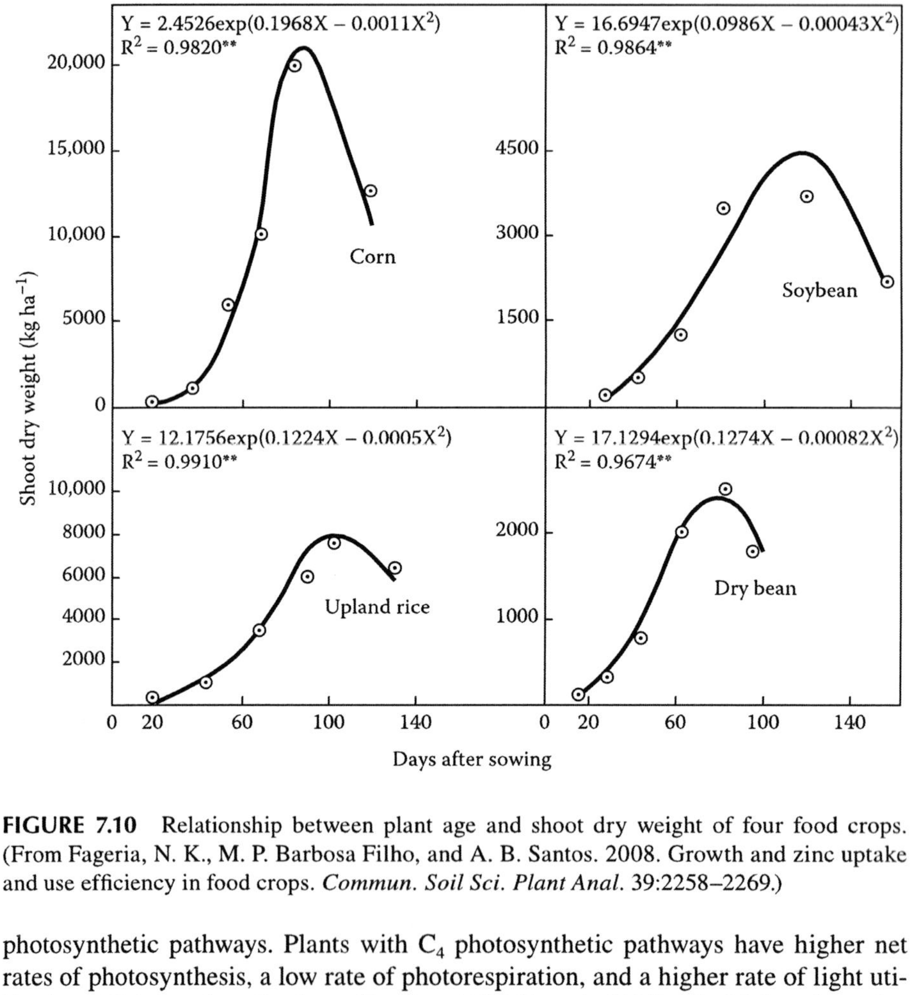

FIGURE 7.10 Relationship between plant age and shoot dry weight of four food crops. (From Fageria, N. K., M. P. Barbosa Filho, and A. B. Santos. 2008. Growth and zinc uptake and use efficiency in food crops. *Commun. Soil Sci. Plant Anal.* 39:2258–2269.)

photosynthetic pathways. Plants with C_4 photosynthetic pathways have higher net rates of photosynthesis, a low rate of photorespiration, and a higher rate of light utilization efficiency than C_3 plants (Ludlow 1985; Fageria 1992). The higher shoot dry weight of cereals (rice and corn) compared with legumes (soybeans and dry beans) was also associated with a lower respiration rate of cereals compared with legumes (Shinano et al. 1991). It has also been noted that low productivity of legumes compared with cereals is associated with higher protein and lipid contents in legume seeds. In legumes, higher quantities of these compounds require larger amounts of energy, which may be responsible for lower grain yield (Yamaguchi 1978).

The concentration and uptake of zinc in lowland rice during the growth cycle were also studied (Table 7.3). Zinc concentration decreased significantly in a quadratic fashion with advanced plant age. Zinc concentration was higher in straw compared with grain at physiological maturity. The decrease in Zn concentration with advanced plant age was related to an increase in straw weight with advanced plant age. However, Zinc uptake increased significantly in a quadratic fashion with advanced plant age. It was 19 g ha^{-1} at 22 days after sowing and increased to 546 g ha^{-1} at physiological maturity. In the case of upland rice, the increase in Zn

TABLE 7.3
Zinc Concentration and Uptake in Straw and Grain of Lowland Rice during Growth Cycle

Plant Age/Growth Stage	Zn Concentration (mg kg^{-1})	Zn Uptake (g ha^{-1})
22 (IT)	52.9	19.0
35 (AT)	43.3	48.8
71 (IP)	33.1	195.6
97 (B)	38.8	410.1
112 (F)	42.2	555.4
140 (PM)	56.8	546.1
140 (grain)	35	35
F-test	*	*
CV (%)	10	15

Regression Analysis

Plant age vs. Zn conc. in straw $(Y) = 68.18 - 0.87X + 0.0057X^2$, $R^2 = 0.80$*

Plant age vs. Zn uptake in straw $(Y) = -149.81 + 6.45X - 0.0083X^2$, $R^2 = 0.92$*

Note: IT, initiation of tillering; AT, active tillering; IP, initiation of primordial floral; B, booting; F, flowering; PM, physiological maturity.

*Significant at the 1% probability level.

uptake with advanced plant age was related to a similar increase in straw weight as discussed earlier in this chapter.

7.6 USE EFFICIENCY

Zinc use efficiency in the shoot (kg shoot dry weight g^{-1} Zn accumulated in shoot) was significantly influenced in four crop species, except dry beans (Figure 7.11). It was linear in corn and exponentially quadratic in upland rice in relation to plant age. In corn, Zn use efficiency varied from 26 to 69 kg g^{-1} at 18 to 119 days of plant age. In upland rice, it varied from 25 to 40 kg g^{-1} at 19 to 130 days of plant age. In soybeans and dry beans, Zn use efficiency was similar to responses with advanced plant age. In soybeans, the values of Zn use efficiency varied from 33 to 63 kg g^{-1} at 27 and 158 days of plant age. Similarly, in dry beans, the values of Zn use efficiency were 24 to 51 kg g^{-1} at 15 to 96 days of plant age. Hence, the Zn use efficiency in food crops can be classified in the order of corn > soybeans > dry beans > upland rice. In corn, higher Zn use efficiency in shoot dry weight production was associated with high dry matter production. Zinc use efficiency data for food crops are scarce; therefore, the results of this study cannot be compared with any published work.

The grain yield of the four food crops was in the order of corn > upland rice > dry bean > soybean (Figure 7.12). Cereals had a higher grain yield than legumes. The zinc concentration in corn grain was 18 mg kg^{-1}; in upland rice grain, 30 mg kg^{-1}; in dry bean grain, 39 mg kg^{-1}; and in soybean grain, 55 mg kg^{-1}. Fageria, Baligar, and Jones (2011) and Fageria (1989) reported more or less similar concentrations of Zn

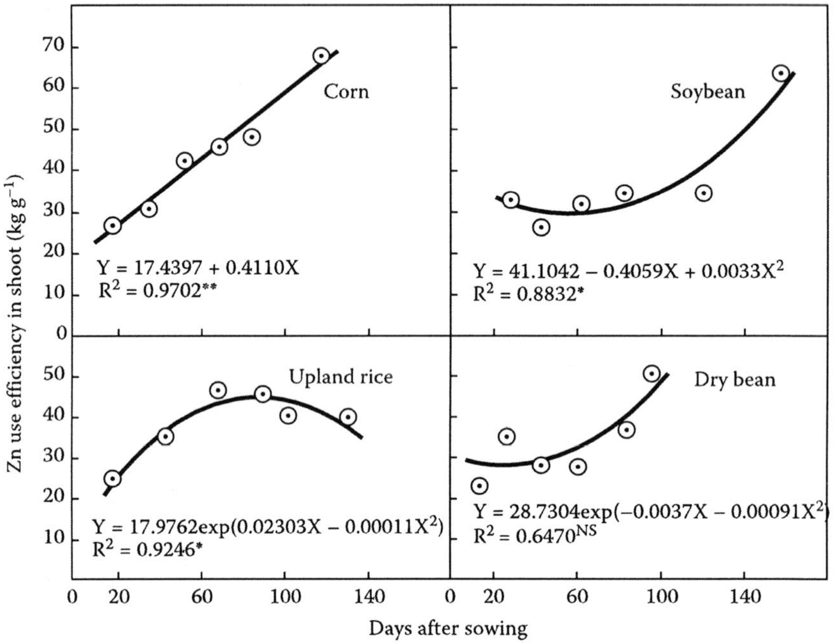

FIGURE 7.11 Relationship between plant age and zinc uptake in the shoots of four food crops. (From Fageria, N. K., M. P. Barbosa Filho, and A. B. Santos. 2008. Growth and zinc uptake and use efficiency in food crops. *Commun. Soil Sci. Plant Anal.* 39:2258–2269.)

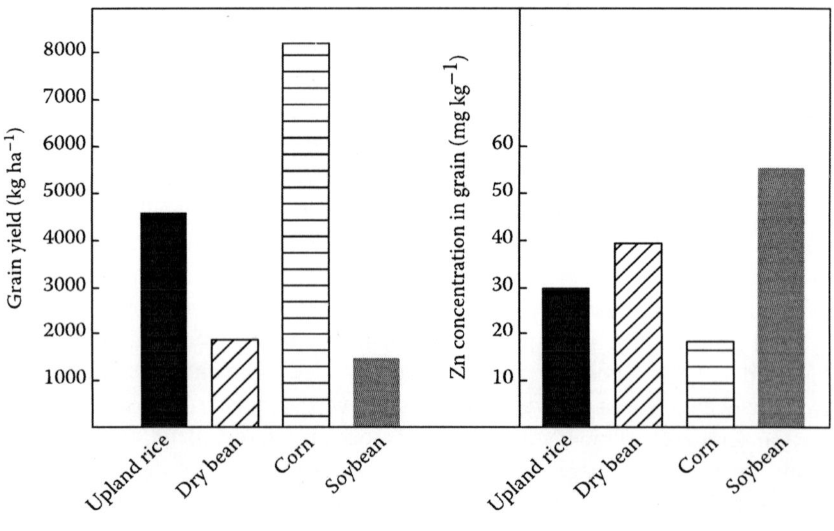

FIGURE 7.12 Grain yield and zinc concentration in the grain of four food crops. (From Fageria, N. K., M. P. Barbosa Filho, and A. B. Santos. 2008. Growth and zinc uptake and use efficiency in food crops. *Commun. Soil Sci. Plant Anal.* 39:2258–2269.)

in the grain of upland rice and dry beans. Zinc concentration in legume crops was higher than that in cereals. High grain Zn concentration is considered a desirable quality that could increase the nutritional value of the grain for humans (Graham, Ascher, and Hynes 1992; Cakmak et al. 1996). Also, high Zn concentration in seed is also a desirable trait for seedling vigor and grain yield of the next crop when re-sown on Zn-deficient soil (Graham, Ascher, and Hynes 1992; Graham and Rengel 1993; Grewal and Graham 1999). Graham, Ascher, and Hynes (1992) also reported that high zinc content in grain is under genetic control, is not tightly linked to agronomic zinc efficiency traits, and may have to be selected for independently.

Zinc uptake in the grain of the four crops was in the order of corn > upland rice > soybeans > dry beans (Figure 7.13). The higher Zn uptake in corn and upland rice was associated with higher grain yield. In soybeans, higher Zn uptake compared with dry beans was associated with high Zn concentration in the seeds of soybeans. Zinc use efficiency in grain production was higher for corn and lower for dry beans and soybeans. The higher Zn efficiency of cereals compared with legumes was associated with higher grain yields of corn and upland rice. Fageria (1992) reported that higher crop yield is associated with higher nutrient use efficiency.

The Zn recovery efficiency in upland rice genotypes was studied (Table 7.4). Zinc recovery efficiency (ZnRE) varied from 8.3 to 23.1% with an average of 13% across 10 genotypes (Table 7.4). Information on Zn recovery efficiency by lowland rice genotypes is limited; therefore, these results cannot be compared with any published work. However, Mortvedt (1994) reported that crop recovery of applied micronutrients is relatively low (5–10%) compared with macronutrients (10–50%). He further states that such low recovery of applied micronutrients is due to their uneven distribution in soil because of low application rates, reaction with soil to form unavailable products, and low mobility in soil.

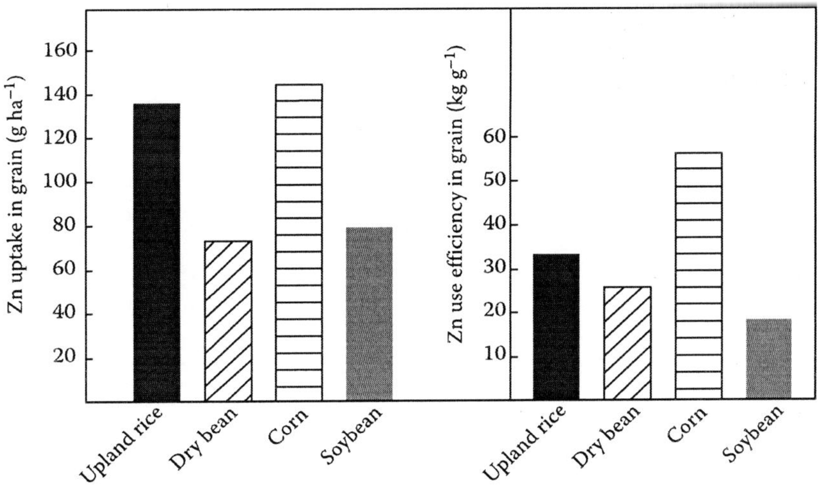

FIGURE 7.13 Zinc uptake and use efficiency in the grain of four food crops. (From Fageria, N. K., M. P. Barbosa Filho, and A. B. Santos. 2008. Growth and zinc uptake and use efficiency in food crops. *Commun. Soil Sci. Plant Anal.* 39:2258–2269.)

TABLE 7.4
Zinc Recovery Efficiency in Upland Rice Genotypes

Genotype	Zinc Recovery Efficiency (%)
Bonança	15.0ab
Caiapó	14.6ab
Canastra	12.4ab
Carajas	13.3ab
Carisma	11.8ab
CNA 8540	10.4b
CNA 8557	8.5b
Guarani	23.1a
Maravilha	13.1ab
IR42	8.3b
Average	13.1
F-test	
Zn level	NS
Genotype	*
Zn × G	NS
CV (%)	32

Source: Fageria, N. K., and V. C. Baligar. 2005. Growth components and zinc recovery efficiency of upland rice genotypes. *Pesquisa Agropecuaria Brasileira*. 40:1211–1215.

Note: Means in the same column followed by the same letter are statistically not different at the 5% probability level by the Tukey's test.

*,NS Significant at the 5% probability level and nonsignificant, respectively.

Zinc use efficiency can be defined in several ways (Fageria, Barbosa Filho, and Santos 2008). Fageria (2009) presented the following equations for calculating these efficiencies:

Agronomic efficiency (AE) = (Grain yield of Zn fertilized pots – grain yield of Zn unfertilized pots)/(quantity of Zn applied) (7.1)

Physiological efficiency (PE) = (Grain and straw yield of Zn fertilized pots – grain and straw yield of Zn unfertilized pots)/(uptake of Zn in grain and straw of Zn fertilized pots – uptake of Zn in grain and straw of unfertilized pots) (7.2)

Agrophysiological efficiency (APE) = (Grain yield of Zn fertilized pots – grain yield of Zn unfertilized pots)/(uptake of Zn in grain and straw of Zn fertilized pots – uptake of Zn in grain and straw of unfertilized pots) (7.3)

TABLE 7.5
Zinc Use Efficiency in Lowland Rice as Influenced by Zn Application Rate

Zn Rate ($mg\ kg^{-1}$)	Agronomic Efficiency ($\mu g\ \mu g^{-1}$)	Physiological Efficiency ($\mu g\ \mu g^{-1}$)	Agrophysiological Efficiency ($\mu g\ \mu g^{-1}$)	Apparent Recovery Efficiency (%)	Utilization Efficiency ($\mu g\ \mu g^{-1}$)
5	815	10326	4881	16.49	1742
10	450	12756	6457	6.86	867
20	266	13072	6440	4.41	535
40	109	10276	4519	2.42	245
80	58	6739	3985	1.45	98
120	36	2106	1033	3.51	74
Average	289	9213	4553	5.86	594
F-test	**	*	NS	**	**

Source: Fageria, N. K., A. B. Santos, and T. Cobucci. 2011. Zinc nutrition of lowland rice. *Commun. Soil Sci. Plant Anal.* 42:1719–1727.

*,**,NS Significant at the 5 and 1% probability levels and nonsignificant, respectively.

Apparent recovery efficiency (ARE) = (Uptake of Zn in grain and straw of Zn fertilized pots – uptake of Zn in grain and straw of unfertilized pots)/(quantity of Zn applied) (7.4)

Utilization efficiency (EU) = PE × ARE. (7.5)

Zn efficiencies were calculated for lowland rice (Table 7.5). AE, PE, ARE, and EU were significantly decreased with increasing Zn rate in the soil. However, APE had a significant quadratic increase with increasing Zn levels. Higher nutrient use efficiency at lower levels is common due to efficient utilization of nutrients at lower levels (Fageria 1992). The decrease in nutrient use efficiency with increasing nutrient rate is also related to a progressive decrease in dry matter or grain yield with increasing nutrient rate in nutrient-deficient soils or where crop response to applied nutrient is obtained (Fageria, Baligar, and Jones 2011). Genc, McDonald, and Graham (2002) also reported higher Zn use efficiency at the lowest Zn level in barley genotypes. ARE of Zn across the Zn rates was about 6%. Information on Zn use efficiency is limited; therefore, results cannot be compared with published works. However, Mortvedt (1994) reported that crop recovery of micronutrients is relatively low (5–10%) compared with that of macronutrients (10–50%) because of poor distribution from low rates applied, fertilizer reactions with soil to form unavailable products, and low mobility in soil. Results of ARE are within this range (Table 7.5).

7.7 HARVEST INDEX

The zinc harvest index (ZnHI; Zn uptake in grain/Zn uptake in grain plus straw) is significantly influenced by soil Zn levels and genotypes and by interaction between soil Zn levels and genotypes (Table 7.6). The ZnHI indicates the quantity

TABLE 7.6
Zinc Harvest Index in 10 Upland Rice Genotypes

	Zinc Harvest Index	
Genotype	**0 mg Zn kg^{-1}**	**10 mg Zn kg^{-1}**
Bonança	0.60abc	0.25ab
Caiapó	0.62ab	0.25ab
Canastra	0.64ab	0.35a
Carajas	0.76a	0.32ab
Carisma	0.60abc	0.31ab
CNA 8540	0.68a	0.37a
CNA 8557	0.46bc	0.30ab
Guarani	0.70a	0.22b
Maravilha	0.57abc	0.26ab
IR42	0.40c	0.21b
Average	0.60	0.28
F-test		
Zn level	*	
Genotype	*	
Zn × XG	*	
CV (%)	12	

Source: Adapted from Fageria, N. K., and V. C. Baligar. 2005. Growth components and zinc recovery efficiency of upland rice genotypes. *Pesquisa Agropecuaria Brasileira*. 40:1211–1215.

Note: Means in the same column followed by the same letter are not statistically different at the 5% probability level by the Tukey's test.

*Significant at the 1% probability levels by F-test.

of Zn translocated to grain from the amount of Zn absorbed by the roots. In Table 7.6, the ZnHI varies from 0.40–0.76 at the low Zn level to 0.21–0.37 at the high Zn level. Average values across the genotypes were 0.60 and 0.28, respectively. These genotypes recorded higher ZnHI at low soil Zn levels than at high soil Zn levels; such differences are due to higher Zn accumulation in the shoot at the higher Zn level.

7.8 MANAGEMENT PRACTICES

Production potentials of many of the world's soils are decreased by low supplies of mineral nutrients because of adverse physical and chemical constraints (Fageria, Baligar, and Clark 2002). Major chemical (salinity, acidity, elemental deficiencies and toxicities, low organic matter) and physical (bulk density, hardpan layers, structure and texture, surface sealing and crusting, water holding capacity, water logging, drying, aeration) constraints affect transformation (mineralization, immobilization), fixation (adsorption, precipitation), and leaching or surface runoff of indigenous and added fertilizer nutrients (Fageria, Baligar, and Clark 2002; Fageria and Stone

2008). Strategies to improve micronutrient uptake in crop plants can be divided into two groups. In the first group, bioavailability of micronutrients can be improved by adopting practices such as the use of adequate rate, source, and methods of application. Fertilizer management practices (source, rate and method of placement, application time) should be optimized based on soil, plant, and climatic factors that reduce mineral losses (leaching, runoff, fixation) (Fageria, Baligar, and Clark 2002). Improvement and consideration of these factors will enhance recovery of added fertilizer minerals (Fageria, Baligar, and Clark 2002). Cakmak (2008) also reported that agronomic biofortification is of great importance in Zn enrichment of food crop seeds. In the second group are the crop species or genotypes within species that are efficient in micronutrient uptake and translocation of a large portion of the micronutrients in the grain (Fageria, Barbosa Filho, and Santos 2008; Fageria 2009). For rice, zinc management practices that can improve Zn use efficiency and yield include effective source, appropriate method and timing of application, adequate rate, and the use of Zn-efficient genotypes. These management practices are discussed in the following sections.

7.8.1 Effective Sources

There are many zinc fertilizers available, and selecting an effective source depends on price and availability in the local market. Various Zn fertilizers and their solubility in water are listed in Table 7.7. The most common commercially manufactured granular Zn fertilizers are Zn sulfates, oxides, oxysulfates, lignosulfunates, and a number of organic chelated materials such as ZnEDTA and ZnHEDTA. Excellent reviews of the manufacturing and properties of Zn fertilizers were given by Tisdale, Nelson, and Beaton (1985), Foth and Ellis (1988), and Martens and

TABLE 7.7
Principal Zinc Carriers

Common Name	Formula	Zn (%)	Solubility in Water
Zinc sulfate (monohydrate)	$ZnSO_4{\cdot}H_2O$	36	Soluble
Zinc sulfate (heptahydrate)	$ZnSO_4{\cdot}7H_2O$	23	Soluble
Zinc oxide	ZnO	78	Insoluble
Zinc carbonate	$ZnCO_3$	52	Insoluble
Zinc frits	Silicates	Varies	Slightly soluble
Zinc EDTA chelates	$Na_2ZnEDTA$	14	Soluble
Zinc HEDTA chelate	NaZnHEDTA	9	Soluble
Zinc chloride	$ZnCl_2$	47	Soluble

Source: Mortvedt, J. J. 1994. Needs for controlled availability micronutrient fertilizers. *Fert. Res.* 38:213–221; Fageria, N. K., M. F. Moraes, E. P. B. Ferreira, and A. M. Knupp. 2012. Biofortification of trace elements in food crops for human health. *Commun. Soil Sci. Plant Anal.* 43:556–570; Fageria, N. K., V. C. Baligar, and R. B. Clark. 2002. Micronutrients in crop production. *Adv. Agron.* 77:85–268.

Westermann (1991). Zinc sulfate (23% Zn) is the most commonly used Zn fertilizer. Water-soluble sources have higher Zn availability for plants than water-insoluble sources.

The use of water-soluble Zn criteria for selecting a Zn fertilizer becomes more important as the severity of Zn deficiency for the immediate crop increases. The recommended rates of soil-applied Zn are about 20 times higher than the total crop uptake of Zn but are required to obtain adequate distribution of Zn fertilizer granules. Micronutrient chelates are more efficient in supplying elements to the plants, but they are very expensive and their commercial use is not economical, especially in developing countries (Fageria et al. 2012).

7.8.2 Appropriate Methods and Timing of Application

There are several ways to apply micronutrients to crops such as soil application, fertigation, foliar spray, seed treatment, or in combination with crop protection products. Each option has specific advantages and disadvantages depending on the nutrient, the crop, and the soil characteristics. Application of these Zn fertilizers to rice is performed using a variety of methods depending on the production system. Most commonly, relatively high rates of inorganic Zn fertilizers are applied to the soil below the seed before or during seeding. In the water-seeded system practiced in California, a surface broadcast application of either $ZnSO_4$ or Zn lignosulfonate is recommended because the roots of rice seedlings are positioned at or near the soil–water interface (Wells et al. 1993). The primary advantage of soil-applied Zn over other Zn fertilization methods (that use much lower Zn application rates) is the residual benefit. A single Zn fertilizer application should provide adequate Zn for several years before additional Zn fertilizer is needed to optimize grain yields.

Due to the small amount of micronutrients required for crop plants (compared with macronutrients), uniform distribution is a problem. To overcome this problem, micronutrients are generally mixed with fertilizers to supply a predetermined quantity. Dry mixing is a simple method that works well with nongranular materials. However, there are often caking problems. Bulk bending is a form of dry mixing but with granular material. The main problem is segregation of the different components, unless all the materials have similar particle sizes. Segregation generally leads to uneven application. Use of fluid fertilizer is another method of micronutrient application. Micronutrients can be applied with fungicides or insecticides as foliar sprays or with foliar application of micronutrient products. They are generally available as prepared mixtures to prevent reactions that create water-insoluble compounds. Such macronutrient solutions can be provided to the farmers by fertilizer industries, if deficiency of micronutrients is well established in agro-climatic regions or in soils.

In greenhouse studies, Liscano et al. (2000) showed that the water solubility of inorganic Zn fertilizers was highly correlated to Zn uptake by seedling rice. They suggested a minimum of 40–50% of a Zn fertilizer's total Zn content should be water soluble to optimize Zn uptake. Amrani, Westfall, and Peterson (1999) and Gangloff et al. (2002) reported similar results for corn. In general, the water solubility of Zn sulfates and lignosulfonate sources is high, and the water solubility of Zn oxides and Zn oxysulfate sources is low to moderate. However, in most cases, the Zn application rate is more critical than the water-soluble Zn content of the fertilizer, but research data clearly show that tissue Zn concentration and total Zn uptake generally increase as the amount of water-soluble Zn in a fertilizer increases. The primary advantage of soil-applied Zn over other Zn fertilization methods that use much lower Zn application rates is the residual benefit. A single Zn fertilizer application should provide adequate Zn for several years before additional Zn fertilizer is needed to optimize grain yields.

7.8.3 Adequate Rate

Use of adequate rate of nutrients is important for achieving maximum economic yield as well as maximum uptake of nutrients. Because micronutrients are needed in small amounts, the economics of their use is generally highly favorable. Although the economics of micronutrient use is compelling in almost all cases, the problem is to get both the diagnosis and the delivery right. Soil testing is one of the most common methods of identifying nutrient deficiency in a given soil and determining fertilizer application for maximum economic yield. Soil testing is any chemical or physical measurement that is made on soil (Fageria et al. 2012). The main objective of soil chemical testing is to measure nutrient status and lime requirements in order to make fertilizer and lime application recommendations for profitable farming. The use of soil analysis as a fertilizer recommendation method is based on the existence of a functional relationship between the amount of nutrient extracted from soil by chemical methods and crop yield. When soil analysis shows low levels of a particular nutrient in a given soil, application of that nutrient is expected to increase crop yield.

Routine soil testing is a valuable tool that can be used to assess the potential for Zn deficiency in crops. Sims and Johnson (1991) reported that the critical soil Zn concentration range for most crops was between 0.5 and 2.0 mg kg^{-1} for Diethylenetriaminepentaacetic acid (DTPA) and 0.5 and 3.0 mg kg^{-1} for Mehlich 1. Most research indicates that the critical soil test Zn concentrations for rice fall within the ranges suggested by Sims and Johnson (1991). Fageria (1989) reported that 1.0 mg Zn kg^{-1} of soil extracted by the Mehlich 1 method was the critical concentration for lowland rice. Critical DTPA extractable soil Zn concentration of 0.8 mg kg^{-1} has been reported for Indian soils for lowland rice (Tiwari and Dwivedi 1994), whereas 0.7 mg Zn kg^{-1} (Sedberry et al. 1978) and 0.5 mg Zn kg^{-1} (Hill et al. 1992) have been suggested for rice grown in the United States. Fageria (2000a) reported that 4 mg Zn kg^{-1} was an adequate level and 35 mg Zn kg^{-1} was a toxic level of Zn extracted by DTPA solution for upland rice grown on Brazilian Oxisol at a pH of about 6.2.

Sedberry et al. (1980) and Wells (1980) indicated that soil pH of silt loam soils was the best predictor of rice response to Zn fertilization. However, their research was conducted on soils that had not previously received applications of Zn fertilizer, that were uniformly low in Zn, and where micronutrients were not commonly measured in routine soil analysis. Thus, for a number of years, Zn fertilizer recommendations were based exclusively on soil texture and soil pH, which triggered the recommendation of using Zn fertilizer on nearly every rice crop grown in rotation on alkaline soils.

In Arkansas, Zn fertilizer recommendations for flooded rice are now based on the soil pH, texture, and Mehlich 3 extractable Zn (Wilson et al. 2001). Zinc fertilizer is recommended for rice grown on silt and sandy loam soils having a pH > 6.0 and Mehlich 3 extractable Zn < 3.5 mg kg^{-1}. Zinc deficiencies are seldom observed on undisturbed clay soils in the United States. Precision land leveling often exposes Zn-deficient subsoils, and Zn deficiency is occasionally observed on leveled soils of all textures. Application of 5–7 kg Zn ha^{-1} as zinc sulfate was found to correct Zn deficiency in lowland as well as upland rice grown on Inceptisols and Oxisols in Brazil (Fageria and Barbosa Filho 1994). Slightly higher rates of 11 kg Zn ha^{-1} are typically recommended for soil application in the United States (Fageria, Slaton, and Baligar 2003).

7.8.4 Use of Efficient Genotypes

Exploring genetic variability is an important strategy for improving micronutrient contents in the staple food crops, including rice. Plant species and cultivars differ in their susceptibility to micronutrient deficiencies and toxicities (Fageria, Baligar, and Clark 2002), and these differences are related to the various morphological, physiological, and biochemical processes of plants (Baligar et al. 2001). Rice, sorghum, and maize have a high sensitivity to Zn deficiency, whereas barley, wheat, oat, and rye are less sensitive (Clark 1990). Relative sensitivity of some plant species to low Zn was maize > tomato > wheat > alfalfa > tall wheatgrass > soybean (Fageria, Baligar, and Clark 2002), and maize > cowpea > soybean > millet > rice > sorghum > wheat (Rashid and Fox 1992). Upland and lowland rice genotypes varied for uptake and utilization of Zn (Fageria and Baligar 2005; Fageria et al. 2012). Differences among genotypes for micronutrient uptake and utilization occur because of many physiological and biochemical factors. Adapting and developing plant genotypes to soils that are deficient in micronutrients or toxic may become a challenging but potentially productive area to pursue in the future.

The steady increases in major crop yields during the last half-century have been achieved through genetic improvement and advanced management practices. Selection of improved genotypes adapted to wide ranges of climate differences has contributed greatly to the overall gain in crop productivity during this time. In spite of these advances, mean yields of major crops are two- to four-fold below recorded maximum potentials (Baligar and Fageria 1997). Newly developed genotypes of rice, maize, wheat, and soybeans have been more efficient in absorption and utilization of nutrients compared with older cultivars (Clark and Duncan 1991; Fageria 1992). Zinc concentration in the grain of 10 upland rice genotypes and Zn recovery efficiency varied significantly among genotypes (Fageria and Baligar 2005).

The responses to Zn fertilization of 20 upland rice genotypes grown on Brazilian Oxisol were studied. Zinc level × genotype interactions were significant

TABLE 7.8
Grain Yield and Panicle Number of 20 Upland Genotypes as Influenced by Zinc Fertilization

	Grain Yield (g plant^{-1})		Panicle Number (plant^{-1})	
Genotype	0 mg Zn kg^{-1}	20 mg Zn kg^{-1}	0 mg Zn kg^{-1}	20 mg Zn kg^{-1}
BRA 01506	7.81def	13.51a	3.66ab	5.33abc
BRA 01596	9.47abcd	12.70ab	4.00ab	5.66abc
BRA 01600	7.66ef	11.46abc	3.66ab	4.66abc
BRA 025535	7.92def	10.61abcdef	4.33ab	6.33ab
BRA 02601	7.88def	10.92abcd	4.33ab	6.33ab
BRA 032033	7.94def	11.74abc	4.66ab	7.00a
BRA 032039	7.76def	12.07abc	4.00ab	6.00abc
BRA 032048	8.03cdef	9.83bcdefg	3.66ab	6.00abc
BRA 032051	7.89def	10.11bcdef	4.00ab	4.00bc
BRA 042094	7.39f	6.98g	3.33b	4.66abc
BRA 042156	9.73abc	11.45abc	3.33b	3.66bc
BRA 042160	8.71bcdef	10.49abcdef	4.00ab	4.00bc
BRA 052015	9.21abcde	9.98abcdefg	3.66ab	3.66bc
BRA 052023	8.08cdef	7.95defg	5.00ab	4.00bc
BRA 052033	10.32ab	7.69efg	5.66a	4.33abc
BRA 052034	7.54ef	7.88defg	4.33ab	5.33abc
BRA 052045	7.29f	7.56fg	5.33ab	6.00abc
BRA 052053	10.35ab	10.66abcde	3.66ab	3.33c
BRS Primavera	10.83a	10.10bcdef	4.00ab	3.66bc
BRS Sertaneja	9.91ab	9.53cdefg	3.66ab	4.00bc
Average	8.58b	10.16a	4.11b	4.90a
F-test				
Zn level (Zn)	**		*	
Genotype (G)	**		**	
Zn × G	**		**	
CVZn (%)	13.94		27.68	
CVG (%)	8.21		17.46	

Note: Means followed by the same letter in the same column are not significantly different at 5% probability level. Means of the average values are compared at low and high Zn levels.

*,**Significant at the 5 and 1% probability levels, respectively.

for grain yield and panicle number (Table 7.8), indicating different responses of genotypes in relation to grain yield and panicle number with changes in Zn levels. At low Zn levels, grain yield varied from 7.29 g plant^{-1} produced by the genotype BRA 052045 to 10.83 g plant^{-1} produced by the cultivar BRS Primavera, with an average value of 8.58 g plant^{-1}. At high Zn levels, grain yield varied from 6.98 g plant^{-1} produced by the genotype BRA 042094 to 13.51 g plant^{-1} produced by the genotype BRA 01506, with an average value of 10.16 g plant^{-1}. Significant variation in upland rice genotypes in grain yield at low and high Zn levels was reported by Fageria, Baligar, and Clark (2002) and Fageria and Baligar (2005). Overall, increase in grain yield was 18% at the high Zn level compared with the low Zn level. Response of upland rice to Zn fertilization has been reported by Fageria and Baligar (2005).

Panicle number varied from 3.33 plant^{-1} produced by the genotypes BRA 042156 and BRA 042156 to 5.66 plant^{-1} produced by the genotype BRA 052033, with an average value of 4.11 plant^{-1} at the low Zn level (see Table 7.8). At the high Zn level, panicle number varied from 3.33 plant^{-1} produced by the genotype BRA 052053 to 7.0 plant^{-1} produced by the genotype BRA 032033, with an average value of 4.90 plant^{-1}. Overall, the increase in panicle number was 19% at the high Zn level compared with the low Zn level. Significant variation in panicle numbers among upland rice genotypes has been reported by Fageria and Baligar (2005). Figures 7.14 through 7.16 show different panicle numbers in six upland rice genotypes and confirm that panicle numbers were higher under high Zn levels compared with low Zn levels.

Maximum root length was significantly influenced by genotype treatment, and it varied from 19.00 to 25.33 cm (Table 7.9). Root dry weight also was significantly influenced by Zn and genotype, and Zn × genotype interaction was also significant. Hence, there was a variation among genotypes in relation to root dry weight as the Zn level changed. Zinc fertilization produced significantly higher root dry weight compared with the control or without Zn fertilization treatment. This is an important conclusion because upland rice is subject to drought stress in Brazilian Oxisol

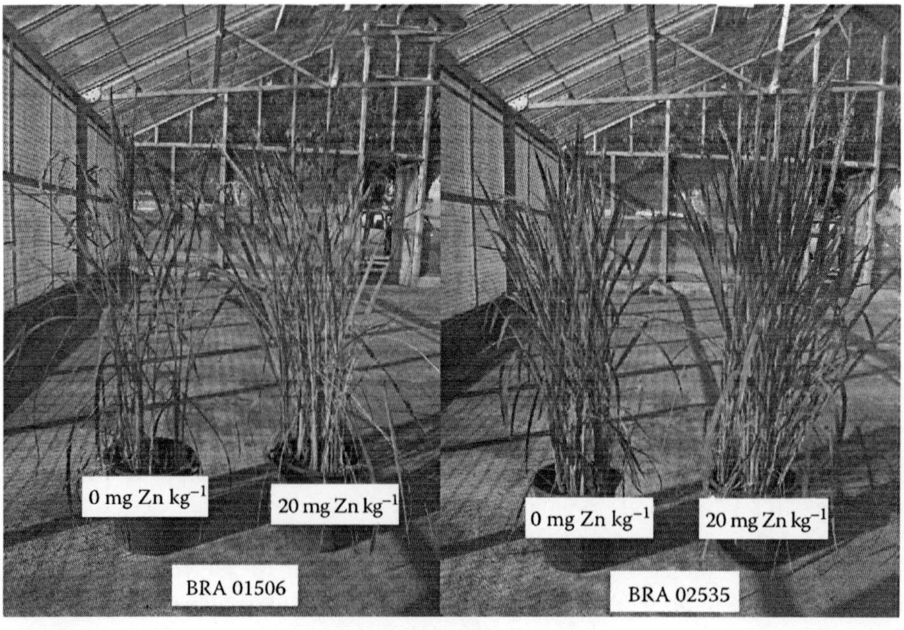

FIGURE 7.14 Response of two upland rice genotypes to zinc fertilization.

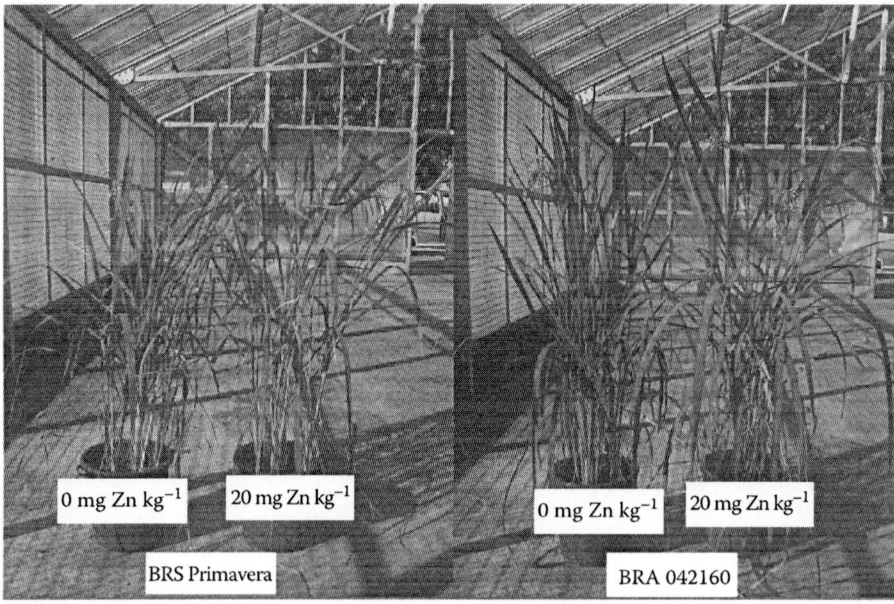

FIGURE 7.15 Response of two upland rice genotypes to zinc fertilization.

FIGURE 7.16 Response of two upland rice genotypes to zinc fertilization.

TABLE 7.9
Maximum Root Length and Root Dry Weight of 20 Upland Rice Genotypes as Influenced by Zinc Fertilization

Genotype	Root Length (cm)	Root Dry Weight (g plant^{-1})	
		0 mg Zn kg^{-1}	20 mg Zn kg^{-1}
BRA 01506	21.83ab	2.61abcde	2.92bcdef
BRA 01596	19.83ab	1.60e	2.06f
BRA 01600	21.33ab	1.65de	2.34ef
BRA 025535	25.33a	2.33abcde	4.29abcd
BRA 02601	21.33ab	3.05ab	5.33a
BRA 032033	21.66ab	1.97bcde	4.67ab
BRA 032039	22.33ab	2.22bcde	4.36abc
BRA 032048	20.33ab	1.95bcde	4.15abcd
BRA 032051	21.50ab	1.93bcde	2.01f
BRA 042094	22.16ab	1.83cde	2.25ef
BRA 042156	20.33ab	1.72cde	1.78f
BRA 042160	21.16ab	2.73abcd	2.54def
BRA 052015	21.83ab	1.64de	2.02f
BRA 052023	21.00ab	3.39a	2.91bcdef
BRA 052033	19.33b	2.51abcde	2.70cdef
BRA 052034	21.50ab	2.21bcde	2.75cdef
BRA 052045	22.16ab	2.79abc	3.87abcde
BRA 052053	19.00b	1.69cde	1.45f
BRS Primavera	23.50ab	1.83cde	1.52f
BRS Sertaneja	22.33ab	1.69cde	2.75cdef
Average	21.48	2.17b	2.94a
F-test			
Zn level (Zn)	NS	**	
Genotype (G)	*	**	
Zn × G	NS	**	
CVZn (%)	16.94	34.03	
CVG (%)	12.43	17.57	

Note: Means followed by the same letter in the same column are not significantly different at 5% probability level. Means of the average values are compared at low and high Zn levels.

*,**,NS Significant at the 5% and 1% probability levels and nonsignificant, respectively.

(Fageria, Baligar, and Clark 2006), and adequate Zn levels may improve root growth and consequently increase uptake of water and nutrients. Figures 7.17 through 7.19 show root growth of six upland rice genotypes, which was higher at higher Zn levels compared with lower Zn levels. Root weight is genetically controlled and also influenced by environmental factors, especially by mineral nutrition (Fageria, Baligar, and Clark 2006; Fageria and Moreira 2011).

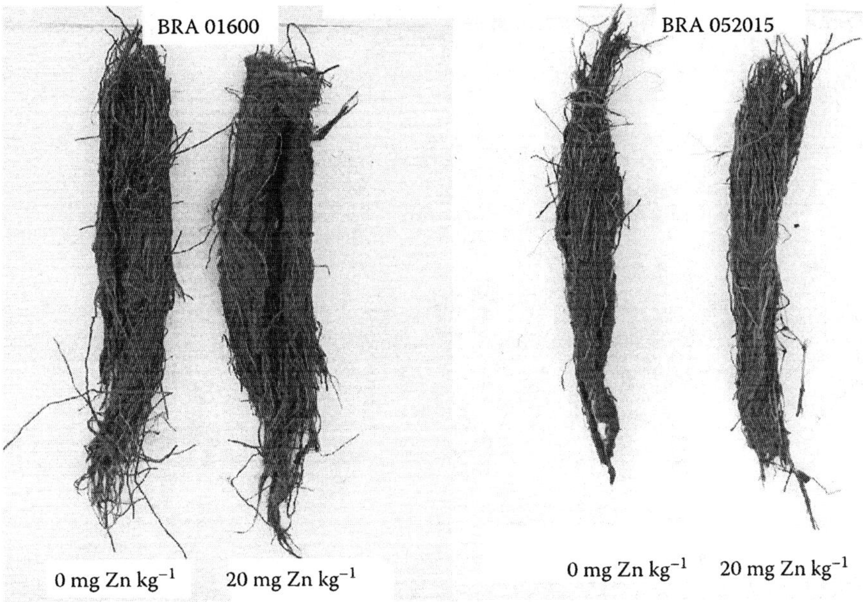

FIGURE 7.17 Root growth of two upland rice genotypes at two Zn levels.

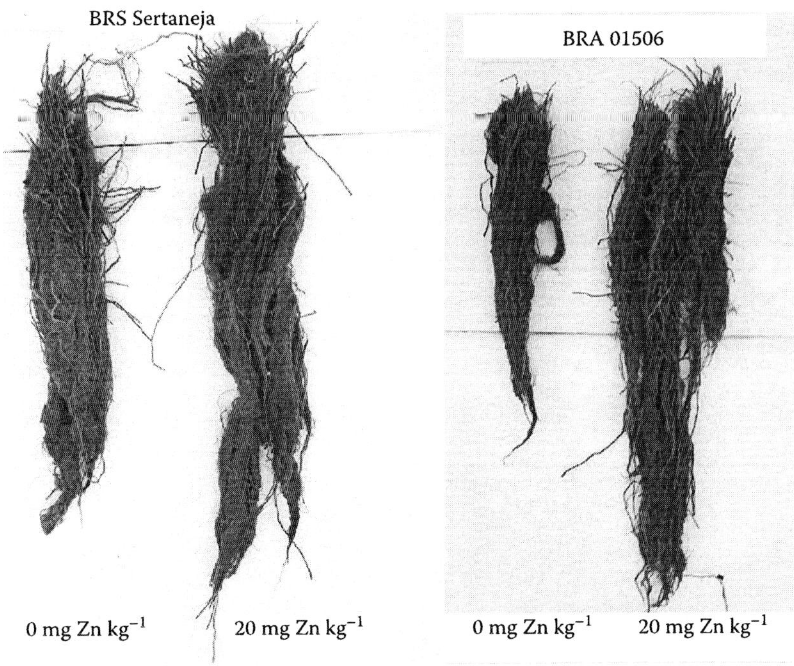

FIGURE 7.18 Root growth of two upland rice genotypes at two Zn levels.

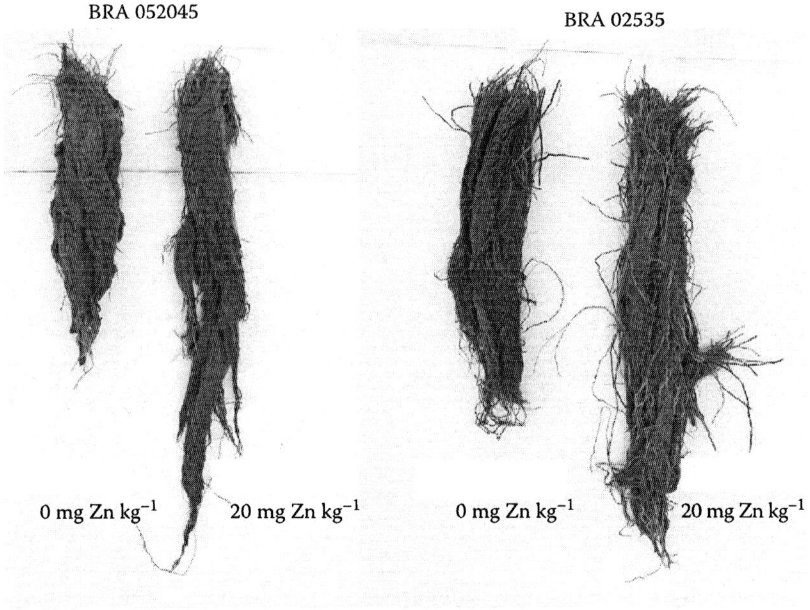

FIGURE 7.19 Root growth of two upland rice genotypes at two Zn levels.

7.9 CONCLUSIONS

Zinc deficiency is widely reported in upland and lowland rice in most rice-growing regions of the world. Zinc deficiency in rice is related to low levels of Zn in highly weathered tropical Oxisols and Ultisols, the use of lime in these soils, and the use of modern high-yielding cultivars that require high Zn rates. In lowland rice, increase in pH due to flooding is also responsible for Zn deficiency. In addition, Zn loss by soil erosion and leaching also contributes to Zn deficiency in rice. Zinc is required for activation of many enzymes that play significant roles in many physiological and biochemical plant processes. Zinc is not a mobile nutrient, and its deficiency symptoms first appear in the younger plant leaves. Typical Zn deficiency symptoms are leaves of a rusty brown color. Top and root growth is also reduced due to insufficient Zn in the soil.

Visual symptoms, as well as soil and plant tissue tests, can be used to identify a Zn deficiency or sufficiency in rice crops. The adequate Zn level in plant tissue varies among growth stages and plant ages. With advanced plant age, Zn concentration (content per unit tissue weight) decreases in plant tissues. This is due to the dilution effects of increased dry matter with advancing age. However, a Zn concentration of about >20 mg kg^{-1} is considered adequate for upland and lowland rice. The adequate soil level of Zn depends on soil type, soil pH, genotype, and extractor used for Zn extraction from the soil. Overall, 1–5 mg Zn kg^{-1} soil is adequate for upland and lowland rice production under most soil and climatic conditions.

Zinc deficiency can be corrected with the addition of 5–10 kg Zn ha^{-1} in most soils and cultivars planted. Zinc sulfate is the most efficient or desirable chemical fertilizer used to correct Zn deficiency because of its availability, cost, and its solubility in water. Zinc mobility is low in the medium to high textured soils, and its application below the seed is more effective method of application in Zn-deficient soils. Genetic variability in Zn uptake and utilization is widely reported among upland and lowland rice genotypes. Therefore, planting Zn-efficient genotypes is an important strategy for correcting Zn deficiency and for reducing production costs and environmental pollution.

REFERENCES

Adriano, D. C. 1986. *Trace Elements in the Terrestrial Environment*. New York: Spring-Verlag.

Alloway, B. J. 2008. Micronutrients and crop production: An introduction. In *Micronutrient Deficiencies in Global Crop Production*, ed. B. J. Alloway, pp. 1–39. New York: Springer.

Alvarez, J. M., M. I. Rico, and A. Obrador. 1997. Leachability and distribution of zinc applied to an acid soil as controlled-release zinc chelates. *Commun. Soil Sci. Plant Anal.* 28:1579–1590.

Amrani, M., D. G. Westfall, and G. A. Peterson. 1999. Influence of water solubility of granular zinc fertilizers on plant uptake and growth. *Plant Nutr.* 22:1815–1827.

Arnesen, A. K. M. and Singh, B. R. 1998. Plant uptake and DTPA-extractability of Cd, Cu, Ni and Zn in a Norwegian alum shale soil as affected by previous addition of dairy and pig manures and peat. *Can. J. Soil Sci.* 78:531–539.

Baligar, V. C., and N. K. Fageria. 1997. Nutrient use efficiency in acid soils: Nutrient management and plant use efficiency. In *Plant-Soil Interactions at Low pH: Sustainable Agriculture and Forestry Production*, eds. A. C. Moniz, A. M. C. Furlani, N. K. Fageria, C. A. Rosolem, and H. Cantarells, pp. 75–95. Campinas, Brazil: Brazilian Soil Science Society.

Baligar, V. C., N. K. Fageria, and Z. L. He. 2001. Nutrient use efficiency in plants. *Commun. Soil Sci. Plant Anal.* 32:921–950.

Barber, S. A. 1995. *Soil Nutrient Bioavailability: A Mechanistic Approach*, 2nd edition. New York: Wiley.

Brown, P. H. 2008. Micronutrient use in agriculture in the United States of America: Current practices, trends and constraints. In *Micronutrient Deficiencies in Global Crop Production*, ed. B. J. Alloway, pp. 267–286. New York: Springer.

Cakmak, I. 2008. Zinc deficiency in wheat in Turkey. In *Micronutrient Deficiencies in Global Crop Production*, ed. B. J. Alloway, pp. 181–200. New York: Springer.

Cakmak, I., A. Yilmaz, M. Kalayci, H. Ekiz, B. Touun, B. Erenoglu, and H. J. Braaun. 1996. Zinc deficiency as a critical problem in wheat production in central Anatolia. *Plant Soil.* 180:165–172.

Cheaney, R. L., and P. R. Jennings. 1975. *Field Problems of Rice in Latin America*. Cali, Colombia: Centro Internacional de Agricultura Tropical.

Chlopecka, A., and D. C. Adriano. 1996. Mimicked in situ stabilization of metals in a cropped soil: Bioavailability and chemical form of zinc. *Environ. Sci. Technol.* 30:3294–3303.

Clark, R. B. 1990. Physiology of cereals for mineral nutrient uptake, use, and efficiency. In *Crops as Enhancers of Nutrient Use*, eds. V. C. Baligar and R. R. Duncan, pp. 131–209. San Diego, CA: Academic Press.

Clark, R. B., and R. R. Duncan. 1991. Improvement of plant mineral nutrition through breeding. *Field Crops Res.* 27:219–240.

De Datta, S. K. 1981. *Principles and Practices of Rice Production*. New York: John Wiley.

Epstein, E., and A. J. Bloom. 2005. *Mineral Nutrition of Plants: Principles and Perspectives*, 5th edition. Sunderland, MS: Sinauer.

Fageria, N. K. 1984. *Fertilization and Mineral Nutrition of Tice*. Rio de Janeiro: EMBRAPA-CNPAF.

Fageria, N. K. 1989. *Tropical Soils and Physiological Aspects of Crop Production*. Brasilia: EMBRAPA-DPU.

Fageria, N. K. 1992. *Maximizing Crop Yields*. New York: Marcel Dekker.

Fageria, N. K. 2000a. Adequate and toxic levels of zinc for rice, common bean, corn, and wheat production in Cerrado soil. *Revista Brasileira de Engenharia Agrícola e Ambiental*. 4:390–395.

Fageria, N. K. 2000b. Upland rice response to soil acidity in Cerrado soil. *Pesq. Agropec. Bras.* 35:2303–2307.

Fageria, N. K. 2001. Screening method of lowland rice genotypes for zinc uptake efficiency. *Scientia Agricola*. 58:623–626.

Fageria, N. K. 2004. Influence of dry matter and length of roots on growth of five field crops at varying soil zinc and copper levels. *J. Plant Nutr.* 27:1517–1523.

Fageria, N. K. 2009. *The Use of Nutrients in Crop Plants*. Boca Raton, FL: CRC Press.

Fageria, N. K., and V. C. Baligar. 2005. Growth components and zinc recovery efficiency of upland rice genotypes. *Pesquisa Agropecuaria Brasileira*. 40:1211–1215.

Fageria, N. K., V. C. Baligar, and R. B. Clark. 2002. Micronutrients in crop production. *Adv. Agron.* 77:85–268.

Fageria, N. K., V. C. Baligar, and R. B. Clark. 2006. *Physiology of Crop Production*. New York: The Haworth Press.

Fageria, N. K., V. C. Baligar, and C. A. Jones. 2011. *Growth and Mineral Nutrition of Field Crops*, 3rd edition. Boca Raton, FL: CRC Press.

Fageria, N. K., V. C. Baligar, and Y. C. Li. 2008. The role of nutrient efficient plants in improving crop yields in the twenty-first century. *J. Plant Nutr.* 31:1121–1157.

Fageria, N. K., and M. P. Barbosa Filho. 1994. *Nutritional Deficiency in Rice: Identification and Correction*. Goiania, Brazil: EMBRAPA Arroz e Feijão.

Fageria, N. K., M. P. Barbosa Filho, and J. R. P. Carvalho. 1982. Response of upland rice to phosphorus fertilization on an Oxisol of central Brazil. *Agron. J.* 74:51–56.

Fageria, N. K., M. P. Barbosa Filho, and A. B. Santos. 2008. Growth and zinc uptake and use efficiency in food crops. *Commun. Soil Sci. Plant Anal.* 39:2258–2269.

Fageria, N. K., M. F. Moraes, E. P. B. Ferreira, and A. M. Knupp. 2012. Biofortification of trace elements in food crops for human health. *Commun. Soil Sci. Plant Anal.* 43:556–570.

Fageria, N. K., and A. Moreira. 2011. The role of mineral nutrition on root growth of crop plants. *Adv. Agron.* 110:251–331.

Fageria, N. K., A. B. Santos, and T. Cobucci. 2011. Zinc nutrition of lowland rice. *Commun. Soil Sci. Plant Anal.* 42:1719–1727.

Fageria, N. K., N. A. Slaton, and V. C. Baligar. 2003. Nutrient management for improving lowland rice productivity and sustainability. *Adv. Agron.* 80:63–152.

Fageria, N. K., and L. F. Stone. 2008. Micronutrient deficiency problems in South America. In *Micronutrient Deficiencies in Global Crop Production*, ed. B. J. Alloway, pp. 245–266. New York: Springer.

Fageria, N. K., F. J. P. Zimmermann, and V. C. Baligar. 1995. Lime and phosphorus interactions on growth and nutrient uptake by upland rice, wheat, common bean, and corn in an Oxisol. *J. Plant Nutr.* 18:2519–2532.

Foth, H. D., and B. G. Ellis. 1988. *Soil Fertility*. New York: John Wiley.

Gangloff, W. J., D. G. Westfall, G. A. Peterson, and J. J. Mortvedt. 2002. Relative availability coefficients of organic and inorganic Zn fertilizers. *J. Plant Nutr.* 25:259–274.

Genc, Y., G. K. McDonald, and R. D. Graham. 2002. Critical deficiency concentration of zinc in barley genotypes differing in zinc efficiency and its relations to growth responses. *J. Plant Nutr.* 25:545–560.

Gilmour, J. T. 1977. Micronutrient status of the rice plant. I. Plant and soil solution concentrations as a function of time. *Plant Soil.* 46:549–557.

Graham, R. D. 1984. Breeding for nutritional characteristics in cereals. *Adv. Plant Nutr.* 1:57–102.

Graham, R. D. 2008. Micronutrient deficiencies in crops and their global significance. In *Micronutrient Deficiencies in Global Crop Production*, ed. B. J. Alloway, pp. 41–61. New York: Springer.

Graham, R. D., J. S. Ascher, and S. C. Hynes. 1992. Selecting zinc efficient genotypes for soils of low zinc status. *Plant Soil.* 146:241–250.

Graham, R. D., and Z. Rengel. 1993. Genotypic variation in zinc uptake and utilization by plants. In *Zinc in Soils and Plants*, ed. A. D. Robson 107–118. Dordrecht, The Netherlands: Kluwer.

Grewal, H. S., and R. D. Graham. 1999. Residual effects of subsoil zinc and oilseed rape genotype on the grain yield and distribution of zinc in wheat. *Plant Soil.* 207:29–36.

Harter, R. D. 1991. Micronutrient adsorption-desorption reactions in soils. *In Micronutrients in Agriculture*, 2nd edition, eds. J. J. Mortvedt, F. R. Fox, L. M. Shuman, and R. M. Welch, pp. 59–87. Madison, WI: SSSA.

Hazra, G. C., and B. Mandal. 1996. Desorption of adsorbed zinc in soils in relation to soil properties. *J. Indian Soc. Soil Sci.* 44:233–237.

Hill, J. E., S. R. Roberts, D. M. Brandon, S. C. Scardaci, J. F. Williams, C. M. Wick, W. M. Canevari, and B. L. Weir. 1992. *Rice Production in California*. Univ. of California Div. of Agri. and Nat. Res. Pub. 21498.

Karlen, D. L., R. L. Flannery, and E. J. Sadler. 1988. Aerial accumulation and partitioning of nutrients by corn. *Agron. J.* 80:232–242.

Kittrick, J. A. 1976. Control of Zn^{2+} in the soil solution by sphalerite. *Soil Sci. Soc. Am. Proc.* 40:314–317.

Lindsay, W. L. 1979. Zinc in soils and plant nutrition. *Adv. Agron.* 24:147–181.

Liscano, J. F., C. E. Wilson, Jr., R. J. Norman, and N. A. Slaton. 2000. *Zinc Availability to Rice from Seven Granular Fertilizers*. Ark. Agric. Exp. Stn Res. Bull. No. 963. Fayetteville, AR.

Ludlow, M. M. 1985. Photosynthesis and dry matter production in C_3 and C_4 pasture plants with special emphasis on tropical C_3 legumes and C_4 grasses. *Aus. J. Plant Physiol.* 12:557–572.

Mandal, B., G. C. Hazra, and L. N. Mandal. 2000. Soil management influences on zinc desorption for rice and maize nutrition. *Soil Sci. Soc. Am. J.* 64:1699–1705.

Martens, D. C., and D. T. Westermann. 1991. Fertilizer applications for correcting micronutrient deficiencies. In *Micronutrients in Agriculture*, 2nd edition, eds. J. J. Mortvedt, F. R. Cox, L. M. Shuman, and R. M. Welch, pp. 549–592. Madison, WI: SSSA.

Mengel, K., A. Kirkby, H. Kosegarten, and T. Appel. 2001. *Principles of Plant Nutrition*, 5th edition. Dordrecht, The Netherlands: Kluwer.

Mikkelsen, D. S., and S. Kuo. 1976. Zinc fertilization and behavior in flooded soils. In *The Fertility of Paddy Soils and Fertilizer Application for Rice*, ed. Food and Fertilizer Technology Center, pp. 170–196. Taipei, Taiwan: Food and Fertilizer Technology Center.

Moore, P. A., and W. H. Patrick, Jr. 1988. Effect of zinc deficiency on alcohol dehydrogenase activity and nutrient uptake in rice. *Agron. J.* 80:882–885.

Mortvedt, J. J. 1994. Needs for controlled availability micronutrient fertilizers. *Fert. Res.* 38:213–221.

Pedrazzini, F. R., and K. L. McKee. 1984. Effect of flooding on activities of soil dehydrogenases and alcohol dehydrogenase in rice (*Oryza sativa* L.) roots. *Soil Sci. Plant Nutr.* 30:359–366.

Qiao, K., and G. Ho. 1997. The effects of clay amendment and composting on metal speciation in digested sludge. *Wat. Res.* 31:951–964.

Quang, V. D., and J. E. Dufey. 1995. Effect of temperature and flooding duration on phosphate sorption in an acid sulphate soil from Vietnam. *European J. Soil Sci.* 46:641–647.
Rashid, A., and R. L. Fox. 1992. Evaluating internal zinc requirements of grain crops by seed analysis. *Agron. J.* 84:469–474.
Sajwan, K. S., and W. L. Lindsay. 1986. Effects of soil redox on zinc deficiency in paddy rice. *Soil Sci. Soc. Am. J.* 50:1264–1269.
Sedberry, J. E., F. J. Peterson, F. E. Wilson, D. B. Mengel, P. E. Schilling, and R. H. Brupbacher. 1980. Influence of soil reaction and applications of zinc on yields and zinc contents of rice plants. *Commun. Soil Sci. Plant Anal.* 11:283–295.
Sedberry, J. E., P. G. Schilling, F. E. Wilson, and F. J. Peterson. 1978. *Diagnosis and Correction of Zinc Problems in Rice Production*. Baton Rouge, LA: Louisiana State University. Agri. Exp. Sta. Bul. 708.
Sah, R. N., and D. S. Mikkelsen. 1986. Sorption and bioavailability of phosphorus during the drainage period of flooded-drained soils. *Plant Soil.* 92:265–278.
Sajwan, K. S., and W. L. Lindsay. 1986. Effects of soil redox on zinc deficiency in paddy rice. *Soil Sci. Soc. Am. J.* 50:1264–1269.
Shimada, N. 1995. Deficiency and excess of micronutrient elements. In *Science of Rice Plant: Physiology*, Vol. 2, eds, T. Matsuo, K. Kumazawa, R. Ishii, K. Ishihara, and H. Hirata, pp. 412–419. Tokyo: Food and Agricultural Policy Research Center.
Shinano, T., M. Osaki, K. Komatsu, and T. Tadano. 1991. Comparison of production efficiency of the harvesting organs among field crops. I. Growth efficiency of the harvesting organs. *Soil Sci. Plant Nutr.* 39:269–280.
Shuman, L. M. 1995. Effects of nitrilotriacetic acid on metal adsorption isotherms for two soils. *Soil Sci.* 160:92–100.
Sillanpaa, M. 1982. *Micronutrients and Nutrient Status of Soils: A Global Study*. Rome: FAO.
Sillanpaa, M., and P. L. G. Vlek. 1985. Micronutrients and the agroecology of tropic and Mediterranean regions. *Fert. Res.* 7:151–167.
Sims, J. T., and G. V. Johnson. 1991. Micronutrient soil tests. In *Micronutrients in Agriculture*, 2nd edition, eds. J. J. Mortvedt, F. R. Cox, L. M. Shuman, and R. M. Welch, pp. 427–476. Madison, WI: SSSA.
Sinclair, A. H., and A. C. Edwards. 2008. Micronutrient deficiency problems in agricultural crops in Europe. In *Micronutrient Deficiencies in Global Crop Production*, ed. B. J. Alloway, pp. 225–266. New York: Springer.
Singh, D., R. G. McLaren, and K. C. Cameron. 1997. Desorption of native and added zinc from a range of New Zealand soils in relation to soil properties. *Aust. J. Soil Res.* 35:1253–1266.
Singh, M. V. 2008. Micronutrient deficiencies in crop and soils in India. In *Micronutrient Deficiencies in Global Crop Production*, ed. B. J. Alloway, pp. 93–123. New York: Springer.
Tisdale, S. L., W. L. Nelson, and J. D. Beaton. 1985. *Soil Fertility and Fertilizers*, 4th edition. New York: Macmillan.
Tiwari, K. N., and B. S. Dwivedi. 1994. Fertilizer Zn needs of rice (*Oryza sativa* L.) as influenced by native soil Zn in Udic Ustochrepts of the Indo-Ganetic plains. *Trop. Agric (Trinidad).* 71:17–21.
Torun, B., A. Yazici, I. Gultekin, and I. Cakmak. 2003. Influence of gyttja on shoot growth and shoot concentrations on zinc and boron of wheat cultivars grown on zinc deficient and boron toxic soil. *J. Plant Nutr.* 26:869–881.
Waals, J. H. V., and M. C. Laker. 2008. Micronutrient deficiencies in crops in Africa with emphasis on southern Africa. In *Micronutrient Deficiencies in Global Crop Production*, ed. B. J. Alloway, pp. 201–224. New York: Springer.
Welch, R. M. 1993. Zinc concentrations and forms in plants for humans and animals. In *Zinc in Soil and Plants*, ed. A. D. Robson, pp. 183–195. Dordrecht, The Netherlands: Kluwer.

Wells, B. R. 1980. *Zinc Nutrition of Rice Growing on Arkansas Soils*. Fayetteville, AR: University of Arkansas.

Wells, B. R., B. A. Huey, R. J. Norman, and R. S. Helms. 1993. Rice. In *Nutrient Deficiencies and Toxicities in Crop Plants*, ed. W. F. Bennett, pp. 15–19. St. Paul, MN: American Phytopathological Society.

Wilson, C. E. Jr., N. A. Slaton, R. J. Norman, and D. M. Miller. 2001. Efficient use of fertilizer. In *Rice Production Handbook*, ed. N. A. Slaton, pp. 51–74. Ark Coop. Ext. Serv. Misc. Publ. No. 192. Little Rock, AR.

Yamaguchi, J. 1978. Respiration and the growth efficiency in relation to crop productivity. *J Faculty Agric Hokkaido Univ.* 59:59–129.

Yoshida, S. 1981. *Fundamentals of Rice Crop Science*. Los Baños, Philippines: IRRI.

Zou, C., X. Gao, R. Shi, X. Fan, and F. Zhang. 2008. Micronutrient deficiencies in crop production in China. In *Micronutrient Deficiencies in Global Crop Production*, ed. B. J. Alloway, pp.127–148. New York: Springer.

8 Copper

8.1 INTRODUCTION

The importance of copper (Cu) for the growth of higher plants was first discovered in 1931 by Lipman and MacKinney (Fageria Baligar, and Jones 2011). Copper deficiency has been observed in an ever-increasing range of crops and environments, and the use of copper fertilizers has increased accordingly (Graham and Nambiar 1981). In the twenty-first century, Cu deficiency in annual crops, including rice, has been reported in many regions (Fageria, Baligar, and Clark 2002). However, variations in soil conditions, climate, crop genotypes, and management result in marked variations in the deficiency problem (Alloway 2008; Fageria and Stone 2008; Graham 2008). Alloway (2008) found that copper deficiency is more predominant in Europe and Australia, where cereals are most affected. Fageria and colleagues reported that micronutrient deficiency in food crops is very common in different parts of the world due to low natural levels of micronutrients in the soils; the use of high-yielding cultivars; liming of acidic soils; interactions among macro- and micronutrients, sandy and calcareous soils; the increased use of high analysis fertilizers having low amounts of micronutrients; and the decreased use of animal manures, composts, and crop residues (Fageria, Baligar, and Clark 2002; Fageria, Slaton, and Baligar 2003; Fageria, Baligar, and Jones 2011; Fageria et al. 2012). Copper deficiencies in the United States have been reported in Florida, Wisconsin, Michigan, and New York, where high value crops are intensively grown on Histosols (Tisdale, Nelson, and Beaton 1985). Graham and Nambiar (1981) noted that copper deficiency is usually best demonstrated in the field by measuring the increase in grain yield following copper application.

Most Cu deficiencies are found in organic soils, but deficiencies also are found in mineral soils high in organic matter and in sandy soils. Soil organic matter causes Cu deficiency in many organic soils (Mortvedt 1994). Copper deficiencies were first recognized in Western Australia in 1939; by 1948, they were found to be widespread on sandy and gravelly soils and on calcareous coastal soils (Holloway, Graham, and Stacey 2008). Similarly, Brennan and Best (1999) reported that Cu deficiency has been reported throughout Australia, but the major areas of potential deficiency include 8–9 million ha in Western Australia, the Eyre Peninsula, the upper southeast of South Australia, and Western Victoria. Fageria (2001a) has reported upland rice copper deficiency in Brazilian Oxisols. Copper deficiency in these soils is related to the use of lime for raising soil pH (Fageria 2000, 2002a, 2002b; Fageria and Baligar 2008). Other factors that can cause Cu deficiency in crop plants include the use of large nitrogen and phosphorus applications, which lead to Cu dilution in plant tissues (Alloway 2008). Data in Table 8.1 show that Cu uptake in upland rice straw decreased significantly with an increase in pH. In Brazilian Oxisol, Cu uptake was also decreased in upland rice with the addition

TABLE 8.1
Uptake of Cu in the Straw of Upland Rice as Influenced by pH and Zn Rate in Brazilian Oxisol

Soil pH in H_2O	Cu Uptake (μg $plant^{-1}$)	Zn Rate (mg kg^{-1})	Cu Uptake (mg kg^{-1})
4.6	75	0	13
5.7	105	5	14
6.2	78	10	12
6.4	64	20	12
6.6	61	40	12
6.8	51	80	11
		120	10
R^2	0.89*	R^2	0.80*

Source: Adapted from Fageria, N. K. 2000. Upland rice response to soil acidity in Cerrado soil. *Pesq. Agropec. Bras.* 35:2303–2307.
*Significant at the 5% probability level by F-test.

TABLE 8.2
Influence of pH on Mehlich 1 Extractable Cu in Inceptisol of Brazil

Soil pH in H_2O	Cu in Soil (mg kg^{-1})
4.9	6.3
5.9	6.3
6.4	5.8
6.7	5.0
7.0	3.3
R^2	0.94*

Source: Adapted from Fageria, N. K., and V. C. Baligar. 1999. Growth and nutrient concentrations of common bean, lowland rice, corn, soybean, and wheat at different soil pH on an Inceptisol. *J. Plant Nutr.* 22:1495–1507.
*Significant at the 1% probability level by F-test.

of Zn. Similarly, Mehlich 1 extractable Cu also decreased significantly with increasing soil pH after harvest of lowland rice grown on Brazilian Inceptisol (Table 8.2).

Copper deficiency can also occur because of adverse reactions between copper and other nutrients in plant absorption processes (Fageria and Gheyi 1999; Fageria, Baligar, and Clark 2006). Bowen (1969, 1981) reported that Cu absorption from solutions is strongly inhibited by zinc and vice versa (Loneragan 1975). Bowen (1981) also noted that copper and zinc are absorbed by the same mechanism, which is different from that of boron or manganese. Many other interactions have been reported in Cu absorption. For example, calcium may competitively inhibit copper uptake under some conditions (Cathala and Salsac 1975), and synergistic effects between copper and manganese have been reported (Younts 1964). Copper uptake was reported to decrease with the addition of iron (Graham and Nambiar 1981) and with the addition of molybdenum (Mackay, Chipman, and Gupta 1966). Based on published information, much progress has been made in diagnosing and correcting copper deficiency in crop plants, including rice; this topic has been reviewed in comprehensive reports on the copper nutrition of rice.

8.2 CYCLE IN SOIL–PLANT SYSTEM

The copper cycle in the soil–plant system involves the addition of copper in soil, its transformation in the root zone, its uptake by plants and losses due to leaching, and Cu adsorption on soil colloids and in soil's organic matter. Concentrations of total Cu in soils range from 2 to 100 mg kg^{-1} (mean of 30 mg kg^{-1}; Mortvedt 2000). However, its concentration may fall to 1–2 mg kg^{-1} in deficient soils (Tisdale, Nelson, and Beaton 1985). Copper concentration in the soil solution is usually very low, in the range of 10^{-8} to 10^{-6} M (0.6–63 ppb) (Tisdale, Nelson, and Beaton 1985). Depending on pH, different forms of copper can exist in the soil. At values below pH 6.9, divalent Cu^{2+} is the dominant form. Above pH 6.9, $Cu(OH)_2^0$ is the principal solution, and $CuOH^+$ assumes some importance near pH 7.0 (Tisdale, Nelson, and Beaton 1985). Hydrolysis reactions of copper ions can be expressed in the following equations (Lindsay 1979):

$$Cu^{2+} + H_2O \leftrightarrow CuOH^+ + H^+ \tag{8.1}$$

$$Cu^{2+} + 2H_2O \leftrightarrow Cu(OH)_2^0 + 2H^+ \tag{8.2}$$

$$Cu^{2+} + 3H_2O \leftrightarrow Cu(OH)_3^- + 3H^+ \tag{8.3}$$

$$Cu^{2+} + 4H_2O \leftrightarrow Cu(OH)_4^{2-} + 4H^+ \tag{8.4}$$

$$2Cu^{2+} + 2H_2O \leftrightarrow Cu_2(OH)_2^{2+} + 2H^+ \tag{8.5}$$

Copper is mostly found in silt and clay fractions of soil and is usually present in carbonate fractions in alkaline soils and in Fe oxide fractions in acidic soils (Shuman 1991). Sulfides are the predominant copper minerals in the earth's crust, with strong covalent bonds formed between reduced copper (Cu^+) and sulfide (S^{2-}) anions. Chalcopyrite ($CuFeS_2$) is the most widely occurring copper mineral. Chalcocite (Cu_2S) and bornite ($CuFeS_4$) are other important copper-containing sulfide minerals (Tisdale, Nelson, and Beaton 1985).

Crops grown on soils developed from sand, sandstones, acidic igneous rocks, and calcareous materials often exhibit Cu deficiency, but deficiencies are not generally found in plants grown in clays and in soils formed from basic rocks (Jarvis 1981). In the United States, soils formed from weathered bedrocks have high Cu, whereas soils formed in the lower Atlantic Coastal Plains have low Cu (Kubota 1983).

Organic, peat, and muck soils generally have low amounts of labile Cu (Oplinger and Ohlrogge 1974). When Histosols are brought under cultivation, plants commonly exhibit Cu deficiency, which has been termed a "reclamation disease" (Welch et al. 1991). Care should be taken to avoid copper deficiency whenever new areas of organic soils are brought into production. The high capacity of organic soils to fix micronutrients, especially Cu, has focused attention on the significance that organic matter may have in the soil. In mineral soils, a significant positive relationship has been reported between organic matter and Cu (Hodgson 1963). Complexation of Cu with organic matter (OM) occurs mainly at solution pH values above 6.5 (Barber 1995), and increased Cu complex formation has been reported with increased pH, decreased ionic strength, and increased OM/Cu ratios (Sanders and Bloomfield 1980). Inorganic Cu commonly complexes with hydroxyls and carbonates when the soil solution pH is >7.0 (McBride 1981). Breakdown of crop residues by soil microbes may release significant amounts of Cu, but natural complexing substances produced during OM decomposition could complex Cu into unavailable forms (Moraghan and Mascagni 1991). Flooding soil also decreased Cu availability to rice (Fageria, Baligar, and Clark 2002). Therefore, Cu deficiency may be expected in lowland or flooded rice grown on soils with low levels of Cu.

The movement of Cu in soil is very limited. It is the least mobile micronutrient in soil compared with Zn and Mn (Hodgson 1963). An increase in soil pH decreases Cu availability in plants, but its availability decrease is lower than Zn (Hodgson 1963). Fageria, Baligar, and Clark (2002) reported that solubility of Cu^{2+} is very much dependent on soil pH and decreases 100-fold for each unit of pH increase. For example, increasing pH from 4 to 7 increased Cu adsorption (Cavallaro and McBride 1984); Cu was adsorbed by inorganic soil components and occluded by soil hydroxide and oxides (Martens and Westermann 1991). Increases in soil pH above 6.0 induce hydrolysis of hydrated Cu, which can lead to stronger Cu adsorption by clay minerals and OM.

8.3 FUNCTIONS

Copper deficiency reduces growth, causes rice spikelet sterility, and consequently reduces yield. In addition to lower yield, Cu deficiency can reduce grain quality, producing shriveled grains and reduced cereal grain viability (Alloway 2008). Significant yield losses of up to 20% (without prior symptoms) are recognized as a noticeable result of Cu deficiency in cereals (Graham and Nambiar 1981). Pollen sterility and impaired carbohydrate metabolism (reduced starch formation) are two of the major causes of the yield reduction associated with hidden Cu deficiency in cereals grown on soils with low levels of available Cu (Jewell, Murray, and Alloway 1988; Alloway 2008). Alloway (2008) summarized Cu functions in crop plants, including its activation of several enzymes, with roles in photosynthesis, respiration, protein and carbohydrate metabolism, lignifications, and pollen formation. Copper-deficient plants are more susceptible to diseases and are also delayed in heading and maturation.

Fageria (2002a, 2002b) studied the influence of Cu on root growth of upland rice, dry beans, corn, wheat, and soybeans grown on Brazilian Oxisol (Table 8.3). Root growth of upland rice and wheat was significantly increased with copper fertilization of

TABLE 8.3
Root Dry Weight of Five Crop Species under Different Copper Levels

Cu Levels (mg kg^{-1})	Upland Rice (g 4 plants^{-1})	Common Bean (g 4 plants^{-1})	Corn (g 4 plants^{-1})	Wheat (g 4 plants^{-1})	Soybean (g 4 plants^{-1})
0	1.50	1.26	6.56	0.53	1.07
2	1.40	1.67	6.93	0.60	1.10
4	1.43	2.37	7.00	0.50	1.00
8	1.23	2.50	7.97	0.48	1.10
16	1.43	1.33	6.00	0.47	0.97
32	1.37	1.33	5.13	0.47	0.67
64	1.16	1.16	5.90	0.43	0.80
96	0.31	0.13	1.90	0.17	0.96
			Regression Analysis		
β_0	1.37	1.84	6.87	0.52	1.11
β_1	0.0068	−0.0044	−0.0038	0.00009	−0.014
β_2	−0.00018	−0.00013	−0.00045	−0.00004	0.00013
R^2	0.94**	0.64^{NS}	0.80*	0.88**	0.79*

Source: Fageria, N. K. 2002a. Micronutrients influence on root growth of upland rice, common bean, corn, wheat, and soybean. *J. Plant Nutr.* 25:613–622.

^{*,**,NS}Significant at the 5 and 1% probability levels and nonsignificant, respectively.

the soil. Fageria (1992) reported the beneficial effect of Cu on roots of upland rice grown in nutrient solution. However, root growth of common bean, corn, and soybean was decreased with the application of copper. Ninety percent of maximum dry weight of upland rice and wheat roots was obtained at 5 and 1 mg Cu kg^{-1} soil, respectively. A toxic level of copper relating to 10% reduction in root dry weight after reaching maximum was 50 mg Cu kg^{-1} for upland rice and 28 mg Cu kg^{-1} for wheat. Baligar, Fageria, and Elrashidi (1998) reported toxicity on root morphology and growth of cereals and legumes at 60–125 mg Cu kg^{-1} of soil. It is believed that the function of copper in a soil containing large amounts of organic materials is to neutralize the adverse effects of toxic substances produced from decomposition of the OM by combining with these substances and thus becoming unavailable to plants (Ishizuka 1978). Many researchers have reported that copper deficiency reduces root development more than shoot development and therefore increases the shoot/root ratio (Chaudhry and Loneragan 1970; King 1974). Even mild to moderate copper deficiency may delay anthesis and maturity (Graham 1976; Nambiar 1976a). Copper-deficient plants are more susceptible to diseases than are copper-sufficient plants (Primavesi and Primavesi 1970).

8.4 DEFICIENCY SYMPTOMS

Copper is an immobile nutrient in plants. Its deficiency first appears in younger leaves or growing points when its concentration is not adequate in the soil solution. Rice plants with Cu deficiency have smaller leaves, retarded tillering, and stunted growth. The leaves of copper-deficient plants often do not unroll, and their laminae become white-tipped and withered with a strong tendency to collapse sharply (wither-tip; Graham and Nambiar 1981). Copper deficiency symptoms in rice include the bending of leaves from the middle. This bending or collapse of leaves has been attributed to tissue structure damage caused by retarded lignification and incomplete formation of vascular structure (Graham and Nambiar 1981). Graham and Nambiar (1981) also reported that copper-deficient leaves have fewer and smaller chloroplasts, and that their stomata are smaller with narrow apertures and with thin walled, damaged guard cells. The newly emerging leaves fail to unroll, and the entire leaf (or half the leaf) appears needlelike, with the basal end developing normally (Fageria 2009). Expanding leaves emerging from the whorl may remain tightly curled. Pale brown necrosis develops near the tips of younger leaves, and leaf tips and margins wither and die to form a twisted spiral.

Leaf tips of very severely Cu-deficient plants may resemble those affected by Ca deficiency; tissue remains stuck together as leaves emerge from the whorl. Leaf tips of rice plants are discolored and distorted from Cu deficiency. An important feature of copper deficiency in the field is that significant losses in grain yield of up to 20% can occur without prior symptoms. Head bending, melanism, and (if there are healthy plants for comparison) a slight delay in maturity may be noticed (Graham and Nambiar 1981). Copper deficiency symptoms in upland and lowland rice leaves are shown in Figures 8.1 through 8.4. Symptoms of wither-tip of leaves are very clear in Figure 8.3.

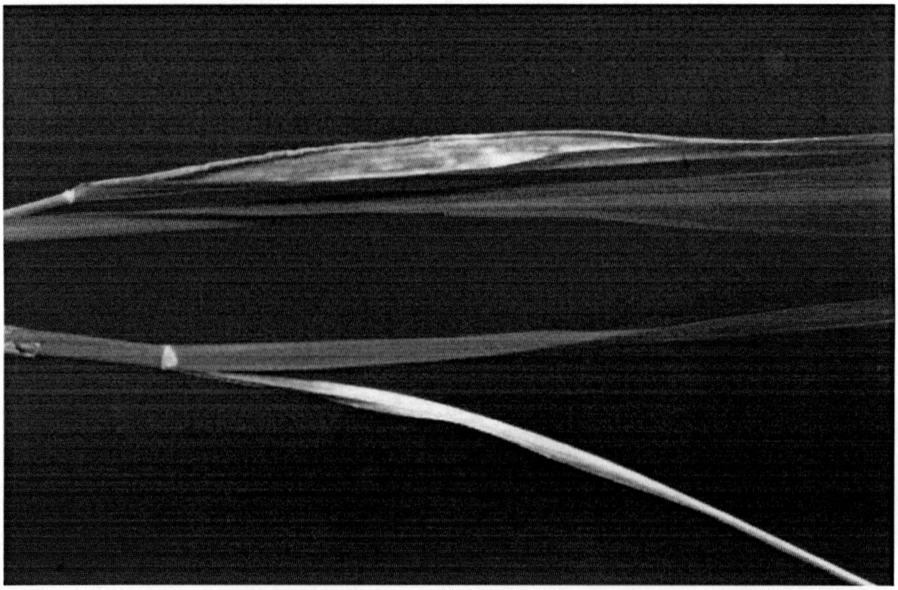

FIGURE 8.1 (See color insert.) Copper deficiency symptoms in upland rice leaves developed in nutrient solution. (From Fageria, N. K., and M. P. Barbosa Filho. 1994. *Nutritional Deficiency in Rice: Identification and Correction*. Goiania, Brazil: EMBRAPA Arroz e Feijão.)

FIGURE 8.2 (See color insert.) Copper deficiency symptoms in upland rice plants developed in nutrient solution. (From Fageria, N. K., and M. P. Barbosa Filho. 1994. *Nutritional Deficiency in Rice: Identification and Correction*. Goiania, Brazil: EMBRAPA Arroz e Feijão.)

8.5 UPTAKE IN PLANT TISSUE

Micronutrient uptake, including Cu, depends on nutrient concentrations at root surfaces, root absorption capacity, and plant demand. Micronutrient acquisition includes dynamic processes in which nutrients must be continuously replenished in the soil solution from the soil solid phase and transported to roots as uptake proceeds. Nutrient transport to roots, absorption by roots, and translocation from roots to shoots occur simultaneously, which means that rate changes of one process will ultimately influence other processes involved in uptake (Fageria, Baligar, and Clark 2002). In soil systems, nutrients move to plant roots by mass flow, diffusion, and root interception (Barber 1995). Mass flow is the passive transport of nutrients to roots as water moves through soil, and occurs when solutes are transported to roots with a convective flow of water from soil. The amount of nutrients supplied to roots depends on water flow rates to roots and the average nutrient content of the water. Amounts of nutrients reaching roots by this process depend on concentrations of nutrients in the soil solution and rates of water transport to and into roots. Mass flow could meet plant micronutrient requirements for Cu provided sufficient nutrient concentrations are in the soil solution, but this is not usually the case (Fageria, Baligar, and Clark 2002).

Copper is taken up by the plants in only very small quantities. The copper content of most plants is generally between 2 and 20 mg kg^{-1} in the dry plant material (Mengel et al. 2001). About 5–8 mg Cu kg^{-1} dry plant tissue may be

FIGURE 8.3 (See color insert.) Lowland rice leaves with copper deficiency grown on Brazilian Inceptisol.

considered the critical level for most field crops (Fageria, Baligar, and Jones 2011). Many researchers have reported that Cu concentrations in the shoots of cereals—including wheat, barley, oats, rye, corn, rice, and sorghum—normally range from 1.0 to 12 mg kg^{-1} (Chaudhry et al. 1973; Oplinger and Ohlrogge 1974; Gladstones, Loneragan, and Simmons 1975). Concentrations of copper in

FIGURE 8.4 (See color insert.) Upland rice leaves with copper deficiency grown on Brazilian Oxisol.

the grain normally range from 0.8 to 6.0 mg kg^{-1} (Gladstones, Loneragan, and Simmons 1975; Nambiar 1976b). Available information does not indicate any relationship between the copper concentrations of straw and grain (Graham and Nambiar 1981). Concentrations of copper in tillers are generally lower than in the main culms (Nambiar 1976b; Loneragan, Snowball, and Robson 1980).

Loneragan (1975) and Loneragan, Snowball, and Robson (1976) proposed the analysis of young leaves for diagnosis of copper deficiency following a careful study of copper's mobility within deficient and sufficient plants. Nambiar (1976b) also proposed the use of studying the younger leaves; although their copper concentrations were generally lower than in older leaves or stems, the range between deficient and sufficient was wider. Moreover, genotypic differences in the critical levels were smaller if young leaves were used (Graham and Nambiar 1981).

Plant age is one of the important factors determining copper concentration in plant tissues. Copper concentration in straw of upland rice varied from 22 mg kg^{-1} at the start of tillering to 5 mg kg^{-1} at physiological maturity (Table 8.4). The decrease was in a quadratic fashion with the advancement of plant age (Table 8.5). Gladstones, Loneragan, and Simmons (1975) and Loneragan (1975) also reported that copper concentration is relatively high in young wheat plants (two- to three-leaf stage) and decreases rapidly until flowering, with further reductions being relatively small. Rate of decline is strongly related to initial content, that is, the higher the initial concentration, the more rapid the rate of decline (Loneragan, Snowball, and Robson 1976, 1980).

Copper uptake in straw was 1.93 g ha^{-1} at the beginning of the tillering growth stage and increased to 55.88 g ha^{-1} at flowering. At the physiological maturity, the uptake of Cu decreased to 34.78 g ha^{-1} due to translocation of this element to the grains. The uptake of Cu followed an exponential quadratic function in the shoot or straw of upland rice with the advancement of plant age (see Table 8.5). The increase in the uptake of Cu in the straw of upland rice was related to an increase in dry matter with the advancement of plant age. The concentration and

TABLE 8.4
Concentration and Uptake of Copper in Straw and Grain of Upland and Lowland Rice during Growth Cycle

	Upland Rice			Lowland Rice	
Plant Age in Days	Concentration (mg kg^{-1})	Uptake (g ha^{-1})	Plant Age in Days	Concentration (mg kg^{-1})	Uptake (g ha^{-1})
19 (IT)	22	1.93	22 (IT)	16	5.59
43 (AT)	14	14.85	35 (AT)	13	15.10
69 (IP)	8	28.34	71 (IP)	5	28.43
90 (B)	7	43.89	97 (B)	3	26.64
102 (F)	6	55.88	112 (F)	2	27.83
130 (PM)	5	34.78	140 (PM)	2	20.73
130 (grain)	20	57	140 (grain)	17	104.65

Source: Fageria, N. K., L. F. Stone, and A. B. Santos. 2003. *Soil Fertility Management for Lowland Rice.* Santo Antonio de Goiás, Brazil: EMBRAPA Arroz e Feijão.

Note: IT, initiation of tillering; AT, active tillering; IP, initiation of primordia floral; B, booting; F, flowering; PM, physiological maturity.

TABLE 8.5
Relationship between Plant Age and Concentration and Uptake of Copper in Upland and Lowland Rice

Variable	Regression Equation	R^2
Plant age vs. Cu concentration in straw of upland rice	$Y = 28.78 - 0.41X + 0.0018X^2$	0.98*
Plant age vs. Cu uptake in straw of upland rice	$Y = 0.42 \exp.(0.096X - 0.00048X^2)$	0.96*
Plant age vs. Cu concentration in straw of lowland rice	$Y = 23.09 - 0.35X + 0.0014X^2$	0.98*
Plant age vs. Cu uptake in straw of lowland rice	$Y = 14.64 + 0.87X - 0.0044X^2$	0.90*

Source: Fageria, N. K. 2001b. Response of upland rice, dry bean, corn, and soybean to base saturation in Cerrado soil. *Rev. Bras. Eng. Amb.* 5:416–424; Fageria, N. K., L. F. Stone, and A. B. Santos. 2003. *Soil Fertility Management for Lowland Rice.* Santo Antonio de Goiás, Brazil: EMBRAPA Arroz e Feijão.

Note: Regression equations were calculated using Table 8.4 data.

*Significant at the 1% probability level.

TABLE 8.6
Adequate Level of Copper in Plant Tissues of Crop Plants

Crop Species	Growth Stage and Plant Part Analyzed	Adequate Cu Level (mg kg^{-1})
Wheat	Heading, leaf blade	>2.2
Barley	Stem extension, whole tops	4.8–6.8
Rice	Tillering, uppermost mature leaves	5–20
Corn	30–45 days after emergence, whole tops	7–20
Sorghum	Seedling, whole tops	8–15
Sorghum	Bloom, 3rd blade below panicle	2–7
Soybean	Prior to pod set, upper fully developed trifoliate	10–30
Dry bean	Early flowering, uppermost blade	5–15
Peanut	Early pegging, upper stems and leaves	10–50

Source: Adapted from Fageria, N. K., V. C. Baligar, and C. A. Jones. 2011. *Growth and Mineral Nutrition of Field Crops*, 3rd edition. Boca Raton, FL: CRC Press.

uptake pattern of Cu in lowland rice was similar to that of upland rice. However, values were different due to variations in straw yield and grain yield.

Table 8.6 presents data regarding adequate Cu levels in plant tissues of principal crop species, including rice. Similarly, in Table 8.7, data display adequate and toxic levels of copper in principal crop species grown on Brazilian Oxisols. Overall, the range between adequate and toxic levels in five crop species was 10–15 mg Cu kg^{-1} shoot dry weight. This means that the copper range between adequate and toxic levels is very narrow, and care should be taken when applying copper fertilizers to correct copper deficiency in crop plants.

TABLE 8.7
Adequate and Toxic Levels of Cu in the Plant Tissues of Five Crop Species

Crop Species	Adequate Level (mg kg^{-1})	Toxic Level (mg kg^{-1})
Upland rice	15	26
Dry bean	6	10
Corn	7	11
Soybean	7	10
Wheat	14	17
Average	10	15

Source: Adapted from Fageria, N. K., M. P. Barbosa Filho, and L. F. Stone. 2003. Adequate and toxic levels of micronutrients in soil and plants for annual crops. Paper presented at the *XXIX Brazilian Soil Science Congress*, Ribeirão Preto, São Paulo, Brazil.

Note: Adequate level was calculated at 90% of the maximum dry weight of shoot, and toxic level was calculated at 10% reduction in shoot dry weight after achieving maximum weight. Plants were harvested at four weeks after sowing, and soil pH at harvest was about 6.0 in water in all five experiments.

8.6 USE EFFICIENCY

Crop recovery values for micronutrients generally range from only 5 to 10% (Mortvedt 1994). Some reasons for the low micronutrient efficiency of fertilizers are poor distribution of the low rates applied to soils, fertilizer reactions with soil that form unavailable reaction products, and low mobility in the soil for reaching plant roots (Martens and Lindsay 1990). Copper use efficiency (dry weight of shoot or grain/copper uptake in shoot or grain) of upland rice and dry beans during the crop growth cycle is presented in Table 8.8. In upland rice and dry beans, copper use efficiency significantly increased in a quadratic fashion (Table 8.8). The variation in copper use efficiency due to plant age was about 98% in rice and about 96% in dry beans. At harvest, copper use efficiency in rice grain was 84.4 kg rice produced with the accumulation of 1 g copper. In dry beans, 85.3 kg grain was produced with the accumulation of 1 g copper. This means that the copper use efficiency in grain production of upland rice and of dry beans was more or less similar.

8.7 HARVEST INDEX

The copper harvest index, or CuHI (Cu translocation to grain/copper uptake in grain plus shoot), was maximum for dry beans and minimum for corn (Table 8.9). The lowest CuHI (in corn) was due to the highest grain and straw yield compared with other crops. From a nutritional point of view, the crops can be classified on the basis of CuHI as dry bean > rice > soybean > corn (Fageria 2009).

TABLE 8.8
Copper Use Efficiency in Shoot and Grain of Upland Rice and Dry Bean during Crop Growth Cycle

Plant Age in Days	Upland Rice (kg g^{-1})	Plant Age in Days	Dry Bean (kg g^{-1})
19	45.3	15	84.5
43	70.9	29	125.8
68	120.7	43	103.5
90	140.3	62	120.9
102	141.7	84	135.7
130	187.7	96	216.2
130 (grain)	84.4	96 (grain)	85.3

Regression Analysis

Plant age (X) vs. CuUE in rice shoot $(Y) = 16.6851 + 1.4727X - 0.0014X^2$, $R^2 = 0.9791$**
Plant age (X) vs. CuUE in dry bean shoot $(Y) = 7.0336 + 2.7758X - 0.0108X^2$, $R^2 = 0.9289$*

Source: Fageria, N. K. 2009. *The Use of Nutrients in Crop Plants*. Boca Raton, FL: CRC Press.
Note: CuUE (kg g^{-1}) = (Shoot or grain dry weight in kg/uptake of Cu in shoot or grain in g). CuUE, copper use efficiency.
*,**Significant at the 5 and 1% probability levels, respectively.

TABLE 8.9
Copper Uptake in Shoot and Grain and Copper Harvest Index in Upland Rice, Dry Bean, Corn, and Soybean Grown on Brazilian Oxisol

Crop Species	Uptake in Shoot (g ha^{-1})	Uptake in Grain (g ha^{-1})	Total (g ha^{-1})	Copper Harvest Index
Upland rice	34.78	56.64	91.41	0.62
Dry bean	8.01	22.41	30.42	0.74
Corn	53.32	13.75	67.07	0.21
Soybean	53.15	30.79	83.94	0.37

Source: Adapted from Fageria, N. K. 2001b. Response of upland rice, dry bean, corn, and soybean to base saturation in Cerrado soil. *Rev. Bras. Eng. Amb.* 5:416–424.
Note: Copper harvest index = (Cu uptake in grain/Cu uptake in grain plus shoot).

8.8 MANAGEMENT PRACTICES

Adopting appropriate crop management practices is important to maximize crop yields, reduce the cost of fertilizer application, and to maximize nutrient use efficiency. Some of the important management practices for maximizing copper use efficiency in crop plants are effective source, appropriate method and timing of application, adequate rate, and planting copper-efficient genotypes.

TABLE 8.10
Principal Copper Carriers

Common Name	Formula	Cu Content (%)
Copper sulfate pentahydrate	$CuSO_4 \cdot 5H_2O$	25
Copper sulfate monohydrate	$CuSO_4 \cdot H_2O$	35
Cupric oxide	CuO	75
Copper chloride	$CuCl_2$	17
Copper frits	Frits	40–50
Copper EDTA chelate	$Na_2CuEDTA$	13
Copper HEDTA chelate	NaCuHEDTA	9
Cuprous oxide	Cu_2O	89

Source: Fageria, N. K. 1989. *Tropical Soils and Physiological Aspects of Crops*. Brasilia: EMBRAPA; Tisdale, S. L., W. L. Nelson, and J. D. Beaton. 1985. *Soil Fertility and Fertilizers*, 4th edition. New York: Macmillan; Foth, H. D., and B. G. Ellis. 1988. *Soil Fertility*. New York: John Wiley; Fageria, N. K. 2009. *The Use of Nutrients in Crop Plants*. Boca Raton, FL: CRC Press.

8.8.1 Effective Sources

Using an effective nutrient source is a prerequisite for correcting crop plant deficiencies in nutrient-deficient soils. Fertilizer sources that are water soluble are more efficient in supplying Cu to crop plants than those that are insoluble. Principal copper carriers are given in Table 8.10. Among these sources, all are water soluble, except cupric oxide (CuO). Copper sulfate is the most commonly used fertilizer for correcting Cu deficiency in crop plants.

8.8.2 Appropriate Methods and Timing of Application

Copper deficiency in crop plants can be corrected by applying Cu in soil as broadcast or band or by foliar application. Soil application of Cu through an appropriate source is better than foliar application. Foliar application of Cu can be used in an emergency situation. For correcting Cu deficiency with banded application, lower rates of Cu are required than with broadcast $CuSO_4$. Rates of band-placed $CuSO_4$ are as low as 1.1 kg Cu ha^{-1} for correcting Cu deficiency in vegetables and as high as 6.6 kg Cu ha^{-1} for highly sensitive crops (Martens and Westermann 1991). If copper deficiency occurs during crop growth, foliar application of $CuSO_4$ can alleviate this problem. Mengel (1980) recommended foliar application of 2.2 kg Cu as $CuSO_4$ in 280 L water ha^{-1} for correcting Cu deficiencies in annual crops.

8.8.3 Adequate Rate

An appropriate rate of copper application is essential for economic and environmental reasons. The best method to determine the appropriate rate is to gather data on crop response to applied copper fertilizers. Data on crop responses to applied copper fertilizers are limited under field as well as greenhouse conditions. However, Fageria (2001a) determined crop response curves to applied copper fertilization for principal food crops, including upland rice (Figure 8.5). There was a quadratic increase in dry weight of upland rice, common beans, corn, and wheat with an increasing copper rate, but at a higher copper rate, dry weight decreased for all four crop species. In the case of soybeans, there was a quadratic decrease in shoot dry weight with increasing copper rate in the growth medium. Adequate copper rate, calculated on the basis of 90% maximum yield, was 3 mg kg^{-1} for upland rice, 2 mg kg^{-1} for common beans, 3 mg kg^{-1} for corn, and 12 mg kg^{-1} for wheat. For soybeans, there was no need for copper application; the original soil level of 1 mg Cu kg^{-1} extracted by Mehlich 1 solution was sufficient to produce maximum plant growth. Martens and Westermann (1991) reported that application of 1.7–7.25 mg Cu kg^{-1} of soil as broadcast $CuSO_4$ could correct a copper deficiency in most annual crops. Similarly, Follett, Murphy, and Donahue (1981) reported that 1–7 kg Cu ha^{-1} (as copper sulfate) is sufficient for most crop plants. Differences in broadcast rates of Cu required for correcting a Cu deficiency reflect variations in soil properties, plant requirements, and concentrations of extractable soil Cu. Copper applied to soil has a prolonged residual effect (three to five years depending on cropping intensity). Hence, soil application of copper is generally recommended.

Critical toxic levels of applied Cu were 51 mg kg^{-1} for upland rice, 37 mg kg^{-1} for common beans, 48 mg kg^{-1} for corn, 15 mg kg^{-1} for soybeans, and 51 mg kg^{-1} for wheat. Upland rice and wheat crops were more tolerant to copper toxicity, and soybeans were most susceptible to toxicity. Practically no data are available regarding toxic rate of application for annual crops grown on Oxisols of central Brazil.

Soil testing is an important criterion for determining copper fertilizer rates for crop plants (Fageria, Baligar, and Jones 2011). The diethylene triamine pentaacetic acid (DTPA) extracting solution is most commonly used solution for soil test of copper. However, in Brazil, Mehlich 1 extracting solution is used in routine micronutrient analysis for economic reasons. Adequate soil test levels of copper were established at 2 mg kg^{-1} for upland rice, 1.5 mg kg^{-1} for common beans, 2.5 mg kg^{-1} for corn, 1 mg kg^{-1} for soybeans, and 10 mg kg^{-1} for wheat, using Mehlich 1 extracting solution (Fageria 2001b). When DTPA extracting solution was used, adequate levels of copper were 1 mg kg^{-1} for upland rice, 0.5 mg kg^{-1} for common beans, 1.5 mg kg^{-1} for corn, 0.5 mg kg^{-1} for soybeans, and 8.5 mg kg^{-1} for wheat (Figure 8.6). Adequate soil test levels of 0.1–10 mg kg^{-1} for Mehlich 1 extracting solution have been reported for annual crops (Sims and Johnson 1991). Using DTPA extracting solution, Galrão (1999) reported that 0.6 mg Cu kg^{-1} was an adequate level in the soil for soybean crops grown on Brazilian Oxisols.

8.8.4 Use of Efficient Genotypes

Genotype differences have been reported for copper utilization by crop plants (Baligar, Fageria, and He 2001; Fageria, Baligar, and Clark 2002; Fageria, Castro, and Baligar 2004). Data in Table 8.11 show that copper use efficiency of upland rice genotypes varied significantly among genotypes and soil acidity levels. The differences in nutrient uptake and utilization may be associated with better root geometry; the plants'

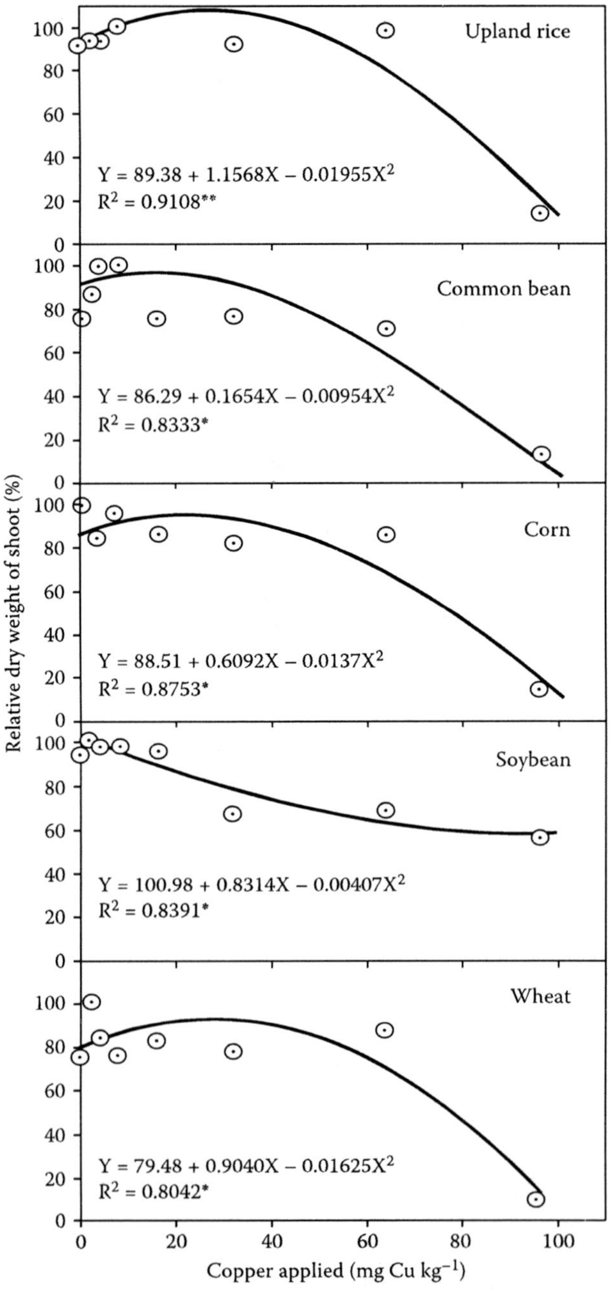

FIGURE 8.5 Relationship between copper applied in soil and relative grain or dry weight of five crop species. (From Fageria, N. K. 2001a. Adequate and toxic levels of copper and manganese in upland rice, common bean, corn, soybean, and wheat grown on an Oxisol. *Commun. Soil Sci. Plant Anal.* 32:1659–1676.)

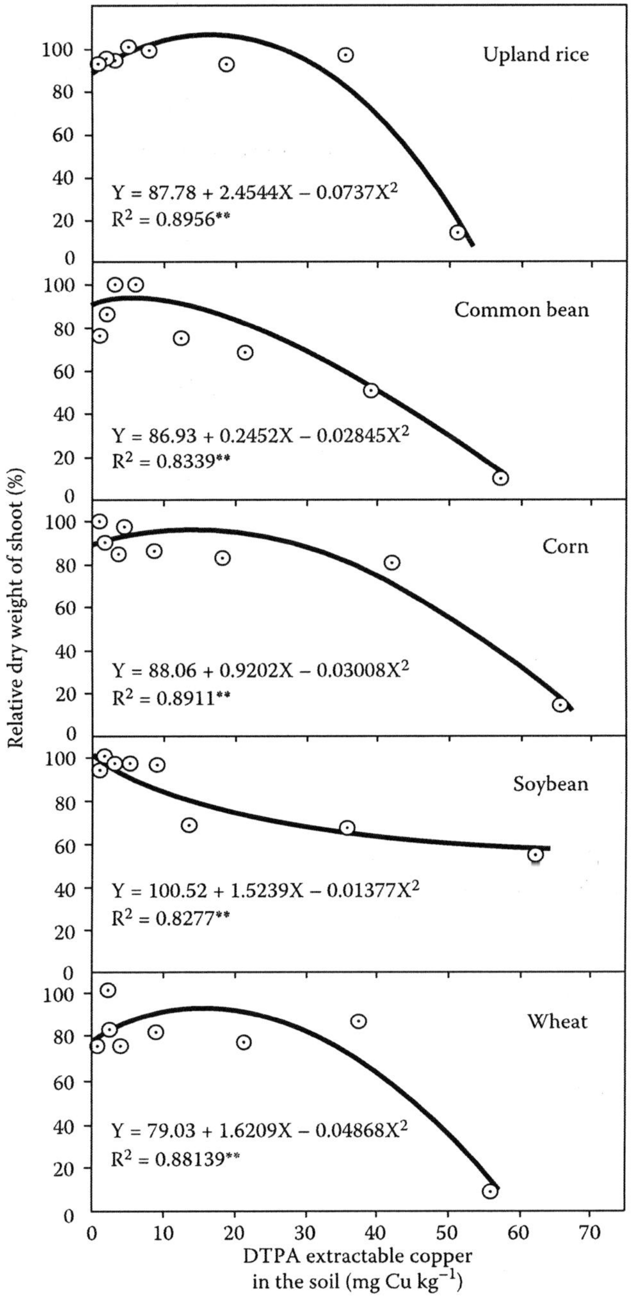

FIGURE 8.6 Relationship between DTPA extractable Cu and relative grain or dry weight of five crop species. (From Fageria, N. K. 2001a. Adequate and toxic levels of copper and manganese in upland rice, common bean, corn, soybean, and wheat grown on an Oxisol. *Commun. Soil Sci. Plant Anal.* 32:1659–1676.)

TABLE 8.11
Copper Use Efficiency (mg Grain per μg Nutrient Accumulated in the Plant) by 10 Upland Rice Genotypes at Two Acidity Levels in Brazilian Oxisol

	Cu Use Efficiency (mg Grain μg^{-1} Cu Uptake)	
Upland Rice Genotype	**High Acidity (Soil pH 4.5)**	**Low Acidity (Soil pH 6.4)**
CRO 97505	150	197
CNAs 8983	207	182
CNAs 8938	191	184
CNAs 8989	138	176
CNAs 8824	86	122
CNAs 8952	96	152
CNAs 8950	79	141
CNAs 8540	76	144
Primavera	117	192
Carisma	127	114
Average	127	160
F-test		
Acidity (A)	*	
Genotype (G)	**	
A × G	*	
CV (%)	18	

Source: Adapted from Fageria, N. K., E. M. Castro, and V. C. Baligar. 2004. Response of upland rice genotypes to soil acidity. In *The Red Soils of China: Their Nature, Management and Utilization*, eds. M. J. Wilson, Z. He, and X. Yang, pp. 219–237. Dordrecht, The Netherlands: Kluwer.

Note: Means in the same column followed by the same letter are not significantly different at the 5% probability level by Tukey's test.

*,**Significant at the 5 and 1% probability levels, respectively. Means in the same column followed by the same letter are not significantly different at the 5% probability level by Tukey's test.

ability to take up sufficient nutrients from lower or subsoil concentrations; their ability to solubilize nutrients in the rhizosphere; better transport, distribution, and utilization within plants; and balanced source–sink relationships (Graham 1984; Baligar, Fageria, and He 2001; Fageria, Baligar, and Li 2008).

8.9 CONCLUSIONS

Copper deficiency has been reported in many rice-growing regions of the world. Liming is the main cause of Cu deficiency in rice grown in acidic soil; liming raises the pH and reduces Cu availability. In flooded rice, Cu deficiency is also related to increased pH of acidic soils. Copper activates many enzymes in rice plants that are important in photosynthesis and in protein and carbohydrate metabolism. Due to copper's immobility in plants, deficiency symptoms first appear in younger leaves. In rice, pollen sterility (leading to failure to set grain), incomplete grain filling, lack of shoot vigor, and reduced root growth are all signs of copper deficiency. Because of the immobility of copper ions in the soil, it is supplied to plant roots mainly by interception. Copper's mobility and its availability to plants are controlled by factors such as pH, cation exchange capacity, OM content, and the presence of Fe and Al hydrous oxides. Of the applied chemical fertilizers, copper recovery efficiency in soil is less than 10%. Therefore, residual effects of Cu fertilizers can be impacted by cropping intensity and yield level. Copper concentration in the rice shoot or straw varies from 2 to 22 mg kg^{-1} depending on the growth stage and the copper status of the soil. Copper concentration is relatively high in the young shoot and decreases until flowering, after which further Cu reductions are relatively small. In plant tissue, the range between Cu deficiency and toxicity is narrow. Excess amounts of P, Zn, Fe, and Mn decrease copper uptake in rice. Soil testing is considered one of the most popular methods of identifying Cu deficiency in crop plants. Determining adequate critical levels of Cu depends on soil type and on the extracting solution used. However, in most rice-producing soils, it can vary from 0.5 to 3 mg kg^{-1} soil. Copper deficiency can be corrected by applying 2–10 kg Cu ha^{-1} in copper-deficient rice-growing soils. Copper sulfate ($CuSO_4 \cdot 5H_2O$, containing about 25% Cu) is the most common chemical fertilizer source. Copper deficiency can also be corrected by foliar application of 0.2% of a water-soluble Cu source. Because the recommended Cu rate is low, it is mixed with macronutrient fertilizers for application to the soil. This practice assures uniform application. The use of Cu-efficient genotypes is one of the most promising strategies for lowering the cost of rice production and reducing environmental pollution.

REFERENCES

Alloway, B. J. 2008. Micronutrients and crop production: An introduction. In *Micronutrient Deficiencies in Global Crop Production*, ed. B. J. Alloway, pp. 1–39. New York: Springer.

Baligar, V. C., N. K. Fageria, and M. A. Elrashidi. 1998. Toxicity and nutrient constraints on root growth. *Hort Sci.* 33:960–965.

Baligar, V. C., N. K. Fageria, and Z. L. He. 2001. Nutrient use efficiency in plants. *Commun. Soil Sci. Plant Anal.* 32:921–950.

Barber, S. A. 1995. *Soil Nutrient Bioavailability: A Mechanistic Approach*, 2nd edition. New York: Wiley.

Bowen, J. E. 1969. Absorption of copper, zinc and manganese by sugarcane leaf tissue. *Plant Physiol.* 44:255–261.

Bowen, J. E. 1981. Kinetics of active uptake of boron, zinc, copper and manganese in barley and sugarcane. *J. Plant Nutr.* 3:215–224.

Brennan, R. F., and E. Best. 1999. Copper. In *Soil Analysis: An Interpretation Manual*, eds. K. I. Sparrow, L. A. Reuter, and D. J. Reuter, pp. 303–307. Melbourne, Australia: CSIRO.

Cathala, N., and L. Salsac. 1975. Copper absorption by maize (*Zea mays* L.) and sunflower. (*Helianthus annuus* L.) roots. *Plant Soil.* 42:65–83.

Cavallaro, N., and M. B. McBride. 1984. Zinc and copper status and fixation by an acid soil clay: Effect of selective dissolutions. *Soil Sci. Soc. Am. J.* 48:1050–1054.

Chaudhry, F. M., and J. F. Loneragan. 1970. Effects of nitrogen, copper, zinc fertilizers on the copper and zinc nutrition of wheat plants. *Aust. J. Agric. Res.* 21:865–879.

Chaudhry, F. M., M. Sharif, A. Latif, and R. H. Zureshi. 1973. Zinc-copper antagonism in the nutrition of rice (*Oryza sativa* L.). *Plant Soil.* 38:573–580.

Fageria, N. K. 1989. *Tropical Soils and Physiological Aspects of Crops*. Brasilia: EMBRAPA.
Fageria, N. K. 1992. *Maximizing Crop Yields*. New York: Marcel Dekker.
Fageria, N. K. 2000. Upland rice response to soil acidity in Cerrado soil. *Pesq. Agropec. Bras.* 35:2303–2307.
Fageria, N. K. 2001a. Adequate and toxic levels of copper and manganese in upland rice, common bean, corn, soybean, and wheat grown on an Oxisol. *Commun. Soil Sci. Plant Anal.* 32:1659–1676.
Fageria, N. K. 2001b. Response of upland rice, dry bean, corn, and soybean to base saturation in Cerrado soil. *Rev. Bras. Eng. Amb.* 5:416–424.
Fageria, N. K. 2002a. Micronutrients influence on root growth of upland rice, common bean, corn, wheat, and soybean. *J. Plant Nutr.* 25:613–622.
Fageria, N. K. 2002b. Influence of micronutrients on dry matter yield and interaction with other nutrients in annual crops. *Pesq. Agropec. Bras.* 37:1765–1772.
Fageria, N. K. 2009. *The Use of Nutrients in Crop Plants*. Boca Raton, FL: CRC Press.
Fageria, N. K., and V. C. Baligar. 1999. Growth and nutrient concentrations of common bean, lowland rice, corn, soybean, and wheat at different soil pH on an Inceptisol. *J. Plant Nutr.* 22:1495–1507.
Fageria, N. K., and V. C. Baligar. 2008. Ameliorating soil acidity of tropical Oxisols by liming for sustainable crop production. *Adv. Agronomy.* 99:345–399.
Fageria, N. K., V. C. Baligar, and R. B. Clark. 2002. Micronutrients in crop production. *Adv. Agron.* 77:189–272.
Fageria, N. K., V. C. Baligar, and R. B. Clark. 2006. *Physiology of Crop Production*. New York: The Haworth Press.
Fageria, N. K., V. C. Baligar, and C. A. Jones. 2011. *Growth and Mineral Nutrition of Field Crops*, 3rd edition. Boca Raton, FL: CRC Press.
Fageria, N. K., V. C. Baligar, and Y. C. Li. 2008. The role of nutrient efficient plants in improving crop yields in the twenty-first century. *J. Plant Nutr.* 31:1121–1157.
Fageria, N. K., and M. P. Barbosa Filho. 1994. *Nutritional Deficiency in Rice: Identification and Correction.* Goiania, Brazil: EMBRAPA Arroz e Feijão.
Fageria, N. K., M. P. Barbosa Filho, and L. F. Stone. 2003. Adequate and toxic levels of micronutrients in soil and plants for annual crops. Paper presented at the *XXIX Brazilian Soil Science Congress*, Ribeirão Preto, São Paulo, Brazil.
Fageria, N. K., E. M. Castro, and V. C. Baligar. 2004. Response of upland rice genotypes to soil acidity. In *The Red Soils of China: Their Nature, Management and Utilization*, eds. M. J. Wilson, Z. He, and X. Yang, pp. 219–237. Dordrecht, The Netherlands: Kluwer Academic Press.
Fageria, N. K., and H. R. Gheyi. 1999. *Efficient Crop Production*. Campina Grande, Brazil: Federal University of Paraiba.
Fageria, N. K., M. F. Moraes, E. P. B. Ferreira, and A. M. Knupp. 2012. Biofortification of trace elements in food crops for human health. *Commun. Soil Sci. Plant Anal.* 43:556–570.
Fageria, N. K., N. A. Slaton, and V. C. Baligar. 2003. Nutrient management for improving lowland rice productivity and sustainability. *Adv. Agron.* 80:63–152.
Fageria, N. K., and L. F. Stone. 2008. Micronutrient deficiency problems in South America. In *Micronutrient Deficiencies in Global Crop Production*, ed. B. J. Alloway, pp. 245–266. New York: Springer.
Fageria, N. K., L. F. Stone, and A. B. Santos. 2003. *Soil Fertility Management for Lowland Rice*. Santo Antonio de Goiás, Brazil: EMBRAPA Arroz e Feijão.
Follett, R. H., L. S. Murphy, and R. L. Donahue. 1981. *Fertilizers and Soil Amendments*. Englewood Cliffs, NJ: Prentice Hall.
Foth, H. D., and B. G. Ellis. 1988. *Soil Fertility*. New York: John Wiley.
Galrão, E. Z. 1999. Methods of copper application and evaluation of its availability for soybean grown on a Cerrado red yellow Latosol. *R. Bras. Ci. Solo.* 23:265–272.

Gladstones, J. S., J. F. Loneragan, and W. J. Simmons. 1975. Mineral elements in temperate crop and pasture plants. III. Copper. *Aust. J. Agric. Res.* 26:113–126.

Graham, R. D. 1976. Anomalous water relations in copper deficient wheat plants. *Aust. J. Plant Physiol.* 3:229–236.

Graham, R. D. 1984. Breeding for nutritional characteristics in cereals. In *Advances in Plant Nutrition,* Vol. 1, eds. P. B. Tinker and A. Lauchi, pp. 57–102. New York: Praeger.

Graham, R. D. 2008. Micronutrient deficiencies in crops and their global significance. In *Micronutrient Deficiencies in Global Crop Production*, ed. B. J. Alloway, pp. 41–61. New York: Springer.

Graham, R. D., and E. K. S. Nambiar. 1981. Advances in research on copper deficiency in cereals. *Aust. J. Agric. Res.* 32:1009–10037.

Hodgson, J. F. 1963. Chemistry of the micronutrient elements in soils. *Adv. Agron.* 15:119–159.

Holloway, R. E., R. D. Graham, and S. P. Stacey. 2008. Micronutrient deficiencies in Australian field crops. In *Micronutrient Deficiencies in Global Crop Production*, ed. B. J. Alloway, pp. 63–92. New York: Springer

Ishizuka, Y. 1978. *Nutrient Deficiencies of Crops*. Taipei, Taiwan: Food and Fertilizer Technology Center.

Jarvis, S. C. 1981. Copper concentrations in plants and their relationship to soil properties. In *Copper in Soils and Plants*, eds. J. F. Loneragan, A. D. Robson, and R. D. Graham, pp. 265–286. Sydney, Australia: Academic Press.

Jewell, A. W., B. G. Murray, and B. J. Alloway. 1988. Light and electron microscope studies on pollen development in barley (*Hordeum vulgare* L.) grown under copper-sufficient and deficient conditions. *Plant Cell Environ.* 11:273–281.

King, P. M. 1974. Copper deficiency symptoms in wheat. *J Agric. South Aust.* 77:96–105.

Kubota, J. 1983. Copper status of United States soils and forage plants. *Agron. J.* 75:913–918.

Lindsay, W. L. 1979. *Chemical Equilibrium in Soils*. New York: John Wiley.

Loneragan, J. F. 1975. The availability and absorption of trace elements in soil-plant systems and their relation to movement and concentrations of trace elements in plants. In *Trace Elements in Soil-Plant-Animal Systems*, eds. D. J. D. Nicholas and A. R. Egan, pp. 109–134. New York: Academic Press.

Loneragan, J. F., K. Snowball, and A. D. Robson. 1976. Remobilisation of nutrients and its significance in plant nutrition. In *Transport and Transfer Processes in Plants*, eds. I. F. Wardlaw and J. B. Passioura, pp. 463–469. New York: Academic Press.

Loneragan, J. F., K. Snowball, and A. D. Robson. 1980. Copper supply in relation to content and redistribution of copper among organs of the wheat plant. *Ann. Bot.* 45:621–632.

Mackay, D. C., E. W. Chipman, and U. C. Gupta. 1966. Copper and molybdenum nutrition of crops grown on acid *Spagnum* peat soil. *Soil Sci. Soc. Am. Proc.* 30:755–759.

Martens, D. C., and W. L. Lindsay. 1990. Testing soils for copper, iron, manganese and zinc. In *Soil Testing and Plant Analysis*, 3rd edition, ed. R. L. Westerman, pp. 229–264. Madison, WI: SSSA.

Martens, D. C., and D. T. Westermann. 1991. Fertilizer applications for correcting micronutrient deficiency. In *Micronutrients in Agriculture*, 2nd edition, eds. J. J. Mortvedt, F. R. Cox, L. M. Shuman, and R. M. Welch, pp. 549–592. Madison, WI: Soil Science Society of America.

McBride, M. B. 1981. Forms and distribution of copper in solid and solution phases of soil. In *Copper in Soils and Plants*, eds. J. F. Loneragan, A. D. Robson, and R. D. Graham, pp. 25–45. Sydney, Australia: Academic Press.

Mengel, D. B. 1980. *Role of Micronutrients in Efficient Crop Production*. Indiana Coop. Ext. Serv. Ay-239. West Lafayate, IN: Purdue University.

Mengel, K, E. A. Kirkby, H. Kosegarten, and T. Appel. 2001. *Principles of Plant Nutrition*, 5th edition. Dordrecht, The Netherlands: Kluwer.

Moraghan, J. T., and H. J. Mascagni, Jr. 1991. Environmental and soil factors affecting micronutrient deficiencies and toxicities. In *Micronutrients in Agriculture*, 2nd edition, eds. J. J. Mortvedt, F. R. Cox, L. M. Shuman, and R. M. Welch, pp. 371–425. Madison, WI: Soil Science Society of America.

Mortvedt, J. J. 1994. Needs for controlled availability micronutrient fertilizers. *Fert. Res.* 38:213–221.

Mortvedt, J. J. 2000. Bioavailability of micronutrients. In *Handbook of Soil Science*, ed. M. E. Sumner, pp. D71–D88. Boca Raton, FL: CRC Press.

Nambiar, E. K. S. 1976a. Genetic differences in the copper nutrition of cereals. I. Differential responses to genotypes to copper. *Aust. J. Agric. Res.* 27:453–463.

Nambiar, E. K. S. 1976b. Genetic differences in copper nutrition of cereals. II. Genotypic differences in response to copper in relation to copper, nitrogen and other mineral contents of plants. *Aust. J. Agric. Res.* 27:465–477.

Oplinger, E. S., and A. J. Ohlrogge. 1974. Response of corn and soybeans to field applications of copper. *Agron. J.* 66:568–571.

Primavesi, A. M., and A. Primavesi. 1970. Effect of the trace element copper on rice (*Oryza sativa* L.). *Agrochimica.* 14:490–495.

Sanders, J. R., and C. Bloomfield. 1980. The influence of pH, ionic strength and reactant concentration on copper complexing by humified organic matter. *J. Soil Sci.* 31:53–63.

Shuman, L. M. 1991. Chemical forms of micronutrients. In *Micronutrients in Agriculture*, 2nd edition, eds. J. J. Mortvedt, F. R. Cox, L. M. Shuman, and R. M. Welch, pp. 113–144. Madison, WI: Soil Science Society of America.

Sims, J. T., and G. V. Johnson. 1991. Micronutrient soil tests. In *Micronutrients in Agriculture*, 2nd edition, ed. J. J. Mortvedt, pp. 427–476. Madison, WI: SSSA.

Tisdale, S. L., W. L. Nelson, and J. D. Beaton. 1985. *Soil Fertility and Fertilizers*, 4th edition. New York: Macmillan.

Welch, R. M., W. H. Allaway, W. A. House, and J. Kubota. 1991. Geographic distribution of trace element problems. In *Micronutrients in Agriculture*, 2nd edition, eds. J. J. Mortvedt, F. R. Cox, L. M. Shuman, and R. M. Welch, pp. 31–57. Madison, WI: SSSA.

Younts, S. E. 1964. Response of wheat to rates, dates of application and sources of copper and to other micronutrients. *Agron. J.* 56:266–269.

9 Manganese

9.1 INTRODUCTION

Manganese (Mn) is an essential micronutrient for the growth and development of crop plants. Natural background concentrations of Mn in soil range from 0.5 to 5000 mg kg^{-1}, and the oxidation state ranges from 0 to +7, with +2, +3, and +4 the most common oxidation states in natural environments (Hernandez-Soriano et al. 2012). In soils, Mn solubility is determined by two major variables, pH and redox potential. In well-aerated alkaline soils, Mn may not be sufficiently available for healthy plant growth (Graham, Davies, and Ascher 1985; Curtin, Martin, and Scott 2008). Mn^{3+} and Mn^{4+} hydroxides precipitate at pH above 5.5 and under aerobic conditions (Hernandez-Soriano et al. 2012). The Mn^{2+} form is highly soluble and is the thermodynamically most stable form in soils at low pH (Porter et al. 2004) or under reducing conditions such as may happen in flooded rice (Foy 1984; Weil, Foy, and Coradetti 1997).

Manganese deficiency has been documented in some rice-growing regions, but it is not common in most soils that are used for rice production. Nutritional disorders related to Mn occur less frequently than those associated with Fe, Cu, and B. Rice grown on Oxisols, Ultisols, and Histosols are most likely to experience disorders related to Mn nutrition (Fageria 2001; Fageria, 2002a,b; Fageria, Slaton, and Baligar 2003; Fageria, Baligar, and Clark 2006). In Oxisols and Ultisols, liming to raise pH may cause Mn deficiency. Data presented in Table 9.1 show that Mn uptake by upland rice grown on Brazilian Oxisol decreased significantly when soil pH was raised from 4.6 to 6.8. The variability in Mn uptake was 99% with increasing soil pH. Similarly, Mn concentration in lowland rice grown on Brazilian Inceptisol also decreased significantly when soil pH was raised from 4.9 to 7.0. The variability in Mn concentration was 89% with the change in soil pH. Lindsay (1979) reported that Mn solubility decreased 100-fold for each unit increase of soil pH in the range of 4–9.

Soil pH influences solubility, soil solution concentration, ionic form, and mobility of micronutrients in soil and, consequently, the acquisition of micronutrient elements by plants (Fageria, Baligar, and Jones 2011). As a rule, the availability of B, Cu, Fe, Mn, and Zn usually increases and Mo decreases as soil pH decreases. These nutrients are usually adsorbed onto sesquioxide soil surfaces. Table 9.2 summarizes important changes in micronutrient concentrations and acquisition by plants as influenced by soil pH.

Soil organic matter (OM) content is another important factor affecting availability of Mn to crop plants. Soil OM content has been related to increased, decreased, and no effects on Mn availability to crop plants (Reisenauer 1988). Within soil fractions, exchangeable and organically bound forms of Mn are important. Higher Mn accumulations in surface soil horizons have been reported, indicating

TABLE 9.1
Influence of pH on Uptake of Manganese by Upland and Lowland Rice

Soil pH in H_2O	Mn Uptake by Upland Rice (μg plant^{-1})	Soil pH in H_2O	Mn Conc. in Lowland Rice Shoot (mg kg^{-1})
4.6	11165	4.9	1433
5.7	5006	5.9	833
6.2	4309	6.4	757
6.4	3607	6.7	767
6.6	2762	7.0	663
6.8	2359		
R^2	0.99*	R^2	0.89*

Source: Adapted from Fageria, N. K., and V. C. Baligar. 1999. Growth and nutrient concentrations of common bean, lowland rice, corn, soybean, and wheat at different soil pH on an Inceptisol. *J. Plant Nutr.* 22:1495–1507; Fageria, N. K. 2000. Upland rice response to soil acidity in Cerrado soil. *Pesq. Agropec. Bras.* 35:2303–2307.

*Significant at the 1% probability level.

TABLE 9.2
Influence of Soil pH on Micronutrient Concentration and Uptake

Micronutrient	Influence on Concentration/Uptake
Boron	Increasing soil pH favors adsorption of boron, and this element generally becomes less available to plants. There is often a dramatic drop in boron availability and uptake at pH levels above 6.0.
Copper	Solubility of Cu^{2+} is very pH dependent and decreases 100-fold for each unit increase, and plant uptake also decreases.
Iron	Fe^{3+} and Fe^{2+} activities in soil solution decrease 1000-fold and 100-fold, respectively, for each unit increase in pH. In most oxidized soils, uptake of iron by crop plants is decreased with increasing soil pH.
Manganese	The principal ionic species in soil solution is Mn^{2+}, and its concentration decreases 100-fold for each unit increase in pH. In extremely acidic soils, Mn^{2+} solubility can be sufficiently great to cause toxicity problems in sensitive crop species.
Molybdenum	Above pH 4.2, the MoO_4^{2-} species is dominant. Concentration of this species increases with increasing soil pH and, hence, uptake to plants also increases. Water-soluble Mo increases six-fold as pH increases from 4.7 to 7.5. Replacement of adsorbed Mo by OH^- ions is responsible for the increase in water-soluble Mo as pH increases.
Zinc	Zinc solubility is highly pH dependent and decreases 100-fold for each unit increase in pH; uptake by plants also decreases as a consequence.
Chlorine	The Cl^- anion is bound vary slightly by most soils in the mildly acidic to neutral pH range and becomes negligible at pH 7.0. Appreciable amounts can be adsorbed, however, with increasing acidity, particularly by Oxisols and Ultisols, which are dominated by kaolinitic clay. In general, increasing soil pH increases Cl uptake by plants.

Source: Compiled from Fageria, N. K., V. C. Baligar, and R. B. Clark. 2002. Micronutrients in crop production. *Adv. Agron.* 77:189–272.

FIGURE 9.1 Response of lowland rice to Mn application in Brazilian Inceptisol.

that Mn may be closely associated with OM (McDaniel and Buol 1991). The positive correlations between OM and Mn indicate that Mn has a strong affinity for OM and that higher Mn concentrations in surface soil (compared with lower layers) are likely due to higher OM in surface horizons (Zhang et al. 1997). The sites of Mn retention have also been associated with calcium carbonate in pH 8 calcareous soils (Karimian and Gholamalizadeh Ahangar 1998). Manganese, particularly Mn^{2+}, forms complexes with fulvic and humic acids and humins, and with ligands, such as organic, amino, and sugar acids, hydroxamates, phenolics, siderophores, and other organic compounds produced by various organisms in soil solution (Stevenson 1986; Tate 1987; Marschner 1995). Hydrated Mn^{2+} forms complexes with carboxyl groups of OM, which helps explain observations that Mn binds weakly to OM compared with Fe, Cu, and Zn (Bloom 1981). Manganese availability in soils high in OM may also decrease because of formation of unavailable Mn complexes. Unavailable Mn complexes form in peat or muck soils (Fageria, Baligar, and Clark 2002).

Low levels of Mn have been reported by Fageria and Baligar (1999) and Fageria, Stone, and Santos (2003) in Brazilian lowlands known as "Varzea" where rice is mainly grown during the rainy season. An example of the response of lowland rice to Mn application to a Varzea soil is presented in Figure 9.1. Soil for this experiment was collected from Varzea in the central Brazilian state of Tocantins. Manganese deficiency in crop plants may also be due to negative interactions between Mn and other elements (Fageria 2009). Fageria (2009) has reported such negative interactions between Mn, Fe, Zn, Ca, and Mg. Foy (1984) found that the toxicity of a given level of soluble Mn in the growth medium, or even within the plant, depends on

TABLE 9.3
Influence of Magnesium on Mn Concentration in the Shoot of Upland Rice

Mg Level in Soil ($cmol_c\ kg^{-1}$)	Mn Concentration ($mg\ kg^{-1}$)
0.30	693
1.05	520
1.15	588
1.33	468
3.52	415
6.22	333
Regression	Linear*

Source: Adapted from Fageria, N. K., and C. M. R. Souza. 1991. Upland rice, common bean, and cowpea response to magnesium application on an Oxisol. *Commun. Soil Sci. Plant Anal.* 22:1805–1816.
*Significant at the 1% probability level.

interactions between Mn and several other mineral elements, particularly Fe and Si. Data in Table 9.3 show increasing Mg levels in Brazilian Oxisol significantly decreased the Mn concentration in an upland rice shoot. The objective of this chapter is to discuss the role of Mn in rice production and the management practices that are appropriate for its efficient use.

9.2 CYCLE IN SOIL–PLANT SYSTEM

The Mn cycle in the soil–plant system involves addition, uptake, and loss during the crop growth cycle. The major Mn addition sources are organic manures, chemical fertilizers, the release from parent materials, and microbial biomass. The principal causes of Mn depletion are uptake by crop plants, loss through soil erosion, leaching in sandy soils, and adsorption on organic compounds or microbial biomass (Fageria 2009). Manganese uptake by plants is influenced by many factors that determine its solubility and transport behavior in soil–plant systems. Manganese uptake is metabolically mediated, and its uptake increases from pH 4 to 6 (Maas, Moore, and Mason 1969). Above pH 6, oxidation of Mn^{2+} to Mn^{4+} occurs and Mn^{2+} uptake is reduced. Soil pH and redox potentials control Mn supply to roots by mass flow and diffusion. Deficiency of Mn usually occurs when soil pH is >6.2, but Mn^{2+} may be sufficient in some soils even though pH is ≥7.5 (Barber 1995). The prevailing source of Mn at root surfaces is Mn^{2+}. Manganese forms complexes with organic compounds (trihydroxamic acid, sideramines) of microbial and plant origin, which increase Mn mobility in soil (Clarkson 1988). Three major sources of Mn exist in soils that are primarily responsible for Mn supply to roots (Marschner 1995). These include exchangeable Mn, organically complexed Mn, and Mn oxides. The proportion of these Mn forms varies with soil type, soil pH, and OM. As soil pH decreases, the proportion of exchangeable

Mn increases dramatically, although proportions of Mn oxides and Mn bound to Mn and Fe oxides decrease. In soils low in available Fe, root reductase activity is stimulated because of acidification of the rhizosphere; this may lead to higher Mn mobility and uptake. Greater ranges in foliage Mn were noted for different species of plants growing in the same soil compared with Cu, Fe, or Zn (Gladstones and Loneragan 1970). These differences were attributed to species ability to acidify the rhizosphere soil rather than to Mn requirement (Fageria, Baligar, and Clark 2002).

The chemistry of Mn in soils is complex because three oxidation states are involved. These states are Mn^{2+}, Mn^{3+}, and Mn^{4+}, with Mn^{2}+ being the primary form absorbed by plants. Manganese forms hydrated oxides with mixed valency states (Lindsay 1979). Knowledge of Mn hydrolysis reactions is important in understanding solubility of its minerals and its uptake by plants. Lindsay (1979) reported these reactions, which are as follows:

$$Mn^{2+} + H_2O \leftrightarrow MnOH^{+} + H^{+} \quad (9.1)$$

$$Mn^{2+} + 2H_2O \leftrightarrow Mn(OH)_2^{0} + 2H^{+} \quad (9.2)$$

$$Mn^{2+} + 3H_2O \leftrightarrow Mn(OH)_3^{-} + 3H^{+} \quad (9.3)$$

$$Mn^{2+} + 4H_2O \leftrightarrow Mn(OH)_4^{2-} + 4H^{+} \quad (9.4)$$

$$2Mn^{2+} + H_2O \leftrightarrow Mn(OH)^{3+} + H^{+} \quad (9.5)$$

$$2Mn^{2+} + 3H_2O \leftrightarrow Mn_2(OH)_3^{+-} + 3H^{+} \quad (9.6)$$

$$Mn^{3+} + 3H_2O \leftrightarrow Mn(OH)^{2+-} + 3H^{+} \quad (9.7)$$

In submerged soils or in flooded rice, Mn^{4+} is reduced to Mn^{2+} due to oxygen depletion. The Mn^{2+} concentration in the soil solution increases when redox potential (Eh) values decrease (Fageria, Slaton, and Baligar 2003). The reduction of Mn^{4+} occurs after NO_3^- reduction and before Fe^{3+} reduction. Patrick and Reddy (1978) classified soils based on redox potential as aerated or well-drained soils, +700 to +500 mV; moderately reduced, +400 to +200 mV; reduced, +100 to –100 mV; and highly reduced, –100 to –300 mV. Under reduced soil conditions, rice uptake of Mn increases. Reduction of Mn can be either chemical or microbiological, although microbiological reduction is likely to predominate in flooded rice soils that are at pH 5.5–6.0 (Patrick and Reddy 1978). The reduction of Mn can be explained from the following reduction equation:

$$MnO_2 + 4H^{+} + 2e^{-} \leftrightarrow Mn^{2+} + 2H_2O \quad (9.8)$$

Rice can tolerate very high concentrations of Mn, and toxicity of this element is not a problem for flooded rice. Fe^{2+} toxicity in flooded rice is avoided with higher Mn concentrations due to a strong opposition in the absorption process between

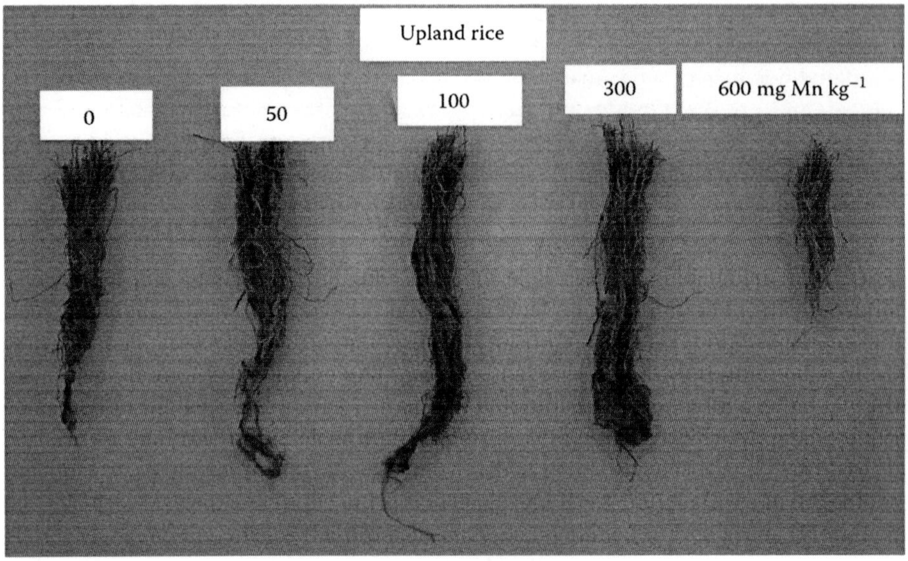

FIGURE 9.2 Influence of Mn on root growth of upland rice.

these two elements. When flooded rice soils have Fe toxicity problems, application of Mn is one strategy for reducing the toxicity.

9.3 FUNCTIONS

Manganese plays an important role in many physiological and biochemical functions in plants. These functions include the formation of chlorophyll, the activation of respiratory enzymes, participation in several important metabolic reactions, and the acceleration of germination and maturity (Fageria 2001). More than 60% of Mn contained in higher plant leaves is found in chloroplast. Along with Fe and Cu, Mn performs a vital role in the electron transport system (Obata 1995). Manganese also functions as a cofactor to activate enzymes such as dehydrogenases and hydrolyses in the glycolysis system and in the citric acid cycle and RNA polymerases in the chloroplasts. Manganese activates the protease enzyme contained in rice seeds (Horiguchi and Kitagishi 1976). The most well-known and extensively studied function of Mn in green plants is its involvement in photosynthetic O_2 evolution (Marschner 1995). In addition, Mn supplied at an adequate rate can improves root growth of rice (Figure 9.2).

9.4 DEFICIENCY SYMPTOMS

Because manganese is immobile in plants, Mn deficiency symptoms initially appear in the younger leaves. Manganese-deficient plants are chlorotic and develop an irregular yellow mottling between the leaf veins. Long chlorotic streaks may be interspersed with smaller dark brown or red lesions that join to form long, dark brown, or

FIGURE 9.3 (See color insert.) Manganese deficiency in upland rice in nutrient solution. (From Fageria, N. K., and M. P. Barbosa Filho. 1994. *Nutritional Deficiency in Rice: Identification and Correction*. Goiania, Brazil: EMBRAPA.)

red lesions. In the beginning, Mn deficiency symptoms are similar to those of iron. As the iron deficiency advances, leaves become white, whereas Mn-deficient leaves develop small, discrete dark reddish brown or black lesions within the chlorotic tissues (Fageria and Barbosa Filho 1994). In the case of Mn toxicity, yellow spots generally develop between leaf veins, extend to the interveinal areas, and eventually turn brown as the toxicity develops (Shimada 1995). However, Mn toxicity rarely occurs in upland and lowland rice. Figures 9.3 and 9.4 show Mn deficiency symptoms in rice leaves developed in solution culture.

9.5 UPTAKE IN PLANT TISSUE

Manganese deficiency in rice occurs when the Mn concentration in the plant tissue is less than 20 mg kg^{-1} (Wells et al. 1993). The critical tissue concentration of Mn in most plants ranges from 10 to 20 mg kg^{-1} in mature leaves and is surprisingly consistent regardless of the plant species, cultivar, or the prevailing environmental conditions (Marschner 1995). Fageria, Baligar, and Jones (2011) reported whole plant Mn concentrations of 30–600 mg kg^{-1} at tillering were sufficient.

Rice can tolerate tissue levels of more than 2500 mg Mn kg^{-1} without adverse effects on either growth or grain yield (Wells et al. 1993). Fageria (2001) reported critical shoot concentration for Mn toxicity in upland rice is about 4600 mg kg^{-1}. Cheng and Quallette (1971) reported a critical, toxic tissue concentration for rice of 7000 mg Mn kg^{-1}. Yoshida (1981) reported that in many cases, a high Mn content in

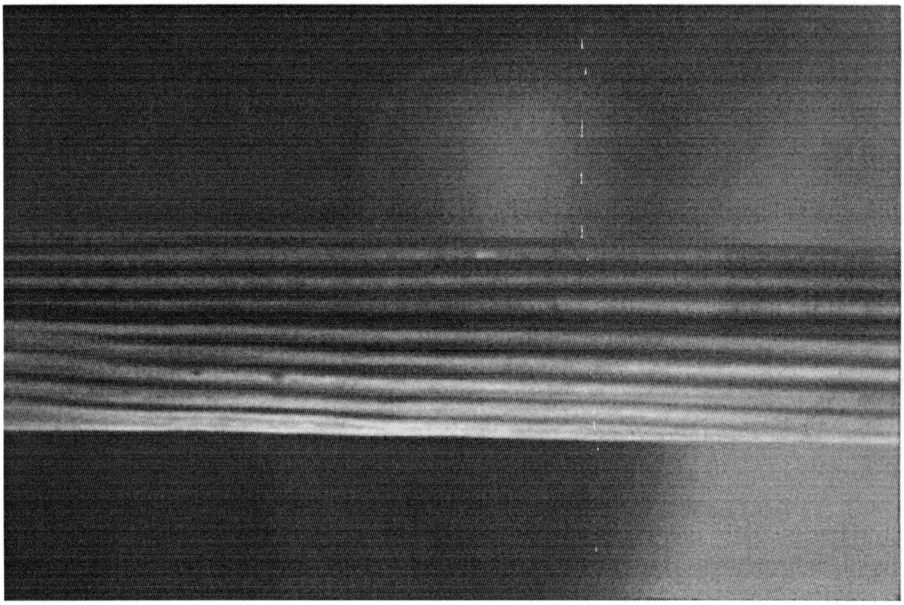

FIGURE 9.4 (See color insert.) Manganese deficiency symptoms in rice leaf. (From Fageria, N. K., and M. P. Barbosa Filho. 1994. *Nutritional Deficiency in Rice: Identification and Correction*. Goiania, Brazil: EMBRAPA.)

rice tissue is frequently associated with high yields, possibly indicating that high Mn content in the soil is associated with various favorable soil conditions.

The concentration and uptake of Mn in straw and grain of upland and lowland rice during the growth cycle were determined (Table 9.4). Concentration in straw or shoot decreased significantly with advanced plant age (Table 9.5). However, uptake increased significantly in a quadratic fashion with advanced plant age. The decrease in Mn concentration with advanced plant age was related to an increase in dry matter of shoot or straw (also with advanced plant age) that caused dilution of Mn in the plant tissue (Fageria 2004). Manganese concentration results in straw or shoot can be used to diagnose Mn deficiency or sufficiency in rice during the growth cycle. Similarly, Mn uptake data can be used to determine the amount of Mn removed from a given soil by upland and lowland rice during the growth cycle in order to maintain soil fertility for sustainable rice production.

9.6 USE EFFICIENCY

Nutrient use efficiency (dry matter or grain yield/unit nutrient uptake) is an important index to determine how a nutrient is utilized in dry matter or grain yield production by a crop species under agroecological conditions. Table 9.6 shows Mn use efficiency in straw and grain of upland and lowland rice. Manganese use efficiency decreased in straw and grain at flowering and in straw at physiological maturity, which might be related to translocation of Mn to grain. Similarly, the Mn use efficiency was higher

TABLE 9.4
Manganese Concentration and Uptake in Upland and Lowland Rice Straw and Grain during Growth Cycle

	Upland Rice			Lowland Rice	
Age in Days	Conc. (mg kg^{-1})	Uptake (g ha^{-1})	Age in Days	Conc. (mg kg^{-1})	Uptake (g ha^{-1})
19 (IT)	250	23	22 (IT)	784	284
43 (AT)	217	234	35 (AT)	847	965
68 (IP)	145	573	71 (IP)	580	3336
90 (B)	150	896	97 (B)	470	4781
102 (F)	212	1542	112 (F)	438	5979
130 (PM)	205	1319	140 (PM)	578	5602
130 (grain)	30	284	140 (grain)	58	369

Source: Adapted from Fageria, N. K. 2004. Dry matter yield and shoot nutrient concentrations of upland rice, common bean, corn, and soybean grown in rotation on an Oxisol. *Commun. Soil Sci. Plant Anal.* 35:961–974; Fageria, N. K., L. F. Stone, and A. B. Santos. 2003. *Soil Fertility Management for Lowland Rice.* Santo Antonio de Goiás, Brazil: EMBRAPA.

Note: IT, initiation of tillering; AT, active tillering; IP, initiation of primordia floral; B, booting; F, flowering; PM, physiological maturity.

TABLE 9.5
Relationship between Plant Age and Mn Concentration and Uptake in Upland Rice and Lowland Rice

Variable	Regression Equation	R^2
Plant age vs. Mn concentration in upland straw	$Y = 313.44 - 3.58X + 0.022X^2$	0.67*
Plant age vs. Mn uptake in upland rice straw	$Y = 4.48\exp(0.104X - 0.00046X^2)$	0.98**
Plant age vs. Mn concentration in lowland rice straw	$Y = 1075.17 - 10.94X + 0.051X^2$	0.85*
Plant age vs. Mn uptake in lowland rice straw	$Y = -2101.65 + 105.35X - 0.35X^2$	0.98**

Source: Adapted from Fageria, N. K. 2004. Dry matter yield and shoot nutrient concentrations of upland rice, common bean, corn, and soybean grown in rotation on an Oxisol. *Commun. Soil Sci. Plant Anal.* 35:961–974; Fageria, N. K., L. F. Stone, and A. B. Santos. 2003. *Soil Fertility Management for Lowland Rice.* Santo Antonio de Goiás, Brazil: EMBRAPA.

Note: Regression equations were calculated from data of Table 9.4.

*,**Significant at the 5 and 1% probability levels, respectively.

TABLE 9.6
Manganese Use Efficiency in Upland Rice and Lowland Rice Straw and Grain

Plant Age in Days	MnUE (kg g^{-1}) in Upland Rice	Plant Age in Days	MnUE (kg g^{-1}) in Lowland Rice
19 (IT)	4.6	22 (IT)	1.28
43 (AT)	5.2	35 (AT)	1.23
68 (IP)	10.4	71 (IP)	1.90
90 (B)	7.2	97 (B)	2.19
102 (F)	5.3	112 (F)	2.33
130 (PM)	5.4	140 (PM)	2.12
130 (grain)	32.3	140 (grain)	17.52

Source: Adapted from Fageria, N. K. 2009. *The Use of Nutrients in Crop Plants*. Boca Raton, FL: CRC Press. Lowland rice from Fageria, unpublished data.

Note: MnUE (kg g^{-1}) = (Straw or grain weight in kg/Uptake of Mn in straw or grain in g). MnUE, manganese use efficiency; IT, initiation of tillering; AT, active tillering; IP, initiation of primordia floral; B, booting; F, flowering; PM, physiological maturity.

TABLE 9.7
Manganese Uptake and Harvest Index in Principal Field Crops

Crop Species	Mn Uptake in Shoot (g ha^{-1})	Mn Uptake in Grain (g ha^{-1})	Total	Mn Harvest Index (%)
Upland rice	1318.51	284.03	1602.55	18
Corn	452.16	82.21	534.36	15
Dry beans	73.19	26.60	99.79	49
Dry beans	31.10	48.7	79.8	61
Soybeans	117.40	120.10	237.50	51

Source: Fageria, N. K. 2009. *The Use of Nutrients in Crop Plants*. Boca Raton, FL: CRC Press.

Note: Mn harvest index = (Mn uptake in grain/Mn uptake in grain plus shoot). Values are averages of two years of field trials.

in grain than in straw, which might be related to higher uptake in straw compared with grain.

9.7 HARVEST INDEX

Nutrient harvest index is also important in determining nutrient distribution in shoot and grain. If more of a nutrient is exported to grain, it means a higher depletion of that nutrient from soil. However, this may be a good characteristic in relation to human consumption. The Mn uptake and harvest index values for four crops are presented in Table 9.7. Exportation of Mn to grain was higher in legumes than in cereals.

9.8 MANAGEMENT PRACTICES

Management practices that impact Mn efficiency in rice are effective source, appropriate method and timing of application, adequate rate, use of acidic fertilizers in the band and neutral salts, and use of Mn-efficient genotypes.

9.8.1 Effective Sources

Effective source is related to the solubility of Mn fertilizers in water. Manganese sulfate and manganese chloride are water-soluble sources and can be used effectively in correcting Mn deficiency in rice. Important manganese sources are cited in Table 9.8. Of these, manganese oxide is insoluble in water, and its efficiency is low compared with manganese sulfate and manganese chloride sources.

9.8.2 Appropriate Methods and Timing of Application

Manganese is a metallic micronutrient that is strongly sorbed by soil clays, and its movement in soil is very low (Mortvedt 1994). Therefore, it does not reach the root surface by mass flow. Its application in the furrow or band near the root is desirable for efficient uptake. Band application with acid-forming fertilizers results in increased efficiency of applied Mn because the rate of oxidation of the divalent Mn to the unavailable tetravalent form (MnO_2) is decreased (Mortvedt 1994). As far as application timing is concerned, it should be applied at sowing. Foliar application can also be effective if the crop has a sufficient leaf area to absorb the applied soluble fertilizers. Snyder, Jones, and Coale (1990) reported that Mn deficiency could occur in seedling rice grown in drained Histosols with pH values near or above 7. Affected seedlings contain <20 mg Mn kg^{-1} tissue; the deficiency can be prevented

TABLE 9.8
Principal Manganese Carriers

Common Name	Formula	Mn Content (%)
Manganese sulfate	$MnSO_4 \cdot 3H_2O$	26–28
Manganese oxide	MnO	68–70
Manganese oxide	MnO_2	63
Manganese EDTA chelate	MnEDTA	12
Manganese frit	Frit	10–25
Manganese carbonate	$MnCO_3$	31
Manganese chloride	$MnCl_2$	17
Manganese polyflavonoid	MnPF	8
Manganese methoxyphenyl-propane	MnPP	5

Source: Tisdale, S. L., W. L. Nelson, and J. D. Beaton. 1985. *Soil Fertility and Fertilizers*, 4th edition. New York: Macmillan; Foth, H. D., and B. G. Ellis. 1988. *Soil Fertility*. New York: John Wiley; Fageria, N. K. 1989. *Tropical Soils and Physiological Aspects of Crops*. Brasilia: EMBRAPA; Fageria, N. K. 2009. *The Use of Nutrients in Crop Plants*. Boca Raton, FL: CRC Press.

by drilling $MnSO_4$ at seeding to provide approximately 15 kg Mn ha^{-1} as $MnSO_4$ and near maximum grain yield. There are low residual effects of applied Mn fertilizers because of ease of oxidation to the unavailable forms (MnO_2 or Mn^{4+}); hence, annual applications are required.

9.8.3 Adequate Rate

Field trials relating Mn rate to grain yield are necessary to determine the adequate rate of Mn application in different agroecological regions. Such trials should be conducted at multiple sites and repeated for at least two years. Three years of field data give more precise results. Fageria, Stone, and Santos (2003) determined the adequate rate of Mn for upland rice in a greenhouse experiment (Table 9.9). The adequate rate was 2 mg kg^{-1} and the toxic rate was 560 mg kg^{-1}, indicating a wide range between adequate and toxic levels.

Soil tests of Mn rate in crops can be applied if calibration data are available. Extractable Mn has been used successfully in many studies, particularly for small grains, to predict crop yield responses to Mn fertilization (Sims and Johnson 1991). The critical Mn level established by soil testing is an important criterion for diagnosis of Mn disorders in crop plants. Sims and Johnson (1991) reported that critical Mn level for Mehlich 1, Mehlich 3, 0.0M HCl, or 0.03M H_3PO_4 extracting solutions varied from 4 to 20 mg kg^{-1}, relative to 1.4 mg kg^{-1} for diethylene triamine pentaacetic acid (DTPA) (0.005M DTPA + 0.01M $CaCl_2$ + 0.1M triethanolamine [TEA]), adjusted to pH 7.3 (Lindsay and Norvell 1978). Lindsay and Cox (1985) reported the critical Mn level in soil for DTPA extracting solution in the range of 1–5 mg kg^{-1} for tropical soils.

TABLE 9.9
Adequate and Toxic Levels of Manganese for Annual Crops Applied to Brazilian Oxisols

Crop Species[a]	Adequate Mn Rate (mg kg^{-1})	Toxic Mn Rate (mg kg^{-1})
Upland rice	2	560
Dry beans	12	112
Corn	12	400
Soybeans	No response	72
Wheat	No response	10

Source: Adapted from Fageria, N. K., L. F. Stone, and A. B. Santos. 2003. *Soil Fertility Management for Lowland Rice*. Santo Antonio de Goiás, Brazil: EMBRAPA.

Note: Adequate rate was calculated at 90% of the maximum dry weight of shoot and toxic level was calculated at 10% reduction in shoot dry weight after achieving maximum weight.

[a] Plants were harvested annually from Brazilian Oxisols at physiological maturity, and soil pH at harvest was about 6.0 in water in all five experiments.

TABLE 9.10
Adequate and Toxic Level of Soil Manganese Extracted by Mehlich 1 and DTPA Extracting Solutions for Five Annual Crops Grown on Brazilian Oxisols

	Mehlich 1 Extracting Solution		DTPA Extracting Solution	
Crop Species[a]	Adequate Mn Level (mg kg^{-1})	Toxic Mn Level (mg kg^{-1})	Adequate Mn Level (mg kg^{-1})	Toxic Mn Level (mg kg^{-1})
Upland rice	8	168	4	80
Dry beans	8	128	6	88
Corn	8	400	4	336
Soybeans	8	92	4	56
Wheat	8	44	3	40

Source: Adapted from Fageria, N. K., L. F. Stone, and A. B. Santos. 2003. *Soil Fertility Management for Lowland Rice*. Santo Antonio de Goiás, Brazil: EMBRAPA.

Note: Adequate level was calculated at 90% of the maximum dry weight of shoot, and toxic level was calculated at 10% reduction in shoot dry weight after achieving maximum weight.

[a] Plants were harvested annually at physiological maturity from Brazilian Oxisols, and soil pH at harvest was about 6.0 in water in all five experiments.

Fageria, Stone, and Santos (2003) determined the adequate and toxic levels of soil Mn extracted by Mehlich 1 and DTPA extracting solutions for five annual crops grown on Brazilian Oxisols (Table 9.10). The Mehlich 1 extractable adequate Mn level was 8 mg kg^{-1} for upland rice, dry beans, corn, soybeans, and wheat. The toxic level of Mehlich 1 extractable Mn for these crops was 168 mg kg^{-1} for upland rice, 128 mg kg^{-1} for dry beans, 400 mg kg^{-1} for corn, 92 mg kg^{-1} for soybeans, and 44 mg kg^{-1} for wheat. The adequate level of Mn by DTPA extracting solution varied from 3 to 6 mg kg^{-1} depending on crop species. It was maximum for dry beans and minimum for wheat (Table 9.10). The toxic level of Mn by DTPA extracting solution was 80 mg kg^{-1} for upland rice, 88 mg kg^{-1} for dry beans, 336 mg kg^{-1} for corn, 56 mg kg^{-1} for soybeans, and 40 mg kg^{-1} for wheat. These results indicate that among the aforementioned five crop species, corn was most tolerant of Mn toxicity and wheat was most susceptible to Mn toxicity. In addition, adequate as well as toxic level of Mn was higher for Mehlich 1 than the DTPA extracting solution. Furthermore, the adequate level of Mn was similar for the five crop species. Overall, adequate level for Mehlich 1 extracting solution was 8 mg kg^{-1}, and for DTPA it was 4 mg kg^{-1}.

9.8.4 Use of Acidic Fertilizers in the Band and Neutral Salts

Decreasing the soil pH in a localized zone by band application of acid-forming N or P fertilizers has resulted in increased Mn uptake and yields (Jackson and Carter 1976). Increased levels of ammonium acetate-extractable Mn have also resulted from application of acid N or P fertilizers (Voth and Christenson 1980). Petrie and Jackson (1984) showed that band application of acid-forming fertilizers was necessary to correct Mn deficiency and produce maximum barley and oat yields on alkaline, high OM mineral soils in the Klamath Lake area of south central Oregon. Hossner and Richards (1968) noted higher Mn uptake from $MnSO_4$ applied with monoammonium phosphate and acidic ammonium polyphosphate than with monocalcium phosphate, the predominant form in superphosphate. Soil band application of acid-forming phosphate fertilizers alone often solubilized sufficient soil Mn to correct soil deficiency (Randall, Schulte, and Corey 1975).

Some neutral salts (KCl, NaCl, and $CaCl_2$) have been shown to increase bioavailability of soil Mn (Jackson, Westerman, and Moore 1966; Krishnamurti and Huang 1992), an effect that was hypothesized as due to an Mn^{4+}, Mn^{3+}, Cl^- redox reaction (Westermann, Jackson, and Moore 1971). For California soils, Khattak, Jarrell, and Page (1989) advocated ion exchange as the dominant mechanism to explain the salt-induced Mn release. Krishnamurti and Huang (1992) reported that increasing KCl concentration increased Mn release from selected soils of different taxonomic orders. They concluded that ionic strength effect coupled with complexation was mainly responsible for the enhanced Mn release by KCl from the soils in the common pH range. Application of chloride (Cl) has also been shown to increase the availability and uptake of native soil Mn. Application of Cl as KCl or $CaCl_2$ increased the leaf Mn concentration in common beans and sweet corn grown on acidic soils; the increased uptake was such that Mn toxicity was evident on common beans (Jackson, Westerman, and Moore 1966). Application of KCl resulted in greater increases in $Mg(NO_3)_2$—extractable Mn than did K_2SO_4 (Westermann, Jackson, and Moore 1971). The increase in extractable Mn was attributed to anoxidation–reduction couple involving Mn and Cl (Petrie and Jackson 1984).

9.8.5 Use of Efficient Genotypes

Response of crop species to applied Mn varied significantly. Manganese uptake and utilization also varied among crop species and among genotypes within species. Data in Table 9.11 show differences in dry matter production by 10 upland rice genotypes at two Mn levels. Shoot dry weight varied from 46.37 to 91.23 g 4 plants^{-1} at the lower Mn level (depending on genotype) with an average value of 62.42 g 4 plants^{-1}. The variation was 97% between Carajas, the genotype producing the lowest shoot dry weight, and IR42, the genotype producing the highest shoot weight. At the higher Mn level, shoot dry weight varied from 51.73 to 76.27 g 4 plants^{-1}, with an average value of 63.42 g 4 plants^{-1}.

Based on Mn use efficiency and grain yield at low Mn level, genotypes were classified into four groups (Figure 9.5). Fageria and Baligar (1993) suggested this type of classification for the nutrient use efficiency of crop genotypes. The first group was efficient and responsive (ER) genotypes. These genotypes produced above average yields at the low Mn level, and Mn use efficiency was higher than average. Genotypes Carisma, CNA8540, and IR42 fall into this group. The second classification was efficient and nonresponsive (ENR) genotypes. These genotypes produced more than the average yield of the 10 genotypes at the low Mn level, but their response to Mn application was lower than the average. The genotypes CNA8557 and Maravilha fell into this group. The third group of genotypes was nonefficient and responsive (NER). The genotypes that produced less than the average grain yield

TABLE 9.11
Shoot Dry Matter Yield of 10 Upland Rice Genotypes at Two Mn Levels Applied to Brazilian Oxisol

	Shoot Dry Weight (g 4 plants^{-1})	
Genotype	**0 mg Mn kg^{-1}**	**20 mg Mn kg^{-1}**
Bonança	51.43cde	60.47abc
Caipó	68.03bc	76.27a
Canastra	65.20bcd	65.87abc
Carajas	46.37e	57.93bc
Carisma	61.00bcde	54.33c
CNA8540	55.70bcde	55.47c
CNA8557	64.57bcd	68.20abc
Guarani	48.83de	51.73c
Maravilha	71.80b	68.23abc
IR42	91.23a	75.70ab
Average	62.42	63.42

Source: Fageria, N. K., M. P. Barbosa Filho, and A. Moreira. 2008. Screening upland rice genotypes for manganese use efficiency. *Commun. Soil Sci. Plant Anal.* 39:2873–2882.

Note: Means in the same column followed by the same letter are statistically not different at the 5% probability level by the Tukey's test.

of the 10 genotypes at low Mn level but responded to Mn application above the average are classified in this group. The only genotype tested for this study that fell into the NER group was Caipó. The fourth group of genotypes produced less than average yield at low Mn level, and their response to applied Mn was also less than average. This group of genotypes was classified as nonefficient and nonresponsive (NENR). These genotypes were Bonança, Canastra, Carajas, and Guarani.

From a practical point of view, the genotypes that fall into the ER group are the most desirable because they can produce more at a low Mn level and they also respond well to applied Mn. These genotypes can produce a reasonably good yield regardless of technology available (Fageria and Baligar 1993). The second most desirable group is made up of the ENR genotypes. They can be planted under a low Mn level and still produce more than an average yield. The NER genotypes sometimes can be used in breeding programs for their Mn-responsive characteristics. The least desirable genotypes are the NENR type. These results indicate that upland rice genotypes differ in Mn use efficiency. Inter- and intraspecific variations in Mn nutrition have been recognized among cereal species and genotypes (Fageria and Baligar 1993; Fageria, Baligar, and Li 2008; Fageria, Baligar, and Jones 2011); this suggests that it may be possible to develop cultivars that are efficient at low-nutrient levels or that are capable of using Mn more efficiently when it is applied as fertilizer.

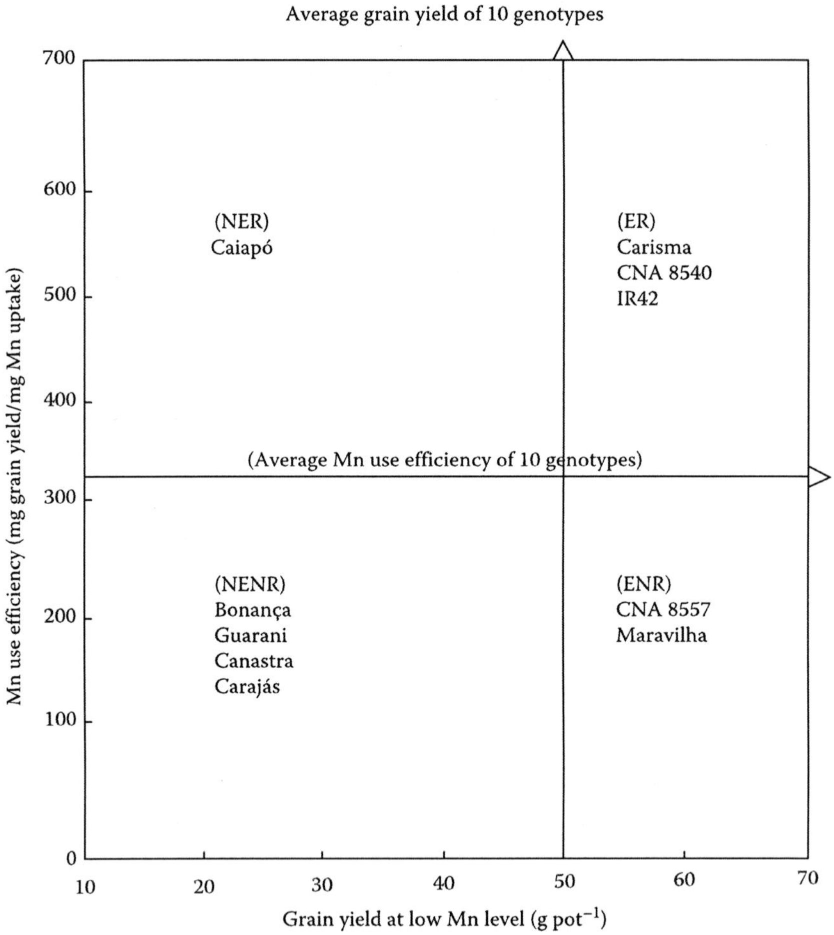

FIGURE 9.5 Classification of upland rice genotypes for Mn use efficiency. (From Fageria, N. K., M. P. Barbosa Filho, and A. Moreira. 2008. Screening upland rice genotypes for manganese use efficiency. *Commun. Soil Sci. Plant Anal.* 39:2873–2882.)

Difference in Mn use efficiency among different plant species or genotypes within species is poorly understood (Graham 1988; Huang, Webb, and Graham 1993, 1994, 1996; Rengel 1999). However, Graham (1988) reported that better internal utilization, lower physiological requirements, better root geometry, and improved absorption rate may be associated with higher Mn use efficiency. Rengel et al. (1996) also suggested that excretion by roots of Mn-efficient genotype substances can solubilize unavailable Mn in the rhizosphere and improve Mn uptake. It is possible that the genetic control of Mn efficiency is expressed through the composition of root exudates encouraging a more favorable balance of Mn reducers to Mn oxidizers in the rhizosphere (Rengel 1999). In Mn-inefficient genotypes, a large number of Mn-oxidizing microbes may be present than in Mn-efficient genotypes (Rengel 1999). The rhizosphere of wheat genotypes contains an increased proportion of Mn reducers under Mn-deficient, as opposed to Mn-sufficient, conditions (Rengel 1999). Rengel (1997) concluded that the increase in the rate of Mn reducers and Mn oxidizers in the rhizosphere of wheat genotypes is one of the mechanisms underlying differential tolerance to Mn deficiency in some, but not all, Mn-efficient genotypes.

9.9 CONCLUSIONS

Manganese deficiency is reported in many rice-growing regions of the world. The main causes are using lime in acidic soils to raise the pH (which decreases the Mn concentration in the soil solution), using high-yielding cultivars that require a high Mn rate, negative interactions with some macro- and micronutrients, and adsorption by OM. Some Histosols are naturally low in Mn. Manganese participates in many physiological and biochemical processes in plants, including the evolution of O_2 in photosynthesis, nitrate assimilation, and chlorophyll formation, and it is a component of several enzyme systems. Manganese is a relatively immobile nutrient in plants, and its deficiency first appears in the younger leaves. Deficiency symptoms include stunting of the plant and interveinal chlorosis of the younger leaves.

In flooded rice, Mn^{4+} is reduced to Mn^{2+} increasing its concentration and availability to rice. Manganese concentration decreases with advanced plant age due to the dilution effect of increased dry matter. This concentration is higher in straw than in grain. A critical level of Mn in rice tissue is about 20 mg kg^{-1} of dry matter. Rice can tolerate a very high level of Mn in the plant tissue (more than 2500 mg kg^{-1}). In soil, critical Mn levels are tied to extracting solution, but they may vary from 4 to 8 mg kg^{-1}. A deficiency can be corrected by applying soluble Mn fertilizers. Manganese sulfate is the fertilizer most commonly used to correct Mn deficiency in rice grown on Mn-deficient soils. The broadcast rate of Mn may vary from 10 to 50 kg ha^{-1}, depending on soil pH, OM content, and genotype planted. The band application rate may be one-third that of broadcast application. Band application with acid-forming fertilizers results in increased efficiency of applied Mn because the rate of oxidation of the divalent Mn to the unavailable tetravalent form (MnO_2) is decreased. The residual effect of applied Mn fertilizers is low because of oxidation, necessitating annual applications.

REFERENCES

Barber, S. A. 1995. *Soil Nutrient Bioavailability: A Mechanistic Approach*, 2nd edition. New York: John Wiley.

Bloom, P. R. 1981. Metal organic matter interactions in soil. In *Chemistry in the Soil Environment*, ed. M. Stelly, pp. 129–149. Madison, WI: ASA and ASSS.

Cheng, B. T., and G. T. Quellette. 1971. Manganese availability in soils. *Soils Fert.* 34:589–595.

Clarkson, D. T. 1988. The uptake and translocation of manganese by plants roots. In *Manganese in Soils and Plants*, eds. R. D. Graham, R. J. Hannam, and N. C. Uren, pp. 101–111. Dordrecht, The Netherlands: Kluwer.

Curtin, D., R. J. Martin, and C. L. Scott. 2008. Wheat (*Triticum aestivum*) response to micronutrients (Mn, Cu, Zn, B) in Canterbury, New Zealand. *N.Z. J. Crop Hortic. Sci.* 36:169–181.

Fageria, N. K. 1989. *Tropical Soils and Physiological Aspects of Crops*. Brasilia: EMBRAPA.
Fageria, N. K. 2000. Upland rice response to soil acidity in Cerrado soil. *Pesq. Agropec. Bras.* 35:2303–2307.
Fageria, N. K. 2001. Adequate and toxic levels of copper and manganese in upland rice, common bean, corn, soybean, and wheat grown on an Oxisol. *Commun. Soil Sci. Plant Anal.* 32:1659–1676.
Fageria, N. K. 2002a. Influence of micronutrients on dry matter yield and interaction with other nutrients in annual crops. *Pesq. Agropec. Bras.* 37:1765–1772.
Fageria, N. K. 2002b. Micronutrients influence on root growth of upland rice, common bean, corn, wheat, and soybean. *J. Plant Nutr.* 25:613–622.
Fageria, N. K. 2004. Dry matter yield and shoot nutrient concentrations of upland rice, common bean, corn, and soybean grown in rotation on an Oxisol. *Commun. Soil Sci. Plant Anal.* 35:961–974.
Fageria, N. K. 2009. *The Use of Nutrients in Crop Plants*. Boca Raton, FL: CRC Press.
Fageria, N. K., and V. C. Baligar. 1993. Screening crop genotypes for mineral stresses. In *Proceedings of the Workshop on Adaptation of Plants to Soil Stress*, August 1–4 1993. INTSORMIL publication No. 94–2, pp. 142–159. Lincoln, NE: University of Nebraska.
Fageria, N. K., and V. C. Baligar. 1999. Growth and nutrient concentrations of common bean, lowland rice, corn, soybean, and wheat at different soil pH on an Inceptisol. *J. Plant Nutr.* 22:1495–1507.
Fageria, N. K., V. C. Baligar, and R. B. Clark. 2002. Micronutrients in crop production. *Adv. Agron.* 77:189–272.
Fageria, N. K., V. C. Baligar, and R. B. Clark. 2006. *Physiology of Crop Production*. New York: The Haworth Press.
Fageria, N. K., V. C. Baligar, and C. A. Jones. 2011. *Growth and Mineral Nutrition of Field Crops*, 3rd edition. Boca Raton, FL: CRC Press.
Fageria, N. K., V. C. Baligar, and Y. C. Li. 2008. The role of nutrient efficient plants in improving crop yields in the twenty-first century. *J. Plant Nutr.* 31:1121–1157.
Fageria, N. K., and M. P. Barbosa Filho. 1994. *Nutritional Deficiency in Rice: Identification and Correction*. Goiania, Brazil: EMBRAPA.
Fageria, N. K., M. P. Barbosa Filho, and A. Moreira. 2008. Screening upland rice genotypes for manganese use efficiency. *Commun. Soil Sci. Plant Anal.* 39:2873–2882.
Fageria, N. K., N. A. Slaton, and V. C. Baligar. 2003. Nutrient management for improving lowland rice productivity and sustainability. *Adv. Agron.* 80:63–152.
Fageria, N. K., and C. M. R. Souza. 1991. Upland rice, common bean, and cowpea response to magnesium application on an Oxisol. *Commun. Soil Sci. Plant Anal.* 22:1805–1816.
Fageria, N. K., L. F. Stone, and A. B. Santos. 2003. *Soil Fertility Management for Lowland Rice*. Santo Antonio de Goiás, Brazil: EMBRAPA.
Follett, R. H., L. S. Murphy, and R. L. Donahue. 1981. *Fertilizers and Soil Amendments*. Englewood Cliffs, NJ: Prentice-Hall.
Foth, H. D., and B. G. Ellis. 1988. *Soil Fertility*. New York: John Wiley.
Foy, C. D. 1984. Physiological effects of hydrogen, aluminum and manganese toxicities in acid soils. In *Soil Acidity and Liming*, 2nd edition, ed. F. Adams, pp. 57–97. Madison, WI: ASA.
Gladstones, J. S., and J. F. Loneragan. 1970. Nutrient elements in herbage plants in relation to soil adaptation and animal nutrition. In *Proceedings of the XI International Grassland Congress*, pp. 350–354. Brisbane, Australia: University of Queensland Press.
Graham, R. D. 1988. Genotype differences in tolerance to manganese deficiency. In *Manganese in Soils and Plants,* eds. R. D. Graham, R. J. Hannam, and N. C. Uren, pp. 261–276. Dordrecht, The Netherlands: Kluwer.
Graham, R. D., W. Davies, and J. Ascher. 1985. The critical concentration of manganese in field-grown wheat. *Aust. J. Agric. Res.* 36:145–155.

Hernandez-Soriano, M. C., F. Degryse, E. Lombi, and E. Smolders. 2012. Manganese toxicity in barley is controlled by solution manganese and soil manganese speciation. *Soil Sci. Soc. Am. J.* 76:399–407.

Horiguchi, T., and K. Kitagishi. 1976. Studies on rice seed protease: Metal ion activation of rice seed peptidase. *J. Sci. Soil Manure.* 22:73–80.

Hossner, L. R., and G. E. Richards. 1968. The effect of phosphorus source on the movement and uptake of band applied manganese. *Soil Sci. Soc. Am. Proc.* 32:83–85.

Huang, C., M. J. Webb, and R. D. Graham. 1993. Effect of pH on Mn absorption by barley genotypes in a chelate-buffered nutrient solution. *Plant Soil.* 155/156:437–440.

Huang, C., M. J. Webb, and R. D. Graham. 1994. Manganese efficiency is expressed in barley growing in soil system but not in a solution culture. *J. Plant Nutr.* 17:83–95.

Huang, C., M. J. Webb, and R. D. Graham. 1996. Pot size affects expression of Mn efficiency in barley. *Plant Soil* 78:205–208.

Jackson, T. L., and G. E. Carter. 1976. Nutrient uptake by Russet Burbank potatoes as influenced by fertilization. *Agron. J.* 68:9–12.

Jackson, T. L., D. T. Westerman, and D. P. Moore. 1966. The effect of chloride and lime on the manganese uptake by bush beans and sweet corn. *Soil Sci. Soc. Am. Proc.* 30:70–73.

Karimian, N., and A. Gholamalizadeh Ahangar. 1998. Manganese retention by selected calcareous soils as related to soil properties. *Commun. Soil Sci. Plant Anal.* 29:1061–1070.

Khattak, R. A., W. M. Jarrell, and A. L. Page. 1989. Mechanism of native manganese release in salt treated soils. *Soil Sci. Soc. Am. J.* 53:701–705.

Krishnamurti, G. S. R., and P. M. Huang. 1992. Dynamics of potassium chloride induced manganese release in different soil orders. *Soil Sci. Soc. Am. J.* 56:1115–1123.

Lindsay, W. L. 1979. *Chemical Equilibrium in Soils*. New York: John Wiley.

Lindsay, W. L., and F. R. Cox. 1985. Micronutrient soil test for the tropics. *Fert. Res.* 7:169–200.

Lindsay, W. L., and W. A. Norvell. 1978. Development of a DTPA soil test for zinc, iron, manganese and copper. *Soil Sci. Soc. Am. J.* 42:421–428.

Maas, E. V., D. P. Moore, and B. J. Mason. 1969. Influence of calcium and manganese on manganese absorption. *Plant Physiol.* 44:796–800.

Marschner, H. 1995. *Mineral Nutrition of Higher Plants*. San Diego, CA: Academic Press.

McDaniel, P. A., and Buol, S. W. 1991. Manganese distribution in acid soils of the North Carolina Piedmont. *Soil Sci. Soc. Am. J.* 55:152–158.

Mortvedt, J. J. 1994. Needs for controlled availability of micronutrient fertilizers. *Fert. Res.* 38:213–221.

Obata, H. 1995. Physiological functions of micro essential elements. In *Science of Rice Plant: Physiology*, Vol. 2, eds. T. Matsuo, K. Kumazawa, R. Ishii, K. Ishihara, and H. Hirata, pp. 402–419. Tokyo: Food and Agricultural Policy Research Center.

Patrick, W. H. Jr., and C. N. Reddy. 1978. Chemical changes in rice soils. In *Soils and Rice*, ed. International Rice Research Institute, pp. 361–379. Los Banos, Philippines: IRRI.

Petrie, S. G., and T. L. Jackson. 1984. Effect of fertilization on soil pH and manganese concentration. *Soil Sci. Soc. Am. J.* 48:315–348.

Porter, G. S., J. B. Bakita-Locke, N. V. Hue, and D. Strand. 2004. Manganese solubility and phytotoxicity affected by soil moisture, oxygen levels, and green manure addition. *Commun. Soil Sci. Plant Anal.* 35:99–116.

Randall, G. W., E. E. Schulte, and R. B. Corey. 1975. Soil Mn availability to soybeans as affected by mono and diammonium phosphate. *Agron. J.* 67:705–709.

Reisenauer, H. M. 1988. Determination of plant-available soil manganese. In *Manganese in Soils and Plants*, eds. R. D. Graham, R. J. Hannam, and N. C. Uren, pp. 87–98. Dordrecht, The Netherlands: Kluwer.

Rengel, Z. 1997. Root exudation and microflora populations in rhizosphere of crop genotypes differing in tolerance to micronutrient deficiency. *Plant Soil.* 196:255–260.

Rengel, Z. 1999. Physiological mechanisms underlying differential nutrient efficiency of crop genotypes. In *Mineral Nutrition of Crops: Fundamental Mechanisms and Implications*, ed. Z. Rengel, 227–265. New York: The Haworth Press.

Rengel, Z., R. Guterridge, P. Hirsch, and D. Hornby. 1996. Plant genotype, micronutrient fertilization and take-all infection influence bacterial populations in the rhizosphere of wheat. *Plant Soil*. 183:269–277.

Shimada, N. 1995. Deficiency and excess of micronutrient elements. In *Science of Rice Plant: Physiology*, Vol. 2, eds. T. Matsuo, K. Kumazawa, R. Ishii, K. Ishihara, and H. Hirata, pp. 412–419. Tokyo: Food and Agricultural Policy Research Center.

Sims, J. T., and G. V. Johnson. 1991. Micronutrient soil test. In *Micronutrients in Agriculture*, 2nd edition, ed. J. J. Mortvedt, 417–476. Madison, WI: SSSA.

Snyder, G. H., D. B. Jones, and F. J. Coale. 1990. Occurrence and correction of manganese deficiency in Histosol grown rice. *Soil Sci. Soc. Am. J.* 54:1634–1638.

Stevenson, F. J. 1986. *Cycles of Soil Carbon, Nitrogen, Phosphorous, Sulfur, and Micronutrients*. New York: John Wiley.

Tate, R. L. 1987. *Soil Organic Matter: Biological and Ecological Effects*. New York: John Wiley.

Tisdale, S. L., W. L. Nelson, and J. D. Beaton. 1985. *Soil Fertility and Fertilizers*, 4th edition. New York: Macmillan.

Voth, R. D., and D. R. Christenson. 1980. Effect of fertilizer reaction and placement on availability of manganese. *Agron. J.* 72:769–773.

Weil, R. R., C. D. Foy, and C. A. Coradetti. 1997. Influence of soil moisture regimes on subsequent manganese toxicity in two cotton genotypes. *Agron. J.* 89:1–8.

Wells, B. R., B. A. Huey, R. J. Norman, and R. S. Helms. 1993. Rice. In *Nutrient Deficiencies and Toxicities in Crop Plants*, ed. W. F. Bennett, pp. 15–19. St. Paul, MN: American Phytopathological Society.

Westermann, D. T., T. L., Jackson, and D. P. Moore. 1971. Effect of potassium salts on extractable soil manganese. *Soil Sci. Soc. Am. Proc.* 35:43–46.

Yoshida, S. 1981. *Fundamentals of Rice Crop Science*. Los Banos, Philippines: IRRI.

Zhang, M., A. K. Alva, Y. C. Li, and D. V. Calvert. 1997. Fractionation of iron, manganese, aluminum and phosphorus in selected sandy soils under citrus production. *Soil Sci. Soc. Am. J.* 61:794–801.

10 Iron

10.1 INTRODUCTION

Iron (Fe) is an essential micronutrient for the growth and development of crop plants. Iron deficiency is a complex disorder, and it occurs in response to multiple soil, environmental, and genetic factors (Wiersma 2005). Factors that can contribute to Fe deficiency in plants include low Fe supply from soil; high lime and P application; high levels of heavy metals such as Zn, Cu, and Mn; low and high temperatures; high levels of nitrate nitrogen; poor aeration; unbalanced cation ratios; and root infection by nematodes (Fageria, Baligar, and Wright 1990; Fageria 2009). Iron deficiency has been observed in Brazilian upland rice when it is grown in rotation with dry beans and soybeans on Oxisols and Ultisols (Fageria, Baligar, and Jones 2011). Oxisols and Ultisols are highly weathered soils that have low fertility and that are acidic in reaction (Fageria and Baligar 2008). Liming is an essential and predominant practice for correcting acidity of these soils. Therefore, iron deficiency observed in upland rice grown on these soils is due to increases in soil pH, rather than low levels of iron.

Bolan, Adriano, and Curtin (2003) reported that with the exception of Mo, the availability of most other trace elements in plants decreases with liming due to decreases in the concentration of these elements in the soil solution. The decrease in Fe uptake with increasing pH is referred to as lime-induced chlorosis. The effect of pH > 6.0 in lowering free metal ion activities in soils has been attributed to the increase in pH-dependent surface charge on oxides of Fe, Al, and Mn (Stahl and James 1991) or the precipitation of metal hydroxides (Lindsay 1979). The solubility of Fe decreases by ~1000-fold for each unit increase of soil pH in the range of 4–9 compared with ~100-fold decreases in the activity of Mn, Cu, and Zn (Lindsay 1979). With increasing pH, iron is precipitated and its availability is reduced according to the following equation (Fageria, Guimaraes, and Portes 1994):

$$Fe^{3+} + 3OH^{-} \Leftrightarrow Fe(OH)_3 \text{ (precipitated)} \quad (10.1)$$

Data in Table 10.1 show iron uptake by upland rice at different soil pH levels. The uptake of iron significantly decreased with increasing soil pH. Similarly, extractable soil iron also decreased with increasing soil pH in a Brazilian Inceptisol where lowland rice is mainly grown (Table 10.2). Figure 10.1 shows that growth of upland rice decreased with increasing soil pH of 4.6–6.8, and iron deficiency was observed in the plants at higher pH.

Iron-deficient soils are often sandy, although deficiencies have been found on fine-textured soils, mucks, and peats (Brown 1961). In addition, iron deficiency is a potential problem on most calcareous soils (Marschner and Romheld 1995; Lucena and Chaney 2007). Calcareous soils are widespread throughout the world

TABLE 10.1
Iron Uptake by Upland Rice under Different Soil pH

Soil pH in H_2O	Fe Uptake (μg $plant^{-1}$)
4.6	4541
5.7	1856
6.2	1978
6.4	1634
6.6	1663
6.8	1569
R^2	0.97*

Source: Adapted from Fageria, N. K. 2000. Upland rice response to soil acidity in Cerrado soil. *Pesq. Agropec. Bras.* 35:2303–2307.

*Significant at the 1% probability level.

TABLE 10.2
Influence of pH on Iron Extracted by Mehlich 1 Extraction Solution of Brazilian Inceptisol

pH in H_2O	Fe (mg kg^{-1})
4.9	297
5.9	231
6.4	187
6.7	158
7.0	148

Regression Analysis

pH vs. Fe in the soil $(Y) = 673.50 - 75.69X$, $R^2 = 0.55$*

Source: Adapted from Fageria, N. K., and V. C. Baligar. 1999. Growth and nutrient concentrations of common bean, lowland rice, corn, soybean, and wheat at different soil pH on an Inceptisol. *J. Plant Nutr.* 22:1495–1507.

Note: Iron extracted by Mehlich 1 extraction solution of Brazilian Inceptisol.

*Significant at the 1% probability level.

(Lombi et al. 2004). The United Nations Food and Agriculture Organization (FAO) estimated the extent of calcareous soils at 800 million ha worldwide, mainly concentrated in areas with arid or Mediterranean climates (Land, FAO, and Plant Nutrition Management 2000). These soils are important in terms of agricultural production in many areas of the world.

Rice is considered very susceptible to Fe deficiency, especially when grown under upland conditions (Fageria, Slaton, and Baligar 2003; Fageria, Baligar, and

FIGURE 10.1 (See color insert.) Iron deficiency symptoms in upland rice grown on Brazilian Oxisol. (From Fageria, N. K. 2009. *The Use of Nutrients in Crop Plants*. Boca Raton, FL: CRC Press.)

Jones 2011). Rice yield losses caused by Fe deficiency measured in research plots range from 10 to 50% (Snyder and Jones 1988, 1991). Ferrous (Fe^{2+}) is preferentially absorbed by plant root systems and generally is present in high concentrations in reduced soils (Marschner 1995). Thus, Fe deficiency does not commonly occur in flooded rice due to the increase in Fe availability associated with the anaerobic soil conditions used for flooded rice production. However, Fe deficiency does occur in seedling rice before flooding in some rice-growing regions of the world (Yoshida 1981; Snyder and Jones 1988). Mori et al. (1991) suggested that seedling rice is highly susceptible to Fe deficiency because rice roots produce relatively low amounts of iron-chelating phytosiderophores compared with other grass species. Soil conditions that limit Fe availability, coupled with plant limitations for obtaining Fe, are the primary reasons why Fe deficiency occurs.

Iron toxicity is a serious problem in flooded rice (Fageria 1984; Fageria, Baligar, and Wright 1990). In many parts of the world (Africa, South America, and Asia) where rice is grown on acidic soils having great potential for rice production, Fe toxicity is or will be a serious problem (Fageria 1984, 1989; Fageria, Baligar, and Wright 1990). In lowland or flooded rice, iron concentration generally increases because of reduction of Fe^{3+} to Fe^{2+} (Fageria et al. 2011). The increase in pH of acidic soils is mainly determined by reduction of iron and manganese oxides that consume H^+ ions. These reduction processes are shown in the following equations (Fageria et al. 2011):

$$Fe_2O_3 + 6H^+ + 2e^- \leftrightarrow 2Fe^{2+} + 3H_2O \quad (10.2)$$

$$MnO_2 + 4H^+ + 2e^- \leftrightarrow Mn^{2+} + 2H_2O \quad (10.3)$$

It is clear that iron deficiency in upland rice and iron toxicity in flooded or lowland rice are major problems. The objective of this chapter is to discuss iron nutrition of upland and lowland rice grown on aerobic and flooded soils.

10.2 CYCLE IN SOIL–PLANT SYSTEM

Knowledge of the iron cycle in the soil–plant system is fundamental for improving iron efficiency in plants and, consequently, rice yield. In addition, transformation (or solubilization), uptake by plants, oxidation–reduction, immobilization, and loss are important components of the iron cycle in the soil–plant system. Figure 10.2 shows a simplified cycle of micronutrients, including iron, in flooded rice soils or reduced soils. In reduced soils, Fe^{2+} is the major species of iron; in oxidized soil, Fe^{3+} is the dominant species. Hydrolysis of Fe^{3+} and Fe^{2+} species is shown here (Lindsay 1979):

$$Fe^{3+} + H_2O \leftrightarrow FeOH^{2+} + H^+ \qquad (10.4)$$

$$Fe^{3+} + 2H_2O \leftrightarrow Fe(OH)_2^{+} + 2H^+ \qquad (10.5)$$

$$Fe^{3+} + 3H_2O \leftrightarrow Fe(OH)_3^{0} + 3H^+ \qquad (10.6)$$

$$Fe^{3+} + 4H_2O \leftrightarrow Fe(OH)_4^{-} + 4H^+ \qquad (10.7)$$

$$2Fe^{3+} + 2H_2O \leftrightarrow Fe_2(OH)_2^{4+-} + 2H^+ \qquad (10.8)$$

$$Fe^{2+} + H_2O \leftrightarrow FeOH^{+} + H^+ \qquad (10.9)$$

$$Fe^{2+} + 2H_2O \leftrightarrow Fe(OH)_2^{0} + 2H^+ \qquad (10.10)$$

$$Fe^{2+} + 3H_2O \leftrightarrow Fe(OH)_3^{-} + 3H^+ \qquad (10.11)$$

$$Fe^{2+} + 4H_2O \leftrightarrow Fe(OH)_4^{2-} + 4H^+ \qquad (10.12)$$

$$3Fe^{3+} + 4H_2O \leftrightarrow Fe_3(OH)_4^{2+-} + 4H^+ \qquad (10.13)$$

Iron minerals commonly found in soils include goethite (FeOOH), hematite (Fe_2O_3), pyrite (FeS), siderite ($FeCO_3$), and magnetite (Fe_3O_4). Therefore, Fe availability for plants is affected by soil properties such as pH and redox that influence the solubility of Fe-containing minerals (Fageria, Slaton, and Baligar 2003). The Fe^{3+} concentration in the soil solution decreases with an increase in soil pH. For each unit increase in soil pH, there is a 1000-fold decrease in the solubility of Fe^{3+} and a 100-fold decrease in the solubility of Fe^{2+} (Tisdale, Nelson, and Beaton 1985). The amounts of Fe^{2+} increase rapidly at redox potentials below 200 mV. Ponnamperuma (1976) found that the soil Fe^{2+} concentration increased to a peak ranging from 0.1 to 600 mg kg^{-1} for several soils shortly after submergence and then declined. Diffusion and mass flow are believed to be the two mechanisms responsible for the movement of Fe from soil to the root surface.

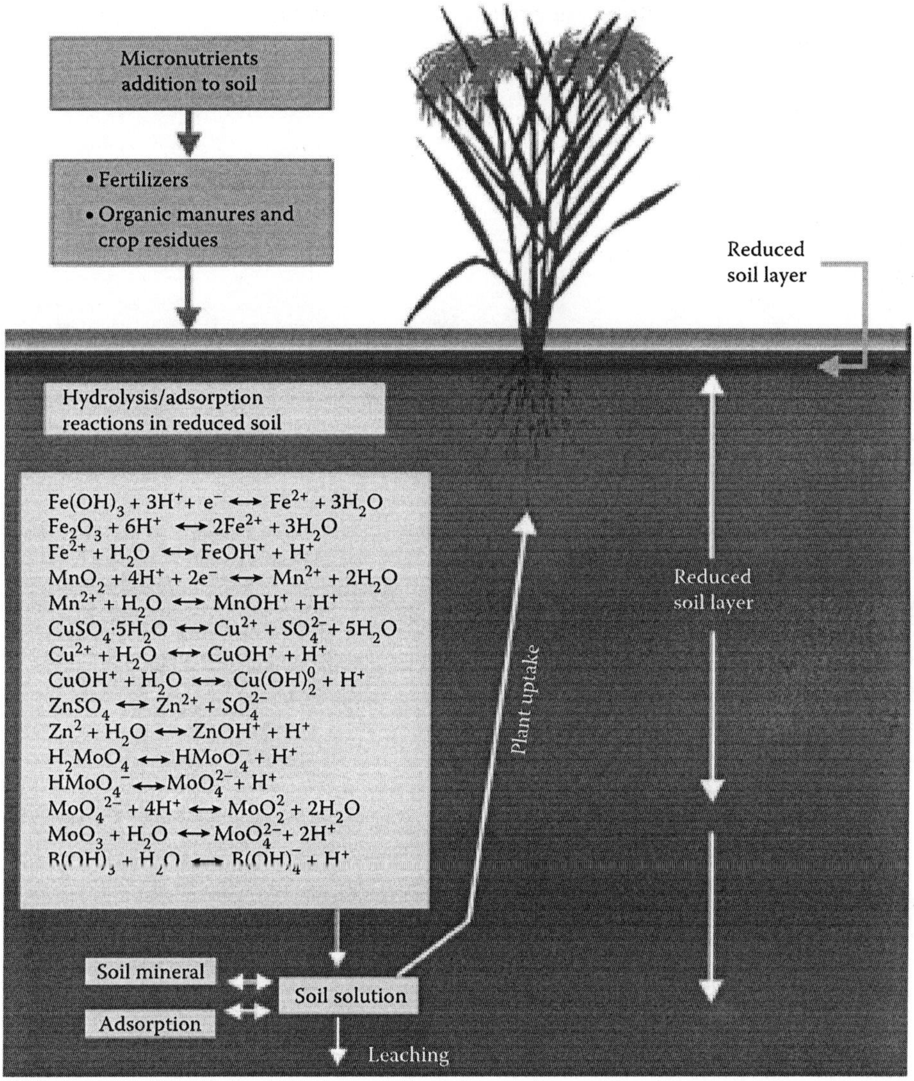

FIGURE 10.2 Micronutrient cycle in flooded rice.

The chemistry of flooded soils is dominated by Fe more than by any other redox element. The major reason for this dominance is the large amount of soil Fe that can undergo reduction, which usually exceeds the total amount of other redox elements by a factor of 10 or more (Patrick and Reddy 1978). Under submerged soil conditions, Fe^{3+} is reduced to Fe^{2+} by respiring microorganisms. Although variable in composition, ferric oxyhydroxides in aerated soils can be represented by the formula $Fe(OH)_3$, which can undergo reduction (Patrick and Reddy 1978):

$$Fe(OH)_3 + 3H^+ + e^- \leftrightarrow Fe^{2+} + 3H_2O \tag{10.14}$$

Although there is general agreement that the reduction of ferric compounds occurs as a result of the respiration of facultative anaerobic bacteria, it has not been demonstrated conclusively that the reduction is brought about by enzymatic transfer of electrons directly to Fe^{3+}, or that the reduction is an indirect chemical reaction between bacterial metabolites and Fe^{3+}. In either case, it is likely that complexing of Fe with organic chelates plays an important role in making the Fe solution more reactive (Patrick and Reddy 1978; Fageria, Slaton, and Baligar 2003).

10.3 FUNCTIONS

Iron is essential for the synthesis of chlorophyll. It is involved in photosynthesis and electron transfer. Iron is involved in respiratory enzyme systems as a part of cytochrome and hemoglobin and also in many other enzyme systems (Bennett 1993). As an electron carrier, iron is involved in oxidation–reduction reactions (Follett, Murphy, and Donahue 1981). It is also a structural component of substances involved in reactions such as the reduction of O_2 to H_2O during respiration (Tisdale, Nelson, and Beaton 1985). Iron is required in protein synthesis and is a constituent of hemoprotein (Follett, Murphy, and Donahue 1981). It is also a part of iron–sulfur proteins, which act as enzymes, and is involved in electron transfer (Marschner 1995). Iron also improves the root systems of rice (Figure 10.3) and the growth and leaf areas of rice (Figure 10.4).

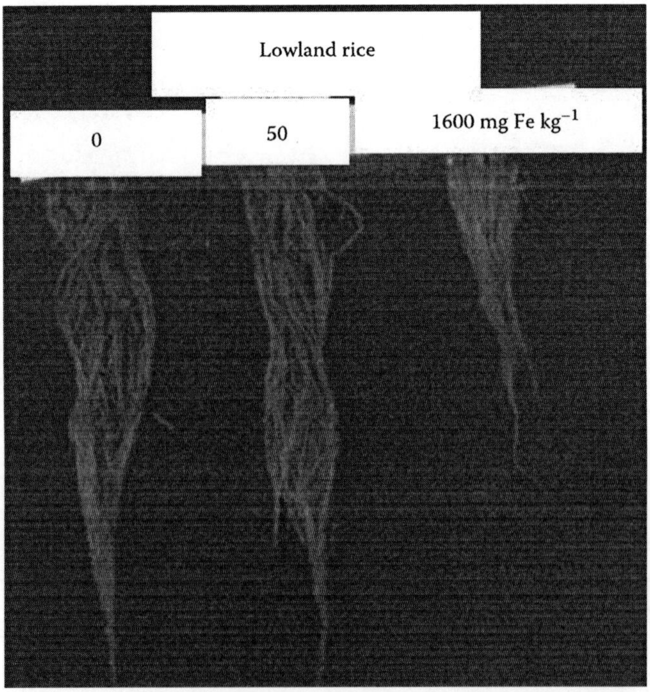

FIGURE 10.3 Influence of iron on root growth of lowland rice. (From Fageria, N. K. 2009. *The Use of Nutrients in Crop Plants*. Boca Raton, FL: CRC Press.)

FIGURE 10.4 Response of upland rice to iron fertilization.

10.4 DEFICIENCY SYMPTOMS

Deficiency symptoms are among the important tools for diagnosing nutrient deficiency or sufficiency in crop plants. There is no cost to the producer, such as for analysis of soil tests and for tissue analysis. However, great experience is required by the analyst because of possible confusion with diseases, drought, and insect infection. The mobility of nutrients within the plant and the position of the leaf on which the deficiency symptoms appear are interrelated (Yoshida 1981). Because iron is immobile in the plant, its deficiency symptoms first appear in the younger leaves. At the onset of Fe deficiency, symptoms begin as an interveinal chlorosis of the youngest leaves, giving plants a striped appearance. Further progression of Fe deficiency causes seedlings to have a uniform pale yellow to bleached white appearance due to a reduction in the chlorophyll formation processes. Iron deficiency reduces seedling dry matter production, leaf chlorophyll content, panicle number per unit area, and grain yield (Snyder and Jones 1988). Figure 10.5 shows iron deficiency in upland rice leaves. Figure 10.6 shows Fe deficiency in upland rice with a high P level. Iron deficiency symptoms are also shown in crops grown on a Brazilian Oxisol with pH about 7.0 in water (Figure 10.7).

10.5 UPTAKE IN PLANT TISSUE

Plant uptake of nutrients is expressed in concentration (content per unit dry matter). The unit of concentration is percentage or gram per kilogram ($\% \times 10 = g\ kg^{-1}$) for macronutrients and milligram per kilogram for micronutrients. Uptake of nutrients is concentration multiplied by dry matter. The unit of uptake is kilogram per

FIGURE 10.5 (See color insert.) Iron deficiency in younger leaves of upland rice leaves grown on Oxisol with pH about 6.5 in water.

FIGURE 10.6 (See color insert.) Iron deficiency in upland rice with a high P level (400 mg P kg^{-1}).

FIGURE 10.7 (See color insert.) Iron deficiency in upland rice grown on Brazilian Oxisol having pH 7.0 in water.

hectare for macronutrients and gram per hectare for micronutrients. Representative Fe concentrations and uptakes of upland and lowland rice grown on Oxisol and Inceptisol of central Brazil are listed in Table 10.3. The concentration of Fe in the whole aboveground tissue is rather high during early tillering and then decreases during the growing season. The decrease in concentration with advanced plant age was related to an increase in dry matter that diluted the Fe concentration. Similarly, uptake in both the rice cultures (upland and lowland) followed the quadratic increase with advanced plant age (Table 10.4). This was due to an increase in plant dry matter with advanced plant age. Similar results were reported by Fageria, Slaton, and Baligar (2003) and Fageria (2004) in lowland and upland rice.

In plant tissue, iron has a relatively wide sufficiency concentration range in plant tissue between the proposed critical concentrations for Fe deficiency and Fe toxicity. At the tillering stage, the sufficiency range of Fe concentrations in the leaf blades ranges from 70 to 300 mg kg^{-1} (Wells et al. 1993). Iron deficiency or toxicity occurs at concentrations below or above this sufficiency range. Fageria, Baligar, and Jones (2011) also reported a similar Fe sufficiency concentration range but noted the concentration varied depending on the plant age and part analyzed. Fageria, Slaton, and Baligar (2003) suggested that the sufficient Fe concentration range of the Y-leaf was 75–150 mg kg^{-1} during vegetative growth. The sufficient Fe concentration of the whole shoot was somewhat lower at 60–100 mg kg^{-1}. Gilmour (1977) found that whole plant and bottom leaf Fe concentrations were significantly greater than that of the Y-leaf up to 69 days after seeding. Similar to other nonmobile elements, the concentration of Fe usually increases as leaf age increases. Tissue Fe concentrations are useful only to establish whether the

TABLE 10.3
Iron Concentration and Uptake in Upland and Lowland Rice Straw and Grain during Growth Cycle

	Upland Rice			Lowland Rice	
Plant Age in Days	Conc. (mg kg^{-1})	Uptake (g ha^{-1})	Plant Age in Days	Conc. (mg kg^{-1})	Uptake (g ha^{-1})
19 (IT)	1325	89	22 (IT)	419	134
43 (AT)	300	322	35 (AT)	374	411
68 (IP)	157	544	71 (IP)	175	1055
90 (B)	82	491	97 (B)	155	1403
102 (F)	72	457	112 (F)	179	2202
130 (PM)	103	654	140 (PM)	448	3499
130 (grain)	25	117	140 (grain)	81	265

Source: Adapted from Fageria, N. K. 2004. Dry matter yield and shoot nutrient concentrations of upland rice, common bean, corn, and soybean grown in rotation on an Oxisol. *Commun. Soil Sci. Plant Anal.* 35:961–974; Fageria, N. K., N. A. Slaton, and V. C. Baligar. 2003. Nutrient management for improving lowland rice productivity and sustainability. *Adv. Agron.* 80:63–152.

Note: IT, initiation of tillering; AT, active tillering; IP, initiation of panicle primordia floral; B, booting; F, flowering; PM, physiological maturity.

TABLE 10.4
Relationship between Concentration and Uptake of Iron in Upland and Lowland Rice

Variable	Regression Equation	R^2
Plant age vs. Fe conc. in upland rice straw	$Y = 1851.69 - 38.62X + 0.20X^2$	0.91**
Plant age vs. Fe uptake in upland rice straw	$Y = -47.78 + 9.63X - 0.035X^2$	0.86*
Plant age vs. Fe conc. in lowland rice straw	$Y = 720.98 - 14.01X + 0.085X^2$	0.94**
Plant age vs. Fe uptake in lowland rice straw	$Y = 66.11 \exp(4.75X - 0.00014X^2)$	0.96**

Note: The regression equations in this table were calculated from the data in the Table 10.2.
*,**Significant at the 5 and 1% probability levels, respectively.

total Fe concentration falls within the suggested sufficiency range. Perhaps, the best method for determining Fe deficiency is studying tissue analysis for active Fe.

Iron toxicity is believed to occur when leaf blade total Fe concentrations exceed 300 mg kg^{-1} (Tanaka, Leo, and Navasero 1966). Fageria, Slaton, and Baligar (2003) indicated this critical concentration was specifically for the Y-leaf. The importance of leaf age, plant part sampled, and sample cleanliness cannot be overemphasized for the diagnosis of Fe toxicity. The presence of toxic concentrations of Fe may be the result of another nutrient deficiency. For example, Zn-deficient whole aboveground rice seedlings commonly contain Fe concentrations > 300 mg kg^{-1}. However, clean

Y-leaf tissue is seldom above this threshold. Whole plant samples of nutritionally healthy plants may easily exceed 300 mg Fe kg^{-1} from contamination of Fe precipitates on the rice stems unless the tissues are thoroughly washed in a mild acid solution before drying and analysis.

10.6 USE EFFICIENCY

Iron use efficiency values, defined as the shoot dry matter (straw yield) or grain yield per unit of Fe uptake in shoot (straw) or grain, for upland and lowland rice are presented in Table 10.5. The Fe use efficiency in upland and lowland rice increased quadratically and was higher in grain compared with shoot or straw. The Fe use efficiency was higher in grain compared with shoot. This type of response may be associated with the increase in dry matter of shoot quadratically with advanced plant age (Fageria 2009). Fageria, Castro, and Baligar (2004) determined iron use efficiency in upland rice under low and high acidity (Table 10.6). The Fe use efficiency was significantly affected by acidity and genotype treatments. Similarly, acidity × genotype interaction was also significant, indicating variation in Fe use efficiency among genotypes with changes in acidity level. The acidity × genotype interaction also suggests that it is necessary to evaluate upland rice genotypes at more than one acidity level. All genotypes produced higher grain yields per unit of Fe uptake or accumulated at low acidity level compared with high acidity level.

TABLE 10.5
Iron Use Efficiency (kg g^{-1}) in Straw and Grain of Upland and Lowland Rice during Growth Cycle

Plant Age in Days	Upland Rice	Plant Age in Days	Lowland Rice
19 (IT)	0.8	22 (IT)	2.56
43 (AT)	3.5	35 (AT)	2.70
68 (IP)	7.0	71 (IP)	5.82
90 (B)	13.9	97 (B)	7.25
102 (F)	14.8	112 (F)	5.66
130 (PM)	10.6	140 (PM)	3.79
130 (grain)	46.8	140 (grain)	12.6

Regression Analysis

Plant age vs. Fe use efficiency upland rice $(Y) = -6.34 + 0.33X - 0.0015X^2$, $R^2 = 0.85$*

Plant age vs. Fe us efficiency $(Y) = -1.49 + 0.18X - 0.00099X^2$, $R^2 = 0.89$*

Source: From Fageria, N. K. 2009. *The Use of Nutrients in Crop Plants*. Boca Raton, FL: CRC Press; Fageria's unpublished data.

Note: Fe use efficiency (kg g^{-1}) = (Straw or grain weight in kg/ Uptake of Fe in straw or grain in g). IT, initiation of tillering; AT, active tillering; IP, initiation of panicle primordia floral; B, booting; F, flowering; PM, physiological maturity.

*Significant at the 5% probability level.

TABLE 10.6
Iron Use Efficiency (mg Grain per μg Fe Uptake in Straw Plus Grain) in Upland Rice Genotypes at Two Acidity Levels

Genotype	High Acidity (pH 4.5)	Low Acidity (pH 6.4)
CRO 97505	2.63a	4.65ab
CNAs 8983	2.38ab	6.23a
CNAs 8938	1.39ab	3.09bc
CNAs 8989	1.54ab	6.56a
CNAs 8824	0.92b	4.69ab
CNAs 8952	1.17ab	5.78a
CNAs 8950	1.18ab	4.67ab
CAN 8540	0.85b	2.92c
BRS Primavera	2.03ab	5.76a
BRS Carisma	1.02b	1.85c
Average	1.51	4.62
F-test		
Acidity (A)	*	
Genotype	*	
A × G	*	
CV (%)	25	

Source: Adapted from Fageria, N. K., E. M. Castro, and V. C. Baligar. 2004. Response of upland rice genotypes to soil acidity. In *The Red Soils of China: Their Nature, Management and Utilization*, eds. M. J. Wilson, Z. He, and X. Yang, pp. 219–237. Dordrecht, The Netherlands: Kluwer.

Note: Means in the same column followed by the same letter are not significantly different at the 1% probability level by Tukey's test.

*Significant at the 1% probability level.

10.7 HARVEST INDEX

Overall, iron harvest index (Fe uptake in grain/Fe uptake in grain plus shoot) was higher in cereals compared with legumes (Table 10.7). This means that for human nutrition, cereals such as corn and lowland rice are superior in furnishing iron compared with legumes such as dry beans and soybeans.

10.8 MANAGEMENT PRACTICES

Correcting iron deficiency is very difficult because it is caused by changes in soil's chemical conditions and not by low levels of iron (Foth and Ellis 1988). If soluble iron is added to soil, it is very quickly precipitated and becomes unavailable to plants. However, adopting appropriate soil and crop management practices can reduce iron deficiency problems in upland rice. These practices are the use of an effective source, appropriate method and timing of application, and the use of efficient genotypes.

TABLE 10.7
Iron Uptake and Harvest Index in Principal Field Crops

Crop Species	Fe Uptake in Shoot ($g\ ha^{-1}$)	Fe Uptake in Grain ($g\ ha^{-1}$)	Total	Fe Harvest Index
Upland rice	654	117	771	0.15
Lowland rice	980	449	1429	0.31
Corn	206	206	412	0.50
Dry bean	268	144	412	0.35
Dry bean	1010	275	1285	0.21
Soybean	778	190	968	0.20

Source: Fageria, N. K. 2009. *The Use of Nutrients in Crop Plants*. Boca Raton, FL: CRC Press.
Note: Fe harvest index = (Fe uptake in grain/Fe uptake in grain plus shoot).

TABLE 10.8
Principal Iron Carriers

Common Name	Formula	Fe Content (%)	Solubility in Water
Ferrous sulfate (monohyd.)	$FeSO_4{\cdot}H_2O$	33	Soluble
Ferrous sulfate (heptahyd.)	$FeSO_4{\cdot}7H_2O$	19	Soluble
Ferric sulfate	$Fe_2(SO_4){\cdot}4H_2O$	23	Soluble
Iron chelate	NaFeEDTA	5–14	Soluble
Iron chelate	NaFeHEDTA	5–9	Soluble
Iron chelate	NaFeDTPA	6	Soluble
Iron frits	Fritted glass	2–6	Slightly soluble

Source: Adapted from Fageria, N. K., M. P. Barbosa Filho, A. Moreira, and C. M. Guimaraes. 2009. Foliar fertilization of crop plants. *J. Plant Nutr.* 32:1044–1064.

10.8.1 Effective Sources

Iron deficiency can be corrected by foliar application of an appropriate source. There are several sources of Fe fertilizers. The important criterion in selecting an Fe source is its cost and solubility in water. Iron chelates are good sources and also soluble in water, but their cost is very high compared with other options. Important Fe sources and their solubility are given in Table 10.8.

10.8.2 Appropriate Methods and Timing of Application

Iron sulfate used on calcareous soils quickly reacts with $CaCO_3$ to form Fe oxides that are less available for plant uptake. However, some researchers have reported that band application of ferrous sulfate can correct iron chlorosis in annual crops.

At present, the most economical and attractive strategies for correcting iron deficiency in crop plants are foliar application and the use of efficient cultivars.

Application of 1–2% ferrous sulfate ($FeSO_4{\cdot}7H_2O$ or $FeSO_4{\cdot}H_2O$) solution can correct Fe deficiency in most crop plants (Fageria 2009). If the deficiency is severe, more than one application is sometimes necessary. One practical difficulty in foliar Fe application is that normally Fe deficiency occurs in the early growth stage of crops, and plants may not have sufficient leaf area for absorbing the applied element. Fageria et al. (2009) discussed some of the important aspects of foliar fertilization. These aspects are summarized in this section.

10.8.2.1 Mechanisms of Uptake of Foliar-Applied Nutrients

Green leaves are organs whose most important function is photosynthesis. However, absorption of inorganic and organic materials can also take place through the surfaces of leaves (Franke 1967). Nutrient absorption by leaves may be different from that by roots because the cell wall of a leaf is covered by a cuticle, which does not possess the root structure. Franke (1967) reported that cuticular membranes are permeable to organic and inorganic ions and undissociated molecules. The penetration of ions is determined by the kind of charge, adsorbability, and ion radius. Under normal conditions, uptake of ions constitutes an accumulation against a concentration gradient in leaves as well as in roots. The energy required for active absorption can be derived from respiratory metabolism, or, as in green leaves, from photosynthetic processes (Franke 1967). Light quality and intensity improves the rate of ion absorption by leaves (Franke 1967). Franke (1967) suggested that ion uptake by leaves may be completed in three stages. In the first stage, substances applied to the leaf surface penetrate the cuticle and the cellulose wall via limited or free diffusion. In the second stage, these substances, having penetrated the free space, are adsorbed into the surface of the plasma membrane by some form of binding. In the third stage, the absorbed substances are taken up into the cytoplasm, requiring metabolically derived energy for this process.

Previous research has shown that a foliar-applied nutrient passes through the cuticular wax, the cuticle, the cell wall, and the membrane in that order (Middleton and Sanderson 1965; Franke 1967). Sometimes the nutrient will pass through these various layers, but at other times it may pass through the spaces between these layers, which are typical for inorganic ions (Dybing and Currier 1961). However, ions are absorbed by a leaf's stomata (Eichert, Goldbach, and Burkhardt 1998; Eichert and Burkhardt 2001). When the stoma is open, foliar absorption is often easier (Burkhardt et al. 1999).

Remobilization of mineral nutrients is important during ontogenesis of a plant. For example, if a nutrient is not capable of travelling from the sprayed tissues to those developing after the spray treatment, the spray treatment should be repeated every time a new flush of growth appears (Papadakis, Sotiropoulos, and Therios 2007). In other words, if a nutrient is immobilized after its foliar application, the positive effects of the spray would be limited only to the sprayed tissues, and deficiency symptoms will appear in the shoots growing after the spraying. Macronutrient mobility in plant tissue is reasonable, except for Ca and S. But, most micronutrient mobility in plant tissues is poor. For example, Gettier, Martens, and Brumback (1985) reported that two or more foliar sprays may be required within the growing season for soybeans because Mn is poorly remobilized, and its mobility in the phloem is low. According to Guzman et al. (1990), Fe is immobile in tomatoes, cucumbers, and navy beans, but mobile in muskmelons. However, in all these species, Mn acts as an immobile nutrient. Marschner (1995) also reported that mobility of Fe and Mn in the plant phloem is low or intermediate, respectively. Garnett and Graham (2005) reported that Fe shows a high reproductive mobility in wheat where the remobilization evidence for Fe is much greater than for Mn. Therefore, it can be concluded that a large difference exists among nutrients and plant species in remobilization in plant tissues.

10.8.2.2 Advantages and Disadvantages of Foliar Fertilization

Foliar fertilization provides more rapid utilization of nutrients and permits the correction of observed deficiencies in less time than would be required by soil application. It has been observed that crops respond to soil-applied fertilizers in five to six days, if climatic conditions are favorable. On the contrary, crop responses to foliar application of nutrients can be seen in three to four days. The soil-applied nutrient has a lengthy influence on plant growth, but plant response to foliar application is often only temporary. This means that in case of severe nutrient deficiency, several foliar applications are necessary. Foliar application is most successful for micronutrients, whereas soil application is effective for macro- and micronutrients.

For immobilized nutrients in the soils, such as iron, foliar application is more effective and economical than soil application. At the early growth stage, when plant roots are not well developed, foliar fertilization is more advantageous in absorption compared with soil application. For foliar application, in order to maximize spray interception, an appropriate leaf area index (LAI) is a prerequisite. In wheat, a leaf LAI of 2–4 seems adequate (Thorne 1955; Gooding and Davies 1992). Foliage burning may be necessary if salt solution concentration is higher than leaves can tolerate. But the chance of this is remote in the case of soil application. In foliar fertilization, wind is a major cause of variability in spray deposition. Therefore, on a windy day, care should be taken when spraying nutrient solution. Such problems do not occur with soil fertilization.

Gooding and Davis (1992) reported that there are several potential benefits of providing N as urea solution to cereals via the foliage. These include reduced N losses through denitrification and leaching compared with N fertilizer applications to the soil, the ability to provide N when root activity is impaired (e.g., in saline or dry conditions), and uptake late in the season that increases grain N concentration. The importance of foliar fertilization may lie in the localization and regulation of the enzyme systems involved in nitrogen assimilation. It is known that molybdenum ions are important components of cofactors of the key enzymes of assimilatory nitrogen metabolism––nitrogen fixation and nitrate uptake and reduction (Gupta and Lipsett 1981; Campbell 1999; Hristozkova et al. 2007). Foliar sprays of urea have reduced the severity of certain diseases (Gooding et al. 1988), which may result in yield benefit. Additionally, sprays of fertilizer provide opportunities to apply other agrochemical in the same operation as tank mixes allowing saving in labor, machinery and energy cost (Gooding and Davies 1992). It is widely assumed that high rates of foliar uptake are also dependent on high relative humidity because rapid drying can lead to crystallization on the leaf surface (Gamble and Emino 1987).

Like soil application, foliar application is also less effective when soil moisture is limited. Using present-day spraying technology, foliar application of N can have benefits over soil treatment for increasing grain protein content and the bread-making quality of wheat when applied at an appropriate timing, such as at and after antithesis (Gooding and Davies 1992). Foliar application of a nutrient solution after flowering may result in severe discoloration of rice spikelets (author's personal observation) and is not recommended. Tom, Miller, and Bowman (1981) also reported that applying fertilizer solution high in urea content to rice after flowering has resulted in severe discoloration of the lemma and palea and in desiccation. Soil application of N after flowering may not create such a problem. Foliar application will not only increase nutrient uptake efficiency and decrease cost of production but also reduce runoff of soil applied P, which is responsible for eutrophication of many lakes and streams (Sharpley et al. 1994). Micronutrients are required in small amounts, and foliar application of these nutrients is more uniform than soil application.

10.8.2.3 Day Timing of Foliar Fertilization

Day timing of foliar fertilization is important for efficient absorption and also to avoid leaf injury from applied fertilizer materials. For efficient absorption of foliar fertilization, leaf stomata should be open and the temperature should not be too high to cause burning of plant foliage. Afternoon when the air temperature is low (after 2–3 PM) is the best time for foliar fertilization. Another factor that may affect foliar fertilization is wind, which can drift the spray solution. Therefore, windy days should be avoided for foliar spray. It will take three or four hours for applied nutrients to be absorbed by plant foliage. Therefore, there should not be any rain at least three to four hours after application of the nutrient solution. When applying the nutrient solution as a spray, some sticky material should be added to the solution so that the spray clings to the plant foliage.

Foliar application should be made when the plant is not in water stress—either too wet or too dry (Denelan 1988). Nutrients are best applied when the plant is cool and filled with water (turgid) (Girma et al. 2007). Applications that are misapplied or applied too late in the season may not be effective. The most critical time to apply is when the crop is under a given nutrient stress. Stress periods occur during growth activity. Usually, this is when the plant is changing from a vegetative to a reproductive stage (Cantisano 2000; Fageria et al. 2009).

10.8.3 Adequate Rate

Application of iron fertilizers in soil may not be very effective due to their conversion into an insoluble form. However, band use of soluble sources can improve Fe uptake by plants to a limited extent. Data are scarce for rice but have been reported for other annual crops. For example, Godsey et al. (2003) reported that application of 81 kg ha^{-1} $FeSO_4{\cdot}H_2O$ in seed row was the most consistent treatment for correcting Fe deficiency in corn. Similarly, Hergert et al. (1996) noted that corn yield was increased by nearly 3.6 Mg ha^{-1} on two soils with pH greater than 8.2 by placing 85 kg ha^{-1} $FeSO_4{\cdot}7H_2O$ directly in the seed row. Mathers (1970) indicated that 112 kg ha^{-1} $FeSO_4{\cdot}7H_2O$ banded 20 cm beneath the soil surface and directly under the seed increased the grain yield of sorghum by 1.0 Mg ha^{-1} compared with the control treatment. However, some researchers have experienced mixed results with the addition of $FeSO_4{\cdot}7H_2O$. Yield responses varied greatly from year to year, and some research has indicated no significant response to the addition of $FeSO_4{\cdot}7H_2O$ (Olson 1950; Mortvedt and Giordano 1970).

10.8.4 Use of Efficient Genotypes

Plants differ in their uptake of Fe, and it has been suggested that monocotyledonous species are less Fe efficient than dicotyledonous species (Brown and Jones 1976). The ability of plant species to extract Fe from soils varies widely. Kashirad and Marschner (1974) showed that sunflower plants reduced the pH of solution and thus overcame Fe chlorosis. However, corn plants did not reduce the pH and failed to utilize precipitated ferric compounds to overcome Fe chlorosis. Mortvedt (1991) suggested that the use of hydrated polymers as a matrix for Fe fertilizers may provide a novel method for improving the efficiency of Fe uptake by plants in calcareous soils. Root Fe^{3+}-reducing activity of soybean genotypes was correlated with genotypic resistance to Fe chlorosis measured in field nurseries and was used as a method for identifying chlorosis-resistant (Fe-efficient) genotypes (Rengel 1999). Soybean genotypes vary in the magnitude and timing of Fe stress responses; resistant genotypes reduce Fe^{3+} sooner and in larger quantities than susceptible genotypes (Jolley et al. 1992; Ellsworth et al. 1998). Rye, a species very tolerant of Fe deficiency in calcareous soils, exudes mugineic acid, hydroxymugineic acid (HMA), and diaminemugineic acid (DMA) into the rhizosphere. It has been suggested that the capacity of rye to exude HMA may be related to the superior tolerance of rye to Fe, as well as to Zn deficiency when compared with wheat, which exudes only DMA (Cakmak et al. 1996; Rengel 1999).

Aqueous solutions of $FeSO_4$, $Fe_2(SO_4)_3$, or other inorganic Fe salts could be absorbed by polymers to form gels that are band applied to soil. As some polymers are only slowly biodegraded in soil, it is possible that these hydrogels could provide available Fe for crops for a long period of time. Technologies to correct Fe deficiencies in fields, vegetables, and tree crops were reviewed by Mortvedt (1991). Similarly, controlled availability of micronutrients, including iron, is discussed by Mortvedt (1994). Crop species efficient or inefficient in iron use are cited in Table 10.9. Similarly, upland rice genotypes classified as efficient or inefficient in iron uptake are listed in Table 10.10.

10.9 IRON TOXICITY IN LOWLAND RICE

In many parts of the world, iron toxicity is very common in flooded rice and is associated with inherent soil properties (Yoshida 1981; Fageria and Rabelo 1987; Olaleye et al. 2001; Fageria, Slaton, and Baligar 2003; Fageria, Stone, and Santos 2003; Fageria, Barbosa Filho, and Guimaraes 2008). Iron toxicity that occurs primarily on acid sulfate soils has been reported as a significant problem in rice-growing areas of Southeast Asia and Africa (Ottow et al. 1983). Iron toxicity has also been reported in flooded rice in the Brazilian states of Santa Catarina and Rio Grande do Sul (Fageria, Stone, and Santos 2003). It is thought to be caused by excessive Fe in the soil solution or to be induced by deficiencies of other nutrients. Direct toxicity is defined as excessive Fe absorption by the plant resulting from high soil

TABLE 10.9
Iron Use Efficiency of Crop Species

Efficient	Moderately Efficient	Inefficient
Wheat	Corn	Sorghum
Sunflower	Oats	Soybeans
Dry beans	Barley	Upland rice
Amaranthus	Alfalfa	Peanuts
	Potato	Millet

Source: Adapted from Fageria, N. K., C. M. Guimaraes, and T. A. Portes. 1994. Iron deficiency in upland rice. *Lav. Arrozeira*. 47:3–5.

TABLE 10.10
Classification of Upland Rice Genotypes for Iron Use Efficiency

Genotype	Grain Yield (kg ha^{-1})	Rating	Classification
CNA 7460	2679	3	Highly efficient
CNA 6895	2890	3	Highly efficient
CNA 6843-1	2570	5	Moderately efficient
Aaguaia	2539	7	Moderately efficient
CNA 7286	2398	7	Moderately efficient
CNA 7475	2375	7	Moderately efficient
CNA 7449	2218	9	Inefficient

Source: Fageria, N. K., C. M. Guimaraes, and T. A. Portes. 1994. Iron deficiency in upland rice. *Lav. Arrozeira*. 47:3–5.

solution concentrations of Fe (Howeler 1973). Indirect toxicity has been blamed on low soil fertility. Specifically, low soil concentrations of Ca, Mg, K, and P have been cited as the common factors in soils expressing Fe toxicity symptoms (Benckiser et al. 1984). The name Fe toxicity is somewhat misleading because, in many cases, the Fe toxicity is actually an affect rather than a cause. Howeler (1973) suggested that rice roots become coated with Fe oxide, which reduces the plants' ability to absorb sufficient quantities of other plant nutrients that are already present in low concentrations. The production of H_2S, FeS, or both in highly reduced flooded soils may contribute to Fe toxicity. Hydrogen sulfide and FeS reduce the oxidizing capacity of the rice root system, thereby increasing the susceptibility of rice plants to Fe toxicity (Inada 1966b). Application of $(NH_4)_2SO_4$ fertilizer has been noted to increase the incidence of Fe toxicity (Inada 1966a). Acidic soil conditions are likely to produce either type of Fe^{2+} toxicity. Severe Fe toxicity can cause significant rice yield reductions (Genon et al. 1994). In general, solution Fe concentrations of flooded soils are very high because of anaerobic soil conditions, but rice is well adapted to this flooded environment and normally is able to regulate Fe uptake.

10.9.1 Iron Uptake Mechanism

In oxidized soils, iron generally is in the ferric (Fe^{3+}) form and is tied up with oxides and hydroxides. The solubility of Fe^{3+} iron is very low. In addition, the major form of iron uptake by plants is the ferrous (Fe^{2+}) ion (Lindsay and Schwab 1982). Therefore, iron has to be reduced to the Fe^{2+} form for uptake by crop plants. Otherwise, the Fe^{3+} form of iron has to be transported by chelating agents inside the roots for uptake and utilization, meaning there are two mechanisms evolved by plants in response to Fe deficiency in the growth medium (Romheld and Marschner 1986; Rogers and Guerinot 2002; Epstein and Bloom 2005). In the first mechanism, the release of protons (H^+) by plant roots is involved; this lowers the pH of the rhizosphere. The lower pH of the rhizosphere may solubilize or reduce Fe^{3+} to Fe^{2+}. The reduced iron is then transported across the plasma membrane by a Fe^{2+}-specific transport system. This type of mechanism is mainly operated in the dicots and nongraminaceous monocots (Epstein and Bloom 2005).

The second mechanism used by crop plants for iron uptake is the release of phytosiderophores (iron carriers) by plant roots. These phytosiderophores form a complex with Fe^{3+} ion without reducing it to Fe^{2+}, and this Fe^{3+} siderophore complex is then transported across the root cell plasma membranes (Epstein and Bloom 2005) (Figure 10.8). Takagi, Nomoto, and Takemoto (1984) showed that the release of chelating compounds or phytosiderophores was specific for grasses or graminaceous species but not for dicot plants. The chelating compounds are characterized as nonprotein amino acids, and mugineic and avenic acids. The pathways for biosynthesis of these acids are still not known, but these amino acids are structurally related to nicotianamine, a compound primarily found in the shoots of green plants. It has been shown to induce regreening of chlorotic mutants of tomatoes, possibly through mobilizing Fe for intercellular and intracellular transport (Kochain 1991).

10.9.2 Factors Inducing Iron Toxicity

Factors that affect iron solubility and concentration in the rice rhizosphere may influence availability. If this influence in favor of higher iron uptake is higher than plant demand, iron toxicity may be induced. Plant soil conditions important to the solubility and availability of iron have not been fully defined, largely because solubility and availability relationships for iron are very complex (Fageria, Baligar, and Wright 1990; Fageria, Barbosa Filho, and Guimaraes 2008). However, principal iron-toxicity-inducing factors are release of iron from parent material to soil solution, reduction in oxidation–reduction potential, increase in ionic strength, low soil fertility, low soil pH, and interaction with other nutrients.

10.9.2.1 Release of Iron from Parent Material in Soil Solution

Iron is a major component of most soils. Iron minerals commonly found in soil include geothite (FeOOH), hematite (Fe_2O_4), pyrite (FeS), siderite ($FeCO_3$), and magnetite (Fe_3O_4) (Fageria, Slaton, and Baligar 2003). Weathering of parent materials releases significant amounts of nutrients, including iron, in the soil solution.

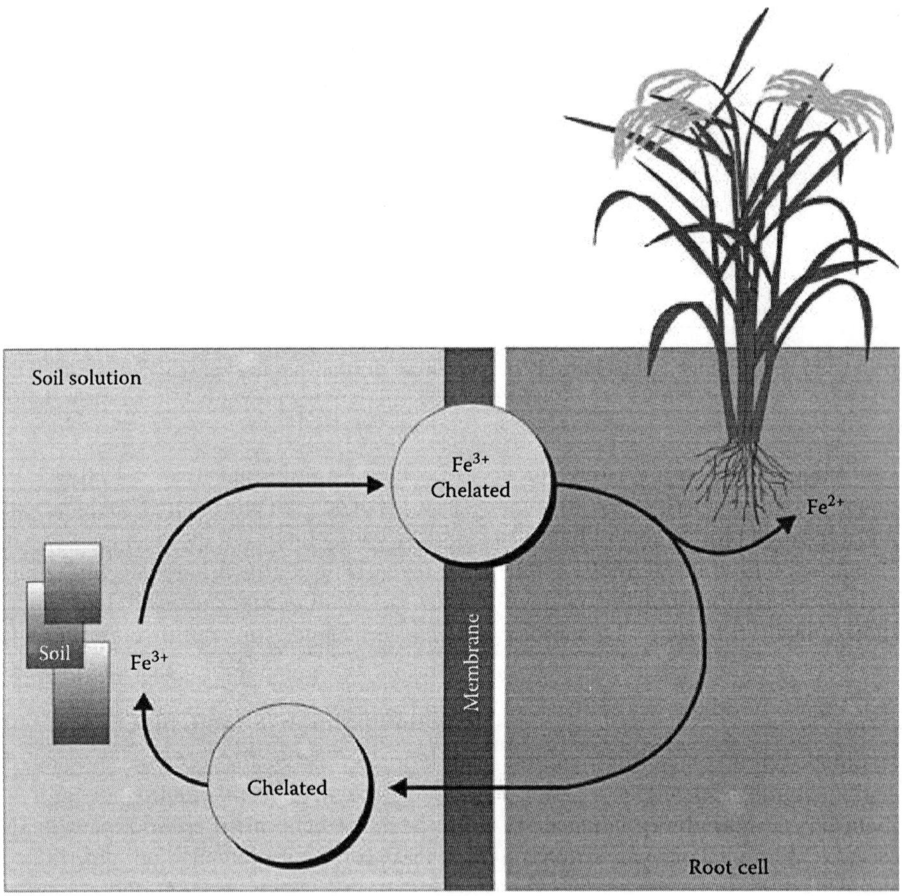

FIGURE 10.8 Iron uptake mechanism in rice plants. (From Fageria, N. K. 2009. *The Use of Nutrients in Crop Plants*. Boca Raton, FL: CRC Press.)

When the Fe^{2+} concentration of reduced or submerged soils is high, uptake may exceed plant demand and may be toxic to rice plants. In West Africa, bronzing occurs in rice plants grown in inland valley swamps, irrigated lowlands, and hydromorphic lands (International Institute of Tropical Agriculture 1983). A study in Nigeria concluded that the occurrence of bronzing was positively correlated with the ferrous iron concentration in the soil, suggesting that the disorder was caused by iron toxicity (Kosaki and Juo 1986).

10.9.2.2 Oxidation–Reduction Potential

Oxidation–reduction (or redox) potential has significant influence on the chemistry of iron and other nutrients in the submerged soils. It is the best single indicator of anaerobiosis in flooded soil and allows reasonable predictions to be made concerning the behavior of several essential plant nutrients (Patrick and Mikkelsen 1971). Oxidation–reduction is a chemical reaction in which electrons are transferred

TABLE 10.11
Principal Reduction Processes in Flooded Soils

$2NO_3^- + 12H^+ + 2e^- \Leftrightarrow N_2 + 8H_2O$
$MnO_2 + 4H^+ + 2e^- \Leftrightarrow Mn^{2+} + 2H_2O$
$SO_4^{2-} + 10H^+ + 8e^- \Leftrightarrow H_2S + 4H_2O$
$CO_2 + 8H^+ + 8e^- \Leftrightarrow CH_4 + 2H_2O$
$2H^+ + 2e^- \Leftrightarrow H_2$

Source: Adapted from Ponnamperuma, F. N. 1977. *Screening Rice for Tolerance to Mineral Stresses*. IRRI Research Paper Series No. 6. Los Baños, Philippines: IRRI.

from a donor to an acceptor. The electron donor loses electrons and increases its oxidation number (or is oxidized); the acceptor gains electrons and decreases its oxidation number (or is reduced). The source of electrons for biological reductions is organic matter (Ponnamperuma 1972). Oxidation–reduction affects the valence of iron and, therefore, its uptake by plants. The Fe^{3+} ion is reduced to Fe^{2+} due to oxidation–reduction processes, and it increases its uptake. On the other hand, when Fe^{2+} is oxidized to Fe^{3+}, its concentration is reduced and its uptake by plants is also reduced. The reduction of Fe^{3+} to Fe^{2+} is expressed by the following equation (Ponnamperuma 1977):

$$Fe(OH)_3 + 3H^+ + e^- \Leftrightarrow Fe^{2+} + 3H_2O \tag{10.15}$$

In addition to the reduction of Fe^{3+} to Fe^{2+}, Mn^{4+} is reduced to Mn^{2+}, NO_3^- to N_2, SO_4^{2-} to H_2S, CO_2 to CH_4, and H^+ to H_2 in submerged soils. These reduction processes are presented in Table 10.11. All these reduction processes influence Fe^{2+} concentration in the soil solution and consequently its uptake by rice plants. Reduction processes in the submerged or flooded soils are influenced by magnitude of oxidation–reduction, which is measured in millivolt. The critical redox potentials for Fe reduction and its consequent dissolution are between +300 and +100 mV at pH 6 and 7, and −100 mV at pH 8, although at pH 5, appreciable reduction occurs at +300 mV (Gotoh and Patrick 1976). Oxidation–reduction or potential reduction values for oxidized and submerged soils and reduction processes are given in Table 10.12.

10.9.2.3 Soil pH

Soil pH is an important chemical property, which determines iron solubility and its uptake by plants. In addition, soil pH is often a highly changeable property because of the dynamic nature of various soil processes and the interactions of these processes with plants and microorganisms (Adams 1984). As pH increases, Fe is converted to less soluble forms, principally to the oxide Fe_2O_3 and iron availability decrease. The reaction responsible for the reduced solubility of Fe with increasing pH is well understood. It results in the precipitation of $Fe(OH)_3$ because the concentration of

TABLE 10.12
Range of Oxidation–Reduction Potential Values in Oxidized and Submerged Soils and at Which Reduction Processes Occur

Soil Moisture/Reduction Processes	Redox Potential (mV)
Well-oxidized soils	+700 to +500
Moderately reduced soils	+400 to +200
Reduced soils	+100 to −100
Highly reduced soils	−100 to −300
NO_3^- to N_2	+280 to +220
Mn^{4+} to Mn^{2+}	+280 to +220
Fe^{3+} to Fe^{2+}	+180 to +150
SO_4^{2-} to S^{2-}	−120 to −180
CO_2 to CH_4	−200 to −280
O_2 to H_2O	+380 to +320
Absence of free O_2	+350

Source: Patrick, W. H. Jr. 1966. Apparatus for controlling the oxidation-reduction potential of waterlogged soils. *Nature.* 212:1278–1279; Patrick, W. H., Jr., and C. N. Reddy. 1978. Chemical changes in rice soils. In *Soils and Rice,* eds. IRRI, pp. 361–379. Los Baños, Philippines: IRRI; and Marschner, H. 1995. *Mineral Nutrition of Higher Plants*, 2nd edition. New York: Academic Press.

OH^- ions is increased as indicated by the following reaction (Fageria, Baligar, and Wright 1990):

$$Fe^{3+} + OH^- \Leftrightarrow Fe(OH)_3 \qquad (10.16)$$

The $Fe(OH)_3$ is chemically equivalent to the hydrated oxide, $FeO_3{\cdot}3H_2O$. Acidification shifts the equilibrium, causing a greater release of Fe^{3+} as a soluble ion. This means that iron toxicity is more severe in acidic soils than in alkaline soils. The overall effect of submergence is to increase the pH of acidic soils and to depress the pH of sodic and calcareous soils (Ponnamperuma 1972). The magnitude of pH increase or decrease depends on soil type, organic matter content, soil fertility, crop rotation, and rice cultivar. Soils high in organic matter content and reducible iron attain a pH of about 6.5 within a few weeks of submergence. Acidic soils low in organic matter or active iron slowly attain pH values that are less than 6.5 (Ponnamperuma 1972). The increase in pH of acidic soils due to submergence is associated with a reduction process, and a decrease in pH of alkaline soils is due to accumulation of CO_2.

All the reduction reactions consume H^+ ions; this means a decrease in acidity and an increase in OH^- ions. Most soils contain more Fe^{3+} hydrates than any other oxidant; the increase in pH of acidic soils is largely associated with iron reduction (Ponnamperuma 1972). The increase in pH of acidic soils after submergence has beneficial effects on rice growth, such as reducing Al^{3+} and Fe^{2+} toxicity and increasing P availability. Bremen and Moormann (1978) reported that iron toxicity has been observed only in flooded soils with a pH below 5.8 when aerobic, and a pH below 6.5 when anaerobic. Also, it has been reported that in most flooded soils that have an aerobic pH below 5.8, the concentration of iron reaches high levels for only a short period. But young acid sulfate and acidic soils low in reducible iron often have iron concentrations exceeding 500 mg kg^{-1} for prolonged periods.

10.9.2.4 Ionic Strength

Ionic strength is defined as the measure of the electrical environment of ions in a solution. Ionic strength can be calculated using the following formula (Fageria, Barbosa Filho, and Guimaraes 2008):

$$\text{Ionic strength} = 1/2\Sigma M_i Z_i^2 \quad (10.17)$$

where M is the molarity of the ion, Z_i is the total charge of the ion (regardless of sign), and Σ is a symbol meaning the "sum of."

Ionic strength of submerged soil increases with the release of macro- and micronutrients in the soil solution (Patrick and Mikkelsen 1971). The increase in ionic strength raises Fe^{2+} uptake by rice plants and Fe toxicity. In acidic soils, the increase in ionic strength is associated with a reduction of Fe^{2+} and Mn^{2+} and solubilization of P. In alkaline soils, the increase in ionic strength is associated with displacement of cations by ferrous iron produced through reduction reactions (Patrick and Mikkelsen 1971).

10.9.2.5 Low Soil Fertility

An adequate level of essential nutrients in soils is an important factor for improving crop productivity, maintaining ecosystem stability, and maintaining a healthy environment. When there is a deficiency of any essential element, iron uptake by rice under reduced or waterlogged soils increases and may reach a toxic level. Nutrient deficiencies or low concentrations of nutrients, such as potassium, phosphorus, calcium, and magnesium, may increase the uptake of Fe^{2+} by rice and cause iron toxicity (Tadano and Tanaka 1970; Tadano 1976). Yamauchi (1989) reported that the application of potassium sulfate reduced the severity of bronzing and increased the dry matter production of rice plants grown in the field. The concentration and accumulation of potassium in the shoots increased when the bronzing severity decreased, and the iron concentration was decreased by the dilution effect caused by the increased dry matter (Yamauchi 1989).

The oxidizing capacity of the rice root, which causes oxidation and precipitation of part of the ferrous iron, is important in depressing the uptake of Fe^{2+} present in high concentrations in the root zone (Tadano 1975). Trolldenier (1977) reported that low potassium and phosphorus apparently aggravate iron toxicity through decreased oxidizing capacity of the roots. By contrast, nitrogen deficiency does not result in higher iron uptake (Yoshida 1981). Trolldenier (1977) reported that higher N levels might even stimulate the uptake of excess iron. High salinity due to sodium chloride or magnesium chloride also increases iron uptake and may thus aggravate iron toxicity (Bremen and Moormann 1978).

10.9.2.6 Interaction with Other Nutrients

Iron interaction with other nutrients may be antagonistic, synergistic, or neutral depending on the growth response of plants. If the growth response is greater with two combined factors as compared with the sum of their individual effects, it is a positive interaction, and when the combined effects are less, the interaction is negative (Sumner and Farina 1986; Fageria, Baligar, and Wright 1990). Nutrient interaction is also measured in terms of influence of one nutrient on the uptake of other nutrients. Antagonistic interactions with other nutrients reduce Fe availability to rice and vice versa. The antagonistic or negative interactions may result from processes that occur either outside the root or within the root. Those taking place in the external root environment are usually precipitation or similar reactions that reduce the chemical availability of the nutrient. The interactions that influence absorption or utilization processes alter the effectiveness of a nutrient by reducing its physiological availability (Fageria, Baligar, and Wright 1990).

In rice plants, one example of an antagonistic interaction with iron is manganese (Olsen and Watanable 1979). Application of Mn in the soils significantly reduces uptake of iron by rice. Fageria, Barbosa Filho, and Carvalho (1981) reported that uptake of P, K, Ca, and Mg decreased in the nutrient solution with the increasing Fe concentration in the range of 0–160 mg L^{-1}. Regression equations showing the relationship between the Fe concentration in the nutrient solution and the uptake of P, K, Ca, and Mg is presented in Table 10.13. The decrease in uptake of P and K was quadratic with increasing Fe concentration from 0 to 160 mg L^{-1} in the solution. However, the uptake of Ca and Mg decreased linearly with increasing concentration of Fe in the nutrient solution (Table 10.13).

10.9.3 Diagnostic Techniques for Iron Toxicity

Diagnostic techniques for nutrient toxicity refer to the methods for identifying nutrient toxicities or imbalances in the soil–plant system. Nutritional toxicity can occur when excess nutrients in the growth medium are absorbed by plant roots or inhibit

TABLE 10.13
Relationship between Fe Concentration in Nutrient Solution and Uptake of P, K, Ca, and Mg in the Shoot of 60-Day-Old Rice Plants

Regression Equation	R^2
Fe Conc. (X) vs. uptake of P (Y) = 0.5838 − 0.0078X + 0.000036X^2	0.7577*
Fe Conc. (X) vs. uptake of K (Y) = 2.5467 − 0.0154X + 0.000045X^2	0.9826**
Fe Conc. (X) vs. uptake of Ca (Y) = 0.1449 − 0.00053X	0.8337*
Fe Conc. (X) vs. uptake of Mg (Y) = 0.3267 − 0.00074X	0.5973*

Source: Adapted from Fageria, N. K., M. P. Barbosa Filho, and J. R. P. Carvalho. 1981. Influence of iron on growth and absorption of P, K, Ca and Mg by rice plant in nutrient solution. *Pesq. Agropec. Bras.* 16:483–488.

Note: 60-day-old rice plants.

*,**Significant at the 5 and 1% probability levels, respectively.

absorption of other essential nutrients. Three methods to assess iron toxicity for lowland rice are (1) visual symptoms, (2) soil testing, and (3) plant analysis. These three approaches are widely used separately or collectively as iron toxicity diagnostic aids. A brief discussion of these techniques is given in the following sections.

10.9.3.1 Visual Symptoms

When the supply of a particular nutrient is at an excess level in the soil or when plant roots absorb more of a nutrient than a plant needs, plants show certain growth disorders. These disorders may be expressed as reduced height, reduced tillering, leaf discoloration, and reduced root growth. Toxicity symptoms of an element always start in the old leaves. However, if toxicity persists for a long duration, symptoms may spread to the whole plant's foliage. The first symptom is the appearance of many small rust-colored spots on the tips of the lower leaves. These spots enlarge and progress down the leaf in rows between the veins of the blade. The mid-vein usually remains green and unaffected for several weeks after the appearance of the problem. Iron toxicity symptoms in rice plants are shown in Figure 10.9 where leaves are brown in color. Iron toxicity can seriously affect root and shoot growth (Figures 10.10 and 10.11). Rice roots commonly have a black coating of FeS, which can also be used to help diagnose Fe toxicity. The degree of leaf bronzing has been suggested as a good measure of the severity of Fe toxicity in flooded rice (Fageria, Slaton, and Baligar 2003).

Data in Table 10.14 show that plant height decreased when Fe concentration was increased more than 5 mg L^{-1} at 20 days of plant growth and at 40 mg L^{-1} at 60 days

FIGURE 10.9 (See color insert.) Iron toxicity symptoms in flooded rice. (From Fageria, N. K., L. F. Stone, and A. B. Santos. 2003. Soil fertility management for irrigated rice. Santo Antônio de Goiás, Brazil: EMBRAPA Arroz e Feijão.)

FIGURE 10.10 Lowland rice growth in nutrient solution at different iron concentrations. (From Fageria, N. K. 1984. *Fertilization and Mineral Nutrition of Rice*. Rio de Janeiro, Brazil: EMBRAPA/CNPAF.)

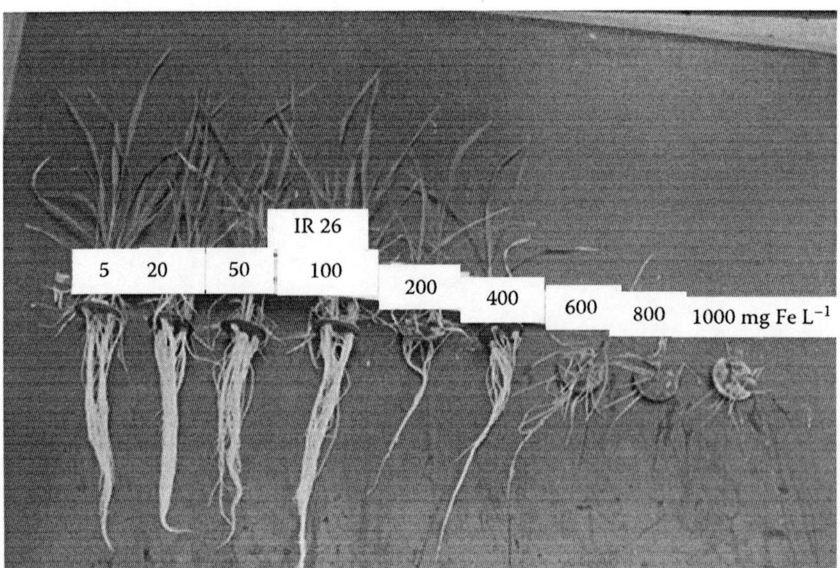

FIGURE 10.11 Roots and top growth of lowland rice (cv. IR 26) at different iron concentrations. (From Fageria, N. K. 1984. *Fertilization and Mineral Nutrition of Rice*. Rio de Janeiro, Brazil: EMBRAPA/CNPAF.)

TABLE 10.14
Plant Height and Top and Root Dry Weight at 20 and 60 Days of Plant Age as Influenced by Fe Concentration Nutrient Solution

	Plant Height (cm)		Top Dry Wt. (g plant^{-1})		Root Dry Wt. (g plant^{-1})	
Fe Conc. (mg L^{-1})	20 days	60 days	20 days	60 days	20 days	60 days
0.0	28	61	0.16	2.75	0.03	0.64
2.5	46	97	0.44	6.91	0.09	1.19
5.0	52	108	0.52	5.89	0.13	1.05
10.0	50	111	0.49	6.31	0.10	1.07
20.0	44	112	0.36	5.69	0.08	1.18
40.0	46	110	0.44	5.90	0.11	1.24
80.0	38	83	0.21	3.23	0.07	0.84
160.0	18	36	0.10	0.17	0.02	0.12
Average	40	90	0.33	4.61	0.08	0.92

Source: Adapted from Fageria, N. K., M. P. Barbosa Filho, and J. R. P. Carvalho. 1981. Influence of iron on growth and absorption of P, K, Ca and Mg by rice plant in nutrient solution. *Pesq. Agropec. Bras.* 16:483–488.

of plant growth. Top dry weight decrease was at 10 mg Fe L^{-1} in the nutrient solution at 20 days of plant growth, and decrease in top dry weight at 60 days plant growth occurred at 80 mg Fe L^{-1}. Dry weight of roots reduced when the Fe concentration was 10 mg L^{-1} at 20 days of plant growth, and at 60 days of plant growth, root dry weight decreased at 80 mg Fe L^{-1}. Toxic levels of iron in nutrient solution may vary from 20 to 500 mg L^{-1} (Ishizuka 1961; Tanaka, Leo, and Navasero 1966). This variability in the toxic level of iron may be associated with cultivar planted, soil fertility levels, and climatic conditions, such as temperature and solar radiation (Bremen and Moormann 1978). Rice plants develop resistance to Fe toxicity with advanced plant age (Fageria, Barbosa Filho, and Carvalho 1981). Toxicity symptoms normally occur over an area and not on an individual plant. If a symptom is found on a single plant, it may be due to disease, insect injury, or a genetic variation. Earlier symptoms are often more useful than later mature symptoms.

Visual symptoms are the least costly nutritional disorder diagnostic technique compared with other methods. However, a lot of experience is necessary on the part of the observer because toxicity symptoms may be confused with insect and disease infestation, herbicide damage, soil salinity, and inadequate drainage problems. The use of visible symptoms has the advantage of direct field application without the need for costly equipment or laboratory support services, such as is the case with soil and plant analysis. A disadvantage is that sometimes it is too late to correct toxicity of iron because the disorder is identified when it is too severe to produce visible symptoms. For some disorders, considerable yield loss may have already occurred by the time visible symptoms appear. Several publications are available in which nutritional disorders have been described and illustrated with color photographs, including iron toxicity for rice (Yoshida 1981; Bennett 1993; Fageria, Stone, and Santos 2003). Readers may refer to these publications to get acquainted with iron toxicity symptoms in lowland rice.

10.9.3.2 Soil Testing

In a broad sense, soil testing is any chemical or physical measurement that is made on a soil (Fageria and Baligar 2005). Soil testing is an important tool in high-yield farming, but it produces the best results only when used in conjugation with other good farming practices. "There is good evidence that the competent use of soil tests can make a valuable contribution to the more intelligent management of the soil." This statement was made by the U.S. National Soil Test Workgroup in its 1951 report and is still applicable today (Fageria and Baligar 2005).

In soil analysis, collecting soil samples properly from the field, drying, and preparing for analysis are important steps. Furthermore, selecting an appropriate extraction solution is also important for soil analysis. For micronutrient analysis of soil (Fe, Zn, Cu, Mn), diethylene triamine pentaacetic acid (DTPA) extracting solution developed by Lindsay and Norvell (1978) is most common. This solution consists of 0.005M DTPA, 0.01M $CaCl_2$, and 0.01M triethanolamine (TEA) buffered at pH 7.3. The use of Mehlich 1 (0.05 M HC1+ 0.0125 M H_2SO_4) extracting solution for Zn, Cu, Mn, and Fe analysis in soils is a standard practice in Brazilian soil analysis laboratories because of economic reasons. Because Mehlich 1 extracting solution is used for P and K analysis, micronutrient analysis can be done on the same soil solution.

Knowledge of iron deficiency and toxicity levels in soil is an important tool for interpretation of soil test results and diagnoses of iron disorders in rice. Sims and Johnson (1991) reported that for most crops, the critical deficiency soil Fe concentration range was 2.5–5.0 mg kg^{-1} of DTPA extractable Fe, but this is also influenced by soil pH. In field and pot experiments, the degree of bronzing in a given variety showed a highly significant correlation (r = 0.90**) with yield (Bremen and Moormann 1978). Iron toxicity in rice plants, as indicated by bronzing of leaves, was reported when soluble iron in soil solution was more than 300–500 mg kg^{-1} by DTPA extracting solution (Ponnamperuma, Bradfield, and Peech 1955; Tanaka, Leo, and Navasero 1966). However, Bremen and Moorman (1978) reported that bronzing symptoms generally appear when iron concentration in soil solution is in the range of 300–400 mg kg^{-1} by DTPA extracting solution. Barbosa Filho, Fageria, and Stone (1983) found that iron toxicity in lowland rice occurred when Mehlich 1 extracted iron in the soil was in the range of 420–730 mg kg^{-1}. This means that determining the iron toxicity level in the soil also depends on the solution used to extract iron from soil. Values for the Mehlich 1 extracting solution are higher than those of the DTPA solution.

10.9.3.3 Plant Analysis

Plant analysis is based on the concept that the concentration of an essential element in a plant or a part of a plant indicates the soil's ability to supply that nutrient. This means it is directly related to the quantity in the soil that is available to the plant (Fageria and Baligar 2005). For annual crops, the primary objective of plant analysis is to identify nutritional problems or to determine or monitor the nutrient status during the growing season. If deficiency or toxicity is identified early in the growth stage of a crop, it is possible to correct it during the current season. Otherwise, steps should be taken to correct it in the next crop. Like soil analysis, plant analysis also involves plant sampling, plant tissue preparation, analysis, and interpretation of analytical results. All of these steps are important for a meaningful plant analysis program.

Many factors such as soil, climate, plant type, and their interaction affect absorption of nutrients. However, concentrations of the essential nutrients are maintained within rather narrow limits in plant tissues. Such consistency may be due to the operation of delicate feedback systems that enable plants to respond in a homeostatic fashion to environmental fluctuations (Fageria and Baligar 2005). For the interpretation of plant analysis results, a critical nutrient concentration concept was developed. This concept is widely used in the interpretation of plant analysis results for diagnosis of nutritional disorders.

Adequate Fe concentration in plant tissue is in the range of 70–300 mg kg^{-1} (Wells, et al. 1993). Iron deficiency or toxicity occurs at concentrations below or above this sufficiency range. Fageria, Barbosa Filho, and Carvalho (1981) reported that toxic levels of Fe in whole plant tops were 680 mg kg^{-1} at 20 days of plant age and 850 mg kg^{-1} at 40 days of plant growth. However, toxic concentration also depends on rice cultivars (Fageria et al. 1984).

10.9.4 Physiological Disorders Related to Iron Toxicity

Iron toxicity causes some physiological disorders in rice plants. These are known as bronzing, akagare type I, and akicochi (Yoshida 1981; Tadano 1995). Bronzing is a discoloration of rice leaves. Many small brown spots appear on the leaves, and these symptoms start on the tips of lower leaves and spread to the basal parts. In severe cases, the brown discoloration even appears on the upper leaves. The bronzing symptoms vary with cultivars and may be purplish orange, yellowish brown, reddish brown, brown, or purplish brown (Tadano 1995).

Akagare type I disorder has been caused by iron toxicity reported in rice in Japan, South Korea, and Sri Lanka (Tadano 1995). The symptoms of this disorder include leaves that turn dark green and small reddish brown spots that appear around the tips of older leaves. Under severe iron toxicity, this spreads all over the plant leaves, and leaves die from the tips. The roots of affected plants turn light brown or dark reddish brown, or are blackened depending on soil types (Tadano 1995). Yoshida (1981) reported that the main cause of akagare disorder in rice-growing soils is K deficiency.

Akicochi disorder in rice caused by iron toxicity appears as small brown spots on the older leaves and may spread to the entire plant foliage under severe iron toxicity. Akicochi disorder symptoms are similar to helminthosporium (a fungus disease of rice). Akicochi disorder mainly occurs when soils are deficient in silicon and potassium. The presence of adequate amounts of silicon and potassium in the soil solution increases the oxidizing power of rice roots and decreases the excessive uptake of iron (Tadano 1976; Yoshida 1981). Application of silicates is considered beneficial when the silica content in rice straw is lower than 11% (Yoshida 1981). Similarly, application of K is desirable when its content in rice soils is lower than 60 mg kg^{-1} (Fageria 1984).

10.9.5 Management Practices to Ameliorate Iron Toxicity

Effective measures to ameliorate Fe^{2+} toxicity include periodic surface drainage to oxidize reduced Fe^{2+}, liming acidic soils, use of adequate amounts of essential nutrients, and planting iron-tolerant cultivars/genotypes. The Fe-excluding ability of rice plants is lowered by deficiencies of P, K, Ca, and Mg (Obata 1995). In particular, K deficiency readily induces Fe toxicity (Fageria, Stone, and Santos 2003). Among these management practices, the use of tolerant cultivars is the most economical and environmentally sound practice. Genetic variability in rice cultivars for Fe^{2+} toxicity has been reported (Fageria et al. 1984; Fageria and Rabelo 1987; Fageria, Baligar, and Wright 1990). Data in Table 10.15 show differences in the shoot dry weight of eight lowland rice cultivars cultivated in solution. Based on the shoot dry weights shown in Table 10.15, the iron tolerance index (ITE) was calculated, and cultivars were classified as tolerant (T), moderately tolerant MT), or susceptible (S) to iron toxicity (Table 10.16). Cultivars BG 90-2, Suvale 1, and Paga Divida were tolerant at three Fe concentrations. Also, cultivars IR 36, IR 22, and IR 26 were susceptible to three Fe concentrations. Figure 10.12 shows tolerance to iron toxicity in relation to root growth for two lines of irrigated rice. Tadano (1976) reported that there might be three mechanisms that are responsible for iron toxicity variability in rice cultivars. These mechanisms are (1) oxidation of Fe^{2+} in the rhizosphere, (2) exclusion of Fe^{2+} at the root surface, and (3) retention of Fe in the root tissues, which prevents translocation of Fe from root to shoot. The introduction of modern high-yielding cultivars has often led to an increase in the incidence of bronzing apparently because the traditional cultivars had a better tolerance for iron toxicity and other adverse soil conditions (Bremen and Moormann 1978).

TABLE 10.15
Shoot Dry Weight (g 4 plants^{-1}) of Lowland Rice Cultivars as Influenced by Iron Concentration in Nutrient Solution

	Fe Concentration (mg L^{-1})			
Cultivar	2	40	60	80
BG 90-2	4.50	2.47	1.50	1.36
Suvale 1	5.88	2.12	1.78	0.63
Paga Divida	4.66	2.35	1.68	1.48
CICA 8	3.17	1.21	0.39	0.18
Bluebelle	2.00	1.44	1.19	0.76
IR 36	1.85	0.40	0.58	0.26
IR 22	1.46	0.67	0.25	0.19
IR 26	1.26	0.18	0.15	0.05
Average	3.10	1.36	0.94	0.61

Source: Adapted from Fageria, N. K., M. P. Barbosa Filho, J. R. P. Carvalho, P. H. N. Rangel, and V. A. Cutrim. 1984. Preliminary screening of rice cultivars for tolerance to iron toxicity. *Pesq. Agropec. Bras.* 19:1271–1278.

TABLE 10.16
Iron Tolerance Index (ITE) of Lowland Rice Cultivars

	Fe Concentration (mg L^{-1})		
Cultivar	**40**	**60**	**80**
BG 90-2	2.64 (T)	2.32 (T)	3.24 (T)
Suvale 1	2.96 (T)	3.59 (T)	1.96 (T)
Paga Divida	2.60 (T)	2.69 (T)	3.65 (T)
CICA 8	0.91 (MT)	0.42 (S)	0.30 (S)
Bluebelle	0.68 (MT)	0.82 (MT)	0.80 (MT)
IR 36	0.18 (S)	0.37 (S)	0.25 (S)
IR 22	0.23 (S)	0.13 (S)	0.15 (S)
IR 26	0.05 (S)	0.06 (S)	0.03 (S)

Note: Iron tolerance index (ITE) = (Shoot dry weight at 2 mg Fe L^{-1} level/average shoot dry weight of eight cultivars at 2 mg Fe L^{-1}) × (Shoot dry weight at 40, 60, or 80 mg Fe L^{-1} levels/average shoot dry weight of eight cultivars at 40, 60, or 80 mg Fe L^{-1} levels). Shoot dry weight values were taken from Table 7. The species having ITE > 1 were classified as tolerant, species having ITE between 0.5 and 1 were classified as moderately tolerant, and species having ITE < 0.5 were classified as susceptible to soil acidity. These efficiency indices rating are arbitrary.

10.10 CONCLUSIONS

Iron is required by rice in large quantities compared with other micronutrients, except chlorine. Iron deficiency is reported in upland rice grown on acidic soils and lowland rice seedlings in Histosols. In upland rice, iron deficiency is mostly related to increased pH of acidic soils, and deficiency in Histosols is related to naturally low levels of iron in these soils. In many parts of the world, iron toxicity is reported in flooded rice due to increased concentrations of Fe^{2+} from reduced soil conditions. The changes in plant available iron resulting from flooding are due to biological oxidation–reduction processes brought into play by the exclusion of oxygen from the flooded soil. Several ferric compounds, such as hydrated ferric oxide and ferric phosphate, are reduced under the anaerobic conditions established in a waterlogged soil by respiring microorganisms. The reduced form of iron is much more soluble than the oxidized form. Iron is immobile in the plant, and its deficiency symptoms in upland rice first appear in the younger leaves. Iron plays many physiological and biochemical roles in plants. Soil testing and plant analysis can be used to identify iron deficiency or sufficiency in rice. The iron level sufficiency in soils is in the range of 5–10 mg kg^{-1} depending on the extracting solution used. In Brazil, Mehlich 1 extracting solution (0.5 N HCl + 0.5 N H_2SO_4) is the commonly used extracting solution for micronutrient analysis. DTPA solution is commonly used in micronutrient analysis in many laboratories in various countries. Adequate tissue concentration of Fe varies according to the plant growth stage and the plant part analyzed. However, 70–300 mg Fe kg^{-1} dry weight is considered an adequate level for rice in most agroecological regions.

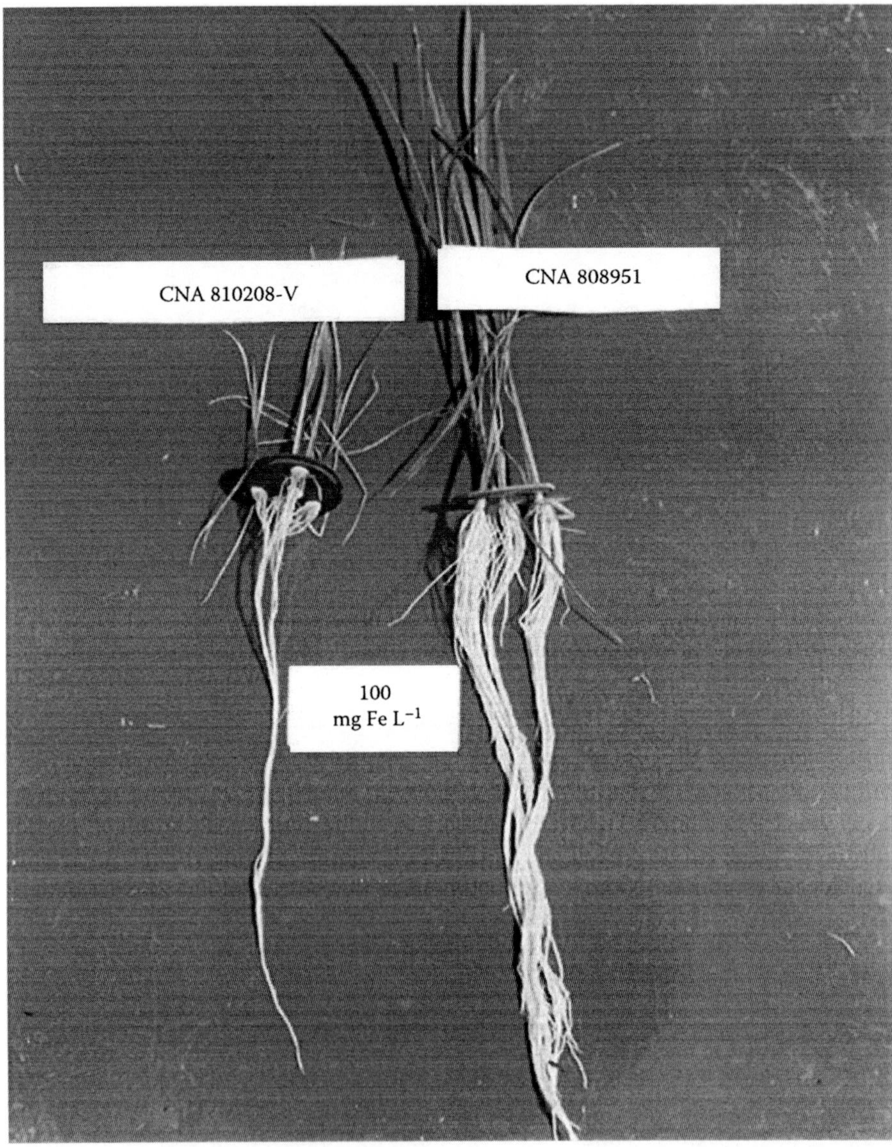

FIGURE 10.12 Root and top growth of two lowland rice lines at 100 mg Fe L^{-1} nutrient solution. (From Fageria, N. K., M. P. Barbosa Filho, and C. M. Guimaraes. 2008. Iron toxicity in lowland rice. *J. Plant Nutr.* 31:1676–1697.)

Iron deficiency can be corrected by foliar application of a ferrous sulfate solution of about 2%, provided there is sufficient foliage to absorb the applied nutrient solution. The foliar spray of chelates FeEDDHA and FeHEDTA is satisfactory for many soil conditions. In addition, the use of higher levels of formulated fertilizers can lower the soil pH and decrease iron deficiency in aerobic rice.

Use of iron-efficient genotypes is another method of correcting iron deficiency in upland rice.

Iron toxicity is one of the main yield-limiting factors for lowland rice. Flooding lowland rice soil creates reduced conditions because of the absence of oxygen for plants and microorganisms in the rhizosphere. The reduced conditions bring physiochemical transformation in the waterlogged soils. One of the most significant transformations in flooded soils is the conversion or reduction of Fe^{3+} to Fe^{2+} form. This transformation may permit higher or excess uptake of Fe^{2+} by growing plants and may be toxic to them. Iron toxicity is attributed to low redox potential of flooded soils, low soil pH, and low soil fertility as well as accumulation of harmful organic acids or hydrogen sulfides. The physicochemical measurement that can differentiate a submerged soil from a well-drained soil is the oxidation–reduction or redox potential. Well-aerated soils have a redox potential of more than +500 mV, whereas waterlogged soils have redox potential as negative as –300 mV, if the reduction processes are sufficiently intense. Effective measures to ameliorate iron toxicity include periodic surface drainage, liming acidic soils, adopting good fertilizer management practices, and the use of iron-tolerant rice cultivars. Among these management practices, planting iron-tolerant cultivars is the most economic and environmentally sound practice for reducing iron toxicity problems in lowland rice.

REFERENCES

Adams, F. 1984. Crop response to lime in the southern United States. In *Acidity and Liming*, 2nd edition, ed. F. Adams, pp. 212–265. Madison, WI: American Society of Agronomy.

Barbosa Filho, M. P., N. K. Fageria, and L. F. Stone. 1983. Water management and liming in relation to grain yield and iron toxicity. *Pesq. Agropec. Bras.* 18:903–910.

Benckiser, G., J. C. G. Ottow, I. Watanbe, and S. Santiago. 1984. The mechanism of excessive iron uptake (iron toxicity) of wetland rice. *J. Plant Nutr.* 7:177–185.

Bennett, W. F. 1993. Plant nutrient utilization and diagnostic plant symptoms. In *Nutrient Deficiencies and Toxicities in Crop Plants*, ed. W. F. Bennett, pp. 1–7. St. Paul, MN: American Phytopathology Society.

Bolan, N. S., D. C. Adriano, and D. Curtin. 2003. Soil acidification and liming interactions with nutrient and heavy metal transformation and bioavailability. *Adv. Agron.* 78:215–272.

Bremen, N. V., and F. R. Moormann. 1978. Iron-toxic soils. In *Soils and Rice*, ed. IRRI, 781–800. Los Bãnos, Philippines: IRRI.

Brown, J. C. 1961. Iron chlorosis in plants. *Adv. Agron.* 13:329–369.

Brown, J. C., and W. E. Jones. 1976. A technique to determine iron efficiency in plants. *Soil Sci. Soc. Am. J.* 40:398–405.

Burkhardt, J., S. Dreitz, H. E. Goldbach, and T. Eichert. 1999. Stomatal uptake as an important factor for foliar fertilization. In *Technology and Application of Foliar Fertilizers: Proceedings of the Second International Workshop on Foliar Fertilization*, ed. Soil and Fertilizer Society of Thailand, 63–72. Bangkok: Soil and Fertilizer Society of Thailand.

Cakmak, I., L. Ozturk, S. Karanlik, H. Marschner, and H. Ekiz. 1996. Zinc-efficient wild grasses enhance release of phytosiderophores under zinc deficiency. *J. Plant Nutr.* 19:551–563.

Campbell, W. H. 1999. Nitrate reductase structure, function and regulation: Bridging the gap between biochemistry and physiology. *Ann. Rev. Plant Physiol. Plant Mol. Biol.* 50:277–303.

Cantisano, A. 2000. What to use for foliar feeding. In *Growing for the Market*, ed. L. Byczynski, 4–6. Lawrence, KS: Fairplain.

Denelan, P. 1988. Foliar feeding. *Mother Earth News.* 111:58–61.

Dybing, C. D., and H. B. Currier. 1961. Foliar penetration of chemicals. *Plant Physiol.* 36:169–174.

Eichert, T., and J. Burkhardt. 2001. Quantification of stomatal uptake of ionic solutes using a new model system. *J. Exp. Bot.* 52:771–781.

Eichert, T., H. E. Goldbach, and J. Burkhardt. 1998. Evidence for the uptake of large anions through stomatal pores. *Botanica Acta.* 111:461–466.

Ellsworth, J. W., V. D. Jolley, D. S. Nuland, and A. D. Blaylock. 1998. Use of hydrogen release or a combination of hydrogen release and iron reduction for selecting iron-efficient dry bean and soybean cultivars. *J. Plant Nutr.* 21:2639–2651.

Epstein, E., and A. J. Bloom. 2005. *Mineral Nutrition of Plants: Principles and Perspectives*, 2nd edition. Sunderland, MA: Sinauer.

Fageria, N. K. 1984. *Fertilization and Mineral Nutrition of Rice*. Rio de Janeiro, Brazil: EMBRAPA/CNPAF.

Fageria, N. K. 1989. *Tropical Soils and Physiological Aspects of Crop Production*. Brasilia: EMBRAPA-DPU.

Fageria, N. K. 2000. Upland rice response to soil acidity in Cerrado soil. *Pesq. Agropec. Bras.* 35:2303–2307.

Fageria, N. K. 2004. Dry matter yield and shoot nutrient concentrations of upland rice, common bean, corn, and soybean grown in rotation on an Oxisol. *Commun. Soil Sci. Plant Anal.* 35:961–974.

Fageria, N. K. 2009. *The Use of Nutrients in Crop Plants*. Boca Raton, FL: CRC Press.

Fageria, N. K., and V. C. Baligar. 1999. Growth and nutrient concentrations of common bean, lowland rice, corn, soybean, and wheat at different soil pH on an Inceptisol. *J. Plant Nutr.* 22:1495–1507.

Fageria, N. K., and V. C. Baligar. 2005. Nutrient availability. In *Encyclopedia of Soils in the Environment*, ed. D. Hillel, pp. 63–72. San Diego, CA: Elsevier.

Fageria, N. K., and V. C. Baligar. 2008. Ameliorating soil acidity of tropical Oxisols by liming for sustainable crop production. *Adv. Agron.* 99:345–399.

Fageria, N. K., V. C. Baligar, and C. A. Jones. 2011. *Growth and Mineral Nutrition of Field Crops*, 3rd edition. Boca Raton, FL: CRC Press.

Fageria, N. K., V. C. Baligar, and R. J. Wright. 1990. Iron nutrition of plants: An overview on the chemistry and physiology of its deficiency and toxicity. *Pesq. Agropec. Bras.* 25:553–570.

Fageria, N. K., M. P. Barbosa Filho, and J. R. P. Carvalho. 1981. Influence of iron on growth and absorption of P, K, Ca and Mg by rice plant in nutrient solution. *Pesq. Agropec. Bras.* 16:483–488.

Fageria, N. K., M. P. Barbosa Filho, J. R. P. Carvalho, P. H. N. Rangel, and V. A. Cutrim. 1984. Preliminary screening of rice cultivars for tolerance to iron toxicity. *Pesq. Agropec. Bras.* 19:1271–1278.

Fageria, N. K., M. P. Barbosa Filho, and C. M. Guimaraes. 2008. Iron toxicity in lowland rice. *J. Plant Nutr.* 31:1676–1697.

Fageria, N. K., M. P. Barbosa Filho, A. Moreira, and C. M. Guimaraes. 2009. Foliar fertilization of crop plants. *J. Plant Nutr.* 32:1044–1064.

Fageria, N. K., G. D. Carvalho, A. B. Santos, E. P. B. Ferreira, and A. M. Knupp. 2011. Chemistry of lowland rice soils and nutrient availability. *Commun. Soil Sci. Plant Anal.* 42:1913–1933.

Fageria, N. K., E. M. Castro, and V. C. Baligar. 2004. Response of upland rice genotypes to soil acidity. In *The Red Soils of China: Their Nature, Management and Utilization*, eds. M. J. Wilson, Z. He, and X. Yang, pp. 219–237. Dordrecht, The Netherlands: Kluwer.

Fageria, N. K., C. M. Guimaraes, and T. A. Portes. 1994. Iron deficiency in upland rice. *Lav. Arrozeira.* 47:3–5.
Fageria, N. K., and N. A. Rabelo. 1987. Tolerance of rice cultivars to iron toxicity. *J. Plant Nutr.* 10:653–661.
Fageria, N. K., N. A. Slaton, and V. C. Baligar. 2003. Nutrient management for improving lowland rice productivity and sustainability. *Adv. Agron.* 80:63–152.
Fageria, N. K., L. F. Stone, and A. B. Santos. 2003. Soil fertility management for irrigated rice. Santo Antônio de Goiás, Brazil: EMBRAPA Arroz e Feijão.
Follett, R. H., L. S. Murphy, and R. L. Donahue. 1981. *Fertilizers and Soil Amendments.* Englewood Cliffs, NJ: Prentice-Hall.
Foth, H. D., and B. G. Ellis. 1988. *Soil Fertility.* New York: John Wiley.
Franke, W. 1967. Mechanisms of foliar penetration of solutions. *Anu. Rev. Plant Physiol.* 18:281–300.
Gamble, P. E., and E. R. Emino. 1987. Morphological and anatomical characterization of leaf burn in corn induced from foliar applied nitrogen. *Agron. J.* 79:92–96.
Garnett, T. P., and R. D. Graham. 2005. Distribution and remobilization of iron and copper in wheat. *Ann. Bot.* 95:817–826.
Genon, J. G., N. Hepcee, J. E. Duffy, B. Delvaux, and P. A. Hennebert. 1994. Iron and other chemical soil constraints to rice in highlands swamps of Burundi. *Plant Soil.* 166:109–111.
Gettier, S. W., D. C. Martens, and T. B. Brumback, Jr. 1985. Timing of foliar manganese application for correction of manganese deficiency in soybeans. *Agron. J.* 77:627–629.
Gilmour, J. T. 1977. Micronutrients status of the rice plant. II. Micronutrient uptake rate as a function of time. *Plant Soil.* 46:549–557.
Girma, K., K. L. Martin, K. W. Freeman, J. Mosali, R. K. Teal, W. R. Raun, S. M. Moges, et al. 2007. Determination of optimum rate and growth for foliar applied phosphorus in corn. *Commun. Soil Sci. Plant Anal.* 38:1137–1154.
Godsey, C. B., J. P. Schmidt, A. J. Schlegel, R. K. Taylor, C. R. Thompson, and R. J. Gehl. 2003. Correcting iron deficiency in corn with seed row applied iron sulfate. *Agron. J.* 95:160 166.
Gooding, M. J., and W. P. Davies. 1992. Foliar urea fertilization of cereals: A review. *Fer. Res.* 32:209–222.
Gooding, M. J., P. S. Kettlewell, and W. P. Davies. 1988. Disease suppression by late season urea sprays on winter wheat and interaction with fungicide. *J. Fertilizer Issues* 5:19–23.
Gotoh, S., and W. H. Patrick, Jr. 1976. Transformation of iron in a waterlogged soil as influenced by redox potential and pH. *Soil Sci. Soc. Am. Proc.* 38:66–71.
Gupta, U. C., and J. Lipsett. 1981. Molybdenum in soil, plants, and animals. *Adv. Agron.* 34:73–115.
Guzman, M., J. L. Valenzuela, A. Sanchez, and L. Romero. 1990. A method for diagnosis the status of horticulture crops. II. Micronutrients. *Phyton. Int. J. Exp. Bot.* 51:43–56.
Hergert, G. W., P. T. Nordquist, J. L. Petersen, and B. A. Skates. 1996. Fertilizer and crop management practices for improving maize yields on high pH soils. *J. Plant Nutr.* 19:1223–1233.
Howeler, R. H. 1973. Iron-induced oranging disease of rice in relation to physico-chemical changes in a flooded Oxisol. *Soil Sci. Soc. Am. Proc.* 37:898–903.
Hristozkova, M., M. Geneva, I. Stancheva, and G. Georgiev. 2007. Nitrogen assimilatory and amino acid content in inoculated foliar fertilizer pea plants grown at reduced molybdenum concentration. *J. Plant Nutr.* 30:1409–1419.
IITA (International Institute of Tropical Agriculture). 1983. *Annual Report for 1982.* Ibadan, Nigeria: IITA.

Inada, K. 1966a. Studies on the bronzing disease of rice plants in Ceylon. I. Effect of field treatment on bronzing occurrence and changes in leaf respiration induced by disease. *Trop. Agric. (Ceylon).* 122:19–29.

Inada, K. 1966b. Studies on the bronzing disease of rice plants in Ceylon. II. Cause of the occurrence of bronzing. *Trop. Agric. (Ceylon).* 125:31–46.

Ishizuka, Y. 1961. Effect of iron, manganese and copper on plant. *J. Soc. Soil Manure.* 32:97–100.

Jolley, V. D., D. J. Fairbanks, W. B. Stevens, R. E. Terry, and J. H. Orf. 1992. Using root-reduction capacity for genotype evaluation of iron efficiency in soybean. *J. Plant Nutr.* 15:1679–1690.

Kashirad, A., and H. Marschner. 1974. Effect of pH and phosphate on iron nutrition of sunflower and corn plants. *Agrochimica.* 18:497–508.

Kochain, L. V. 1991. Mechanisms of micronutrient uptake and translocation in plants. In *Micronutrients in Agriculture*, 2nd edition, eds. J. J. Mortvedt, F. R. Fox, L. M. Shuman, and R. M. Welch, pp. 229–296. Madison, WI: SSSA.

Kosaki, T., and A. S. R. Juo. 1986. Iron toxicity of rice in inland valleys: A case from Nigeria. In *The Wetland and Rice in Subsaharan Africa*, eds. A. S. R. Juo and J. A. Lowe, 167–174. Ibadan, Nigeria: IITA.

Land, FAO and Plant Nutrition Management. 2000. Prosoil-problem soil database. FAO, Rome, Italy. http://www.fao.org/ag/AGL/agll/prosoil/default.htm. Accessed December 16, 2003.

Lindsay, W. L. 1979. *Chemical Equilibria in Soils.* New York: Wiley-Interscience.

Lindsay, W. L., and W. A. Norvell. 1978. Development of a DTPA soil test for zinc, iron, manganese and copper. *Soil Sci. Soc. Am. J.* 42:421–428.

Lindsay, W. L., and A. P. Schwab. 1982. The chemistry of iron in soils and its availability to plants. *J. Plant Nutr.* 5:821–840.

Lombi, E., M. J. McLaughlin, C. Johnston, R. D. Armstrong, and R. E. Holloway. 2004. Mobility and lability of phosphorus from granular and fluid monoammonium phosphate differs in a calcereous soil. *Soil Sci. Soc. Am. J.* 68:682–689.

Lucena, J. J., and R. L. Chaney. 2007. Response of cucumber plants to low doses of different synthetic iron chelates in hydroponics. *J. Plant Nutr.* 30:795–809.

Marschner, H. 1995. *Mineral Nutrition of Higher Plants*, 2nd edition. New York: Academic Press.

Marschner, H., and V. Romheld. 1995. Strategies of plants for acquisition of iron. In *Iron Nutrition in Soils and Plants*, ed. J. Abadia, pp. 375–388. Dordrecht, The Netherlands: Kluwer.

Mathers, A. C. 1970. Effect of ferrous sulfate and sulfuric acid on grain sorghum yields. *Agron. J.* 62:555–556.

Middleton, L. J., and J. Sanderson. 1965. The uptake of inorganic ions by plant leaves. *J. Exp. Bot.* 16:197–215.

Mori, S., N. Nishazawa, H. Hayashi, M. Chino, E. Yoshimura, and J. Ishihara. 1991. Why are young rice plants highly susceptible to iron deficiency? *Plant Soil.* 130:143–156.

Mortvedt, J. J. 1991. Correcting iron deficiencies in annual and perennial plants. Present technologies and future prospects. *Plant Soil.* 130:273–279.

Mortvedt, J. J. 1994. Needs for controlled availability micronutrient fertilizers. *Fert. Res.* 38:213–221.

Mortvedt, J. J., and P. M. Giordano. 1970. Crop response to iron sulfate applied with fluid polyphosphate fertilizers. *Fert. Sol.* 14:22–27.

Obata, H. 1995. Physiological functions of micro essential elements. In *Science of Rice Plant: Physiology*, Vol. 2, eds. T. Matsu, K. Kumazawa, R. Ishii, K. Ishihara, and H. Hirata, pp. 402–419. Tokyo: Food and Agricultural Policy Research Center.

Olaleye, A. O., F. O. Tabi, A. O. Ogunkunle, B. N. Singh, and K. L. Sahrawat. 2001. Effect of toxic iron concentrations on the growth of lowland rice. *J. Plant Nutr.* 24:441–457.
Olsen, S. R., and F. S. Watanable. 1979. Interaction of added gypsum in alkaline soils with uptake of iron, molybdenum, manganese, and zinc by sorghum. *Soil Sci. Soc. Am. J.* 43:125–130.
Olson, R. V. 1950. Effects of acidification, iron oxide addition, and other soil treatments on sorghum chlorosis and iron absorption. *Soil Sci. Soc. Am. Proc.* 15:97–101.
Ottow, J. C. G., G. Benckiser, I. Watanabe, and S. Santiago. 1983. Multiple nutritional soil stress as the prerequisite for iron toxicity of wetland rice (*Oryza sativa* L.). *Trop. Agric. (Trinidad).* 60:102–106.
Papadakis, I. E., T. E. Sotiropoulos, and I. N. Therios. 2007. Mobility of iron and manganese within two citrus genotypes after foliar applications of iron sulfate and manganese. *J. Plant Nutr.* 30:1385–1396.
Patrick, W. H. Jr. 1966. Apparatus for controlling the oxidation-reduction potential of waterlogged soils. *Nature.* 212:1278–1279.
Patrick, W. H. Jr., and D. S. Mikkelsen. 1971. Plant nutrient behavior in flooded soil. In *Fertilizer Technology and Use*, 2nd edition, ed. R. C. Dinauer, pp. 187–215. Madison, WI: SSA.
Patrick, W. H., Jr., and C. N. Reddy. 1978. Chemical changes in rice soils. In *Soils and Rice,* eds. IRRI, pp. 361–379. Los Baños, Philippines: IRRI.
Ponnamperuma, F. N. 1972. The chemistry of submerged soils. *Adv. Agron.* 24:29–96.
Ponnamperuma, F. N. 1976. Screening rice for tolerance to mineral stresses. In *Plant Adaption to Mineral Stress in Problem Soils*, ed. M. J. Wright, pp. 341–353. Ithaca, NY: Cornell University.
Ponnamperuma, F. N. 1977. *Screening Rice for Tolerance to Mineral Stresses.* IRRI Research Paper Series No. 6. Los Baños, Philippines: IRRI.
Ponnamperuma, F. N., R. Bradfield, and M. Peech. 1955. Physiological disease of rice attributable to iron toxicity. *Nature.* 175:265.
Rengel, Z. 1999. Physiological mechanisms underlying differential nutrient efficiency of crop genotypes. In *Mineral Nutrition of Crops: Fundamental Mechanisms and Implications*, ed. Z. Rengel, pp. 227–265. New York: The Haworth Press.
Rogers, E. E., and M. L. Guerinot. 2002. Iron acquisition in plants. In *Molecular and Cellular Iron Transport*, ed. D. M. Templeton, pp. 359–393. New York: Marcel Dekker.
Romheld, V., and H. Marschner. 1986. Evidence for a specific uptake system for iron phytosiderophores in roots of grasses. *Plant Physiol.* 80:175–180.
Sims, J. T., and G. V. Johnson. 1991. Micronutrient soil tests. In *Micronutrients in Agriculture*, 2nd edition, eds. J. J. Mortvedt, F. R. Fox, L. M. Shuman, and R. M. Welch, pp. 427–476. Madison, WI: SSSA.
Snyder, G. H., and D. B. Jones. 1988. Prediction and prevention of iron-related rice seedling chlorosis on Everglades Histosols. *Soil Sci. Soc. Am. J.* 52:1043–1046.
Snyder, G. H., and D. B. Jones. 1991. Post-emergence treatment of iron-related rice-seedling chlorosis. *Plant Soil.* 138:313–317.
Stahl, R. S., and B. R. James. 1991. Zinc sorption by B horizon soils as a function of pH. *Soil Sci. Soc. Am. J.* 55:1592–1597.
Sumner, M. E., and M. P. W. Farina. 1986. Phosphorus interactions with other nutrients and lime in field cropping systems. *Ad. Soil Sci.* 5:201–236.
Tadano, T. 1975. Devices of rice roots to tolerate high iron concentrations in growth media. *JARQ.* 9:34–39.
Tadano, T. 1976. Studies on the methods to prevent iron toxicity in lowland rice. *Mem. Fac. Ag.* 10:22–88.
Tadano, T. 1995. Akagare disease. In *Science of Rice Plant: Physiology*, Vol. 2, eds. T. Matsu, K. Kumazawa, R. Ishii, K. Ishihara, and H. Hirata, pp. 939–953. Tokyo: Food and Agricultural Policy Research Center.

Tadano, T., and A. Tanaka. 1970. Studies on the iron nutrition of rice plants: Iron absorption affected by potassium status of the plants. *J. Sci. Soil Man.* 41:142–148.

Takagi, S., K. Nomoto, and T. Takemoto. 1984. A physiological aspect of mugenic acid: A possible phytosiderophore of graminaceous plants. *J. Plant Nutr.* 7:469–477.

Tanaka, A., R. Leo, and S. A. Navasero. 1966. Some mechanism involved in the development of iron toxicity symptoms in the rice plant. *Soil Sci. Plant Nutr.* 12:158–162.

Tisdale, S. L., W. L. Nelson, and J. D. Beaton. 1985. *Soil Fertility and Fertilizers*, 4th edition. New York: Macmillan.

Thorne, G. N. 1955. The effect on yields and leaf area of wheat of applying nitrogen as a top-dressing in April or in sprays at ear emergence. *J Agric. Sci.* 46:449–456.

Tom, W. O., T. C. Miller, and D. H. Bowman. 1981. Foliar fertilization of rice after midseason. *Agron. J.* 73:411–414.

Trolldenier, G. 1977. Mineral nutrition and reduction processes in the rhizosphere of rice. *Plant Soil.* 47:193–202.

Wells, B. R., B. A. Huey, R. J. Norman, and R. S. Helms. 1993. Rice. In *Nutrient Deficiencies and Toxicities in Crop Plants*, ed. W. F. Bennett, pp. 15–19. St. Paul, MI: American Phytopathology Society.

Wiersma, J. V. 2005. High rates of Fe-EDDHA and seed iron concentration suggest partial solutions to iron deficiency in soybean. *Agron. J.* 97:924–934.

Yamauchi, M. 1989. Rice bronzing in Nigeria caused by nutrient imbalances and its control by potassium sulfate application. *Plant Soil.* 117:275–286.

Yoshida, S. 1981. *Fundamentals of Rice Crop Science.* Los Baños, Philippines: IRRI.

11 Boron

11.1 INTRODUCTION

The importance of boron (B) as an essential micronutrient for plant growth was discovered in 1923 by Katherine Warington of Oxford, United Kingdom (Marschner 1995). Boron's value was first uncovered for legumes, and later for nonleguminous plants, by A. L. Sommer and C. B. Lipman of the University of California (Epstein and Bloom 2005). Boron is present as undissociated boric acid, or $B(OH)_3$, in soil solutions at pH values less than 7; it disassociates to $B(OH)_4^-$ only at higher pH values. Boron is the only micronutrient present over a wide pH range as a neutral molecule rather than as an ion (Epstein and Bloom 2005) and is the only nonmetal among the micronutrient elements.

Boron deficiency has been reported in many crops worldwide, including rice (Gupta 1979; Fageria 2009). Figure 11.1 shows the response to B fertilization of lowland or flooded rice grown on Brazilian Inceptisol. Some early surveys have shown that B deficiency was common throughout the United States (Berger 1962). It is frequently applied in semiarid tropical regions, on calcareous and sandy soils (Page and Paden 1954; Woodruff, Moore, and Musen 1987; Sahrawat et al. 2008; Ziaeyan and Rajaie 2009; Nelson and Meinhart 2011). Over the past 50 years, a number of research papers have addressed various aspects of B deficiency related to geographic distribution, chemical reactions and equilibriums, functions in plants, mechanisms of plant uptake, soil and tissue testing, and fertilization (Mortvedt et al. 1991).

Increasingly, research and educational literature on soils and crop nutrition reflect the accumulating knowledge on B deficiency and its wider implications (Yau and Ryan 2008). In addition to deficiency, B toxicity is also reported in many dry regions of the world, especially in alkaline soils (Cartwright, Zarcinas, and Spoucer 1986; Wayne 1986; Yau and Saxena 1997; Yau and Ryan 2008). Boron is unique as a micronutrient in that the threshold between deficiency and toxicity is narrow (Mortvedt et al. 1991). Therefore, care should be taken when correcting B deficiency in crop plants in order to avoid over fertilization, which may create toxicity. Figure 11.2 shows that in upland rice, B toxicity appeared when the B application rate was increased from 20 to 40 mg kg^{-1}. At 40 mg B kg^{-1}, upland rice plants were nearly dead, but the same level of B in the lowland rice did not produce B toxicity symptoms (as shown in Figure 11.1). Therefore, it can be concluded that B toxicity also depends on the rhizosphere environment. The environment of lowland or flooded rice is different from that of upland rice. Flooding increases soil pH and improves uptake of P, Ca^{2+}, Mg^{2+}, Fe^{2+}, Mn^{2+}, Mo, and Si (Fageria, Carvalho, et al. 2011); these conditions might have avoided or reduced toxicity in the B application range of 0–40 mg kg^{-1}. These chemical changes in the flooded rice rhizosphere occur because of physical reactions between the soil and water and also because of biological activities of anaerobic

FIGURE 11.1 Response of flooded rice to boron fertilization.

FIGURE 11.2 Response of upland rice to boron fertilization in Brazilian Oxisol. (From Gerloff, G. C. 1987. Intact-plant screening for tolerance of nutrient-deficiency. In *Genetic Aspects of Plant Mineral Nutrition*, eds. W. H. Gableman and B. C. Loughman, pp. 55–68. The Hague, The Netherlands: Martinus Nijhoff.)

microorganisms. The objective of this chapter is to discuss B nutrition of upland and lowland rice, with special emphasis on correcting B deficiency.

11.2 CYCLE IN SOIL–PLANT SYSTEM

The B cycle in the soil–plant system is similar to that of other micronutrients, and it involves addition, transformation, or solubilization; uptake by plants; and losses by leaching. However, the B cycle is simpler than that of other micronutrients. Boron does not undergo oxidation–reduction reactions or volatilization reactions in soils (Goldberg 1997). Boron concentration in soil solution is generally controlled by B adsorption reactions as is the amount of water-soluble B available for plant uptake. Boron sorption of aluminum and iron oxide minerals plays a significant role in soils and helps explain the need for B fertilizers on many tropical soils (Mortvedt 1994). The main factors that affect the availability of B for crop plants are pH, soil texture, soil moisture, and temperature.

Soluble B content in soils is significantly correlated with solution pH (Elrashidi and O'Connor 1982; Goldberg 1997). Similarly, B adsorption maxima is positively correlated with solution pH (Evans 1987). Boron adsorption by soils increases as a function of solution pH in the range of 3–9 (Keren, Bingham, and Rhoades 1985; Barrow 1989; Goldberg 1997). Competing ions such as silicate, sulfate, phosphate, and oxalate decrease the magnitude of B adsorption on oxides (Goldberg and Glaubig 1988; Goldberg 1997). Increasing soil pH decreases B availability to plants; therefore, the use of lime in acidic soils may lead to B deficiency. The reduction in B availability from increasing soil pH by liming results from B adsorption by iron and aluminum hydroxides. The mechanism of B adsorption by aluminum and iron oxide is a ligand exchange with reactive surface hydroxyl groups; anions become specifically adsorbed on mineral surfaces (Goldberg, Forster, and Heick 1993; Goldberg 1997).

Boron adsorption is very pH dependent. Maximum adsorption by $Al(OH)_3$ and $Fe(OH)_3$ occurs in the soil pH range of 7–9; this corresponds to the soil pH range of lowest B availability (Tisdale, Nelson, and Beaton 1985). Data in Table 11.1 show that uptake of B in a shoot of lowland rice decreased when soil pH increased more

TABLE 11.1
Concentration of B in the Shoot of Lowland Rice at Different Soil pH Levels

Soil pH in H_2O	B Concentration (mg kg^{-1})
4.9	11
5.9	17
6.4	10
6.7	10
7.0	11

Source: Fageria, N. K., and V. C. Baligar. 1999. Growth and nutrient concentrations of common bean, lowland rice, corn, soybean, and wheat at different soil pH on an Inceptisol. *J. Plant Nutr.* 22:1495–1507.

than 5.9 by liming. Organic matter also adsorbs B and acts as a reservoir to replenish soil solution B upon crop removal or loss via leaching. Replacement by other anions or mineralization of the organic matter releases B (Foth and Ellis 1988). Clay content also influences B adsorption. Barber (1995) reported that B adsorption by fine-textured soils is two to three times greater than by coarse-textured soils.

The earlier discussion about B availability is applicable only to upland rice or aerobic soils. Very limited information is available about the chemistry of B in flooded soils of irrigated rice (Fageria, Barbosa Filho, and Stone 2003). The concentration of B in the soil solution is believed to remain more or less constant following soil submergence (Ponnamperuma 1975). In flooded soils, with pH buffered around neutrality, H_3BO_3 is the dominant component in the soil solution. As with upland soils, the adsorption of B by Fe and Al oxides (Sims and Bingham 1968) seems to be an important mechanism in governing B solubility in flooded soils (Patrick and Reddy 1978). Extraction with hot water is the most reliable evaluation of soil boron availability (Tisdale, Nelson, and Beaton 1985).

11.3 FUNCTIONS

Boron plays a significant role in many physiological and biochemical processes in the growth and development of plants. These processes are cell wall synthesis, lignification, carbohydrate transport, nutrient uptake, cell wall structure, cell division, carbohydrate metabolism, ribonucleic acid (RNA) metabolism, respiration, indole acetic acid (IAA) metabolism, salinity stress, phenol metabolism, and membrane function (Gary, Sharma, and Kona 1979; Mozafar 1989; Marschner 1995; Cakmak and Romheld 1997; Fageria 2009; Fageria, Baligar, and Jones 2011; Nelson and Meinhardt 2011). In cereals, B is essential for pollen germination, pollen grain development, and seed reproduction (Vaughan 1977; Lordkaew et al. 2010; Nelson and Meinhardt 2011). Boron deficiency may cause small and twisted panicles. In some crops, B reduces the severity of disease (Donald and Porter 2009; Thomidis and Exadaktylou 2010).

11.4 DEFICIENCY SYMPTOMS

Boron is an immobile nutrient in plants. Therefore, its deficiency symptoms first appear in plant's younger leaves or growing parts. The tips of emerging leaves become white and rolled (similar to the symptoms of calcium deficiency). In severe cases, the growing points may die, but new tillers continue to be produced (De Datta 1981). Obata (1995) noted that B deficiency also retarded a plant's root elongation. The B requirement for a plant's vegetative growth—especially grasses—is very low, but the need for B increases with seed production (Marschner 1995). This is one possible reason why B deficiency symptoms have not been documented on rice leaves in commercial fields. In rice, B deficiency may be expressed solely in the form of reduced grain yields due to floret sterility; this may be blamed mistakenly on poor environmental conditions during anthesis. This hypothesis is yet to be proven in replicated field trials. However, there is some preliminary evidence that supports this theory.

Okuda, Hori, and Ida (1961) observed that the panicles of B-deficient rice plants failed to exsert from the boot. Likewise, Dunn (1978) noted that B deficiency produced symptoms that were similar to those associated with the physiological disorder

FIGURE 11.3 (See color insert.) Boron deficiency symptoms in rice plants. (From Fageria, N. K., and M. P. Barbosa Filho. 1994. *Nutritional Deficiency in Rice: Identification and Correction*. Goiânia, Brazil: EMBRAPA Arroz e Feijão.)

straighthead that has been induced in greenhouse studies by arsenic toxicity. Very little research has been conducted to verify or refute B deficiency as a possible cause of straighthead. Figures 11.3 and 11.4 show B deficiency in upland rice plants and leaves. These symptoms were developed in the nutrient solution without addition of B.

11.5 UPTAKE IN PLANT TISSUE

Boron is absorbed by roots as undissociated boric acid [$B(OH)_3$ or H_3BO_3], and it is not clear whether uptake is active or passive (Marschner 1995; Mengel et al. 2001). Nevertheless, B uptake by rice appears to be passive under a normal B supply and active under a low B supply (Yu and Bell 1998); B uptake is the result of passive assimilation of undissociated boric acid (Hu and Brown 1997). With a high B supply, passive uptake and active excretion of B have been noted (Yu and Bell 1998). Boron distribution in plant tissue appears to be governed primarily by transpiration because B is highly mobile and moves with water. Boron is supplied to roots primarily by mass flow (Fageria, Baligar, and Clark 2002).

Boron uptake can be expressed as concentration (content per unit dry weight) or accumulation (concentration × dry weight). Boron concentration as well as accumulation varies widely among plant species. Rice, together with wheat and barley, has a lower requirement for B than do nongramineous crops (Obata 1995). The B requirement is higher during the reproductive growth stages than during vegetative growth due its important function in grain formation. Due to the lack of published research on B nutrition for rice, very little data are available on the critical tissue

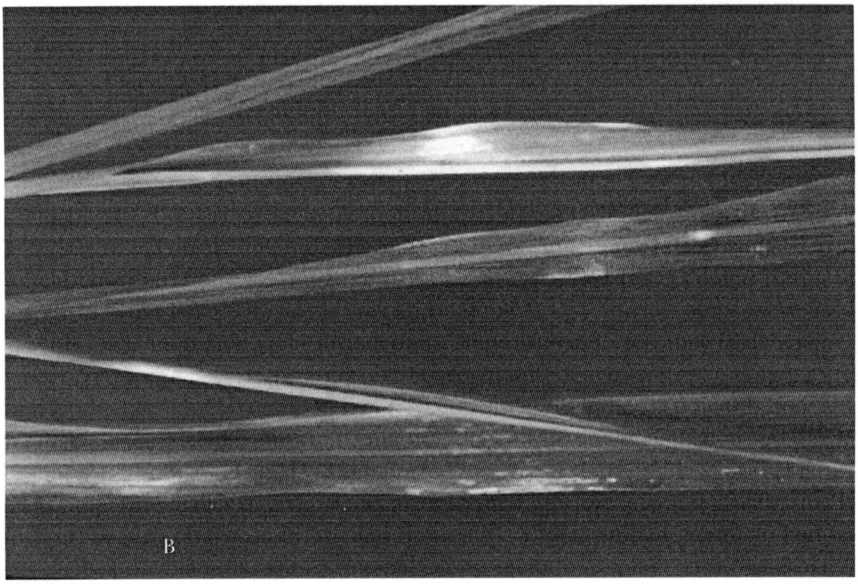

FIGURE 11.4 (See color insert.) Boron deficiency symptoms in rice leaves. (From Fageria, N. K., and M. P. Barbosa Filho. 1994. *Nutritional Deficiency in Rice: Identification and Correction*. Goiânia, Brazil: EMBRAPA Arroz e Feijão.)

B concentrations required for the production of maximum rice yields. Dobermann and Fairhurst (2000) suggested the optimum B concentration range for the Y-leaf of tillering rice as 6–15 mg kg^{-1}. Fageria, Baligar, and Jones (2011) suggested that a critical concentration of 8 mg B kg^{-1} in rice straw at maturity. Yu and Bell (1998) reported 18.5 mg B kg^{-1} in rice leaves and 8.9 mg B kg^{-1} in rice stems were associated with maximum rice yield production. They also reported that B deficiency in rice occurred when B concentrations in the top mature leaves were <7.3 mg kg^{-1}. In a review of the literature, Yu and Bell (1998) concluded that the sufficient B concentration range of rice varied from 5 to 67 mg kg^{-1} depending on the plant age and the plant part analyzed. Foth and Ellis (1988) reported that general guidelines for evaluating B content in mature leaves of crop plants are <15 mg kg^{-1} deficient and 20–200 mg kg^{-1} sufficient. Fageria, Barbosa Filho, and Stone (2003) determined adequate and toxic levels of B in the plant tissue of principal crops, including upland rice (Table 11.2).

Boron concentration and uptake in upland and lowland rice during the growth cycle is presented in Table 11.3. Similarly, the association between rice plant age and B concentration and uptake in the straw of upland and lowland rice is given in Table 11.4. Data in Tables 11.3 and 11.4 suggest that the B concentration of upland and lowland rice decreased significantly with advanced plant age, whereas uptake in upland and lowland rice significantly increased with advanced plant age. The decrease in B concentration and the increase in B uptake with advanced plant age were related to increase in dry matter yield (Fageria 2004). In the last sampling, the decrease in B uptake was related to the translocation of B to grain.

TABLE 11.2
Adequate and Toxic Levels of B in Tissues of Principal Field Crops

Crop Species	Adequate Level (mg kg^{-1})	Toxic Level (mg kg^{-1})
Upland rice	10	20
Dry bean	24	135
Corn	20	68
Soybean	75	155
Wheat	13	144

Source: Adapted from Fageria, N. K., M. P. Barbosa Filho, and L. F. Stone. 2003. Adequate and toxic levels of micronutrients in soil and plants for annual crops. Paper presented at the XXIX Brazilian Soil Science Congress, Ribeirão Preto, São Paulo, Brazil.

Note: Adequate level was calculated at 90% of the maximum dry weight of shoot, and toxic level was calculated at the 10% reduction in shoot dry weight after achieving maximum weight. Plants were harvested at four weeks after sowing; soil pH at harvest was about 6.0 in water in all five experiments.

TABLE 11.3
Boron Concentration and Uptake in Upland and Lowland Rice Straw and Grain during Crop Growth Cycle

	Upland Rice			Lowland Rice	
Plant Age in Days	Conc. (mg kg^{-1})	Uptake (g ha^{-1})	Plant Age in Days	Conc. (mg kg^{-1})	Uptake (g ha^{-1})
19	9.50	0.80	22	7.8	2.8
43	9.17	10.36	35	7.5	8.6
68	9.17	43.61	71	7.1	41.1
90	8.83	56.67	97	6.7	70.6
102	8.50	72.28	112	6.9	92.5
130	8.00	50.05	140	7.3	69.4
130 (grain)	6.0	29.52	140 (grain)	5.3	35.0

Source: Adapted from Fageria, N. K. 2001. Response of upland rice, dry bean, corn and soybean to base saturation in Cerrado soil. *Rev. Bras. Eng. Agric. Amb.* 5:416–424; Fageria, N. K., N. A. Slaton, and V. C. Baligar. 2003. Nutrient management for improving lowland rice productivity and sustainability. *Adv. Agron.* 80:63–152.

TABLE 11.4
Relationship between Plant Age and Concentration and Uptake of B in the Straw of Upland and Lowland Rice

Variable	Regression Equation	R^2
Plant age vs. B conc. in upland rice straw	$Y = 9.83 - 0.013X$	0.92*
Plant age vs. B uptake in upland rice straw	$Y = 0.09\exp(0.113 - 0.00065X^2)$	0.98*
Plant age vs. B conc. in lowland rice straw	$Y = 8.52 - 0.036X + 0.00019X^2$	0.94*
Plant age vs. B uptake in lowland rice straw	$Y = 0.53\exp(0.088X - 0.00038X^2)$	0.99*

Note: Data used from Table 11.3 for calculating regression equations.
*Significant at the 1% probability level.

11.6 USE EFFICIENCY

Boron use efficiency (BUE) is an important index for crop species under certain agroecological conditions. BUE for upland rice grown on Brazilian Oxisols is presented in Table 11.5. At harvest, BUE was higher in grain compared with shoot. The higher BUE in grain was associated with a lower B accumulation compared with shoot (Fageria 2004, 2009).

11.7 HARVEST INDEX

Boron harvest index (BHI) is an important index for determining the distribution of absorbed B to grain and straw in crop plants. BHI (proportion of B accumulation in grain/proportion of B accumulation in grain plus shoot) was less than 50% in four annual crops, as shown in Table 11.6. However, it was higher in legumes (dry beans and soybeans) than in cereals (upland rice and corn). The higher B exportation in legume grain demonstrates the superiority of these crops over cereals for human nutrition. Total B uptake in cereals (grain plus straw) was higher compared with legumes and was associated with higher yields of cereals compared with legumes (Fageria 2004).

11.8 MANAGEMENT PRACTICES

Soil and plant management practices can be adopted to improve B use efficiency and consequently rice yields. These practices include effective source, appropriate method and timing of application, adequate rate, and use of B-efficient genotypes of upland and lowland rice. A discussion of these practices follows.

11.8.1 Effective Sources

Fertilizer source is very important for correcting nutrient deficiencies in crop plants when a given nutrient's concentration is low in a soil. When considering a fertilizer

TABLE 11.5
Boron Use Efficiency in the Shoot and Grain of Upland Rice during Growth Cycle

Plant Age in Days	Boron Use Efficiency (kg g^{-1})
19	121.9
43	137.2
68	102.6
90	120.3
102	113.0
130	131.5
130 (grain)	177.1

Source: Adapted from Fageria, N. K. 2009. *The Use of Nutrients in Crop Plants*. Boca Raton, FL: CRC Press.

Note: $\text{Boron use efficiency (kg g}^{-1}) = \frac{\text{Shoot or grain yield in kg}}{\text{B uptake in shoot or grain in g}}$.

TABLE 11.6
Boron Uptake and Harvest Index in Principal Field Crops

Crop Species	B Uptake in Shoot (g ha^{-1})	B Uptake in Grain (g ha^{-1})	Total	B Harvest Index (%)
Upland rice	53.05	29.52	82.57	36
Corn	103.12	42.62	145.75	29
Dry bean	19.74	13.94	33.68	41
Soybean	22.35	20.46	42.80	48

Source: Fageria, N. K. 2009. *The Use of Nutrients in Crop Plants*. Boca Raton, FL: CRC Press.

Note: Boron harvest index = (B uptake in grain/B uptake in grain plus shoot). Values are average of two years of field trials.

source for crop use, efficiency and cost are always taken into consideration. Principal B fertilizers are listed in Table 11.7. The most common B crop plant fertilizers are borax, boric acid, and sodium tetra borate.

11.8.2 Appropriate Methods and Timing of Application

Appropriate method and timing of application are important considerations for improving B use efficiency in crop plants. Because only a small amount of B is required per hectare (about 2 kg ha^{-1}), uniform distribution is a problem. Therefore, B should be blended with N, P, and K fertilizers. It can be broadcast or applied

TABLE 11.7
Principal Boron Fertilizers, Their Formulae, and B Contents

Common Name	Formula	B Content (%)
Boric acid	H_3BO_3	17
Borax	$Na_2B_4O_7 \cdot 10H_2O$	11
Sodium borate (anhydrous)	$Na_2B_4O_7$	20
Sodium pentaborate	$Na_2B_{10}O_{16} \cdot 10H_2O$	18
Sodium tetraborate	Na_2B_4O	20
Sodium tetraborate pentahydrate	$Na_2B_4O_7 \cdot 5H_2O$	14
Boron frits	Fritted glass	10–17
Boron oxide	B_2O_3	31
FTE 11		11
FTE 115		11
FTE 171		2
FTE 181		2
176 E		2
176 F		6
501		6

Source: Adapted from Fageria, N. K. 2009. *The Use of Nutrients in Crop Plants*. Boca Raton, FL: CRC Press; Tisdale, S. L., W. L. Nelson, and J. D. Beaton. 1985. *Soil Fertility and Fertilizers*, 4th edition. New York: Macmillan; Foth, H., and B. G. Ellis. 1988. *Soil Fertility*. New York: John Wiley.

in a band at sowing. Both of these methods are effective due to the high water solubility of B fertilizers. However, band application near the seed or root system is preferable for efficient uptake. Band application generally requires a smaller amount than broadcast application. In emergency situations, foliar application of B fertilizers can be utilized. Foliar application may be combined with fungicides and insecticides to avoid the cost of an extra application (Tisdale, Nelson, and Beaton 1985).

11.8.3 Adequate Rate

Using an adequate rate is very important for improving plant B uptake and utilization efficiency. Adequate rate should be determined by conducting field experiments in a specific agroecological region. Fageria, Stone, and Santos (2003) determined the B requirements of lowland rice grown on Brazilian Inceptisol (Table 11.8). A maximum lowland rice grain yield was achieved with a band application of 2 kg B ha^{-1} at sowing time.

Soil analysis sometimes serves as a basis for identifying potentially B-deficient soils and has been used to make fertilizer recommendations for some crops (Martens and Westermann 1991; Sims and Johnson 1991). However, B deficiency of rice is yet to be recognized as a serious yield-limiting factor, and critical soil test B concentrations for rice have not been developed. Research conducted with

TABLE 11.8
Response to Boron Fertilization in Lowland Rice

B Rate (kg ha^{-1})	Grain Yield (kg ha^{-1})
0	4337
2	5123
4	4690
8	4795
F-test	*
CV (%)	5

Source: Adapted from Fageria, N. K., L. F. Stone, and A. B. Santos. 2003. *Soil Fertility Management of Irrigated Rice.* Santo Antônio de Goiás, Brazil: EMBRAPA Arroz e Feijão.
*Significant at the 5% probability level.

other crops suggests that if positive rice yield responses to B fertilization are found, fertilizer recommendations potentially can be based on routine soil analysis (Fageria, Barbosa Filho, and Stone 2003).

Extraction with the hot-water procedure is generally considered the standard process for evaluating crop response to B. Sensitive crops are likely to respond to B fertilization if the soil level of hot-water-soluble B is less than 0.5 mg kg^{-1} (Sims and Johnson 1991). A particular problem with B is the narrow margin between deficiency and toxicity. Sensitive crops can be affected by B toxicity at soil levels over 5.0 mg kg^{-1}, so B fertilizer must be used with caution. Sims and Johnson (1991) reported critical B soil levels of 0.1–2.0 mg kg^{-1}, depending on yield level, soil type, pH, and organic matter content. Because B is a mobile nutrient in soil, tests should be done just before the rice crop is sown in order to verify B status of the soil.

11.8.4 Use of Efficient Genotypes

Use of B-efficient rice genotypes is an important strategy for producing higher yield in B-deficient soils. This strategy is more favorable when absorbed B is not available to plants. Use of vigorous root system or higher root length genotypes is important to take up B from lower soil layers when B is leached at lower surface layers. Fageria and Moreira (2011) and Fageria (2013) reported significant variations in root weight and maximum root length of upland and lowland rice genotypes. Similarly, large differences exist among crop species or genotypes of the same species in the uptake and utilization of B (Mengel et al. 2001; Fageria, Baligar, and Clark 2002, 2006; Epstein and Bloom 2005).

The differences among crop genotypes in B absorption and utilization have been attributed to genetics, physiological/biochemical mechanisms, responses to climate variables, tolerance to pest and diseases, and responses to agronomic management practices (Fageria, Baligar, and Clark 2002). Genetic variations in plant micronutrient acquisition have been reviewed (Gerloff and Gableman 1983; Graham 1984; Gerloff 1987; Marschner 1995; Fageria, Baligar, and Clark 2002). The development of genotypes/cultivars effective in the acquisition and use of micronutrients and with the desired agronomic characteristics is vital for improving yields and achieving genotypic adaption to diversified environmental conditions and increased resistance to pests (Graham 1984; Fageria, Baligar, and Clark 2002). Plant and external factors affecting micronutrient use by plants and mechanisms and processes influencing genotypic differences in micronutrient efficiency have been summarized in Tables 11.9 and 11.10.

TABLE 11.9
Plant and External Factors That Affect Micronutrient Use Efficiency in Plants

Plant Factors	External Factors
Genetic control	*Agronomic management practices*
Species/cultivars/genotype	Liming, crop rotation, incorporate crop residue, cover crops
Physiological	*Soil*
Root length; density of main, lateral, and root hairs	Aeration/reducing conditions
Higher shoot yield, harvest index, internal demand	pH
Higher physiological efficiency	Organic matter levels and forms
Higher nutrient uptake and utilization	Temperature
Excretion of H^+, OH^-, and HCO	Moisture status
	Texture/structure
	Compaction
Biochemical	*Fertilizers*
Enzymes: rhodotorulic acid (Fe), ferroxamine b (Fe), ascorbic acid oxidase (Cu), carbonic acid anhydrate (Zn)	Source
	Timing, depth, method of placement, and application
Metallothionein (trace elements)	Use of slow release form
Proline, aspharagine pinitol (salinity)	Elements
Absissic acid, proline (drought).	Toxicities in acid soils (Al, Mn, pH) and saline soils (B, Cl, Na)
Root exudates (citric, malic, transaconitic acids)	
Phytosiderophores	Deficiencies in acid soils (Cu, Zn, Mn, Mo) and alkaline soils (Zn, Fe, Mn, Cu)
Others	*Others*
Tolerance to stress (drought, acidity, alkalinity)	Arbuscular mycorrhizae, beneficial soil microbes
Tolerance/resistance to diseases/pests	Control of weeds, diseases, and insects
Arial temperature, light quality, humidity	

Source: Compiled from Baligar, V. C., and N. K. Fageria. 1997. Nutrient use efficiency in acid soils: Nutrient management and plant use efficiency. In *Plant-Soil Interactions at Low pH: Sustainable Agriculture and Forestry Production*, eds. A. C. Moniz, M. C. Furlani, R. E. Schaffer, N. K. Fageria, C. A. Rosolem, and H. Canatarella, pp. 75–95. Campinas/Viçosa, Brazil: Brazilian Soil Science Society; Fageria, N. K. 1992. *Maximizing Crop Yields*. New York: Marcel Dekker; Fageria, N. K., V. C. Baligar, and R. B. Clark. 2002. Micronutrients in crop production. *Adv. Agron.* 77:185–268.

TABLE 11.10
Soil and Plant Mechanisms and Processes and Other Factors That Influence Genotypic Differences in Nutrient Efficiency in Plants Grown under Nutrient Stress Conditions

A. *Nutrient acquisition*
  1. Diffusion and mass flow (buffer capacity, ionic concentration, ionic properties, tortuosity, soil moisture, bulk density, temperature)
  2. Root morphological factors (number, length, root hair density, root extension, root density)
  3. Physiological (root: shoot, root microorganisms such as mycorrhizal fungi, nutrient status, water uptake, nutrient influx and efflux, rate of nutrient transport in roots and shoots, affinity to uptake [K_m], threshold concentration [C_{min}])
  4. Biochemical (enzyme secretion as phosphatase, chelating compounds, phytosiderophore), proton exudate; organic acid production, such as citric, trans-aconitic, and malic acid exudates

B. *Nutrient movement in root*
  1. Transfer across endodermis and transport within root
  2. Compartmentalization/binding within roots
  3. Rate of nutrient release to xylem

C. *Nutrient accumulation and remobilization in shoot*
  1. Demand at cellular level and storage in vacuoles
  2. Retransport from older to younger leaves and from vegetative to reproductive parts
  3. Rate of chelates in xylem transport

D. *Nutrient utilization and growth*
  1. Metabolism at reduced tissue concentration of nutrient
  2. Lower element concentration in supporting structure, particularly the stem
  3. Elemental substitution (e.g., Na for K function)
  4. Biochemical (nitrate reductase for N use efficiency, glutamate dehydrogenase for N metabolism, peroxidase for Fe efficiency, pyruvate kinase for K deficiency, metallothionein for metal toxicities)

E. *Other factors*
  1. Soil factors
     a. Soil solution (ionic equilibria, solubility precipitation, competing ions, organic ions, pH, phytotoxic ions)
     b. Physicochemical properties of soil (organic matter, pH, aeration, structure, texture, compaction, and soil moisture)
  2. Environmental effects
     a. Intensity and quality of light (solar radiation)
     b. Temperature
     c. Moisture (rainfall, humidity, and drought)
  3. Plant diseases, insects, and allelopathy

Source: Compiled from Baligar, V. C., and N. K. Fageria. 1997. Nutrient use efficiency in acid soils: Nutrient management and plant use efficiency. In *Plant-Soil Interactions at Low pH: Sustainable Agriculture and Forestry Production*, eds. A. C. Moniz, M. C. Furlani, R. E. Schaffer, N. K. Fageria, C. A. Rosolem, and H. canatarella, pp. 75–95. Campinas/Viçosa, Brazil: Brazilian Soil Science Society; Fageria, N. K. 1992. *Maximizing Crop Yields*. New York: Marcel Dekker; Fageria, N. K., V. C. Baligar, and R. B. Clark. 2002. Micronutrients in crop production. *Adv. Agron*. 77:185–268; Gerloff, G. C. 1987. Intact-plant screening for tolerance of nutrient-deficiency. In *Genetic Aspects of Plant Mineral Nutrition*, eds. W. H. Gableman and B. C. Loughman, pp. 55–68. The Hague, The Netherlands: Martinus Nijhoff.

11.9 CONCLUSIONS

Boron deficiency has been reported in upland and lowland rice-growing regions of the world. Boron is the most widely applied micronutrient in crop production and is required by plants for many physiological and biochemical reactions. In cereals, B is essential for pollen germination, pollen grain development, and seed reproduction. In rice, B deficiency may result in small and twisted panicles. Boron also reduces the severity of diseases. The chemistry of B in the soil–plant system is simple because it is not involved in the oxidation and reduction reactions or in the volatilization reactions in soils. Generally, less than 5% of the total soil B is available to plants. Undissociated boric acid (H_3BO_3) is the predominant component in soil solutions at pH values ranging from 5 to 9. Soil factors affecting B availability for plants are pH, texture, moisture, temperature, organic matter, and clay mineralogy. Major B-adsorbing surfaces in soils are hydroxides of iron, aluminum, magnesium, clay minerals, calcium carbonate, and organic matter. Boron uptake by plants is decreased in a quadratic fashion with increasing pH in the range of 3–9. Hence, liming acid soils decreases B uptake, especially in upland rice.

Important B fertilizers that can be applied to correct B deficiency are borax, boric acid, and sodium tetraborate. Generally, about 2 kg B ha^{-1} is sufficient in order to correct B deficiency in most crops, including rice. However, there may be some exceptions depending on soil type, crop management practices, rainfall, liming, and organic matter content. When blended with commercial fertilizers, B is economical for correction of a deficiency. It can be broadcast, banded, or applied as a foliar spray. Boron is also mixed with N, P, and K fertilizers and applied to the soil with these nutrients. As an anion in the soil, boron is easily leached from upper to lower horizons, and the residual effect is very low. Therefore, application of B each year or for each crop may be required. Boron can also be used in a foliar application with a concentration range of 0.1–0.25% solution. For foliar application, B can be mixed with insecticides or fungicides in order to reduce application costs. Use of B-efficient genotypes is an important strategy in correcting B deficiency in crop plants, including rice. Variation in B uptake and utilization in rice genotypes has been reported.

REFERENCES

Baligar, V. C., R. R. Duncan, and N. K. Fageria. 1990. Soil-plant interaction on nutrient use efficiency in plants: An overview. In *Crops as Enhancers of Nutrient Use*, eds. V. C. Baligar and R. R. Duncan, pp. 351–373. San Diego, CA: Academic Press.

Baligar, V. C., and N. K. Fageria. 1997. Nutrient use efficiency in acid soils: Nutrient management and plant use efficiency. In *Plant-Soil Interactions at Low pH: Sustainable Agriculture and Forestry Production*, eds. A. C. Moniz, M. C. Furlani, R. E. Schaffer, N. K. Fageria, C. A. Rosolem, and H. Canatarella, pp. 75–95. Campinas/Viçosa, Brazil: Brazilian Soil Science Society.

Barber, S. A. 1995. *Soil Nutrient Bioavailability: A Mechanistic Approach*, 2nd edition. New York: John Wiley.

Barrow, N. J. 1989. Testing a mechanistic model. X. The effect of pH and electrolyte concentration on borate sorption by a soil. *J. Soil Sci.* 40:427–435.

Berger, K. C. 1962. Micronutrient deficiencies in the United States. *J. Agric. Food. Chem.* 10:178–181.

Cakmak, I., and V. Romheld. 1997. Boron deficiency induced impairments of cellular functions in plants. *Plant Soil.* 193:71–83.

Cartwright, B., B. A. Zarcinas, and L. R. Spoucer. 1986. Boron toxicity in South Australian barley crops. *Aust. J. Agric. Res.* 37:351–359.

De Datta, S. K. 1981. *Principles and Practices of Rice Production.* New York: John Wiley.

Dobermann, A., and T. Fairhurst. 2000. *Rice: Nutritional Disorders and Nutrient Management.* Singapore/Los Baños, Philippines: Potash and Phosphate Institute of Canada/IRRI.

Donald, C., and I. Porter. 2009. Integrated control of clubroot. *J. Plant Growth Regl.* 28:289–303.

Dunn, R. J. 1978. A study of boron and arsenic as straighthead inducing agents of rice. MS thesis. University of Arkansas.

Elrashidi, M. A., and G. A. O'Connor. 1982. Boron sorption and desorption in soils. *Soil Sci. Soc. Am. J.* 46:27–31.

Epstein, E., and A. J. Bloom. 2005. *Mineral Nutrition of Plants: Principles and Perspectives*, 2nd edition. Sunderland, MA: Sinauer.

Evans, L. J. 1987. Retention of B by agricultural soils from Ontario. *Can. J. Soil Sci.* 67:33–42.

Fageria, N. K. 1992. *Maximizing Crop Yields.* New York: Marcel Dekker.

Fageria, N. K. 2001. Response of upland rice, dry bean, corn and soybean to base saturation in Cerrado soil. *Rev. Bras. Eng. Agric. Amb.* 5:416–424.

Fageria, N. K. 2004. Dry matter yield and shoot nutrient concentrations of upland rice, common bean, corn, and soybean grown in rotation of an Oxisol. *Commun. Soil Sci. Plant Anal.* 35:961–974.

Fageria, N. K. 2009. *The Use of Nutrients in Crop Plants.* Boca Raton, FL: CRC Press.

Fageria, N. K. 2013. *The Role of Plant Roots in Crop Production.* Boca Raton, FL: CRC Press.

Fageria, N. K., and V. C. Baligar. 1999. Growth and nutrient concentrations of common bean, lowland rice, corn, soybean, and wheat at different soil pH on an Inceptisol. *J. Plant Nutr.* 22:1495–1507.

Fageria, N. K., V. C. Baligar, and R. B. Clark. 2002. Micronutrients in crop production. *Adv. Agron.* 77:185–268.

Fageria, N. K., V. C. Baligar, and R. B. Clark. 2006. *Physiology of Crop Production.* New York: The Haworth Press.

Fageria, N. K., V. C. Baligar, and C. A. Jones. 2011. *Growth and Mineral Nutrition of Crop Plants*, 3rd edition. Boca Raton, FL: CRC Press.

Fageria, N. K., and M. P. Barbosa Filho. 1994. *Nutritional Deficiency in Rice: Identification and Correction.* Goiânia, Brazil: EMBRAPA Arroz e Feijão.

Fageria, N. K., M. P. Barbosa Filho, and L. F. Stone. 2003. Adequate and toxic levels of micronutrients in soil and plants for annual crops. Paper presented at the XXIX Brazilian Soil Science Congress, Ribeirão Preto, São Paulo, Brazil.

Fageria, N. K., G. D. Carvalho, A. B. Santos, E. P. B Ferreira, and A. M. Knupp. 2011. Chemistry of lowland rice soils and nutrient availability. *Commun. Soil Sci. Plant Anal.* 42:1913–1933.

Fageria, N. K., and A. Moreira. 2011. The role of mineral nutrition on root growth of crop plants. *Adv. Agron.* 110:251–330.

Fageria, N. K., N. A. Slaton, and V. C. Baligar. 2003. Nutrient management for improving lowland rice productivity and sustainability. *Adv. Agron.* 80:63–152.

Fageria, N. K., L. F. Stone, and A. B. Santos. 2003. *Soil Fertility Management of Irrigated Rice.* Santo Antônio de Goiás, Brazil: EMBRAPA Arroz e Feijão.

Foth, H., and B. G. Ellis. 1988. *Soil Fertility.* New York: John Wiley.

Gary, O., A. Sharma, and G. Kona. 1979. Effect of boron on pollen vitality and yield of rice plants (*Oryza sativa* L., var. Jaya). *Plant Soil.* 52:951–954.

Gerloff, G. C. 1987. Intact-plant screening for tolerance of nutrient-deficiency. In *Genetic Aspects of Plant Mineral Nutrition*, eds. W. H. Gableman and B. C. Loughman, pp. 55–68. The Hague, The Netherlands: Martinus Nijhoff.

Gerloff, G. C., and W. H. Gableman. 1983. Genetic basis of inorganic plant nutrition. In *Inorganic Plant Nutrition: Encyclopedia of Plant Physiology*, Vol. 15B, eds. A. Lauchli and R. L. Bieleski, pp. 453–480. New York: Springer-Verlag.

Goldberg, S. 1997. Reactions of boron with soils. *Plant Soil.* 193:35–48.

Goldberg, S., H. S. Forster, and E. L. Heick. 1993. Boron adsorption mechanisms on oxides, clay minerals and soils inferred from ionic strength effects. *Soil Sci. Soc. Am. J.* 57:704–708.

Goldberg, S., and R. A. Glaubig. 1988. Boron and silicon adsorption on an aluminum oxide. *Soil Sci. Soc. Am. J.* 52:87–91.

Graham, R. D. 1984. Breeding for nutrition characteristics in cereals. *Adv. Plant Nutr.* 1:57–102.

Gupta, U. C. 1979. Boron nutrition of crops. *Adv. Agron.* 31:273–307.

Hu, H., and P. H. Brown. 1997. Absorption of boron by plant roots. *Plant Soil.* 193:49–58.

Keren, R., F. T. Bingham, and J. D. Rhoades. 1985. Plant uptake of boron as affected by boron distribution between liquid and solid phases in soil. *Soil Sci. Soc. Am. J.* 49:297–302.

Lordkaew, S., B. Dell, S. Jamjod, and B. Rerkasem. 2010. Boron deficiency in maize. *Plant Soil.* 342:207–220.

Marschner, H. 1995. *Mineral Nutrition of Higher Plants*. London: Academic Press.

Martens, D. C., and D. T. Westermann. 1991. Fertilizer applications for correcting micronutrient deficiencies. In *Micronutrients in Agriculture*, 2nd edition, eds. J. J. Mortvedt, F. R. Cox, L. M. Shuman, and R. M. Welch, pp. 549–592. Madison, WI: SSSA.

Mengel K., E. A. Kirkby, H. Kosegarten, and T. Appel. 2001. *Principles of Plant Nutrition*, 5th edition. Dordrecht, The Netherlands: Kluwer.

Mortvedt, J. J. 1994. Needs for controlled-availability micronutrient fertilizers. *Fert. Res.* 38:213–221.

Mortvedt, J. J., F. R. Fox, L. M. Shuman, and R. M. Welch. 1991. *Micronutrients in Agriculture*, 2nd edition. Madison, WI: SSSA.

Mozafar, A. 1989. Boron effects on mineral nutrients of maize. *Agron. J.* 81:285–290.

Nelson, K. A., and C. G. Meinhardt. 2011. Foliar boron and pyraclostrobin effects on corn yield. *Agron. J.* 103:1352–1358.

Obata, H. 1995. Physiological functions of micro essential elements. In *Science of Rice Plant: Physiology,* Vol. 2, eds. T. Matsuo, K. Kumazawa, R. Ishii, K. Ishihara, and H. Hirata, pp. 402–419. Tokyo: Food and Agricultural Policy Research Center.

Okuda, A. S. Hori, and S. Ida. 1961. Boron nutrition in higher plants. I. A method of growing boron deficient plants. *J. Sci. Soil Manure.* 32:153–157.

Page, N. R., and R. W. Paden. 1954. Boron supplying power of several South Carolina soils. *Soil Sci.* 77:427–434.

Patrick, W. H., Jr., and C. N. Reddy. 1978. Chemical changes in rice soils. In *Soils and Rice*, ed. IRRI, pp. 361–379. Los Baños, Philippines: IRRI.

Ponnamperuma, F. N. 1975. Micronutrient limitations in acid tropical rice soils. In *Soil Management in Tropical America*, eds. E. Bornemisza and A. Alvarado, pp. 330–347. Raleigh, NC: North Carolina State University.

Sahrawat, K. L., T. J. Rego, S. P. Wani, and G. Pardhasaradhi. 2008. Sulfur, boron, and zinc fertilization effects on grain and straw quality of maize and sorghum grown in semi-arid tropical region of India. *J. Plant Nutr.* 31:1578–1584.

Sims, J. R., and F. T. Bingham. 1968. Retention of boron by layer silicate, and soil materials. II. Sesquioxides. *Soil Sci. Soc. Am. Proc.* 32:364–369.

Sims, J. T., and G. V. Johnson. 1991. Micronutrient soil test. In *Micronutrients in Agriculture*, 2nd edition, eds. J. J. Mortvedt, F. R. Cox, L. M. Shuman, and R. M. Welch, pp. 427–476. Madison, WI: SSSA.

Thomidis, T., and E. Exadaktylou. 2010. Effect of boron on the development of brown rot (*Monilinia laxa*) on peaches. *Crop Prot.* 29:572–576.

Tisdale, S. L., W. L. Nelson, and J. D. Beaton. 1985. *Soil Fertility and Fertilizers*, 4th edition. New York: Macmillan.

Vaughan, A. K. F. 1977. The relation between the concentration of boron in the reproductive and vegetative organs of maize plants and their development. *Rhod. J. Agric. Res.* 15:163–170.

Wayne, R. 1986. Boron toxicity in southern cereals. *Rural Res.* 130:25–27.

Woodruff, J. R., F. W. Moore, and H. L. Musen. 1987. Potassium, boron, nitrogen and lime effects on corn yield and earleaf nutrient concentrations. *Agron. J.* 79:520–524.

Yau, S. K., and J. Ryan. 2008. Boron toxicity tolerance in crops: A viable alternative to soil amelioration. *Crop Sci.* 48:854–865.

Yau, S. K., and M. C. Saxena. 1997. Variation in growth, development and yield of durum wheat in response to high soil boron. I. Boron effects. *Aust. J. Agric. Res.* 48:945–949.

Yu, X., and P. F. Bell. 1998. Nutrient deficiency symptoms and boron uptake mechanisms of rice. *J. Plant Nutr.* 21:2077–2088.

Ziaeyan, A. H., and M. Rajaie. 2009. Combined effect of zinc and boron on yield and nutrients accumulation in corn. *Int. J. Plant Prod.* 3:35–44.

12 Molybdenum

12.1 INTRODUCTION

In 1938, the importance of molybdenum (Mo) for crop plants was established by Arnon and Stout working with tomatoes in water culture or solution culture (Marschner 1995; Fageria, Baligar, and Jones 2011). The Mo-deficient tomato plants developed mottling of the leaves and evolution of the laminae, showing that some specific metabolic processes in the plant had been affected by the deficiency. Since then, in some regions, spectacular increases in crop growth have been obtained with Mo fertilization, and every year new areas benefit from Mo. Upland rice plant response to Mo fertilization applied to Brazilian Oxisol has been observed (Figure 12.1). Similarly, lowland rice grown on Brazilian Inceptisol also responded to Mo fertilization (Figure 12.2). In India, Subba Rao and Adinarayana (1995) reported an average grain yield response of 130–880 kg ha^{-1} (5–35%) for rice grown in different soils to which Mo had been applied. Anderson (1956) reviewed literature on Mo fertilizer and reported the marked response of subterranean clover (*Trifolium subterraneum* L.) and alfalfa (*Medicago sativa* L.) to Mo in Australia. The commercial use of Mo on nearby deficient pastures followed almost immediately (Anderson 1956). Graham (2008) reported that 15% of the world's soils are Mo deficient. Similarly, among 190 soils worldwide, it was reported that 3% had an acute Mo deficiency and 12% had a latent deficiency (Sillanpaa 1982). Zou et al. (2008) reported that about 47% of the agricultural land in China, mainly in the east, is Mo deficient.

The importance of Mo for nitrogen fixation and for nitrate reduction is well known. Similarly, for many soils, the effects of lime and other alkaline materials can now be explained in terms of their effect on Mo availability. A number of plant diseases, such as cauliflower "whiptail" and citrus "yellow spot," that had baffled pathologists, are now known to be caused by Mo deficiency (Anderson 1956). Discoveries of Mo-deficient areas have explained some of the past failures of crops and pastures. In some regions, particularly in Australia and New Zealand, the correction of Mo deficiency has made possible the agricultural development of extensive areas of land that had previously been neglected. These transformations usually can be achieved with as little as 32–63 g Mo ha^{-1} (Anderson 1956). In addition to low natural Mo levels in the soils, Mo deficiency can also occur due to the antagonistic effects of other essential plant nutrients on Mo absorption by plants. SO_4^{2-}, Cu, Mn, and Zn have been reported to negatively affect Mo absorption if they are present in excess amounts in the soil solution. On the contrary, P reportedly has synergistic effects on Mo absorption. The nitrate form of nitrogen apparently encourages Mo uptake by plants, yet ammoniacal sources act oppositely (Tisdale, Nelson, and Beaton 1985).

FIGURE 12.1 Upland rice response to Mo fertilization.

FIGURE 12.2 Response of lowland rice grown on Brazilian Inceptisol to Mo fertilization.

In the United States, Mo deficiency occurs largely in the acidic, sandy soils of the Atlantic and Gulf coasts, although similar responses of crops to Mo have been reported in California and the Pacific Northwest, Nebraska, and the Great Lakes states (Tisdale, Nelson, and Beaton 1985). Soils in eastern Canada are acidic, coarsely textured, and highly leached in character. These soils also may be deficient in Mo (Tisdale, Nelson, and Beaton 1985). Molybdenum deficiency has been reported in numerous annual crops in India, China, and many African countries (Alloway 2008). Singh (2008) reported that Mo deficiency occurs in annual crops in India, including rice. Plants exhibiting Mo deficiency usually are grown in broad areas of acidic, well-drained soils, and in soils formed from parent materials low in Mo (Fageria, Baligar, and Clark 2002). Alloway (2008) reported that Mo deficiency is most likely to occur in acidic and severely leached soils and is mainly a problem in brassicas and legumes (such as peanuts, subterranean clover, and soybeans) but other crops, including wheat and sunflowers, also can be affected. However, the objective of this chapter is to discuss Mo deficiencies with respect to rice, including functions, deficiency symptoms, uptake, and management practices for improving efficient Mo use and, consequently, crop yields.

12.2 CYCLE IN SOIL–PLANT SYSTEM

Molybdenum is the least abundant micronutrient in the lithosphere (Mortvedt 2000), and soil concentrations range from 0.2 to 5 mg kg^{-1} (mean of 2 mg kg^{-1}). Table 12.1 displays data related to micronutrients in soils and rocks. Molybdenum is the only micronutrient where availability normally increases along with increases in soil pH. Figure 12.3 compares the availability of Mo with other micronutrients as a function of soil pH. It is clear from Figure 12.4 that Mo availability increases in tandem with a soil pH increase. Below pH 7, the increase in Mo availability is reported to be 100-fold per unit increase in soil pH (Tisdale, Nelson, and Beaton 1985). Goldberg (2009) also reported that Mo adsorption decreased with increasing pH in the range of 4–8. An adsorption maximum was found near pH 4. He further reported that Mo adsorption as a function of solution pH was independent of solution salinity from pH 4 to 8.

TABLE 12.1
Micronutrients in Earth's Crust, Basic Rocks, and Soils

Micronutrient (mg kg^{-1})	Earth's Crust	Basic Rocks	Soils
Mo	1	1.4	2
Zn	65	130	58
Cu	45	140	20
Fe	50,000	86,000	38,000
Mn	1000	2000	850
B	3	10	10

Source: Compiled from Hodgson, J. F. 1963. Chemistry of micronutrients in soils. *Adv. Agron.* 15:119–159.

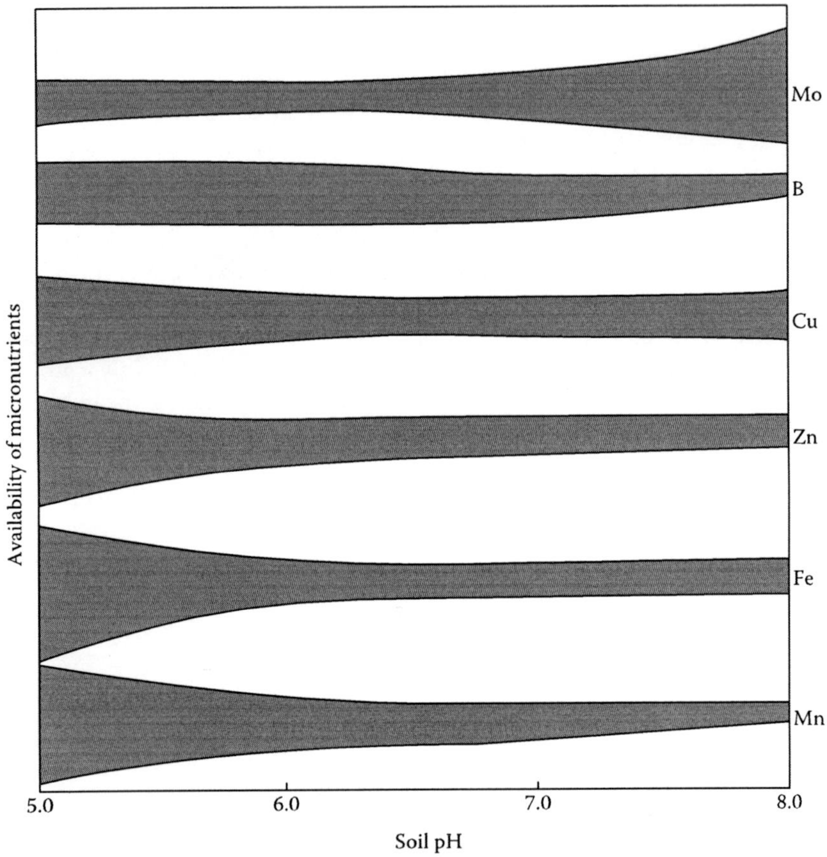

FIGURE 12.3 Hypothetical diagram showing micronutrient availability as a function of soil pH.

The active form of Mo is normally MoO_4^{2-}, which tends to polymerize when in solution. This condition is enhanced by acidification, which could partially explain the low availability of Mo in acidic soils (Fageria, Baligar, and Clark 2002). Solubility of $CaMoO_4$ and H_2MoO_4 (molybdic acid) increases with increases in soil pH. The major solution species of Mo in decreasing order of importance are $MoO_4^{2-} > HMoO_4^- > H_2MoO_4^0 > MoO_2(OH)^+ > MoO_2^{2+}$ (Lindsay 1979; Tisdale, Nelson, and Beaton 1985). In the pH range of 3–5, the first three species contribute significantly to total Mo in solution. The latter two ions generally can be ignored in soils (Lindsay 1979). Plants absorb Mo mainly as MoO_4^{2-}, which is transported to plant roots by mass flow. Diffusion becomes the dominant means for Mo transfer to plant roots when the concentration is less than 4 ppb (Tisdale, Nelson, and Beaton 1985).

Figure 12.4 shows the distribution of Mo species as a function of soil pH. Molybdenum sorption on Fe oxides increases with decreases in soil pH in the range of 4.5–7.8 (Hodgson 1963). Adsorption of Mo on Al and Fe oxides reached a

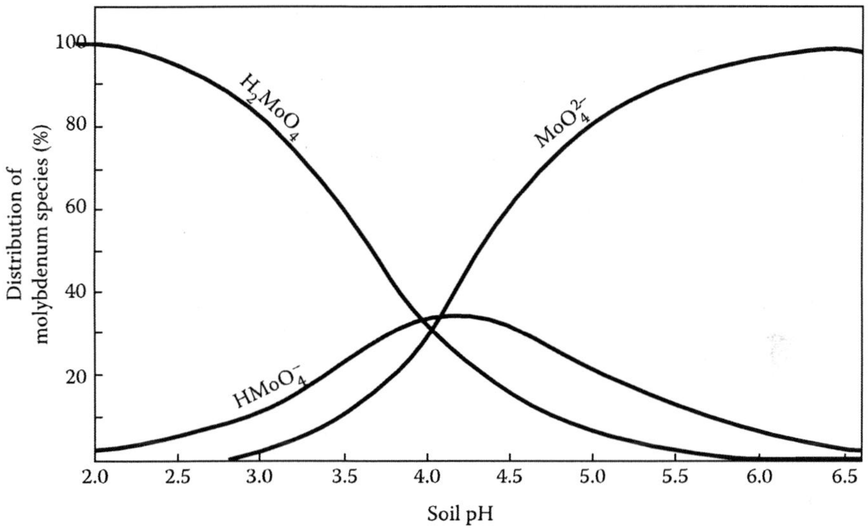

FIGURE 12.4 Relationship between soil pH and distribution of molybdenum species in the soil solution. (Modified from Harter, R. D. 1991. Micronutrient adsorption-desorption reactions in soils. In *Micronutrients in Agriculture*, 2nd edition, eds. J. J. Mortvedt, F. R. Cox, L. M. Schuman, and R. M. Welch, pp. 59–87. Madison, WI: SSSA.)

maximum at pH < 5 and decreased as pH increased >5, with little or no adsorption at pH 8 (Goldberg, Foster, and Godfrey 1996). Soil pH had pronounced effects on Mo adsorption between 3 and 10.5 with virtually no adsorption at pH 8 (Goldberg and Foster 1998). Adsorption of Mo on hydrous Fe and Al oxides decreases as soil pH increases; the addition of lime to the soil normally increases Mo solubility and acquisition by plants (Williams and Thornton 1972). In addition, maximum Mo adsorption on Al and Fe oxides was at pH 4–5, but adsorption was greatest at pH 3.5 with humic acid and decreased as soil pH increased (Bibak and Borggaard 1994). Different mechanisms were apparent for Mo adsorption with humic acid versus Al/Fe oxides, which involved complex formation between carboxyl and phenolic groups. Harmful effects occasionally arise for legumes grown in acidic soils because Mo deficiency may be more dominant than Al toxicity (Bohn, NcNeal, and O'Connor 1979). In some cases, lime and Mo applications may be needed to provide adequate Mo to plants (Lindsay 1991).

Molybdenum dissociation and hydrolysis reactions in soil can be depicted by the following equations (Lindsay 1979):

Molybdic acid dissociation:

$$H_2MoO_4^0 \leftrightarrow H^+ + HMoO_4^- \quad (12.1)$$

$$HMoO_4^- \leftrightarrow H^+ + MoO_4^{2-} \quad (12.2)$$

$$H_2MoO_4^0 \leftrightarrow 2H^+ + MoO_4^{2-} \quad (12.3)$$

$$H_2oO_4(c) \leftrightarrow 2H^+ + MoO_4^{2-} \quad (12.4)$$

Hydrolysis reactions:

$$MoO_4^{2-} + 3H^+ \leftrightarrow MoO_2(OH)^+ + H_2O \quad (12.5)$$

$$MoO_4^{2-} + 4H^+ \leftrightarrow MoO_4^{2-} + 2H_2O \quad (12.6)$$

12.3 FUNCTIONS

Mo participates in many metabolic activities in plants and is an essential constituent of enzymes, including nitrate reductase and sulfite oxidase. Molybdenum-deficient plants result from a reduced conversion of nitrate N to ammonium forms. Deficiency of Mo in plants interferes with protein metabolism; the deficient plants are often pale green with symptoms resembling nitrogen deficiency (Anderson 1956). Molybdenum controls many diseases in plants; it has been reported to have a suppressive effect on zoospores of *Phytophthora spp.* (Halsall 1977) and on the inhibition of nematodes (Haque and Mukhopadhyaya 1983). A putative role of Molybdenum in disease abatement is a decrease in leaf nitrate concentrations resulting from its recognized role in activating the nitrate reductase enzyme (Graham and Stangoulis 2007).

12.4 DEFICIENCY SYMPTOMS

In crop plants, visual symptoms of nutrient deficiency provide a convenient and low-cost means of identifying the problem. When a plant's micronutrient supply is deficient, in addition to crop yields and quality being affected, there may also be visible symptoms of physiological stress, especially in cases of severe deficiency. Although plant species differ in symptoms of micronutrient deficiencies displayed, the most common are stunted plant growth, discoloration, and necrotic spots on the leaves (Alloway 2008). In the case of Mo deficiency in rice, plant growth is stunted and leaf area is reduced, which may influence photosynthetic activities. The leaves of Mo-deficient plants are pale yellow green in color; Mo deficiency symptoms are similar to those of N and S deficiency.

Because Mo is moderately mobile, a deficiency may appear first in the older or younger parts of the plants (Clark 1982; Mengel et al. 2001; Alloway 2008). In the United States, Foth and Ellis (1988) reported that field deficiency of Mo has been observed most often on vegetable crops and that the younger leaves are the most often affected.

With rice, first a yellowing of younger leaves due to Mo deficiency in solution culture was observed (Figure 12.5). The light green to yellow color was uniform in the leaves, and in extreme cases, the leaves became cupped or curled in the margins, with yellowing between the veins and a twisted and distorted appearance. Tissue concentration of 0.1–0.5 mg kg^{-1} is associated with Mo deficiencies. The data in Table 12.2 show the relative susceptibility of principal food crops to Mo deficiency; rice's

FIGURE 12.5 (See color insert.) Molybdenum deficiency symptoms in rice leaves. (From Fageria, N. K., and M. P. Barbosa Filho. 1994. *Nutritional Deficiency in Rice: Identification and Correction*. Goiania, Brazil: EMBRAPA Arroz e Feijão.)

TABLE 12.2
Relative Susceptibility of Principal Food Crops to Molybdenum Deficiency

Crop	Susceptibility
Rice	Low
Barley	Low
Corn	High
Oat	Low/medium
Potato	Low
Pea	Medium
Sorghum	Low
Soybean	Medium
Sugarbeet	Medium
Wheat	Low
Tomato	Medium
Rye	Low

Source: Adapted from Alloway, B. J. 2008. Micronutrients and crop production: An introduction. In *Micronutrient Deficiencies in Global Crop Production*, ed. B. J. Alloway, pp. 1–39. New York: Springer.

susceptibility is low. Colored photographs of Mo deficiency have been reproduced and presented by Wallace (1961) and by Fageria and Barbosa Filho (1994).

12.5 UPTAKE IN PLANT TISSUE

Although Mo is required by plants, rice needs a very small amount. This very low concentration of Mo in healthy rice plants means that the physiological demand is also low. Singh (2008) reviewed the literature on Mo nutrition for plants and reported that the uptake of Mo by intensive cropping systems ranged from 12 to 32 g ha^{-1}. It is absorbed by plants as molybdate. The Mo uptake is metabolically coupled; roots' uptake rate of Mo is not controlled by the internal concentration and is merely proportional to the external concentration (Franco and Munns 1981). Molybdenum is located primarily in the phloem and vascular parenchyma and is moderately mobile in the plant. This characteristic should be taken into account when Mo is applied, particularly when in combination with the liming of acidic soils (Mortvedt 1981; Marschner 1995). The critical level of Mo may vary from 0.1 to 1 mg kg^{-1} depending on plant species and plant part analyzed (Marschner 1995). Deficiency is usually under 0.2 mg kg^{-1} dry matter (Mengel et al. 2001). A unique feature of Mo nutrition is the wide variation between the critical deficiency and toxicity levels. These levels may differ by a factor of up to 10^4 (i.e., 0.1–1000 mg kg^{-1}) (Marschner 1995). Romheld and Marschner (1991) reported that Mo can be applied in a much higher concentration of up to 200–1000 mg kg^{-1} without adverse effects on plant growth.

Bennett (1993) reported that 0.1–0.5 mg Mo kg^{-1} dry plant tissue is sufficient for normal growth of crop plants, and Adriano (1986) found that Mo toxicity in crop plants did not occur until Mo concentration exceeded 500 mg kg^{-1}. Adequate levels of Mo in principal crop tissues, including rice, are given in Table 12.3. Data in Table 12.3 show that adequate Mo concentration in legume plant tissue is much higher than in cereals.

Molybdenum uptake is especially important in the foraging diet of animals. Molybdenum concentration in forages above 5–10 mg kg^{-1} dry weight is high enough to induce toxicity in animals, particularly in ruminants (cattle, sheep, and horses), that is known as *molybdenosis* or *teart* (Marschner 1995). Molybdenosis is actually caused by an imbalance of Mo and copper in ruminants' diets. Fertilizers containing gypsum or sulfur can be used to lower Mo uptake. The sulfate in gypsum is more effective than the equivalent amount of sulfate in superphosphate (Pasricha et al. 1977). Water culture experiments have shown that uptake can be depressed by SO_4^{2-} (Stout et al. 1951; Mengel et al. 2001).

12.6 MANAGEMENT PRACTICES

Appropriate management practices, such as liming acidic soils and using an adequate source, method, and rate of application, can improve Mo uptake and efficiency by crop plants. A soil test is one of the best criteria for determining an adequate rate of Mo application. The use of Mo-efficient genotypes is another important management practice utilized to improve Mo's uptake and application. These management practices are discussed in the next section.

TABLE 12.3
Adequate Level of Molybdenum in Plant Tissue of Principal Crop Species

Crop Species	Plant Part Analyzed	Adequate Level (mg kg^{-1})
Alfalfa	Whole tops prior to bloom	1–5
Barley	Whole tops at heading	0.3–0.5
Barley	Whole tops at boot stage	0.1–0.2
Dry beans	Whole tops, 8 weeks old	0.4–0.6
Dry beans	Whole tops, 56 days after sowing	0.4–0.8
Corn	Ear leaf, prior to silking	0.1–2.0
Corn	Ear leaf at silk	0.6–1.0
Wheat	Whole tops at boot stage	1–2
Wheat	Leaf blade at stem extension	0.05–0.1
Rice	Uppermost mature leaves at tillering	0.5–2
Soybeans	Upper fully developed trifoliate, prior to pod set	1–5
Peanut	Upper stem and leaves at early pegging	1–5

Source: Adapted from Gupta, U. C., and J. Lipsett. 1981. Molybdenum in soils, plants and animals. *Adv. Agron.* 34:73–115; Voss, R. D. 1993. Corn. In *Nutrient Deficiencies and Toxicities in Crop Plants*, ed. W. F. Bennett, pp. 11–14. St. Paul, MN: American Phytopathological Society; Fageria, N. K. 2009. *The Use of Nutrients in Crop Plants*. Boca Raton, FL: CRC Press; Fageria, N. K., V. C. Baligar, and C. A. Jones. 2011. *Growth and Mineral Nutrition of Field Crops,* 3rd edition. Boca Raton, FL: CRC Press.

12.6.1 Liming Acid Soils

With the exception of Mo, availability of essential micronutrients decreases with increasing pH. Because Mo becomes more available with increasing pH, liming will correct a deficiency in acidic soils if the soil contains enough of the nutrient. Where the total Mo content of the soil is low, lime may have no significant impact. Anderson (1956) reported a marked response to either Mo or lime on a soil containing 10 mg Mo kg^{-1} and responses to Mo but not to lime on a soil that contained less than 1 mg Mo kg^{-1}. Adequate pH values, which can be defined as the pH needed to obtain the maximum economic yield of a crop, depend on soil type, plant species, or genotypes within species. The soluble Mo fraction of submerged acidic soils generally increases, presumably as a result of decreased adsorption of MoO_4 under the higher pH conditions associated with waterlogging in flooded rice (Ponnamperuma 1985). Molybdenum deficiency never occurs in neutral and calcareous soils (Moraghan and Mascagni 1991). Anderson (1956) reported that lime increased the amount of water-soluble Mo in the soil and that in most cases, this was associated with an increase in the Mo content of the plants.

12.6.2 Effective Sources

Molybdenum deficiency can also be corrected by applying Mo fertilizers. Ammonium and sodium molybdates are soluble compounds used as sources of Mo fertilizers.

TABLE 12.4
Principal Molybdenum Carriers

Common Name	Formula	Mo Content (%)
Sodium molybdate	$Na_2MoO_4{\cdot}2H_2O$	39
Ammonium molybdate	$(NH_4)_6Mo_7O_{24}{\cdot}2H_2O$	54
Molybdenum trioxidenum	MoO_3	66
Molybdenum frits	Fritted glass	1–30
Molybdenum sulfide	MoS_2	60

Source: Adapted from Tisdale, S. L., W. L. Nelson, and J. D. Beaton. 1985. *Soil Fertility and Fertilizers*, 4th edition. New York: Macmillan; Foth, H., and B. G. Ellis. 1988. *Soil Fertility*. New York: John Wiley; Fageria, N. K. 1989. *Tropical Soils and Physiological Aspects of Crops*. Brasilia: EMBRAPA; Fageria, N. K. 2009. *The Use of Nutrients in Crop Plants*. Boca Raton, FL: CRC Press.

Molybdenum is quite effective when used as sodium molybdate, ammonium molybdate, or the less soluble Mo trioxide, and it appears that about the same level of application is needed for each of these forms (Anderson 1956). Principal Mo carriers are listed in Table 12.4.

12.6.3 Appropriate Methods and Timing of Application

Mo can be successfully applied in a variety of ways. It is rarely applied alone in a separate operation but generally is mixed with materials used on crops or pastures. Soil, foliar, or seed application of Mo, when the soil pH measures are between 4.9 and 6.0, can correct Mo deficiency in crop plants. Soil application (500 g Mo ha^{-1}), foliar application (100 g Mo ha^{-1}), and seed treatments (10 g Mo ha^{-1}) are all effective in correcting Mo deficiency (Franco and Day 1980; Fageria 2009). Martens and Westermann (1991) reported that application rates of 100–500 g Mo ha^{-1} to the soil correct a deficiency.

Molybdenum is not easily lost via leaching, and its recovery is less than 15%. At an adequate rate, the Mo application may have lengthy residual effects. Therefore, its application is not needed frequently. Residual effects of Mo fertilizer can last two to three years, depending on soil type and crop intensity (Gupta and Lipsett 1981). However, Scott (1963) reported that on some soils in New Zealand, a new application of Mo is required after five to six years. Anderson (1956) also reported that an adequate rate of Mo remains effective for many years.

12.6.4 Adequate Rate

Mo deficiency is one of the simplest and least expensive of all deficiencies corrected by the addition of fertilizers because of the very low levels of Mo required (Anderson 1956). The amount of Mo required per unit area may vary with the crop, the soil, the method of application, and other factors. However, an adequate rate of Mo can be determined by a soil test. A soil test is a valid criterion for determining crop responses to applied Mo. For the comparison of soil test results, the definition of an extracting solution is very important. The acid ammonium oxalate (AAO) solution, first proposed by Grigg (1953) as a soil test extractant, is perhaps still the most commonly used for Mo (Sims and Johnson 1991). AAO solution is composed of $(NH_4)_2C_2O_4 \cdot H_2O + H_2C_2O_4$, and a detailed description of soil Mo determination by this solution is given by Reisenauer (1965). Ammonium acetate (NH_4Ac) is also used as an extractant to evaluate Mo availability in soils (Zou et al· 2008). Critical Mo values ranging from 0.2 to 0.5 mg kg^{-1} have been noted for most crops (Fageria 1992). Similarly, Sims and Johnson (1991) reported critical levels of Mo in the range of 0.1–0.3 mg kg^{-1} for crop plants. If a Mo soil test is lower than the adequate level, application of about 500 g Mo ha^{-1} with an appropriate source can correct a Mo deficiency in rice.

12.6.5 Use of Efficient Genotypes

The utilization of Mo-efficient genotypes is an important strategy for maximizing Mo use efficiency in crop plants. Clark (1984) reviewed literature on genotype differences in Mo uptake and use efficiency and concluded that differences for efficient use of Mo among genotypes exist for most crop plants. In general, monocotyledonous plants require less Mo and are less sensitive to Mo deficiency than are dicotyledonous plants (Mengel et al. 2001). Seed-borne Mo may be largely responsible for many differences among plants that are susceptible to Mo deficiency (Clark 1984). Zou et al. (2008) reported genotypic differences in Mo absorption and utilization for cereals and legumes. These differences may be caused by variations in Mo uptake capacities and phloem transportation (Yu, Hu, and Wang 2003).

Alloway (2008) reviewed the literature and cited several possible explanations for the differences in micronutrient efficiency among genotypes. These are (1) the volume and length of roots; (2) the presence, or not, of proteoid roots; (3) root-induced changes in the rhizosphere; (4) increased absorption through vesicular mycorrhizae, if present; (5) release of root exudates to facilitate uptake; (6) efficiency of micronutrient utilization once absorbed into plants; (7) recycling of elements within the growing plant's tissues; and (8) tolerance of inhibiting factors (i.e., bicarbonate ions inhibiting Zn uptake in rice).

12.7 CONCLUSIONS

Throughout the world, large agricultural areas are Mo deficient. Molybdenum deficiency has been reported in rice grown on sandy soils and on highly leached, acidic soils. In Brazil, the response of upland rice grown on Oxisols and lowland or flooded rice grown on Inceptisols was observed. Molybdenum is the only micronutrient where uptake by plants increases with soil pH. Molybdenum concentration in plant tissues is usually fairly low, normally below 1 mg kg^{-1}. In most plant tissues, adequate Mo supply is about 0.5 mg kg^{-1}. Excess Mo can be taken into plants without causing toxicity problems. In acidic soils, Mo is strongly adsorbed by iron and aluminum oxides. Hence, in some cases, Mo deficiency can be corrected with liming of acidic soils. Molybdenum can be added along with other nutrients, such as N,P, and K.

A soil application of about 200–500 g Mo ha^{-1} is sufficient to correct a deficiency in most crop plants, including rice. Ammonium molybdate and sodium molybdate are the most common forms of Mo fertilizer. The foliar application rate may be about 200 g Mo ha^{-1} for annual crops such as rice. Use of Mo-efficient genotypes is one important strategy for maximizing Mo use efficiency, reducing the cost of crop production, and avoiding environmental pollution.

REFERENCES

Adriano, D. 1986. *Trace Elements in the Terrestrial Environment*. New York: Springer-Verlag.

Alloway, B. J. 2008. Micronutrients and crop production: An introduction. In *Micronutrient Deficiencies in Global Crop Production*, ed. B. J. Alloway, pp. 1–39. New York: Springer.

Anderson, A. J. 1956. Molybdenum as a fertilizer. *Adv. Agron.* 8:163–202.

Bennett, W. F. 1993. Plant nutrient utilization and diagnostic plant symptoms. In *Nutrient Deficiencies and Toxicities in Crop Plants*, ed. W. F. Bennett, pp. 1–7. St. Paul, MN: American Phytopathological Society.

Bibak, A., and O. K. Borggaard. 1994. Molybdenum adsorption by aluminum and iron and humic acid. *Soil Sci.* 153:323–336.

Bohn, H., B. NcNeal, and G. O'Connor. 1979. *Soil Chemistry*. New York: Wiley.

Clark, R. B. 1982. Plant response to mineral element toxicity and deficiency. In *Breeding Plants for Less Favorable Environments*, eds. M. N. Christiansen and C. F. Lewis, pp. 71–142. New York: John Wiley.

Clark, R. B. 1984. Physiological aspects of calcium, magnesium, and molybdenum deficiencies in plants. In *Soil Acidity and Liming*, 2nd edition, ed. F. Adams, pp. 99–170. Madison, WI: ASA, CSSA, and SSSA.

Fageria, N. K. 1989. *Tropical Soils and Physiological Aspects of Crops*. Brasilia: EMBRAPA.

Fageria, N. K. 1992. *Maximizing Crop Yields*. New York: Marcel Dekker.

Fageria, N. K. 2009. *The Use of Nutrients in Crop Plants*. Boca Raton, FL: CRC Press.

Fageria, N. K., V. C. Baligar, and R. B. Clark. 2002. Micronutrients in crop production. *Adv. Agron.* 77:185–268.

Fageria, N. K., V. C. Baligar, and C. A. Jones. 2011. *Growth and Mineral Nutrition of Field Crops,* 3rd edition. Boca Raton, FL: CRC Press.

Fageria, N. K., and M. P. Barbosa Filho. 1994. *Nutritional Deficiency in Rice: Identification and Correction.* Goiania, Brazil: EMBRAPA Arroz e Feijão.

Follett, R. H., L. S. Murphy, and R. L. Donahue. 1981. *Fertilizers and Soil Amendments.* Englewood Cliffs, NJ: Prentice-Hall.

Foth, H., and B. G. Ellis. 1988. *Soil Fertility*. New York: John Wiley.

Franco, A. A., and J. M. Day. 1980. Effects of lime and molybdenum on nodulation and nitrogen fixation of *Phaseolus vulgaris* L. in acid soils of Brazil. *Turrialba.* 30:99–105.

Franco, A. A., and D. N. Munns. 1981. Response of *Phaseolus vulgaris* L. to molybdenum under acid conditions. *Soil Sci. Soc. Am. J.* 45:1144–1148.

Goldberg, S. 2009. Influence of solution salinity on molybdenum adsorption by soils. *Soil Sci.* 174:9–13.

Goldberg, S., and H. S. Foster. 1998. Factors affecting molybdenum adsorption by soils and minerals. *Soil Sci.* 163:109–114.

Goldberg, S., H. S. Foster, and C. L. Godfrey. 1996. Molybdenum adsorption on oxides, clay minerals and soils. *Soil Sci. Soc. Am. J.* 60:425–432.

Graham, R. D. 2008. Micronutrient deficiencies in crops and their global significance. An introduction. In *Micronutrient Deficiencies in Global Crop Production*, ed. B. J. Alloway, pp. 41–61. New York: Springer.

Graham, R. D., and J. C. R. Stangoulis. 2007. Molybdenum and plant disease. In *Mineral Nutrition and Plant Disease*, eds. L. E. Datnoff, W. H. Elmer, and D. M. Huber, pp. 203–205. St. Paul, MN: American Phytopathological Society.

Grigg, J. L. 1953. Determination of the available molybdenum of soils. *NZ J. Sci. Technol.* 34:405–414.

Gupta, U. C., and J. Lipsett. 1981. Molybdenum in soils, plants and animals. *Adv. Agron.* 34:73–115.

Halsall, D. M. 1977. Effects of certain cations on the formation and infectivity of *Phytophthora* zoospores: 2. Effects of copper, boron, cobalt, manganese, molybdenum and zinc ions. *Can. J. Microbiol.* 23:1002–1010.

Haque, M. S., and M. C. Mukhopadhyaya. 1983. Influence of some micro-nutrients on *Rotylenchulus reniformis*. *Indian J. Nematol.* 13:115–116.

Harter, R. D. 1991. Micronutrient adsorption-desorption reactions in soils. In *Micronutrients in Agriculture*, 2nd edition, eds. J. J. Mortvedt, F. R. Cox, L. M. Schuman, and R. M. Welch, pp. 59–87. Madison, WI: SSSA.

Hodgson, J. F. 1963. Chemistry of micronutrients in soils. *Adv. Agron.* 15:119–159.

Lindsay, W. L. 1979. *Chemical Equilibrium in Soils*. New York: Wiley.

Lindsay, W. L. 1991. Inorganic equilibrium affecting micronutrients in soil. In *Micronutrients in Agriculture*, 2nd edition, eds. J. J. Mortvedt, F. R. Cox, L. M. Shuman, and R. M. Welch, pp. 89–112. Madison, WI: SSSA.

Marschner, H. 1995. *Mineral Nutrition of Higher Plants*, 2nd edition. New York. Academic Press.

Martens, D. C., and D. T. Westermann. 1991. Fertilizer application for correcting micronutrient deficiencies. In *Micronutrients in Agriculture*, 2nd edition, ed. J. J. Mortvedt, pp. 549–592. Madison, WI: SSSA.

Mengel, K., E. A. Kirkby, H. Kosegarten, and T. Appel. 2001. *Principles of Plant Nutrition*, 5th edition. Dordrecht, The Netherlands: Kluwer.

Moraghan, J. T., and H. J. Mascagni, Jr. 1991. Environmental and soil factors affecting micronutrient deficiencies and toxicities. In *Micronutrient in Agriculture*, 2nd edition, eds. J. J. Mortvedt, P. M. Giordano, and W. L. Lindsay, pp. 371–425. Madison, WI: SSSA.

Mortvedt, J. J. 1981. Nitrogen and molybdenum uptake and dry matter relationship in soybeans and forage legumes in response to applied molybdenum on acid soil. *J. Plant Nutr.* 3:245–256.

Mortvedt, J. J. 2000. Bioavailability of micronutrients. In *Handbook of Soil Science*, ed. M. E. Sumner, pp. D71–D88. Boca Raton, FL: CRC Press.

Pasricha, N. S., V. K. Nayyar, N. S. Randhawa, and M. K. Sinha. 1977. Influence of sulphur fertilization on suppression of molybdenum uptake by berseem (*Trifolium alexandrinum*) and oats (*Avena sativa*) grown on a molybdenum-toxic soil. *Plant Soil.* 46:245–250.

Ponnamperuma, F. N. 1985. Chemical kinetics of wetland rice soils relative to soil fertility. In *Wetland Soils: Characterization, Classification, and Utilization*, ed. International Rice Research Institute, pp. 71–89. Los Baños, Philippines: IRRI.

Reisenauer, H. M. 1965. Molybdenum. In *Methods of Soil Analysis*, Part 2, ed. C. A. Black, pp. 1050–1058. Madison, WI: ASA.

Romheld, V., and H. Marschner. 1991. Functions of micronutrients in plants. In *Micronutrients in Agriculture*, 2nd edition, eds. J. J. Mortvedt, F. R. Cox, L. M. Shuman, and R. M. Welch, pp. 297–328. Madison, WI: SSSA.

Scott, R. S. 1963. Long-term studies of molybdenum applied to pasture. III. Rates of molybdenum application in relation to pasture production. *N.Z. J. Agric. Res.* 6:567–577.

Sillanpaa, M. 1982. *Micronutrients and the Nutrient Status of Soils: A Global Study*. Soil Bulletin 48. Rome: Food and Agricultural Organization of the United Nations, p. 444.

Sims, J. T., and G. V. Johnson. 1991. Micronutrient soil test. In *Micronutrients in Agriculture*, 2nd edition, eds. J. J. Mortvedt, P. M. Giordano, and W. L. Lindsay, pp. 427–476. Madison, WI: SSSA.

Singh, M. V. 2008. Micronutrient deficiencies in crops and soils in India. In *Micronutrient Deficiencies in Global Crop Production*, ed. B. J. Alloway, pp. 93–125. New York: Springer.

Stout, P. R., W. R. Meagher, G. A. Pearson, and C. M. Johnson. 1951. Molybdenum nutrition of crop plants. I. The influence of phosphate and sulfate on the absorption of molybdenum from soils and solution cultures. *Plant Soil.* 3:51–87.

Subba Rao, V. V., and V. Adinarayana. 1995. Molybdenum research and agricultural production. In *Micronutrient Research and Agricultural Production*, ed. H. L. S. Tandon, pp. 60–77. New Delhi: FFDCO.

Tisdale, S. L., W. L. Nelson, and J. D. Beaton. 1985. *Soil Fertility and Fertilizers*, 4th edition. New York: Macmillan.

Voss, R. D. 1993. Corn. In *Nutrient Deficiencies and Toxicities in Crop Plants*, ed. W. F. Bennett, pp. 11–14. St. Paul, MN: American Phytopathological Society.

Wallace, T. 1961. *The Diagnosis of Mineral Deficiencies in Plants*, 2nd edition. New York: Chemical Publications.

Williams, C., and I. Thornton. 1972. The effect of soil additives on the uptake of molybdenum and selenium from soils from different environments. *Plant Soil.* 36:395.

Yu, M., C. Hu, and Y. Wang. 2003. Response of different winter wheat cultivars to Mo deficiency. *J. Huazhong Agric Univ.* 3:34–38.

Zou, C., X. Gao, R. Shi, X. Fan, and F. Zhang. 2008. Micronutrient deficiencies in crop production in China. In *Micronutrient Deficiencies in Global Crop Production*, ed. B. J. Alloway, pp. 127–148. New York: Springer.

13 Chlorine

13.1 INTRODUCTION

Of the earth's 92 natural elements, chlorine (Cl) is ranked as the eighteenth most abundant (Graedel and Keene 1996), and thus far it is the most prevalent anion found in Martian meteorites (Sawyer et al. 2000). Chlorine, or more correctly the chloride (Cl^-), is classified as a micronutrient because the chlorine requirement for optimal growth is between 340 and 1200 mg kg^{-1} (Marschner 1995). However, its uptake by crop plants is equal to that of macronutrients (2000–20,000 mg kg^{-1}) (Fageria, Baligar, and Clark 2002; Fageria 2009). Because of these two contrasting properties, chlorine is known as a unique element. It also is a nonmetal micronutrient like boron (B). In 1954, the importance of chlorine for plants was established using a tomato plant for the purpose of testing (Broyer et al. 1954). Later on, the same group of scientists proved the importance of this element for corn, dry beans, alfalfa, barley, and sugarbeets. All these studies were conducted in nutrient solutions. Chlorine and nickel (Ni) are among the latest micronutrients discovered to be essential for higher plants.

The potential role of Cl^- in crop production was not seriously considered until the 1970s, when research in the Philippines (Von Uexkull 1972), Europe (Russell 1978), and the northwestern United States (Powelson and Jackson 1978) reported that Cl^- could play an important role in crop production. Although chlorine has not been studied intensively, positive responses to this nutrient have been documented for corn (Heckman 1995, 1998) and small grain crops in limited situations (Fixen et al. 1986; Engel, Eckhoff, and Berg 1994). More importantly, reports of chloride deficiency in more than 11 commercial crops have been published (Gausman, Cummingham, and Struchtemeyer 1958; Fixen 1993; Engel et al. 1997). The first mention of chloride in crop management was a recommendation that NaCl be used as a top-dressing on barley to prevent lodging, a condition that may have been precipitated by root disease (Tottingham 1919). Chloride deficiency was also found in sugarbeets (Ulrich and Ohki 1956) and in eight other plant species (Johnson et al. 1957). Chloride uptake and accumulation in plants is decreased by high concentrations of NO_3^- and SO_4^{2-} (Tisdale, Nelson, and Beaton 1985). The antagonism between these anions is related to carrier site competition at the root surface. Because there is limited available literature on the chlorine nutrition of rice, most studies are based on general information about annual crops.

13.2 CYCLE IN SOIL–PLANT SYSTEM

The main source of micronutrients for plant growth is soil. Therefore, knowledge of the trace element cycle in soil–plant systems, including that of chlorine, is very important in order to understand uptake processes and deficiency or sufficiency problems (Welch et al. 1991). Most micronutrients are associated with soil's solid phase.

Lake, Kirk, and Lester (1984) reported that generally, <10% are in soluble and exchangeable forms. However, redistribution among forms, due to changes in soil properties brought about by natural or anthropogenic causes, makes the so-called capacity factor important for micronutrients (Shuman 1991).

Chlorine is an abundant element in the lithosphere and atmosphere and is subject to rapid recycling (Mengel et al. 2001). In aqueous solutions, it occurs mainly as Cl^-. It is loosely held by soil minerals and is subject to leaching in well-drained soils. In humid climates, where rainfall is high, its leaching from surface to subhorizon is significant. However, chlorine may accumulate in arid climates where evaporation is higher than precipitation. It may also accumulate in coastal soils due to high levels of Cl^- in seawater. Chlorine is commonly added to soil through manures and fertilizers (KCl), and irrigation water, and by rainfall and sea spray (Needham 1983).

Generally, chloride is bound tightly in mildly acidic to neutral pH soils and becomes negligible at pH 7.0. Appreciable amounts can be adsorbed with increasing acidity, particularly by Oxisols and Ultisols, which are dominated by kaolinitic clay. Increasing soil pH generally increases Cl^- uptake by plants (Fageria, Baligar, and Clark 2002). Therefore, liming may improve the uptake of Cl^- in acidic soils. When chloride is added to the soil, Cl replaces more OH^- ions than H_2O (Wang, Guo, and Dong 1989). The release of OH^- ions during the specific adsorption of chloride has a negative impact on the removal of free iron oxides (Xu et al. 2000). This hydroxyl ion release, caused by specific adsorption of Cl, increases the pH value in chloride solution (Zhang et al. 1989). Chlorine is one of the most soluble micronutrients and is added to soils in considerable amounts by rainfall. Chlorine's incidental addition to soils in fertilizers and in other ways helps prevent the deficiency of chlorine in field conditions (Brady and Weil 2002).

13.3 FUNCTIONS

Chlorine is an essential micronutrient for higher plants because it is required for the water-splitting reaction of photosynthesis (Clarke and Eaton-Rae 2000). Chloride is also an osmotically active solute in the vacuole and is involved in both turgor and osmoregulation (Valencia et al. 2008). In the cytoplasm, chloride may regulate the activities of key enzymes (Philip and Broadley 2001). The enzymes asparagine synthetase (Rognes 1980), amylase (Metzler 1979), and ATPase (Churchill and Sze 1984) appear to require Cl for optimal activity. In those plants where asparagine is the major compound in the long distance transport of soluble nitrogen, chloride may also play a role in nitrogen metabolism (Marschner 1995; Xu et al. 2000).

The active uptake of Cl in plant roots and its relatively low biochemical activity are two important properties that make Cl particularly well suited to serve as a key osmotic solute in plants (Maas 1986; Xu et al. 2000). The accumulation of chloride by plants contributes greatly to an increase in cell hydration and turgor pressure, both of which are essential for cell elongation (Maas 1986; Xu et al. 2000). Chlorine is required for the fractional assembly of the photosystem II cluster comprising four Mn atoms (Merchant and Dreyfuss 1998). An adequate level of Cl^- in soil was found to enhance the evolution of O_2 as well as photophosphorylation (Bove et al. 1963).

In some plant species, chlorine may influence photosynthesis indirectly through its effect on stomatal regulation of the guard cells. However, high tissue chloride (Cl^-)

concentration can be toxic to crop plants and may restrict crop production in saline regions (Valencia et al. 2008). Functions of chlorine include regulation of cation balance. Chlorine inhibits nitrification and increases NH^{4+} and Mn availability. The slow rate of nitrification inhibition induced by chloride fertilization, particularly in slightly acidic soils, might help to increase N use efficiency in rice fields by preventing N losses due to denitrification in flooded soils (Xu et al. 2000). It also increases fluorescent pseudomonads and Mn-reducing microbes (Elmer 2007). Chlorine also helps in reducing certain diseases in plants. Elmer (2007) and Trolldenier (1985) found that most of the reported benefits of chloride application in crop production have been made in environments where considerable environmental stress or disease pressure was present. Such observations suggest that chloride fertilization may improve defense mechanisms against stress factors and may explain the lack of response when disease or stress factors are absent. Data in Table 13.1 show the impact of chloride fertilization on several plant diseases, including stem rot and sheath blight in rice.

An important impact of Cl^-, when added in adequate quantities, is the inhibition of nitrification (Golden et al. 1981); this has been repeatedly linked to disease suppression (Huber and Wilhelm 1988). Huber and Wilhelm (1988) reported that nitrification inhibition can decrease soil pH (greater uptake of NH_4^+ and lower NO_3^-) and may be responsible for a greater uptake of micronutrients—especially Mn, which decreases disease infestation (Graham and Webb 1991). In addition, NH_4^+ uptake may have other effects on host physiology that lead to disease resistance (Huber and Watson 1974). Higher uptake of NH_4^+ compared to NO_3^- may decrease blast disease in upland rice.

TABLE 13.1
Plant Diseases Suppressed by Chloride Fertilization

Crop Species	Common Name	Scientific Name
Rice	Stem rot	*Sclerotium oryzae*
	Sheath blight	*Rhizoctonia solani*
Winter wheat	Take-all root rot	*Gaeumannomyces graminis var. tritici*
	Tanspot	*Pyrenophora trichostoma*
	Stripe rust	*Puccinia striiformis*
	Septoria leaf blotch	*Septoria tritici* and *S. avenae* f.sp. *triticea*
	Leaf rust	*Puccinia recondita* f.sp. *tritici*
Spring wheat	Common root rot	*Cochliobolus sativus*
Barley	Common root rot	*Cochliobolus sativus*
	Fusarium root rot	*Fusarium* spp.
	Spot blotch	*Cochliobolus sativus* f. *teres*
Durum wheat	Common root rot	*Cochliobolus sativus*
Corn	Stalk rot	*Gibberella roseum* f.sp. *cerealis*
Pearl millet	Downy mildew	*Sclerospora graminicola*
Potatoes	Hollow heart	Physiological disorder
	Brown center	Physiological disorder

Source: Adapted from Fixen, P. E., R. H. Gelderman, J. R. Grewing, and B. G. Farber. 1987. Calibration and implementation of a soil Cl test. *J. Fertil. Issues.* 4:91–97.

13.4 DEFICIENCY SYMPTOMS

Chlorine deficiency symptoms have not been reported in rice. However, deficiency symptoms have been observed in other crop species (Broyer et al. 1954; Ulrich and Ohki 1956). Epstein and Bloom (2005) reported that symptoms of chlorine deficiency in plants include a blue-green color and shiny appearance of younger leaves. These authors also noted that younger leaves of chlorine-deficient plants show wilting symptoms and dangle down. As the deficiency progresses, a characteristic bronzing appears on the leaves, followed by chlorosis and necrosis (Johnson et al. 1957; Ozanne, Woolley, and Broyer 1957). Plants that are severely deficient become spindly, and their growth is stunted (Epstein and Bloom 2005). In chlorine-deficient plants, leaf area is reduced; this ultimately affects photosynthesis (Mengel et al. 2001). Chlorine deficiency also reduces root growth and root hair formation (Fageria, Baligar, and Clark 2002).

13.5 UPTAKE IN PLANT TISSUE

The rate of Cl^- uptake by plant species is relatively high compared to other micronutrients. This uptake rate depends on concentrations in the soil solution and is also determined by crop species (Mengel et al. 2001). Uptake occurs in aqueous solutions mainly as Cl^-, which is highly mobile and easily taken up by plants (Romheld and Marschner 1991). Chlorine uptake is metabolically controlled, and uptake takes place against an electrochemical potential gradient (based upon the active transport definition first proposed by Ussing in 1949). Presumably, the gradient is mediated by a Cl^-/H^+ cotransport across the plasmalemma because low pH promotes Cl^- uptake (Gerson and Poole 1972). Chloride uptake occurs through specific ion channels that are responsible for the plasmalemma and tonoplast transport (Hedrich and Schroeder 1989; Mengel et al. 2001).

The Cl^- ion retains its negative charge after absorption by roots, whereas SO_4^{2-} and NO_3^- are partly or completely reduced during metabolism in plants (Xu et al. 2000). Physiological mechanisms for the control of Cl accumulation in plant cells operate at the cell or organ level (Cram 1988). Changes in root temperature and Cl concentration in the soil solution affect Cl absorption by plants (Cram 1983, 1988), and uptake of Cl is affected by forms of N. If N is present in the form of ammonium, higher uptake of cations leads to uptake of more anions in order to maintain the electrical neutrality of the process (Harward et al. 1956). As a result, when Cl is present in the soil solution, ammonium uptake increases the salt sensitivity of crop plants (Speer, Brune, and Kaiser 1994). Plants fertilized with NH_4^+ usually contain much more Cl in the tissue than do plants fertilized with NO_3^-. Uptake of Cl^- (concentration × dry weight) is a function of the dry matter or grain yield of a crop. De Datta and Mikkelsen (1985) reported that in lowland rice, Cl accumulation was 54 kg ha^{-1} in 8.3 Mg straw ha^{-1} and was 41 kg ha^{-1} in 9.8 Mg grain ha^{-1}. Therefore, lowland rice accumulated 95 kg Cl ha^{-1} in straw and grain combined. Rice is relatively tolerant of chloride (Xu et al. 2000). When Cl concentration in a rice shoot was lower than 3 or 3000 mg kg^{-1} dry weight, irrigation water containing 50–150 mg Cl m^{-3} increased rice yield (Yin, Sun, and Liu 1989). A negative effect of Cl was not observed either for rice yield or for quality when the Cl content in mature straw was 12–13 mg g^{-1} or 12,000–13,000 mg kg^{-1} dry weight (Zhu and Yu 1991; Huang, Rao, and Liao 1995). Rice can tolerate chloride as high as 400–800 mg kg^{-1} in the soil (Zhu and Yu 1991) or 300–500 g m^{-3} in the irrigation water (Yin, Sun, and Liu 1989). Fixen (1993) reported that because the biochemical functions of Cl require no more than 100 mg kg^{-1}, Cl is classified as a micronutrient. However, much higher concentrations (in the range of 2000 to 20,000 mg kg^{-1}) are normally present in plants, indicating that Cl, unlike other micronutrients, is relatively nontoxic at higher concentrations.

Xu et al. (2000) reported that wheat crop can remove 18 and 61 kg ha^{-1} at low and high Cl content in the soil, respectively. Of the total Cl uptake by crops, the distribution of Cl was 2.15% in spring wheat grain, 1.34% in soybeans, and 1.62% in rice (Xu et al. 2000). Tsukada and Takeda (2008) analyzed Cl content of polished rice, rice bran, hull, straw, and roots (Table 13.2). Data in Table 13.2 show that a rice crop removed about 27 kg Cl ha^{-1} in the shoot or aboveground plant parts. The maximum concentration of Cl was in the roots. Because roots are difficult to remove from the soils, accumulation of Cl in this organ is not possible. However, straw accumulated the maximum amount of chlorine in the aboveground plant parts. Hence, incorporation of straw can recycle chlorine for the succeeding crops.

The nutrient concentration of a plant, or a plant part, is a reflection of the soil's available micronutrient status; this status is not always easily determined by a soil test (Jones 1991). Therefore, nutrient concentration in plants is an important diagnostic criterion for deficiency or sufficiency. The average concentration of Cl in plants ranges from 1 to 20 g kg^{-1} and, therefore, is in the range of macronutrients (Romheld and Marschner 1991). Compared with the high tissue concentrations of Cl found in most plants, the requirement for optimal growth generally is much lower (150–300 mg kg^{-1}). Martens and Westermann (1991) reported that the critical Cl concentration is about 70–200 mg kg^{-1} in the dry matter for most plants.

Several methods are available for Cl determination in plant tissues. Among them, the more classic procedures have been described by Williams (1979). A gravimetric and two volumetric procedures are provided in the Association of Official Analytical Chemists (AOAC) manual (Williams 1984). Jones (1991) reported that

TABLE 13.2
Concentration, Uptake, and Distribution of Chlorine in Rice Plant Parts

Plant Part	Conc. (mg kg^{-1}) Dry Wt.	Uptake (kg ha^{-1})	Percent of Total Uptake
Polished rice	140	0.65	0.12
Rice bran	670	0.33	0.06
Hull	600	0.65	0.12
Straw	3800	25	4.7
Roots	4800	–	–
Aboveground parts		27	5.0

Source: Adapted from Tsukada, H., and A. Takeda. 2008. Concentration of chlorine in rice plant components. *J. Radioanal. Nucl. Chem.* 278:387–390.

in order to minimize interferences with either electrode methods, Cl is determined either in a 0.5 M HNO_3 extract of the plant tissue or after the tissue is dry ashed and the ash is solubilized in dilute HNO_3. The chlorine in plant tissue is also determined by ion chromatography; a detailed discussion of this procedure is given by Kalbasi and Tabatabai (1985) as well as by Grunau and Swiader (1986). Deficient, adequate, and toxic Cl levels in principal crop species, including rice, are presented in Table 13.3.

TABLE 13.3
Critical, Adequate, and Toxic Levels of Chloride in the Plant Tissue of Principal Crop Species

Crop Species	Plant Part Analyzed	Cl Concentration ($mg\ kg^{-1}$)			Reference
		Deficient	Adequate	Toxic	
Rice	Shoot	3.0	Not given	7.0–8.0	Yin, Sun, and Liu 1989
Rice	Mature straw	Not given	5.1–10.0	>14.0	Huang, Rao, and Liao 1995
Wheat	Heading shoot	1.2–4.0	>4.0	Not given	Engel et al. 1997
Barley	Heading shoot	1.2–4.0	>4.0	Not given	Engel et al. 1997
Alfalfa	Shoot	0.65	0.9–2.7	6.1	Ozanne, Woolley, and Broyer 1957; Eaton 1966
Corn	Ear leaves	Not given	1.1–10.0	>33	Parker, Gaines, and Gascho 1985
Corn	Shoot	0.05–0.11	Not given	Not given	Johnson et al. 1957
Peanut	Shoot	Not given	<3.9	>4.6	Wang, Guo, and Dong 1989
Potato	Petioles	0.71–1.42	18.0	44.8	Whitehead 1985
Soybean	Leaves	Not given	0.3–1.5	16.7–24.3	Parker, Gaines, and Gascho 1986
Sugar beet	Petioles	<5.7	>7.2	>50.8	Zhou and Zhang 1992

Source: Eaton, F. M. 1966. Chlorine. In *Diagnostic Criteria for Plants and Soils*, ed. H. D. Chapman, pp. 98–135. Riverside, CA: University of California; Engel, R. E., P. L. Bruckner, D. E. Mathre, and S. K. Z. Brumfield. 1997. A chloride deficient leaf spot syndrome of wheat. *Soil Sci. Soc. Am. J.* 61:176–184; Huang, Y., Y. P. Rao, and T. J. Liao. 1995. Migration of chloride in soil and plant. *J. Southwest Agric. Univ.* (in Chinese) 17:259–263; Johnson, C. M., P. R. Stout, T. C. Boyer, and A. B. Carlton. 1957. Comparative chlorine requirements of different plant species. *Plant Soil.* 8:337–353; Ozanne, P. G., J. T. Woolley, and T. C. Broyer. 1957. Chlorine and bromine in the nutrition of higher plants. *Aust. J. Biol. Sci.* 10:66–79; Parker, M. B., T. P. Gaines, and G. J. Gascho. 1985. Chloride effects on corn. *Commun. Soil Sci. Plant Anal.* 16:1319–1333; Parker, M. B., T. P. Gaines, and G. J. Gascho. 1986. The chloride toxicity problem in soybean in Georgia. In *Special Bulletin on Chloride and Crop Production*, ed. T. L. Jackson, No. 2, pp. 100–108. Atlanta, GA: Potash & Phosphate Institute; Wang, D. Q., B. C. Guo, and X. Y. Dong. 1989. Toxicity effects of chloride on crops. *Chin. J. Soil Sci.* 30:258–261; Whitehead, D. C. 1985. Chlorine deficiency in red clover grown in solution culture. *J. Plant Nutr.* 8:193–198; Yin, M. J., J. J. Sun, and C. S. Liu. 1989. Contents and distribution of chloride and effects of irrigation water of different chloride levels on crops. *Soil Fertil.* (in Chinese) 1:3–7; Zhou, B. K., and X. Y. Zhang. 1992. Effects of chloride on growth and development of sugarbeet. *Soil Fertil.* (in Chinese) 3:41–43.

13.6 MANAGEMENT PRACTICES

Management practices that can improve Cl^- uptake and use efficiency by plants include effective source, appropriate method and timing of application, adequate rate, and use of efficient genotypes. These involve soil, crop, and fertilizer management and should be adopted together (rather than in isolation) for maximizing crop yields.

13.6.1 EFFECTIVE SOURCES

When selecting a micronutrient source, the important factors that should be taken into account are cost; market availability; N, P, and K fertilizer compatibility; convenience of application; water solubility; and agronomic effectiveness. Agronomic effectiveness of a *micronutrient source* is defined as the degree of crop response per unit of applied micronutrient (Mortvedt 1991). The source that produces the maximum yield at the lowest rate, or the highest yield at the same rate, is considered most effective. Relative agronomic effectiveness varies with rate, which emphasizes the need for including multiple rates (Mortvedt 1991). Important chloride sources are presented in Table 13.4. Fixen et al. (1986) and Mohr (1992) reported that all common Cl^- sources were equally effective in crop production.

13.6.2 APPROPRIATE METHODS AND TIMING OF APPLICATION

Soil and foliar application are the common methods of applying micronutrients. The micronutrient sources should be mixed with N–P–K fertilizers for field application. This method may result in a uniform application and may reduce application cost. Some chemical reactions can occur when micronutrients are mixed with N–P–K fertilizers during manufacturing and storage. Therefore, care should be taken when selecting micronutrient sources that are to be combined with N–P–K fertilizers. When mixed with N–P–K sources, micronutrients should be applied immediately

TABLE 13.4
Principal Chlorine Carriers

Common Name	Formula	Cl (%)
Potassium chloride	KCl	47
Ammonium chloride	NH_4Cl	66
Calcium chloride	$CaCl_2$	65
Magnesium chloride	$MgCl_2$	74
Sodium chloride	$NaCl_2$	60
Zinc chloride	$ZnCl_2$	48
Manganese chloride	$MnCl_2$	56

Source: Fageria, N. K. 2009. *The Use of Nutrients in Crop Plants*. Boca Raton, FL: CRC Press.

and not stored for a longer duration. Fertilizers, including micronutrients, should be applied in bands or furrows in order to obtain maximum crop production efficiency. Considerable placement flexibility appears to exist for Cl fertilizers. Comparisons of preplant broadcast, band, and early spring top-dress applications for cereals have given very similar results (Fixen et al. 1986; Mohr 1992).

13.6.3 Adequate Rate

Using an adequate rate of nutrients is fundamental for sustainable crop production. In addition, this also reduces costs, minimizes environmental pollution, and maximizes crop yields. The main criteria used in determining the adequate rate of a given nutrient are field trial data relating to nutrient rate versus grain yield, soil test calibration data relating to nutrient concentration in the soil, and grain yield and nutrient uptake by straw and grain for a determined yield level. Data related to chlorine rate and grain yield of rice under field conditions or field trials are limited. Tisdale, Nelson, and Beaton (1985) reported that band application of 40–45 kg Cl ha^{-1} can reduce take-all root rot disease of wheat. Similarly, these authors reported that if Cl is applied as broadcast, 80–140 kg ha^{-1} is required in order to depress effects of crop plant diseases.

As far as soil testing is concerned, Cl^- can be extracted with water or a dilute electrolyte because chloride salts are highly water soluble and Cl^- is not strongly adsorbed. Fixen, Gelderman, and Denning (1988) have described in detail the extraction of Cl^- using either 0.01 M $Ca(NO_3)_2$, 0.5 M K_2SO_4, or a CaO-saturated solution corresponding to analysis by colorimetric, potentiometric, or ion chromatographic procedures, respectively. Little research has been conducted on the use of Cl soil tests, presumably because deficiencies of this element are rare. Sims and Johnson (1991) reported that suitable extraction and analytical procedures are available for routine testing. Calibration and correlation data are lacking to the degree that test interpretation is difficult at best. Because Cl is highly leachable in the soil profile, the soil sampling depth should be given special consideration. In rice-growing soil, a large amount of Cl is washed by water to a depth of 40–60 cm after one growing season (Huang, Rao, and Liao 1995). On North America's Great Plains, sampling down to 60 cm has been recommended for spring wheat and barley (Fixen et al. 1987). Soil test data are not available for rice. However, Fixen et al. (1987) reported soil test results of <17 mg kg^{-1} for wheat response to Cl application as low; they recommended 75 kg Cl ha^{-1} and 17–34 mg kg^{-1} as medium, and 65 kg Cl ha^{-1} and >34 mg kg^{-1} as high, which do not need Cl fertilization. James, Weaver, and Reeder (1970) reported that a Cl level in the soil of about 13 mg kg^{-1} may be deficient for potatoes.

13.6.4 Use of Efficient Genotypes

Within species, variation in the absorption and utilization of chlorine among crops and genotypes has been reported by Fageria (2009). Similarly, Fixen (1993) noted that significant cultivar differences in chloride response potential appear to exist for several crops. Hence, use of chlorine-efficient rice genotypes is an important strategy for improving rice yield in chlorine-deficient soils. However, data for rice are limited, and more research is required to identify genotype variation in Cl uptake and utilization.

13.7 CONCLUSIONS

Chlorine is an abundant element in nature, and its deficiency for crop production is rarely observed in most agroclimatic conditions. Its main functions are osmoregulation (cell elongation and stomatal opening) and charge compensation (i.e., counter anion in cation transport) in higher plants. It also plays a positive role in enzyme activities such as in water-splitting enzymes. An adequate rate of chlorine application may suppress plant diseases, especially rice blast. For plants, chlorine deficiency symptoms include leaf wilting, leaflet curling, bronzing and chlorosis (similar to Mn deficiency), and severely inhibited root growth. Chlorine uptake by roots is an active process and requires energy. Plant and soil analyses appear to be helpful in predicting crop response to chlorine. Rice plants remove about 6.5 kg Cl in grain plus straw to produce one metric ton of grain. In rice plants, most accumulated Cl stays in the straw, and a small part is translocated to the grains. Rice is relatively tolerant to Cl accumulation in the tops. It can tolerate Cl content of about 13 mg g^{-1} or 13,000 mg kg^{-1} dry weight without an adverse effect on grain yield or quality. The critical level of Cl in the plants varies among genotypes and yield levels, but it may be in the range of 70–300 mg kg^{-1}. Critical Cl soil test levels vary among genotypes and soil types. However, published results show that with a Cl level lower than 13 mg kg^{-1}, crop response to applied Cl is expected. In chlorine-deficient soils, using about 80 kg K ha^{-1} with potassium chloride is sufficient to fulfill the Cl requirements of upland as well as lowland rice. The use of Cl-efficient genotypes is a very attractive strategy for improving rice yield in Cl-deficient soils. However, there is limited research data regarding various rice genotypes and their uptake and utilization of Cl.

REFERENCES

Bove, J. M., C. Bove, F. R. Whatley, and D. I. Arnon. 1963. Chloride requirement for oxygen evolution in photosynthesis. *Z. Naturforsch.* 18:683–688.

Brady, N. C., and R. R. Weil. 2002. *The Nature and Properties of Soils*, 13th edition. Upper Saddle, NJ: Prentice Hall.

Broyer, T. C., A. B. Carlton, C. M. Johnson, and P. R. Stout. 1954. Chlorine: A micronutrient element for higher plants. *Plant Physiol.* 29:526–532.

Churchill, K. A., and H. Sze. 1984. Anion sensitive, H^+ pumping ATPase of oat roots. *Plant Physiol.* 76:490–497.

Clarke, S. M., and J. J. Eaton-Rye. 2000. Amino acid deletions loop of the chlorophyll binding protein CP47 alter the chloride requirement and/or prevent the assembly of photosystem II. *Plant Mol. Biol.* 44:591–601.

Cram, W. J. 1983. Chloride accumulation as a homeostatic system: Set points perturbations. The physiological significance of influx isotherms, temperature effects and the influences of plant growth substances. *J. Exp. Bot.* 34:181–1502.

Cram, W. J. 1988. Transport of nutrient ions across cell membranes in vivo. *Adv. Plant Nutr.* 3:1–54.

De Datta, S. K., and D. S. Mikkelsen. 1985. Potassium nutrition of rice. In *Potassium in Agriculture*, ed. R. D. Munson, pp. 665–699. Madison, WI: ASA.

Eaton, F. M. 1966. Chlorine. In *Diagnostic Criteria for Plants and Soils*, ed. H. D. Chapman, pp. 98–135. Riverside, CA: University of California.
Elmer, W. H. 2007. Chlorine and plant disease. In *Mineral Nutrition and Plant Disease*, eds. L. E. Datnoff, W. H. Elmer, and D. M. Huber, pp. 189–202. St. Paul, MN: American Phytopathological Society.
Engel, R. E., P. L. Bruckner, D. E. Mathre, and S. K. Z. Brumfield. 1997. A chloride deficient leaf spot syndrome of wheat. *Soil Sci. Soc. Am. J.* 61:176–184.
Engel, R. E., J. Eckhoff, and R. Berg. 1994. Grain yield, kernel weight, and disease responses of winter wheat cultivars to chloride fertilization. *Agron. J.* 86:891–896.
Epstein, E., and A. J. Bloom. 2005. *Mineral Nutrition of Plants: Principles and Perspectives*, 2nd edition. Sunderland, MA: Sinauer.
Fageria, N. K. 2009. *The Use of Nutrients in Crop Plants*. Boca Raton, FL: CRC Press.
Fageria, N. K., V. C. Baligar, and R. B. Clark. 2002. Micronutrients in crop production. *Adv. Agron.* 77:185–268.
Fixen, P. E. 1993. Crop responses to chloride. *Adv. Agron.* 50:107–150.
Fixen, P. E., R. H. Gelderman, and J. L. Denning. 1988. Chloride tests. In *Recommended Chemical Soil Test Procedures for the North Central Region,* ed. W. Dahnke, pp. 26–29. North Dakota Agricultural Experimental Station, North Central Region Publication 221 (revised).
Fixen, P. E., R. H. Gelderman, J. R. Gerwing, and F. A. Cholick. 1986. Response of spring wheat, barley, and oats to chloride in potassium chloride fertilizers. *Agron. J.* 78:664–668.
Fixen, P. E., R. H. Gelderman, J. R. Grewing, and B. G. Farber. 1987. Calibration and implementation of a soil Cl test. *J. Fertil. Issues.* 4:91–97.
Gausman, H. W., C. E. Cummingham, and R. A. Struchtemeyer. 1958. Effects of chloride and sulfate on ^{32}P uptake by potatoes. *Agron. J.* 50:90–91.
Gerson, D. F., and R. J. Poole. 1972. Chloride accumulation by mung bean root tips. A low affinity active transport system at the plasmalemma. *Plant Physiol.* 50:603–607.
Golden, D. C., S. Sivasubramanian, S. Sanderman, and M. A. Wijedasa. 1981. Inhibitory effects of commercial potassium chloride on the nitrification rates of added ammonium sulfate in an acid red yellow podzolic soil. *Plant Soil.* 59:147–151.
Graedel, T. E., and W. C. Keene. 1996. The budget and cycle of earth's natural chlorine. *Pure Appl. Chem.* 68:1689–1697.
Graham, R. D., and M. J. Webb. 1991. Micronutrients and disease resistance and tolerance in plants. In *Micronutrients in Agriculture*, eds. J. J. Mortvedt, F. R. Cox, L. M. Shuman, and R. M. Welch, pp. 329–370. Madison, WI: SSSA.
Grunau, J. A., and J. M. Swiader. 1986. Application of ion chromatography to anion analysis in vegetable leaf extracts. *Commun. Soil Sci. Plant Anal.* 17:321–335.
Harward, M. E., W. A. Jackson, J. R. Piland, and D. D. Mason. 1956. The relationship of chloride and sulfate ions to forms of nitrogen in the nutrition of Irish potatoes. *Soil Sci. Soc. Am. Proc.* 20:231–236.
Heckman, J. R. 1995. Corn responses to chloride in maximum yield research. *Agron. J.* 87:415–419.
Heckman, J. R. 1998. Corn stalk rot suppression and grain yield response to chloride. *J. Plant Nutr.* 21:149–155.
Hedrich, R., and J. I. Schroeder. 1989. The physiology of ion channels and electrogenic pumps in higher plants. *Annu. Rev. Plant Physiol.* 40:539–569.
Huang, Y., Y. P. Rao, and T. J. Liao. 1995. Migration of chloride in soil and plant. *J. Southwest Agric. Univ.* (in Chinese) 17:259–263.
Huber, D. M., and R. D. Watson. 1974. Nitrogen form and plant disease. *Annu. Rev. Phytopathol.* 12:139–165.
Huber, D. M., and N. S. Wilhelm. 1988. The role of manganese in resistance to plant disease. In *Manganese in Soils and Plants*, eds. R. D. Graham, R. J. Hannam, and N. C. Uren, pp. 155–173. Dordrecht, The Netherlands: Kluwer Academic.

James, D. W., W. H. Weaver, and R. L. Reeder. 1970. Chloride uptake by potatoes and the effects of potassium chloride, nitrogen and phosphorus fertilization. *Soil Sci.* 109:48–52.

Johnson, C. M., P. R. Stout, T. C. Boyer, and A. B. Carlton. 1957. Comparative chlorine requirements of different plant species. *Plant Soil.* 8:337–353.

Jones, J. B. Jr. 1991. Plant tissue analysis in micronutrients. In *Micronutrients in Agriculture*, eds. J. J. Mortvedt, F. R. Cox, L. M. Shuman, and R. M. Welch, pp. 477–521. Madison, WI: SSSA.

Kalbasi, M., and M. A. Tabatabai. 1985. Simultaneous determination of nitrate, chloride, sulfate, and phosphate in plant materials by ion chromatography. *Commun. Soil Sci. Plant Anal.* 16:787–800.

Lake, D. L., P. W. W. Kirk, and J. N. Lester. 1984. Fractionation, characterization and speciation of heavy metals in sewage sludge and sludge-amended soils. A review. *J. Environ. Qual.* 13:175–183.

Maas, E. V. 1986. Physiological response to chloride. In *Special Bulletin on Chloride and Crop Production,* No. 2, ed. T. L. Jackson, pp. 4–20. Atlanta, GA: Potash & Phosphate Institute.

Marschner, H. 1995. *Mineral Nutrition of Higher Plants*, 2nd edition. San Diego, CA: Academic Press.

Martens, D. C., and D. T. Westermann. 1991. Fertilizer applications for correcting micronutrient deficiencies. In *Micronutrients in Agriculture*, eds. J. J. Mortvedt, F. R. Cox, L. M. Shuman, and R. M. Welch, pp. 549–590. Madison, WI: SSSA.

Mengel, K., E. A. Kirkby, H. Kosegarten, and T. Appel. 2001. *Principles of Plant Nutrition*, 5th edition. Dordrecht, The Netherlands: Kluwer Academic.

Merchant, S., and B. W. Dreyfuss. 1998. Posttranslational assembly of photosynthetic metalloproteins. *Annu. Rev. Plant Physiol. Plant Mol. Biol.* 49:25–51.

Metzler, D. E. 1979. *Biochemistry: The Chemical Reactions of Living Cells.* New York: Academic press.

Mohr, R. M. 1992. The effect of chloride fertilization on growth and yield of barley and spring wheat. MS thesis, University of Manitoba.

Mortvedt, J. J. 1991. Micronutrient fertilizer technology. In *Micronutrients in Agriculture*, eds. J. J. Mortvedt, F. R. Cox, L. M. Shuman, and R. M. Welch, pp. 523–536. Madison, WI: SSSA.

Needham, P. 1983. The occurrence and treatment of mineral disorders in the field. In *Diagnosis of Mineral Disorders in Plants: Principles*, Vol. 1., eds. C. Bould, E. J. Hewitt, and P. Needham, pp. 131–170. London: Her Majesty's Stationary Office.

Ozanne, P. G., J. T. Woolley, and T. C. Broyer. 1957. Chlorine and bromine in the nutrition of higher plants. *Aust. J. Biol. Sci.* 10:66–79.

Parker, M. B., T. P. Gaines, and G. J. Gascho. 1985. Chloride effects on corn. *Commun. Soil Sci. Plant Anal.* 16:1319–1333.

Parker, M. B., T. P. Gaines, and G. J. Gascho. 1986. The chloride toxicity problem in soybean in Georgia. In *Special Bulletin on Chloride and Crop Production*, ed. T. L. Jackson, No. 2, pp. 100–108. Atlanta, GA: Potash & Phosphate Institute.

Philip, J., and R. Broadley. 2001. Chloride in soils and its uptake and movement within the plant: A review. *Ann. Bot.* 88:967–988.

Powelson, R. L., and T. L. Jackson. (1978). Suppression of take-all (*Gaeumannomyces graminis*) root rot of wheat with fall applied chloride fertilizers. Proceedings of 29th Annual Northwest Fertilizer Conference, Beaverton, Oregon, 175–182.

Rognes, S. E. 1980. Anion regulation of lupin asparagine synthetase: Chloride activation of the glutamine utilizing reaction. *Phytochemistry.* 19:2287–2293.

Romheld, V., and H. Marschner. 1991. Function of micronutrients in plants. In *Micronutrients in Agriculture*, eds. J. J. Mortvedt, F. R. Cox, L. M. Shuman, and R. M. Welch, pp. 297–326. Madison, WI: SSSA.

Russell, G. E. 1978. Some effects of applied sodium and potassium on yellow rust in winter wheat. *Ann. Appl. Biol.* 90:163–168.

Sawyer, D. J., M. D. McGehee, J. Canepa, and C. B. Moore. 2000. Water soluble ions in the Nakhla Martian meteorite. *Meteorit. Planet. Sci.* 35:743–748.

Shuman, L. M. 1991. Chemical forms of micronutrients in soils. In *Micronutrients in Agriculture*, eds. J. J. Mortvedt, F. R. Cox, L. M. Shuman, and R. M. Welch, pp. 113–142. Madison, WI: SSSA.

Sims, J. T., and G. V. Johnson. 1991. Micronutrient soil tests. In *Micronutrients in Agriculture*, eds. J. J. Mortvedt, F. R. Cox, L. M. Shuman, and R. M. Welch, pp. 427–475. Madison, WI: SSSA.

Speer, M., A. Brune, and W. M. Kaiser. 1994. Replacement of nitrate by ammonium as the nitrogen source increases the salt sensitivity of pea plants. I. Ion concentration in root and leaves. *Plant Cell Environ.* 17:1215–1221.

Tisdale, S. L., W. L. Nelson, and J. D. Beaton. 1985. *Soil Fertility and Fertilizers*, 4th edition. New York: Macmillan.

Tottingham, W. E. 1919. A preliminary study of the influence of chloride on the growth of certain agricultural plants. *J. Am. Soc. Agron.* 11:1–32.

Trolldenier, G. 1985. Effect of potassium chloride vs. potassium sulphate fertilization at different soil moisture on take-all of wheat. *Phytopathol. Z.* 112:56–62.

Tsukada, H., and A. Takeda. 2008. Concentration of chlorine in rice plant components. *J. Radioanal. Nucl. Chem.* 278:387–390.

Ulrich, A., and K. Ohki. 1956. Chlorine: Bromine and sodium as nutrients for sugar beet plants. *Plant Physiol.* 31:171–181.

Ussing, H. H. 1949. Transport of ions across cellular membranes. *Physiol. Rev.* 29:127–155.

Valencia, R., P. Chen, T. Ishibashi, and M. Conatser. 2008. A rapid and effective method for screening salt tolerance in soybean. *Crop Sci.* 48:1773–1779.

Von Uexkull, H. R. 1972. Response of coconuts to potassium chloride in the Philippines. *Oleagineux.* 27:13–19.

Wang, D. Q., B. C. Guo, and X. Y. Dong. 1989. Toxicity effects of chloride on crops. *Chin. J. Soil Sci.* 30:258–261.

Welch, R. M., W. H. Allaway, W. A. House, and J. Kubota. 1991. Geographic distribution of trace element problems. In *Micronutrients in Agriculture*, eds. J. J. Mortvedt, F. R. Cox, L. M. Shuman, and R. M. Welch, pp. 31–56. Madison, WI: SSSA.

Whitehead, D. C. 1985. Chlorine deficiency in red clover grown in solution culture. *J. Plant Nutr.* 8:193–198.

Williams, S. 1984. *Official Methods of Analysis of the Association of Official Analytical Chemists*. Arlington, VA: AOAC.

Williams, W. J. 1979. *Handbook of Anion Determination*. London: Butterworths.

Xu, G., H. Magen, J. Tarchitzky, and U. Kafkafi. 2000. Advances in chloride nutrition of plants. *Adv. Agron.* 68:97–150.

Yin, M. J., J. J. Sun, and C. S. Liu. 1989. Contents and distribution of chloride and effects of irrigation water of different chloride levels on crops. *Soil Fertil.* (in Chinese) 1:3–7.

Zhang, X. N., G. Y. Zhang, A. Z. Zhao, and T. R. Yu. 1989. Surface eletro-chemical properties of B horizon of a Rhodic Ferralsol, China. *Geordema.* 44:275–286.

Zhou, B. K., and X. Y. Zhang. 1992. Effects of chloride on growth and development of sugar-beet. *Soil Fertil.* (in Chinese) 3:41–43.

Zhu, Q. S., and B. S. Yu. 1991. Critical tolerance of chloride of rice and wheat on three types of soils. *Wubei Agric. Sci.* 5:22–26.

14 Nickel

14.1 INTRODUCTION

Nickel (Ni) was the last to be added to the list of essential nutrients for plant growth. Its importance for higher plants was established in the 1980s (Welch 1981; Eskew, Welch, and Cary 1983), using the soybean as a test plant. Although Ni deficiency in wheat, potatoes, and beans was observed in 1945 (Alloway 2008), its importance was not conclusively demonstrated until 1987 (Brown 2007). Micronutrient deficiency in soils is due to low natural levels. In addition, micronutrient deficiencies are due not only to low contents of these elements in soils but more often to their unavailability to growing plants (Brady and Weil 2002). Soil types and properties commonly associated with Ni deficiency are sandy texture, high soil pH (>7.0), high $CaCO_3$ content (>15%), and calcareous soils.

Nickel deficiency in rice is rarely reported in most growing regions. Several authors reviewed the micronutrient deficiency problems in Australia (Holloway, Graham, and Stacey 2008), India (Singh 2008), China (Zou et al. 2008), the Near East (Rashid and Ryan 2008), Africa (Waals and Laker 2008), Europe (Sinclair and Edwards 2008), South America (Fageria and Stone 2008), and the United States (Brown 2008), but none mentioned Ni deficiency problems in crop production. However, Alloway (2008) reported that the global incidence of micronutrient deficiencies is likely to increase due to the intensification of cropping systems. In addition, Fageria et al. (2002, 2012) reported that micronutrient deficiency in food crops is very common throughout the world due to low natural levels of micronutrients in the soils; the use of high-yielding cultivars; liming of acidic soils; interactions among macro- and micronutrients; sandy and calcareous soils; increased use of high analysis fertilizers with low amounts of micronutrients; and the decreased use of animal manures, composts, and crop residues.

Holloway, Graham, and Stacey (2008) reported that interactions between macro- and micronutrients have such a huge impact on the bioavailability of micronutrients that they are still among the least understood aspects of plant nutrition. Wood and Reilly (2007) reported that the incidence and severity of Ni deficiency in world agriculture will increase as cultural practices lead to greater metal accumulation due to excessive fertilizer usage or to the addition of excessive amounts of metals to potting mixes in an attempt to maximize growth. Zornoza, Robles, and Martin (1999) reported that simultaneous supplies of NO_3-N and NH_4-N reduced Ni toxicity in sunflower, whereas addition of Ni enhanced growth. Low Ni plants became nitrogen (N) deficient from lack of urease activity, with a high accumulation of urea but low tissue N (Gerendas and Sattelmacher 1997; Fageria, Baligar, and Clark 2002). In many cases, this deficiency will necessitate regular application of micronutrients, including Ni, either to the soil, foliage, or seed; ideally, these applications should be based on regular soil or plant tissue testing (Alloway 2008).

Principles for the use of micronutrients were well developed in the latter part of the twentieth century, and micronutrients contributed significantly to twentieth-century agriculture. In the same period, new information came to light that will challenge the agronomy of micronutrient use well into the twenty-first century (Graham 2008). The role of micronutrients, including Ni, is expected to increase in the twenty-first century in order to maintain cropping system sustainability and increase food production. Hence, information is required regarding the Ni cycle in soil–plant systems, functions, deficiency symptoms, uptake mechanisms, and management practices needed to improve the uptake and use efficiency of Ni in crop production. These aspects of Ni are discussed in this chapter.

14.2 CYCLE IN SOIL–PLANT SYSTEM

Nickel is added to soil–plant systems mainly through weathering of parent materials, liming acid soils, and the use of sewage sludge and municipality compost. Similarly, depletion from soil–plant systems involves soil erosion, sorption by soil colloids, immobilization by microbial mass, and uptake by plants. Nickel deficiency is rarely observed in crop plants. However, as a result of industrial and mining activities, Ni contamination of soil is a serious problem. Because Ni is highly toxic to plants and animals, its fate and mobility in soil are of great concern. High Ni content in soil accelerates its absorption by plants in favorable conditions (Mishra and Kar 1971). Nickel sorption of soil minerals can result in adsorbed (outer- and inner-sphere complexes) and precipitated phases (Scheidegger, Lamble, and Sparks 1997; Scheckel and Sparks 2001). The diethylene triamine pentaacetic acid (DTPA) extracting solution (Lindsay and Norvell 1978) consisting of 0.005 M DTPA with 0.01 M $CaCl_2$ and 0.1 M triethanolamine (TEA) at pH 7.3 is a good method for Ni determination in soil (Echevarria et al. 1998).

The sorption of Ni into soil surfaces controls the Ni distribution in soil and aquatic systems (Yamaguchi, Scheinost, and Sparks 2001). The concentration of Ni in soil averages 5–500 mg kg^{-1}, with a range up to 53,000 mg kg^{-1} in contaminated soil near metal refineries and in dried sludges (USEPA 1990). Agricultural soils contain approximately 3–1000 mg Ni kg^{-1} (World Health Organization 1991). Similarly, Daroub and Snyder (2007) reported that most soils contain less than 100 Ni mg kg^{-1}. Mengel et al. (2001) also reported that most soils usually contain less than 100 mg Ni kg^{-1}, well below the level at which Ni toxicity occurs. Soil pH is one of the most important properties that determines Ni uptake by plants. Such as with most other micronutrients, Ni uptake is higher in acidic soils or low pH soils and decreases with increasing soil pH (Alloway 2008). Hence, Ni toxicity can be alleviated by liming. The application of K can also reduce Ni toxicity, but P has the reverse effect.

14.3 FUNCTIONS

Brady and Weil (2002) and Daroub and Snyder (2007) reported that Ni is essential for urease, hydrogenases, and methyl reductase and for urea and ureide metabolism, in order to avoid toxic levels of these nitrogen fixation products in legumes. In addition,

Ni deficiency results in a variety of physiological effects in plants (Brown, Welch, and Madison 1990; Mengel et al. 2001). Nickel is part of the plant enzyme urease—the enzyme that catalyzes the degradation of urea to carbon dioxide and ammonia (Dixon et al. 1975). The reaction can be written as follows:

$$CO(NH_2)_2 + H_2O \Leftrightarrow 2NH_3 + CO_2 \tag{14.1}$$

Urease activity in garden peas was significantly increased with the addition of 100 μmol Ni^{2+} L^{-1} compared with a control treatment (Singh et al. 2004). The increase in urease activity in peas with the addition of Ni parallels the response of soybeans regarding the importance of Ni for expression of urease (Klucas, Hanus, and Russell 1983). The Ni is not required for the synthesis of the enzyme protein but is essential for the structure and functioning of the enzyme (Dixon, Blakeley, and Zerner 1980; Klucas, Hanus, and Russell 1983; Winkler et al. 1983; Marschner 1995). Perhaps the most striking effects of Ni deficiency have been described for cereals such as wheat, barley, and oats (Brown, Welch, and Madison 1990). In these species, Ni deficiency results in growth reduction, premature senescence, decreased tissue Fe levels, inhibited grain development, and grain inviability (Brown et al. 1987b).

Nickel is essential for plants supplied with urea. Nickel deficiency causes severe disruption in N metabolism as well as in other metabolic processes (Brown, Welch, and Madison 1990). Nickel-deficient plants accumulate toxic levels of urea in their leaf tips because of reduced urease activity (Daroub and Snyder 2007). Nickel stimulates proline biosynthesis in plants, which is responsible for osmotic balance in plant tissues (Salt et al. 1995; Singh et al. 2004). In studies with animals, Ni has also been found to replace Co (Underwood 1971). Nickel has been used as a systemic fungicide to control many plant diseases (Rowell 1968), and salts of Ni control cereal rusts, rice blast, and sheath spot disease as well as cotton wilt (Mishra and Kar 1971). In addition, the germination of seeds of several species was stimulated by treatment with Ni salts (Welch 1981). When the supply of Ni is inadequate, hydrogenase activity in soybean nodules diminishes (Klucas, Hanus, and Russell 1983; Dalton, Evans, and Hanus 1985). Nickel assists with the utilization of nitrogen translocated from roots to tops via guanidines or ureides that are subsequently used for anabolic reactions in growing tissues (Welch 1981).

14.4 DEFICIENCY SYMPTOMS

Nickel deficiency may be due to low levels of this element in the soil, or it may be caused by interaction with other elements. Nickel deficiency symptoms are characterized by marginal chlorosis of leaves, premature senescence, and diminished seed sets (Epstein and Bloom 2005). In pecan (*Carya illinoinensis*) leaves, the expression "mouse ear" is used to describe Ni deficiency symptoms (Malavolta and Moraes 2007). Brown et al. (1987a) observed chlorosis of young leaves, reduced leaf area, and less upright growth of leaves as symptoms of Ni deficiency in wheat, barley, and oats. Brown et al. (1987a, 1987b) also observed premature senescence in Ni-deficient oat plants. Nickel deficiency also depresses growth, inhibits grain development, and causes grain inviability, and it can reduce iron levels in plant tissues (Wood and

Reilly 2007). The necrosis of leaf tips appears to be a key defining characteristic of Ni deficiency in all higher plant families and is a reliable diagnostic trait for identifying this deficiency (Wood and Reilly 2007). In graminaceous species, Ni deficiency symptoms include chlorosis similar to that caused by iron deficiency, including interveinal chlorosis and patchy necrosis in the youngest leaves (Brown 2007).

If Ni accumulates in high concentrations in the soil, Ni toxicity is expected. Excess Ni causes several physiological disturbances, such as the yellowing of leaves or chlorosis. When Ni is present in toxic levels in plants, anthocyanin accumulation in leaves has been observed (Someya et al. 2007). The fundamental cause of chlorosis produced by excess Ni has been attributed to induced iron deficiency because application of iron salts to the chlorotic plants restores the green color (Mishra and Kar 1971). Mishra and Kar (1971) reported that chlorosis is severe at a Ni:Fe ratio value above 6 and usually negligible when the ratio value is below 1. Reduction in root growth has been commonly observed in plants subjected to heavy metal toxicity (Woolhouse 1983). Alteration of plasma membrane integrity, through lipid peroxidation, is another indication of heavy metal toxicity (Pandolfini, Gabbrielli, and Comparini 1992; Baccouch, Chaoui, and El Ferjani 2001).

14.5 UPTAKE IN PLANT TISSUE

The absorption rate of Ni is high in most crop species. It is absorbed in the form of divalent cations (Ni^{2+}) and may compete with other divalent cations such as Ca^{2+}, Mg^{2+}, Fe^{2+}, and Zn^{2+}; its concentration in plant tissue is in the range of 0.1–1.0 mg kg^{-1} dry weight (Marschner 1995). However, Brown (2007) reported that the Ni concentration in leaves of plants grown in uncontaminated soil ranges from 0.05 to 5.0 mg kg^{-1} dry weight. Brown (2007) and Gerendas et al. (1999) reported that the adequate range of Ni appears to fall between 0.01 and 10 mg kg^{-1} dry weight, which is an extremely wide range compared with that for other elements. Its critical toxic level may be >10 mg kg^{-1} in sensitive crop species and >50 mg kg^{-1} in tolerant species (Welch 1981; Bollard 1983; Marschner 1995). In tolerant crop species, Ni may be combined with organic acids, which may contribute to the crops' tolerance of high levels of Ni. However, other mechanisms may also be involved (Woolhouse 1983).

14.6 MANAGEMENT PRACTICES

Production potentials of many of the world's soils are decreased by low supplies of mineral nutrients from physical and chemical constraints of adverse soil (Fageria, Baligar, and Clark 2002). Major chemical (salinity, acidity, elemental deficiencies and toxicities, low organic matter) and physical (bulk density, hardpan layers, structure and texture, surface sealing and crusting, water holding capacity, water logging, drying, aeration) constraints affect the transformation (mineralization, immobilization), fixation (adsorption, precipitation), and leaching or surface runoff of indigenous and added fertilizer nutrients (Fageria, Baligar, and Clark 2002; Fageria and Stone 2008).

Strategies that can be adopted to improve micronutrient uptake, including Ni, in crops can be divided into two groups. In the first group, bioavailability of micronutrients can be improved by adopting practices such as the use of adequate rate, source, and methods of application. Fertilizer management practices (source, rate, method of placement, and application time) should be optimized based on soil, plant, and climatic factors that reduce mineral losses (leaching, runoff, fixation; Fageria, Baligar, and Clark 2002). Improvement and consideration of these factors will enhance recovery of added fertilizer minerals (Fageria, Baligar, and Clark 2002). In the second group, practices such as the use of crop species or genotypes within species that are efficient in micronutrient uptake and translocation of large part of it in the grain (Fageria 2009).

14.6.1 Effective Sources

Nickel is ubiquitous in soil, and most P fertilizers contain sufficient Ni for plant productivity, so Ni is not usually applied to soils (Fageria, Baligar, and Clark 2002). In addition, there are several sources of Ni that can be used to correct deficiencies of this nutrient, if required. Important sources of Ni are presented in Table 14.1.

14.6.2 Appropriate Methods and Timing of Application

In crop plants, soil and foliar application of Ni can alleviate a deficiency. If the Ni level in the soil is low, application can take place (along with the addition of N, P, and K fertilizers) when crops are planted. Nickel, applied as a foliar spray, can be absorbed by plant foliage through the cuticle (Chamel and Neumann 1987); hence, timely application of Ni is highly effective in correcting a deficiency (Wood and Reily 2007). Wood, Reilly, and Nyczepir (2004a) reported that a Ni deficiency is easily corrected by timely foliar application of Ni carriers. Wood, Reilly, and Nyczepir (2004b) suggested that one or two spray applications of Ni at a concentration of 10–100 mg L^{-1} during the early canopy expansion stage, or when sufficient leaf area is achieved, correct Ni deficiencies and ensure normal growth of plants.

14.6.3 Liming Acidic Soils

Nickel uptake increases with an increase in soil pH. Therefore, if the growth medium is deficient in Ni, liming can correct the problem. Fageria (2009) discussed liming versus Ni as methods of correcting deficiencies of this micronutrient.

TABLE 14.1
Nickel Carriers

Carrier	Formula	Ni Content (%)	Solubility in H_2O
Nickel chloride	$NiCl_2 \cdot 6H_2O$	25	Soluble
Nickel nitrate	$Ni(NO_3) \cdot 6H_2O$	20	Soluble
Nickel oxide	NiO	79	Insoluble

Source: Adapted from Fageria, N. K., V. C. Baligar, and R. B. Clark. 2002. Micronutrients in crop production. *Adv. Agron.* 77:185–268.

14.6.4 Use of Adequate Rate of Fertilizers

Most of the work on Ni is directed toward correcting its toxicity rather than addressing deficiencies because Ni deficiencies are rarely observed in crop plants. Brown (2007) reported that under all normal field conditions, it is unlikely that the application of Ni fertilizer will be required. Use of an adequate rate of N, K, Ca, Mg, and Mo in soils reduces the toxic effects of Ni on crop plants (Mishra and Kar 1971). Hunter and Vergnano (1952, 1953) reported that the application of N and K fertilizers could correct Ni toxicity in crop plants. Similarly, Crooke and Inkson (1955) found that in sand culture experiments with oats, the symptoms of Ni toxicity can be reduced by the application of N, K, Ca, or Mg to the culture solution. Mishra and Kar (1971) noted that application of Mo to the soil and foliage decreased the severity of chlorosis and other toxic effects of Ni. This corrective treatment for Ni toxicity resulted from the antagonistic effect of Mo on Ni. Phosphorus fertilization also tends to enhance the toxic effect of Ni (Halstead, Finn, and McLean 1969; Mishra and Kar 1971).

14.6.5 Use of Efficient Genotypes

It is well known that a spectrum of genetically controlled adaptations to soils low in available micronutrients exists in the germplasm of major staple crops, including rice (Graham 2008). Data are not available about rice genotype and uptake and use efficiency for Ni. Therefore, research is needed to identify differences in Ni uptake and utilization for rice genotypes.

14.7 CONCLUSIONS

Nickel was the last addition in the list of nutrients essential for higher plants. The major physiological functions of Ni are urea and ureide metabolism, N fixation, reproductive growth, iron absorption, and seed viability. Deficiency of Ni in crop plants, including rice, is rarely observed. However, Ni deficiency may occur in crop plants grown in soils with excessive heavy metals such as Ca, Mg, Mn, Fe, Cu, and Zn. Deficiency symptoms are dwarfed foliage with blunted apices and a necrotic margin at the leaf or leaflet tip. Nickel salts sprayed in an appropriate concentration (10–100 mg L^{-1}) can correct this deficiency and also can control cereal rusts and other plant diseases. However, Ni toxicity, in various crop species' studies, is frequently observed in greenhouse experiments. Increasing applications of sewage sludge to agricultural soils and the continuing release of industrial wastes cause a redistribution of heavy metals in the environment. Nickel toxicity is also commonly found in higher plants grown on serpentine soils. In recent years, soil and plant tests for heavy metals, including Ni, have made significant advances due in part to improvements in analytical procedures. However, plant and soil calibration data are still not sufficient for heavy metal deficiency or toxicity diagnostic techniques. Adequate Ni concentration in plant tissue is in the range of 0.1–5 mg kg^{-1} dry weight depending on crop species.

Micronutrients, including Ni, are vitally important for maintaining and increasing food production for a growing world population. Despite the sizable body of research undertaken in the twentieth century identifying the critical nature and the extent of serious micronutrient deficiency in various parts of the world, much remains to be done to understand precisely where and when Ni is needed and its availability in soil–plant systems, and to determine appropriate management strategies to improve Ni uptake and its efficient use in crop plants, including rice.

REFERENCES

Alloway, B. J. 2008. Micronutrients and crop production: An introduction. In *Micronutrient Deficiencies in Global Crop Production*, ed. B. J. Alloway, pp. 1–39. New York: Springer.

Baccouch, S., A. Chaoui, and E. El Ferjani. 2001. Nickel toxicity induces oxidative damage in *Zea mays* roots. *J. Plant Nutr.* 24:1085–1097.

Bollard, E. G. 1983. Involvement of unusual elements in plant growth and nutrition. In *Inorganic Plant Nutrition, Encyclopedia Plant Physiology*, New Series Vol. 15M, eds. A. Lauchli and R. L. Bieleski, pp. 695–744. New York: Springer Verlag.

Brady, N. C., and R. R. Weil. 2002. *The Nature and Properties of Soils*, 13th edition. Upper Saddle River, NJ: Prentice Hall.

Brown, P. H. 2007. Nickel. In *Handbook of Plant Nutrition*, eds. A. V. Barker and D. J. Pilbeam, pp. 395–410. Boca Raton, FL: CRC Press.

Brown, P. H. 2008. Micronutrient use in agriculture in the United States of America: Current practices, trends and constraints. In *Micronutrient Deficiencies in Global Crop Production*, ed. B. J. Alloway, pp. 267–286. New York: Springer.

Brown, P. H., R. M. Welch, and E. E. Cary. 1987a. Nickel: A micronutrient essential for higher plants. *Plant Physiol.* 85:801–803.

Brown, P. H., R. M. Welch, E. E. Cary, and R. T. Checkai. 1987b. Beneficial effects of nickel on plant growth. *J. Plant Nutr.* 10:2125–2135.

Brown, P. H., R. M. Welch, and J. T. Madison. 1990. Effect of nickel deficiency on soluble anion, amino acid, and nitrogen levels. *Plant Soil.* 125:19–27.

Chamel, A., and P. Neumann. 1987. Foliar absorption of nickel: Determination of its cuticular behavior using isolated cuticles. *J. Plant Nutr.* 10:99–111.

Crooke, W. M., and R. H. E. Inkson. 1955. Relation between nickel toxicity and major nutrient supply. *Plant Soil.* 6:1–15.

Dalton, D. A., H. J. Evans, and F. J. Hanus. 1985. Stimulation by nickel of soil microbial urease activity and urease and hydrogenase activities in soybeans grown in a low nickel soil. *Plant Soil.* 88:245–285.

Daroub, S. M., and G. H. Snyder. 2007. The chemistry of plant nutrients in soil. In *Mineral Nutrition and Plant Disease*, ed. L. E. Datnoff, W. H. Elmer, and D. M. Huber, pp. 1–7. St. Paul, MN: American Phytopathological Society.

Dixon, N. E., R. L. Blakeley, and B. Zerner. 1980. Jack bean urease (EC 3.5.1.5). III. The involvement of active-site nickel ion in inhibition by β-mercaptoethanol, phosphoramidate, and fluoride. *Can. J. Biochem.* 58:481–488.

Dixon, N. E., C. Gazzola, R. L. Blakeley, and B. Zerner. 1975. Jack bean urease (EC 3.5.1.5). A metalloenzyme. A simple biological role for nickel? *J. Am. Chem. Soc.* 97:4131–4133.

Echevarria, G., J. L. Morel, J. C. Fardeau, and E. Leclerc-Cessac. 1998. Assessment of phytoavailability of nickel in soils. *J. Environ. Q.* 27:1064–1070.

Epstein, E., and A. J. Bloom. 2005. *Mineral Nutrition of Plants: Principles and Perspectives*, 2nd edition. Sunderland, MA: Sinauer Associates.

Eskew, D. L., R. M. Welch, and E. E. Cary. 1983. Nickel: An essential micronutrient for legumes and possibly higher plants. *Science.* 222:621–623.

Fageria, N. K. 2009. *The Use of Nutrients in Crop Plants*. Boca Raton, FL: CRC Press.

Fageria, N. K., V. C. Baligar, and R. B. Clark. 2002. Micronutrients in crop production. *Adv. Agron.* 77:185–268.

Fageria, N. K., M. F. Moraes, E. P. B. Ferreira, and A. M. Knupp. 2012. Biofortification of trace elements in food crops for human health. *Commun. Soil Sci. Plant Anal.* 43:556–570.
Fageria, N. K., and L. F. Stone. 2008. Micronutrient deficiency problems in South America. In *Micronutrient Deficiencies in Global Crop Production*, ed. B. J. Alloway, pp. 245–266. New York: Springer.
Gerendas, J., J. C. Polacco, S. K. Freyermuth, and B. Sattelmachr. 1999. Significance of nickel for plant growth and metabolism. *J. Plant Nutr. Soil Sci.* 162:241–256.
Gerendas, J., and B. Sattelmacher. 1997. Significance of Ni supply for growth, urease activity, and the concentration of urea, amino acids, and mineral nutrients of urea-grown plants. *Plant Soil.* 190:153–162.
Graham, R. D. 2008. Micronutrient deficiencies in crops and their global significance. In *Micronutrient Deficiencies in Global Crop Production*, ed. B. J. Alloway, pp. 41–61. New York: Springer.
Halstead, R. L., B. J. Finn, and A. J. McLean. 1969. Extractability of nickel added to the soils and its concentration in plants. *Can J. Soil Sci.* 49:335–342.
Holloway, R. E., R. D. Graham, and S. P. Stacey. 2008. Micronutrient deficiencies in Australian field crops. In *Micronutrient Deficiencies in Global Crop Production*, ed. B. J. Alloway, pp. 63–92. New York: Springer.
Hunter, J. G., and O. Vergnano. 1952. Nickel toxicity in plants. *Ann. Appl. Biol.* 39:279–281.
Hunter, J. G., and O. Vergnano. 1953. Trace element toxicities in oat plants. *Ann. Appl. Biol.* 40:761–777.
Klucas, R. V., F. J. Hanus, and S. A. Russell. 1983. Nickel. A micronutrient element for hydrogen-dependent growth of *Rhizobium japonicum* and for expression of urease activity in soybean leaves. *Proc. Nat. Acad. Sci. USA.* 80:2253–2257.
Lindsay, W. L., and W. A. Norvell. 1978. Development of a DTPA soil test for zinc, iron, manganese and copper. *Soil Sci. Soc. Am.* 42:421–428.
Malavolta, E., and M. F. Moraes. 2007. Nickel-from toxic to essential nutrient. *Better Crops.* 91:26–27.
Marschner, H. 1995. *Mineral Nutrition of Higher Plants*, 2nd edition. New York: Academic Press.
Mengel, K., E. A. Kirkby, H. Kosegarten, and T. Appel. 2001. *Principles of Plant Nutrition*, 5th edition. Dordrecht, The Netherlands: Kluwer Academic.
Mishra, D., and M. Kar. 1971. Nickel in plant growth and metabolism. *Bot. Rev.* 40:395–452.
Pandolfini, T., R. Gabbrielli, and C. Comparini. 1992. Nickel toxicity and peroxidase activity in seedlings of *Triticum aestivum* L. *Plant Cell Environ.* 15:719–725.
Rashid, A., and J. Ryan. 2008. Micronutrient constraints to crop production in the near East. In *Micronutrient Deficiencies in Global Crop Production*, ed. B. J. Alloway, pp. 149–180. New York: Springer.
Rowell, J. B. 1968. Chemical control of the cereal rusts. *Annu. Rev. Phytopathol.* 54:999–1008.
Salt, D. E., M. Blaylock, P. B. A. Kumar Nanda, V. Dushenkov, B. O. Ensley, L. Chet, and I. Raskin. 1995. I. Phytoremediation: A novel strategy for removal of toxic metals from the environment using plants. *Biotechnology.* 13:468–478.
Scheckel, K. G., and D. L. Sparks. 2001. Dissolution kinetics of nickel surface precipitates on clay mineral and oxide surfaces. *Soil Sci. Soc. Am. J.* 65:685–694.
Scheidegger, A. M., G. M. Lamble, and D. L. Sparks. 1997. Spectroscopic evidence for the formation of mixed cation hydroxide phases upon metal sorption on clays and aluminum oxides. *J. Colloid Interface Sci.* 186:118–128.
Sinclair, A. H., and A. C. Edwards. 2008. Micronutrient deficiency problems in agricultural crops in Europe. In *Micronutrient Deficiencies in Global Crop Production*, ed. B. J. Alloway, pp. 225–266. New York: Springer.
Singh, M. V. 2008. Micronutrient deficiencies in crop and soils in India. In *Micronutrient Deficiencies in Global Crop Production*, ed. B. J. Alloway, pp. 93–123. New York: Springer.

Singh, S., A. M. Kayastha, R. K. Asthana, and S. P. Singh. 2004. Response of garden pea to nickel toxicity. *J. Plant Nutr.* 27:1543–1560.

Someya, N., Y. Sato, I. Yamaguchi, H. Hamamoto, Y. Ichiman, K. Akutsu, H. Sawada, et al. 2007. Alleviation of nickel toxicity in plants by a rhizobacterium strain is not dependent on its siderophore production. *Commun. Soil Sci. Plant Anal.* 38:1155–1162.

Underwood, E. J. 1971. *Trace Elements in Human and Animal Nutrition.* New York: Academic Press.

USEPA (United States Environmental Protection Agency). 1990. *Project Summary Health Assessment Document for Nickel.* EPA/600/S8-83/012. Washington, DC: Office of Health and Environmental Assessment.

Waals, J. H. V., and M. C. Laker. 2008. Micronutrient deficiencies in crops in Africa with emphasis on southern Africa. In *Micronutrient Deficiencies in Global Crop Production*, ed. B. J. Alloway, pp. 201–224. New York: Springer.

Welch, R. M. 1981. The biological significance of nickel. *J. Plant Nutr.* 3:345–356.

Winkler, R. G., J. C. Polacco, D. L. Eskew, and R. M. Welch. 1983. Nickel is not required for apo-urease synthesis in soybean seeds. *Plant Physiol.* 72:262–263.

Wood, B. W., and C. C. Reilly. 2007. Nickel and plant disease. In *Mineral Nutritional and Plant Disease*, eds. L. E. Datnoff, W. H. Elmer, and D. M. Huber, pp. 215–231. St. Paul, MN: American Phytopathological Society.

Wood, B. W., C. C. Reilly, and A. P. Nyczepir. 2004a. Mouse-ear of pecan: II. Influence of nutrient applications. *HortScience.* 39:95–100.

Wood, B. W., C. C. Reilly, and A. P. Nyczepir. 2004b. Mouse-ear of pecan: A nickel deficiency. *HortScience.* 39:1238–1242.

Woolhouse, H. W. 1983. Toxicity and tolerance in response of plants to metals. In *Encyclopedia of Plant Physiology: Physiological Plant Ecology* III, 12C, eds. O. L. Lange, P. S. Nobel, C. B. Osmond, and H. Ziegler, pp. 245–300. New York: Springer Verlag.

World Health Organization. 1991. *International Program on Chemical Safety. Environmental Health Criteria 108: Nickel.* Geneva, Switzerland: WHO.

Yamaguchi, N. U., A. C. Scheinost, and D. L. Sparks. 2001. Surface-induced nickel hydroxide precipitation in the presence of citrate and salicylate. *Soil Sci. Soc. Am. J.* 65:729–736.

Zornoza, P., S. Robles, and N. Martin. 1999. Alleviating of nickel toxicity by ammonium supply to sunflower plants. *Plant Soil.* 208:221–226.

Zou, C., X. Gao, R. Shi, X. Fan, and F. Zhang. 2008. Micronutrient deficiencies in crop production in China. In *Micronutrient Deficiencies in Global Crop Production*, ed. B. J. Alloway, pp. 127–148. New York: Springer.

15 Silicon

15.1 INTRODUCTION

Plant nutrients are divided into two groups known as essential and beneficial elements. The essential nutrients required by plants to complete their normal life cycles include carbon (C), hydrogen (H), oxygen (O), nitrogen (N), phosphorus (P), potassium (K), calcium (Ca), magnesium (Mg), sulfur (S), iron (Fe), manganese (Mn), zinc (Zn), copper (Cu), boron (B), molybdenum (Mo), chlorine (Cl), and nickel (Ni) (Fageria, Baligar, and Jones 2011). Beneficial elements are required by some plants to enhance their growth. These elements are not considered essential according to the definition established by Arnon and Stout (1939). The beneficial elements are silicon (Si), cobalt (Co), aluminum (Al), selenium (Se), sodium (Na), and vanadium (V). The role of these essential nutrients in crop production/rice production has been discussed in the previous chapters in this volume. Of the beneficial elements, silicon has more (or specific) benefits for rice production compared with other beneficial elements.

In scientific literature, silicon is also referred to as silica (SiO_2). It is second most abundant element of the earth's crust, and soil contains approximately 32% silicon by weight (Lindsay 1979). Silicon is considered to be a "quasi-essential" element for most living organisms (Hirota et al. 2010). It acts as a component of the outer skeleton of diatomaceous protozoans (Azam, Hemmingsen, and Volcani 1974), as a trace element to help animal bone and tooth development (Chumlea 2007), and it enhances plant tissue strength and disease resistance (Snyder, Matichenkov, and Datnoff 2007). The beneficial effects of silicon on the growth of sugar beets were reported as early as 1840 by Justius von Leibig, who used sodium silicate as a silicon fertilizer. A field study at the Rothamsted experimental station (UK), which began in 1956 and continues today, has demonstrated sodium silicate's marked effect on grass productivity (Snyder, Matichenkov, and Datnoff 2007). Since 1840, the beneficial effects of silicon application for the growth of rice, wheat, barley, corn, and sugarcane have been studied in greenhouse and field experiments (Snyder, Matichenkov, and Datnoff 2007). Silicon deficiencies in crop plants have been reported in Australia, the United States, Sri Lanka, India, Japan, Puerto Rico, and South Africa (Tisdale, Nelson, and Beaton 1985). The objective of this chapter is to present an overview of silicon as a nutrient producing higher rice yields.

15.2 CYCLE IN SOIL–PLANT SYSTEM

Knowledge of the silicon cycle in the soil–plant system is very important for understanding its uptake behavior by plants and for adopting appropriate management practices to improve its efficiency. Figure 15.1 depicts the main silicon components in the soil–plant system of rice. In a usual pH range of soils, $H_4SiO_4^0$ or $Si(OH)_4$

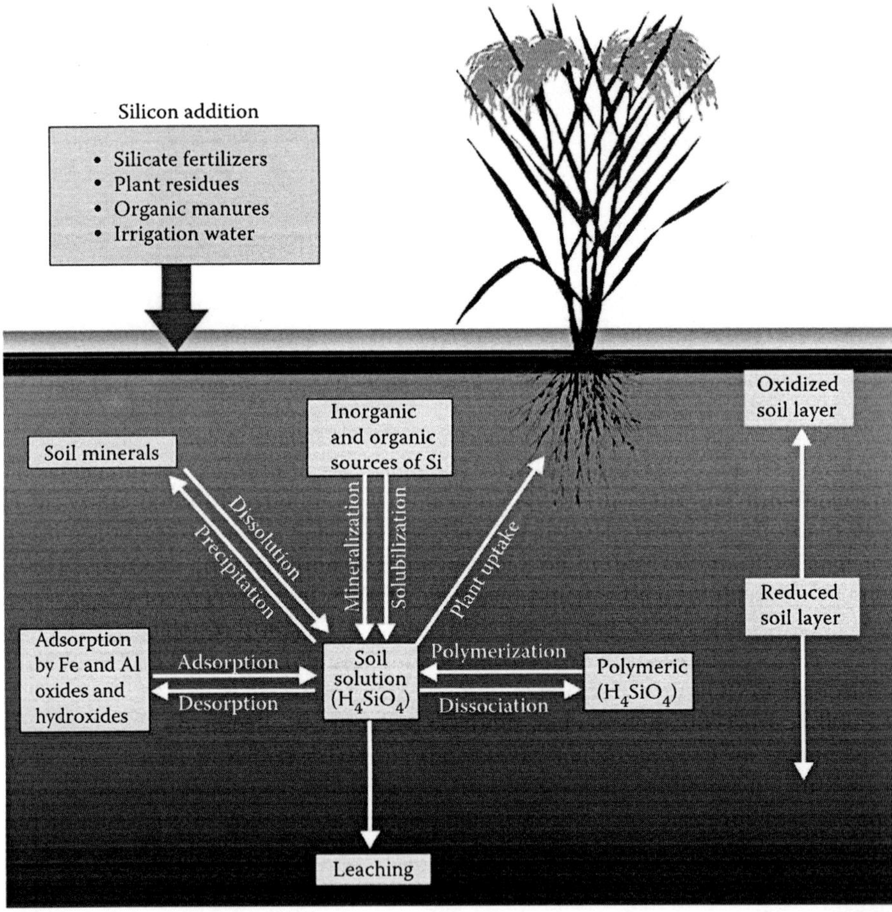

FIGURE 15.1 Silicon cycle in soil–plant system of rice.

is the principal silicate species in soil solution (Tisdale, Nelson, and Beaton 1985). The concentration of Si in the soil solution varies from 3.0 to 56 mg kg^{-1}, which is in the same order of magnitude as potassium, calcium, and other micronutrients (Epstein 1991; Bogdan and Schenk 2008; Fleck et al. 2011). However, Hallmark et al. (1982) reported a concentration of Si in soil solution in the 1–40 mg L^{-1} range and that it seems to be controlled more by chemical kinetics than by thermodynamics. Similarly, Tisdale, Nelson, and Beaton (1985) found that Si levels in solution from 3 to 37 mg kg^{-1} have been reported for a wide range of normal soils.

Silicon is absorbed by plants in the form of monosilicic (H_4SiO_4) or orthosilicic acid, which is chemically very active (Daroub and Snyder 2007). In addition to monosilicic acid, polymerized forms (polysilicic acids) are also an integral part of soil solution. Polysilicic acids have a higher Si concentration (>65 mg L^{-1}) in soil solution. Major sources of silicon include primary silicate minerals, secondary aluminosilicates, and several forms of silica (SiO_2). Silica occurs as six distinct minerals—quartz, tridymite, cristobalite, coesite, stishovite, and opal. The first three of these minerals plus opal comprise the major weight and volume percentage of most soils. Quartz is overwhelmingly the most common mineral species in most soils and can often make up 90–95% of all sand and silt fractions (Tisdale, Nelson, and Beaton 1985).

The Si sorption capacity of soils increases with greater soil pH. Therefore, Si availability for plants decreases with increasing soil pH. However, Mengel et al. (2001) and Drees et al. (1989) reported that the concentration of soluble Si, mainly $Si(OH)_4$, remains at a constant level over a broad pH range (2–9) until deprotonation begins at pH 9 with a steep solubility increase in the form of $Si(OH)_3O$. Acidic soils tend to contain higher Si concentrations in the form of nondissociated silicic acid (H_4SiO_4). In the soil solution, silicon's concentration is in the range of 100–600 mmol m^{-3} and generally is higher than that of phosphate (Epstein 1999). Liming has been found to decrease Si availability and uptake by a number of crop plants (Mengel et al. 2001).

Monosilicic acid can combine with Fe, Mn, and Al ions in soil solution. The higher the ratios of Si/Al or Si/Fe, the greater the Si uptake by rice (Jones and Handreck 1967). The anion of monosilicic acid [$Si(OH)_3$] can replace P anions ($H_2PO_4^-$ and HPO_4^{2-}) from Ca, Mg, Al, and Fe phosphates, making P more readily available to plants (Daroub and Snyder 2007). Therefore, beneficial effects of Si in P uptake can be found in acidic as well as in alkaline soils. This is because P is mainly fixed by Al and Fe oxides at lower pH (acid range < 7.0); at higher pH (alkaline range > 7.0), it is mainly fixed by Ca and Mg carbonates. Silicon content increases in flooded soils, possibly due to an increase of organic acids that dissolve silica under the reducing conditions that are characteristic of submerged soils (Datnoff, Rodrigues, and Seebold 2007). Ponnamperuma (1965) also reported that the concentration of silicon in the soil solution generally increases with flooding and that soils with a high organic matter content have the greatest increases.

Addition of P to the soil improves Si content. This may be due to a release of Si by P ions on the clay minerals. This reaction was also found to be reversible by the addition of neutral silicate (Reifenberg and Buckwold 1954). This is believed to be the underlying mechanism for the enhanced availability of P in soils after the application of silicates. Despite the large amount of Si present in mineral soils, its deficiency can occur due to uptake by some crops, such as rice. Rice can uptake Si at rates of roughly 230–470 kg ha^{-1}; as a result of intensive cropping, Si can be removed from the soil solution faster than it can be replenished naturally (Savant, Datnoff, and Snyder 1997; Savant, Snyder, and Datnoff 1997; Datnoff, Rodrigues, and Seebold 2007). Silicon deficiency occurs more often in Oxisols and Ultisols, which are used for the cultivation of upland rice in Asia, Africa, and Latin America, than in other soil orders (Datnoff, Rodrigues, and Seebold 2007). Heavy rainfall in regions where these two types of soils occur can cause high degrees of weathering, leaching, acidification, and desilification (Singer and Munns 1987; Savant, Datnoff, and Snyder 1997; Savant, Snyder, and Datnoff 1997). In addition, Histosols can also be deficient in Si because of their high organic matter content and low mineral content (Singer and Munns 1987).

The Si solubility of soil is low. Knowledge of Si's equilibrium reaction (dissolution and sorption) is important in order to understand its availability to plants. Equilibrium reactions of various silicate minerals and silicate ions are presented in the following equations (Lindsay 1979).

15.2.1 Reaction of Silicate Minerals with Water

$$SiO_2 \text{ (silicate glass)} + 2H_2O \leftrightarrow H_4SiO_4^0 \quad (15.1)$$

$$SiO_2 \text{ (amorp)} + 2H_2O \leftrightarrow H_4SiO_4^0 \quad (15.2)$$

$$SiO_2 \text{ (coesite)} + 2H_2O \leftrightarrow H_4SiO_4^0 \quad (15.3)$$

$$SiO_2 \text{ (soil)} + 2H_2O \leftrightarrow H_4SiO_4^0 \quad (15.4)$$

$$SiO_2 \text{ (tridymite)} + 2H_2O \leftrightarrow H_4SiO_4^0 \quad (15.5)$$

$$SiO_2 \text{ (cristobalite)} + 2H_2O \leftrightarrow H_4SiO_4^0 \quad (15.6)$$

$$SiO_2 \text{ (quartz)} + 2H_2O \leftrightarrow H_4SiO_4^0 \quad (15.7)$$

15.2.2 Silicate Ions

$$H_4SiO_4^0 \leftrightarrow H_3SiO_4^- + H^+ \quad (15.8)$$

$$H_4SiO_4^0 \leftrightarrow H_2SiO_4^- + 2H^+ \quad (15.9)$$

$$H_4SiO_4^0 \leftrightarrow HSiO_4^- + 3H^+ \quad (15.10)$$

$$H_4SiO_4^0 \leftrightarrow SiO_4^{4-} + 4H^+ \quad (15.11)$$

$$4H_4SiO_4^0 \leftrightarrow H_6Si_4O_{12}^{2-} + 2H^+ + 4H_2O \quad (15.12)$$

15.3 FUNCTIONS

Silica participates in many physiological and biochemical functions that are responsible for improving growth and development and, consequently, higher rice yields. Silica improves the rice plant's water economy. Because the rice plant deposits most of its absorbed silica on the leaf blade surface, the cuticular transpiration is significantly reduced. Takahashi (1995) reported that with the addition of silica, rice production efficiency (dry matter production per unit of water uptake) was enhanced by 30% with straw and by 450% with unhulled grain. Silica reduces the drooping of leaf blades, which is responsible for higher solar radiation use and higher photosynthetic capacity of plants. Yoshida (1981) reported that the maintenance of erect leaves as the result of silicate application can easily account for a 10% increase in the photosynthesis of the canopy and, consequently, a similar increase in yield. Silica increases the leaf area index, which is responsible for higher photosynthetic activity in rice plants. Silica also promotes oxidation power in plants and increases the ammonium tolerance of rice plants. When Si is absorbed in adequate amounts, plant lodging is reduced when a higher rate of N is applied (Takahashi 1995). Increased silicon absorption decreases transpiration losses, perhaps through the cuticle, and increases the plants' tolerance for decreased osmotic potential in the growth medium (Yoshida 1981).

In addition to these functions, silica is reported to increase resistance in rice to fungus diseases, especially blast (*Magnaporthe grisea*). Blast is a serious disease in South American rice (Prabhu et al. 2006). The blast resistance mechanism in rice associated with silica deposition in the epidermal tissue mechanically protects hyphae invasion; silica physiologically promotes ammonium assimilation and restrains the increase in soluble nitrogen compounds, including amino acids and amide, which are instrumental for the propagation of hyphae (Takahashi 1995). Data in Table 15.1 show that many crop plant diseases (including those of rice) can be controlled with the addition of Si. In rice, silicon can positively affect the activity of some enzymes involved in photosynthesis and can reduce leaf senescence (Kang 1980; Epstein 1991, 1999; Datnoff, Rodrigues, and Seebold 2007). Silicon can reduce uptake of Fe^{2+} ions in flooded rice, thereby avoiding iron toxicity, which is a problem in many flooded-rice-producing areas worldwide (Fageria et al. 2008). Okuda and Takahashi (1961) reported that the addition of silicon to nutrient solution increases the oxidation power of rice roots; this leads to an oxidation of Fe^{2+}

TABLE 15.1
Influence of Silicon on Plant Diseases of Principal Food Crops

Crop Species	Disease	Pathogen	Effect of Si
Barley	Powdery mildew	*Erysiphe graminis* f.sp. hordei	Decreases
	Black point	*Alternaria spp.*	Decreases
Corn	Stalk rot	*Pythium aphanidermatum, Fusarium moniliforme*	Decreases
Rice	Blast	*Magnaporthe grisea*	Decreases
	Brown spot	*Cochliobolus miyabeanus*	Decreases
	Sheath blight	*Thanatephorus cucumeris*	Decreases
	Leaf scald	*Monographella albescens*	Decreases
	Stem rot	*Magnaporthe salvinii*	Decreases
	Grain discoloration	Many fungal species	Decreases
Sorghum	Anthracnose	*Colletotrichum graminicola*	Decreases
Sugarcane	Rust	*Puccinia melanocephala*	No effect
Wheat	Powdery mildew	*Blumeria graminis*	Decreases
	Brown rust	*Puccinia recondita*	No effect
	Foot rot	*Fusarium spp.*	Decreases
	Leaf spot	*Phaeosphaeria nodorum*	Decreases
Pea	Leaf spot	*Mycosphaerella pinodes*	Decreases
Soybean	Stem canker	*Diaporthe phaseolorum*	Decreases

Source: Compiled from Datnoff, L. E., F. A. Rodrigues, and K. W. Seebold. 2007. Silicon and plant disease. In *Mineral Nutrition and Plant Disease*, eds. L. E. Datnoff, W. H. Elmer, and D. M. Huber, pp. 233–246. St. Paul, MN: American Phytopathological Society.

and Mn^{2+}, subsequent precipitation on the root surface, and a reduction in rice plant uptake of Fe and Mn. In a recent study, however, Fleck et al. (2011) reported that Si nutrition of rice plants reduced the oxidation power of roots and enhanced the development of Casparian bands in the exodermis and endodermis, as well as lignin depositions in the sclerenchyma. These changes are probably the reason for reduced radial oxygen loss (the diffusion of oxygen from the root to the anaerobic rhizosphere) and might be useful for those plants growing in anaerobic soils and coping with unfavorable conditions. Datnoff, Rodrigues, and Seebold (2007) reported that the most significant effect of Si, besides improving plant fitness and increasing agricultural productivity, is that it restricts grazing and parasitism.

15.4 DEFICIENCY SYMPTOMS

Silicon is easily translocated in plant parts; therefore, deficiency symptoms first appear in the older leaves. Silicon deficiency produces minute circular white leaf spots (freckles) that are more severe on older leaves (Gascho, Anderson, and Bowen 1993). In rice, typical symptoms of Si deficiency are necrosis of the older leaves and wilting associated with a higher rate of transpiration (Mitsui and Takatoh 1963). Silicon deficiency reduces tillering in rice. Symptoms of Si deficiency in rice plants are soft, droopy leaves; increased lodging; reduction in grain yield; and increased incidence of brown spot (*Helminthosporium oryzae*; Wells et al. 1993). Yamauchi and Winslow (1989) reported that the fungal infestation and characteristic coloration of the grains of upland rice could be cured by silica application. Rice responds to Si and often exhibits deficiency symptoms when the Si concentration in the leaf or stem drops below 5% (Wells et al. 1993). Park (1975) reported that silicon content in straw has a quadratic association with grain yield of rice. Maximum grain yield was obtained with a SiO_2 content of about 130 mg Si g^{-1} of straw dry weight.

15.5 UPTAKE IN PLANT TISSUE

Silicon is absorbed by plants as monosilicic acid or its anion, orthosilicic acid [$H_4SiO_4 = Si(OH)_4$] (Snyder, Matichenkov, and Datnoff 2007). The uptake of Si by plants may be passive or active. Jones and Handreck (1967) reported that uptake was passive after observing that values for the plants' uptake agreed closely with calculated values derived from data of soil solution concentration and water uptake by transpiration. Okuda and Takahashi (1965) found that the concentration of Si in the sap of the rice plant can be several hundred-fold greater than in the outer solution. Therefore, rice plant uptake of silicon is against the concentration gradient, indicating that energy is involved in the uptake process. Barber and Shone (1966) also reported that Si uptake by barley roots required metabolic energy and was sensitive to metabolic inhibitors and variations in temperature.

Tamai and Ma (2003) reported that Si uptake by rice roots is a transporter-mediated process, with a proteinaceous transporter demonstrating a low affinity for silicic acid and a peptide consisting entirely of cysteines. After root absorption, monosilicic acid is translocated rapidly into the plant's leaves in the transpiration stream (Ma 2003). Silicon is concentrated in the epidermal tissue as a fine layer of silicon-cellulose membrane and is associated with pectin and calcium ions (Snyder, Matichenkov, and Datnoff 2007). By this means, the double-cuticular layer can protect and mechanically strengthen plant structure (Snyder, Matichenkov, and Datnoff 2007).

Silicon content in plant tissues can vary from 1% to more than 10% depending on crop plant species (Epstein 1991; Datnoff, Rodrigues, and Seebold 2007). Plant species may be divided into Si accumulators and non accumulators. They are considered Si accumulators when the concentration of Si is greater than 1% of dry weight (Epstein 1999). Wetland species such as rice, horsetails (*Equisetum arvense*), and members of the Pinaceae (pine family of conifers), all of which contain 50–75 mg Si g^{-1} dry matter, are accumulator species (Mengel et al. 2001). Dicots, such as tomatoes and soybeans, with less than 0.1% Si in their biomass, are relatively poor accumulators of Si, compared with monocots (Datnoff, Rodrigues, and Seebold 2007). Dry land grasses, such as wheat, oat, rye, barley, sorghum, corn, and sugarcane, contain about 1% Si in their biomass, yet aquatic grasses contain up to 5% (Jones and Handreck 1967; Epstein 1991; Datnoff, Rodrigues, and Seebold 2007). Lowland rice usually contains between 4.6 and 7.0% silicon in the straw (Tisdale, Nelson, and Beaton 1985). Most Si accumulated by rice is deposited in leaves (71%), followed by hulls (13%), roots (10%), and culms (6%) (Chen 1990; Epstein 1991).

Ma, Miyake, and Takahashi (2001) reported that Si concentration in rice plants is 6.6%, in wheat 3.8%, in corn 2.2%, in dry beans 1.2%, in tomatoes 0.1%, and in morning glories 0.1%. The amount of silicon absorbed by a crop species depends on its productivity. On average, the largest amounts of silicon are absorbed by sugarcane (300–700 kg Si ha^{-1}), rice (150–300 kg Si ha^{-1}), and wheat (50–150 kg Si ha^{-1}) (Snyder, Matichenkov, and Datnoff 2007). Savant, Snyder, and Datnoff (1997) reported that a rice crop producing 5 Mg ha^{-1} grain yield accumulates about 250 Si kg ha^{-1}, 125 kg K ha^{-1}, 100 kg N ha^{-1}, and 12 kg P ha^{-1}. Nair, Mishra, and Patnaik (1982) reported that in India, 12 rice cultivars (90–140 days duration) grown on an Inceptisol removed 205–611 kg Si ha^{-1} when grain yields ranged from 4.6 to 8.4 Mg ha^{-1}. These values of Si accumulation are equal to macronutrient accumulation such as that of N and K.

Prakash et al. (2011) determined Si concentration and uptake in the grain and straw of lowland rice under different N–P–K, Si, and pesticide (insecticide + fungicide) spray treatments (Table 15.2). They concluded that application of silicic acid at 4 ml L^{-1} along with a half dose of the recommended pesticide as a foliar spray increased concentration and uptake of Si in the grain and straw compared with a control treatment. Overall, concentration of Si in the straw was about three times higher compared with grain. Similarly, uptake of Si in the straw was 4.2 times higher compared with the grain across the seven treatments. Eighty-one percent of the Si absorbed (grain + straw) was retained in the straw and only 19% was translocated to the grain.

15.6 USE EFFICIENCY

Data related to silicon use efficiency in rice are limited. However, this author calculated silicon use efficiency (SUE) in lowland rice grain and straw from the data of Prakash et al. (2011; Table 15.3). In rice grain, SUE varied from 83 to 111 kg grain

TABLE 15.2
Concentration and Uptake of Silicon in the Grain and Straw of Lowland Rice as Influenced by Foliar Spray of Silicic Acid (SA) Treatments

	Si Concentration (%)		Si Uptake (kg ha^{-1})	
Treatment	**Grain**	**Straw**	**Grain**	**Straw**
T_1 – Recommended N–P–K	0.9	2.6	45.5	188.8
$T_2 - T_1$ + 2 ml L^{-1} SA	1.0	3.0	59.3	248.8
$T_3 - T_1$ + 4 ml L^{-1} SA	1.0	3.2	63.8	285.7
$T_4 - T_1$ + 8 ml L^{-1} SA	0.9	2.9	49.3	214.4
$T_5 - T_2$ + ½ dose pesticide	1.1	3.2	66.2	280.3
$T_6 - T_3$ + ½ dose pesticide	1.2	3.3	80.1	320.0
$T_7 - T_4$ + ½ dose pesticide	0.9	3.1	48.8	227.1
Average	1.0	2.9	59.0	252.2
SEM±	0.1	0.1	4.5	4.8
CD (5%)	0.2	0.2	13.9	14.9

Source: Adapted from Prakash, N. B., N. Chandrashekar, C. Mahendra, S. U. Patil, G. N. Thippesshappa, and H. M. Laane. 2011. Effect of foliar spray of soluble silicic acid on growth and yield parameters of wetland rice in hilly and coastal zone soils of Karnataka, South India. *J. Plant Nutr.* 34:1883–1893.

TABLE 15.3
Silicon Use Efficiency (SUE) in the Grain and Straw of Lowland Rice as Influenced by Foliar Spray of Silicic Acid (SA) Treatments

Treatment	**SUE in Grain (kg kg^{-1})**	**SUE in Straw (kg kg^{-1})**
T_1 – Recommended N–P–K	111	38
$T_2 - T_1$ + 2 ml L^{-1} SA	100	33
$T_3 - T_1$ + 4 ml L^{-1} SA	100	31
$T_4 - T_1$ + 8 ml L^{-1} SA	111	34
$T_5 - T_2$ + ½ dose pesticide	91	31
$T_6 - T_3$ + ½ dose pesticide	83	30
$T_7 - T_4$ + ½ dose pesticide	111	32
Average	101	33

Source: Values of SUE in grain and straw were calculated from the data of Prakash, N. B., N. Chandrashekar, C. Mahendra, S. U. Patil, G. N. Thippesshappa, and H. M. Laane. 2011. Effect of foliar spray of soluble silicic acid on growth and yield parameters of wetland rice in hilly and coastal zone soils of Karnataka, South India. *J. Plant Nutr.* 34:1883–1893.

Note: $\text{SUE (kg kg}^{-1}) = \dfrac{\text{Grain or straw yield in kg}}{\text{Uptake of Si in grain or straw in kg}}.$

produced per kg Si uptake, with an average value of 101 kg grain per kg Si uptake. In the straw, the SUE varied from 30 to 38 kg dry matter produced per kg Si uptake, with an average value of 33 kg straw produced per kg Si uptake. Overall, the SUE in the straw was about three times lower than that of grain. The lower SUE in the straw was associated with a higher straw Si uptake than grain uptake.

15.7 SILICON HARVEST INDEX

The silicon harvest index (SHI) is an important tool for determining the amount of Si in grain and straw (Table 15.4). SHI varied from 0.17 to 0.20 with an average value of 0.18. Therefore, it can be concluded that about 82% of accumulated Si was retained in straw and only about 18% was translocated to grain. Incorporation of straw after the rice crop harvest is an important strategy for improving or maintaining the soil's Si level.

15.8 MANAGEMENT PRACTICES

Silicon is considered a quasi-essential nutrient for rice growth, and maintaining an adequate level of this element in the soil is important for improving rice yield. In order to achieve higher rice yields, management practices that can be adopted are using an effective source, using an appropriate method, timing of application, adequate rate, and planting Si-efficient rice genotypes.

TABLE 15.4
Silicon Harvest Index in the Lowland Rice as Influenced by Foliar Spray of Silicic Acid (SA) Treatments

Treatment	Silicon Harvest Index
T_1 – Recommended N–P–K	0.19
$T_2 - T_1$ + 2 ml L^{-1} SA	0.19
$T_3 - T_1$ + 4 ml L^{-1} SA	0.18
$T_4 - T_1$ + 8 ml L^{-1} SA	0.19
$T_5 - T_2$ + ½ dose pesticide	0.19
$T_6 - T_3$ + ½ dose pesticide	0.20
$T_7 - T_4$ + ½ dose pesticide	0.17
Average	0.18

Source: Values were calculated from the data in Table 15.2. Prakash, N. B., N. Chandrashekar, C. Mahendra, S. U. Patil, G. N. Thippesshappa, and H. M. Laane. 2011. Effect of foliar spray of soluble silicic acid on growth and yield parameters of wetland rice in hilly and coastal zone soils of Karnataka, South India. *J. Plant Nutr.* 34:1883–1893.

Note: $SHI = \frac{\text{Si uptake in grain in kg}}{\text{Si uptake in grain plus straw in kg}}$.

15.8.1 Effective Sources

Using an effective silicon source is an important strategy in managing rice-producing soils. Important carriers of silicon are wollastonite ore or calcium silicate slag (mostly $CaAl_2Si_2O_8$ with some $CaSiO_3$, a byproduct of electric furnaces), containing 18–21% Si. In addition, calcium silicate ($CaSiO_3$) containing 31% Si and sodium metasilicate ($NaSiO_3$) containing 23% Si are important Si sources (Tisdale, Nelson, and Beaton 1985; Snyder, Jones, and Gascho 1986). Chinese scientists tested new silicate fertilizers containing 230 g water-soluble Si kg^{-1} fertilizer for rice with promising results (Liang et al. 1994). In addition to wollastonite, Korean scientists have used different slags such as slowly cooled slag (13% Si), air-cooled slag (13% Si), quenched slag (16% Si), and fused phosphate (11.5% Si) in field experiments (Savant, Snyder, and Datnoff 1997). In India, lignite fly ash from a thermal power plant containing 23% Si (141 mg kg^{-1} available Si) has been used as a Si source for rice (Raghupathy 1993). Because of its high Si requirement, rice plants generally respond well to applied Si sources. For rice in Japan, the use of silicate sources is very common, and Si has been regarded as an agronomically essential element (Takahashi, Ma, and Miyake 1990). Takahashi, Ma, and Miyake (1990) reported that in Japan, Si management in rice cultivation includes Si fertilization and plant Si recycling.

15.8.2 Appropriate Methods and Timing of Application

When fertilizers or soil amendments are applied in larger amounts, broadcast application and mixing in the soil are most effective in furnishing nutrients to a given crop or changing soil properties in favor of higher yields. Because silicate sources are applied in larger amounts (>1 Mg ha^{-1}) (Savant, Snyder, and Datnoff 1997), they should be applied as broadcast and mixed in the soil before sowing the rice crop. Tisdale, Nelson, and Beaton (1985) also reported that broadcasting followed by mixing into the soil before planting is an acceptable method of application. Savant, Snyder, and Datnoff (1997) reviewed the literature on method and timing of silicon sources and concluded that broadcast was a better method than band application.

Prakash et al. (2011) reported that foliar spraying of silicon with pesticides is an important strategy for increasing rice yield. These authors conducted field experiments in the state of Karnataka, India, evaluating several foliar Si application treatments. Grain and straw yield increased with foliar spraying of silica at various levels (Table 15.5). Foliar spraying of silicic acid at 2 and 4 ml L^{-1} increased grain and straw yield over the control, and further increased levels (8 ml L^{-1}) decreased the yield. The increase in grain and straw yield may be related to a decrease in the percent of spikelet sterility, an increased rate of photosynthesis, an increased number of productive tillers, and a reduction in the number of insect pests and diseases (Munir et al. 2003; Singh and Singh 2005; Singh, Dwivedi, and Shukla 2006; Prakash et al. 2011).

15.8.3 Adequate Rate

Adequate rate of silicon depends on soil tests for Si, type of soil, chemical properties, and yield level of rice. However, based on field experiment results in Japan, Korea, and Taiwan, Savant, Snyder, and Datnoff (1997) reported that application of 1–2 Mg ha^{-1}

TABLE 15.5
Grain and Straw Yield of Lowland Rice as Influenced by Foliar Spray of Silicic Acid (SA)

Treatment	Grain Yield (kg ha^{-1})	Straw Yield (kg ha^{-1})
T_1 – Recommended N–P–K	5057	7261
$T_2 - T_1$ + 2 ml L^{-1} SA	5932	8294
$T_3 - T_1$ + 4 ml L^{-1} SA	6380	8929
$T_4 - T_1$ + 8 ml L^{-1} SA	5474	7392
$T_5 - T_2$ + ½ dose pesticide	6022	8759
$T_6 - T_3$ + ½ dose pesticide	6679	9697
$T_7 - T_4$ + ½ dose pesticide	5424	7326
SEM±	178	81
CD (%)	547	248

Source: Adapted from Prakash, N. B., N. Chandrashekar, C. Mahendra, S. U. Patil, G. N. Thippesshappa, and H. M. Laane. 2011. Effect of foliar spray of soluble silicic acid on growth and yield parameters of wetland rice in hilly and coastal zone soils of Karnataka, South India. *J. Plant Nutr.* 34:1883–1893.

of calcium silicate slag may be adequate for lowland rice. In Japan, silicate slag application at an optimum rate of 1.5–2.0 Mg ha^{-1} is widely used in degraded paddy soils (Savant, Snyder, and Datnoff 1997). However, Korndorfer and Lepsch (2001) recommended higher fertilizer rates for rice based on soil test results in Florida Histosols. (Results of these authors are discussed in the sections that follow on soil and plant testing.) Yoshida (1981) reported that a rice yield increase of 10% with the addition of silica is common in Japan. This author further noted that a 30% yield increase was reported in Japan when a rice crop was infested with blast disease.

15.8.3.1 Soil Test

When determining an adequate silicon rate for a given crop, important criteria to consider for maximum economic results are field studies relating to silica rate versus grain yield, soil tests, and plant tissue tests. In soil tests, the extracting solution should be observed first. Imaizumi and Yoshida (1958) compared the amounts of Si absorbed from soils by rice plants and the amount of Si extracted from the soils by hot HCl, ammonium oxalate (pH 3.0), 2% Na_2CO_3 saturated water, 0.002 N H_2SO_4, and ammonium acetate (pH 4.0). These authors concluded that dilute acid-soluble Si was better correlated with Si absorption by plants and that dilute alkaline soluble Si did not reflect the Si-supplying power of the soil. Snyder (1991) developed a soil test based on Si extraction by 0.5 M acetic acid in order to identify the need for Si fertilization in Histosols planted with rice in the Everglades Agricultural Area in Florida. This test is widely used to define critical Si levels in soils amended with calcium silicate (Savant, Snyder, and Datnoff 1997). On the basis of soil test levels of Si, Korndorfer and Lepsch (2001) made the following recommendations for silicate fertilizers: when a Si soil test was <6 mg kg^{-1}, the level in the soil was considered low and 7.5 Mg ha^{-1} $CaSiO_3$ was recommended; when a Si soil test was in the range of 6–24 mg kg^{-1}, the level in the soil was considered medium and the fertilizer recommendation was 5.6 Mg $CaSiO_3$ ha^{-1}; and when a Si soil test was >24 mg kg^{-1}, the level in the soil was considered high and no silicate fertilizer was recommended. At the high level of a soil test, these authors reported a relative yield of 95%. However, a 100% relative yield was obtained in a test of about 40 mg Si kg^{-1} soil (Korndorfer and Lepsch 2001).

Another criterion for determining adequate rate of silicate fertilizers is the critical level of a given nutrient for a given crop. Imaizumi and Yoshida (1958) reported that Si application can have a positive effect on rice yield when the Si content of the sodium acetate buffer pH 4.0 is less than 49 mg kg^{-1} soil. However, Liang et al. (1994) reported that the sodium acetate buffer-extractable Si content of calcareous soils in China ranged from 71 to 181 mg Si kg^{-1} and that rice yields still responded to the application of silicate fertilizers. For tropical rice soils in Malaysia and Thailand, Kawaguchi (1966) used 33 mg Si kg^{-1} soil, and for those in Sri Lanka, Takijima, Wijayaratna, and Seneviratne (1970) used 38 mg Si kg^{-1} as tentative criteria for the acetate buffer-extractable Si for describing Si deficiency.

15.8.3.2 Plant Tissue Test

Plant tissue tests primarily are used to diagnose nutrient deficiencies or sufficiencies in crop plants. The main problem of plant tissue test interpretation is that nutrient concentration varies with plant age because of the dilution effect. In mineral nutrition, the dilution effect is defined as a decrease in nutrient concentration due to an increase in dry weight of plants with the advancement of plant age. In rice, straw usually is analyzed at harvest because it is related to grain yield (Fageria 2009; Fageria, Baligar, and Jones 2011), and the results can be easily compared with those obtained in previous studies (Savant, Snyder, and Datnoff 1997).

Based upon a large number of experiments conducted to correlate the Si content of straw and grain yield of rice to Si fertilization in Japan, Korea, and Taiwan, Lian (1976) reported that straw's critical Si level for no yield responded to slag application. For straw, it was > 6.1% Si in Japan and Korea and 5.1% Si in Taiwan. In Sri Lanka and India, *indica* rice cultivars may respond to Si application when Si content in straw is below 3.7% (Nair and Aiyer 1968; Takijima, Wijayaratna, and Seneviratne 1970). Snyder, Jones, and Gascho (1986) reported that for straw, more than 3% Si (6.4% SiO_2) is needed for a good yield of rice grown on organic soils (Histosols) in the Everglades Agricultural Area in Florida. Korndorfer and Lepsch (2001) recommended $CaSiO_3$ on the basis of Si concentration in the rice straw in Florida's Histosols. These recommendations were as follows: When Si in the straw was < 17 g kg^{-1}, the level in the plant was considered low and 5.6 Mg $CaSiO_3$ ha^{-1} was recommended; when Si concentration in the straw was in the range of 17–34 g kg^{-1}, the level in the plant was considered medium and 4.3 Mg $CaSiO_3$ was recommended; and when the Si concentration in the straw was > 34 g kg^{-1}, the level in the plant was considered high and no silicate fertilizer was recommended.

The Si content of rice plant tissue can be determined by several methods. Kilmer (1965) suggested that Si content in rice straw can be determined by high temperature fusion with NaOH, followed by development of a silicomolybdous chromophore and subsequent colorimetry determination. Plant tissue can be analyzed for Si gravimetrically by freeing Si from the tissue matrix through wet digestion procedures, followed by weighing it as Si (Yoshida et al. 1976; Elliott et al. 1988). Colorimetric Si determination in aliquots of nitric acid digestion of rice plant tissue has been suggested (CRRI 1974). Hydrofluoric acid may be used to solubilize Si in plant material or ash, with subsequent determination of Si by atomic absorption spectrophotometry or colorimetry (Van der Vorm 1980; Novozamsky, van Eck, and Houba 1984).

15.8.4 Recycling Silicon Content of Plant Residues

Recycling Si content of plant residues is an important strategy from an economic and soil quality point of view. Plant residues not only supply Si but also add to soil's organic matter content. Recycling of Si content in rice straw for succeeding crops is important because rice straw content has a significant amount of Si at harvest. On average, rice straw content has about 4% Si and rice hull content about 8% Si (Savant, Snyder, and Datnoff 1997). These two products can add significant amounts of Si to the succeeding crops. Diamond, Sri Adiningsih, and Sudjadi (1986) reported that Si content in an Indonesian Oxisol increased from 50 to 150 mg kg^{-1} with the incorporation of 5.0 Mg ha^{-1} of rice straw. Rice straw decomposition may be a problem due to a high C:N ratio (>60). Liu and Ho (1960) reported that making rice straw and hull compost by an aerobic or anaerobic process could be a good source of Si because the plants' Si (associated with cellulose and hemicelluloses) would be released as it degraded during composting. The deposition of silicon content in rice plant parts is summarized in Table 15.6.

TABLE 15.6
Silicon Content in Rice Plant Parts and Observation of Its Deposition

Plant Part	Si Content (%)	Observation
Root	0.9	No localization, uniformally distributed in all roots
Stem	2.4	Accumulation in the epidermis, sclerenchyma plus bundle sheath, and the cell walls in the parenchyma
Leaf blade	5.6	Maximum accumulation in the interspace between the cuticle and the epidermal cells
Leaf sheath	4.7	Maximum accumulation in the epidermis and along the cell walls in the parenchyma
Hull	7.1	Maximum deposition in the interspace between the cuticle and the epidermal cells

Source: Yoshida, S., Y. Ohnishi, and K. Kitagishi. 1962. Histochemistry of silicon in rice plant. II. Localization of silicon within plant tissues. III. The presence of cuticle-silica double layer in the epidermal tissue. *Soil Sci. Plant Nutr.* 8:30–51; Yoshida, S. 1975. *The Physiology of Silicon in Rice.* Technical Bulletin No. 25. Taipei, Taiwan: Food and Fertilization Technology Center; Savant, N. K., G. H. Snyder, and L. E. Datnoff. 1997. Silicon management and sustainable rice production. *Adv. Agron.* 58:1245–1252.

15.8.5 Use of Efficient Genotypes

The use of silicon-efficient rice genotypes in soils deficient in Si is an important strategy for reducing crop production costs as well as environmental pollution. Variation in rice genotypes' Si uptake and utilization has been reported by Deren (2001) and Deren, Datnoff, and Snyder (1992). Deren (2001) reported that significant variations in Si concentration among rice genotypes were identified and that ranking of genotypes over environments was fairly stable. This author also reported that *indica* rice appeared to be less efficient in acquiring Si than *japonica*. Yuan and Cheng (1977) grew *indica* and *japonica* rice cultivars in nutrient solutions containing a fixed 100 mg Si L^{-1} concentration. Total plant Si concentration ranged from 117 to 171 mg g^{-1}, demonstrating that some genotypes are better at accumulating Si than others. Deren, Datnoff, and Snyder (1992) screened rice cultivars for Si uptake using different Si levels (0, 2, and 5 Mg ha^{-1}). Silica concentration in plant tissue increased as expected with an increasing rate of Si fertilization, but within each Si treatment and within the control, cultivars varied for Si concentration. These authors also reported that temperate *japonica* genotypes had the greatest concentration of Si in all treatments, and the single *indica* entry had consistently low Si.

Genetic studies of Si in rice are limited. Majumder, Rakshit, and Borthakur (1985) created a seven-parent diallel cross to investigate the inheritance of Si uptake in rice. Genotypes ranged in leaf Si concentration from 11 to 70 mg g^{-1} at 60 days and 32 to 85 mg g^{-1} at harvest. Variation was largely additive and some heterosis was observed (Deren 2001). Disease resistance is a major component of most crop breeding programs that include rice. Deren (2001) frequently found that genotypes with greater disease resistance did in fact have greater Si concentrations. Blast is a serious rice disease worldwide, and its interaction with Si has been studied widely. In India, Rabindra et al. (1981) found blast resistance in rice cultivars was related to Si content. In Japan, the association between blast and Si concentration in rice genotypes has also been reported by Tanaka (1965) and Suzuki (1965). Osuna-Canizalez, De Datta, and Bonman (1991) reported that in the Philippines, Si fertilization reduced the severity of blast disease in lowland rice. Similarly, Seebold (1998) reported that upland rice cultivars' inherent disease resistance to blast could be enhanced with Si fertilization. Prabhu et al. (2001) reported that in Brazilian Oxisols, grain discoloration in upland rice was related to Si content in the plant tissues; this was reduced with the addition of Si fertilizer (compared with a control treatment). Correa-Victoria et al. (2001) found that rice leaf blast severity and neck blast incidence in Colombian Oxisol were reduced from about 26% and 53% in nonamended plots to 15% in Si-amended plots. These authors also reported that rice yield increased by about 40% with the addition of Si fertilizer.

In Africa, Winslow (1992) and Wilson, Okada, and Correa-Vitoria (1997) reported that blast and husk discoloration (*Bipolaris oryzae* and other organisms) decreased in rice with the addition of Si under upland conditions. Deren (2001) reported that genotypic differences in Si uptake were associated with rice ecotype or species. Tropical *japonica* rice had a 93% greater Si concentration than *indica* ecotypes or species. It was hypothesized that the *japonica* species evolved in the Si-deficient uplands and had developed mechanisms to attain greater Si concentrations, whereas *indica* evolved in lowlands where Si was more available (Deren 2001).

Many researchers identified disease and insect resistance mechanisms related to Si concentration in rice tissues. Yoshida, Ohnishi, and Kitagishi (1962) reported that soluble Si is deposited as a gel under the cuticle of plant cells, forming a double layer of cuticle and Si. This double layer acts as a physical barrier to pathogens and pests and likely contributes to resistance against disease, insects, and nematodes as well as reducing transpiration. The *indica* and *japonica* rice genotypes evaluated by Yuan and Cheng (1977) had cell wall Si contents that ranged from 180 to 211 mg g^{-1}, which were greater than the concentrations in the whole plant.

In plant tissues, varietal resistance to nematodes and insects is also related to Si concentration. Swain and Prasad (1991) reported that Si concentration in rice roots was associated with resistance to root-knot nematodes (*Meloidogyne spp.*). Similarly, Patanakamjorn and Pathak (1967) and Djamin and Pathak (1967) reported that Asiatic rice borer (*Chilo suppressalis*) infestations decreased with increasing Si concentration in the rice tissues. These authors suggested that selecting rice cultivars with high Si concentration is an important strategy as compared with Si fertilization.

15.9 CONCLUSIONS

Silicon is the second most abundant element in the earth's crust. Silicon deficiency has been reported in many rice-growing regions of the world. Although not essential, Si is a beneficial element because it supports the healthy growth and development of many crop species, particularly rice. Generally, the beneficial effects of Si are obvious in plants that encounter biotic and abiotic stresses; Si decreases plants' susceptibility to certain diseases and insects. It is mobile in plant tissue and its deficiency symptoms first appear in older leaves. Silicon is absorbed by plants in the form of monosilicis acid (H_4SiO_4). Uptake of Si is affected by pH, iron, and aluminum oxides; flooding; and soil fertility. Uptake of silicon may be via active and passive processes. Active uptake of Si requires energy, and its accumulation in rice tissues equals that of macronutrients such as N and K. Uptake of Si is much higher in cereals than in legumes. Average Si concentration in rice plants is about 5%, and the major part (>70%) is retained in the straw. SUE is defined as dry matter production per unit of Si uptake and is about three times lower in straw than in grain. This lower value of Si use in straw is related to higher uptake in straw compared to grain. Calcium silicates are major sources of Si fertilizers. An adequate rate of silicate fertilizer should be defined on the basis of soil and plant tissue tests. Soil and plant tissue test calibration studies are limited for rice. Broadcasting, followed by mixing into the soil before sowing, is an acceptable method of application. Data related to Si genetics in rice are limited. However, genotypic differences have been reported in Si uptake and utilization.

REFERENCES

Arnon, D. I., and P. R. Stout. 1939. The essentiality of certain elements in minute quantity for plants with special reference to copper. *Plant Physiol.* 14:371–375.

Azam, F., B. B. Hemmingsen, and B. F. Volcani. 1974. Role of silicon in diatom metabolism. V. Silicic acid transport and metabolism in the heterotrophic diatom *Nitzschia Alba*. *Arch. Microbiol.* 97:103–114.

Barber, D. A., and M. G. T. Shone. 1966. The absorption of silica from aqueous solutions by plants. *J. Exp. Bot.* 17:569–578.

Bogdan, K., and M. K. Schenk. 2008. Arsenic in rice (*Oryza sativa* L.) related to dynamics of arsenic and silicic acid in paddy soils. *Environ. Sci. Tech.* 42:7885–7890.

Chen, Y. 1990. Characteristics of silicon uptake and accumulation in rice. *J. Guizhou Agric. Sci.* 6:37–40.

Chumlea, W. C. 2007. Silica, a mineral of unknown but emerging health importance. *J. Nutr. Health Aging.* 11:93.

Correa-Victoria, F. J., L. E. Datnoff, K. Okada, D. K. Friesen, J. J. Sanz, and G. H. Snyder. 2001. Effects of silicon fertilization on disease development and yield of rice in Colombia. In *Silicon in Agriculture*, eds. L. E. Datnoff, G. H. Snyder, and G. H. Korndorfer, pp. 313–322. New York: Elsevier.

CRRI (Central Rice Research Institute). 1974. *Annual report*, 1973. Orissa, India: Cuttack.

Daroub, S. H., and G. H. Snyder. 2007. The chemistry of plant nutrients in soil. In *Mineral Nutrition and Plant Disease*, eds. L. E. Datnoff, W. H. Elmer, and D. M. Huber, pp. 1–7. St. Paul, MN: American Phytopathological Society.

Datnoff, L. E., F. A. Rodrigues, and K. W. Seebold. 2007. Silicon and plant disease. In *Mineral Nutrition and Plant Disease*, eds. L. E. Datnoff, W. H. Elmer, and D. M. Huber, pp. 233–246. St. Paul, MN: American Phytopathological Society.

Deren, C. W. 2001. Plant genotype, silicon concentration, and silicon related responses. In *Silicon in Agriculture*, eds. L. E. Datnoff, G. H. Snyder, and G. H. Korndorfer, pp. 149–158. New York: Elsevier.

Deren, C. W., L. E. Datnoff, and G. N. Snyder. 1992. Variable silicon content of rice cultivars grown on Everglades Histosols. *J. Pant Nutr.* 15:2363–2368.

Diamond, R. B., S. J. Sri Adiningsih, and M. Sudjadi. 1986. Maintaining yields of lowland rice with reduced fertilizer applications in Indonesia. Paper presented at the INSFFER workshop. Hangzhow, China.

Djamin, A., and M. D. Pathak. 1967. Role of silica in resistance to Asiatic rice borer, *Chilo suppressalis* (Walker), in rice varieties. II. *Riso.* 28:235–253.

Drees, L. R., L. P. Wilding, N. E. Smeck, and A. L. Senkayi. 1989. Silica in soils: Quartz and disordered silica polymorphs. In *Minerals in Soil Environments*, eds. D. E. Kissel and W. S. Wadison, pp. 913–974. Madison, WI: SSSA.

Elliott, C. L., G. H. Snyder, and D. B. Jones. 1988. Rapid gravimetric determination of Si in rice straw. *Commun. Soil Sci. Plant Anal.* 19:1118–1119.

Epstein, E. 1991. The anomaly of silicon in plant biology. *Proc. Natl. Acad. Sci.* 30:207–210.

Epstein, E. 1999. Silicon. *Ann. Rev. Plant Physiol. Plant Mol. Biol.* 50:641–664.

Fageria, N. K. 2009. *The Use of Nutrients in Crop Plants*. Boca Raton, FL: CRC Press.

Fageria, N. K., V. C. Baligar, and C. A. Jones. 2011. *Growth and Mineral Nutrition of Field Crops*, 3rd edition. Boca Raton, FL: CRC Press.

Fageria, N. K., A. B. Santos, M. P. Barbosa Filho, and C. M. Guimaraes. 2008. Iron toxicity in lowland rice. *J. Plant Nutr.* 31:1676–1697.

Fleck, A. T., T. Nye, C. Repenning, F. Stahl, M. Zahn, and M. K. Schenk. 2011. Silicon enhances suberization and lignification in roots of rice (*Oryza sativa*). *J. Exp. Bot.* 62:2001–2011.

Gascho, G. J., D. L. Anderson, and J. E. Bowen. 1993. Sugarcane. In *Nutrient Deficiencies & Toxicities in Crop Plants*, ed. W. F. Bennett, pp. 37–42. St. Paul, MN: American Phytopathology Society.

Hallmark, C. T., L. P. Wilding, and N. E. Smeck. 1982. Silicon. *Agron. Monogr.* 9:263–265.

Hirota, R., Y. Hata, T. Ikeda, T. Ishida, and A. Kuroda. 2010. The silicone layer supports acid resistance bacillus cereus spores. *J. Bacteriol.* 192:111–116.

Imaizumi, K., and S. Yoshida. 1958. Edaphological studies on silicon supplying power of paddy soils. *Bull. Natl. Inst. Agric. Sci. B.* 8:261–304.
Jones, L. H. P., and K. A. Handreck. 1967. Silica in soils, plants and animals. *Adv. Agron.* 19:107–149.
Kang, Y. K. 1980. Silicon influence on physiological activities in rice. PhD dissertation, University of Arkansas, Fayetteville.
Kawaguchi, K. 1966. Tropical paddy soils. *Jpn. Agric. Res. Q.* 1:7–11.
Kilmer, V. J. 1965. Silicon. In *Methods of Soil Analysis: Chemical and Microbiological Properties*, part 2, ed. C. A. Black, pp. 959–962. Madison, WI: ASA.
Korndorfer, G. H., and I. Lepsch. 2001. Effect of silicon on plant growth and crop yield. In *Silicon in Agriculture*, eds. L. E. Datnoff, G. H. Snyder, and G. H. Korndorfer, pp. 133–1145. New York: Elsevier.
Lian, S. 1976. Silica fertilization of rice. In *The Fertility of Paddy Soils and Fertilizer Application for Rice,* ed. Food and Fertilizer Technology Center, pp. 197–220. Taipei, Taiwan: Food and Fertilizer Technology Center.
Liang, Y. C., T. S. Ma, F. J. Li, and Y. J. Feng. 1994. Silicon availability and response of rice and wheat to silicon in calcareous soils. *Commun. Soil Sci. Plant Anal.* 25:2285–2297.
Lindsay, W. L. 1979. *Chemical Equilibrium in Soils*. New York: John Wiley.
Liu, S. L., and C. H. Ho. 1960. Study in the nature of silicon in rice hull. I. Solubility of the silicon part. *J. Chin. Chem. Soc.* 6:141–153.
Ma, J. F. 2003. Function of silicon in higher plants. *Prog. Mol. Subcell Biol.* 33:127–147.
Ma, J. F., Y. Miyake, and E. Takahashi. 2001. Silicon as a beneficial element for crop plants. In *Silicon in Agriculture*, eds. L. E. Datnoff, G. H. Snyder, and G. H. Korndorfer, pp. 17–39. New York: Elsevier.
Majumder, N. D., S. C. Rakshit, and D. N. Borthakur. 1985. Genetics of silica uptake in selected genotypes of rice (*Oryza sativa* L.). *Plant Soil.* 88:449–453.
Mengel, K., E. A. Kirkby, H. Kosegarten, and T. Appel. 2001. *Principles of Plant Nutrition*, 5th edition. Dordrecht, The Netherlands: Kluwer.
Mitsui, S., and H. Takatoh. 1963. Nutritional study of silicon in graminaceous crops. Part I. *Soil Sci. Plant Nutr.* 9:49–53.
Munir, M., A. C. A. Carlos, G. F. Heilo, and C. C. Juliano. 2003. Nitrogen and silicone fertilization of upland rice. *Scientia Agricola.* 60:1–10.
Nair, P. K., and R. S. Aiyer. 1968. Status of available silica in the rice soils of Kerala state of India. II. Silicon uptake by different varieties of rice in relation to available silica contributed by soil and irrigation water. *Agric. Res. J. Kerala.* 6:88–94.
Nair, P. K., A. K. Mishra, and S. Patnaik. 1982. Silica in rice and flooded rice soils. II. Uptake of silica in relation to growth of rice varieties of different durations grown on an Inceptisol. *Oryza.* 19:82–92.
Novozamsky, I., R. van Eck, and V. J. G. Houba. 1984. A rapid determination of silicon in plant material. *Commun. Soil Sci. Plant Anal.* 15:205–211.
Okuda, A., and E. Takahashi. 1961. Studies on the physiological role of silicon in crop plants. 4. Effect of silicon on the growth of barley, tomato, radish, green onion, Chinese cabbage, and their nutrients uptake. *J. Soil Sci. Soil Manure.* 32:623–626.
Okuda A. and E. Takahashi. 1965. The role of silicon. In *The Mineral Nutrition of the Rice Plant*, ed. International Rice Research Institut, pp. 123–146. Baltimore, Maryland: The Johns Hopkins Press.
Osuna-Canizalez, F. J., S. K. De Datta, and J. M. Bonman. 1991. Nitrogen form and silicone nutrition effects on resistance to blast disease of rice. *Plant Soil.* 135:223–231.
Park, C. S. 1975. The micronutrient problem of Korean agriculture. In *Symposium Commemorating the 30th Anniversary of Korean Liberation*, ed. Nat. Acad. Sci. Rep. Korea, pp. 847–862. Seoul: Nat. Acad. Sci. Rep. Korea.

Patanakamjorn, S., and M. D. Pathak. 1967. Varietal resistance of rice to the Asiatic rice borer, *Chilo suppressalis* (Lepidoptera: Crambidae) and its association with various plant characters. *Ann. Entom. Soc. Am.* 60:287–292.

Ponnamperuma, F. N. 1965. Dynamic aspects of flooded soils and the nutrition of the rice plant. In *The Mineral Nutrition of Rice Plant*, ed. IRRI, pp. 295–328. Baltimore, MD: Johns Hopkins.

Prabhu, A. S., M. P. Barbosa Filho, M. C. Filippi, L. E. Datnoff, and G. H. Snyder. 2001. Silicon from rice disease control perspective in Brazil. In *Silicon in Agriculture*, eds. L. E. Datnoff, G. H. Snyder, and G. H. Korndorfer, pp. 293–311. New York: Elsevier.

Prabhu, A. S., G. B. Silva, L. G. Araujo, and M. C. C. Filippi. 2006. *Blast in Rice: Control, Genetics, Progress and Perspectives*. Santo Antônio de Goiás, Brazil: EMBRAPA Arroz e Feijão.

Prakash, N. B., N. Chandrashekar, C. Mahendra, S. U. Patil, G. N. Thippesshappa, and H. M. Laane. 2011. Effect of foliar spray of soluble silicic acid on growth and yield parameters of wetland rice in hilly and coastal zone soils of Karnataka, South India. *J. Plant Nutr.* 34:1883–1893.

Rabindra, B., B. S. Gowda, K. T. P. Gowda, and H. K. Rajaooa. 1981. Blast disease as influence by silicon in some rice varieties. *Curr. Res.* (Bangalore). 10:82–83.

Raghupathy, B. 1993. Effect of lignite fly ash (LFA) on rice. *Rice Res. Notes.* 18:27–28.

Reifenberg, A., and S. J. Buckwold. 1954. The release of silica from soils by the orthophosphate anion. *J. Soil Sci.* 5:106–115.

Savant, N. K., L. E. Datnoff, and G. H. Snyder. 1997. Depletion of plant-available silicon in soils: A possible cause of declining rice yields. *Commun. Soil Sci. Plant Anal.* 28:1245–1252.

Savant, N. K., G. H. Snyder, and L. E. Datnoff. 1997. Silicon management and sustainable rice production. *Adv. Agron.* 58:1245–1252.

Seebold, K. W. Jr. 1998. The influence of silicon fertilization on the development and control of blast caused by *Magnaporthe grisea* Herber (Barr) in upland rice. PhD dissertation, University of Florida.

Singer, M. J., and D. N. Munns. 1987. *Soils: An Introduction*. New York: Macmillan.

Singh, K. K., and K. Singh. 2005. Effect of N and Si on growth, yield attributes and yield of rice in Alfisols. *Int. Rice Res. Notes.* 12:40–41.

Singh, V. K., B. S. Dwivedi, and A. K. Shukla. 2006. Yields and nitrogen and phosphorus use efficiency as influenced by fertilizer NP additions in wheat under rice-wheat and pigeon pea-wheat system on a Typic Ustochrept soil. *Indian J. Agric. Sci.* 76:92–97.

Snyder, G. H. 1991. *Development of a Silicon Soil Test for Histosol Grown Rice.* Belle Glade EREC Research Report EV–1991–2, Belle Glade, FL: University of Florida.

Snyder, G. H., D. B. Jones, and G. J. Gascho. 1986. Silicon fertilization of rice on Everglades Histosols. *Soil Sci. Soc. Am. J.* 50:1259–263.

Snyder, G. H., V. V. Matichenkov, and L. E. Datnoff. 2007. Silicon. In *Handbook of Plant Nutrition*, eds. A. V. Barker and D. J. Pilbeam, pp. 551–568. Boca Raton, FL: CRC Press.

Suzuki, N. 1965. Nature of resistance to blast. In *The Rice Blast Disease*, ed. IRRI, pp. 277–301. Baltimore, MD: Johns Hopkins.

Swain, B. N., and J. S. Prasad. 1991. Influence of silica content in the roots of rice varieties on the resistance to root-knot nematodes. *Ind. J. Nematol.* 18:360–361.

Takahashi, E. 1995. Uptake mode and physiological functions of silica. In *Science of the Rice Plant: Physiology*, Vol. 2, eds. T. Matsuo, K. Kumazawa, R. Ishii, K. Ishihara, and H. Hirata, pp. 420–433. Tokyo: Food and Agriculture Policy Research Center.

Takahashi, E., J. F. Ma, and Y. Miyake. 1990. The possibility of silicon as an essential element for higher plants. *Comments Agric. Food Chem.* 2:99–122.

Takijima, Y., H. M. S. Wijayaratna, and C. J. Seneviratne. 1970. Nutrient deficiency and physiological disease of lowland rice I Ceylon, III. Effect of silicate fertilizers and dolomite for increasing rice yields. *Soil Sci. Plant Nutr.* 16:11–16.

Tamai, K., and J. F. Ma. 2003. Characterization of silicon uptake by rice roots. *New Phytol.* 158:431–436.

Tanaka, S. 1965. Nutrition of *Piricularia oryzae* in vitro. In *The Rice Blast Disease*, ed. IRRI, pp. 23–34. Baltimore, MD: Johns Hopkins.

Tisdale, S. L., W. L. Nelson, and J. D. Beaton. 1985. *Soil Fertility and Fertilizers*, 4th edition. New York: Macmillan.

Van der Vorm, P. D. J. 1980. Uptake of Si by five plant species as influenced by variations in Si-supply. *Plant Soil.* 56:153–56.

Wells, B. R., B. A. Huey, R. J. Norman, and R. S. Helms. 1993. Rice. In *Nutrient Deficiencies & Toxicities in Crop Plants*, ed. W. F. Bennett, pp. 15–19. St. Paul, MN: American Phytopathology Society.

Wilson, M. D., K. Okada, and F. Correa-Vitoria. 1997. Silicon deficiency and the adaptation of tropical rice genotypes. *Plant Soil.* 188:239–248.

Winslow, M. D. 1992. Silicon, disease resistance, and yield of rice genotypes under upland cultural conditions. *Crop Sci.* 32:1208–1213.

Yamauchi, M. and M. D. Winslow. 1989. Effect of silica and magnesium on yield of upland rice in humid tropics. *Plant Soil.* 113:265–269.

Yoshida, S. 1975. *The Physiology of Silicon in Rice.* Technical Bulletin No. 25. Taipei, Taiwan: Food and Fertilization Technology Center.

Yoshida, S. 1981. *Fundamentals of Rice Crop Science*. Los Baños, Philippines: IRRI.

Yoshida, S., D. A. Forno, J. H. Cook, and K. A. Gomez. 1976. *Laboratory Manual for Physiological Studies of Rice*, 3rd edition, ed. IRRI. Los Banõs, Philippines: IRRI.

Yoshida, S., Y. Ohnishi, and K. Kitagishi. 1962. Histochemistry of silicon in rice plant. II. Localization of silicon within plant tissues. III. The presence of cuticle-silica double layer in the epidermal tissue. *Soil Sci. Plant Nutr.* 8:30–51.

Yuan, H. F., and Y. S. Cheng. 1977. The physiological significance in rice plants: The influence of pH on silicon uptake of rice plants and the presence of silica in the cell wall. *Proc. Nat. Sci. Council* (Taiwan). 10:2 13.

FIGURE 2.7 Upland rice cultivar BRS Primavera growth with adequate rate of P and K but without N (left) and with adequate rate of N + P + K (at right).

FIGURE 2.8 Growth of upland rice at three N levels.

FIGURE 2.13 Two lowland rice genotypes showing N deficiency symptoms in pots that did not receive N fertilization.

FIGURE 3.5 Upland rice growth on Brazilian Oxisol at three P levels.

FIGURE 3.9 Upland rice plants without P and with P.

FIGURE 4.1 Response of upland rice, grown on Brazilian Oxisol, to N, P, and K fertilization.

FIGURE 4.2 Response of upland rice, grown on Brazilian Oxisol, to potassium fertilization.

FIGURE 4.7 Growth of lowland rice, grown on Brazilian Inceptisol with and without K fertilization.

FIGURE 4.8 Potassium deficiency symptoms in the leaves of upland rice plants grown on Brazilian Oxisol.

FIGURE 4.9 Potassium deficiency symptoms in the leaves of lowland rice plants grown on Brazilian Inceptisol.

FIGURE 5.6 Rice plants with calcium deficiency symptoms. (From Fageria, N. K., and M. P. Barbosa Filho. 1994. *Nutrient Deficiency in Rice: Identification and Correction*. Goiania, Brazil: EMBRAPA-SPI/EMBRAPA-CNPAF.)

FIGURE 5.7 Magnesium deficiency symptoms in rice plants. (From Fageria, N. K., and M. P. Barbosa Filho. 1994. *Nutrient Deficiency in Rice: Identification and Correction*. Goiania, Brazil: EMBRAPA-SPI/EMBRAPA-CNPAF.)

FIGURE 5.8 Magnesium deficiency symptoms in rice leaves. (From Fageria, N. K., and M. P. Barbosa Filho. 1994. *Nutrient Deficiency in Rice: Identification and Correction*. Goiania, Brazil: EMBRAPA-SPI/EMBRAPA-CNPAF.)

FIGURE 6.2 Upland rice growth on Brazilian Oxisol at different S levels.

FIGURE 6.5 Sulfur deficiency symptoms in lowland rice plants.

FIGURE 7.5 Zinc deficiency symptoms in upland rice grown on Brazilian Oxisol.

FIGURE 7.6 Zinc deficiency symptoms in upland and lowland rice grown on Brazilian Oxisol and Inceptisol, respectively.

FIGURE 8.1 Copper deficiency symptoms in upland rice leaves developed in nutrient solution. (From Fageria, N. K., and M. P. Barbosa Filho. 1994. *Nutritional Deficiency in Rice: Identification and Correction*. Goiania, Brazil: EMBRAPA Arroz e Feijão.)

FIGURE 8.2 Copper deficiency symptoms in upland rice plants developed in nutrient solution. (From Fageria, N. K., and M. P. Barbosa Filho. 1994. *Nutritional Deficiency in Rice: Identification and Correction*. Goiania, Brazil: EMBRAPA Arroz e Feijão.)

FIGURE 8.3 Lowland rice leaves with copper deficiency grown on Brazilian Inceptisol.

FIGURE 8.4 Upland rice leaves with copper deficiency grown on Brazilian Oxisol.

FIGURE 9.3 Manganese deficiency in upland rice in nutrient solution. (From Fageria, N. K., and M. P. Barbosa Filho. 1994. *Nutritional Deficiency in Rice: Identification and Correction*. Goiania, Brazil: EMBRAPA.)

FIGURE 9.4 Manganese deficiency symptoms in rice leaf. (From Fageria, N. K., and M. P. Barbosa Filho. 1994. *Nutritional Deficiency in Rice: Identification and Correction*. Goiania, Brazil: EMBRAPA.)

FIGURE 10.1 Iron deficiency symptoms in upland rice grown on Brazilian Oxisol. (From Fageria, N. K. 2009. *The Use of Nutrients in Crop Plants*. Boca Raton, FL: CRC Press.)

FIGURE 10.5 Iron deficiency in younger leaves of upland rice leaves grown on Oxisol with pH about 6.5 in water.

FIGURE 10.6 Iron deficiency in upland rice with a high P level (400 mg P kg^{-1}).

FIGURE 10.7 Iron deficiency in upland rice grown on Brazilian Oxisol having pH 7.0 in water.

FIGURE 10.9 Iron toxicity symptoms in flooded rice. (From Fageria, N. K., L. F. Stone, and A. B. Santos. 2003. Soil fertility management for irrigated rice. Santo Antônio de Goiás, Brazil: EMBRAPA Arroz e Feijão.)

FIGURE 11.3 Boron deficiency symptoms in rice plants. (From Fageria, N. K., and M. P. Barbosa Filho. 1994. *Nutritional Deficiency in Rice: Identification and Correction*. Goiânia, Brazil: EMBRAPA Arroz e Feijão.)

FIGURE 11.4 Boron deficiency symptoms in rice leaves. (From Fageria, N. K., and M. P. Barbosa Filho. 1994. *Nutritional Deficiency in Rice: Identification and Correction*. Goiânia, Brazil: EMBRAPA Arroz e Feijão.)

FIGURE 12.5 Molybdenum deficiency symptoms in rice leaves. (From Fageria, N. K., and M. P. Barbosa Filho. 1994. *Nutritional Deficiency in Rice: Identification and Correction*. Goiania, Brazil: EMBRAPA Arroz e Feijão.)

Index

A

B

C

H

I

J

K

L

N

R

S

T

U

V

W

Y

Z